# THE DREAMS THAT STUFF IS MADE OF

# THE DREAMS THAT STUFF IS MADE OF

The Most Astounding Papers on Quantum Physics—and How They Shook the Scientific World

Edited, with an introduction, by Stephen Hawking

RUNNING PRESS
PHILADELPHIA · LONDON

Library of Congress Control Number: 2011936730
ISBN    978-0-7624-3434-3

Cover design by Bill Jones
Interior design by Aptara, Inc.
Edited by David Goldberg, Joel Allred, and Jennifer Kasius
Commentary by Joel Allred

Running Press Book Publishers
2300 Chestnut Street
Philadelphia, PA 19103-4371
Visit us on the web!
www.runningpress.com

Set in 12/16 pt ITCGaramond by Aptara_R, Inc., New Delhi, India

# CONTENTS

## A Note on the Text

The texts in this book are based on translations of the original printed
editions. We have made no attempt to modernize the authors' own
distinct usage, spelling, or punctuation, or to make the texts consistent
with each other in this regard.

# INTRODUCTION

## BY STEPHEN HAWKING

The goal of physical science is to explain what the universe is made of, and how it works. Ever since Kepler, Galileo, and Newton, we have represented our knowledge of natural phenomena through physical laws. These have evolved over time, as we have enlarged the domain of our observations. When, early in the twentieth century, physicists developed the tools to investigate the structure of atoms and their interaction with radiation, they discovered that their picture of Nature, which was based on observations of the objects of everyday life, was inadequate in a fundamental way. This fascinating volume employs original texts to trace the development of the revolutionary new concepts required to explain Nature at and below the scale of atoms. It is a compelling story, a tale of troubling observations and profound flashes of insight, leading to a new worldview in which familiar properties such as position and momentum take on a new meaning, ideas such as the trajectory of a particle have to be abandoned, and the very idea of what is meant by prediction has to be redefined.

It was observations of the light produced by glowing objects, called "black-body radiation," that first challenged the credibility of the old "classical" picture. Not only did the theory based on that picture not match experimental observations, it also predicted that an infinite amount of radiation would be emitted by such bodies. That is an absurd result. In 1899 Max Planck showed that he could derive the correct mathematical description if he made what seemed at the time to be a limited and ad hoc assumption: that, for any frequency of light, there is a fundamental unit of energy. As a result, the energy radiated by the black body at any frequency must be an integer multiple of that fundamental "quantum."

Around the same time, the classical picture also failed to explain the nature of another phenomenon, the photoelectric effect, in which electric current is produced when light strikes metals. In 1905 Albert Einstein employed Planck's idea to account for the mystery. But Einstein's explanation had importance that reached far beyond the photoelectric effect. By employing the quantum to explain a phenomenon unrelated to blackbody radiation, Einstein had shown that Planck's idea had fundamental significance, and was not just a mysterious property of blackbody radiation. Quantum physics was born.

In the ensuing two decades, experimental observations revealed new mysteries, and the quantum always seemed to be the idea needed to solve the puzzle. Ernest Rutherford and Hans Geiger, for example, conducted experiments that seemed to show that the protons in an atom are clumped together at its center, the nucleus, while the electrons orbit around them. But according to classical theory, charged particles travelling in that manner should radiate away their energy, and spiral inward. Why then, are atoms stable?

Niels Bohr employed the quantum idea to explain this. He proposed that the radius of electron orbits, like energy, is quantized. That would mean that electrons can only be at certain discrete allowed distances from the nucleus, and therefore cannot spiral inward. In Bohr's model, when an electron jumps from one allowed orbit to another, it emits or absorbs energy. In this way, he explained the atomic spectrum of hydrogen.

The idea of quantized distances and energy levels in an atom was another indication of the universality of the quantum principle, but quantum theory didn't become a fully developed theory until Werner Heisenberg and Erwin Schrodinger developed their equations in 1926, describing how a quantum system will evolve over time, and under the influence of any force. A few years later, Paul Dirac showed how to modify that theory to include special relativity. Dirac's theory required the existence of a new kind of matter, anti-matter. Quantum theory had predicted the existence of the positron, which was discovered in experiments performed soon thereafter.

The success of quantum theory, and its interpretation, raised many philosophical issues because quantum theory is non-deterministic, meaning that when a system starts in a given state, the results of measurements on its future state cannot in general be precisely predicted. One can calculate the probability of obtaining various results, but, if repeated, experiments that all begin with the same initial state can produce different results. The development of quantum theory meant the end of the idea that science could in principle predict all future events given enough information about the system at present. That bothered many physicists, such as Einstein and Schrodinger, who raised arguments against quantum theory, but their specific objections were eventually shown to be invalid.

Today, thanks to Richard Feynman, we know that quantum theory means that a physical system doesn't have a single history, but rather has many histories, each associated with a different probability. That picture was used to create a theory of quantum electrodynamics, which explains how quantum particles interact with electromagnetic fields, and how they emit and absorb radiation. The predictions of quantum electrodynamics match experimental observations to a degree of accuracy unparalleled in the rest of science.

As this volume traces all these developments we are reminded of Bertrand Russell's words, "We all start from 'naive realism,' i.e., the doctrine that things are what they seem. We think that grass is green, that stones are hard, and that snow is cold. But physics assures us that the greenness of grass, the hardness of stones, and the coldness of snow are not the greenness, hardness, and coldness that we know in our experience, but something very different..." It is these dreams that stuff is made of.

# Chapter One

The question of the nature of light has been a central issue for much of the history of physics. Isaac Newton theorized that light was particle-like—a beam of light was a stream of little particles in much the same way that a stream of water is composed of tiny water molecules. Because of his reputation as one of the great founders of physics, his theory of light was widely accepted. However, in 1801 Thomas Young definitively showed that Newton's particle theory could not be the complete description of light. He demonstrated that light incident on two closely-spaced narrow openings produces an interference pattern on a distant screen. Interference is a wave phenomenon and cannot be explained by a particle theory of light. Another major blow to the particle theory of light came in the 1860s when James Clerk Maxwell united the theories of electricity and magnetism and showed that light is an electromagnetic wave. Thus the wave theory of light was on very good experimental and theoretical grounds.

However, in the early twentieth century, the groundbreaking explanations of two troubling observations changed our understanding of light and began the quantum revolution, in which we find that light and matter have *both* wave- and particle-like properties. The first of these was the explanation of the shape of the black-body radiation spectrum.

We have all seen hot things glow—the red glow from the dying embers of a fire or the heating coils on a stove, the light produced by the tungsten filament in a normal incandescent light bulb, even the brilliant white light emitted from the surface of the sun are all examples of the same phenomenon. We call the light produced by glowing hot objects, *black-body radiation*. We experience it every day and in many different ways. It seems very ordinary, so it is surprising that to understand black-body radiation required a break from classical physics and opened the door to the revolution of quantum mechanics. But this is precisely what happened.

Black-body radiation was found to be only dependent on the temperature of the object. Hotter objects radiate more energy, and the peak of the emission spectrum is toward higher frequencies of light. As an example, consider heating a metal rod. At first it does not seem to glow at all. Of course, it is still emitting, but the radiation is primarily produced in the infrared region of the electromagnetic spectrum, which our eyes cannot see. As it is heated, it begins to glow a dull red as its emission spectrum moves into the visible range. With further heating it becomes bright red, then orange, then yellow, and so on as the peak of its spectrum passes through the visible portion of the electromagnetic spectrum.

The spectrum produced by black-body radiation can be measured, and this was done by several researchers in the latter part of the nineteenth century. However, no physical theory at that point was able to correctly predict how the spectrum would change as the temperature varied. Wilhelm Wien found an empirical relationship that described the spectrum at high frequencies. However, he was not able to derive this relationship from previously discovered physical laws, so it was not grounded well conceptually. In other words, the Wien law worked, but no one knew why. Even worse, in the late 1890s observations of the black-body spectrum were made at low frequencies which completely disagreed with the predictions of the Wien law.

In "On the Law of Distribution of Energy in the Normal Spectrum," Max Planck was able to resolve this discrepancy and derive a mathematical expression that described the spectrum of black-body radiation correctly at all frequencies. To do this, Planck had to make what has turned out to be a revolutionary assumption. He assumed that black-body radiation was produced by a large number of microscopic oscillators and that the total thermal energy of the black-body was not distributed continuously among these, but rather in finite and discrete portions. In other words, the energy was "quantized" in that it was an integer multiple of some small unit of energy. Planck showed this small energy element is proportional to the frequency of the oscillator. The constant of proportionality, which he labeled *h*, is known as Planck's constant—it is a fundamental parameter of quantum mechanics. Its value dictates the scale level at where classical physics fails and a theory of quantum physics is needed.

The second troubling phenomenon that could not be explained by the classical wave theory of light is known as the photoelectric effect. In the early 1900s it had been noticed that when light impacted on metals, an electrical current could be produced. Today this is a well-known and well-used concept. It is, in fact, part of how solar cells produce electricity from sunlight. However, at that time the photoelectric effect was something of a mystery. At first glance, the wave theory of light provides a simple explanation. Light waves impacting on the metal give energy to electrons on the metal's surface, which removes them from the atoms to which they are bound. They are free to move and so can produce an electrical current. According to wave theory, the more intense the light, the more energy the electrons will have. But this is not what was observed. In 1902 Phillip Lenard observed that the energy of the freed electrons is independent of the light's intensity. A more intense light produces more electrons, but an individual electron's energy is not affected by the intensity of light—rather by its color. This was an odd result, because the color (or *frequency*) of light waves should have nothing to do with the waves' energy. Even stranger, there was a cutoff frequency below which no electrons were freed no matter the light's intensity. These strange details required an explanation beyond the wave theory of light.

In "On a Heuristic Viewpoint Concerning the Production and Transformation of Light," Albert Einstein was able to explain the photoelectric effect using Planck's quantization principle. He theorized that the energy in a light ray was not continuously distributed but consisted of a finite number of "energy quanta" that could not be divided. Monochromatic light, then, consists of a large but finite number of particles of light, each with an energy given by the product of Planck's constant and the frequency of the light. The particle of light is now called the *photon*. An individual photon can only be absorbed as a complete unit. When an electron absorbs the energy from a photon, it is the photon's energy, not the overall intensity, that determines how much energy the electron gains. In the photoelectric effect it requires a certain amount of energy to remove an electron from the metal. If the frequency of light is too small, then no photon will have enough energy to remove the electron regardless of how many photons are present (i.e., the intensity of the light). Thus with the assumption of energy quantization Einstein could explain the strange details of the photoelectric effect.

With the publication of Planck's explanation of black-body radiation and Einstein's explanation of the photoelectric effect, the theory of quantum physics was born. Given the revolutionary nature of what was introduced we can ask, "Where did Planck get the inspiration to make this assumption?" In a 1909 lecture at Columbia University entitled "The Atomic Theory of Matter," Plank revealed how his work in statistical mechanics led him to make his brilliant assumption of energy quantization. However, it seems likely that neither Planck nor Einstein truly understood the extent to which physics would be changed as a result of their work, for both resisted some of the implications that accompanied quantum physics once it was fully formed. We shall see in subsequent chapters that our understanding of the universe and reality at a fundamental level was radically altered as a result of the quantum revolution.

# On the Law of Distribution of Energy in the Normal Spectrum

By

Max Planck

*First published in Annalen der Physik,* vol. 4, p. 553 ff (1901)

The recent spectral measurements made by O. Lummer and E. Pringsheim[*], and even more notable those by H. Rubens and F. Kurlbaum[†], which together confirmed an earlier result obtained by H. Beckmann[‡], show that the law of energy distribution in the normal spectrum, first derived by W. Wien from molecular-kinetic considerations and later by me from the theory of electromagnetic radiation, is not valid generally.

In any case the theory requires a correction, and I shall attempt in the following to accomplish this on the basis of the theory of electromagnetic radiation which I developed. For this purpose it will be necessary first to find in the set of conditions leading to Wien's energy distribution law that term which can be changed; thereafter it will be a matter of removing this term from the set and making an appropriate substitution for it.

In my last article[§] I showed that the physical foundations of the electromagnetic radiation theory, including the hypothesis of "natural radiation," withstand the most severe criticism; and since to my knowledge there are no errors in the calculations, the principle persists that the law of energy distribution in the normal spectrum is completely determined when one succeeds in calculating the entropy S of an irradiated, monochromatic, vibrating resonator as a function of its vibrational energy U. Since one then obtains, from the relationship

[*] O. Lummer and E. Pringsheim, *Transactions of the German Physical Society* 2 (1900), p. 163

[†] H. Rubens and F. Kurlbaum, *Proceedings of the Imperial Academy of Science*, Berlin, October 25, 1900, p. 929.

[‡] H. Beckmann, *Inaugural dissertation*, Tübingen 1898. See also H. Rubens, *Weid. Ann.* 69 (1899) p. 582.

[§] M. Planck, *Ann. d. Phys.* 1 (1900), p. 719.

$dS/dU = 1/\theta$, the dependence of the energy U on the temperature $\theta$, and since the energy is also related to the density of radiation at the corresponding frequency by a simple relation*, one also obtains the dependence of this density of radiation on the temperature. The normal energy distribution is then the one in which the radiation densities of all different frequencies have the same temperature.

Consequently, the entire problem is reduced to determining S as a function of U, and it is to this task that the most essential part of the following analysis is devoted. In my first treatment of this subject I had expressed S, by definition, as a simple function of U without further foundation, and I was satisfied to show that this from of entropy meets all the requirements imposed on it by thermodynamics. At that time I believed that this was the only possible expression and that consequently Wein's law, which follows from it, necessarily had general validity. In a later, closer analysis,[†] however, it appeared to me that there must be other expressions which yield the same result, and that in any case one needs another condition in order to be able to calculate S uniquely. I believed I had found such a condition in the principle, which at the time seemed to me perfectly plausible, that in an infinitely small irreversible change in a system, near thermal equilibrium, of N identical resonators in the same stationary radiation field, the increase in the total entropy $S_N = NS$ with which it is associated depends only on its total energy $U_N = NU$ and the changes in this quantity, but not on the energy U of individual resonators. This theorem leads again to Wien's energy distribution law. But since the latter is not confirmed by experience one is forced to conclude that even this principle cannot be generally valid and thus must be eliminated from the theory.[‡]

Thus another condition must now be introduced which will allow the calculation of S, and to accomplish this it is necessary to look more deeply into the meaning of the concept of entropy. Consideration of the untenability of the hypothesis made formerly will help to orient

---

*Compare with equation (8).

[†]M. Planck, *loc. cit.*, pp. 730 ff.

[‡]Moreover one should compare the critiques previously made of this theorem by W. Wien (*Report of the Paris Congress* 2, 1900, p. 40) and by O. Lummer (*loc. cit.*, 1900, p. 92.).

our thoughts in the direction indicated by the above discussion. In the following a method will be described which yields a new, simpler expression for entropy and thus provides also a new radiation equation which does not seem to conflict with any facts so far determined.

## I. CALCULATIONS OF THE ENTROPY OF A RESONATOR AS A FUNCTION OF ITS ENERGY

§ 1. Entropy depends on disorder and this disorder, according to the electromagnetic theory of radiation for the monochromatic vibrations of a resonator when situated in a permanent stationary radiation field, depends on the irregularity with which it constantly changes its amplitude and phase, provided one considers time intervals large compared to the time of one vibration but small compared to the duration of a measurement. If amplitude and phase both remained absolutely constant, which means completely homogeneous vibrations, no entropy could exist and the vibrational energy would have to be completely free to be converted into work. The constant energy U of a single stationary vibrating resonator accordingly is to be taken as time average, or what is the same thing, as a simultaneous average of the energies of a large number N of identical resonators, situated in the same stationary radiation field, and which are sufficiently separated so as not to influence each other directly. It is in this sense that we shall refer to the average energy U of a single resonator. Then to the total energy

$$U_N = NU \tag{1}$$

of such a system of N resonators there corresponds a certain total entropy

$$S_N = NS \tag{2}$$

of the same system, where S represents the average entropy of a single resonator and the entropy $S_N$ depends on the disorder with which the total energy $U_N$ is distributed among the individual resonators.

§ 2. We now set the entropy $S_N$ of the system proportional to the logarithm of its probability W, within an arbitrary additive constant, so that the N resonators together have the energy $E_N$:

$$S_N = k \log W + \text{constant} \qquad (3)$$

In my opinion this actually serves as a definition of the probability W, since in the basic assumptions of electromagnetic theory there is no definite evidence for such a probability. The suitability of this expression is evident from the outset, in view of its simplicity and close connection with a theorem from kinetic gas theory.*

§ 3. It is now a matter of finding the probability W so that the N resonators together possess the vibrational energy $U_N$. Moreover, it is necessary to interpret $U_N$ not as a continuous, infinitely divisible quantity, but as a discrete quantity composed of an integral number of finite equal parts. Let us call each such part the energy element $\varepsilon$; consequently we must set

$$U_N = P\varepsilon \qquad (4)$$

where P represents a large integer generally, while the value of $\epsilon$ is yet uncertain.

The above paragraph in the original German

Now it is evident that any distribution of the P energy elements among the N resonators can result only in a finite, integral, definite number. Every such form of distribution we call, after an expression used by L. Boltzmann for a similar idea, a "complex." If one denotes the resonators by the numbers 1, 2, 3, ... N, and writes these side by side, and if one sets under each resonator the number of energy elements assigned to it by some arbitrary distribution, then one obtains for every complex a pattern of the following form:

| 1 | 2 | 3 | 4 | 5 | 6 | 7 | 8 | 9 | 10 |
|---|---|---|---|---|---|---|---|---|---|
| 7 | 38 | 11 | 0 | 9 | 2 | 20 | 4 | 4 | 5 |

Here we assume N = 10, P = 100. The number R of all possible complexes is obviously equal to the number of arrangements that one can obtain in this fashion for the lower row, for a given N and P. For the sake of clarity we should note that two complexes must be

*L. Boltzmann, *Proceedings of the Imperial Academy of Science*, Vienna, (II) 76 (1877), p. 428.

considered different if the corresponding number patters contain the same numbers but in a different order.

From combination theory one obtains the number of all possible complexes as:

$$R = \frac{N(N+1)(N+2)\cdots\cdot(N+P-1)}{1\cdot 2\cdot 3\cdots\cdot P} = \frac{(N+P-1)!}{(N-1)!P!}$$

Now according to Stirling's theorem, we have in the first approximation:

$$N! = N^N$$

Consequently, the corresponding approximation is:

$$R = \frac{(N+P)^{N+P}}{N^N \cdot P^P}$$

§ 4. The hypothesis which we want to establish as the basis for further calculation proceeds as follows: in order for the N resonators to possess collectively the vibrational energy $U_N$, the probability W must be proportional to the number R of all possible complexes formed by distribution of the energy $U_N$ among the N resonators; or in other words, any given complex is just as probable as any other. Whether this actually occurs in nature one can, in the last analysis, prove only by experience. But should experience finally decide in its favor it will be possible to draw further conclusions from the validity of this hypothesis about the particular nature of resonator vibrations; namely in the interpretation put forth by J. V. Kries* regarding the character of the "original amplitudes, comparable in magnitude but independent of each other." As the matter now stands, further development along these lines would appear to be premature.

§ 5. According to the hypothesis introduced in connection with equation (3), the entropy of the system of resonators under consideration is, after suitable determination of the additive constant:

$$S_N = k \log R$$
$$= k\left\{(N+P)\log(N+P) - N\log N - P\log P\right\} \qquad (5)$$

---

*Joh. v. Kries, *The Principles of Probability Calculation* (Freiburg, 1886), p. 36.

and by considering (4) and (1):

$$S^N = kN \left\{ \left( 1 + \frac{U}{\epsilon} \right) \log \left( 1 + \frac{U}{\epsilon} \right) - \frac{U}{\epsilon} \log \frac{U}{\epsilon} \right\}$$

Thus, according to equation (2) the entropy S of a resonator as a function of its energy U is given by:

$$S = k \left\{ \left( 1 + \frac{U}{\epsilon} \right) \log \left( 1 + \frac{U}{\epsilon} \right) - \frac{U}{\epsilon} \log \frac{U}{\epsilon} \right\} \tag{6}$$

## II. INTRODUCTION OF WIEN'S DISPLACEMENT LAW

§ 6. Next to Kirchoff's theorem of the proportionality of emissive and absorptive power, the so-called displacement law, discovered by and named after W. Wien,[*] which includes as a special case the Stefan-Boltzmann law of dependence of total radiation on temperature, provides the most valuable contribution to the firmly established foundation of the theory of heat radiation, In the form given by M. Thiesen[†] it reads as follows:

$$E \cdot d\lambda = \theta^5 \psi (\lambda\theta) \cdot d\lambda$$

where $\lambda$ is the wavelength, E $d\lambda$ represents the volume density of the "black-body" radiation[‡] within the spectral region $\lambda$ to $\lambda + d\lambda$, $\theta$ represents temperature and $\psi(x)$ represents a certain function of the argument x only.

§ 7. We now want to examine what Wien's displacement law states about the dependence of the entropy S of our resonator on its energy U and its characteristic period, particularly in the general case where the resonator is situated in an arbitrary diathermic medium. For this purpose we next generalize Thiesen's form of the law for the radiation in an arbitrary diathermic medium with the velocity of light $c$. Since we do not have to consider the total radiation, but only the monochromatic radiation, it becomes necessary in order to compare different diathermic media to introduce the frequency $\nu$ instead of the wavelength $\lambda$.

---

[*] W. Wien, *Proceedings of the Imperial Academy of Science*, Berlin, February 9, 1893, p. 55.
[†] M. Thiesen, *Transactions of the German Physical Society* 2 (1900), p. 66.
[‡] Perhaps one should speak more appropriately of a "white" radiation, to generalize what one already understands by total white light.

Thus, let us denote by u dν the volume density of the radiation energy belonging to the spectral region ν to ν + dν; then we write: u dν instead of E dλ; c/ν instead of λ, and cdν/ν² instead of dλ. From which we obtain

$$u = \theta^5 \frac{c}{\nu^2} \cdot \psi \left( \frac{c\theta}{\nu} \right)$$

Now according to the well-known Kirchoff-Clausius law, the energy emitted per unit time at the frequency ν and temperature θ from a black surface in a diathermic medium is inversely proportional to the square of the velocity of propagation $c^2$; hence the energy density U is inversely proportional to $c^3$ and we have:

$$u = \frac{\theta^5}{\nu^2 c^3} \cdot f \left( \frac{\theta}{\nu} \right)$$

where the constants associated with the function $f$ are independent of $c$.

In place of this, if $f$ represents a new function of a single argument, we can write:

$$u = \frac{\nu^3}{c^3} \cdot f \left( \frac{\theta}{\nu} \right) \tag{7}$$

and from this we see, among other things, that as is well known, the radiant energy $u \cdot \lambda^3$ at a given temperature and frequency is the same for all diathermic media.

§ 8. In order to go from the energy density $u$ to the energy U of a stationary resonator situated in the radiation field and vibrating with the same frequency ν, we use the relation expressed in equation (34) of my paper on irreversible radiation processes*:

$$K = \frac{\nu^2}{c^2} U$$

(K is the intensity of a monochromatic linearly, polarized ray), which together with the well-known equation:

$$u = \frac{8\pi K}{c}$$

*M. Planck, *Ann. D. Phys.* 1 (1900), p. 99.

yields the relation:

$$u = \frac{8\pi v^2}{c^3} U \qquad (8)$$

From this and from equation (7) follows:

$$U = v \cdot f\left(\frac{\theta}{v}\right)$$

where now $c$ does not appear at all. In place of this we may also write:

$$\theta = v \cdot f\left(\frac{U}{v}\right)$$

§ 9. Finally, we introduce the entropy S of the resonator by setting

$$\frac{1}{\theta} = \frac{dS}{dU} \qquad (9)$$

We then obtain:

$$\frac{dS}{dU} = \frac{1}{v} \cdot f\left(\frac{U}{v}\right)$$

and integrated:

$$S = f\left(\frac{U}{v}\right) \qquad (10)$$

that is, the entropy of a resonator vibrating in an arbitrary diathermic medium depends only on the variable U/$v$, containing besides this only universal constants. This is the simplest form of Wien's displacement law known to me.

§ 10. If we apply Wien's displacement law in the latter form to equation (6) for the entropy S, we then find that the energy element $\varepsilon$ must be proportional to the frequency $v$, thus:

$$\varepsilon = hv$$

and consequently:

$$S = k\left\{\left(1 + \frac{U}{hv}\right) \log\left(1 + \frac{U}{hv}\right) - \frac{U}{hv} \log \frac{U}{hv}\right\}$$

here h and k are universal constants.

By substitution into equation (9) one obtains:

$$\frac{1}{\theta} = \frac{k}{h\nu} \log\left(1 + \frac{h\nu}{U}\right)$$

$$U = \frac{h\nu}{e^{h\nu/k\theta} - 1} \tag{11}$$

and from equation (8) there then follows the energy distribution law sought for:

$$u = \frac{8\pi h\nu^3}{c^3} \cdot \frac{1}{e^{h\nu/k\theta} - 1} \tag{12}$$

or by introducing the substitutions given in § 7, in terms of wavelength $\lambda$ instead of the frequency:

$$E = \frac{8\pi ch}{\lambda^5} \cdot \frac{1}{e^{ch/k\lambda\theta} - 1} \tag{13}$$

I plan to derive elsewhere the expressions for the intensity and entropy of radiation progressing in a diathermic medium, as well as the theorem for the increase of total entropy in nonstationary radiation processes.

## III. Numerical Values

§ 11. The values of both universal constants h and k may be calculated rather precisely with the aid of available measurements. F. Kurlbaum*, designating the total energy radiating into air from 1 sq cm of a black body at temperature $\iota°$C in 1 sec by $S_\iota$, found that:

$$S_{100} - S_0 = 0.0731 \text{ watt/cm}^2 = 7.31 \times 10^5 \text{erg/cm}^2 \cdot \text{sec}$$

From this one can obtain the energy density of the total radiation energy in air at the absolute temperature 1:

$$\frac{4 \cdot 7.31 \cdot 10^5}{3 \cdot 10^{10}\left(373^4 - 273^4\right)} = 7.061 \cdot 10^{-15} \text{erg/cm}^3 \cdot \text{deg}^4$$

---

*F. Kurlbaum, *Wied. Ann.* 65 (1898), p. 759.

On the other hand, according to equation (12) the energy density of the total radiant energy for $\theta = 1$ is:

$$u^* = \int_0^\infty u\,dv = \frac{8\pi h}{c^3} \int_0^\infty \frac{v^3\,dv}{e^{hvlk} - 1}$$

$$= \frac{8\pi h}{c^3} \int_0^\infty v^3 (e^{-hv/k} + e^{-2hv/k} + e^{-3hv} + \cdots)\,dv$$

and by termwise integration:

$$u^* = \frac{8\pi h}{c^3} \cdot 6 \left(\frac{k}{h}\right)^4 \left(1 + \frac{1}{24} + \frac{1}{34} + \frac{1}{44} + \cdots\right)$$

$$= \frac{48\pi k^4}{c^3 h^3} \cdot 1.0823$$

If we set this equal to $7.061 \cdot 10^{-15}$, then, since $c = 3 \cdot 10^{10}$ cm/sec, we obtain:

$$\frac{k^4}{h^3} = 1.1682 \cdot 10^{15} \tag{14}$$

§ 12. O. Lummer and E. Pringswim* determined the product $\lambda_m \theta$, where $\lambda_m$ is the wavelength of maximum energy in air at temperature 0, to be 2940 micron· degree. Thus, in absolute measure:

$$\lambda_m = 0.294\,\text{cm} \cdot \text{deg}$$

On the other hand, it follows from equation (13), when one sets the derivative of E with respect to $\theta$ equal to zero, thereby finding $\lambda = \lambda_m$

$$\left(1 - \frac{ch}{5k\lambda_m\theta}\right) \cdot e^{ch/k\lambda_m\theta} = 1$$

and from this transcendental equation:

$$\lambda_m \theta = ch/4.9651k$$

consequently:

$$h/k = (4.9561 \cdot 0.294)/3 \cdot 10^{10} = 4.866 \cdot 10^{-11}$$

---

*O. Lummer and Pringsheim, *Transactions of the German Physical Society* 2 (1900), p. 176.

From this and from equation (14) the values for the universal constants become:

$$h = 6.55 \cdot 10^{-27} \mathrm{erg} \cdot \sec \qquad (15)$$

$$k = 1.346 \cdot 10^{-16} \mathrm{erg/deg} \qquad (16)$$

These are the same number that I indicated in my earlier communication.

# On a Heuristic Viewpoint Concerning the Production and Transformation of Light

By

Albert Einstein

*First presented in Bern, Switzerland, March 17, 1905.*

A profound formal distinction exists between the theoretical concepts which physicists have formed regarding gases and other ponderable bodies and the Maxwellian theory of electromagnetic processes in so–called empty space. While we consider the state of a body to be completely determined by the positions and velocities of a very large, yet finite, number of atoms and electrons, we make use of continuous spatial functions to describe the electromagnetic state of a given volume, and a finite number of parameters cannot be regarded as sufficient for the complete determination of such a state. According to the Maxwellian theory, energy is to be considered a continuous spatial function in the case of all purely electromagnetic phenomena including light, while the energy of a ponderable object should, according to the present conceptions of physicists, be represented as a sum carried over the atoms and electrons. The energy of a ponderable body cannot be subdivided into arbitrarily many or arbitrarily small parts, while the energy of a beam of light from a point source (according to the Maxwellian theory of light or, more generally, according to any wave theory) is continuously spread an ever increasing volume.

The wave theory of light, which operates with continuous spatial functions, has worked well in the representation of purely optical phenomena and will probably never be replaced by another theory.

Reprinted with permission from *American Journal of Physics,* 33, 367–374 (1965).
© 1965, American Association of Physics Teachers.

It should be kept in mind, however, that the optical observations refer to time averages rather than instantaneous values. In spite of the complete experimental confirmation of the theory as applied to diffraction, reflection, refraction, dispersion, etc., it is still conceivable that the theory of light which operates with continuous spatial functions may lead to contradictions with experience when it is applied to the phenomena of emission and transformation of light.

It seems to me that the observations associated with blackbody radiation, fluorescence, the production of cathode rays by ultraviolet light, and other related phenomena connected with the emission or transformation of light are more readily understood if one assumes that the energy of light is discontinuously distributed in space. In accordance with the assumption to be considered here, the energy of a light ray spreading out from a point source is not continuously distributed over an increasing space but consists of a finite number of energy quanta which are localized at points in space, which move without dividing, and which can only be produced and absorbed as complete units.

In the following I wish to present the line of thought and the facts which have led me to this point of view, hoping that this approach may be useful to some investigators in their research.

## 1. CONCERNING A DIFFICULTY WITH REGARD TO THE THEORY OF BLACKBODY RADIATION

We start first with the point of view taken in the Maxwellian and the electron theories and consider the following case. In a space enclosed by completely reflecting walls, let there be a number of gas molecules and electrons which are free to move and which exert conservative forces on each other on close approach: i.e. they can collide with each other like molecules in the kinetic theory of gases.* Furthermore, let there be a number of electrons which are bound to widely separated

---

*This assumption is equivalent to the supposition that the average kinetic energies of gas molecules and electrons are equal to each other at thermal equilibrium. It is well known that, with the help of this assumption, Herr Drude derived a theoretical expression for the ratio of thermal and electrical conductivities of metals.

points by forces proportional to their distances from these points. The bound electrons are also to participate in conservative interactions with the free molecules and electrons when the latter come very close. We call the bound electrons "oscillators": they emit and absorb electromagnetic waves of definite periods.

According to the present view regarding the origin of light, the radiation in the space we are considering (radiation which is found for the case of dynamic equilibrium in accordance with the Maxwellian theory) must be identical with the blackbody radiation—at least if oscillators of all the relevant frequencies are considered to be present.

For the time being, we disregard the radiation emitted and absorbed by the oscillators and inquire into the condition of dynamical equilibrium associated with the interaction (or collision) of molecules and electrons. The kinetic theory of gases asserts that the average kinetic energy of an oscillator electron must be equal to the average kinetic energy of a translating gas molecule. If we separate the motion of an oscillator electron into three components at angles to each other, we find for the average energy $\overline{E}$ of one of these linear components the expression

$$\overline{E} = (R/N)\ T,$$

where $R$ denotes the universal gas constant. $N$ denotes the number of "real molecules" in a gram equivalent, and $T$ the absolute temperature. The energy $\overline{E}$ is equal to two-thirds the kinetic energy of a free monatomic gas particle because of the equality the time average values of the kinetic and potential energies of the oscillator. If through any cause—in our case through radiation processes—it should occur that the energy of an oscillator takes on a time-average value greater or less than $\overline{E}$, then the collisions with the free electrons and molecules would lead to a gain or loss of energy by the gas, different on the average from zero. Therefore, in the case we are considering, dynamic equilibrium is possible only when each oscillator has the average energy $\overline{E}$.

We shall now proceed to present a similar argument regarding the interaction between the oscillators and the radiation present in the cavity. Herr Planck has derived* the condition for the dynamics equilibrium in this case under the supposition that the radiation can be considered a completely random process.† He found

$$(\overline{E_\nu}) = \left(L^3/8\pi\nu^2\right)\rho_\nu,$$

where $(\overline{E_\nu})$ is the average energy (per degree of freedom) of an oscillator with eigenfrequency $\nu$, $L$ the velocity of light, $\nu$ the frequency, and $\rho\nu d\nu$ the energy per unit volume of that portion of the radiation with frequency between $\nu$ and $\nu + d\nu$.

If the radiation energy of frequency $\nu$ is not continually increasing or decreasing, the following relations must obtain:

$$(R/N)\,T = \overline{E} = \overline{E}\nu = \left(L^3/8\pi\nu^2\right)\rho_\nu,$$
$$\rho_\nu = (R/N)\left(8\pi\nu^2/L^3\right)T.$$

These relations, found to be the conditions of dynamic equilibrium, not only fail to coincide with experiment, but also state that in our model there can be not talk of a definite energy distribution between ether and matter. The wider the range of wave numbers of the oscillators, the greater will be the radiation energy of the space, and in the limit we obtain

$$\int_0^\infty \rho_\nu\,d\nu = \frac{R}{N}\cdot\frac{8\pi}{L^3}\cdot T\int_0^\infty \nu^2 d\nu = \infty.$$

---

*M. Planck, Ann. Phys. 1, 99 (1900).

†This problem can be formulated in the following manner. We expand the $Z$ component of the electrical force ($Z$) at an arbitrary point during the time interval between $t = 0$ and $t = T$ in a Fourier series in which $A_\nu \geq 0$ and $0 \leq \alpha_\nu \leq 2\pi$: the time $T$ is taken to be very large relative to all the periods of oscillation that are present:

$$Z = \sum_{\nu=1}^{\nu=\infty} A_\nu \sin\left(2\pi\nu\frac{t}{T}+\alpha_\nu\right),$$

If one imagines making this expansion arbitrary often at a given point in space at randomly chosen instants of time, one will obtain various sets of values of $A_\nu$ and $\alpha_\nu$. There then exist for the frequency of occurrence of different sets of values of $A_\nu$ and $\alpha_\nu$ (statistical) probabilities $dW$ of the form:

$$dW = f(a_1, A_2, \ldots, \alpha_1, \alpha_2, \ldots)\,dA_1\,dA_2\ldots d\alpha_1\,d\alpha_2\ldots,$$

The radiation is then as disordered as conceivable if

$$f(A_1, A_2, \ldots, \alpha_1, \alpha_2, \ldots) = F_1(A_1)\,F_2(A_2)\ldots f_1(\alpha_1)\,f_2(\alpha_2)\ldots,$$

i.e., if the probability of a particular value of $A$ or $\alpha$ is independent of other values of $A$ or $\alpha$. The more closely this condition is fulfilled (namely, that the individual pairs of values of $A_\nu$ and $\alpha_\nu$ are dependent upon the emission and absorption processes of specific groups of oscillators) the more closely will radiation in the case being considered approximate a perfectly random state.

## 2. Concerning Planck's Determination
## of the Fundamental Constants

We wish to show in the following that Herr Planck's determination of the fundamental constants is, to a certain extent, independent of his theory of blackbody radiation.

Planck's formula,* which has proved adequate up to this point, gives for $\rho \nu$

$$\rho_\nu = \frac{\alpha \nu^3}{e^{\beta \nu / T} - 1},$$
$$\alpha = 6.10 \times 10^{-56},$$
$$\beta = 4.866 \times 10^{-11}.$$

For large values of $T/\nu$; i.e. for large wavelengths and radiation densities, this equation takes the form

$$\rho_\nu = (\alpha/\beta)\, \nu^2\, T.$$

It is evident that this equation is identical with the one obtained in Sec. 1 from the Maxwellian and electron theories. By equating the coefficients of both formulas one obtains

$$(R/N)\left(8\pi/L^3\right) = (\alpha/\beta)$$

or

$$N = (\beta/\alpha)\left(8\pi R/L^3\right) = 6.17 \times 10^{23}.$$

i.e., an atom of hydrogen weighs $1/N$ grams $= 1.62 \times 10^{-24}$ g. This is exactly the value found by Herr Planck, which in turn agrees with values found by other methods.

We therefore arrive at the conclusion: the greater the energy density and the wavelength of a radiation, the more useful do the theoretical principles we have employed turn out to be: for small wavelengths and small radiation densities, however, these principles fail us completely.

*M. Planck, Ann. Phys. 4, 561 (1901).

In the following we shall consider the experimental facts concerning blackbody radiation without invoking a model for the emission and propagation of the radiation itself.

## 3. CONCERNING THE ENTROPY OF RADIATION

The following treatment is to be found in a famous work by Herr W. Wien and is introduced here only for the sake of completeness.

Suppose we have radiation occupying a volume $v$. We assume that the observable properties of the radiation are completely determined when the radiation density $\rho(\nu)$ is given for all frequencies.* Since radiation of different frequencies are to be considered independent of each other when there is no transfer of heat or work, the entropy of the radiation can be represented by

$$S = v \int_0^\infty \varphi(\rho, \nu)\, d\nu,$$

where $\varphi$ is a function of the variables $\rho$ and $\nu$.

$\varphi$ can be reduced to a function of a single variable through formulation of the condition that the entropy of the radiation is unaltered during adiabatic compression between reflecting walls. We shall not enter into this problem, however, but shall directly investigate the derivation of the function $\varphi$ from the blackbody radiation law.

In the case of blackbody radiation, $\rho$ is such a function of $\nu$ that the entropy is maximum for a fixed value of energy; i.e.,

$$\delta \int_0^\infty \varphi(\rho, \nu)\, d\nu = 0,$$

providing

$$\delta \int_0^\infty \rho\, d\nu = 0.$$

*This assumption is an arbitrary one. One will naturally cling to this simplest assumption as long as it is not controverted experiment.

From this it follows that for every choice of $\delta\rho$ as a function of $\nu$

$$\int_0^\infty \left(\frac{\partial\varphi}{\partial\rho} - \lambda\right) \delta\rho\,d\nu = 0,$$

where $\lambda$ is independent of $\nu$. In the case of blackbody radiation, therefore, $\partial\varphi/\partial\rho$ is independent of $\nu$.

The following equation applies when the temperature of a unit volume of blackbody radiation increases by $dT$

$$dS = \int_{\nu=0}^{\nu=\infty} \left(\frac{\partial\varphi}{\partial\rho}\right) d\rho\,d\nu,$$

or, since $\partial\varphi/\partial\rho$ is independent of $\nu$.

$$dS = (\partial\varphi/\partial\rho)\,dE.$$

Since $dE$ is equal to the heat added and since the process is reversible, the following statement also applies

$$dS = (1/T)\,dE.$$

By comparison one obtains

$$\partial\varphi/\partial\rho = 1/T.$$

This is the law of blackbody radiation. Therefore one can derive the law of blackbody radiation from the function $\varphi$, and, inversely, one can derive the function $\varphi$ by integration, keeping in mind the fact that $\varphi$ vanishes when $\rho = 0$.

## 4. ASYMPTOTIC FROM FOR THE ENTROPY OF MONOCHROMATIC RADIATION AT LOW RADIATION DENSITY

From existing observations of the blackbody radiation, it is clear that the law originally postulated by Herr W. Wien,

$$\rho = \alpha\nu^3 e^{-\beta\nu/T},$$

is not exactly valid. It is, however, well confirmed experimentally for large values of $\nu/T$. We shall base our analysis on this formula, keeping in mind that our results are only valid within certain limits.

This formula gives immediately

$$(1/T) = -(1/\beta v)\ln\left(\rho/\alpha v^3\right)$$

and then, by using the relation obtained in the preceeding section,

$$\varphi(\rho,v) = -\frac{\rho}{\beta v}\left[\ln\left(\frac{\rho}{\alpha v^3}\right) - 1\right].$$

Suppose that we have radiation of energy $E$, with frequency between $v$ and $v + dv$, enclosed in volume $v$. The entropy of this radiation is:

$$S = v\varphi(\rho,v)dv = -\frac{E}{\beta v}\left[\ln\left(\frac{E}{v\alpha v^3 dv}\right) - 1\right].$$

If we confine ourselves to investigating the dependence of the entropy on the volume occupied by the radiation, and if we denote by $S_0$ the entropy of the radiation at volume $v_0$, we obtain

$$S - S_0 = (E/\beta v)\ln(v/v_0).$$

This equation shows that the entropy of a monochromatic radiation of sufficiently low density varies with the volume in the same manner as the entropy of an ideal gas or a dilute solution. In the following, this equation will be interpreted in accordance with the principle introduced into physics by Herr Boltzmann, namely that the entropy of a system is a function of the probability its state.

## 5. MOLECULAR—THEORETIC INVESTIGATION OF THE DEPENDENCE OF THE ENTROPY OF GASES AND DILUTE SOLUTIONS ON THE VOLUME

In the calculation of entropy by molecular–theoretic methods we frequently use the word "probability" in a sense differing from that employed in the calculus of probabilities. In particular "gases of equal probability" have frequently been hypothetically established when one theoretical models being utilized are definite enough to permit a deduction rather than a conjecture. I will show in a separate paper that the so-called "statistical probability" is fully adequate for the treatment of thermal phenomena, and I hope that by doing so I will eliminate a logical difficulty that obstructs the application of Boltzmann's Principle. Here, however, only a general formulation and application to very special cases will be given.

If it is reasonable to speak of the probability of the state of a system, and futhermore if every entropy increase can be understood as a transition to a state of higher probability, then the entropy $S_1$ of a system is a function of $W_1$, the probability of its instantaneous state. If we have two noninteracting systems $S_1$ and $S_2$, we can write

$$S_1 = \varphi_1(W_1),$$
$$S_2 = \varphi_2(W_2).$$

If one considers these two systems as a single system of entropy $S$ and probability $W$, it follows that

$$S = S_1 + S_2 = \varphi(W)$$

and

$$W = W_1 \cdot W_2.$$

The last equation says that the states of the two systems are independent of each other.

From these equation it follows that

$$\varphi(W_1 \cdot W_2) = \varphi_1(W_1) + \varphi_2(W_2)$$

and finally

$$\varphi_1(W_1) = C \ln(W_1) + \text{const},$$
$$\varphi_2(W_2) = C \ln(W_2) + \text{const},$$
$$\varphi(W) = C \ln(W) + \text{const}.$$

The quantity $C$ is therefore a universal constant; the kinetic theory of gases shows its value to be $R/N$, where the constants $R$ and $N$ have been defined above. If $S_0$ denotes the entropy of a system in some initial state and $W$ denotes the relative probability of a state of entropy $S$, we obtain in general

$$S - S_0 = (R/N)\ln W.$$

First we treat the following special case. We consider a number ($n$) of movable points (e.g., molecules) confined in a volume $v_0$. Besides these points, there can be in the space any number of other movable points of any kind. We shall not assume anything concerning the law in accordance with which the points move in this space except that with regard to this motion, no part of the space (and no direction within it) can be distinguished from any other. Further, we take the number of these movable points to be so small that we can disregard interactions between them.

This system, which, for example, can be an ideal gas or a dilute solution, possesses an entropy $S_0$. Let us imagine transferring all $n$ movable points into a volume $v$ (part of the volume $v_0$) without anything else being changed in the system. This state obviously possesses a different entropy ($S$), and now wish to evaluate the entropy difference with the help of the Boltzmann Principle.

We inquire: How large is the probability of the latter state relative to the original one? Or: How large is the probability that at a randomly chosen instant of time all $n$ movable points in the given volume $v_0$ will be found by chance in the volume $v$?

For this probability, which is a "statistical probability", one obviously obtains:

$$W = (v/v_0)^n \, ;$$

By applying the Boltzmann Principle, one then obtains

$$S - S_0 = R\,(n/N)\ln(v/v_0)\,.$$

It is noteworthy that in the derivation of this equation, from which one can easily obtain the law of Boyle and Gay–Lussac as well as the analogous law of osmotic pressure thermodynamically,[*] no assumption had to be made as to a law of motion of the molecules.

---

[*] If $E$ is the energy of the system, one obtains:

$$-d \cdot (E - TS) = pdv = TdS = RT \cdot (n/N) \cdot (dv/v)\,;$$

therefore

$$pv = R \cdot (n/N) \cdot T.$$

## 6. INTERPRETATION OF THE EXPRESSION FOR THE VOLUME DEPENDENCE OF THE ENTROPY OF MONOCHROMATIC RADIATION IN ACCORDANCE WITH BOLTZMANN'S PRINCIPLE

In Sec. 4, we found the following expression for the dependence of the entropy of monochromatic radiation on the volume

$$S - S_0 = (E/\beta v) \ln (v/v_0).$$

If one writes this in the from

$$S - S_0 = (R/N) \ln \left[(v/v_0)^{(N/R)(E/\beta v)}\right].$$

and if one compares this with the general formula for the Boltzmann principle

$$S - S_0 = (R/N) \ln W,$$

one arrives at the following conclusion:

If monochromatic radiation of frequency $v$ and energy $E$ is enclosed by reflecting walls in a volume $v_0$, the probability that the total radiation energy will be found in a volume $v$ (part of the volume $v_0$) at any randomly chosen instant is

$$W = (v/v_0)^{(N/R)(E/\beta v)}.$$

From this we further conclude that: Monochromatic radiation of low density (within the range of validity of Wien's radiation formula) behaves thermodynamically as though it consisted of a number of independent energy quanta of magnitude $R\beta v/N$.

We still wish to compare the average magnitude of the energy quanta of the blackbody radiation with the average translational kinetic energy of a molecule at the same temperature. The latter is $^3/_2(R/N)T$, while, according to the Wien formula, one obtains for the average magnitude of an energy quantum

$$\int_0^\infty \alpha v^3 e^{-\beta v/T} dv \bigg/ \int_0^\infty \frac{N}{R\beta v} \alpha v^3 e^{-\beta v/T} dv = 3(RT/N).$$

If the entropy of monochromatic radiation depends on volume as though the radiation were a discontinuous medium consisting of energy quanta of magnitude $R\beta v/N$, the next obvious step is to investigate whether the laws of emission and transformation of light are also of such a nature that they can be interpreted or explained by considering light to consist of such energy quanta. We shall examine this question in the following.

## 7. Concerning Stokes's Rule

According to the result just obtained, let us assume that, when monochromatic light is transformed through photoluminescence into light of a different frequency, both the incident and emitted light consist of energy quanta of magnitude $R\beta v/N$, where $v$ denotes the relevant frequency. The transformation process is to be interpreted in the following manner. Each incident energy quantum of frequency $v_1$ is absorbed and generates by itself–at least at sufficiently low densities of incident energy quanta—a light quantum of frequency $v_2$; it is possible that the absorption of the incident light quanta can give rise to the simultaneous emission of light quanta of frequencies $v_3$, $v_4$ etc., as well as to energy of other kinds, e.g., heat. It does not matter what intermediate processes give rise to this final result. If the fluorescent substance is not a perpetual source of energy, the principle of conservation of energy requires that the energy of an emitted energy quantum cannot be greater than that of the incident light quantum; it follows that

$$R\,\beta v_2/N \le R\,\beta v_1/N$$

or

$$v_2 \le v_1.$$

This is the well–known Stokes's Rule.

It should be strongly emphasized that according to our conception the quantity of light emitted under conditions of low illumination (other conditions remaining constant) must be proportional to the strength of the incident light, since each incident energy quantum will cause an elementary process of the postulated kind, independently of the action of other incident energy quanta. In particular, there will be no lower limit for the intensity of incident light necessary to excite the fluorescent effect.

According to the conception set forth above, deviations from Stokes's Rule are conceivable in the following cases:

1. when the number of simultaneously interacting energy quanta per unit volume is so large that an energy quantum of emitted light can receive its energy from several incident energy quanta;
2. when the incident (or emitted) light is not of such a composition that it corresponds to blackbody radiation within the range of validity of Wien's Law, that is to say, for example, when the incident light is produced by a body of such high temperature that for the wavelengths under consideration Wien's Law is no longer valid.

The last-mentioned possibility commands especial interest. According to the conception we have outlined, the possibility is not excluded that a "non-Wien radiation" of very low density can exhibit an energy behavior different from that of a blackbody radiation within the range of validity of Wien's Law.

## 8. CONCERNING THE EMISSION OF CATHODE RAYS THROUGH ILLUMINATION OF SOLID BODIES

The usual conception that the energy of light is continuously distributed over the space through which it propagates, encounters very serious difficulties when one attempts to explain the photoelectric phenomena, as has been pointed out in Herr Lenard's pioneering paper.*

According to the concept that the incident light consists of energy quanta of magnitude $R\beta v/N$, however, one can conceive of the ejection of electrons by light in the following way. Energy quanta penetrate into the surface layer of the body, and their energy is transformed, at least in part, into kinetic energy of electrons. The simplest way to imagine this is that a light quantum delivers its entire energy to a single electron: we shall assume that this is what happens. The possibility

---

*P. Lenard, Ann. Phys., 8, 169, 170 (1902).

should not be excluded, however, that electrons might receive their energy only in part from the light quantum.

An electron to which kinetic energy has been imparted in the interior of the body will have lost some of this energy by the time it reaches the surface. Furthermore, we shall assume that in leaving the body each electron must perform an amount of work $P$ characteristic of the substance. The ejected electrons leaving the body with the largest normal velocity will be those that were directly at the surface. The kinetic energy of such electrons is given by

$$R\beta v/N - P.$$

In the body is charged to a positive potential $\Pi$ and is surrounded by conductors at zero potential, and if $\Pi$ is just large enough to prevent loss of electricity by the body, if follows that:

$$\Pi\epsilon = R\beta v/N - P$$

where $\epsilon$ denotes the electronic charge, or

$$\Pi E = R\beta v - P'$$

where $E$ is the charge of a gram equivalent of a monovalent ion and $P'$ is the potential of this quantity of negative electricity relative to the body.[*]

If one takes $E = 9.6 \times 10^3$, then $\Pi \cdot 10^{-8}$ is the potential in volts which the body assumes when irradiated in a vacuum.

In order to see whether the derived relation yields an order of magnitude consistent with experience, we take $P' = 0$, $v = 1.03 \times 10^{15}$ (corresponding to the limit of the solar spectrum toward the ultraviolet) and $\beta = 4.866 \times 10^{-11}$. We obtain $\Pi \cdot 10^7 = 4.3$ volts, a result agreeing in order magnitude with those of Herr Lenard.[†]

If the derived formula is correct, then $\Pi$, when represented in Cartesian coordinates as a function of the frequency of the incident light, must be a straight line whose slope is independent of the nature of the emitting substance.

---

[*] If one assumes that the individual electron is detached from a neutral molecule by light with the performance of a certain amount of work, nothing in the relation derived above need be changed; one can simply consider $P'$ as the sum of two terms.
[†] P. Lenard, Ann. Phys. 8, pp. 163, 185, and Table I, Fig. 2 (1902).

As far as I can see, there is no contradiction between these conceptions and the properties of the photoelectric observed by Herr Lenard. If each energy quantum of the incident light, independently of everything else, delivers its energy of electrons, then the velocity distribution of the ejected electrons will be independent of the intensity of the incident light; on the other hand the number of electrons leaving the body will, if other conditions are kept constant, be proportional to the intensity of the incident light.*

Remarks similar to those made concerning hypothetical deviations from Stokes's Rule can be made with regard to hypothetical boundaries of validity of the law set forth above.

In the foregoing it has been assumed that the energy of at least some of the quanta of the incident light is delivered completely to individual electrons. If one does not make this obvious assumption, one obtains, in place of the last equation:

$$\Pi E + P' \le R\beta v.$$

For fluorescence induced by cathode rays, which is the inverse process to the one discussed above, one obtains by analogous considerations:

$$\Pi E + P' \ge R\beta v.$$

In the case, of the substances investigated by Herr Lenard, $PE^{\dagger}$ is always significantly greater than $R\beta v$, since the potential difference, which the cathode rays must traverse in order to produce visible light, amounts in some cases to hundreds and in others to thousands of volts.[‡] It is therefore to be assumed that the kinetic energy of an electron goes into the production of many light energy quanta.

## 9. Concerning the Ionization of Gases by Ultraviolet Light Solid Bodies

We shall have to assume that, the ionization of a gas by ultraviolet light, an individual light energy quantum is used for the ionization of

---

*P. Lenard, Ref. 9, p. 150 and p. 166–168.
†Should be $\Pi E$ (translator's note).
‡P. Lenard, Ann. Phys., 12, 469 (1903).

an individual gas molecule. From this is follows immediately that the work of ionization (i.e., the work theoretically needed for ionization) of a molecule cannot be greater than the energy of an absorbed light quantum capable of producing this effect. If one denotes by $J$ the (theoretical) work of ionization per gram equivalent, then it follows that:

$$R\beta v \geq J.$$

According to Lenard's measurements, however, the largest effective wavelength for air is approximately $1.9 \times 10^{-5}$ cm: therefore:

$$R\beta v = 6.4 \cdot 10^{12} \, \text{erg} \geq J.$$

An upper limit for the work of ionization can also be obtained from the ionization potentials of rarefied gases according to J. Stark* the smallest observed ionization potentials for air (at platinum anodes) is about 10 V.[†] One therefore obtains $9.6 \times 10^{12}$ as an upper limit for $J$, which is nearly equal to the value found above.

There is another consequence the experimental testing of which seems to me to be of great importance. If every absorbed light energy quantum ionizes a molecule, the following relation must obtain between the quantity of absorbed light $L$ and the number of gram molecules of ionized gas $j$:

$$j = L/R\beta v.$$

If our conception is correct, this relationship must be valid for all gases which (at the relevant frequency) show no appreciable absorption without ionization.

Bern, 17 March 1905
Received 18 March 1905

*J. Stark, Die Electrizitët in Gasen (Leipzig, 1902, p. 57)
[†]In the interior of gases the ionization potential for negative ions is, however, five times greater.

# "The Atomic Theory of Matter"

## By
## Max Planck

The problem with which we shall be occupied in the present lecture is that of a closer investigation of the atomic theory of matter. It is, however, not my intention to introduce this theory with nothing further, and to set it up as something apart and disconnected with other physical theories, but I intend above all to bring out the peculiar significance of the atomic theory as related to the present general system of theoretical physics; for in this way only will it be possible to regard the whole system as one containing within itself the essential compact unity, and thereby to realize the principal object of these lectures.

Consequently it is self evident that we must rely on that sort of treatment which we have recognized in last week's lecture as fundamental. That is, the division of all physical processes into reversible and irreversible processes. Furthermore, we shall be convinced that the accomplishment of this division is only possible through the atomic theory of matter, or, in other words, that irreversibility leads of necessity to atomistics.

I have already referred at the close of the first lecture to the fact that in pure thermodynamics, which knows nothing of an atomic structure and which regards all substances as absolutely continuous, the difference between reversible and irreversible processes can only be defined in one way, which a priori carries a provisional character and does not withstand penetrating analysis. This appears immediately evident when one reflects that the purely thermodynamic definition of

Published in *Eight Lectures on Theoretical Physics* (Mineola: Dover Publications 1998). First published in *Eight Lectures on Theoretical Physics, Delivered at Columbia University in 1909*, as Publication Number 3 of the *Ernest Kempton Adams Fund For Physical Research*.

irreversibility which proceeds from the impossibility of the realization of certain changes in nature, as, e.g., the transformation of heat into work without compensation, has at the outset assumed a definite limit to man's mental capacity, while, however, such a limit is not indicated in reality. On the contrary: mankind is making every endeavor to press beyond the present boundaries of its capacity, and we hope that later on many things will be attained which, perhaps, many regard at present as impossible of accomplishment. Can it not happen then that a process, which up to the present has been regarded as irreversible, may be proved, through a new discovery or invention, to be reversible? In this case the whole structure of the second law would undeniably collapse, for the irreversibility of a single process conditions that of all the others.

It is evident then that the only means to assure to the second law real meaning consists in this, that the idea of irreversibility be made independent of any relationship to man and especially of all technical relations.

Now the idea of irreversibility harks back to the idea of entropy; for a process is irreversible when it is connected with an increase of entropy. The problem is hereby referred back to a proper improvement of the definition of entropy. In accordance with the original definition of Clausius, the entropy is measured by means of a certain reversible process, and the weakness of this definition rests upon the fact that many such reversible processes, strictly speaking all, are not capable of being carried out in practice. With some reason it may be objected that we have here to do, not with an actual process and an actual physicist, but only with ideal processes, so-called thought experiments, and with an ideal physicist who operates with all the experimental methods with absolute accuracy. But at this point the difficulty is encountered: How far do the physicist's ideal measurements of this sort suffice? It may be understood, by passing to the limit, that a gas is compressed by a pressure which is equal to the pressure of the gas, and is heated by a heat reservoir which possesses the same temperature as the gas, but, for example, that a saturated vapor shall be transformed through isothermal compression in a reversible manner to a liquid without at any time a part of the vapor being condensed, as in certain thermodynamic considerations is supposed, must certainly appear doubtful. Still more striking, however, is the liberty as regards thought experiments, which in physical chemistry is granted the theorist. With his semi-permeable membranes, which in reality are only realizable under certain special conditions and then only with a certain approximation, he separates in a reversible manner, not only all possible varieties of molecules, whether or not they are in stable or unstable conditions, but he also separates the oppositely charged ions from one another and from the undissociated molecules, and he is disturbed, neither by the enormous electrostatic forces which resist such a separation, nor by the circumstance that in reality, from the beginning of the separation, the molecules become in part dissociated while the ions in part again combine. But such ideal processes are necessary throughout in order to make possible the comparison of the entropy of the undissociated molecules with the entropy of the dissociated molecules; for the law of thermodynamic equilibrium does not permit in general of derivation in any other way, in case one wishes to retain pure thermodynamics as a basis. It must be considered remarkable that all these ingenious thought processes have so well found confirmation of their results in experience, as is shown by the examples considered by us in the last lecture.

If now, on the other hand, one reflects that in all these results every reference to the possibility of actually carrying out each ideal process has disappeared—there are certainly left relations between directly measurable quantities only, such as temperature, heat effect, concentration, etc.—the presumption forces itself upon one that perhaps the introduction as above of such ideal processes is at bottom a round-about method, and that the peculiar import of the principle of increase of entropy with all its consequences can be evolved from the original idea of irreversibility or, just as well, from the impossibility of perpetual motion of the second kind, just as the principle of conservation of energy has been evolved from the law of impossibility of perpetual motion of the first kind.

This step: to have completed the emancipation of the entropy idea from the experimental art of man and the elevation of the second law thereby to a real principle, was the scientific life's work of Ludwig Boltzmann. Briefly stated, it consisted in general of referring back the idea of entropy to the idea of probability. Thereby is also explained, at the same time, the significance of the above (p. 17) auxiliary term used by me; "preference" of nature for a definite state. Nature prefers the more probable states to the less probable, because in nature processes take place in the direction of greater probability. Heat goes from a body at higher temperature to a body at lower temperature because the state of equal temperature distribution is more probable than a state of unequal temperature distribution.

Through this conception the second law of thermodynamics is removed at one stroke from its isolated position, the mystery concerning the preference of nature vanishes, and the entropy principle reduces to a well understood law of the calculus of probability.

The enormous fruitfulness of so "objective" a definition of entropy for all domains of physics I shall seek to demonstrate in the following lectures. But today we have principally to do with the proof of its admissibility; for on closer consideration we shall immediately perceive that the new conception of entropy at once introduces a great number of questions, new requirements and difficult problems. The first requirement is the introduction of the atomic hypothesis into the system of physics. For, if one wishes to speak of the probability of a physical state, i.e., if he wishes to introduce the probability for a given state as a definite quantity into the calculation, this can only be brought about, as in cases of all probability calculations, by referring the state back to a variety of possibilities; i.e., by considering a finite number of a priori equally likely configurations (complexions) through each of which the state considered may be realized. The greater the number of complexions, the greater is the probability of the state. Thus, e.g., the probability of throwing a total of four with two ordinary six-sided dice is found through counting the complexions by which the throw with a total of four may be realized. Of these there are three complexions:

with the first die, 1, with the second die, 3,
with the first die, 2, with the second die, 2,
with the first die, 3, with the second die, 1.

On the other hand, the throw of two is only realized through a single complexion. Therefore, the probability of throwing a total of four is three times as great as the probability of throwing a total of two.

Now, in connection with the physical state under consideration, in order to be able to differentiate completely from one another the complexions realizing it, and to associate it with a definite reckonable number, there is obviously no other means than to regard it as made up of numerous discrete homogeneous elements—for in perfectly continuous systems there exist no reckonable elements—and hereby the atomistic view is made a fundamental requirement. We have, therefore, to regard all bodies in nature, in so far as they possess an entropy, as constituted of atoms, and we therefore arrive in physics at the same conception of matter as that which obtained in chemistry for so long previously.

But we can immediately go a step further yet. The conclusions reached hold, not only for thermodynamics of material bodies, but also possess complete validity for the processes of heat radiation, which are thus referred back to the second law of thermodynamics. That radiant heat also possesses an entropy follows from the fact that a body which emits radiation into a surrounding diathermanous medium experiences a loss of heat and, therefore, a decrease of entropy. Since the total entropy of a physical system can only increase, it follows that one part of the entropy of the whole system, consisting of the body and the diathermanous medium, must be contained in the radiated heat. If the entropy of the radiant heat is to be referred back to the notion of probability, we are forced, in a similar way as above, to the conclusion that for radiant heat the atomic conception possesses a definite meaning. But, since radiant heat is not directly connected with matter, it follows that this atomistic conception relates, not to matter, but only to energy, and hence, that in heat radiation certain energy elements play an essential rôle. Even though this conclusion appears so singular and even though in many circles today vigorous objection is strongly urged against it, in the long run physical research will not be able to withhold its sanction from it, and the less, since it is confirmed by experience in quite a satisfactory manner. We shall return to this point in the lectures on heat radiation. I desire here only to mention that the novelty involved by the introduction of atomistic conceptions into the theory of heat radiation is by no means so revolutionary as, perhaps, might appear at the first glance. For there is, in my opinion at least, nothing which makes necessary the consideration of the heat processes in a complete vacuum as atomic, and it suffices to seek the atomistic features at the source of radiation, i.e., in those processes which have their play in the centres of emission and absorption of radiation. Then the Maxwellian electrodynamic differential equations can retain completely their validity for the vacuum, and, besides, the discrete elements of heat radiation are relegated exclusively to a domain which is still very mysterious and where there is still present plenty of room for all sorts of hypotheses.

Returning to more general considerations, the most important question comes up as to whether, with the introduction of atomistic conceptions and with the reference of entropy to probability, the content of the principle of increase of entropy is exhaustively comprehended, or whether still further physical hypotheses are required in order to secure the full import of that principle. If this important question had been settled at the time of the introduction of the atomic theory into thermodynamics, then the atomistic views would surely have been spared a large number of conceivable misunderstandings and justifiable attacks. For it turns out, in fact—and our further considerations will confirm this conclusion—that there has as yet nothing been done with atomistics which in itself requires much more than an essential generalization, in order to guarantee the validity of the second law.

We must first reflect that, in accordance with the central idea laid down in the first lecture (p. 7), the second law must possess validity as an objective physical law, independently of the individuality of the physicist. There is nothing to hinder us from imagining a physicist—we shall designate him a "microscopic" observer—whose senses are so sharpened that he is able to recognize each individual atom and to follow it in its motion. For this observer each atom moves exactly in accordance with the elementary laws which general dynamics lays down for it, and these laws allow, so far as we know, of an inverse performance of every process. Accordingly, here again the question is neither one of probability nor of entropy and its increase. Let us imagine, on the other hand, another observer, designated a "macroscopic" observer, who regards an ensemble of atoms as a homogeneous gas, say, and consequently applies the laws of thermodynamics to the mechanical and thermal processes within it. Then, for such an observer, in accordance with the second law, the process in general is an irreversible process. Would not now the first observer be justified in saying: "The reference of the entropy to probability has its origin in the fact that irreversible processes ought to be explained through reversible processes. At any rate, this procedure appears to me in the highest degree dubious. In any case, I declare each change of state which takes place in the ensemble of atoms designated a gas, as reversible, in opposition to the macroscopic observer." There is not the slightest thing, so far as I know, that one can urge against the validity of these statements. But do we not thereby place ourselves in the painful position of the judge who declared in a trial the correctness of the position of each separately of two contending parties and then, when a third contends that only one of the parties could emerge from the process victorious, was obliged to declare him also correct? Fortunately we find ourselves in a more favorable position. We can certainly mediate between the two parties without its being necessary for one or the other to give up his principal point of view. For closer consideration shows that the whole controversy rests upon a misunderstanding—a new proof of how necessary it is before one begins a controversy to come to an understanding with his opponent concerning the subject of the quarrel. Certainly, a given change of state cannot be both reversible and irreversible. But the one observer connects a wholly different idea with the phrase "change of state" than the other. What is then, in general, a change of state? The state of a physical system cannot well be otherwise defined than as the aggregate of all those physical quantities, through whose instantaneous values the time changes of the quantities, with given boundary conditions, are uniquely determined. If we inquire now, in accordance with the import of this definition, of the two observers as to what they understand by the state of the collection of atoms or the gas considered, they will give quite different answers. The microscopic observer will mention those quantities which determine the position and the velocities of all the individual atoms. There are present in the simplest case, namely, that in which the atoms may be considered as material points, six times as many quantities as atoms, namely, for each atom the three coordinates and the three velocity components, and in the case of combined molecules, still more quantities. For him the state and the progress of a process is then first determined when all these various quantities are individually given. We shall designate the state defined in this way the "micro-state." The macroscopic observer, on the other hand, requires fewer data. He will say that the state of the homogeneous gas considered by him is determined by the density, the visible velocity and the temperature at each point of the gas, and he will expect that, when these quantities are given, their time variations and, therefore, the progress of the process, to be completely determined in accordance with the two laws of thermo-dynamics, and therefore accompanied by an increase in entropy. In this connection he can call upon all the experience at his disposal, which will fully confirm his expectation. If we call this state the "macro-state," it is clear that the two laws: "the micro-changes of state are reversible" and "the macro-changes of state are irreversible," lie in wholly different domains and, at any rate, are not contradictory.

But now how can we succeed in bringing the two observers to an understanding? This is a question whose answer is obviously of fundamental significance for the atomic theory. First of all, it is easy to see that the macro-observer reckons only with mean values; for what he calls density, visible velocity and temperature of the gas are, for the micro-observer, certain mean values, statistical data, which are derived from the space distribution and from the velocities of the atoms in an appropriate manner. But the micro-observer cannot operate with these mean values alone, for, if these are given at one instant of time, the progress of the process is not determined throughout; on the contrary: he can easily find with given mean values an enormously large number of individual values for the positions and the velocities of the atoms, all of which correspond with the same mean values and which, in spite of this, lead to quite different processes with regard to the mean values. It follows from this of necessity that the micro-observer must either give up the attempt to understand the unique progress, in accordance with experience, of the macroscopic changes of state—and this would be the end of the atomic theory—or that he, through the introduction of a special physical hypothesis, restrict in a suitable manner the manifold of micro-states considered by him. There is certainly nothing to prevent him from assuming that not all conceivable micro-states are realizable in nature, and that certain of them are in fact thinkable, but never actually realized. In the formularization of such a hypothesis, there is of course no point of departure to be found from the principles of dynamics alone; for pure dynamics leaves this case undetermined. But on just this account any dynamical hypothesis, which involves nothing further than a closer specification of the micro-states realized in nature, is certainly permissible. Which hypothesis is to be given the preference can only be decided through comparison of the results to which the different possible hypotheses lead in the course of experience.

In order to limit the investigation in this way, we must obviously fix our attention only upon all imaginable configurations and velocities of the individual atoms which are compatible with determinate values of the density, the velocity and the temperature of the gas, or in other words: we must consider all the micro-states which belong to a determinate macro-state, and must investigate the various kinds of processes which follow in accordance with the fixed laws of dynamics from the different micro-states. Now, precise calculation has in every case always led to the important result that an enormously large number of these different micro-processes relate to one and the same macro-process, and that only proportionately few of the same, which are distinguished by quite special exceptional conditions concerning the positions and the velocities of neighboring atoms, furnish exceptions. Furthermore, it has also shown that one of the resulting macro-processes is that which the macroscopic observer recognizes, so that it is compatible with the second law of thermodynamics.

Here, manifestly, the bridge of understanding is supplied. The micro-observer needs only to assimilate in his theory the physical hypothesis that all those special cases in which special exceptional conditions exist among the neighboring configurations of interacting atoms do not occur in nature, or, in other words, that the micro-states are in elementary disorder. Then the uniqueness of the macroscopic process is assured and with it, also, the fulfillment of the principle of increase of entropy in all directions.

Therefore, it is not the atomic distribution, but rather the hypothesis of elementary disorder, which forms the real kernel of the principle of increase of entropy and, therefore, the preliminary condition for the existence of entropy. Without elementary disorder there is neither entropy nor irreversible process.* Therefore, a single atom can never possess an entropy; for we cannot speak of disorder in connection

---

*To those physicists who, in spite of all this, regard the hypothesis of elementary disorder as gratuitous or as incorrect, I wish to refer the simple fact that in every calculation of a coefficient of friction, of diffusion, or of heat conduction, from molecular considerations, the notion of elementary disorder is employed, whether tacitly or otherwise, and that it is therefore essentially more correct to stipulate this condition instead of ignoring or concealing it. But he who regards the hypothesis of elementary disorder as self-evident, should be reminded that, in accordance with a law of H. Poincaré, the precise investigation concerning the foundation of which would here lead us too far, the assumption of this hypothesis for all times is unwarranted for a closed space with absolutely smooth walls,—an important conclusion, against which can only be urged the fact that absolutely smooth walls do not exist in nature.

with it. But with a fairly large number of atoms, say 100 or 1,000, the matter is quite different. Here, one can certainly speak of a disorder, in case that the values of the coordinates and the velocity components are distributed among the atoms in accordance with the laws of accident. Then it is possible to calculate the probability for a given state. But how is it with regard to the increase of entropy? May we assert that the motion of 100 atoms is irreversible? Certainly not; but this is only because the state of 100 atoms cannot be defined in a thermodynamic sense, since the process does not proceed in a unique manner from the standpoint of a macro-observer, and this requirement forms, as we have seen above, the foundation and preliminary condition for the definition of a thermodynamic state.

If one therefore asks: How many atoms are at least necessary in order that a process may be considered irreversible?, the answer is: so many atoms that one may form from them definite mean values which define the state in a macroscopic sense. One must reflect that to secure the validity of the principle of increase of entropy there must be added to the condition of elementary disorder still another, namely, that the number of the elements under consideration be sufficiently large to render possible the formation of definite mean values. The second law has a meaning for these mean values only; but for them, it is quite exact, just as exact as the law of the calculus of probability, that the mean value, so far as it may be defined, of a sufficiently large number of throws with a six-sided die, is $3\frac{1}{2}$.

These considerations are, at the same time, capable of throwing light upon questions such as the following: Does the principle of increase of entropy possess a meaning for the so-called Brownian molecular movement of a suspended particle? Does the kinetic energy of this motion represent useful work or not? The entropy principle is just as little valid for a single suspended particle as for an atom, and therefore is not valid for a few of them, but only when there is so large a number that definite mean values can be formed. That one is able to see the particles and not the atoms makes no material difference; because the progress of a process does not depend upon the power of an observing instrument. The question with regard to useful work plays no rôle in this connection; strictly speaking, this possesses, in general, no objective physical meaning. For it does not admit of an answer without reference to the scheme of the physicist or technician who proposes to make use of the work in question. The second law, therefore, has fundamentally nothing to do with the idea of useful work (cf. first lecture, p. 15).

But, if the entropy principle is to hold, a further assumption is necessary, concerning the various disordered elements,—an assumption which tacitly is commonly made and which we have not previously definitely expressed. It is, however, not less important than those referred to above. The elements must actually be of the same kind, or they must at least form a number of groups of like kind, e.g., constitute a mixture in which each kind of element occurs in large numbers. For only through the similarity of the elements does it come about that order and law can result in the larger from the smaller. If the molecules of a gas be all different from one another, the properties of a gas can never show so simple a law-abiding behavior as that which is indicated by thermodynamics. In fact, the calculation of the probability of a state presupposes that all complexions which correspond to the state are a priori equally likely. Without this condition one is just as little able to calculate the probability of a given state as, for instance, the probability of a given throw with dice whose sides are unequal in size. In summing up we may therefore say: the second law of thermodynamics in its objective physical conception, freed from anthropomorphism, relates to certain mean values which are formed from a large number of disordered elements of the same kind.

The validity of the principle of increase of entropy and of the irreversible progress of thermodynamic processes in nature is completely assured in this formularization. After the introduction of the hypothesis of elementary disorder, the microscopic observer can no longer confidently assert that each process considered by him in a collection of atoms is reversible; for the motion occurring in the reverse order will not always obey the requirements of that hypothesis. In fact, the motions of single atoms are always reversible, and thus far one may say, as before, that the irreversible processes appear reduced to a reversible process, but the phenomenon as a whole is nevertheless irreversible, because upon reversal the disorder of the numerous individual elementary processes would be eliminated. Irreversibility is inherent, not in the individual elementary processes themselves, but solely in their irregular constitution. It is this only which guarantees the unique change of the macroscopic mean values.

Thus, for example, the reverse progress of a frictional process is impossible, in that it would presuppose elementary arrangement of interacting neighboring molecules. For the collisions between any two molecules must thereby possess a certain distinguishing character, in that the velocities of two colliding molecules depend in a definite way upon the place at which they meet. In this way only can it happen that in collisions like directed velocities ensue and, therefore, visible motion.

Previously we have only referred to the principle of elementary disorder in its application to the atomic theory of matter. But it may also be assumed as valid, as I wish to indicate at this point, on quite the same grounds as those holding in the case of matter, for the theory of radiant heat. Let us consider, e.g., two bodies at different temperatures between which exchange of heat occurs through radiation. We can in this case also imagine a microscopic observer, as opposed to the ordinary macroscopic observer, who possesses insight into all the particulars of electromagnetic processes which are connected with emission and absorption, and the propagation of heat rays. The microscopic observer would declare the whole process reversible because all electrodynamic processes can also take place in the reverse direction, and the contradiction may here be referred back to a difference in definition of the state of a heat ray. Thus, while the macroscopic observer completely defines a monochromatic ray through direction, state of polarization, color, and intensity, the microscopic observer, in order to possess a complete knowledge of an electromagnetic state, necessarily requires the specification of all the numerous irregular variations of amplitude and phase to which the most homogeneous heat ray is actually subject. That such irregular variations actually exist follows immediately from the well known fact that two rays of the same color never interfere, except when they originate in the same source of light. But until these fluctuations are given in all particulars, the micro-observer can say nothing with regard to the progress of the process. He is also unable to specify whether the exchange of heat radiation between the two bodies leads to a decrease or to an increase of their difference in temperature. The principle of elementary disorder first furnishes the adequate criterion of the tendency of the radiation process, i.e., the warming of the colder body at the expense of the warmer, just as the same principle conditions the irreversibility of exchange of heat through conduction. However, in the two cases compared, there is indicated an essential difference in the kind of the disorder. While in heat conduction the disordered elements may be represented as associated with the various molecules, in heat radiation there are the numerous vibration periods, connected with a heat ray, among which the energy of radiation is irregularly distributed. In other words: the disorder among the molecules is a material one, while in heat radiation it is one of energy distribution. This is the most important difference between the two kinds of disorder; a common feature exists as regards the great number of uncoordinated elements required. Just as the entropy of a body is defined as a function of the macroscopic state, only when the body contains so many atoms that from them definite mean values may be formed, so the entropy principle only possesses a meaning with regard to a heat ray when the ray comprehends so many periodic vibrations, i.e., persists for so long a time, that a definite mean value for the intensity of the ray may be obtained from the successive irregular fluctuating amplitudes.

Now, after the principle of elementary disorder has been introduced and accepted by us as valid throughout nature, the fundamental question arises as to the calculation of the probability of a given state, and the actual derivation of the entropy therefrom. From the entropy all the laws of thermodynamic states of equilibrium, for material substances, and also for energy radiation, may be uniquely derived. With regard to the connection between entropy and probability, this is inferred very simply from the law that the probability of two independent configurations is represented by the product of the individual probabilities:

$$W = W_1 \cdot W_2,$$

while the entropy $S$ is represented by the sum of the individual entropies:

$$S = S_1 + S_2.$$

Accordingly, the entropy is proportional to the logarithm of the probability:

$$S = k \log W. \tag{1}$$

$k$ is a universal constant. In particular, it is the same for atomic as for radiation configurations, for there is nothing to prevent us assuming that the configuration designated by 1 is atomic, while that designated by 2 is a radiation configuration. If $k$ has been calculated, say with the aid of radiation measurements, then $k$ must have the same value for atomic processes. Later we shall follow this procedure, in order to utilize the laws of heat radiation in the kinetic theory of gases. Now, there remains, as the last and most difficult part of the problem, the calculation of the probability $W$ of a given physical configuration in a given macroscopic state. We shall treat today, by way of preparation for the quite general problem to follow, the simple problem: to specify the probability of a given state for a single moving material point, subject to given conservative forces. Since the state depends upon 6 variables: the 3 generalized coordinates $\varphi_1$, $\varphi_2$, $\varphi_3$, and the three corresponding velocity components $\dot{\varphi}_1$, $\dot{\varphi}_2$, $\dot{\varphi}_3$, and since all possible values of these 6 variables constitute a continuous manifold, the probability sought is, that these 6 quantities shall lie respectively within certain infinitely small intervals, or, if one thinks of these 6 quantities as the rectilinear orthogonal coordinates of a point in an ideal six-dimensional space, that this ideal "state point" shall fall within a given, infinitely small "state domain." Since the domain is infinitely small, the probability will be proportional to the magnitude of the domain and therefore proportional to

$$\int d\varphi_1 \cdot d\varphi_2 \cdot d\varphi_3 \cdot d\dot{\varphi}_1 \cdot d\dot{\varphi}_2 \cdot d\dot{\varphi}_3.$$

But this expression cannot serve as an absolute measure of the probability, because in general it changes in magnitude with the time, if each state point moves in accordance with the laws of motion of material points, while the probability of a state which follows of necessity from another must be the same for the one as the other. Now, as is well known, another integral quite similarly formed, may be specified in place of the one above, which possesses the special property of not changing in value with the time. It is only necessary to employ, in addition to the general coordinates $\varphi_1$, $\varphi_2$, $\varphi_3$, the three so-called momenta $\psi_1$, $\psi_2$, $\psi_3$, in place of the three velocities $\dot{\varphi}_1$, $\dot{\varphi}_2$, $\dot{\varphi}_3$, as the determining coordinates of the state. These are defined in the following way:

$$\psi_1 = \left(\frac{\partial H}{\partial \dot{\varphi}_1}\right)_\phi, \quad \psi_2 = \left(\frac{\partial H}{\partial \dot{\varphi}_2}\right)\phi, \quad \dot{\varphi}_3 = \left(\frac{\partial H}{\partial \dot{\varphi}_3}\right)_\phi,$$

wherein $H$ denotes the kinetic potential (Helmholz). Then, in Hamiltonian form, the equations of motion are:

$$\psi_1 = \frac{d\psi_1}{dt} = -\left(\frac{\partial E}{\partial \varphi_1}\right)_\psi, \quad \cdots, \quad \dot{\varphi}_1 = \frac{d\varphi_1}{dt} = \left(\frac{\partial E}{\partial \psi_1}\right)_\phi, \quad \cdots,$$

($E$ is the energy), and from these equations follows the "condition of incompressibility":

$$\frac{\partial \dot{\varphi}_1}{\partial \varphi_1} + \frac{\partial \psi_1}{\partial \psi_1} + \cdots = 0.$$

Referring to the six-dimensional space represented by the coordinates $\varphi_1$, $\varphi_2$, $\varphi_3$, $\psi_1$, $\psi_2$, $\psi_3$, this equation states that the magnitude of an arbitrarily chosen state domain, viz.:

$$\int d\varphi_1 \cdot d\varphi_2 \cdot d\varphi_3 \cdot d\psi_1 \cdot d\psi_2 \cdot d\psi_3$$

does not change with the time, when each point of the domain changes its position in accordance with the laws of motion of material points. Accordingly, it is made possible to take the magnitude of this domain as a direct measure for the probability that the state point falls within the domain.

From the last expression, which can be easily generalized for the case of an arbitrary number of variables, we shall calculate later the probability of a thermodynamic state, for the case of radiant energy as well as that for material substances.

# Chapter Two

The question of whether matter is infinitely divisible has troubled philosophers for millennia. In approximately 450 BCE, the Greek philosopher Democritus speculated that there must be some smallest unit of matter from which every material thing is built. He named this the *atom*, which means "indivisible" in Greek. However, by the end of the nineteenth century, it was known that what we now call atoms are, in fact, divisible. This is not to say that Democritus was wrong, for we have good reasons to believe that elementary particles like electrons are fundamental and indivisible. It simply means what we call an atom was misnamed.

In the latter part of the nineteenth century, atoms were known to be composed of positively charged protons and negatively charged electrons, but it was not known how the protons and electrons were structured in the atom. From 1909 through 1911, Ernest Rutherford and his assistant Hans Geiger conducted experiments to explore this question. The results of their study, are presented in the groundbreaking paper, "The Scattering of $\alpha$ and $\beta$ Particles by Matter and the Structure of the Atom." They bombarded gold foil with $\alpha$ particles in the hope that by watching how the $\alpha$ particles interacted with the atoms in the foil they could determine the structure of the gold atoms. $\alpha$ Particles have a strong positive electrical charge, so they are ideally suited to studying where the positive and negative charges reside inside an atom. The experiment seems straightforward, but what Rutherford and Geiger discovered was totally unexpected. They found that all the protons were very tightly clumped together at the center of the atom, in what we now call the nucleus. The electrons were found surrounding the nucleus. Rutherford speculated that the electrons orbit the nucleus in much the same way that planets orbit the sun. Consequently, Rutherford's model of the atom was called the *planetary model.* The planetary model was completely unexpected because it seems to violate well-known physical laws. For example, we know that electrical charges of the same sign repel each other. All protons are positively charged, so the protons in the nucleus should strongly repel each other and blow the nucleus apart—but they don't! What could be holding it together? Rutherford did not know. He simply postulated that there must be some force holding it together. This force would not be well understood until the 1970s with the advent of quantum chromodynamics.

Another problem with the planetary model of the atom was that the electrons' orbit around the nucleus should be unstable. We know from the theory of electrodynamics that an orbiting charged particle will radiate electromagnetic waves, causing it to lose energy and spiral into the nucleus. But atoms are stable—the electrons in an atom do not do this. Why not? In 1913 Niels Bohr tackled this question in his paper, "On the Constitution of Atoms and Molecules." His answer was the next important step in the development of quantum mechanics. He simply postulated that only discrete distances from the nucleus were allowed for an electron orbiting a nucleus. In other words, he assumed the radius of orbit (and equivalently, the energy) of electrons in atoms to be quantized. It is as if the allowed orbits are steps in a staircase. The electron can be on the third step or the fourth step, for example, but it cannot be in between. Electrons can jump between levels but cannot spiral inward because that would involve being in an intermediate space between levels. With this assumption, Bohr's model could also explain the atomic spectrum of hydrogen. When an electron jumps from a higher to lower energy level it emits electromagnetic radiation with an energy amount given by the difference in energies of the two levels. Since the energy levels are discrete transitions, atomic spectra have distinct, sharp lines.

Bohr's model was immediately recognized as revolutionary, and it won him the 1922 Nobel prize. However, the Bohr model has a number of shortcomings. It only works for atoms with a single electron, and even for single-electron atoms it cannot explain the fine structure of atomic spectra, nor does it provide an explanation of why the electron energy levels should be quantized. It was not until the 1920s, when a more complete theory of quantum mechanics was developed, that we understood why we needed to assume energy level quantization.

# The Scattering of $\alpha$ and $\beta$ Particles by Matter and the Structure of the Atom

## By
### Ernest Rutherford[*]

First appeared in *Philosophical Magazine*, Series 6, vol. 21, May 1911,
p. 669–688

§ 1. It is well known that the $\alpha$ and the $\beta$ particles suffer deflexions from their rectilinear paths by encounters with atoms of matter. This scattering is far more marked for the $\beta$ than for the $\alpha$ particle on account of the much smaller momentum and energy of the former particle. There seems to be no doubt that such swiftly moving particles pass through the atoms in their path, and that the deflexions observed are due to the strong electric field traversed within the atomic system. It has generally been supposed that the scattering of a pencil of $\alpha$ or $\beta$ rays in passing through a thin plate of matter is the result of a multitude of small scatterings by the atoms of matter traversed. The observations, however, of Geiger and Marsden[†] on the scattering of $\alpha$ rays indicate that some of the $\alpha$ particles, about 1 in 20,000 were turned through an average angle of 90 degrees in passing though a layer of gold-foil about 0.00004 cm. thick, which was equivalent in stopping-power of the $\alpha$ particle to 1.6 millimetres of air. Geiger[‡] showed later that the most probable angle of deflexion for a pencil of $\alpha$ particles being deflected through 90 degrees is vanishingly small. In addition, it will be seen later that the distribution of the $\alpha$ particles for various angles of large deflexion does not follow the probability law

---

[*]Communicated by the Author. A brief account of this paper was communicated to the Manchester Literary and Philosophical Society in February, 1911.

[†]Proc. Roy. Soc. lxxxii, p. 495 (1909)

[‡]Proc. Roy. Soc. lxxxiii, p. 492 (1910)

to be expected if such large deflexion are made up of a large number of small deviations. It seems reasonable to suppose that the deflexion through a large angle is due to a single atomic encounter, for the chance of a second encounter of a kind to produce a large deflexion must in most cases be exceedingly small. A simple calculation shows that the atom must be a seat of an intense electric field in order to produce such a large deflexion at a single encounter.

Recently Sir J. J. Thomson* has put forward a theory to explain the scattering of electrified particles in passing through small thicknesses of matter. The atom is supposed to consist of a number N of negatively charged corpuscles, accompanied by an equal quantity of positive electricity uniformly distributed throughout a sphere. The deflexion of a negatively electrified particle in passing through the atom is ascribed to two causes – (1) the repulsion of the corpuscles distributed through the atom, and (2) the attraction of the positive electricity in the atom. The deflexion of the particle in passing through the atom is supposed to be small, while the average deflexion after a large number $m$ of encounters was taken as [the square root of] $m \cdot \theta$, where $\theta$ is the average deflexion due to a single atom. It was shown that the number N of the electrons within the atom could be deduced from observations of the scattering was examined experimentally by Crowther[†] in a later paper. His results apparently confirmed the main conclusions of the theory, and he deduced, on the assumption that the positive electricity was continuous, that the number of electrons in an atom was about three times its atomic weight.

The theory of Sir J. J. Thomson is based on the assumption that the scattering due to a single atomic encounter is small, and the particular structure assumed for the atom does not admit of a very large deflexion of diameter of the sphere of positive electricity is minute compared with the diameter of the sphere of influence of the atom.

*Camb. Lit. & Phil Soc. xv pt. 5 (1910)
[†]Crowther, Proc. Roy. Soc. lxxxiv. p. 226 (1910)

Since the $\alpha$ and $\beta$ particles traverse the atom, it should be possible from a close study of the nature of the deflexion to form some idea of the constitution of the atom to produce the effects observed. In fact, the scattering of high-speed charged particles by the atoms of matter is one of the most promising methods of attack of this problem. The development of the scintillation method of counting single $\alpha$ particles affords unusual advantages of investigation, and the researches of H. Geiger by this method have already added much to our knowledge of the scattering of $\alpha$ rays by matter.

§ 2. We shall first examine theoretically the single encounters* with an atom of simple structure, which is able to produce large deflections of an $\alpha$ particle, and then compare the deductions from the theory with the experimental data available.

Consider an atom which contains a charge $\pm Ne$ at its centre surrounded by a sphere of electrification containing a charge $\pm Ne$ [N.B. in the original publication, the second plus/minus sign is inverted to be a minus/plus sign] supposed uniformly distributed throughout a sphere of radius R. $e$ is the fundamental unit of charge, which in this paper is taken as $4.65 \times 10^{-10}$ E.S. unit. We shall suppose that for distances less than $10^{-12}$ cm. the central charge and also the charge on the alpha particle may be supposed to be concentrated at a point. It will be shown that the main deductions from the theory are independent of whether the central charge is supposed to be positive or negative. For convenience, the sign will be assumed to be positive. The question of the stability of the atom proposed need not be considered at this stage, for this will obviously depend upon the minute structure of the atom, and on the motion of the constituent charged parts.

In order to form some idea of the forces required to deflect an alpha particle through a large angle, consider an atom containing a positive charge $Ne$ at its centre, and surrounded by a distribution of negative electricity $Ne$ uniformly distributed within a sphere of radius R. The electric force X and the potential V at a distance $r$ from the

---

*The deviation of a particle throughout a considerable angle from an encounter with a single atom will in this paper be called 'single' scattering. The deviation of a particle resulting from a multitude of small deviations will be termed 'compound' scattering.

centre of an atom for a point inside the atom, are given by

$$X = Ne \left( \frac{1}{r^2} - \frac{r}{R^3} \right)$$

$$V = Ne \left( \frac{1}{r} - \frac{3}{2R} + \frac{r^2}{2R^3} \right).$$

Suppose an $\alpha$ particle of mass $m$ and velocity $u$ and charge E shot directly towards the centre of the atom. It will be brought to rest at a distance $b$ from the centre given by

$$\frac{1}{2} mu^2 = Ne E \left( \frac{1}{b} - \frac{3}{2R} + \frac{b^2}{2R^3} \right).$$

It will be seen that $b$ is an important quantity in later calculations. Assuming that the central charge is 100 $e$, it can be calculated that the value of $b$ for an $\alpha$ particle of velocity $2.09 \times 10^9$ cms. per second is about $3.4 \times 10^{-12}$ cm. In this calculation $b$ is supposed to be very small compared with R. Since R is supposed to be of the order of the radius of the atom, viz. $10^{-8}$ cm., it is obvious that the $\alpha$ particle before being turned back penetrates so close to the central charge, that the field due to the uniform distribution of negative electricity may be neglected. In general, a simple calculation shows that for all deflexions greater than a degree, we may without sensible error suppose the deflexion due to the field of the central charge alone. Possible single deviations due to the negative electricity, if distributed in the form of corpuscles, are not taken into account at this stage of the theory. It will be shown later that its effect is in general small compared with that due to the central field.

Consider the passage of a positive electrified particle close to the centre of an atom. Supposing that the velocity of the particle is not appreciably changed by its passage through the atom, the path of the particle under the influence of a repulsive force varying inversely as the square of the distance will be an hyperbola with the centre of the atom S as the external focus. Suppose the particle to enter the atom in the direction PO (Fig. 1), and that the direction of motion on escaping the atom is OP'. OP and OP' make equal angles with the

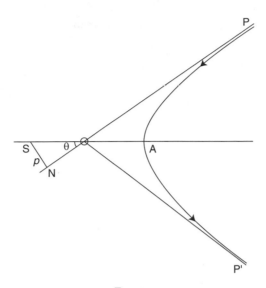

FIG. 1

line SA, where A is the apse of the hyperbola. $p =$ SN = perpendicular distance from centre on direction of initial motion of particle.

Let angle POA $= \theta$.

Let V = velocity of particle on entering the atom, $v$ its velocity at A, then from consideration of angular momentum

$$pV = SA \cdot v.$$

From conservation of energy

$$(1/2)mV^2 = (1/2)mv^2 - (NeE/SA),$$
$$v^2 = V^2(1 - (b/SA)).$$

Since the eccentricity is sec $\theta$,

$$SA = SO + OA = p \operatorname{cosec} \theta (1 + \cos \theta)$$
$$= p \cot \theta / 2$$
$$p^2 = SA(SA - b) = p \cot \theta / 2 (p \cot \theta / 2 - b),$$

therefore $b = 2p \cot \theta$.

The angle of deviation $\theta$ of the particles in $\pi - 2\theta$ and

$$\cot \theta/2 = (2p/b)^* \qquad (1)$$

This gives the angle of deviation of the particle in terms of $b$, and the perpendicular distance of the direction of projection from the centre of the atom.

For illustration, the angle of deviation $\phi$ for different values of $p/b$ are shown in the following table:—

| $p/b$..... | 10 | 5 | 2 | 1 | 0.5 | 0.25 | 0.125 |
|---|---|---|---|---|---|---|---|
| $\phi$...... | 5°.7 | 11°.4 | 28° | 53° | 90° | 127° | 152° |

## § 3. PROBABILITY OF SINGLE DEFLEXION THROUGH ANY ANGLE

Suppose a pencil of electrified particles to fall normally on a thin screen of matter of thickness $t$. With the exception of the few particles which are scattered through a large angle, the particles are supposed to pass nearly normally through the plate with only a small change of velocity. Let $n$ = number of atoms in unit volume of material. Then the number of collisions of the particle with the atom of radius R is $\pi R^2 nt$ in the thickness $t$.

The probability $m$ of entering an atom within a distance $p$ of its center is given by

$$m = \pi p^2 nt.$$

Chance $dm$ of striking within radii $p$ and $p + dp$ is given by

$$dm = 2\pi pnt \cdot dp = (\pi/4)\, ntb^2 \cot \phi/2 \operatorname{cosec}^2 \phi/2\, d\phi \qquad (2)$$

since

$$\cot \phi/2 = 2p/b$$

The value of $dm$ gives the fraction of the total number of particles which are deviated between the angles $\phi$ and $\phi + d\phi$.

---

*A simple consideration shows that the deflexion is unaltered if the forces are attractive instead of repulsive.

The fraction $p$ of the total number of particles which are deflected through an angle greater than $\phi$ is given by

$$p = (\pi/4)\, ntb^2 \cot^2 \phi/2. \tag{3}$$

The fraction $p$ which is deflected between the angles $\phi_1$ and $\phi_2$ is given by

$$p = (\pi/4)\, ntb^2 \left(\cot^2 \phi_1/2 - \cot^2 \phi_2/2\right). \tag{4}$$

It is convenient to express the equation (2) in another form for comparison with experiment. In the case of the $\alpha$ rays, the number of scintillations appearing on the *constant* area of the zinc sulphide screen are counted for different angles with the direction of incidence of the particles. Let $r =$ distance from point of incidence of $\alpha$ rays on scattering material, then if Q be the total number of particles falling on the scattering material, the number y of $\alpha$ particles falling on unit area which are deflected through an angle $\phi$ is given by

$$y = Qdm/2\pi r^2 \sin \phi \cdot d\phi = \left(ntb^2 \cdot Q \cdot \operatorname{cosec}^4 \phi/2\right)/16r^2 \tag{5}$$

Since $b = 2NeE/mu^2$, we see from this equation that the number of $\alpha$ particles (scintillations) per unit area of zinc sulphide screen at a given distance $r$ from the point of Incidence of the rays is proportional to

(1) $\operatorname{cosec}^4 \phi/2$ or $1/\phi^4$ if $\phi$ be small;

(2) thickness of scattering material $t$ provided this is small;

(3) magnitude of central charge Ne;

(4) and is inversely proportional to $(mu^2)^2$, or to the fourth power of the velocity if $m$ be constant.

In these calculations, it is assumed that the $\alpha$ particles scattered through a large angle suffer only one large deflexion. For this to hold, it is essential that the thickness of the scattering material should be so small that the chance of a second encounter involving another large deflexion is very small. If, for example, the probability of a single deflexion $\phi$ in passing through a thickness $t$ is 1/1000, the probability of two successive deflexions each of value $\phi$ is $1/10^6$, and is negligibly small.

The angular distribution of the $\alpha$ particles scattered from a thin metal sheet affords one of the simplest methods of testing the general correctness of this theory of single scattering. This has been done recently for $\alpha$ rays by Dr. Geiger*, who found that the distribution for particles deflected between 30° and 150° from a thin gold-foil was in substantial agreement with the theory. A more detailed account of these and other experiments to test the validity of the theory will be published later.

## § 4. ALTERATION OF VELOCITY IN AN ATOMIC ENCOUNTER

It has so far been assumed that an $\alpha$ or $\beta$ particle does not suffer an appreciable change of velocity as the result of a single atomic encounter resulting in a large deflexion of the particle. The effect of such an encounter in altering the velocity of the particle can be calculated on certain assumptions. It is supposed that only two systems are involved, viz., the swiftly moving particle and the atom which it traverses supposed initially at rest. It is supposed that the principle of conservation of momentum and of energy applies, and that there is no appreciable loss of energy or momentum by radiation.

Let $m$ be mass of the particle,

$v_1$ = velocity of approach,
$v_2$ = velocity of recession,
$M$ = mass of atom,
$V$ = velocity communicated to atom as result of encounter.

Let OA (Fig. 2) represent in magnitude and direction the momentum $mv_1$ of the entering particle, and OB the momentum of the receding particle which has been turned through an angle AOB = $\phi$. Then BA represents in magnitude and direction the momentum MV of the recoiling atom.

$$(MV)^2 = (mv_1)^2 + (mv_2)^2 - 2m^2 v_1 v_2 \cos \phi \qquad (1)$$

By conservation of energy

$$MV^2 = mv_1^2 - mv_2^2 \qquad (2)$$

*Manch. Lit. & Phil. Soc. 1910.

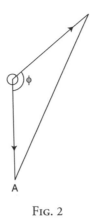

FIG. 2

Suppose $M/m = K$ and $v_2 = pv_1$, where $p < 1$. From (1) and (2).

$$(K + 1)\rho^2 - 2\rho \cos \phi = k - 1,$$

or

$$\rho = \frac{\cos \phi}{K + 1} + \frac{1}{K + 1}\sqrt{K^2 - \sin^2 \phi}.$$

Consider the case of an $\alpha$ particle of atomic weight 4, deflected through an angle of 90° by an encounter with an atom of gold of atomic weight 197.

Since $K = 49$ nearly,

$$p = \sqrt{\frac{K - 1}{K + 1}} = 0.979$$

or the velocity of the particle is reduced only about 2 per cent. by the encounter.

In the case of aluminium $K = 27/4$ and for $\phi = 90°$ $p = 0.86$.

It is seen that the reduction of velocity of the $\alpha$ particle becomes marked on this theory for encounters with the lighter atoms. Since the range of an $\alpha$ particle in air or other matter is approximately proportional to the cube of the velocity, it follows that an $\alpha$ particle of range 7 cms. has its range reduced to 4.5 cms. after incurring a single deviation of 90° in traversing an aluminium atom. This is of a magnitude to be easily detected experimentally. Since the value of K is very large for an encounter of a $\beta$ particle with an atom, the reduction of velocity on this formula is very small.

Some very interesting cases of the theory arise in considering the changes of velocity and the distribution of scattered particles when the $\alpha$ particle encounters a light atom, for example a hydrogen or helium atom. A discussion of these and similar cases is reserved until the question has been examined experimentally.

## § 5. Comparison of single and compound scattering

Before comparing the results of theory with experiment, it is desirable to consider the relative importance of single and compound scattering in determining the distribution of the scattered particles. Since the atom is supposed to consist of a central charge surrounded by a uniform distribution of the opposite sign through a sphere of radius R, the chance of encounters with the atom involving small deflexions is very great compared with the change of a single large deflexion.

This question of compound scattering has been examined by Sir J. J. Thomson in the paper previously discussed (§ 1). In the notation of this paper, the average deflexion $\phi_1$ due to the field of the sphere of positive electricity of radius R and quantity N$e$ was found by him to be

$$\phi_1 = \frac{\pi}{4} \cdot \frac{NeE}{mu^2} \cdot \frac{1}{R}.$$

The average deflexion $\phi_2$ due to the N negative corpuscles supposed distributed uniformly throughout the sphere was found to be

$$\phi_2 = \frac{16}{5} \frac{eE}{mu^2} \cdot \frac{1}{R}\sqrt{\frac{3N}{2}}.$$

The mean deflexion due to both positive and negative electricity was taken as

$$\left(\phi_1^2 + \phi_2^2\right)^{1/2}$$

In a similar way, it is not difficult to calculate the average deflexion due to the atom with a central charge discussed in this paper.

Since the radial electric field X at any distance $r$ from the centre is given by

$$X = Ne \left( \frac{1}{r^2} - \frac{r}{R^3} \right),$$

it is not difficult to show that the deflexion (supposed small) of an electrified particle due to this field is given by

$$\theta = \frac{b}{p} \left( 1 - \frac{p^2}{R^2} \right)^{3/2},$$

Where $p$ is the perpendicular from the center on the path of the particles and $b$ has the same value as before. It is seen that the value of $\theta$ increases with diminution of $p$ and becomes great for small value of $\phi$.

Since we have already seen that the deflexions become very large for a particle passing near the center of the atom, it is obviously not correct to find the average value by assuming $\theta$ is small.

Taking R of the order $10^{-8}$ cm., the value of $p$ for a large deflexions is for $\alpha$ and $\beta$ particles of the order $10^{-11}$ cm. Since the chance of an encounter involving a large deflexion is small compared with the chance of small deflexions, a simple consideration shows that the average small deflexion is practically unaltered if the large deflexions are omitted. This is equivalent to integrating over that part of the cross section of the atom where the deflexions are small and neglecting the small central area. It can in this way be simply shown that the average small deflexion is given by

$$\phi_1 = \frac{3\pi}{8} \frac{b}{R}.$$

This value of $\phi_1$ for the atom with a concentrated central charge is three times the magnitude of the average deflexion for the same value of $Ne$ in the type of atom examined by Sir J. J. Thomson. Combining the deflexions due to the electric field and to the corpuscles, the average deflexion is

$$(\phi_1^2 + \phi_2^2)^{1/2} \text{ or } \frac{b}{2R} \left( 5.54 + \frac{15.4}{N} \right)^{1/2}.$$

It will be seen later that the value of N is nearly proportional to the atomic weight, and is about 100 for gold. The effect due to scattering of the individual corpuscles expressed by the second term of the equation is consequently small for heavy atoms compared with that due to the distributed electric field.

Neglecting the second term, the average deflexion per atom is $3\pi b/8R$. We are now in a position to consider the relative effects on the distribution of particles due to single and to compound scattering. Following J. J. Thomson's argument, the average deflexion $\theta$ after passing through a thickness $t$ of matter is proportional to the square root of the number of encounters and is given by

$$\theta_t = \frac{3\pi b}{8R}\sqrt{\pi R^2 \cdot n \cdot t} = \frac{3\pi b}{8}\sqrt{\pi n t},$$

where $n$ as before is equal to the number of atoms per unit volume.

The probability $p_1$ for compound scattering that the deflexion of the particle is greater than $\phi$ is equal to $e^{-\phi^2/\theta_t^2}$.

Consequently

$$\phi^2 = -\frac{9\pi^3}{64}b^2 n t \log p_1.$$

Next suppose that single scattering alone is operative. We have seen (§ 3) that the probability $p_2$ of a deflexion greater than $\phi$ is given by

$$p = (\pi/4)\, b^2 \cdot n \cdot t(\cot^2 \phi/2).$$

By comparing these two equations

$$p_2 \log p_1 = -0.181\phi^2 \cot^2 \phi/2,$$

$\phi$ is sufficiently small that

$$\tan \phi/2 = \phi/2,$$
$$p_2 \log p_1 = -0.72$$

If we suppose that

$$p_2 = 0.5, \text{ then } p_1 = 0.24$$

If

$$p_2 = 0.1, \text{ then } p_1 = 0.0004$$

It is evident from this comparison, that the probability for any given deflexion is always greater for single than for compound scattering. The difference is especially marked when only a small fraction of the particles are scattered through any given angle. It follows from this result that the distribution of particles due to encounters with the atoms is for small thicknesses mainly governed by single scattering. No doubt compound scattering produces some effect in equalizing the distribution of the scattered particles; but its effect becomes relatively smaller, the smaller the fraction of the particles scattered through a given angle.

## § 6. COMPARISON OF THEORY WITH EXPERIMENTS

On the present theory, the value of the central charge $Ne$ is an important constant, and it is desirable to determine its value for different atoms. This can be most simply done by determining the small fraction of $\alpha$ or $\beta$ particles of known velocity falling on a thin metal screen, which are scattered between $\phi$ and $\phi + d\phi$ where $\phi$ is the angle of deflexion, The influence of compound scattering should be small when this fraction is small.

Experiments in these directions are in progress, but it is desirable at this stage to discuss in the light of the present theory the data already published on scattering of $\alpha$ and $\beta$ particles,

The following points will be discussed:—

(a) The 'diffuse reflexion' of $\alpha$ particles, i.e. the scattering of $\alpha$ particles through large angles (Geiger and Marsden.)

(b) The variation of diffuse reflexion with atomic weight of the radiator (Geiger and Marsden.)

(c) The average scattering of a pencil of $\alpha$ rays transmitted through a thin metal plate (Geiger.)

(d) The experiments of Crowther on the scattering of $\beta$ rays of different velocities by various metals.

(a) In the paper of Geiger and Marsden (*loc. cit.*) on the diffuse reflexion of $\alpha$ particles falling on various substances it was shown that about 1/8000 of the $\alpha$ particles from radium C falling on a thick plate of platinum are scattered back in the direction of the incidence. This fraction is deduced on the assumption that the $\alpha$ particles are uniformly scattered in all directions, the observation being made for a deflexion of about 90°. The form of experiment is not very suited for accurate calculation, but from the data available it can be shown that the scattering observed is about that to be expected on the theory if the atom of platinum has a central charge of about 100 $e$.

In their experiments on this subject, Geiger and Marsden gave the relative number of $\alpha$ particles diffusely reflected from thick layers of different metals, under similar conditions. The numbers obtained by them are given in the table below, where $z$ represents the relative number of scattered particles, measured by the of scintillations per minute on a zinc sulphide screen.

| Metal | Atomic weight | $z$ | $z/A^{3/2}$ |
|---|---|---|---|
| Lead | 207 | 62 | 208 |
| Gold | 197 | 67 | 242 |
| Platinum | 195 | 63 | 232 |
| Tin | 119 | 34 | 226 |
| Silver | 108 | 27 | 241 |
| Copper | 64 | 14.5 | 225 |
| Iron | 56 | 10.2 | 250 |
| Aluminium | 27 | 3.4 | 243 |
| | | | Average 233 |

On the theory of single scattering, the fraction of the total number of $\alpha$ particles scattered through any given angle in passing through a thickness $t$ is proportional to $n \cdot A^2 t$, assuming that the central charge is proportional to the atomic weight A. In the present case, the thickness of matter from which the scattered $\alpha$ particles are able to emerge and affect the zinc sulphide screen depends on the metal. Since Bragg has shown that the stopping power of an atom for an $\alpha$ particle is proportional to the square root of its atomic weight, the value of $nt$ for different elements is proportional to $1/$[square root of] A. In this case $t$ represents the greatest depth from which the scattered $\alpha$ particles emerge. The number $z$ of $\alpha$ particles scattered back from a thick layer is consequently proportional to $A^{3/2}$ or $z/A^{3/2}$ should be a constant.

To compare this deduction with experiment, the relative values of the latter quotient are given in the last column. Considering the difficulty of the experiments, the agreement between theory and experiment is reasonably good.*

The single large scattering of $\alpha$ particles will obviously affect to some extent the shape of the Bragg ionization curve for a pencil of $\alpha$ rays. This effect of large scattering should be marked when the $\alpha$ rays have traversed screens of metals of high atomic weight, but should be small for atoms of light atomic weight.

(c) Geiger made a careful determination of the scattering of $\alpha$ particles passing through thin metal foils, by the scintillation method, and deduced the most probable angle through which the $\alpha$ particles are deflected in passing through known thickness of different kinds of matter.

A narrow pencil of homogeneous $\alpha$ rays was used as a source. After passing through the scattering foil, the total number of $\alpha$ particles are deflected through different angles was directly measured. The angle for which the number of scattered particles was a maximum was taken as the most probable angle. The variation of the most probable angle with thickness of matter was determined, but calculation from these data is somewhat complicated by the variation of velocity of the $\alpha$ particles in their passage through the scattering material. A consideration of the curve of distribution of the $\alpha$ particles given in the paper (*loc.cit.* p. 498) shows that the angle through which half the particles are scattered is about 20 per cent greater than the most probable angle.

We have already seen that compound scattering may become important when about half the particles are scattered through a given

---

*The effect of change of velocity in an atomic encounter is neglected in this calculation.

angle, and it is difficult to disentangle in such cases the relative effects due to the two kinds of scattering. An approximate estimate can be made in the following ways:—From (§ 5) the relation between the probabilities $p_1$ and $p_2$ for compound and single scattering respectively is given by

$$p_2 \log p_1 = -0.721.$$

The probability $q$ of the combined effects may as a first approximation be taken as

$$q = \left(p_1^2 + p_2^2\right)^{1/2}.$$

If $q = 0.5$, it follows that

$$p_1 = 0.2 \text{ and } p_2 = 0.46$$

We have seen that the probability $P_2$ of a single deflexion greater than $\phi$ is given by

$$p_2 = (\pi/4)n \cdot t \cdot b^2 \left(\cot^2 \phi/2\right).$$

Since in the experiments considered $\phi$ is comparatively small

$$\frac{\phi\sqrt{p_2}}{\sqrt{\pi n t}} = b = \frac{2NeE}{mu^2}.$$

Geiger found that the most probable angle of scattering of the $\alpha$ rays in passing through a thickness of gold equivalent in stopping power to about 0.76 cm. of air was $1° 40'$. The angle $\phi$ through which half the $\alpha$ particles are tuned thus corresponds to $2°$ nearly.

$$t = 0.00017 \text{ cm.}; n = 6.07 \times 10^{22};$$

$$u \left(\text{average value}\right) = 1.18 \times 10^9.$$

$$E/m = 1.5 \times 10^4 \text{ E.S units}; e = 4.65 \times 10^{-10},$$

Taking the probability of single scattering $= 0.46$ and substituting the above value in the formula, the value of N for gold comes out to be 97.

For a thickness of gold equivalent in stopping power to 2.12 cms, of air, Geiger found the most probable angle to be $3° 40'$. In this case, $t = 0.00047$, $\phi = 4°.4$, and average $u = 1.7 \times 10^9$, and N comes out to be 114.

Geiger showed that the most probable angle of deflexion for an atom was nearly proportional to its atomic weight. It consequently follows that the value for N for different atoms should be nearly proportional to their atomic weights, at any rate for atomic weights between gold and aluminum.

Since the atomic weight of platinum is nearly equal to that of gold, it follows from these considerations that the magnitude of the diffuse reflexion of $\alpha$ particles through more than 90° from gold and the magnitude of the average small angle scattering of a pencil of rays in passing through gold-foil are both explained on the hypothesis of single scattering by supposing the atom of gold has a central charge of about 100 e.

(d) *Experiments of a Crowther on scattering of $\alpha$ rays.*—We shall now consider how far the experimental results of Crowther on scattering of $\beta$ particles of different velocities by various materials can be explained on the general theory of single scattering. On this theory, the fraction of $\beta$ particles $p$ turned through an angel greater than $\phi$ is given by

$$p = (\pi/4)\, n \cdot t \cdot b^2 \left(\cot^2 \phi/2\right).$$

In most of Crowther's experiments $\phi$ is sufficiently small that tan $\phi/2$ may be put equal to $\phi/2$ without much error. Consequently

$$\phi^2 = 2\pi n \cdot t \cdot b^2 \quad \text{if} \quad p = 1/2$$

On the theory of compound scattering, we have already seen that the chance $p_1$ that the deflexion of the particles is greater than $\phi$ is given by

$$\phi^2 / \log p_1 = -\frac{9\pi^3}{64} n \cdot t \cdot b^2.$$

Since in the experiments of Crowther the thickness $t$ of matter was determined for which $p_1 = 1/2$,

$$\phi^2 = 0.96\pi n t b^2.$$

For the probability of 1/2, the theories of single and compound scattering are thus identical in general form, but differ by a numerical constant. It is thus clear that the main relations on the theory of compound scattering of Sir J. J. Thomson, which were verified experimentally by Crowther, hold equally well on the theory of single scattering.

For example, it $t_m$ be the thickness for which half the particles are scattered through an angle $\phi$, Crowther showed that $\phi$/[square root of] $t_m$ and also $mu^2$/E times [square root of] $t_m$ were constants for a given material when $\phi$ was fixed. These relations hold also on the theory of single scattering. Notwithstanding this apparent similarity in form, the two theories are fundamentally different. In one case, the effects observed are due to cumulative effects of small deflexion, while in the other the large deflexions are supposed to result from a single encounter. The distribution of scattered particles is entirely different on the two theories when the probability of deflexion greater than $\phi$ is small.

We have already seen that the distribution of scattered $\alpha$ particles at various angles has been found by Geiger to be in substantial agreement with the theory of single scattering, but can not be explained on the theory of compound scattering alone. Since there is every reason to believe that the laws of scattering of $\alpha$ and $\beta$ particles are very similar, the law of distribution of scattered $\beta$ particles should be the same as for $\alpha$ particles for small thicknesses of matter. Since the value of $mu^2$/E for $\beta$ particles is in most cases much smaller than the corresponding value for the $\alpha$ particles, the chance of large single deflexions for $\beta$ particles in passing through a given thickness of matter is much greater than for $\alpha$ particles. Since on the theory of single scattering the fraction of the number of particles which are undeflected through this angle is proportional to $kt$, where $t$ is the thickness supposed small and $k$ a constant, the number of particles which are undeflected through this angle is proportional to $1 - kt$. From considerations based on the theory of compound scattering, Sir J.J. Thomson deduced that the probability of deflexion less than $\Phi$ is proportional to $1 - e^{\mu/t}$ is where $\mu$ is a constant for any given value of $\phi$.

The correctness of this latter formula was tested by Crowther by measuring electrically the fraction I/I$_0$ of the scattered $\beta$ particles

which passed through a circular opening subtending an angle of 36°
with the scattering material. If

$$I/I_0 = 1 - 1 - e^{\mu/t},$$

the value of I should decrease very slowly at first with increase of
$t$. Crowther, using aluminium as scattering material, states that the
variation of $I/I_0$ was in good accord with this theory for small values of
$t$. On the other hand, if single scattering be present, as it undoubtedly
is for $\alpha$ rays, the curve showing the relation between $I/I_0$ and $t$ should
be nearly linear in the initial stages. The experiments of Marsden[*]
on scattering of $\beta$ rays, although not made with quite so small a
thickness of aluminium as that used by Crowther, certainly support
such a conclusion. Considering the importance of the point at issue,
further experiments on this question are desirable.

From the table given by Crowther of the value $\phi/$[square root of]
$t_m$ for different elements for $\beta$ rays of velocity $2.68 \times 10^{-10}$ cms.
per second, the value of the central charge $Ne$ can be calculated on
the theory of single scattering. It is supposed, as in the case of the
$\alpha$ rays, that for given value of $\phi/$[square root of] $t_m$ the fraction of
the $\beta$ particles deflected by single scattering through an angle greater
than $\phi$ is 0.46 instead of 0.5

The value of N calculated from Crowther's data are given below.

| Element | Atomic weight | $\phi/$[square root of] $t_m$ | N |
|---------|---------------|-------------------------------|---|
| Aluminium | 27 | 4.25 | 22 |
| Copper | 63.2 | 10.0 | 42 |
| Silver | 108 | 29 | 138 |
| Platnium | 194 | 29 | 138 |

It will be remembered that the values of N for gold deduced
from scattering of the $\alpha$ rays were in two calculations 97 and 114.
These numbers are somewhat smaller than the values given above

[*] Phil. Mag. xviii. p. 909 (1909)

for platinum (viz. 138), whose atomic weight is not very different from gold. Taking into account the uncertainties involved in the calculation from the experimental data, the agreement is sufficiently close to indicate that the same general laws of scattering hold for the $\alpha$ and $\beta$ particles, notwithstanding the wide differences in the relative velocity and mass of these particles.

As in case of the $\alpha$ rays, the value of N should be most simply determined for any given element by measuring the small fraction of the incident $\beta$ particles scattered through a large angle. In this way, possible errors due to small scattering will be avoided.

The scattering data for the $\beta$ rays, as well as for the $\alpha$ rays indicate that the central charge in an atom is approximately proportional to its atomic weight. This falls in with the experimental deductions of Schmidt.* In his theory of absorption of $\beta$ rays, he supposed that in traversing a thin sheet of matter, a small fraction $\alpha$ of the particles are stopped, and a small fraction $\beta$ are reflected or scattered back in the direction of incidence. From comparison of the absorption curves of different elements, he deduced that the value of the constant $\beta$ for different elements is proportional to $nA^2$ where $n$ is the number of atoms per unit volume and A the atomic weight of the element. This is exactly the relation to be expected on the theory of single scattering if the central charge on an atom is proportional to its atomic weight.

## § 7. GENERAL CONSIDERATIONS

In comparing the theory outlined in this paper with the experimental results, it has been supposed that the atom consists of a central charge supposed concentrated at a point, and that the large single deflexions of the $\alpha$ and $\beta$ particles are mainly due to their passage through the strong central field. The effect of the equal and opposite compensation charge supposed distributed uniformly throughout a sphere has been neglected. Some of the evidence in support of these assumptions will now be briefly considered. For concreteness, consider the passage of

* *Annal. d. Phys.* iv. 23. p. 671 (1907)

a high speed $\alpha$ particle through an atom having a positive central charge N$e$, and surrounded by a compensating charge of N electrons. Remembering that the mass, momentum, and kinetic energy of the $\alpha$ particle arc very large compared with the corresponding values of an electron in rapid motion, it does not seem possible from dynamic considerations that an $\alpha$ particle can be deflected through a large angle by a close approach to an electron, even if the latter be in rapid motion and constrained by strong electrical forces. It seems reasonable to suppose that the chance of single deflexions through a large angle due to this cause, if not zero, must be exceedingly small compared with that due to the central charge.

It is of interest to examine how far the experimental evidence throws light on the question of extent of the distribution of central charge. Suppose, for example, the central charge to be composed of N unit charges distributed over such a volume that the large single deflexions are mainly due to the constituent charges and not to the external field produced by the distribution. It has been shown (§ 3) that the fraction of the $\alpha$ particles scattered through a large angle is proportional to $(NeE)^2$, where N$e$ is the central charge concentrated at a point and E the charge on the deflected particles, If, however, this charge is distributed in single units, the fraction of the $\alpha$ particles scattered through a given angle is proportional of $Ne^2$ instead of $N^2e^2$. In this calculation, the influence of mass of the constituent particle has been neglected, and account has only been taken of its electric field. Since it has been shown that the value of the central point charge for gold must be about 100, the value of the distributed charge required to produce the same proportion of single deflexions through a large angle should be at least 10,000. Under these conditions the mass of the constituent particle would be small compared with that of the $\alpha$ particle, and the difficulty arises of the production of large single deflexions at all. In addition, with such a large distributed charge, the effect of compound scattering is relatively more important than that of single scattering. For example, the probable small angle of deflexion of pencil of $\alpha$ particles passing through a thin gold foil would be much greater than that experimentally observed by Geiger (§ b–c). The large and small angle scattering could not then be explained by the assumption of a central charge of the same value. Considering the evidence as a whole, it seems simplest to suppose that the atom contains a central charge distributed through a very small volume, and that the large single deflexions are due to the central charge as a whole, and not to its constituents. At the same time, the experimental evidence is not precise enough to negative the possibility that a small fraction of the positive charge may be carried by satellites extending some distance from the centre. Evidence on this point could be obtained by examining whether the same central charge is required to explain the large single deflexions of $\alpha$ and $\beta$ particles; for the $\alpha$ particle must approach much closer to the center of the atom than the $\beta$ particle of average speed to suffer the same large deflexion.

The general data available indicate that the value of this central charge for different atoms is approximately proportional to their atomic weights, at any rate of atoms heavier than aluminium. It will be of great interest to examine experimentally whether such a simple relation holds also for the lighter atoms. In cases where the mass of the deflecting atom (for example, hydrogen, helium, lithium) is not very different from that of the $\alpha$ particle, the general theory of single scattering will require modification, for it is necessary to take into account the movements of the atom itself (see § 4).

It is of interest to note that Nagaoka* has mathematically considered the properties of the Saturnian atom which he supposed to consist of a central attracting mass surrounded by rings of rotating electrons. He showed that such a system was stable if the attracting force was large. From the point of view considered in his paper, the chance of large deflexion would practically be unaltered, whether the atom is considered to be disk or a sphere. It may be remarked that the approximate value found for the central charge of the atom of gold (100 $e$) is about that to be expected if the atom of gold consisted of 49 atoms of

---

*Nagaoka, Phil. Mag. vii. p. 445 (1904).

helium, each carrying a charge of 2 *e*. This may be only a coincidence, but it is certainly suggestive in view of the expulsion of helium atoms carrying two unit charges from radioactive matter.

The deductions from the theory so far considered are independent of the sign of the central charge, and it has not so far been found possible to obtain definite evidence to determine whether it be positive or negative. It may be possible to settle the question of sign by consideration of the difference of the laws of absorption of the $\beta$ particles to be expected on the two hypothesis, for the effect of radiation in reducing the velocity of the $\beta$ particle should be far more marked with a positive than with a negative center. If the central charge be positive, it is easily seen that a positively charged mass if released from the center of a heavy atom, would acquire a great velocity in moving through the electric field. It may be possible in this way to account for the high velocity of expulsion of $\alpha$ particles without supposing that they are initially in rapid motion within the atom.

Further consideration of the application of this theory to these and other questions will be reserved for a later paper, when the main deductions of the theory have been tested experimentally. Experiments in this direction are already in progress by Geiger and Marsden.

University of Manchester
April 1911

# On the Constitution of Atoms and Molecules

By

Niels Bohr*

First published in

*Philosophical Magazine* Series 6, Volume 26 July 1913, p. 1–25

In order to explain the results of experiments on scattering of $\alpha$ rays by matter Prof. Rutherford[†] has given a theory of the structure of atoms. According to this theory, the atoms consist of a positively charged nucleus surrounded by a system of electrons kept together by attractive forces from the nucleus; the total negative charge of the electrons is equal to the positive charge of the nucleus. Further, the nucleus is assumed to be the seat of the essential part of the mass of the atom, and to have linear dimensions exceedingly small compared with the linear dimensions of the whole atom. The number of electrons in an atom is deduced to be approximately equal to half the atomic weight. Great interest is to be attributed to this atom-model; for, as Rutherford has shown, the assumption of the existence of nuclei, as those in question, seems to be necessary in order to account for the results of the experiments on large angle scattering of the $\alpha$ rays[‡].

In an attempt to explain some of the properties of matter on the basis of this atom-model we meet however, with difficulties of a serious nature arising from the apparent instability of the system of electrons: difficulties purposely avoided in atom-models previously considered, for instance, in the one proposed by Sir J. J. Thomson.[§] According to the theory of the latter the atom consists of a sphere of uniform positive electrification, inside which the electrons move in circular orbits.

---

*Communicated by Prof. E. Rutherford, F.R.S.
[†] E. Rutherford, Phil. Mag. xxi. p. 669 (1911).
[‡] See also Geiger and Marsden, Phil. Mag. April 1913.
[§] J. J. Thomson, Phil. Mag. vii. p. 237 (1904).

The principal difference between the atom-models proposed by Thomson and Rutherford consists in the circumstance the forces acting on the electrons in the atom-model of Thomson allow of certain configurations and motions of the electrons for which the system is in a stable equilibrium; such configurations, however, apparently do not exist for the second atom-model. The nature of the difference in question will perhaps be most clearly seen by noticing that among the quantities characterizing the first atom a quantity appears—the radius of the positive sphere—of dimensions of a length and of the same order of magnitude as the linear extension of the atom, while such a length does not appear among the quantities characterizing the second atom, viz. the charges and masses of the electrons and the positive nucleus; nor can it be determined solely by help of the latter quantities.

The way of considering a problem of this kind has, however, undergone essential alterations in recent years owing to the development of the theory of the energy radiation, and the direct affirmation of the new assumptions introduced in this theory, found by experiments on very different phenomena such as specific heats, photoelectric effect, RÃ¶ntgen &c. The result of the discussion of these questions seems to be a general acknowledgment of the inadequacy of the classical electrodynamics in describing the behaviour of systems of atomic size.* Whatever the alteration in the laws of motion of the electrons may be, it seems necessary to introduce in the laws in question a quantity foreign to the classical electrodynamics, *i.e.* Planck's constant, or as it often is called the elementary quantum of action. By the introduction of this quantity the question of the stable configuration of the electrons in the atoms is essentially changed as this constant is of such dimensions and magnitude that it, together with the mass and charge of the particles, can determine a length of the order of magnitude required.

This paper is an attempt to show that the application of the above ideas to Rutherford's atom-model affords a basis for a theory of the

---

*See f. inst., 'ThÃ©orie du ravonnement et les quanta.' Rapports de la rÃ©union Ã Bruxelles, Nov. 1911. Paris, 1912.

constitution of atoms. It will further be shown that from this theory we are led to a theory of the constitution of molecules.

In the present first part of the paper the mechanism of the binding of electrons by a positive nucleus is discussed in relation to Planck's theory. It will be shown that it is possible from the point of view taken to account in a simple way for the law of the line spectrum of hydrogen. Further, reasons are given for a principal hypothesis on which the considerations contained in the following parts are based.

I wish here to express my thanks to Prof. Rutherford his kind and encouraging interest in this work.

# Part I: Binding of Electrons by Positive Nuclei

## Â§ 1. General Considerations

The inadequacy of the classical electrodynamics in accounting for the properties of atoms from an atom-model as Rutherford's, will appear very clearly if we consider a simple system consisting of a positively charged nucleus of very small dimensions and an electron describing closed orbits around it. For simplicity, let us assume that the mass of the electron is negligibly small in comparison with that of the nucleus, and further, that the velocity of the electron is small compared with that of light.

Let us at first assume that there is no energy radiation. In this case the electron will describe stationary elliptical orbits. The frequency of revolution $\omega$ and the major-axis of the orbit $2a$ will depend on the amount of energy w which must be transferred to the system in order to remove the electron to an infinitely great distance apart from the nucleus. Denoting the charge of the electron and of the nucleus by $-e$ and E respectively and the mass of the electron by $m$ we thus get

$$\omega = \frac{\sqrt{2}}{\pi} \frac{W^{\frac{3}{2}}}{eE\sqrt{m}}, \quad 2\alpha = \frac{eE}{W}. \tag{1}$$

Further, it can easily be shown that the mean value of the kinetic energy of the electron taken for a whole revolution is equal to W. We see that if the value of W is not given there will be no values of $\omega$ and $a$ characteristic for the system in question.

Let us now, however, take the effect of the energy radiation into account, calculated in the ordinary way from the acceleration of the electron. In this case the electron will no longer describe stationary orbits. W will continuously increase, and the electron will approach the nucleus describing orbits of smaller and smaller dimensions, and with greater and greater frequency; the electron on the average gaining in kinetic energy at the same time as the whole system loses energy.

This process will go on until the dimensions of the orbit are of the same order of magnitude as the dimensions of the electron or those of the nucleus. A simple calculation shows that the energy radiated out during the process considered will be enormously great compared with that radiated out by ordinary molecular processes.

It is obvious that the behaviour of such a system will be very different from that of an atomic system occurring in nature. In the first place, the actual atoms in their permanent state seem to have absolutely fixed dimensions and frequencies. Further, if we consider any molecular process, the result seems always to be that after a certain amount of energy characteristic for the systems in question is radiated out, the systems will again settle down in a stable state of equilibrium, in which the distances apart of the particles are of the same order of magnitude as before the process.

Now the essential point in Planck's theory of radiation is that the energy radiation from an atomic system does not take place in the continuous way assumed in the ordinary electrodynamics, but that it, on the contrary, takes place in distinctly separated emissions, the amount of energy radiated out from an atomic vibrator of frequency $v$ in a single emission being equal to $\tau h v$ where $\tau$ is an entire number, and $h$ is a universal constant*.

Returning to the simple case of an electron and a positive nucleus considered above, let us assume that the electron at the beginning of the interaction with the nucleus was at a great distance apart from the nucleus, and bad no sensible velocity relative to the latter. Let us further assume that the electron after the interaction has taken place has settled down in a stationary orbit around the nucleus. We shall, for reasons referred to later, assume that the orbit in question is circular; this assumption will, however, make no alteration in the calculations for systems containing only a single electron.

Let us now assume that, during the binding of the electron, a homogeneous radiation is emitted of a frequency $v$, equal to half the

*See f. inst., M. Planck, *Ann. d. Phys.* xxxi. p. 758 (1910); xxxvii. p. 642 (1912); *Verh. deutsch. Phys. Ges.* 1911, p. 138.

frequency of revolution of the electron in its final orbit; then, from Planck's theory, we might expect, that the amount of energy emitted by the process considered is equal to $\tau h\nu$, where $h$ is Planck's constant and $\tau$ an entire number. If we assume that the radiation emitted is homogeneous, the second assumption concerning the frequency of the radiation suggests itself, since the frequency of revolution of the electron at the beginning of the emission is 0. The question, however, of the rigorous validity of both assumptions, and also of the application made of Planck's* theory will be more closely discussed in § 3.

Putting

$$W = \tau h \frac{\omega}{2}, \qquad (2)$$

we can by help of the formula (1)

$$W = \frac{2\pi^2 m e^2 E^2}{\tau^2 h^2}, \quad \omega = \frac{4\pi^2 m e^2 E^2}{\tau^3 h^3}, \quad 2\alpha = \frac{\tau^2 h^2}{2\pi^2 m e E}$$

If in these expressions we give $\tau$ different values we get -a series of values for W, $\omega$, and $a$ corresponding to a series of configurations of the system. According to the above considerations, we are led to assume that these configurations will correspond to states of the system in which there is no radiation of energy states which consequently will be stationary as long as the system is not disturbed from outside. We see that the value of W' is greatest if $\tau$ has its smallest value 1. This case will therefore correspond to the most stable state of the system, *i.e.* will correspond to the binding of the electron for the breaking up of which the greatest amount of energy is required.

Putting in the above expressions $\tau = l$ and $E = e$, and introducing the experimental values

$$e = 4.7 \times 10\,\text{Å}^{-10}, e/m = 5.31 \times 10^{17}, h = 6.5 \times 10\,\text{Å}^{-27}$$

we get

$$2a = 1.1 \times 10\,\text{Å}^{-8}\,\text{cm.}, \omega = 6.2 \times 10^{15}\,\text{sec}\,\text{Å}^{-1}, W/e = 13\,\text{volt.}$$

We see that these values are of the same order of magnitude as the linear dimensions of the atoms, the optical frequencies, and the ionization-potentials.

The general importance of Planck's theory for the discussion of the behaviour of atomic systems was originally pointed out by Einstein*. The considerations of Einstein have been developed and applied on a number of different phenomena, especially by Stark, Nernst, and Sommerfield [sic]. The agreement as to the order of magnitude between values observed for the frequencies and dimensions of the atoms, and values for these quantities calculated by considerations similar to those given above, has been the subject of much discussion. It was first pointed out by Haas†, in an attempt to explain the meaning and the value of Planck's constant on the basis of J. J. Thomson's atom-model by help of the linear dimensions and frequency of an hydrogen atom.

Systems of the kind considered in this paper, in which the forces between the particles vary inversely as the square of the distance, are discussed in relation to Planck's theory by J. W. Nicholson‡. In a series of papers this author has shown that it seems to be possible to account for lines of hitherto unknown origin in the spectra of the stellar nebulae and that of the solar corona by assuming the presence in these bodies of certain hypothetical elements of exactly indicated constitution. The atoms of these elements are supposed to consist simply of a ring of a few electrons surrounding a positive nucleus of negligibly small dimensions. The ratios between the frequencies corresponding to the lines in question are compared with the ratios between the frequencies corresponding to different modes of vibration of the ring of electrons. Nicholson has obtained a relation to Planck's theory showing that the ratios between the wave-length of different sets of lines of the coronal spectrum can be accounted for with great accuracy by assuming that the ratio between the energy of the system

---

*A. Einstein, *Ann. d. Phys.* xvii. p. 132 (1905); xx. p. 199 (1906); xxii. p. 180 (1907).

†A. E. Haas, *Jahrb. d. Rad. u. El.* vii. p. 261 (1910). See further, A. Schidlof, *Ann. d. Phys.* xxxv. p. 90 (1911); E. Wertheimer, *Phys. Zeitschr.* xii. p. 409 (1911), *Verh. deutsch. Phys. Ges.* 1912, p. 431; F. A. Lindemann, *Verh. deutsch. Phys. Ges.* 1911, pp. 482, 1107; F. Haber, *Verh. deutsch. Phys. Ges.* 1911, p. 1117.

‡J. W. Nicholson, *Month. Not. Roy. Astr. Soc.* lxxii. pp. 49,130, 677, 693, 729 (1912).

and the frequency of rotation of the ring is equal to an entire multiple of Planck's constant. The quantity Nicholson refers to as the energy is equal to twice the quantity which we have denoted above by W. In the latest paper cited Nicholson has found it necessary to give the theory a more complicated form, still, however, representing the ratio of energy to frequency by a simple function of whole numbers.

The excellent agreement between the calculated and observed values of the ratios between the wave-lengths in question seems a strong argument in favour of the validity of the foundation of Nicholson's calculations.

These objections are intimately connected with the problem of the homogeneity of the radiation emitted. In Nicholson's calculations the frequency of lines in a line-spectrum is identified with the frequency of vibration of a mechanical system, in a distinctly indicated state of equilibrium. As a relation from Planck's theory is used, we might expect that the radiation is sent out in quanta; but systems like those considered, in which the frequency is a function of the energy, cannot emit a finite amount of a homogeneous radiation; for, as soon as the emission of radiation is started, the energy and also the frequency of the system are altered. Further, according to the calculation of Nicholson, the systems are unstable for some modes of vibration. Apart from such objections—which may be only formal—it must be remarked, that the theory in the form given does not seem to be able to account for the well-known laws of Miner and Rydberg connecting the frequencies of the lines in the line-spectra of the ordinary elements.

It will now be attempted to show that the difficulties in question disappear if we consider the problems from the point of view taken in this paper. Before proceeding it may be useful to restate briefly the ideas characterizing the calculations on p. 5. The principal assumptions used are:

(1) That the dynamical equilibrium of the systems in the stationary states can be discussed by help of the ordinary mechanics, while the passing of the systems between different stationary states cannot be treated on that basis.

(2) That the latter process is followed by the emission of a *homogeneous* radiation, for which the relation between the frequency and the amount of energy emitted is the one given by Planck's theory.

The first assumption seems to present itself; for it is known that the ordinary mechanics cannot have an absolute validity, but will only hold in calculations of certain mean values of the motion of the electrons. On the other hand, in the calculations of the dynamical equilibrium in a stationary state in which there is no relative displacement of the particles, we need not distinguish between the actual motions and their mean values. The second assumption is in obvious contrast to the ordinary ideas of electrodynamics but appears to be necessary in order to account for experimental facts.

In the calculations on page 5 we further made use of the more special assumptions, viz. that the different stationary states correspond to the emission of a different number of Planck's energy-quanta, and that the frequency of the radiation emitted during the passing of the system from a state in which no energy is yet radiated out to one of the stationary states, is equal to half the frequency of revolution of the electron in the latter state. We can, however (see Â§ 3), also arrive at the expressions (3) for the stationary states by using assumptions of somewhat different form. We shall, therefore, postpone the discussion of the special assumptions, and first show how by the help of the above principal assumptions, and of the expressions (3) for the stationary states, we can account for the line-spectrum of hydrogen.

## Â§ 2. EMISSION OF LINE-SPECTRA.

*Spectrum of Hydrogen.*—General evidence indicates that an atom of hydrogen consists simply of a single electron rotating round a positive nucleus of charge $e$*. The reformation of a hydrogen atom, when the electron has been removed to great distances away from the

---

*See f. inst. N. Bohr, Phil. Mag. xxv. p. 24 (1913). The conclusion drawn in the paper cited is strongly supported by the fact that hydrogen, in the experiments on positive rays of Sir J. J. Thomson, is the only element which never occurs with a positive charge corresponding to the loss of more than one electron (comp. Phil. Mag. xxiv. p. 672 (1912)).

nucleus—*e.g.* by the effect of electrical discharge in a vacuum tube—will accordingly correspond to the binding of an electron by a positive nucleus considered on p. 5. If in (3) we put E $=$ *e*, we get for the total amount of energy radiated out by the formation of one of the stationary states,

$$W_\tau = \frac{2\pi^2 me^4}{h^2\tau^2}.$$

The amount of energy emitted by the passing of the system from a state corresponding to $\tau = \tau_1$ to one corresponding to $\tau = \tau_2$, is consequently

$$W_{\tau_2} - W_{\tau_1} = \frac{2\pi^2 me^4}{h^2}\left(\frac{1}{\tau_2^2} - \frac{1}{\tau_1^2}\right).$$

If now we suppose that the radiation in question is homogeneous, and that the amount of energy emitted is equal to $h\nu$, where $\nu$ is the frequency of the radiation, we get

$$W_{\tau_2} - W_{\tau_1} = h\nu,$$

and from this

$$\nu = \frac{2\pi^2 me^4}{h^3}\left(\frac{1}{\tau_2^2} - \frac{1}{\tau_1^2}\right). \tag{3}$$

We see that this expression accounts for the law connecting lines in the spectrum of hydrogen. If we put $\tau_2 = 2$ and let $\tau_1$ vary, we get the ordinary Balmer series. If we put $\tau_2 = 3$, we get the series in the ultra-red observed by Paschen* and previously suspected by Ritz. If we put $\tau_2 = 1$ and $\tau_2 = 4, 5, \ldots$, we get series respectively in the extreme ultra-violet and the extreme ultra-red, which are not observed, but the existence of which may be expected.

The agreement in question is quantitative as well as qualitative. Putting

$$e = 4.7 \times 10 \text{\AA}^{-10}, e/m = 5.31 \times 10^{17}, h = 6.5 \times 10 \text{\AA}^{-27}$$

---

*F. Paschen, *Ann. d. Phys.* xxvii. p. 565 (1908).

we get

$$\frac{2\pi^2 me^4}{h^3} = 3 \cdot 1.10^{15}.$$

The observed value for the factor outside the bracket in the formula (4) is

$$3.290 \times 10^{15}.$$

The agreement between the theoretical and observed values is inside the uncertainty due to experimental errors in the constants entering in the expression for the theoretical value. We shall in § 3 return to consider the possible importance of the agreement in question.

It may be remarked that the fact, that it has not been possible to observe more than 12 lines of the Balmer series in experiments with vacuum tubes, while 33 lines are observed in the spectra of some celestial bodies, is just what we should expect from the above theory. According to the equation (3) the diameter of the orbit of the electron in the different stationary states is proportional to $\tau_2$. For $\tau = 12$ the diameter is equal to $1.6 \times 10Å^{-6}$ cm., or equal to the mean distance between the molecules in a gas at a pressure of about 7 mm. mercury; for $\tau = 33$ the diameter is equal to $1.2 \times 10Å^{-5}$ cm., corresponding to the mean distance of the molecules at a pressure of about 0.02 mm. mercury. According to the theory the necessary condition for the appearance of a great number of lines is therefore a very small density of the gas; for simultaneously to obtain an intensity sufficient for observation the space filled with the gas must be very great. If the theory is right, we may therefore never expect to be able in experiments with vacuum tubes to observe the lines corresponding to high numbers of the Balmer series of the emission spectrum of hydrogen; it might, however, be possible to observe the lines by investigation of the absorption spectrum of this gas (see § 4).

It will be observed that we in the above way do not obtain other series of lines, generally ascribed to hydrogen; for instance, the series first observed by Pickering[*] in the spectrum of the star ζ Puppis,

*E. C. Pickering, Astrophys. J. iv p. 369 (1896); v. p. 92 (1897).

and the set of series recently found by Fowler* by experiments with vacuum tubes containing a mixture of hydrogen and helium. We shall, however, see that, by help of the above theory, we can account naturally for these series of lines if we ascribe them to helium.

A neutral atom of the latter element consists. according to Rutherford's theory, of a positive nucleus of charge $2e$ and two electrons. Now considering the binding of a single electron by a helium nucleus, we get, putting $E = 2e$ in the expressions (3) on page 5, and proceeding in exactly the same way as above,

$$\nu = \frac{8\pi^2 m e^4}{h^3}\left(\frac{1}{\tau_2^2} - \frac{1}{\tau_1^2}\right) = \frac{2\pi^2 m e^4}{h^3}\left(\frac{1}{\left(\frac{\tau_2}{2}\right)^2} - \frac{1}{\left(\frac{\tau_1}{2}\right)^2}\right)$$

If we in this formula put, $\tau_2 = 1$ or $\tau_2 = 2$, we get series of lines in the extreme ultra-violet. If we put $\tau_2 = 3$, and let $\tau_1$ vary, we get a series which includes 2 of the series observed by Fowler, and denoted by him as the first and second principal series of the hydrogen spectrum. If we put $\tau_2 = 4$, we get the series observed by Pickering in the spectrum of $\zeta$ Puppis. Every second of the lines in this series is identical with a line in the Balmer series of the hydrogen spectrum; the presence of hydrogen in the star in question may therefore account for the fact that these lines are of a greater intensity than the rest of the lines in the series. The series is also observed in the experiments of Fowler, and denoted in his paper as the Sharp series of the hydrogen spectrum. If we finally in the above formula put $\tau_2 = 5, 6, \ldots$, we get series, the strong lines of which are to be expected in the ultra-red.

The reason why the spectrum considered is not observed in ordinary helium tubes may be that in such tubes the ionization not so complete as in the star considered or in the experiments of Fowler, where a strong discharge was sent through a mixture of hydrogen and helium. The condition for the appearance of the spectrum is, according to the above theory, that helium atoms are present in a state in which they have lost both their electrons. Now we must assume the amount of energy to be used in removing the second electron from

*A. Fowler, Month. Not. Roy. Astr. Soc. lxxiii Dec. 1912.

a helium atom is much greater than that to be used in removing the first. Further, it is known from experiments on positive rays, that hydrogen atoms can acquire a negative charge; therefore the presence of hydrogen in the experiments of Fowler may effect that more electrons are removed from some of the helium atoms than would be the case if only helium were present.

*Spectra of other substances.*—In case of systems containing more electrons we must—in conformity with the result of experiments—expect more complicated laws for the line-spectra those considered. I shall try to show that the view taken above allows, at any rate, a certain understanding of the laws observed.

According to Rydberg's theory—with the generalization given by Ritz*—the frequency corresponding to the lines of the spectrum of an element call be expressed by

$$\nu = F_r(\tau_1 \text{ minus } F_s(\tau_2)),$$

where $\tau_1$ and $\tau_2$ are entire numbers, and $F_1$, $F_2$, $F_2$, ... are functions of $\tau$ which approximately are equal to $K/(\tau + a_1)^2$, $K/(\tau + a_2)^2$, ... K is a universal constant, equal to the factor outside the bracket in the formula (4) for the spectrum of hydrogen. The different series appear if we put $\tau_1$ or $\tau_2$ equal to a fixed number and let the other vary.

The circumstance that the frequency can be written as a difference between two functions of entire numbers suggests an origin of the lines in the spectra in question similar to the one we have assumed for hydrogen; *i.e.* that the lines correspond to a radiation emitted during the passing of the system between two different stationary states. For systems containing more than one electron the detailed discussion may be very complicated, as there will be many different configurations of the electrons which can be taken into consideration as stationary states. This may account for the different sets of series in the line spectra emitted from the substances in question. Here I shall only try to show how, by help of the theory, it can be simply explained

---

*W. Ritz, Phys. Zeitschr. ix p. 521 (1908).

that the constant K entering in Rydberg's formula is the same for all substances.

Let us assume that the spectrum in question corresponds to the radiation emitted during the binding of an electron; and let us further assume that the system including the electron considered is neutral. The force, on the electron, when at a great distance apart from the nucleus and the electrons previously bound, will be very nearly the same as in the above case of the binding of an electron by a hydrogen nucleus. The energy corresponding to one of the stationary states will therefore for $\tau$ great be very nearly equal to that given by the expression (3) on p. 5, if we put $E = e$. For $\tau$ great we consequently get

$$\lim \left(\tau^2 \cdot F_1\left(\tau\right)\right) = \lim \left(\tau^2 \cdot F_2\left(\tau\right)\right) = \cdots = 2\pi^2 me^4 / h^3$$

in conformity with Rydberg's theory.

## Â§ 3. GENERAL CONSIDERATIONS CONTINUED

We shall now return to the discussion (see p. 7) of the special assumptions used in deducing the expressions (3) on p. 5 for the stationary states of a system consisting of an electron rotating round a nucleus.

For one, we have assumed that the different stationary states correspond to an emission of a different number of energy-quanta. Considering systems in which the frequency is a function of the energy, this assumption, however, may be regarded as improbable; for as soon as one quantum is sent out the frequency is altered. We shall now see that we can leave the assumption used and still retain the equation (2) on p. 5, and thereby the formal analogy with Planck's theory.

Firstly, it will be observed that it has not been necessary, in order to account for the law of the spectra by help of the expressions (3) for the stationary states, to assume that in any case a radiation is sent out corresponding to more than a single energy-quantum, $h\nu$. Further information on the frequency of the radiation may be obtained by comparing calculations of the energy radiation in the region of slow vibrations based on the above assumptions with calculations based on the ordinary mechanics. As is known, calculations on the latter basis are in agreement with experiments on the energy radiation in the named region.

Let us assume that the ratio between the total amount of energy emitted and the frequency of revolution of the electron for the different stationary states is given by the equation $W = f(\tau) \cdot h\nu$, instead of by the equation (2). Proceeding in the same way as above we get in this case instead of (3)

$$W = \frac{\pi^2 m e^2 E^2}{2h^2 f^2(\tau)}, \qquad \omega = \frac{\pi^2 m e^2 E^2}{2h^3 f^3(\tau)},$$

Assuming as above that the amount of energy emitted during the passing of the system from a state corresponding to $\tau = \tau_1$ to one for which $-r = -r2$ is equal to h$\nu$, we get instead of (4)

$$\nu = \frac{\pi^2 m e^2 E^2}{2h^3} \left( \frac{1}{f^2(\tau_2)} - \frac{1}{f^2(\tau_1)} \right).$$

We see that in order to get an expression of the same form as the Balmer series we must put $f(\tau) = c\tau$.

In order to determine $c$ let us now consider the passing of the system between two successive stationary states, corresponding to $\tau = N$ and $\tau = N - 1$; introducing $f(\tau) = c\tau$, we get for the frequency of the radiation emitted

$$\nu = \frac{\pi^2 m e^2 E^2}{2c^2 h^3} \cdot \frac{2N - 1}{N^2 (N - 1)^2}.$$

For the frequency of revolution of the electron before and after the emission we have

$$\omega_N = \frac{\pi^2 m e^2 E^2}{2c^3 h^3 N^3} \quad \text{and} \quad \omega_{N-1} = \frac{\pi^2 m e^2 E^2}{2c^3 h^3 (N - 1)^3}.$$

If N is great the ratio between the frequency before and after the emission will be very near equal to 1; and according to the ordinary electrodynamics we should therefore expect that the ratio between the frequency of radiation and the frequency of revolution also is very nearly equal to 1. This condition will only be satisfied if $c = 1/2$.

Putting $f(\tau) = \tau/2$, we however, again arrive at the equation (2) and consequently at the expression (3) for the stationary states.

If we consider the passing of the system between two states corresponding to $\tau = N$ and $\tau = N - n$, where n is small compared with N, we get with the same approximation as above putting $f(\tau) = \tau/2$,

$$v = n\omega$$

The possibility of an emission of a radiation of such a frequency may also be interpreted from analogy with the ordinary elecrodynamics, as in electron rotating round a nucleus in an elliptical orbit will emit a radiation which according to Fourier's theorem can be resolved into homogeneous components, the frequencies of which are $n\omega$, if $\omega$ is the frequency of revolution of the electron.

We are thus led to assume that the interpretation of the equation (2) is not that the different stationary states correspond to an emission of different numbers of energy-quanta, but that the frequency of the energy emitted during the passing of the system from a state in which no energy is yet radiated out to one of the different stationary states, is equal to different multiples of $\omega/2$ where $\omega$ is the frequency of revolution of the electron in the state considered. From this assumption we get exactly the same expressions as before for the stationary states, and from these by help of the principal assumptions on p. 7 the same expression for the law of the hydrogen spectrum. Consequently we may regard our preliminary considerations on p. 5 only as a simple form of representing the results of the theory.

Before we leave the discussion of this question, we shall for a moment return to the question of the significance of the agreement between the observed and calculated values of the constant entering in the expressions (4) for the Balmer series of the hydrogen spectrum. From the above consideration it will follow that, taking the starting-point in the form of the law of the hydrogen spectrum and assuming that the different lines correspond to a homogeneous radiation emitted during the passing between different stationary states, we shall arrive at exactly the same expression for the constant in question as that given by (4), if we only assume (1) that th, radiation is sent out in quanta $h\nu$ and (2) that the frequency of the radiation emitted during the passing of the system between successive stationary states will coincide with the frequency of revolution of the electron in the region of slow vibrations.

As all the assumptions used in this latter way of representing the theory are of what we may call a qualitative character, we are justified in expecting—if the whole way of considering is a sound one—an absolute agreement between the values calculated and observed for the constant in question, and not only an approximate agreement. The formula (4) may therefore be of value in the discussion of the results of experimental determinations of the constants $e$, $m$, and $h$.

While, there obviously can be no question of a mechanical foundation of the calculations given in this paper, it is, however possible to give a very simple interpretation of the result of the calculation on p. 5 by help of symbols taken from the mechanics. Denoting the angular momentum of the electron round the nucleus by M, we have immediately for a circular orbit $\pi M = T/\omega$ where $\omega$ is the frequency of revolution and T the kinetic energy of the electron; for a circular orbit we further have $T = W$ (see p. 3) and from (2), p. 5 we consequently get

$$M = \tau M_o, \text{ where}$$

$$M_o = h/2\pi = 1.04 \times 10\,\text{Å}^{-27}$$

If we therefore assume that the orbit of the electron in the stationary states is circular, the result of the calculation on p. 5 can be expressed by the simple condition: that the angular momentum of the electron round the nucleus in a stationary state of the system is equal to an entire multiple of a universal value, independent of the charge on the nucleus. The possible importance of the angular momentum in the discussion of atomic systems in relation to Planck's theory is emphasized by Nicholson*.

*J. W. Nicholson *loc. cit.* p. 679.

The great number of different stationary states we do not observe except by investigation of the emission and absorption of radiation. In most of the other physical phenomena, however, we only observe the atoms of the matter in a single distinct, state, *i.e.* the state of the atoms at low temperature. From the preceding considerations we are immediately led to the assumption that the "permanent" state is the one among the stationary states during the formation of which the greatest amount of energy is emitted. According to the equation (3) on p. 5, this state is the one corresponds to $\tau = 1$.

## Â§ 4. ABSORPTION OF RADIATION

In order to account for Kirchhoff's law it is necessary to introduce assumptions on the mechanism of absorption of radiation hich correspond to those we have used considering the emission. Thus we must assume that a system consisting of a nucleus and in electron rotating round it under certain circumstances can absorb a radiation of a frequency equal to the frequency of the homogeneous radiation emitted during the passing of the system between different stationary states. Let us consider the radiation emitted during the passing of the system between two stationary states $A_1$ and $A_2$ corresponding to values for $\tau$ equal to $\tau_1$ and $\tau_2$, $\tau_1 > \tau_2$. As the necessary condition for an emission of the radiation in question was the presence of systems in the state $A_1$, we must assume that the necessary condition for an absorption of the radiation is the presence of systems in the state $A_2$.

These considerations seem to be in conformity with experiments on absorption in gases. In hydrogen gas at ordinary conditions for instance there is no absorption of a radiation of a frequency corresponding to the line-spectrum of this gas; such an absorption is only observed in hydrogen gas in a luminous state. This is what we should expect according to the above. We have on p. 9 assumed that the radiation in question was emitted during the passing of the systems between stationary states corresponding to $\tau$ [greater than or equal to] 2. The state of the atoms in hydrogen gas at ordinary conditions should, however, correspond to $\tau = 1$; furthermore, hydrogen atoms at ordinary conditions combine into molecules, i.e. into systems in which the electrons have frequencies different from those in the atoms (see Part III.). From the circumstance that certain substances in a non-luminous state, as, for instance, sodium vapour, absorb radiation corresponding to lines in the line-spectra of the substances, we may, on the other hand, conclude that the lines in question are emitted during the passing of the system between two states, one of which is the permanent state.

How much the above considerations differ from an interpretation based on the ordinary electrodynamics is perhaps most clearly shown by the fact that we have been forced to assume that a system of electrons will absorb a radiation of a frequency different from the frequency of vibration of the electrons calculated in the ordinary way. It may in this connexion be of interest to mention a generalization of the considerations to which we are led by experiments on the photo-electric effect, and which may be able to throw some light on the problem in question. Let us consider a state of the system in which the electron is free, i.e. in which the electron possesses kinetic energy sufficient to remove to infinite distances from the nucleus. If we assume that the motion of the electron is governed by the ordinary mechanics and that there is no (sensible) energy radiation, the total energy of the system—as in the above considered stationary states will be constant. Further, there will be perfect continuity between the two kinds of states, as the difference between frequency and dimensions of the systems in successive stationary states will diminish without limit if $\tau$ increases. In the following considerations we shall for the sake of brevity refer to the two kinds of states in question as "mechanical," states; by this notation only emphasizing the assumption that the motion of the electron in both cases can be accounted for by the ordinary mechanics.

Tracing the analogy between the two kinds of mechanical states, we might now expect the possibility of an absorption of radiation, not only corresponding to the passing of the system between two different stationary states, but also corresponding to the passing between one of the stationary states and a state in which the electron is free; and as above, we might expect that the frequency of this radiation was determined by the equation $E = h\nu$, where E is the difference between the total energy of the system in the two states. As it will be seen, such an absorption of radiation is just what is observed in experiments on ionization by ultra-violet light and by Rĭntgen rays. Obviously, we get in this way the same expression for the kinetic energy of an electron ejected from an atom by photo-electric effect as that deduced by Einstein*, *i.e.* $T = h\nu - W$, where T is the kinetic energy of the electron ejected, and W the total amount of energy emitted during the original binding of the electron.

The above considerations may further account for the result of some experiments of R.W. Wood[†] on absorption of light by sodium vapour. In these experiments, an absorption corresponding to a very great number of lines in the principal series of the sodium spectrum is observed, and in addition a continuous absorption which begins at the head of the series and extends to the extreme ultra-violet. This is exactly what we should expect according to the analogy in question, and, as we shall see, a closer consideration of the above experiments allows us to trace the analogy still further. As mentioned on p. 9 the radii of the orbits of the electrons will for stationary states corresponding to high values for $\tau$ be very great compared with ordinary atomic dimensions. This circumstance was used as an explanation of the non-appearance in experiments with vacuum-tubes of lines corresponding to the higher numbers in the Balmer series of the hydrogen spectrum. This is also in conformity with experiments on the emission spectrum of sodium; in the principal series of the emission spectrum of this substance rather few lines are observed. Now in Wood's experiments the pressure was not very low, and the states corresponding to high values for $\tau$ could therefore not appear; yet in the absorption spectrum about 50

*A. Einstein, *Ann. d. Phys.* xvii. p. 146 (1905).
[†]R. W. Wood, Physical Optics p. 513 (1911).

lines were detected. In the experiments in question we consequently observe an absorption of radiation which is not accompanied by a complete transition between two different stationary states. According to the present theory we must assume that this absorption is followed by an emission of energy during which the systems pass back to tile original stationary state. If there are no collisions between the different systems this energy will be emitted as a radiation of the same frequency as that absorbed, and there will be no true absorption but only a scattering of the original radiation; a true absorption will not occur unless the energy in question is transformed by collisions into kinetic energy of free particles. In analogy we may now from the above experiments conclude that a bound electron—also in cases in which there is no ionization—will have an absorbing (scattering) influence on a homogeneous radiation, as soon as the frequency of the radiation is greater than $W/h$, where W is the total amount of energy emitted during the binding of the electron. This would be highly in favour of a theory of absorption as the one sketched above, as there can in such a case be no question of a coincidence of the frequency of the radiation and a characteristic frequency of vibration of the electron. It will further be seen that the assumption, that there will be an absorption (scattering) of any radiation corresponding to a transition between two different mechanical states, is in perfect analogy with the assumption generally used that a free electron will have an absorbing (scattering) influence on light of any frequency. Corresponding considerations will hold for the emission of radiation.

In analogy to the assumption used in this paper that the emission of line-spectra is due to the re-formation of atoms after one or more of the lightly bound electrons are removed, we may assume that the homogeneous RÃ¶ntgen radiation is emitted during the settling down of the systems after one of the firmly bound electrons escapes, *e.g.* by impact of cathode particles*. In the next part of this paper, dealing with the constitution of atoms, we shall consider the question more

---

*Compare J.J. Thomson, Phil. Mag, xxiii. p. 456 (1912).

closely and try to show that a calculation based on this assumption is in quantitative agreement with the results of experiments: here we shall only mention briefly a problem with which we meet in such a calculation.

Experiments on the phenomena of X-rays suggest that not only the emission and absorption of radiation cannot be treated by the help of the ordinary electrodynamics, but not even the result of a collision between two electrons of which the one is bound in an atom. This is perhaps most clearly shown by some very instructive calculations on the energy of $\beta$-particles emitted from radioactive substances recently published by Rutherford*. These calculations strongly suggest that an electron of great velocity in passing through an atom and colliding with the electrons bound will loose energy in distinct finite quanta. As is immediately seen, this is very different from what we might expect if the result of the collisions was governed by the usual mechanical laws. The failure of the classical mechanics in such a problem might also be expected beforehand from the absence of anything like equipartition of kinetic energy between free electrons and electrons bound in atoms. From the point of view of the "mechanical" states we see, however, that the following assumption—which is in accord with the above analogy—might be able to account for the result of Rutherford calculation and for the absence of equipartition of kinetic energy: two colliding electrons, bound or free, will, after the collision as well as before, be in mechanical states. Obviously, the introduction of such an assumption would not make any alteration necessary in the classical treatment of a collision between two free particles. But, considering a collision between a free and a bound electron, it would follow that the bound electron by the collision could not acquire a less amount of energy than the difference in energy corresponding to successive stationary states, and consequently that the free electron which collides with it could not lose a less amount.

*E. Rutherford, Phil. Mag. xxiv. pp. 453 & 893 (1912).

The preliminary and hypothetical character of the above considerations needs not to be emphasized. The intention, however, has been to show that the sketched generalization of the theory of the stationary states possibly may afford a simple basis of representing a number of experimental facts which cannot be explained by help of the ordinary electrodynamics, and that the assumptions used do not seen, to be inconsistent with experiments on phenomena for which a satisfactory explanation has been given by the classical dynamics and the wave theory of light.

## Â§ 5. THE PERMANENT STATE OF AN ATOMIC SYSTEM

We shall now return to the main object of this paper—the discussion of the "permanent" state of a system consisting of nuclei and bound electrons. For a system consisting of a nucleus and an electron rotating round it, this state is, according to the above, determined by the condition that the angular momentum of the electron round the nucleus is equal to $h/2\pi$.

On the theory of this paper the only neutral atom which contains a single electron is the hydrogen atom. The permanent state of this atom should correspond to the values of $a$ and $\omega$ calculated on p. 5. Unfortunately, however, we know very little of the behavior of hydrogen atoms on account of the small dissociation of hydrogen molecules at ordinary temperatures. In order to get a closer comparison with experiments, it is necessary to consider more complicated systems.

Considering systems in which more electrons are bound by a positive nucleus, a configuration of the elections which presents itself as a permanent state is one in which the electrons are arranged in a ring round the nucleus. In the discussion of this problem on the basis of the ordinary electrodynamics, we meet—apart from the question of the energy radiation—with new difficulties due to the question of the stability of the ring. Disregarding for a moment this latter difficulty, we shall first consider the dimensions and frequency of the systems in relation to Planck's theory of radiation.

Let us consider a ring consisting of $n$ electrons rotating round a nucleus of charge E, the electrons being arranged at equal angular intervals around the circumference of a circle of radius $a$.

The total potential energy of the system consisting of the electrons and the nucleus is

where

For the radial force exerted on an electron by the nucleus and the other electrons we get.

Denoting the kinetic energy of an electron by T and neglecting the electromagnetic forces due to the motion of the electrons (see Part II), we get, putting the centrifugal force on an electron equal to the radial force,

or

From this we get for the frequency of revolution

The total amount of energy W necessary transferred to the system in order to remove the electrons to infinite distances apart from the nucleus and from each other is equal to the total kinetic energy of the electrons.

We see that the only difference in the above formula and those holding for the motion of a single electron in a circular orbit round a nucleus is the exchange of E for E—$es_n$. It is also immediately seen that corresponding to the motion of an electron in an elliptical orbit round a nucleus, there will be a motion of the $n$ electrons in which each rotates in an elliptical orbit with the nucleus in the focus, and the $n$ electrons at any moment are situated at equal angular intervals on a circle with the nucleus as the center. The major axis and frequency of the orbit of the single electrons will for this motion be given by the expressions (1) on p. 3 if we replace E by E—$es_n$ and W by W/$n$.

Let us now suppose that the system of $n$ electrons rotating in a ring round a nucleus is formed in a way analogous to the one assumed for a single electron rotating round a nucleus. It will thus be assumed that the electrons, before the binding by the nucleus, were at a great distance apart from the latter and possessed no sensible velocities, and also that during the binding a homogeneous radiation is emitted. As in the case of a single electron, we have here that the total amount of energy emitted during the formation of the system is equal to the final kinetic energy of the electrons. If we now suppose that during the formation of the system the electrons at any moment are situated at equal angular intervals on the circumference of a circle with the nucleus in the centre, from analogy with the considerations on p. 5 we are here led to assume the existence of a series of stationary configurations in which the kinetic energy per electron is equal to $\tau h\,(\omega/2)$, where $\tau$ is an entire number, $h$ Planck's constant, and $\omega$ the frequency of revolution. The configuration in which the greatest amount of energy is emitted is, as before, the one in which $\tau = 1$. This configuration we shall assume to be the permanent state of the system if the electrons in this state are arranged in a single ring. As for the case of a single electron, we get that the angular momentum of each of the electrons is equal to $h/2\pi$. It may be remarked that instead of considering the single electrons we might have considered the ring as an entity. This would, however, lead to the same result, for in this case the frequency of revolution $\omega$ will be replaced by the frequency $n\omega$ of the radiation from the whole ring calculated from the ordinary electrodynamics, and T by the total kinetic energy $n$T.

There may be many other stationary states corresponding to other ways of forming the system. The assumption of the existence of such states seems necessary in order to account for the line-spectra of systems containing more than one electron (p. 11); it is also suggested by the theory of Nicholson mentioned on p. 6, to which we shall return in a moment. The consideration of the spectra, however, gives, as far as I can see, no indication of the existence of stationary states in which all the electrons are arranged in a ring and which correspond to greater values for the total energy emmitted than the one we above have assumed to be the permanent state.

Further, there may be stationary configurations of a system of $n$ electrons and a nucleus of charge E in which all the electrons are not arranged in a single ring. The question, however, of the existence of such stationary configurations is not essential for our determination of the permanent state, as long as we assume that the electrons in this state of the system are arranged in a single ring. Systems corresponding to more complicated figurations will be discussed on p. 24.

Using the relation $T = h(\omega/2)$ we get, by help of the above expressions for T and $\omega$, values for $a$ and $\omega$ corresponding to the permanent state of the system which only differ from those given by the equations (3) on p. 5, by exchange of E for E—$es_n$.

The question of stability of a ring of electrons rotating round a positive charge is discussed in great detail by Sir J.J. Thomson*. An adaption of Thomson's analysis for the case here considered of a ring rotating around a nucleus of negligibly small linear dimensions is given by Nicholson†. The investigation of the problem in question naturally divides in two parts: one concerning the stability for displacements of the electrons in the plane of the ring; one concerning displacements perpendicular to this plane. As Nicholson's calculations show, the answer to the question of stability differs very much in the two cases in question. While the ring for the latter displacements in general is stable if the number of electrons is not great; the ring is in no case considered by Nicholson stable for displacements of the first kind.

According, however, to the point of view taken in this paper, the question of stability for displacements of the electrons in the plane of the ring is most intimately connected with the question of the mechanism of the binding of the electrons, and like the latter cannot be treated on the basis of the ordinary dynamics. The hypothesis of which we shall make use in the following is that the stability of a ring of electrons rotating round a nucleus is secured through the above condition of the universal constancy of the angular momentum, together with the further condition that the configuration of the

---

*Loc. cit.
†Loc. cit.

particles is the one by the formation of which the greatest amount of energy is emitted. As will be shown, this hypothesis is, concerning the question of stability for a displacement of the electrons perpendicular to the plane of the ring, equivalent to that used in ordinary mechanical calculations.

Returning to the theory of Nicholson on the origin of lines observed in the spectrum of the solar corona, we shall now see that he difficulties mentioned on p. 7 may be only formal. In the first place, from the point of view considered above the objection as to the instability of the systems for displacements of the electrons in the plane of the ring may not be valid. Further, the objection as to the emission of the radiation in quanta will not have reference to the calculations in question, if we assume that in the coronal spectrum we are not dealing with a true emission but only with a scattering of radiation. This assumption seems probable if we consider the conditions in the celestial body in question; for on account of the enormous rarefaction of the matter there may be comparatively few collisions to disturb the stationary states and to cause a true emission of light corresponding to the transition between different stationary states: on the other hand there will in the solar corona be intense illumination of light of all frequencies which may excite the natural vibrations of the systems in the different stationary states. If the above assumption is correct, we immediately understand the entirely different from for the laws connecting the lines discussed by Nicholson and those connecting the ordinary line-spectra considered in this paper.

Proceeding to consider systems of a more complicated constitution, we shall make use of the following theorem, which can be very simply proved:—

> "In every system consisting of electrons and positive nuclei, in which the nuclei are at rest and the electrons move in circular orbits with a velocity small compared with the velocity of light, the kinetic energy will be numerically equal to half the potential energy."

By help of this theorem we get—as in the previous cases of a single electron or of a ring rotating round a nucleus—that the total amount of energy emitted, by the formation of the systems from a configuration in which the distances apart of the particles are infinitely great and in which the particles have no velocities relative to each other, is equal to the kinetic energy of the electrons in the final configuration.

In analogy with the case of a single ring we are here led to assume that corresponding to any configuration of equilibrium a series of geometrically similar, stationary configurations of the system will exist in which the kinetic figurations of the systems will exist in which the kinetic energy of every electron is equal to the frequency of revolution multiplied by $(\tau/2)h$ where $\tau$ is an entire number and $h$ Planck's constant. In any such series of stationary configurations the one corresponding to the greatest amount of energy emitted will be the one in which $\tau$ for every electron is equal to 1. Considering that the ratio of kinetic energy to frequency for a particle rotating in a circular orbit is equal to $\pi$ times the angular momentum round the centre of the orbit, we are therefore led to the following simple generalization of the hypotheses mentioned on pp. 15 and 22.

"In any molecular system consisting of positive nuclei and electrons in which the nuclei are at rest relative to each other and the electrons move in circular orbits, the angular momentum of every electron round the centre of its orbit will in the permanent state of the system be equal to $h/(2\pi)$, where $h$ is Planck's constant"*.

In analogy with the considerations on p. 23, we shall assume that a configuration satisfying this condition is stable if the total energy of the system is less then in any neighbouring configured satisfying the same condition of the angular momentum of the electrons.

As mentioned in the introduction, the above hypothesis will be used in a following communication as a basis for a theory of the constitution of atoms and molecules. It will be shown that it leads to results which seem to be in conformity with experiments on a number of different phenomena.

---

*In the considerations leading to this hypothesis we have assumed that the velocity of the electrons is small compared with the velocity of light. The limits of validity of this assumption will be discussed in Part II.

The foundation of the hypothesis has been sought entirely in its relation with Planck's theory of radiation; by help of considerations given later it will be attempted to throw some further light on the foundation of it from another point of view.

April 5, 1913

# THE STRUCTURE OF THE ATOM

BY

NIELS BOHR

Nobel Lecture, December 11, 1922

Ladies and Gentlemen. Today, as a consequence of the great honour the Swedish Academy of Sciences has done me in awarding me this year's Nobel Prize for Physics for my work on the structure of the atom, it is my duty to give an account of the results of this work and I think that I shall be acting in accordance with the traditions of the Nobel Foundation if I give this report in the form of a survey of the development which has taken place in the last few years within the field of physics to which this work belongs.

## The general picture of the atom

The present state of atomic theory is characterized by the fact that we not only believe the existence of atoms to be proved beyond a doubt, but also we even believe that we have an intimate knowledge of the constituents of the individual atoms. I cannot on this occasion give a survey of the scientific developments that have led to this result; I will only recall the discovery of the electron towards the close of the last century, which furnished the direct verification and led to a conclusive formulation of the conception of the atomic nature of electricity which had evolved since the discovery by Faraday of the fundamental laws of electrolysis and Berzelius's electrochemical theory, and had its greatest triumph in the electrolytic dissociation theory of Arrhenius. This discovery of the electron and elucidation of its properties was the result of the work of a large number of investigators, among whom Lenard and J. J. Thomson may be particularly mentioned. The latter especially has made very important contributions to our subject by his ingenious attempts to develop ideas about atomic constitution on the basis of the electron theory. The present state of our knowledge of the elements of atomic structure was reached, however, by the discovery of the atomic nucleus, which we owe to Rutherford, whose work on the radioactive substances discovered towards the close of the last century has much enriched physical and chemical science.

According to our present conceptions, an atom of an element is built up of a nucleus that has a positive electrical charge and is the seat of by far the greatest part of the atomic mass, together with a number of electrons, all having the same negative charge and mass, which move at distances from the nucleus that are very great compared to the dimensions of the nucleus or of the electrons themselves. In this picture we at once see a striking resemblance to a planetary system, such as we have in our own solar system. Just as the simplicity of the laws that govern the motions of the solar system is intimately connected with the circumstance that the dimensions of the moving bodies are small in relation to the orbits, so the corresponding relations in atomic structure provide us with an explanation of an essential feature of natural phenomena in so far as these depend on the properties of the elements. It makes clear at once that these properties can be divided into two sharply distinguished classes.

To the first class belong most of the ordinary physical and chemical properties of substances, such as their state of aggregation, colour, and chemical reactivity. These properties depend on the motion of the electron system and the way in which this motion changes under the influence of different external actions. On account of the large mass of the nucleus relative to that of the electrons and its smallness in comparison to the electron orbits, the electronic motion will depend only to a very small extent on the nuclear mass, and will be determined to a close approximation solely by the total electrical charge of the nucleus. Especially the inner structure of the nucleus and the way in which the charges and masses are distributed among its separate particles will have a vanishingly small influence on the motion of the electron system surrounding the nucleus. On the other hand, the structure of the nucleus will be responsible for the second class of properties that are shown in the radioactivity of substances. In the radioactive processes we meet with an explosion of the nucleus, whereby positive or negative particles, the so-called $\alpha$- and $\beta$-particles, are expelled with very great velocities.

Our conceptions of atomic structure afford us, therefore, an immediate explanation of the complete lack of interdependence between the two classes of properties, which is most strikingly shown in the existence of substances which have to an extraordinarily close approximation the same ordinary physical and chemical properties, even though the atomic weights are not the same, and the radioactive properties are completely different. Such substances, of the existence of which the first evidence was found in the work of Soddy and other investigators on the chemical properties of the radioactive elements, are called isotopes, with reference to the classification of the elements according to ordinary physical and chemical properties. It is not necessary for me to state here how it has been shown in recent years that isotopes are found not only among the radioactive elements, but also among ordinary stable elements; in fact, a large number of the latter that were previously supposed simple have been shown by Aston's well-known investigations to consist of a mixture of isotopes with different atomic weights.

The question of the inner structure of the nucleus is still but little understood, although a method of attack is afforded by Rutherford's experiments on the disintegration of atomic nuclei by bombardment with a-particles. Indeed, these experiments may be said to open up a new epoch in natural philosophy in that for the first time the artificial transformation of one element into another has been accomplished. In what follows, however, we shall confine ourselves to a consideration of the ordinary physical and chemical properties of the elements and the attempts which have been made to explain them on the basis of the concepts just outlined.

It is well known that the elements can be arranged as regards their ordinary physical and chemical properties in a *natural system* which displays most suggestively the peculiar relationships between the different elements. It was recognized for the first time by Mendeleev and Lothar Meyer that when the elements are arranged in an order which is practically that of their atomic weights, their chemical and physical properties show a pronounced periodicity. A diagrammatic representation of this so-called Periodic Table is given in Fig. 1, where, however, the elements are not arranged in the ordinary way but in a somewhat modified form of a table first given by Julius Thomsen, who has also made important contributions to science in this domain. In the figure the elements are denoted by their usual chemical symbols, and the different vertical columns indicate the so-called periods. The

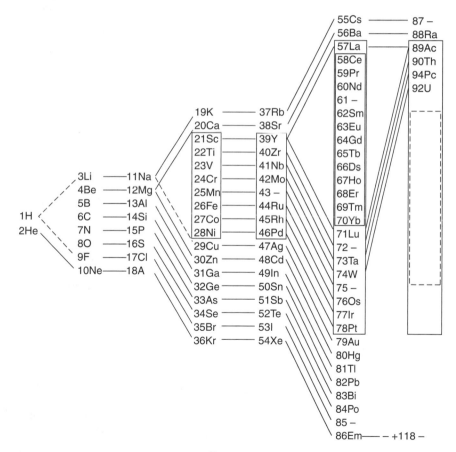

Fig. 1

elements in successive columns which possess homologous chemical and physical properties are connected with lines. The meaning of the square brackets around certain series of elements in the later periods, the properties of which exhibit typical deviations from the simple periodicity in the first periods, will be discussed later.

In the development of the theory of atomic structure the characteristic features of the natural system have found a surprisingly simple interpretation. Thus we are led to assume that the ordinal number of an element in the Periodic Table, the so-called atomic number, is just equal to the number of electrons which move about the nucleus in the neutral atom. In an imperfect form, this law was first stated by Van den Broek; it was, however, foreshadowed by J. J. Thomson's investigations of the number of electrons in the atom, as well as by Rutherford's measurements of the charge on the atomic nucleus. As we shall see, convincing support for this law has since been obtained in various ways, especially by Moseley's famous investigations of the X-ray spectra of the elements. We may perhaps also point out, how the simple connexion between atomic number and nuclear charge offers an explanation of the laws governing the changes in chemical properties of the elements after expulsion of $\alpha$- or $\beta$-particles, which found a simple formulation in the so-called radioactive displacement law.

## Atomic stability and electrodynamic theory

As soon as we try to trace a more intimate connexion between the properties of the elements and atomic structure, we encounter profound difficulties, in that essential differences between an atom and a planetary system show themselves here in spite of the analogy we have mentioned.

The motions of the bodies in a planetary system, even though they obey the general law of gravitation, will not be completely determined by this law alone, but will depend largely on the previous history of the system. Thus the length of the year is not determined by the masses of the sun and the earth alone, but depends also on the conditions that existed during the formation of the solar system, of which we have very little knowledge. Should a sufficiently large foreign body some day traverse our solar system, we might among other effects expect that from that day the length of the year would be different from its present value.

It is quite otherwise in the case of atoms. The definite and unchangeable properties of the elements demand that the state of an atom cannot undergo permanent changes due to external actions. As soon as the atom is left to itself again, its constituent particles must arrange their motions in a manner which is completely determined by the electric charges and masses of the particles. We have the most convincing evidence of this in spectra, that is, in the properties of the radiation emitted from substances in certain circumstances, which can be studied with such great precision. It is well known that the wavelengths of the spectral lines of a substance, which can in many cases be measured with an accuracy of more than one part in a million, are, in the same external circumstances, always exactly the same within the limit of error of the measurements, and quite independent of the previous treatment of this substance. It is just to this circumstance that we owe the great importance of spectral analysis, which has been such an invaluable aid to the chemist in the search for new elements, and has also shown us that even on the most distant bodies of the universe there occur elements with exactly the same properties as on the earth.

On the basis of our picture of the constitution of the atom it is thus impossible, so long as we restrict ourselves to the ordinary mechanical laws, to account for the characteristic atomic stability which is required for an explanation of the properties of the elements.

The situation is by no means improved if we also take into consideration the well-known electrodynamic laws which Maxwell succeeded in formulating on the basis of the great discoveries of Oersted and Faraday in the first half of the last century. Maxwell's theory has not only shown itself able to account for the already known electric and magnetic phenomena in all their details, but has also celebrated its greatest triumph in the prediction of the electromagnetic waves which were discovered by Hertz, and are now so extensively used in wireless telegraphy.

For a time it seemed as though this theory would also be able to furnish a basis for an explanation of the details of the properties of the elements, after it had been developed, chiefly by Lorentz and Larmor, into a form consistent with the atomistic conception of electricity. I need only remind you of the great interest that was aroused when Lorentz, shortly after the discovery by Zeeman of the characteristic changes that spectral lines undergo when the emitting substance is brought into a magnetic field, could give a natural and simple explanation of the main features of the phenomenon. Lorentz assumed that the radiation which we observe in a spectral line is sent out from an electron executing simple harmonic vibrations about a position of equilibrium, in precisely the same manner as the electromagnetic waves in radiotelegraphy are sent out by the electric oscillations in the antenna. He also pointed out how the alteration observed by Zeeman in the spectral lines corresponded exactly to the alteration in the motion of the vibrating electron which one would expect to be produced by the magnetic field.

It was, however, impossible on this basis to give a closer explanation of the spectra of the elements, or even of the general type of the laws holding with great exactness for the wavelengths of lines in these spectra, which had been established by Balmer, Rydberg, and Ritz. After we obtained details as to the constitution of the atom, this difficulty became still more manifest; in fact, so long as we confine ourselves to the classical electrodynamic theory we cannot even understand why we obtain spectra consisting of sharp lines at all. This theory can even be said to be incompatible with the assumption of the existence of atoms possessing the structure we have described, in that the motions of the electrons would claim a continuous radiation of energy from the atom, which would cease only when the electrons had fallen into the nucleus.

## The origin of the quantum theory

It has, however, been possible to avoid the various difficulties of the electrodynamic theory by introducing concepts borrowed from the so-called quantum theory, which marks a complete departure from the ideas that have hitherto been used for the explanation of natural phenomena. This theory was originated by Planck, in the year 1900, in his investigations on the law of heat radiation, which, because of its independence of the individual properties of substances, lent itself peculiarly well to a test of the applicability of the laws of classical physics to atomic processes.

Planck considered the equilibrium of radiation between a number of systems with the same properties as those on which Lorentz had based his theory of the Zeeman effect, but he could now show not only that classical physics could not account for the phenomena of heat radiation, but also that a complete agreement with the experimental law could be obtained if—in pronounced contradiction to classical theory—it were assumed that the energy of the vibrating electrons could not change continuously, but only in such a way that the energy of the system always remained equal to a whole number of so-called energy-quanta. The magnitude of this quantum was found to be proportional to the frequency of oscillation of the particle, which, in accordance with classical concepts, was supposed to be also the frequency of the emitted radiation. The proportionality factor had to be regarded as a new universal constant, since termed Planck's constant, similar to the velocity of light, and the charge and mass of the electron.

Planck's surprising result stood at first completely isolated in natural science, but with Einstein's significant contributions to this subject a few years after, a great variety of applications was found. In the first place, Einstein pointed out that the condition limiting the amount of vibrational energy of the particles could be tested by investigation of the specific heat of crystalline bodies, since in the case of these we have to do with similar vibrations, not of a single electron, but of whole atoms about positions of equilibrium in the crystal lattice.

Einstein was able to show that the experiment confirmed Planck's theory, and through the work of later investigators this agreement has proved quite complete. Furthermore, Einstein emphasized another consequence of Planck's results, namely, that radiant energy could only be emitted or absorbed by the oscillating particle in so-called "**quanta** of radiation", the magnitude of each of which was equal to Planck's constant multiplied by the frequency.

In his attempts to give an interpretation of this result, Einstein was led to the formulation of the so-called "hypothesis of light-quanta", according to which the radiant energy, in contradiction to Maxwell's electromagnetic theory of light, would not be propagated as electromagnetic waves, but rather as concrete light atoms, each with an energy equal to that of a quantum of radiation. This concept led Einstein to his well-known theory of the photoelectric effect. This phenomenon, which had been entirely unexplainable on the classical theory, was thereby placed in a quite different light, and the predictions of Einstein's theory have received such exact experimental confirmation in recent years, that perhaps the most exact determination of Planck's constant is afforded by measurements on the photoelectric effect. In spite of its heuristic value, however, the hypothesis of light-quanta, which is quite irreconcilable with so-called interference phenomena, is not able to throw light on the nature of radiation. I need only recall that these interference phenomena constitute our only means of investigating the properties of radiation and therefore of assigning any closer meaning to the frequency which in Einstein's theory fixes the magnitude of the light-quantum.

In the following years many efforts were made to apply the concepts of the quantum theory to the question of atomic structure, and the principal emphasis was sometimes placed on one and sometimes on the other of the consequences deduced by Einstein from Planck's result. As the best known of the attempts in this direction, from which, however, no definite results were obtained, I may mention the work of Stark, Sommerfeld, Hasenöhrl, Haas, and Nicholson.

From this period also dates an investigation by Bjerrum on infrared absorption bands, which, although it had no direct bearing on atomic structure, proved significant for the development of the quantum theory. He directed attention to the fact that the rotation of the molecules in a gas might be investigated by means of the changes in certain absorption lines with temperature. At the same time he emphasized the fact that the effect should not consist of a continuous widening of the lines such as might be expected from classical theory, which imposed no restrictions on the molecular rotations, but in accordance with the quantum theory he predicted that the lines should be split up into a number of components, corresponding to a sequence of distinct possibilities of rotation. This prediction was confirmed a few years later by Eva von Bahr, and the phenomenon may still be regarded as one of the most striking evidences of the reality of the quantum theory, even though from our present point of view the original explanation has undergone a modification in essential details.

## The quantum theory of atomic constitution

The question of further development of the quantum theory was in the meantime placed in a new light by Rutherford's discovery of the atomic nucleus (1911). As we have already seen, this discovery made it quite clear that by classical conceptions alone it was quite impossible to understand the most essential properties of atoms. One was therefore led to seek for a formulation of the principles of the quantum theory that could immediately account for the stability in atomic structure and the properties of the radiation sent out from atoms, of which the observed properties of substances bear witness. Such a formulation was proposed (1913) by the present lecturer in the form of two postulates, which may be stated as follows:

(1) Among the conceivably possible states of motion in an atomic system there exist a number of so-called *stationary states* which, in spite of the fact that the motion of the particles in

these states obeys the laws of classical mechanics to a considerable extent, possess a peculiar, mechanically unexplainable stability, of such a sort that every permanent change in the motion of the system must consist in a complete transition from one stationary state to another.

(2) While in contradiction to the classical electromagnetic theory no radiation takes place from the atom in the stationary states themselves, a process of transition between two stationary states can be accompanied by the emission of electromagnetic radiation, which will have the same properties as that which would be sent out according to the classical theory from an electrified particle executing an harmonic vibration with constant frequency. This frequency $v$ has, however, no simple relation to the motion of the particles of the atom, but is given by the relation

$$h v = E' - E'',$$

where $h$ is Planck's constant, and $E'$ and $E''$ are the values of the energy of the atom in the two stationary states that form the initial and final state of the radiation process. Conversely, irradiation of the atom with electromagnetic waves of this frequency can lead to an absorption process, whereby the atom is transformed back from the latter stationary state to the former.

While the first postulate has in view the general stability of the atom, the second postulate has chiefly in view the existence of spectra with sharp lines. Furthermore, the quantum-theory condition entering in the last postulate affords a starting-point for the interpretation of the laws of series spectra.

The most general of these laws, the combination principle enunciated by Ritz, states that the frequency v for each of the lines in the spectrum of an element can be represented by the formula

$$v = T'' - T',$$

where $T''$ and $T'$ are two so-called "spectral terms" belonging to a manifold of such terms characteristic of the substance in question.

According to our postulates, this law finds an immediate interpretation in the assumption that the spectrum is emitted by transitions between a number of stationary states in which the numerical value of the energy of the atom is equal to the value of the spectral term multiplied by Planck's constant. This explanation of the combination principle is seen to differ fundamentally from the usual ideas of electrodynamics, as soon as we consider that there is no simple relation between the motion of the atom and the radiation sent out. The departure of our considerations from the ordinary ideas of natural philosophy becomes particularly evident, however, when we observe that the occurrence of two spectral lines, corresponding to combinations of the same spectral term with two other different terms, implies that the nature of the radiation sent out from the atom is not determined only by the motion of the atom at the-beginning of the radiation process, but also depends on the state to which the atom is transferred by the process.

At first glance one might, therefore, think that it would scarcely be possible to bring our formal explanation of the combination principle into direct relation with our views regarding the constitution of the atom, which, indeed, are based on experimental evidence interpreted on classical mechanics and electrodynamics. A closer investigation, however, should make it clear that a definite relation may be obtained between the spectra of the elements and the structure of their atoms on the basis of the postulates.

## The hydrogen spectrum

The simplest spectrum we know is that of hydrogen. The frequencies of its lines may be represented with great accuracy by means of Balmer's formula:

$$v = K \left( \frac{1}{n''^2} - \frac{1}{n'^2} \right),$$

where $K$ is a constant and $n'$ and $n''$ are two integers. In the spectrum we accordingly meet a single series of spectral terms of the form

$Kln^2$, which decrease regularly with increasing term number $n$. In accordance with the postulates, we shall therefore assume that each of the hydrogen lines is emitted by a transition between two states belonging to a series of stationary states of the hydrogen atom in which the numerical value of the atom's energy is equal to $hKln^2$.

Following our picture of atomic structure, a hydrogen atom consists of a positive nucleus and an electron which—so far as ordinary mechanical conceptions are applicable—will with great approximation describe a periodic elliptical orbit with the nucleus at one focus. The major axis of the orbit is inversely proportional to the work necessary completely to remove the electron from the nucleus, and, in accordance with the above, this work in the stationary states is just equal to $hKln^2$. We thus arrive at a manifold of stationary states for which the major axis of the electron orbit takes on a series of discrete values proportional to the squares of the whole numbers. The accompanying Fig. 2 shows these relations diagrammatically. For the sake of simplicity the electron orbits in the stationary states are represented by circles, although in reality the theory places no restriction on the eccentricity of the orbit, but only determines the length of the major

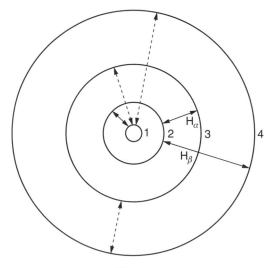

FIG. 2

axis. The arrows represent the transition processes that correspond to the red and green hydrogen lines, $\mathbf{H}_\alpha$ and $\mathbf{H}_\beta$, the frequency of which is given by means of the Balmer formula when we put $n'' = 2$ and $n' = 3$ and 4 respectively. The transition processes are also represented which correspond to the first three lines of the series of ultraviolet lines found by Lyman in 1914, of which the frequencies are given by the formula when $n$ is put equal to 1, as well as to the first line of the infrared series discovered some years previously by Paschen, which are given by the formula if $n''$ is put equal to 3.

This explanation of the origin of the hydrogen spectrum leads us quite naturally to interpret this spectrum as the manifestation of a process whereby the electron is bound to the nucleus. While the largest spectral term with term number corresponds to the final stage in the binding process, the small spectral terms that have larger values of the term number correspond to stationary states which represent the initial states of the binding process, where the electron orbits still have large dimensions, and where the work required to remove an electron from the nucleus is still small. The final stage in the binding process we may designate as the normal state of the atom, and it is distinguished from the other stationary states by the property that, in accordance with the postulates, the state of the atom can only be changed by the addition of energy whereby the electron is transferred to an orbit of larger dimensions corresponding to an earlier stage of the binding process.

The size of the electron orbit in the normal state calculated on the basis of the above interpretation of the spectrum agrees roughly with the value for the dimensions of the atoms of the elements that have been calculated by the kinetic theory of matter from the properties of gases. Since, however, as an immediate consequence of the stability of the stationary states that is claimed by the postulates, we must suppose that the interaction between two atoms during a collision cannot be completely described with the aid of the laws of classical mechanics, such a comparison as this cannot be carried further on the basis of such considerations as those just outlined.

A more intimate connexion between the spectra and the atomic model has been revealed, however, by an investigation of the motion in those stationary states where the term number is large, and where the dimensions of the electron orbit and the frequency of revolution in it vary relatively little when we go from one stationary state to the next following. It was possible to show that the frequency of the radiation sent out during the transition between two stationary states, the difference of the term numbers of which is small in comparison to these numbers themselves, tended to coincide in frequency with one of the harmonic components into which the electron motion could be resolved, and accordingly also with the frequency of one of the wave trains in the radiation which would be emitted according to the laws of ordinary electrodynamics.

The condition that such a coincidence should occur in this region where the stationary states differ but little from one another proves to be that the constant in the Balmer formula can be expressed by means of the relation

$$K = \frac{2\pi^2 e^4 m}{h^3},$$

where $e$ and $m$ are respectively the charge and mass of the electron, while $h$ is Planck's constant. This relation has been shown to hold to within the considerable accuracy with which, especially through the beautiful investigations of Millikan, the quantities $e$, $m$, and $h$ are known.

This result shows that there exists a connexion between the hydrogen spectrum and the model for the hydrogen atom which, on the whole, is as close as we might hope considering the departure of the postulates from the classical mechanical and electrodynamic laws. At the same time, it affords some indication of how we may perceive in the quantum theory, in spite of the fundamental character of this departure, a natural generalization of the fundamental concepts of the classical electrodynamic theory. To this most important question we shall return later, but first we will discuss how the interpretation of the hydrogen spectrum on the basis of the postulates has proved suitable in several ways, for elucidating the relation between the properties of the different elements.

## Relationships between the elements

The discussion above can be applied immediately to the process whereby an electron is bound to a nucleus with any given charge. The calculations show that, in the stationary state corresponding to a given value of the number $n$, the size of the orbit will be inversely proportional to the nuclear charge, while the work necessary to remove an electron will be directly proportional to the square of the nuclear charge. The spectrum that is emitted during the binding of an electron by a nucleus with charge $N$ times that of the hydrogen nucleus can therefore be represented by the formula:

$$\nu = N^2 K \left( \frac{1}{n''^2} - \frac{1}{n'^2} \right).$$

If in this formula we put $N = 2$, we get a spectrum which contains a set of lines in the visible region which was observed many years ago in the spectrum of certain stars. Rydberg assigned these lines to hydrogen because of the close analogy with the series of lines represented by the Balmer formula. It was never possible to produce these lines in pure hydrogen, but just before the theory for the hydrogen spectrum was put forward, Fowler succeeded in observing the series in question by sending a strong discharge through a mixture of hydrogen and helium. This investigator also assumed that the lines were hydrogen lines, because there existed no experimental evidence from which it might be inferred that two different substances could show properties resembling each other so much as the spectrum in question and that of hydrogen. After the theory was put forward, it became clear, however, that the observed lines must belong to a spectrum of helium, but that they were not like the ordinary helium spectrum emitted from the neutral atom. They came from an ionized helium atom which consists of a single electron moving about a nucleus with double charge. In this way there was brought to light a new feature of the relationship between the elements, which corresponds exactly with our present ideas of atomic structure, according to which the physical and chemical properties of an element depend in the first instance only on the electric charge of the atomic nucleus.

Soon after this question was settled the existence of a similar general relationship between the properties of the elements was brought to light by Moseley's well-known investigations on the characteristic X-ray spectra of the elements, which was made possible by Laue's discovery of the interference of X-rays in crystals and the investigations of W. H. and W. L. Bragg on this subject. It appeared, in fact, that the X-ray spectra of the different elements possessed a much simpler structure and a much greater mutual resemblance than their optical spectra. In particular, it appeared that the spectra changed from element to element in a manner that corresponded closely to the formula given above for the spectrum emitted during the binding of an electron to a nucleus, provided $N$ was put equal to the atomic number of the element concerned. This formula was even capable of expressing, with an approximation that could not be without significance, the frequencies of the strongest X-ray lines, if small whole numbers were substituted for $n'$ and $n''$.

This discovery was of great importance in several respects. In the first place, the relationship between the X-ray spectra of different elements proved so simple that it became possible to fix without ambiguity the atomic number for all known substances, and in this way to predict with certainty the atomic number of all such hitherto unknown elements for which there is a place in the natural system. Fig. 3 shows how the square root of the frequency for two characteristic X-ray lines depends on the atomic number. These lines belong to the group of so-called K-lines, which are the most penetrating of the characteristic rays. With very close approximation the points lie on straight lines, and the fact that they do so is conditioned not only by our taking account of known elements, but also by our leaving an open place between molybdenum (42) and ruthenium (44), just as in Mendeleev's original scheme of the natural system of the elements.

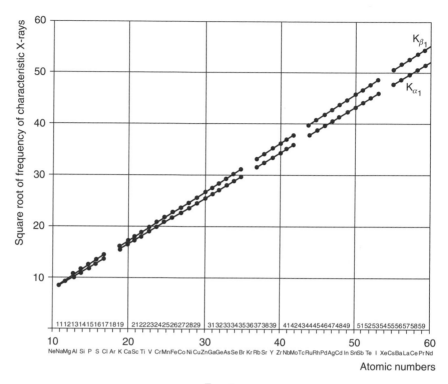

FIG. 3

Further, the laws of X-ray spectra provide a confirmation of the general theoretical conceptions, both with regard to the constitution of the atom and the ideas that have served as a basis for the interpretation of spectra. Thus the similarity between X-ray spectra and the spectra emitted during the binding of a single electron to a nucleus may be simply interpreted from the fact that the transitions between stationary states with which we are concerned in X-ray spectra are accompanied by changes in the motion of an electron in the inner part of the atom, where the influence of the attraction of the nucleus is very great compared with the repulsive forces of the other electrons.

The relations between other properties of the elements are of a much more complicated character, which originates in the fact that we have to do with processes concerning the motion of the electrons in the outer part of the atom, where the forces that the electrons

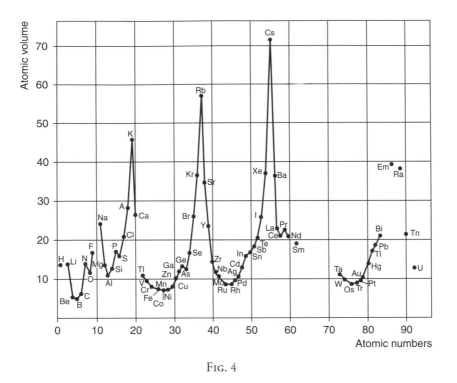

FIG. 4

exert on one another are of the same order of magnitude as the attraction towards the nucleus, and where, therefore, the details of the interaction of the electrons play an important part. A characteristic example of such a case is afforded by the spatial extension of the atoms of the elements. Lothar Meyer himself directed attention to the characteristic periodic change exhibited by the ratio of the atomic weight to the density, the so-called atomic volume, of the elements in the natural system. An idea of these facts is given by Fig. 4, in which the atomic volume is represented as a function of the atomic number. A greater difference between this and the previous figure could scarcely be imagined. While the X-ray spectra vary uniformly with the atomic number, the atomic volumes show a characteristic periodic change which corresponds exactly to the change in the chemical properties of the elements.

Ordinary optical spectra behave in an analogous way. In spite of the dissimilarity between these spectra, Rydberg succeeded in tracing a certain general relationship between the hydrogen spectrum and other spectra. Even though the spectral lines of the elements with higher atomic number appear as combinations of a more complicated manifold of spectral terms which is not so simply co-ordinated with a series of whole numbers, still the spectral terms can be arranged in series each of which shows a strong similarity to the series of terms in the hydrogen spectrum. This similarity appears in the fact that the terms in each series can, as Rydberg pointed out, be very accurately represented by the formula $K/(n + \alpha)^2$, where $K$ is the same constant that occurs in the hydrogen spectrum, often called the Rydberg constant, while $n$ is the term number, and $\alpha$ a constant which is different for the different series.

This relationship with the hydrogen spectrum leads us immediately to regard these spectra as the *last step of a process whereby the neutral atom is built up by the capture and binding of electrons to the nucleus*, one by one. In fact, it is clear that the last electron captured, so long as it is in that stage of the binding process in which its orbit is still large compared to the orbits of the previously bound electrons, will be subjected to a force from the nucleus and these electrons, that differs but little from the force with which the electron in the hydrogen atom is attracted towards the nucleus while it is moving in an orbit of corresponding dimensions.

The spectra so far considered, for which Rydberg's laws hold, are excited by means of electric discharge under ordinary conditions and are often called arc spectra. The elements emit also another type of spectrum, the so-called spark spectra, when they are subjected to an extremely powerful discharge. Hitherto it was impossible to disentangle the spark spectra in the same way as the arc spectra. Shortly after the above view on the origin of arc spectra was brought forward, however, Fowler found (1914) that an empirical expression for the spark spectrum lines could be established which corresponds exactly to Rydberg's laws with the single difference that the constant $K$ is replaced by a constant four times as large. Since, as we have seen, the constant that appears in the spectrum sent out during the binding of an electron to a helium nucleus is exactly equal to **4** *K*, it becomes evident that spark spectra are due to the ionized atom, and that their emission corresponds to *the last step but one in the formation of the neutral atom* by the successive capture and binding of electrons.

## Absorption and excitation of spectral lines

The interpretation of the origin of the spectra was also able to explain the characteristic laws that govern absorption spectra. As Kirchhoff and Bunsen had already shown, there is a close relation between the selective absorption of substances for radiation and their emission spectra, and it is on this that the application of spectrum analysis to the heavenly bodies essentially rests. Yet on the basis of the classical electromagnetic theory, it is impossible to understand why substances in the form of vapour show absorption for certain lines in their emission spectrum and not for others.

On the basis of the postulates given above we are, however, led to assume that the absorption of radiation corresponding to a spectral line emitted by a transition from one stationary state of the atom to a state of less energy is brought about by the return of the atom from the last-named state to the first. We thus understand immediately that in ordinary circumstances a gas or vapour can only show selective absorption for spectral lines that are produced by a transition from a state corresponding to an earlier stage in the binding process to the normal state. Only at higher temperatures or under the influence of electric discharges whereby an appreciable number of atoms are being constantly disrupted from the normal state, can we expect absorption for other lines in the emission spectrum in agreement with the experiments.

A most direct confirmation for the general interpretation of spectra on the basis of the postulates has also been obtained by investigations on the excitation of spectral lines and ionization of atoms by means of impact of free electrons with given velocities. A decided advance in this direction was marked by the well-known investigations of Franck and

Hertz (1914). It appeared from their results that by means of electron impacts it was impossible to impart to an atom an arbitrary amount of energy, but only such amounts as corresponded to a transfer of the atom from its normal state to another stationary state of the existence of which the spectra assure us, and the energy of which can be inferred from the magnitude of the spectral term.

Further, striking evidence was afforded of the independence that, according to the postulates, must be attributed to the processes which give rise to the emission of the different spectral lines of an element. Thus it could be shown directly that atoms that were transferred in this manner to a stationary state of greater energy were able to return to the normal state with emission of radiation corresponding to a single spectral line.

Continued investigations on electron impacts, in which a large number of physicists have shared, have also produced a detailed confirmation of the theory concerning the excitation of series spectra. Especially it has been possible to show that for the *ionization* of an atom by electron impact an amount of energy is necessary that is exactly equal to the work required, according to the theory, to remove the last electron captured from the atom. This work can be determined directly as the product of Planck's constant and the spectral term corresponding to the normal state, which, as mentioned above, is equal to the limiting value of the frequencies of the spectral series connected with selective absorption.

## The quantum theory of multiply-periodic systems

While it was thus possible by means of the fundamental postulates of the quantum. theory to account directly for certain general features of the properties of the elements, a closer development of the ideas of the quantum theory was necessary in order to account for these properties in further detail. In the course of the last few years a more general theoretical basis has been attained through the development of formal methods that permit the fixation of the stationary states for electron motions of a more general type than those WC have hitherto considered. For a simply periodic motion such as we meet in the pure harmonic oscillator, and at least to a first approximation, in the motion of an electron about a positive nucleus, the manifold of stationary states can be simply co-ordinated to a series of whole numbers. For motions of the more general class mentioned above, the so-called *multiply-periodic* motions, however, the stationary states compose a more complex manifold, in which, according to these formal methods, each state is characterized by several whole numbers, the so-called "quantum numbers".

In the development of the theory a large number of physicists have taken part, and the introduction of several quantum numbers can be traced back to the work of Planck himself. But the definite step which gave the impetus to further work was made by Sommerfeld (1915) in his explanation of the fine structure shown by the hydrogen lines when the spectrum is observed with a spectroscope of high resolving power. The occurrence of this fine structure must be ascribed to the circumstance that we have to deal, even in hydrogen, with a motion which is not exactly simply periodic. In fact, as a consequence of the change in the electron's mass with velocity that is claimed by the theory of relativity, the electron orbit will undergo a very slow precession in the orbital plane. The motion will therefore be doubly periodic, and besides a number characterizing the term in the Balmer formula, which *we* shall call the *principal quantum number* because it determines in the main the energy of the atom, the fixation of the stationary states demands another quantum number which we shall call the *subordinate quantum number.*

A survey of the motion in the stationary states thus fixed is given in the diagram (Fig. 5), which reproduces the relative size and form of the electron orbits. Each orbit is designated by a symbol $n_k$, where n is the principal quantum number and $k$ the subordinate quantum number. All orbits with the same principal quantum number have, to a first approximation, the same major axis, while orbits with the same value of $k$ have the same parameter, i.e. the same value for the shortest chord

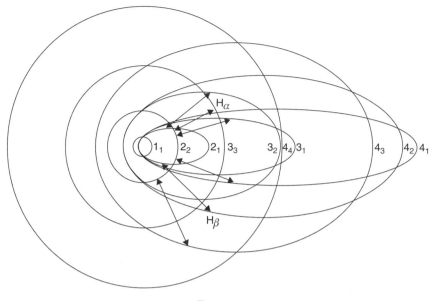

FIG. 5

through the focus. Since the energy values for different states with the same value of $n$ but different values of $k$ differ a little from each other, we get for each hydrogen line corresponding to definite values of $n'$ and $n''$ in the Balmer formula a number of different transition processes, for which the frequencies of the emitted radiation as calculated by the second postulate are not exactly the same. As Sommerfeld was able to show, the components this gives for each hydrogen line agree with the observations on the fine structure of hydrogen lines to within the limits of experimental error. In the figure the arrows designate the processes that give rise to the components of the red and green lines in the hydrogen spectrum, the frequencies of which are obtained by putting $n'' = 2$ and $n' = 3$ or 4 respectively in the Balmer formula.

In considering the figure it must not be forgotten that the description of the orbit is there incomplete, in so much as with the scale used the slow precession does not show at all. In fact, this precession is so slow that even for the orbits that rotate most rapidly the electron performs about 40,000 revolutions before the perihelion has gone round once. Nevertheless, it is this precession alone that is

127

responsible for the multiplicity of the stationary states characterized by the subordinate quantum number. If, for example, the hydrogen atom is subjected to a small disturbing force which perturbs the regular precession, the electron orbit in the stationary states will have a form altogether d&rent from that given in the figure. This implies that the fine structure will change its character completely, but the hydrogen spectrum will continue to consist of lines that are given to a close approximation by the Balmer formula, due to the fact that the approximately periodic character of the motion will be retained. Only when the disturbing forces become so large that even during a single revolution of the electron the orbit is appreciably disturbed, will the spectrum undergo essential changes. The statement often advanced that the introduction of two quantum numbers should be a necessary condition for the explanation of the Balmer formula must therefore be considered as a misconception of the theory.

Sommerfeld's theory has proved itself able to account not only for the fine structure of the hydrogen lines, but also for that of the lines in the helium spark spectrum. Owing to the greater velocity of the electron, the intervals between the components into which a line is split up are here much greater and can be measured with much greater accuracy. The theory was also able to account for certain features in the fine structure of X-ray spectra, where we meet frequency differences that may even reach a value more than a million times as great as those of the frequency differences for the components of the hydrogen lines.

Shortly after this result had been attained, Schwarzschild and Epstein (1916) simultaneously succeeded, by means of similar considerations, in accounting for the characteristic changes that the hydrogen lines undergo in an electric field, which had been discovered by Stark in the year 1914. Next, an explanation of the essential features of the Zeeman effect for the hydrogen lines was worked out at the same time by Sommerfeld and Debye (1917). In this instance the application of the postulates involved the consequence that only certain orientations of the atom relative to the magnetic field were allowable, and this characteristic consequence of the quantum theory has quite recently received a most direct confirmation in the beautiful researches of Stern and Gerlach on the deflexion of swiftly moving silver atoms in a nonhomogenous magnetic field.

## The correspondence principle

While this development of the theory of spectra was based on the working out of formal methods for the fixation of stationary states, the present lecturer succeeded shortly afterwards in throwing light on the theory from a new viewpoint, by pursuing further the characteristic connexion between the quantum theory and classical electrodynamics already traced out in the hydrogen spectrum. In connexion with the important work of Ehrenfest and Einstein these efforts led to the formulation of the so-called *correspondence principle*, according to which the occurrence of transitions between the stationary states accompanied by emission of radiation is traced back to the harmonic components into which the motion of the atom may be resolved and which, according to the classical theory, determine the properties of the radiation to which the motion of the particles gives rise.

According to the correspondence principle, it is assumed that every transition process between two stationary states can be co-ordinated with a corresponding harmonic vibration component in such a way that the probability of the occurrence of the transition is dependent on the amplitude of the vibration. The state of polarization of the radiation emitted during the transition depends on the further characteristics of the vibration, in a manner analogous to that in which on the classical theory the intensity and state of polarization in the wave system emitted by the atom as a consequence of the presence of this vibration component would be determined respectively by the amplitude and further characteristics of the vibration.

With the aid of the correspondence principle it has been possible to confirm and to extend the above-mentioned results. Thus it was possible to develop a complete quantum theory explanation of the Zeeman effect for the hydrogen lines, which, in spite of the essentially

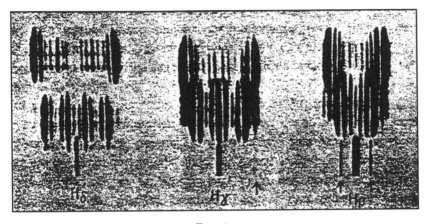

FIG. 6

different character of the assumptions that underlie the two theories, is very similar throughout to Lorentz's original explanation based on the classical theory. In the case of the Stark effect, where, on the other hand, the classical theory was completely at a loss, the quantum theory explanation could be so extended with the help of the correspondence principle as to account for the polarization of the different components into which the lines are split, and also for the characteristic intensity distribution exhibited by the components. This last question has been more closely investigated by Kramers, and the accompanying figure will give some impression of how completely it is possible to account for the phenomenon under consideration.

Fig. 6 reproduces one of Stark's well-known photographs of the splitting up of the hydrogen lines. The picture displays very well the varied nature of the phenomenon, and shows in how peculiar a fashion the intensity varies from component to component. The components below are polarized perpendicular to the field, while those above are polarized parallel to the field.

Fig. 7 gives a diagrammatic representation of the experimental and theoretical results for the line **Hγ**, the frequency of which is given by the Balmer formula with $n'' = 2$ and $n' = 5$. The vertical lines denote the components into which the line is split up, of which the picture on the right gives the components which are polarized parallel

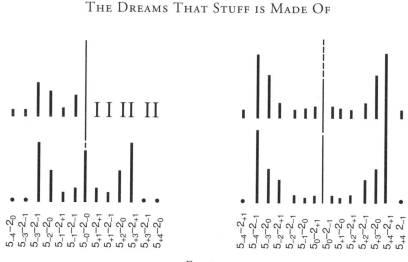

FIG. 7

to the field and that on the left those that are polarized perpendicular to it. The experimental results are represented in the upper half of the diagram, the distances from the dotted line representing the measured displacements of the components, and the lengths of the lines being proportional to the relative intensity as estimated by Stark from the blackening of the photographic plate. In the lower half is given for comparison a representation of the theoretical results from a drawing in Kramers' paper.

The symbol $(n'_{s'} - n''_{s''})$ attached to the lines gives the transitions between the stationary states of the atom in the electric field by which the components are emitted. Besides the principal quantum integer $n$, the stationary states are further characterized by a subordinate quantum integer s, which can be negative as well as positive and has a meaning quite different from that of the quantum number $k$ occurring in the relativity theory of the fine structure of the hydrogen lines, which fixed the form of the electron orbit in the undisturbed atom. Under the influence of the electric field both the form of the orbit and its position undergo large changes, but certain properties of the orbit remain unchanged, and the surbordinate quantum number s is connected with these. In Fig. 7 the position of the components corresponds to the frequencies calculated for the different transitions, and the lengths of the lines are proportional to the probabilities as calculated on the basis of the correspondence principle, by which also the polarization of the radiation is determined. It is seen that the theory reproduces completely the main feature of the experimental results, and in the light of the correspondence principle we can say that the Stark effect reflects down to the smallest details the action of the electric field on the orbit of the electron in the hydrogen atom, even though in this case the reflection is so distorted that, in contrast with the case of the Zeeman effect, it would scarcely be possible directly to recognize the motion on the basis of the classical ideas of the origin of electromagnetic radiation.

Results of interest were also obtained for the spectra of elements of higher atomic number, the explanation of which in the meantime had made important progress through the work of Sommerfeld, who introduced several quantum numbers for the description of the electron orbits. Indeed, it was possible, with the aid of the correspondence principle, to account completely for the characteristic rules which govern the seemingly capricious occurrence of combination lines, and it is not too much to say that the quantum theory has not only provided a simple interpretation of the combination principle, but has further contributed materially to the clearing up of the mystery that has long rested over the application of this principle.

The same viewpoints have also proved fruitful in the investigation of the so-called band spectra. These do not originate, as do series spectra, from individual atoms, but from molecules; and the fact that these spectra are so rich in lines is due to the complexity of the motion entailed by the vibrations of the atomic nuclei relative to each other and the rotations of the molecule as a whole. The first to apply the postulates to this problem was Schwarz-schild, but the important work of Heurhnger especially has thrown much light on the origin and structure of band spectra. The considerations employed here can be traced back directly to those discussed at the beginning of this lecture in connexion with Bjerrum's theory of the influence of molecular rotation on the infrared absorption lines of gases. It is true we no longer think that the rotation is reflected in the spectra in the way claimed by classical electrodynamics, but rather that the line components are due to transitions between stationary states which differ as regards rotational motion. That the phenomenon retains its essential feature, however, is a typical consequence of the correspondence principle.

## The natural system of the elements

The ideas of the origin of spectra outlined in the preceding have furnished the basis for a theory of the structure of the atoms of the elements which has shown itself suitable for a general interpretation of the main features of the properties of the elements, as exhibited in the natural system. This theory is based primarily on considerations of the manner in which the atom can be imagined to be built up by the capture and binding of electrons to the nucleus, one by one. As we have seen, the optical spectra of elements provide us with evidence on the progress of the last steps in this building-up process.

An insight into the kind of information that the closer investigation of the spectra has provided in this respect may be obtained from Fig. 8, which gives a diagrammatic representation of the orbital motion in the stationary states corresponding to the emission of the arc-spectrum of potassium. The curves show the form of the orbits described in the stationary states by the last electron captured in the potassium atom, and they can be considered as stages in the process whereby the 19th electron is bound after the 18 previous electrons have already been bound in their normal orbits. In order not to complicate the figure, no attempt has been made to draw any of the orbits of these inner electrons, but the region in which they move is enclosed by a dotted circle. In an atom with several electrons the orbits will, in general, have a complicated character. Because of the symmetrical nature of the field of force about the nucleus, however, the motion of each single electron can be approximately described as a plane periodic motion on which is superimposed a uniform rotation in the plane of the orbit. The orbit of each electron will therefore be to a first

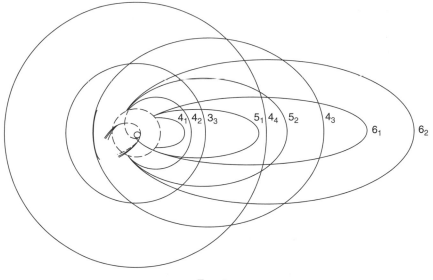

FIG. 8

approximation doubly periodic, and will be fixed by two quantum numbers, as are the stationary states in a hydrogen atom when the relativity precession is taken into account.

In Fig. 8, as in Fig. 5, the electron orbits are marked with the symbol $n_k$, where $n$ is the principal quantum number and $k$ the subordinate quantum number. While for the initial states of the binding process, where the quantum numbers are large, the orbit of the last electron captured lies completely outside of those of the previously bound electrons, this is not the case for the last stages. Thus, in the potassium atom, the electron orbits with subordinate quantum numbers 2 and 1 will, as indicated in the figure, penetrate partly into the inner region. Because of this circumstance, the orbits will deviate very greatly from a simple Kepler motion, since they will consist of a series of successive outer loops that have the same size and form, but each of which is turned through an appreciable angle relative to the preceding one. Of these outer loops only one is shown in the figure. Each of them coincides very nearly with a piece of a Kepler ellipse, and they are connected, as indicated, by a series of inner loops of a complicated character in which the electron approaches the nucleus closely. This holds especially for the orbit with subordinate quantum number 1, which, as a closer investigation shows, will approach nearer to the nucleus than any of the previously bound electrons.

On account of this penetration into the inner region, the strength with which an electron in such an orbit is bound to the atom will—in spite of the fact that for the most part it moves in a field of force of the same character as that surrounding the hydrogen nucleus—be much greater than for an electron in a hydrogen atom that moves in an orbit with the same principal quantum number, the maximum distance of the electron from the nucleus at the same time being considerably less than in such a hydrogen orbit. As we shall see, this feature of the binding process in atoms with many electrons is of essential importance in order to understand the characteristic periodic way in which many properties of the elements as displayed in the natural system vary with the atomic number.

In the accompanying table (Fig. 9) is given a summary of the results concerning the structure of the atoms of the elements to which the author has been led by a consideration of successive capture and binding of electrons to the atomic nucleus. The figures before the different elements are the atomic numbers, which give the total number of electrons in the neutral atom. The figures in the different columns give the number of electrons in orbits corresponding to the values of the principal and subordinate quantum numbers standing at the top. In accordance with ordinary usage we will, for the sake of brevity, designate an orbit with principal quantum number $n$ as an $n$-quantum orbit. The first electron bound in each atom moves in an orbit that corresponds to the normal state of the hydrogen atom with quantum symbol $1_1$. In the hydrogen atom there is of course only one electron; but we must assume that in the atoms of other elements the next electron also will be bound in such a r-quantum orbit of type $1_1$. As the table shows, the following electrons are bound in 2-quantum orbits. To begin with, the binding will result in a $2_1$ orbit, but later electrons will be bound in $2_2$ orbits, until, after binding the first 10 electrons in the atom, we reach a closed configuration of the a-quantum orbits

| | $1_1$ | $2_1 2_2$ | $3_1 3_2 3_3$ | $4_1 4_2 4_3 4_4$ | $5_1 5_2 5_3 5_4 5_5$ | $6_1 6_2 6_3 6_4 6_5 6_6$ | $7_1 7_2$ |
|---|---|---|---|---|---|---|---|
| 1 H | 1 | | | | | | |
| 2 He | 2 | | | | | | |
| 3 Li | 2 | 1 | | | | | |
| 4 Be | 2 | 2 | | | | | |
| 5 B | 2 | 2 (1) | | | | | |
| – – | – | – – | | | | | |
| 10 Ne | 2 | 4 4 | | | | | |
| 11 Na | 2 | 4 4 | 1 | | | | |
| 12 Mg | 2 | 4 4 | 2 | | | | |
| 13 Al | 2 | 4 4 | 2 1 | | | | |
| – – | – | – – | – – | | | | |
| 18 A | 2 | 4 4 | 4 4 | | | | |
| 19 K | 2 | 4 4 | 4 4 | 1 | | | |
| 20 Ca | 2 | 4 4 | 4 4 | 2 | | | |
| 21 Sc | 2 | 4 4 | 4 4 1 | (2) | | | |
| 22 Ti | 2 | 4 4 | 4 4 2 | (2) | | | |
| – – | – | – – | – – – | – | | | |
| 29 Cu | 2 | 4 4 | 6 6 6 | 1 | | | |
| 30 Zn | 2 | 4 4 | 6 6 6 | 2 | | | |
| 31 Ga | 2 | 4 4 | 6 6 6 | 2 1 | | | |
| – – | – | – – | – – – | – – | | | |
| 36 Kr | 2 | 4 4 | 6 6 6 | 4 4 | | | |
| 37 Rb | 2 | 4 4 | 6 6 6 | 4 4 | 1 | | |
| 38 Sr | 2 | 4 4 | 6 6 6 | 4 4 | 2 | | |
| 39 Y | 2 | 4 4 | 6 6 6 | 4 4 1 | (2) | | |
| 40 Zr | 2 | 4 4 | 6 6 6 | 4 4 2 | (2) | | |
| – – | – | – – | – – – | – – – | – | | |
| 47 Ag | 2 | 4 4 | 6 6 6 | 6 6 6 | 1 | | |
| 48 Cd | 2 | 4 4 | 6 6 6 | 6 6 6 | 2 | | |
| 49 In | 2 | 4 4 | 6 6 6 | 6 6 6 | 2 1 | | |
| – – | – | – – | – – – | – – – | – – | | |
| 54 X | 2 | 4 4 | 6 6 6 | 6 6 6 | 4 4 | | |
| 55 Cs | 2 | 4 4 | 6 6 6 | 6 6 6 | 4 4 | 1 | |
| 56 Ba | 2 | 4 4 | 6 6 6 | 6 6 6 | 4 4 | 2 | |
| 57 La | 2 | 4 4 | 6 6 6 | 6 6 6 | 4 4 1 | (2) | |
| 58 Ce | 2 | 4 4 | 6 6 6 | 6 6 6 1 | 4 4 1 | (2) | |
| 59 Pr | 2 | 4 4 | 6 6 6 | 6 6 6 2 | 4 4 1 | (2) | |
| – – | – | – – | – – – | – – – – | – – – | – | |
| 71 Cp | 2 | 4 4 | 6 6 6 | 8 8 8 8 | 4 4 1 | (2) | |
| 72 – | 2 | 4 4 | 6 6 6 | 8 8 8 8 | 4 4 2 | (2) | |
| – – | – | – – | – – – | – – – – | – – – | – | |
| 79 Au | 2 | 4 4 | 6 6 6 | 8 8 8 8 | 6 6 6 | | |
| 80 Hg | 2 | 4 4 | 6 6 6 | 8 8 8 8 | 6 6 6 | 2 | |
| 81 Tl | 2 | 4 4 | 6 6 6 | 8 8 8 8 | 6 6 6 | 2 1 | |
| – – | – | – – | – – – | – – – – | – – – | – – | |
| 86 Em | 2 | 4 4 | 6 6 6 | 8 8 8 8 | 6 6 6 | 4 4 | |
| 87 – | 2 | 4 4 | 6 6 6 | 8 8 8 8 | 6 6 6 | 4 4 | 1 |
| 88 Ra | 2 | 4 4 | 6 6 6 | 8 8 8 8 | 6 6 6 | 4 4 | 2 |
| 89 Ac | 2 | 4 4 | 6 6 6 | 8 8 8 8 | 6 6 6 | 4 4 1 | (2) |
| 90 Th | 2 | 4 4 | 6 6 6 | 8 8 8 8 | 6 6 6 | 4 4 2 | (2) |
| – – | – | – – | – – – | – – – – | – – – | – – – | – |
| 118 ? | 2 | 4 4 | 6 6 6 | 8 8 8 8 | 8 8 8 8 | 6 6 6 | 4 4 |

Fig. 9

in which we assume there are four orbits of each type. This configuration is met for the first time in the neutral neon atom, which forms the conclusion of the second period in the system of the elements. When we proceed in this system, the following electrons are bound in 3-quantum orbits, until, after the conclusion of the third period of the system, we encounter for the first time, in elements of the fourth period, electrons in 4-quantum orbits, and so on.

This picture of atomic structure contains many features that were brought forward by the work of earlier investigators. Thus the attempt to interpret the relations between the elements in the natural system by the assumption of a division of the electrons into groups goes as far back as the work of J. J. Thomson in 1904. Later, this viewpoint was developed chiefly by Kossel (1916), who, moreover, has connected such a grouping with the laws that investigations of X-ray spectra have brought to light.

Also G. R. Lewis and I. Langmuir have sought to account for the relations between the properties of the elements on the basis of a grouping inside the atom. These investigators, however, assumed that the electrons do not move about the nucleus, but occupy positions of equilibrium. In this way, though, no closer relation can be reached between the properties of the elements and the experimental results concerning the constituents of the atoms. Statical positions of equilibrium for the electrons are in fact not possible in cases in which the, forces between the electrons and the nucleus even approximately obey the laws that hold for the attractions and repulsions between electrical charges.

The possibility of an interpretation of the properties of the elements on the basis of these latter laws is quite characteristic for the picture of atomic structure developed by means of the quantum theory. As regards this picture, the idea of connecting the grouping with a classification of electron orbits according to increasing quantum numbers was suggested by Moseley's discovery of the laws of X-ray spectra, and by Sommerfeld's work on the fine structure of these spectra. This has been principally emphasized by Vegard, who some years ago in connexion with investigations of X-ray spectra proposed a grouping of electrons in the atoms of the elements, which in many ways shows a likeness to that which is given in the above table.

A satisfactory basis for the further development of this picture of atomic structure has, however, only recently been created by the study of the binding processes of the electrons in the atom, of which we have experimental evidence in optical spectra, and the characteristic features of which have been elucidated principally by the correspondence principle. It is here an essential circumstance that the restriction on the course of the binding process, which is expressed by the presence of electron orbits with higher quantum numbers in the normal state of the atom, can be naturally connected with the general condition for the occurrence of transitions between stationary states, formulated in that principle.

Another essential feature of the theory is the influence, on the strength of binding and the dimensions of the orbits, of the penetration of the later bound electrons into the region of the earlier bound ones, of which we have seen an example in the discussion of the origin of the potassium spectrum. Indeed, this circumstance may be regarded as the essential cause of the pronounced periodicity in the properties of the elements, in that it implies that the atomic dimensions and chemical properties of homologous substances in the different periods, as, for example, the alkali-metals, show a much greater similarity than that which might be expected from a direct comparison of the orbit of the last electron bound with an orbit of the same quantum number in the hydrogen atom.

The increase of the principal quantum number which we meet when we proceed in the series of the elements, affords also an immediate explanation of the characteristic deviations from simple periodicity which are exhibited by the natural system and are expressed in Fig. 1 by the bracketing of certain series of elements in the later periods. The first time such a deviation is met with is in the 4th period, and the reason for it can be simply illustrated by means of our figure of the orbits of the last electron bound in the atom of potassium, which is the first element in this period. Indeed, in potassium we encounter for the first time in the sequence of the elements a case in which the principal quantum number of the orbit of the last electron bound is, in the normal state of the atom, larger than in one of the earlier stages of the binding process. The normal state corresponds here to a $4_1$ orbit, which, because of the penetration into the inner region, corresponds to a much stronger binding of the electron than a 4-quantum orbit in the hydrogen atom. The binding in question is indeed even stronger than for a 2-quantum orbit in the hydrogen atom, and is therefore more than twice as strong as in the circular $3_3$ orbit which is situated completely outside the inner region, and for which the strength of the binding differs but little from that for a 3-quantum orbit in hydrogen.

This will not continue to be true, however, when we consider the binding of the 19th electron in substances of higher atomic number, because of the much smaller relative difference between the field of force outside and inside the region of the first eighteen electrons bound. As is shown by the investigation of the spark spectrum of calcium, the binding of the 19th electron in the $4_1$ orbit is here but little stronger than in $3_3$ orbits, and as soon as we reach scandium, we must assume that the $3_3$ orbit will represent the orbit of the 19th electron in the normal state, since this type of orbit will correspond to a stronger binding than a $4_1$ orbit. While the group of electrons in 2-quantum orbits has been entirely completed at the end of the 2nd period, the development that the group of 3-quantum orbits undergoes in the course of the 3rd period can therefore only be described as a provisional completion, and, as shown in the table, this electron group will, in the bracketed elements of the 4th period, undergo a stage of further development in which electrons are added to it in 3-quantum orbits.

This development brings in new features, in that the development of the electron group with 4-quantum orbits comes to a standstill, so to speak, until the 3-quantum group has reached its final closed form. Although we are not yet in a position to account in all details for the steps in the gradual development of the 3-quantum electron group, still we can say that with the help of the quantum theory we see at once why it is in the 4th period of the system of the elements that there occur for the first time successive elements with properties that resemble each other as much as the properties of the iron *group;* indeed, we can even understand why these elements show their well-known paramagnetic properties. Without further reference to the quantum theory, Eadenburg had on a previous occasion already suggested the idea of relating the chemical and magnetic properties of these elements with the development of an inner electron group in the atom.

I will not enter into many more details, but only mention that the peculiarities we meet with in the 5th period are explained in much the same way as those in the 4th period. Thus the properties of the bracketed elements in the 5th period as it appears in the table, depend on a stage in the development of the 4-quantum electron group that is initiated by the entrance in the normal state of electrons in $4_3$ orbits. In the 6th period, however, we meet new features. In this period we encounter not only a stage of the development of the electron groups with 5- and 6-quantum orbits, but also the final completion of the development of the 4-quantum electron group, which is initiated by the entrance for the first time of electron orbits of the $4_4$ type in the normal state of the atom. This development finds its characteristic expression in the occurrence of the peculiar family of elements in the 6th period, known as the *rare-earths.* These show, as we know, a still greater mutual similarity in their chemical properties than the elements of the iron family. This must be ascribed to the fact that we have here to do with the development of an electron group that lies deeper in the atom. It is of interest to note that the theory can also naturally account for the fact that these elements, which resemble each other in so many ways, still show great differences in their magnetic properties.

The idea that the occurrence of the rare-earths depends on the development of an inner electron group has been put forward from different sides. Thus it is found in the work of Vegard, and at the same time as my own work, it was proposed by Bury in connexion with considerations of the systematic relation between the chemical properties and the grouping of the electrons inside the atom from the point of view of Langmuir's static atomic model. While until now it has not been possible, however, to give any theoretical basis for such a development of an inner group, we see that our extension of the quantum theory provides us with an unforced explanation. Indeed, it is scarcely an exaggeration to say that if the existence of the rare-earths had not been established by direct chemical investigation, the occurrence of a family of elements of this character within the 6th period of the natural system of the elements might have been theoretically predicted.

When we proceed to the 7th period of the system, we meet for the first time with 7-quantum orbits, and we shall expect to find within this period features that are essentially similar to those in the 6th period, in that besides the first stage in the development of the 7-quantum orbits, we must expect to encounter further stages in the development of the group with 6- or 5-quantum orbits. However, it has not been possible directly to confirm this expectation, because only a few elements are known in the beginning of the 7th period. The latter circumstance may be supposed to be intimately connected with the instability of atomic nuclei with large charges, which is expressed in the prevalent radioactivity among elements with high atomic number.

## X-ray spectra and atomic constitution

In the discussion of the conceptions of atomic structure we have hitherto placed the emphasis on the formation of the atom by successive capture of electrons. Our picture would, however, be incomplete without some reference to the confirmation of the theory afforded by the study of X-ray spectra. Since the interruption of Moseley's fundamental researches by his untimely death, the study of these spectra has been continued in a most admirable way by Prof. Siegbahn in Lund. On the basis of the large amount of experimental evidence adduced by him and his collaborators, it has been possible recently to give a classification of X-ray spectra that allows an immediate interpretation on the quantum theory. In the first place it has been possible, just as in the case of the optical spectra, to represent the frequency of each of the X-ray lines as the difference between two out of a manifold of spectral terms characteristic of the element in question. Next, a direct connexion with the atomic theory is obtained by the assumption that each of these spectral terms multiplied by Planck's constant is equal to the work which must be done on the atom to remove one of its inner electrons. In fact, the removal of one of the inner electrons from the completed atom may, in accordance with the above considerations on the formation of atoms by capture of electrons, give rise to transition processes by which the place of the electron removed is taken by an electron belonging to one of the more loosely bound electron groups of the atom, with the result that after the transition an electron will be lacking in this latter group.

The X-ray lines may thus be considered as giving evidence of stages in a process by which the atom undergoes a *reorganization* after a disturbance in its interior. According to our views on the stability of the electronic configuration such a disturbance must consist in the removal of electrons from the atom, or at any rate in their transference from normal orbits to orbits of higher quantum numbers than those belonging to completed groups; a circumstance which is clearly illustrated in the characteristic difference between selective absorption in the X-ray region, and that exhibited in the optical region.

The classification of the X-ray spectra, to the achievement of which the above-mentioned work of Sommerfeld and Kossel has contributed materially, has recently made it possible, by means of a closer examination of the manner in which the terms occurring in the X-ray spectra vary with the atomic number, to obtain a very direct test of a number of the theoretical conclusions as regards the structure of the atom. In Fig. 10 the abscissæ are the atomic numbers and the ordinates are proportional to the square roots of the spectral terms, while

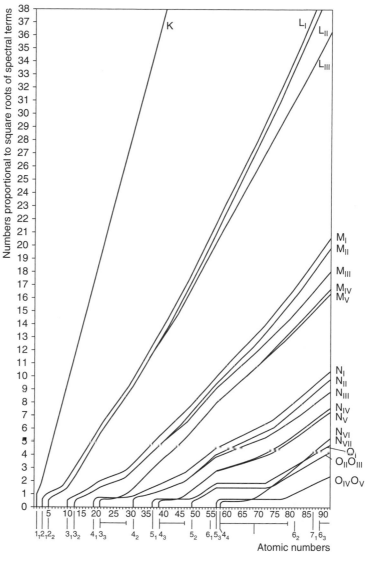

FIG. 10

the symbols K, L, M, N, O, for the individual terms refer to the characteristic discontinuities in the selective absorption of the elements for X-rays; these were originally found by Barkla before the discovery of the interference of X-rays in crystals had provided a means for the closer investigation of X-ray spectra. Although the curves generally

run very uniformly, they exhibit a number of deviations from uni-
formity which have been especially brought to light by the recent
investigation of Coster, who has for some years worked in Siegbahn's
laboratory.

These deviations, the existence of which was not discovered until after the publication of the theory of atomic structure discussed
above, correspond exactly to what one might expect from this theory.
At the foot of the figure the vertical lines indicate where, according
to the theory, we should first expect, in the normal state of the atom,
the occurrence of $n_k$ orbits of the type designated. We see how it has
been possible to connect the occurrence of every spectral term with
the presence of an electron moving in an orbit of a definite type, to
the removal of which this term is supposed to correspond. That in
general there corresponds more than one curve to each type of orbit
$n_k$ is due to a complication in the spectra which would lead us too
far afield to enter into here, and may be attributed to the deviation
from the previously described simple type of motion of the electron
arising from the interaction of the different electrons within the same
group.

The intervals in the system of the elements, in which a further
development of an inner electron group takes place because of the
entrance into the normal atom of electron orbits of a certain type, are
designated in the figure by the horizontal lines, which are drawn between the vertical lines to which the quantum symbols are affixed. It is
clear that such a development of an inner group is everywhere reflected
in the curves. Particularly the course of the N- and O-curves may be
regarded as a direct indication of that stage in the development of
the electron groups with 4-quantum orbits of which the occurrence of
the rare-earths bears witness. Although the apparent complete absence
of a reflection in the X-ray spectra of the complicated relationships
exhibited by most other properties of the elements was the typical and
important feature of Moseley's discovery, we can recognize, nevertheless, in the light of the progress of the last years, an intimate connexion

between the X-ray spectra and the general relationships between the elements within the natural system.

Before concluding this lecture I should like to mention one further point in which X-ray investigations have been of importance for the test of the theory. This concerns the properties of the hitherto unknown element with atomic number 72. On this question opinion has been divided in respect to the conclusions that could be drawn from the relationships within the Periodic Table, and in many representations of the table a place is left open for this element in the rare-earth family. In Julius Thomsen's representation of the natural system, however, this hypothetical element was given a position homologous to titanium and zirconium in much the same way as in our representation in Fig. 1. Such a relationship must be considered as a necessary consequence of the theory of atomic structure developed above, and is expressed in the table (Fig. 9) by the fact that the electron configurations for titanium and zirconium show the same sort of resemblances and differences as the electron configurations for zirconium and the element with atomic number 72. A corresponding view was proposed by Bury on the basis of his above-mentioned systematic considerations of the connexion between the grouping of the electrons in the atom and the properties of the elements.

Recently, however, a communication was published by Dauvillier announcing the observation of some weak lines in the X-ray spectrum of a preparation containing rare-earths. These were ascribed to an element with atomic number 72 assumed to be identical with an element of the rare-earth family, the existence of which in the preparation used had been presumed by Urbain many years ago. This conclusion would, however, if it could be maintained, place extraordinarily great, if not unsurmountable, difficulties in the way of the theory, since it would claim a change in the strength of the binding of the electrons with the atomic number which seems incompatible with the conditions of the quantum theory. In these circumstances Dr. Coster and Prof. Hevesy, who are both for the time working in Copenhagen, took up a short time ago the problem of testing a preparation of zircon-bearing minerals by X-ray spectroscopic analysis. These investigators have been able to establish the existence in the minerals investigated of appreciable quantities of an element with atomic number 72, the chemical properties of which show a great similarity to those of zirconium and a decided difference from those of the rare-earths*.

I hope that I have succeeded in giving a summary of some of the most important results that have been attained in recent years in the field of atomic theory, and I should like, in concluding, to add a few general remarks concerning the viewpoint from which these results may be judged, and particularly concerning the question of how far, with these results, it is possible to speak of an explanation, in the ordinary sense of the word. By a theoretical explanation of natural phenomena we understand in general a classification of the observations of a certain domain with the help of analogies pertaining to other domains of observation, where one presumably has to do with simpler phenomena. The most that one can demand of a theory is that this classification can be pushed so far that it can contribute to the development of the field of observation by the prediction of new phenomena.

When we consider the atomic theory, we are, however, in the peculiar position that there can be no question of an explanation in this last sense, since here we have to do with phenomena which from the very nature of the case are simpler than in any other field of observation, where the phenomena are always conditioned by the combined action of a large number of atoms. We are therefore obliged to be modest in our demands and content ourselves with concepts which are formal in the sense that they do not provide a visual picture of the sort one is accustomed to require of the explanations with which natural philosophy deals. Bearing this in mind I have sought to convey the impression that the results, on the other hand, fulfill, at least in

---

*For the result of the continued work of Coster and Hevesy with the new element, for which they have proposed the name hafnium, the reader may be referred to their letters in *Nature* of January 20, February 10 and 24, and April 7.

some degree, the expectations that are entertained of any theory; in fact, I have attempted to show how the development of atomic theory has contributed to the classification of extensive fields of observation, and by its predictions has pointed out the way to the completion of this classification. It is scarcely necessary, however, to emphasize that the theory is yet in a very preliminary stage, and many fundamental questions still await solution.

# Chapter Three

The following papers are some of the most central to the development of quantum mechanics, and it would be very hard to overstate their importance. In them, the beginnings of quantum theory are transformed from a group of loosely collected ideas into a fully developed theory capable of describing much of the physical world. Without the insight gained from these papers, many of the technological innovations of the last century would have been impossible. In this section, we see in the papers of Erwin Schrodinger and lectures by Werner Heisenberg the establishment of two full and independent quantum theories. This at first seemed a problem: How can two different theories describe the same reality? In 1926 Schrodinger himself demonstrated the equivalency of his and Heisenberg's approaches.

Heisenberg's method is included here as described in a series of lectures he presented in 1929 at the University of Chicago, and is summarized in his Nobel lecture in 1933. In the Heisenberg formulation, quantities that can be observed, for example a particle's position or energy, are represented by square matrices. It is well known from mathematics that a square matrix can be characterized by a set of vectors, called the eigenvectors, and a set of numbers, called the eigenvalues. In Heisenberg's matrix representation of quantum mechanics, the eigenvalues represent all possible values of an experimental observation of the quantity represented by that matrix. For example, the eigenvalues of the matrix corresponding to a particle's position would be every possible location that the particle can have. When an observation is made, a certain eigenvalue is measured, and the corresponding eigenvector represents the state of the system immediately after that observation. The state of the particle is said to have "collapsed" into a state represented by that eigenvector. Thus, the act of measurement itself alters the state of the system. This is one of the most important and fundamental conclusions of quantum mechanics. No matter how cleverly we design our experiments, we will never be able to find a way to take a measurement without altering the system in some way. This is mathematically expressed in Heisenberg's Uncertainty Principle.

In mathematics, matrix multiplication does not always commute. This means that if $A$ and $B$ are matrices, $A$ times $B$ is not always the same as $B$ times $A$. In Heisenberg's quantum mechanics, observable quantities correspond to matrices, and their non-commutation corresponds to the idea that measuring one quantity and then a second will give a different answer than measuring the second quantity first. For example, measuring a particle's position then its speed will result in different values than measuring its speed first then its position. We can understand this result like this: A measurement of position will alter the original state. Taking a measurement of speed afterward will be on this altered state, and so will produce a different answer than if it were taken on the original state. The more precisely we measure the particle's speed, the more we alter the state, and consequently the less we can know about its original velocity. It is impossible to get a complete measurement of both simultaneously. This idea was extremely revolutionary to the physicists at the time. It required them to abandon the classical idea of a particle's trajectory, or the path it takes when it moves. When we throw a ball, we can watch its position and measure its speed at every point along its path. But for a quantum particle this is impossible. The impossibility comes not from our inability to design a clever enough experiment, but seems is inherent in the physical laws of nature as formalized in the Heisenberg Uncertainty Principle.

Less than a year after Heisenberg published his formulation of quantum mechanics, Schrodinger published a completely different method in a series of four papers entitled, "Quantization as an Eigenvalue Problem." His approach was inspired by the wave-particle duality proposed by Albert Einstein and Louis de Broglie. In explaining the photoelectric effect, Einstein showed that electromagnetic waves could act as particles. De Broglie took this a step further and theorized that perhaps matter that is typically thought to be particle-like could act like waves. Schrodinger took this idea and looked for a wave equation that could describe matter. In the first of these four exceptional papers, Schrodinger made a brilliant guess as to what the proper equation should be. This wave equation is now known as the Schrodinger equation. It establishes a way to derive the time evolution of a quantum system that is evolving under the influence of an arbitrary force. It is the quantum mechanical equivalent to Newton's laws, which are the basic equations of classical mechanics. Schrodinger recognized that he could solve for the possible states of an electron in a hydrogen atom using his equation. In the hydrogen atom the only force is the electrostatic force between the electron and the proton in the nucleus. When the solution of his equation for the electrostatic force was coupled with spin, which will be discussed in the next chapter, he was able to re-derive the hydrogen atomic spectrum that Niels Bohr had earlier explained using the planetary model. Bohr's explanation required an assumption of energy quantization without having a good theoretical basis for doing so. Now with Schrodinger's and Heisenberg's quantum theories it was possible to understand why that assumption was needed. All the pieces were coming together to form a quantum theory capable of explaining much of the observable world.

# Excerpts from: The Physical Principles of the Quantum Theory

By

Werner Heisenbeg

Translated into English by Carl Eckart and Frank C. Hoyt

## FOREWORD TO THE ENGLISH EDITION

It is an unusual pleasure to present Professor Heisenberg's Chicago lectures on "The Physical Principles of the Quantum Theory" to a wider audience than could attend them when they were originally delivered. Professor Heinsenberg's leading place in the development of the new quantum mechanics is well recognized by those who have been following its growth. It was in fact he who first saw clearly that in the older forms of quantum theory we were describing our spectra in terms of atomic mechanisms regarding which we could gain no definite knowledge, and who first found a way to interpret (or at least describe) spectroscopic phenomena without assuming the existence of such atomic mechanisms. Likewise, "the uncertainty principle" has become a household phrase throughout our universities, and it is especially fortunate to have this opportunity of learning its significance from one who is responsible for its formulation.

The power of the new quantum mechanics in giving us a better understanding of events on an atomic scale is becoming increasingly evident. The structure of the helium atom, the existence of half-quantum numbers in band spectra, the continuous spatial distribution of photo-electrons, and the phenomenon of radioactive disintegration, to mention only a few examples, are achievements of the new theory which had baffled the old. While the writing of this chapter of the history of physics is doubtless not yet complete, it has progressed to such a stage that we may profitably pause and consider the significance of what has been written. As we make this survey, we are indeed fortunate to have Professor Heinsenberg to guide our thoughts.

ARTHUR H. COMPTON

# Preface

The lectures which I gave at the University of Chicago in the spring of 1929 afforded me the opportunity of reviewing the fundamental principles of quantum theory. Since the conclusive studies of Bohr in 1927 there have been no essential changes in these principles, and many new experiments have confirmed important consequences of the theory (for example, the Raman effect). But even today the physicist more often has a kind of faith in the correctness of the new principles than a clear understanding of them. For this reason the publication of these Chicago lectures in the form of a small book seems justified.

Since the formal mathematical apparatus of the quantum theory is already available in several excellent texts and is more familiar to many than the physical principles, I have placed it at the end of the book, in what is little more than a collection of formulas.* In the text itself I have been at pains to use only elementary formulas and calculations, so far as this is possible.

In the body of the text particular emphasis has been placed on the complete equivalence of the corpuscular and wave concepts, which is clearly reflected in the newer formulations of the mathematical theory. This symmetry of the book with respect to the words "particle" and "wave" shows that nothing is gained by discussing fundamental problems (such as causality) in terms of one rather than the other. I have also attempted to make the distinction between waves in space-time and the Schrödinger waves in configuration space as clear as possible.

On the whole the book contains nothing that is not to be found in previous publications, particularly in the investigations of Bohr. The purpose of the book seems to me to be fulfilled if it contributes somewhat to the diffusion of that *"Kopenhagener Geist der Quantentheorie,"*

---

*TRANSLATORS' NOTE.—In the English edition, Professor Heisenberg's lectures on the mathematical part of the theory have been reproduced in more detail. This seemed advisable since a treatment of the general transformation theory and the quantum theory of wave fields was not available in English at the time the manuscript was prepared. The former has since been treated in several texts (E. U. Condon and P. M. Morse, *Quantum Mechanics;* A. E. Ruark and H. C. Urey, *Atoms, Molecules and Quanta;* both published by McGraw-Hill).
The English text also deviates in several other points from the German, but these are felt to be unessential changes.

if I may so express myself, which has directed the entire development of modern atomic physics.

My thanks are due in the first place to Drs. C. Eckart and F. Hoyt, of the University of Chicago, who have taken on themselves not only the labor of preparing the English translation, but have also contributed essentially to the improvement of the book by working over several sections and giving me the benefit of their advice. I am also indebted to Dr. G. Beck for reading proof of the German edition and for valuable assistance in the preparation of the manuscript.

W. HEISENBERG

LEIPZIG
March 3, 1930

# INTRODUCTORY

## § 1. THEORY AND EXPERIMENT

The experiments of physics and their results can be described in the language of daily life. Thus if the physicist did not demand a theory to explain his results and could be content, say, with a description of the lines appearing on photographic plates, everything would be simple and there would be no need of an epistemological discussion. Difficulties arise only in the attempt to classify and synthesize the results, to establish the relation of cause and effect between them—in short, to construct a theory. This synthetic process has been applied not only to the results of scientific experiment, but, in the course of ages, also to the simplest experiences of daily life, and in this way all concepts have been formed. In the process, the solid ground of experimental proof has often been forsaken, and generalizations have been accepted uncritically, until finally contradictions between theory and experiment have become apparent. In order to avoid these contradictions, it seems necessary to demand that no concept enter a theory which has not been experimentally verified at least to the same degree of accuracy as the experiments to be explained by the theory. Unfortunately it is quite impossible to fulfil this requirement, since the commonest ideas and words would often be excluded. To avoid these insurmountable difficulties it is found advisable to introduce a great wealth of concepts into a physical theory, without attempting to justify them rigorously, and then to allow experiment to decide at what points a revision is necessary.

Thus it was characteristic of the special theory of relativity that the concepts "measuring rod" and "clock" were subject to searching criticism in the light of experiment; it appeared that these ordinary concepts involved the tacit assumption that there exist (in principle, at least) signals that are propagated with an infinite velocity. When it became evident that such signals were not to be found in nature, the task of eliminating this tacit assumption from all logical deductions was undertaken, with the result that a consistent interpretation was found for facts which had seemed irreconcilable. A much more radical departure from the classical conception of the world was brought about by the general theory of relativity, in which only the concept of coincidence in space-time was accepted uncritically. According to this theory, ordinary language (i.e., classical concepts) is applicable only to the description of experiments in which both the gravitational constant and the reciprocal of the velocity of light may be regarded as negligibly small.

Although the theory of relativity makes the greatest of demands on the ability for abstract thought, still it fulfils the traditional requirements of science in so far as it permits a division of the world into subject and object (observer and observed) and hence a clear formulation of the law of causality. This is the very point at which the difficulties of the quantum theory begin. In atomic physics, the concepts "clock" and "measuring rod" need no immediate consideration, for there is a large field of phenomena in which $1/c$ is negligible. The concepts "space-time coincidence" and "observation," on the other hand, do require a thorough revision. Particularly characteristic of the discussions to follow is the interaction between observer and object; in classical physical theories it has always been assumed either that this interaction is negligibly small, or else that its effect can be eliminated from the result by calculations based on "control" experiments. This assumption is not permissible in atomic physics; the interaction between observer and object causes uncontrollable and large changes in the system being observed, because of the discontinuous changes characteristic of atomic processes. The immediate consequence of this circumstance is that in general every experiment performed to determine some numerical quantity renders the knowledge of others illusory, since the uncontrollable perturbation of the observed system alters the values of previously determined quantities. If this perturbation be followed in its quantitative details, it appears that in many cases it is impossible to obtain an exact determination of the simultaneous values of two variables, but rather that there is a lower limit to the accuracy with which they can be known.[*]

The starting-point of the critique of the relativity theory was the postulate that there is no signal velocity greater than that of light. In a similar manner, this lower limit to the accuracy with which certain variables can be known simultaneously may be postulated as a law of nature (in the form of the so-called uncertainty relations) and made the starting-point of the critique which forms the subject matter of the following pages. These uncertainty relations give us that measure of freedom from the limitations of classical concepts which is necessary for a consistent description of atomic processes. The program of the following considerations will therefore be: first, to obtain a general survey of all concepts whose introduction is suggested by the atomic experiments; second, to limit the range of application of these concepts; and third, to show that the concepts thus limited, together with the mathematical formulation of quantum theory, form a self-consistent scheme.

## § 2. THE FUNDAMENTAL CONCEPTS OF QUANTUM THEORY

The most important concepts of atomic physics can be induced from the following experiments:

*(a) Wilson[†] photographs.* The $\alpha$- and $\beta$-rays emitted by radioactive elements cause the condensation of minute droplets when allowed to pass through supersaturated water vapor. These drops are not distributed at random, but are arranged along definite tracks which, in the case of $\alpha$-rays (Fig. 1), are nearly straight lines, in the case of $\beta$-rays, are irregularly curved. The existence of the tracks and their continuity show that the rays may appropriately be regarded as streams of minute particles moving at high speeds. As is well known, the mass and charge of these particles may be determined from the deflection of the rays by electric and magnetic fields.

---

[*] W. Heisenberg, *Zeitschrift für Physik*, **43**, 172, 1927.
[†] *Proceedings of the Royal Society*, A, **85**, 285, 1911; see also *Jahrbuch der Radioaktivität*, **10**, 34, 1913.

FIG. 1   Tracks of $\alpha$-particles in Wilson Chamber

*(b) Diffraction of matter waves (Davisson and Germer;\* Thomson,[†] Rupp[‡]).* After the conception of $\beta$-rays as streams of particles had remained unchallenged for more than fifteen years, another series of experiments was performed which indicated that they could be diffracted and were capable of interference as if they were waves. Typical of these experiments is that of G. P. Thomson, in which a narrow beam of artificial $\beta$-rays of moderate energy is passed through a thin foil of matter. The foil is composed of minute crystals oriented at random, but the atoms in each crystal are regularly arranged. A photographic plate receiving the emergent rays exhibits rings of blackening (Fig. 2), as though the rays were waves and were diffracted by the minute crystals. From the diameters of the rings and the structure of the crystals, the length of these waves may be determined and is found to be $\lambda = h/mv$, where $m$ is the mass and $v$ the velocity of the particles as determined by the above-mentioned

\* *Physical Review*, **30**, 705, 1927; *Proceedings of the National Academy*, **14**, 317, 1928.
[†] *Proceedings of the Royal Society*, A, **117**, 600, 1928; A, **119**, 651, 1928.
[‡] *Annalen der Physik*, **85**, 981, 1928.

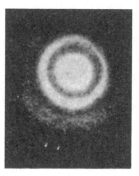

FIG. 2   Diffraction of electrons on passing through a thin foil of matter.

experiments. Similar experiments were performed by Davisson and Germer, Kikuchi,* and Rupp.

(c) *The diffraction of X-rays.* The same dual interpretation is necessary in the case of light and electromagnetic radiation in general. After Newton's objections to the wave theory of light had been refuted and the phenomena of interference explained by Fresnel, this theory dominated all others for many years, until Einstein[†] pointed out that the experiments of Lenard on the photoelectric effect could only be explained by a corpuscular theory. He postulated that the momentum of the hypothetical particles was related to the wave-length of the radiation by the formula $p = h/\lambda$ (cf. § 2b). The necessity for both interpretations is particularly clear in the case of X-rays: If a homogeneous beam of X-rays is passed through a crystalline mass, and the emergent rays received on a photographic plate (Fig. 3), the result is much like the result of G. P. Thomson's experiment, and it may be concluded that X-rays are a form of wave motion, with a determinable wave-length.

(d) *The Compton-Simon* experiment.* When a beam of X-rays passes through supersaturated water vapor, it is scattered by the molecules. Secondary products of the scattering are the "recoil"

*Japanese Journal of Physics, **5**, 83, 1928.
[†]Annalen der Physik, **17**, 145, 1905.
*Physical Review, **25**, 306, 1925.

FIG. 3    Diffraction of X-rays by MgO powder

electrons, which are apparently particles of considerable energy, since they form tracks of condensed droplets as do the $\beta$-rays. These tracks are not very long, however, and occur with random direction. They apparently originate within the region traversed by the primary X-ray beam. Other secondary products of the scattering are the photo-electrons, which again make themselves evident by longer tracks of condensed water droplets. Under suitable conditions these tracks originate at points outside the primary X-ray beam, but the two secondary products are not unrelated. If it be assumed that the X-ray beam consists of a stream of light-particles (photons) and that the scattering process is the collision of a photon with one of the electrons of the molecule, as a result of which the electron recoils in the observed direction, Einstein's postulate regarding the energy and momentum of the photons enables the direction of the photon after the collision to be calculated. This photon then collides with a second molecule, and gives up its remaining energy to an electron (the photoelectron). This assumption has been quantitatively verified (Fig. 4).

FIG. 4 Photograph showing recoil electron and associated photo electron liberated by X-rays. The upper photograph is retouched.

*(e) The collision experiments of Franck and Hertz.* * When a beam of slow electrons with homogeneous velocity passes through a gas, the electronic current as function of the velocity changes discontinuously at certain values of the velocity (energy). The analysis of these experiments leads to the conclusion that the atoms in the gas can only assume discrete energy values (Bohr's postulate). When the energy of the atom is known, one speaks of a "stationary state of the atom." When the kinetic energy of the electron is too small to change the atom from its stationary state to a higher one, the electron makes only elastic collisions with the atoms, but when the kinetic energy suffices for excitation some electrons will transfer their energy to the atom, so the electronic current as a function of the velocity changes rapidly in the critical region. The concept of stationary states, which is suggested by these experiments, is the most direct expression of the discontinuity in all atomic processes.

From these experiments it is seen that both matter and radiation possess a remarkable duality of character, as they sometimes exhibit the properties of waves, at other times those of particles. Now it is obvious that a thing cannot be a form of wave motion and composed

* *Verhandlungen der Deutschen Physikalische Gesellschaft*, **15**, 613, 1913.

of particles at the same time—the two concepts are too different. It is true that it might be postulated that two separate entities, one having all the properties of a particle, and the other all the properties of wave motion, were combined in some way to form "light." But such theories are unable to bring about the intimate relation between the two entities which seems required by the experimental evidence. As a matter of fact, it is experimentally certain only that light sometimes behaves as if it possessed some of the attributes of a particle, but there is no experiment which proves that it possesses all the properties of a particle; similar statements hold for matter and wave motion. The solution of the difficulty is that the two mental pictures which experiments lead us to form—the one of particles, the other of waves— are both incomplete and have only the validity of analogies which are accurate only in limiting cases. It is a trite saying that "analogies cannot be pushed too far," yet they may be justifiably used to describe things for which our language has no words. Light and matter are both single entities, and the apparent duality arises in the limitations of our language.

It is not surprising that our language should be incapable of describing the processes occurring within the atoms, for, as has been remarked, it was invented to describe the experiences of daily life, and these consist only of processes involving exceedingly large numbers of atoms. Furthermore, it is very difficult to modify our language so that it will be able to describe these atomic processes, for words can only describe things of which we can form mental pictures, and this ability, too, is a result of daily experience. Fortunately, mathematics is not subject to this limitation, and it has been possible to invent a mathematical scheme—the quantum theory—which seems entirely adequate for the treatment of atomic processes; for visualization, however, we must content ourselves with two incomplete analogies—the wave picture and the corpuscular picture. The simultaneous applicability of both pictures is thus a natural criterion to determine how far each analogy may be "pushed" and forms an obvious starting-point for the critique of the concepts which have entered atomic theories in the course of their development, for, obviously, uncritical deduction of consequences from both will lead to contradictions. In this way one obtains the limitations of the concept of a particle by considering the concept of a wave. As N. Bohr* has shown, this is the basis of a very simple derivation of the uncertainty relations between co-ordinate and momentum of a particle. In the same manner one may derive the limitations of the concept of a wave by comparison with the concept of a particle.

It must be emphasized that this critique cannot be carried through entirely without using the mathematical apparatus of the quantum theory, for the development of the latter preceded the clarification of the physical principles in the historic sequence. In order to avoid obscuring the essential relationships by too much mathematics, however, it has seemed advisable to relegate this formalism to the Appendix. The exposition of mathematical principles given there does not pretend to be complete, but only to furnish the reader with those formulas which are essential for the argument of the text. References to this Appendix are given as A (16), etc.

---

* *Nature*, **121**, 580, 1928; *Naturwissenschaften*, **16**, 245, 1928.

# CRITIQUE OF THE PHYSICAL CONCEPTS OF THE CORPUSCULAR THEORY OF MATTER

## § 1. THE UNCERTAINTY RELATIONS

The concepts of velocity, energy, etc., have been developed from simple experiments with common objects, in which the mechanical behavior of macroscopic bodies can be described by the use of such words. These same concepts have then been carried over to the electron, since in certain fundamental experiments electrons show a mechanical behavior like that of the objects of common experience. Since it is known, however, that this similarity exists only in a certain limited region of phenomena, the applicability of the corpuscular theory must be limited in a corresponding way. According to Bohr,[*] this restriction may be deduced from the principle that the processes of atomic physics can be visualized equally well in terms of waves or particles. Thus the statement that the position[†] of an electron is known to within a certain accuracy $\Delta x$ at the time $t$ can be visualized by the picture of a wave packet in the proper position with an approximate extension $\Delta x$. By "wave packet" is meant a wavelike disturbance whose amplitude is appreciably different from zero only in a bounded region. This region is, in general, in motion, and also changes its size and shape, i.e., the disturbance spreads. The velocity of the electron corresponds to that of the wave packet, but this latter cannot be exactly defined, because of the diffusion which takes place. This indeterminateness is to be considered as an essential characteristic of the electron, and not as evidence of the inapplicability of the wave picture. Defining momentum as $p_x = \mu v_x$ (where $\mu$ = mass of electron, $v_x$ = $x$-component of velocity), this uncertainty in the velocity causes an uncertainty in

---

[*] N. Bohr, *Nature*, **121**, 580, 1928.

[†] The following considerations apply equally to any of the three space co-ordinates of the electron, therefore only one is treated explicitly.

$p_x$ of amount $\Delta p_x$; from the simplest laws of optics, together with the empirically established law $\lambda = h/p$, it can readily be shown that

$$\Delta x \Delta p_x \geq h. \tag{1}$$

Suppose the wave packet made up by superposition of plane sinusoidal waves, all with wave-lengths near $\lambda_0$. Then, roughly speaking, $n = \Delta x / \lambda_0$ crests or troughs fall within the boundary of the packet. Outside the boundary the component plane waves must cancel by interference; this is possible if, and only if, the set of component waves contains some for which at least $n + 1$ waves fall in the critical range. This gives

$$\frac{\Delta x}{\lambda_0 - \Delta\lambda} \geq n + 1,$$

where $\Delta\lambda$ is the approximate range of wave-lengths necessary to represent the packet. Consequently

$$\frac{\Delta x \Delta\lambda}{\lambda_0^2} \geq 1. \tag{2}$$

On the other hand, the group velocity of the waves (i.e., the velocity of the packet) is by A (85)

$$v_g = \frac{h}{\mu\lambda_0}, \tag{3}$$

so that the spreading of the packet is characterized by the range of velocities

$$\Delta v_g = \frac{h}{\mu\lambda_0^2}\Delta\lambda.$$

By definition $\Delta p_x = \mu \Delta v_g$ and therefore by equation (2),

$$\Delta x \Delta p_x \geq h.$$

This uncertainty relation specifies the limits within which the particle picture can be applied. Any use of the words "position" and "velocity" with an accuracy exceeding that given by equation (1) is just as meaningless as the use of words whose sense is not defined.*

---

*In this connection one should particularly remember that the human language permits the construction of sentences which do not involve any consequences and which therefore have no content at all—in spite of the fact that these sentences produce some kind of

The uncertainty relations can also be deduced without explicit use of the wave picture, for they are readily obtained from the mathematical scheme of quantum theory and its physical interpretation.[†] Any knowledge of the co-ordinate $q$ of the electron can be expressed by a probability amplitude $S(q')$, $|S(q')|^2 dq'$ being the probability of finding the numerical value of the co-ordinate of the electron between $q'$ and $q' + dq'$. Let

$$\bar{q} = \int q' |S(q')|^2 dq' \tag{4}$$

be the average value of $q$. Then $\Delta q$ defined by

$$(\Delta q)^2 = 2 \int (q' - \bar{q})^2 |S(q')|^2 dq' \tag{5}$$

can be called the uncertainty in the knowledge of the electron's position. In an exactly analogous way $|T(p')|^2 dp'$ gives the probability of finding the momentum of the electron between $p'$ and $p' + dp'$; again $\bar{p}$ and $\Delta p$ may be defined as

$$\bar{p} = \int p' |T(p')|^2 dp', \tag{6}$$

$$(\Delta p)^2 = 2 \int (p' - \bar{p})^2 |T(p')|^2 dp'. \tag{7}$$

By equation A(169), the probability amplitudes are related by the equations

$$\left. \begin{array}{l} T(p') = \displaystyle\int S(q') R(q' p') dq', \\[2mm] S(q') = \displaystyle\int T(p') R^*(q' p') dp', \end{array} \right\} \tag{8}$$

where $R(q' p')$ is the matrix of the transformation from a Hilbert space in which $q$ is a diagonal matrix to one in which $p$ is diagonal. From equation A(41) we have

$$\int p(q' q'') R(q'' p') dq'' = \int R(q' p'') p(p'' p') dp'',$$

picture in our imagination; e.g., the statement that besides our world there exists another world, with which any connection is impossible in principle, does not lead to any experimental consequence, but does produce a kind of picture in the mind. Obviously such a statement can neither be proved nor disproved. One should be especially careful in using the words "reality," "actually," etc., since these words very often lead to statements of the type just mentioned.

[†] Kennard, *Zeitschrift für Physik*, **44**, 326, 1927.

and by equation A(42) this is equivalent to

$$\frac{h}{2\pi i}\frac{\partial}{\partial q'}R\left(q'p'\right) = p'R\left(q'p'\right), \tag{9}$$

whose solution is

$$R = ce^{\frac{2\pi i}{h}p'q'}. \tag{10}$$

Normalizing gives $c$ the value $1/\sqrt{h}$. The values of $\Delta p$, $\Delta q$ are thus not independent. To simplify further calculations, we introduce the following abbreviations:

$$\left.\begin{array}{l} x = q' - \bar{q}, \quad y = p' - \bar{p} \\[4pt] s\left(x\right) = S(q')e^{\frac{2\pi i}{h}\bar{p}q'}, \\[4pt] t(y) = T\left(p'\right)e^{-\frac{2\pi i}{h}\bar{q}(p'-\bar{p})}. \end{array}\right\} \tag{11}$$

Then equations (5) and (7) become

$$(\Delta q)^2 = 2\int x^2 \left|s\left(x\right)\right|^2 dx, \tag{5a}$$

$$(\Delta p)^2 = 2\int y^2 \left|t\left(y\right)\right|^2 dy, \tag{7a}$$

while equations (8) become

$$t(y) = \frac{1}{\sqrt{h}}\int s(x)e^{\frac{2\pi i}{h}xy}dx,$$

$$s(x) = \frac{1}{\sqrt{h}}\int t(y)e^{-\frac{2\pi i}{h}xy}dy. \tag{8a}$$

Combining (5a), (7a), and (8a), the expression for $(\Delta p)^2$ may be transformed, giving

$$\frac{1}{2}(\Delta p)^2 = \frac{1}{\sqrt{h}}\int y^2 t^*\left(y\right)dy\int s(x)\,e^{\frac{2\pi i}{h}xy}dx,$$

$$= \frac{1}{\sqrt{h}}\int t^*(y)dy\int s(x)\left(\frac{h}{2\pi i}\frac{d}{dx}\right)^2 e^{\frac{2\pi i}{h}xy}dx,$$

$$= \frac{1}{\sqrt{h}}\left(\frac{h}{2\pi i}\right)^2\int t^*(y)dy\int \frac{d^2 s}{dx^2}e^{\frac{2\pi i}{h}xy}dx,$$

$$= \left(\frac{h}{2\pi i}\right)^2\int s^*\left(x\right)\frac{d^2 s}{dx^2}dx,$$

or

$$\frac{1}{2}(\Delta p)^2 = \frac{h^2}{4\pi^2}\int\left|\frac{ds}{dx}\right|^2 dx. \tag{12}$$

Now

$$\left|\frac{ds}{dx}\right|^2 \geq \frac{1}{(\Delta q)^2}\,|s\,(x)|^2 - \frac{d}{dx}\left(\frac{x}{(\Delta q)^2}\,|s\,(x)|^2\right) - \frac{x^2}{(\Delta q)^4}\,|s\,(x)|^2, \tag{13}$$

as may be proved by rearranging the obvious relation

$$\left|\frac{x}{(\Delta q)^2}\,s\,(x) + \frac{ds}{dx}\right|^2 \geq 0. \tag{13a}$$

Hence it follows from equation (12) that

or

$$\left.\begin{aligned}\frac{1}{2}(\Delta p)^2 &\geq \frac{1}{2}\frac{h^2}{4\pi^2}\frac{1}{(\Delta q)^2},\\[2mm]\Delta p\,\Delta q &\geq \frac{h}{2\pi},\end{aligned}\right\} \tag{14}$$

which was to be proved. The equality can be true in (14) only when the left side of (13a) vanishes, i.e., when

or

$$\left.\begin{aligned}s\,(x) &= c\,e^{-\frac{x^2}{2(\Delta q)^2}},\\[2mm]S\,(q') &= c\,e^{-\frac{(q'-\bar{q})^2}{2(\Delta q)^2}-\frac{2\pi i}{h}\bar{p}q'},\end{aligned}\right\} \tag{15}$$

where $c$ is an arbitrary constant. Thus the Gaussian probability distribution causes the product $\Delta p\,\Delta q$ to assume its minimum value.

It must be emphasized again that this proof does not differ at all in mathematical content from that given at the beginning of this section on the basis of the duality between the wave and corpuscular pictures of atomic phenomena. The first proof, if carried through precisely, would also involve all the equations (4)–(14). Physically, the last proof appears to be more general than the former, which was proved on the assumption that $x$ was a cartesian co-ordinate and applies specifically only to free electrons because of the relation $\lambda = h/\mu v_g$ which enters into the proof. Equation (14), on the other hand, applies to any pair of canonic conjugates $p$ and $q$. This greater generality of

(14) is rather specious, however. As Bohr* has emphasized, if a measurement of its co-ordinate is to be possible at all, the electron must be practically free.

## § 2. ILLUSTRATIONS OF THE UNCERTAINTY RELATIONS

The uncertainty principle refers to the degree of indeterminateness in the possible present knowledge of the simultaneous values of various quantities with which the quantum theory deals; it does not restrict, for example, the exactness of a position measurement alone or a velocity measurement alone. Thus suppose that the velocity of a free electron is precisely known, while the position is completely unknown. Then the principle states that every subsequent observation of the position will alter the momentum by an unknown and undeterminable amount such that after carrying out the experiment our knowledge of the electronic motion is restricted by the uncertainty relation. This may be expressed in concise and general terms by saying that every experiment destroys some of the knowledge of the system which was obtained by previous experiments. This formulation makes it clear that the uncertainty relation does not refer to the past; if the velocity of the electron is at first known and the position then exactly measured, the position for times previous to the measurement may be calculated. Then for these past times $\Delta p \Delta q$ is smaller than the usual limiting value, but this knowledge of the past is of a purely speculative character, since it can never (because of the unknown change in momentum caused by the position measurement) be used as an initial condition in any calculation of the future progress of the electron and thus cannot be subjected to experimental verification. It is a matter of personal belief whether such a calculation concerning the past history of the electron can be ascribed any physical reality or not.

*(a) Determination of the position of a free particle.* As a first example of the destruction of the knowledge of a particle's momentum

* *Loc. cit.*

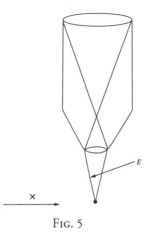

FIG. 5

by an apparatus determining its position, we consider the use of a microscope.* Let the particle be moving at such a distance from the microscope that the cone of rays scattered from it through the objective has an angular opening $\varepsilon$. If $\lambda$ is the wave-length of the light illuminating it, then the uncertainty in the measurement of the $x$-co-ordinate (see Fig. 5) according to the laws of optics governing the resolving power of any instrument is:

$$\Delta x = \frac{\lambda}{\sin \epsilon}. \qquad (16)$$

But, for any measurement to be possible at least one photon must be scattered from the electron and pass through the microscope to the eye of the observer. From this photon the electron receives a Compton recoil of order of magnitude $h/\lambda$. The recoil cannot be exactly known, since the direction of the scattered photon is undetermined within the bundle of rays entering the microscope. Thus there is an uncertainty of the recoil in the $x$-direction of amount

$$\Delta p_x \sim \frac{h}{\lambda} \sin \epsilon, \qquad (17)$$

and it follows that for the motion after the experiment

$$\Delta p_x \Delta x \sim h. \qquad (18)$$

*N. Bohr, *loc. cit.*

Objections may be raised to this consideration; the indeterminateness of the recoil is due to the uncertain path of the light quantum within the bundle of rays, and we might seek to determine the path by making the microscope movable and measuring the recoil it receives from the light quantum. But this does not circumvent the uncertainty relation, for it immediately raises the question of the position of the microscope, and its position and momentum will also be found to be subject to equation (18). The position of the microscope need not be considered if the electron and a fixed scale be simultaneously observed through the moving microscope, and this seems to afford an escape from the uncertainty principle. But an observation then requires the simultaneous passage of at least two light quanta through the microscope to the observer—one from the electron and one from the scale—and a measurement of the recoil of the microscope is no longer sufficient to determine the direction of the light scattered by the electron. And so on *ad infinitum.*

One might also try to improve the accuracy by measuring the maximum of the diffraction pattern produced by the microscope. This is only possible when many photons co-operate, and a calculation shows that the error in measurement of $x$ is reduced to $\Delta x = \lambda / \sqrt{m}$ sin $\epsilon$ when $m$ photons produce the pattern. On the other hand, each photon contributes to the unknown change in the electron's momentum, the result being $\Delta p_x = \sqrt{m} h$ sin $\epsilon / \lambda$ (addition of independent errors). The relation (18) is thus not avoided.

It is characteristic of the foregoing discussion that simultaneous use is made of deductions from the corpuscular and wave theories of light, for, on the one hand, we speak of resolving power, and, on the other hand, of photons and the recoils resulting from their collision with the particle under consideration. This is avoided, in so far as the theory of light is concerned, in the following considerations.

If electrons are made to pass through a slit of width $d$ (Fig. 6), then their co-ordinates in the direction of this width are known at the moment after having passed it with the accuracy $\Delta x = d$. If we assume the momentum in this direction to have been zero before

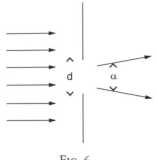

FIG. 6

passing through the slit (normal incidence), it would appear that the uncertainty relation is not fulfilled. But the electron may also be considered to be a plane de Broglie wave, and it is at once apparent that diffraction phenomena are necessarily produced by the slit. The emergent beam has a finite angle of divergence $\alpha$, which is, by the simplest laws of optics,

$$\sin \alpha \sim \frac{\lambda}{d}, \tag{19}$$

where $\lambda$ is the wave-length of the de Broglie waves. Thus the momentum of the electron parallel to the screen is uncertain, after passing through the slit, by an amount

$$\Delta p = \frac{h}{\lambda} \sin \alpha \tag{20}$$

since $h/\lambda$ is the momentum of the electron in the direction of the beam. Then, since $\Delta x = d$,

$$\Delta x \, \Delta p \sim h.$$

In this discussion we have avoided the dual character of light, but have made extensive use of the two theories of the electron.

As a last method of determining position we discuss the well-known method of observing scintillations produced by $\alpha$-rays when they are received on a fluorescent screen or of observing their tracks in a Wilson chamber. The essential point of these methods is that the position of the particle is indicated by the ionization of an atom; it is obvious that the lower limit to the accuracy of such a measurement is given by the linear dimension $\Delta q_s$ of the atom, and also that the momentum of the impinging particle is changed during the act of ionization. Since the momentum of the electron ejected from the atom is measurable, the uncertainty in the change of momentum of the impinging particle is equal to the range $\Delta p_s$ within which the momentum of this electron varies while moving in its unionized orbit. This variation in momentum is again related to the size of the atom by the inequality

$$\Delta p_s \Delta q_s \geq h.$$

Later discussion will show, in fact, that quite generally*

$$\Delta p_s \Delta q_s \sim nh,$$

where $n$ is the quantum number of the stationary state concerned (cf. § 2c below). Thus the uncertainty relation also governs this type of position measurement; here the dualism of treatment is relegated to the background, and the uncertainty relation appears rather to be the result of the Bohr quantum conditions determining the stationary state, but naturally the quantum conditions are themselves manifestations of the duality.

*(b) Measurement of the velocity or momentum of a free particle.* The simplest and most fundamental method of measuring velocity depends on the determination of position at two different times. If the time interval elapsing between the position measurements is sufficiently large, it is possible to determine the velocity before the second was made with any desired accuracy, but it is the velocity after this measurement which alone is of importance to the physicist, and this cannot be determined with exactness. The change in momentum which is necessarily produced by the last observation is subject to such an indeterminateness that the uncertainty relation is again fulfilled, as has been shown in the last section.

Another common method of determining the velocity of charged particles makes use of the Doppler effect. Figure 7 shows the experimental arrangement in its essentials. The component, $p_x$, of the

---

*N. Bohr, *loc. cit.*

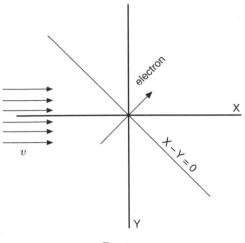

FIG. 7

electron's momentum may be supposed to be known with ideal exactness, its $x$-co-ordinate therefore completely unknown. On the other hand, the $y$-co-ordinate of the electron will be assumed to have been accurately determined, and $p_y$ correspondingly unknown. The problem is therefore to determine the velocity in the $y$-direction, and it is to be shown that the knowledge of the $y$-co-ordinate is destroyed by this measurement to the extent demanded by the uncertainty relation. The light may be supposed incident along the $x$-axis, and the scattered light observed in the $y$-direction. (It is to be noted that the Doppler effect vanishes, under these conditions, if the electron moves along the straight line $x - y = 0$.) The theory of the Doppler effect is in this case identical with that of the Compton effect, and it is only necessary to use the laws of conservation of energy and momentum of the electron and light quantum. Letting $E$ denote the energy of the electron, $\nu$ the frequency of the incident light, and using primes to distinguish the same quantity before and after the collision, we have

$$
\left.
\begin{aligned}
h\nu + E &= h\nu' + E', \\
\frac{h\nu}{c} + p_x &= p'_x, \\
p_y &= \frac{h\nu'}{c} + p'_y,
\end{aligned}
\right\}
\tag{21}
$$

whence

$$\left.\begin{aligned}
h\left(v - v'\right) &= E' - E, \\[4pt]
&= \frac{1}{2\mu}\left[p_x'^2 + p_y'^2 - p_x^2 - p_y^2\right], \\[4pt]
&\sim \frac{1}{\mu}\left[\left(p_x' - p_x\right)p_x + \left(p_y' - p_y\right)p_y\right], \\[4pt]
&= \frac{1}{\mu}\left[\frac{hv}{c}p_x - \frac{hv'}{c}p_y\right], \\[4pt]
&\sim \frac{hv}{\mu c}\left(p_x - p_y\right).
\end{aligned}\right\} \tag{22}$$

Since it is assumed that $p_x$ and $v$ are known, the accuracy of the determination of $p_y$ is conditioned only by the accuracy with which the frequency $v'$ of the scattered light is measured:

$$\Delta p_y' = \frac{\mu c}{v}\Delta v'. \tag{23}$$

To determine $v'$ with this accuracy, it is necessary to observe a train of waves of finite length, which in turn demands a finite time:

$$T = \frac{1}{\Delta v'}.$$

As it is unknown whether the photon collided with the electron at the beginning or at the end of this time interval, it is also unknown whether the electron moved with the velocity $(1/\mu)p_y$ or $(1/\mu)p_y'$ during this time. The uncertainty in the position of the electron which is produced by this cause is thus

$$\Delta y = \frac{1}{\mu}\left(p_y - p_y'\right)T = \frac{hv}{c\mu}T,$$

whence

$$\Delta p_y \Delta y \sim h.$$

A third method of velocity measurement depends on the deflection of charged particles by a magnetic field. For this purpose a beam must be defined by a slit, whose width will be designated by $d$. This ray then enters a homogeneous magnetic field, whose direction is to be taken perpendicular to the plane of Figure 8. The length of that part of the

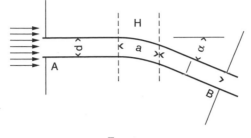

FIG. 8

ray which lies in the region of the field may be $a$; after leaving this region, the ray traverses a field-free region of length $l$ and then passes through a second slit also of width $d$, whose position determines the angle of deflection $\alpha$. The velocity of the particles in the direction of the beam is to be determined from the equation

$$\alpha = \frac{\dfrac{a}{v}He\dfrac{v}{c}}{\mu v} = \frac{a\,He}{\mu v c}. \tag{24}$$

The corresponding errors in measurement are related by

$$\Delta\alpha = \frac{a\,He}{\mu c}\frac{\Delta v}{v^2}.$$

It may be supposed that the position of the particle in the direction of the ray was initially known with great accuracy. This may be achieved, for example, by opening the first slit only during a very brief interval. It will again be shown that this knowledge is lost during the experiment in such a manner that the relation $\Delta p \Delta q \sim h$ is fulfilled after the experiment. To begin with, the accuracy with which the angle $\alpha$ can be determined is obviously $d/(l+a)$, but even this accuracy can only be attained if the natural de Broglie scattering of the ray is less than this. Therefore

$$\Delta\alpha \geq \frac{d}{l+a}, \qquad \Delta\alpha \geq \frac{\lambda}{d},$$

whence

$$(\Delta\alpha)^2 \geq \frac{\lambda}{l+a}.$$

The uncertainty in the position of the particle in the ray after the experiment is equal to the product of the time required to pass through the field and reach the second slit and the uncertainty in the velocity. Thus

$$\Delta q \sim \frac{l+a}{v} \Delta v,$$

whence

$$\Delta q \, \Delta v \sim \frac{l+a}{v} (\Delta v)^2,$$

$$\sim \frac{l+a}{v} \left(\frac{\mu c v}{a \, He}\right)^2 (\Delta \alpha)^2,$$

$$\geq \frac{\lambda}{v} \left(\frac{\mu c v^2}{a \, He}\right)^2.$$

The terms in the parentheses are equal to $v/\alpha$ and $\lambda = h/\mu v$, whence

$$\mu \Delta q \, \Delta v \geq \frac{h}{\alpha^2} \geq h,$$

since equation (24) is valid only for small values of $\alpha$. For large angles of deflection, this derivation requires radical modification. One must remember, among other things, that the experiment as described here would not distinguish between $\alpha = 0$ and $\alpha = 2\pi$.

*(c) Bound electrons.* If it be required to deduce the uncertainty relations for the position, $q$, and momentum, $p$, of bound electrons, two problems must be clearly distinguished. The first assumes that the energy of the system, i.e., its stationary state, is known, and then inquires what accuracy of knowledge of $p$ and $q$ is implied in, or is compatible with, this knowledge of the energy. The second, distinct problem disregards the possibility of determining the energy of the system and merely inquires what the greatest accuracy is with which $p$ and $q$ may simultaneously be known. In this second case, the experiments necessary for the measurement of $p$ and $q$ may produce transitions from one stationary state to another; in the first case, the methods of measurement must be so chosen that transitions are not induced.

We consider the first problem in some detail, and assume an atom in a given stationary state. As Bohr has shown,[*] the corpuscular theory then forces one to conclude that $\Delta p \Delta q$ is in general greater than $h$. For it is obvious that we are concerned with the variation of $p$ and $q$ as the electron moves in its orbit, and it follows from

$$\int p\,dq = nh \tag{25}$$

that

$$\Delta q_s \Delta p_s \sim nh. \tag{26}$$

This may most readily be comprehended from a diagram of the orbit in phase space as given by classical mechanics (Fig. 9). The integral is nothing else than the area inclosed by the orbit, and $\Delta p_s \Delta q_s$ is obviously of the same order of magnitude. The index $s$ which accompanies these uncertainties is to indicate that they are not the absolute minima of these quantities, but are the special values which are assumed by them when the stationary state of the atom is known simultaneously and exactly. This uncertainty is of practical importance, for example, in the discussion of the scintillation method of counting $\alpha$-particles (chap. ii, § 2a). In the classical theory, it would seem strange to consider this as an essential uncertainty, for further experiments could be made without disturbing the orbit. The quantum theory, however, shows that a knowledge of the energy is a "determinate case" (*reiner Fall*),[†] i.e., a case which is represented in the mathematical scheme by a definite wave packet (in configuration space) which does not involve any undetermined constants. This wave packet is the Schrödinger function of the stationary state. If the calculation of pages 16–19 is carried through for this packet, the value of $\Delta p_s \Delta q_s$ is found to be greater in proportion to the number of nodes possessed by the characteristic function. If we consider a function $s$ in equation (12) which possesses $n$ nodes, the calculation would show that

$$\Delta p_s \Delta q_s \sim nh.$$

[*] *Ibid.*
[†] The translators believe that the literal rendering of the German phrase ("pure case") does not at all convey the concept involved.

178

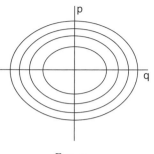

FIG. 9

To pass on to the second problem: The maximum accuracy is obviously given by $\Delta p \Delta q \sim h$ if all knowledge of the stationary states be disregarded. Then the measurements can be carried out by such violent agents that the electron can be regarded as free (acted on only by negligible forces). The momentum of the electron can most readily be measured by suddenly rendering the interaction of the electron with the nucleus and neighboring electrons negligible. It will then execute a straight-line motion and its momentum can be measured in the manner already explained. The disturbance necessary for such a measurement is therefore obviously of the same order of magnitude as the binding energy of the electron.

The relation [eq. (6)] is of importance, as Bohr points out, for the equivalence of classical and quantum mechanics in the limit of large quantum numbers. This is seen when the validity of the concept of an "orbit" is examined. As the highest accuracy attainable is $\Delta p \Delta q \sim h$, the orbit must be the path of a probability packet whose cross-section $(|S(p')|^2 |S(q')|^2)$ is approximately $h$. Such a packet can describe a well-defined, approximately closed path only if the area in-closed by this path is much greater than the cross-section of the wave packet. This, according to equation (26), is possible only in the limit of large quantum numbers; for small $n$, on the other hand, the concept of an orbit loses all significance, in phase space as well as in configuration space. It is thus seen to be essential for this limiting equivalence of the two theories that the factor $n$ occurs on the right side of equation (26).

179

The inapplicability of the concept of an orbit in the region of small quantum numbers can be made clear from direct physical considerations in the following manner: The orbit is the temporal sequence of the points in space at which the electron is observed. As the dimensions of the atom in its lowest state are of the order $10^{-8}$ cm, it will be necessary to use light of wave-length not greater than $10^{-9}$ cm in order to carry out a position measurement of sufficient accuracy for the purpose. A single photon of such light is, however, sufficient to remove the electron from the atom, because of the Compton recoil. Only a single point of the hypothetical orbit is thus observable. One can, however, repeat this single observation on a large number of atoms, and thus obtain a probability distribution of the electron in the atom. According to Born, this is given mathematically by $\psi \psi^*$ (or, in the case of several electrons, by the average of this expression taken over the co-ordinates of the other electrons in the atom). This is the physical significance of the statement that $\psi \psi^*$ is the probability of observing the electron at a given point. This result is stranger than it seems at first glance. As is well known, $\psi$ diminishes exponentially with increasing distance from the nucleus; there is thus always a small but finite probability of finding the electron at a great distance from the center of the atom. The potential energy of the electrons is negative at such a point, but very small. The kinetic energy is always positive; so that the total energy is therefore certainly greater than the energy of the stationary state under consideration. This paradox finds its resolution when the energy imparted to the electron by the photon used in making the position measurement is taken into account. This energy is considerably greater than the ionization energy of the electron, and thus suffices to prevent any violation of the law of conservation of energy, as is readily calculated explicitly from the theory of the Compton effect.

This paradox also serves as a warning against carrying out the "statistical interpretation" of quantum mechanics too schematically. Because of the exponential behavior of the Schrödinger function at infinity, the electron will sometimes be found as much as, say, 1 cm from the nucleus. One might suppose that it would be possible to verify the presence of the electron at such a point by the use of red light. This red light would not produce any appreciable Compton recoil and the foregoing paradox would arise once more. As a matter of fact, the red light will not permit such a measurement to be made; the atom as a whole will react with the light according to the formulas of dispersion theory, and the result will not yield any information regarding the position of a given electron in the atom. This may be made plausible if one remembers that (according to the corpuscular theory) the electron will execute a number of rotations about the nucleus during one period of the red light. The statistical predictions of quantum theory are thus significant only when combined with experiments which are actually capable of observing the phenomena treated by the statistics. In many cases it seems better not to speak of the probable position of the electron, but to say that its size depends upon the experiment being performed.

The orbital concept has a significance when applied to highly excited states of the atom; therefore it must be possible to carry out the determination of the position of the electron with an uncertainty less than the dimension of the atom. It does not follow any longer that the electron will be removed from the atom by the Compton recoil, as may be seen from the following equations. It is necessary that the wave-length of the light, $\lambda$, be much less than $\Delta q_s$, or by equation (26),

$$\frac{h}{\lambda} \gg \frac{\Delta p_s}{n}.$$

The energy imparted to the electron by its recoil is approximately

$$\frac{h}{\lambda}\frac{\Delta p_s}{\mu} \gg \frac{(\Delta p_s)^2}{n\mu} \sim \frac{|E|}{n} \tag{26a}$$

($E$ is the energy of the atom, $\mu$, the mass of the electron); for large values of $n$, this recoil energy is much less than $|E|$, the ionization energy of the electron. On the other hand, this energy will always be great compared to the energy differences between neighboring stationary states in this region of the spectrum, which is also, in general, of the order $|E|/n$. As a matter of fact, from equation ($26a$) it follows at once that

$$h\nu \gg \frac{|E|}{n},$$

so that the frequency of the light used in making the measurement is great compared to the frequency of the electron in its orbit.

The Compton effect has as its consequence that the electron is caused to jump from a state, say $n = 1000$, to some other state for which $n$ is, say, greater than 950 and less than 1050. The particular orbit to which the electron jumps remains essentially indeterminate because of the considerations of chapter ii, § 1$b$. The result of the position measurement is therefore to be represented in the mathematical scheme by a probability packet in configuration space, which is built up of characteristic functions of the states between $n = 950$ and 1050. Its size is determined by the exactitude of the position measurement. This packet describes an orbit analogous to that of a corpuscle of classical mechanics, but, in general, spreads and increases in size with the time. The result of a future measurement of position can therefore only be predicted statistically. The mathematical representation of the physical process changes discontinuously with each new measurement; the observation singles out of a large number of possibilities one of which is the one which has happened. The wave packet which has spread out is replaced by a smaller one which represents the result of this observation. As our knowledge of the system does change discontinuously at each observation its mathematical representation must also change discontinuously; this is to be found in classical statistical theories as well as in the present theory.

The motion and spreading of probability packets has been studied by various authors,* and therefore no mathematical discussion of it need be given here. A simple consideration of Ehrenfest's[†] may be mentioned, however. Consider the motion of a single electron moving in a field of force whose potential is $V(q)$. The wave function satisfies

---

*Kennard, *loc, cit.*; C. G. Darwin, *Proceedings of the Royal Society*, A, **117**, 258, 1927.
[†] P. Ehrenfest, *Zeitschrift für Physik*, **45**, 455, 1927.

[cf. eq. A (80)]

$$-\frac{h^2}{8\pi^2\mu}\nabla^2\psi + eV\psi = -\frac{h}{2\pi i}\frac{\partial\psi}{\partial t}, \tag{27}$$

and the probable value of $q$ is given by equation (4) with $\psi = S$; $q$ is one of the rectangular co-ordinates $x, y, z$. Then differentiating by $t$:

$$\mu\dot{q} = \mu\int q\left(\frac{\partial\psi}{\partial t}\psi^* + \psi\frac{\partial\psi^*}{\partial t}\right)d\tau;$$

on substituting the value of $\partial\psi/\partial t$ and $\partial\psi^*/\partial t$ from (27):

$$\mu\dot{q} = \frac{h}{4\pi}\int q\left(-\psi^*\nabla^2\psi + \psi\nabla^2\psi^*\right)d\tau;$$

integrating by parts:

$$\mu\dot{q} = \frac{h}{4\pi}\int\left(\psi^*\frac{\partial\psi}{\partial q} - \psi\frac{\partial\psi^*}{\partial q}\right)d\tau.$$

This process may be repeated a second time to obtain $\mu\ddot{q}$. As the calculation is lengthy, but simple, we give only the result:

$$\mu\ddot{q} = -e\int\frac{\partial V}{\partial q}\psi\psi^*d\tau. \tag{28}$$

If $\psi$ represents a wave packet whose spatial dimension is small compared to the distance within which $\partial V/\partial q$ changes appreciably, this may be written

$$\mu\ddot{q} = -e\frac{\partial V(\bar{q})}{\partial\bar{q}}. \tag{29}$$

This proves that, so long as the wave packet remains small, its center will move according to the classical equations of motion of the electron.

A remark concerning the rate of spreading of the wave packet may not be out of place at this point. If the classical motion of the system is periodic, it may happen that the size of the wave packet at first undergoes only periodic changes. The number of revolutions which the packet may perform before it spreads completely over the whole region of the atom can be calculated qualitatively as follows: If there were no spreading at all, it would be possible to make a Fourier analysis of the probability density into which only integral multiples of the fundamental frequency of the orbit enter. As a matter of fact, however, the "overtones" of quantum theory are not exactly integral multiples of this fundamental frequency. The time in which the phase of the quantum theoretical overtones is completely shifted from that of the classical overtones will be qualitatively the same as the time required for the spreading of the wave packet. Let $J$ be the action variable of classical theory, then this time will be

$$t \sim \frac{1}{h \dfrac{\partial v}{\partial J}},$$

and the number of revolutions performed in this time is

$$N \sim \frac{v}{h \dfrac{\partial v}{\partial J}}. \tag{30}$$

In the special case of the harmonic oscillator, $N$ becomes infinite—the wave packet remains small for all time. In general, however, $N$ will be of the order of magnitude of the quantum number $n$.

In relation to these considerations, one other idealized experiment (due to Einstein) may be considered. We imagine a photon which is represented by a wave packet built up out of Maxwell waves.* It will thus have a certain spatial extension and also a certain range of frequency. By reflection at a semi-transparent mirror, it is possible to decompose it into two parts, a reflected and a transmitted packet. There is then a definite probability for finding the photon either in one part or in the other part of the divided wave packet. After a sufficient time the two parts will be separated by any distance desired; now if an experiment yields the result that the photon is, say, in the reflected part of the packet, then the probability of finding the photon in the other part of the packet immediately becomes zero. The experiment at the position of the reflected packet thus exerts a kind of action (reduction of the wave packet) at the distant point occupied by the transmitted packet, and one sees that this action is propagated with a

*For a single photon the configuration space has only three dimensions; the Schrödinger equation of a photon can thus be regarded as formally identical with the Maxwell equations.

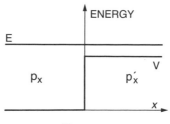

FIG. 10

velocity greater than that of light. However, it is also obvious that this kind of action can never be utilized for the transmission of signals so that it is not in conflict with the postulates of the theory of relativity.

*(d) Energy measurements.* The measurement of the energy of a free electron is identical with the measurement of its velocity, so that most of the possible methods have already been treated. A method not yet discussed for measuring the energy of free electrons is that in which they are caused to move against a retarding field. If the electron passes through the field it is customary to assume the result of classical theory, that its energy $E$ is certainly greater than the energy $V$ corresponding to the highest potential of the field, and if it is reflected, that its energy is smaller than this critical value. Such a conclusion is certainly incorrect in the quantum theory, and a brief discussion of the method will therefore be given here. If the width of the potential barrier is comparable to the de Broglie wave-length, $\lambda$, of the electron, a certain number of electrons will penetrate it even though their energies $E$ are less than the critical value necessary on the classical theory. This number decreases exponentially as the width of the barrier and $V - E$ increase. Conversely, when $E > V$, a certain number will be reflected if the potential changes appreciably in a distance $\lambda$. In any practicable experiment, these conditions are not realizable, and the conclusions of the classical theory can be used without appreciable error. The mathematical treatment of the situation just sketched is important, however, and will therefore be illustrated in the case of an abrupt discontinuity in the potential distribution. The Schrödinger equation for a single electron will be used; this is not identical with

the wave theory of matter, for this latter would take the reaction of the wave on itself into account. The potential distribution is shown in Figure 10. For the incident $\psi$-wave in the region I ($x < 0$), we then readily obtain the expression

$$\psi_i = a e^{\frac{2\pi i}{h}(px - Et)}, \qquad \frac{1}{2\mu} p^2 = E, \qquad p > 0; \qquad (31a)$$

for the wave penetrating into the region II ($x > 0$),

$$\psi_t = a' e^{\frac{2\pi i}{h}(p'x - Et)}, \qquad \frac{1}{2\mu} p'^2 = E - V; \qquad (31b)$$

and for the reflected wave in I,

$$\psi_r = a'' e^{\frac{2\pi i}{h}(-px - Et)}. \qquad (31c)$$

If $p'$ is real, it is to be taken greater than zero; if it is imaginary, total reflection occurs and it is to be taken as positive imaginary, since $\psi_t$ must remain finite as $x \to \infty$. At the discontinuity ($x = 0$), $\psi$ must be continuous and possess a continuous first derivative; hence

$$\left.\begin{array}{c} \psi_i + \psi_r = \psi_t \\[2mm] \dfrac{\partial \psi_i}{\partial x} + \dfrac{\partial \psi_r}{\partial x} = \dfrac{\partial \psi_t}{\partial x}, \end{array}\right\} \quad \text{when } x = 0;$$

or

$$a + a'' = a'$$
$$p(a - a'') = a'p'.$$

Solving these equations for $a'$ and $a''$:

$$\left.\begin{array}{c} a'' = a\dfrac{p - p'}{p + p'}, \\[4mm] a' = a\dfrac{2p}{p + p'}. \end{array}\right\} \qquad (32)$$

The number of electrons that pass through a given cross-section per unit time is given by the square of the absolute magnitude of the wave amplitude multiplied by the momentum provided it is real. Thus, when $E > V$, the intensities of the incident, transmitted and

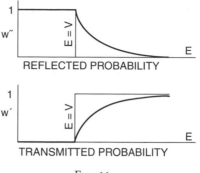

FIG. 11

reflected waves are respectively proportional to

$$
\left.
\begin{aligned}
I_i &= |a|^2\, p; \\[2mm]
I_t &= |a|^2 \left(\frac{2p}{p+p'}\right)^2; \\[2mm]
I_r &= -|a|^2 \left(\frac{p-p'}{p+p'}\right)^2.
\end{aligned}
\right\}
\tag{33}
$$

For imaginary values of $p'$, the wave $\psi_t$ does not represent a current of electrons, but a stationary charge distribution, and $I_t = 0$. As $|a''| = |a|$ in this case, $I_r = -I_i$. In both cases

$$
I_i = I_t - I_r.
$$

The relative probabilities for reflection and penetration of the electron are, by (33) and (31),

$$
\left.
\begin{aligned}
p'' &= \frac{I_r}{I_i} = \left|\frac{\sqrt{E}-\sqrt{E-V}}{\sqrt{E}+\sqrt{E-V}}\right|^2, \\[3mm]
p' &= \frac{I_t}{I_i} = \sqrt{\frac{E-V}{E}}\left|\frac{2\sqrt{E}}{\sqrt{E}+\sqrt{E-V}}\right|^2.
\end{aligned}
\right\}
\tag{34}
$$

These expressions are plotted as solid lines in Figure 11; the curves expected from the classical theory are the dotted lines.

For the elucidation of the physical principles of the quantum theory a consideration of the mesaurement of the energy of atoms is more important than that of free electrons, and this will be given in greater detail than the preceding. As the phase of the electronic motion is the variable which is canonically conjugate to the energy, it follows from the uncertainty principle that this must be completely unknown if the energy is precisely determined. Since the phase of the electronic motion determines the phase of the radiation emitted, it is this latter which is to enter the physical discussion. It will be shown that any experiment which separates atoms that are in the stationary state $n$ from those in $m$ necessarily destroys any pre-existing knowledge of the phase of the radiation corresponding to the transition $n \underset{\leftarrow}{\overset{\rightarrow}{\phantom{m}}} m$.

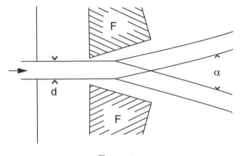

FIG. 12

Let $S$ be a beam of atoms (Fig. 12), of width $d$ in the $x$-direction, which is sent through an inhomogeneous field $F$ (which is not necessarily a magnetic field, as in the experiment of Stern-Gerlach, but may be electric or gravitational). The energy of the atoms in state $m$ will be designated by $E_m$; it will depend on the magnitude of the field $F$ at the center of gravity of the atom, so that the deflecting force of the field in the $x$-direction is $\partial(E_m(F))/\partial x = (dE_m/dF)(dF/dx)$, and is different for atoms in different states. If $T$ be the time required by the atoms to pass through the field, and $p$ the momentum of the atoms in the direction of the beam, the angular deflection of the atoms will be

$$\frac{\partial E_m}{\partial x} \frac{T}{p}.$$

The original beam will thus be divided into several, each containing only atoms in one state; the angular separation $\alpha$ of the two beams containing atoms in states $n$ and $m$, respectively, will then be

$$\alpha = \left( \frac{\partial E_m}{\partial x} - \frac{\partial E_n}{\partial x} \right) \frac{T}{p}.$$

This angle must be greater than the natural scattering of the atomic beams if the two kinds of atoms are to be separated; hence

$$\alpha \geq \frac{\lambda}{d} = \frac{h}{pd}. \tag{35}$$

The Schrödinger function $\psi_n$ contains the periodic factor $e^{\frac{2\pi i}{h} E_n t}$. As $E_n$ is a function of $F$, the frequency and phase of the wave are changed while passing through the field. This change is indeterminate, to a certain extent, since it is impossible to tell in what part of the beam the atom is moving and $F$ varies from point to point. The uncertainty, $\Delta\varphi$, of the phase change of the radiation of frequency $(E_m - E_n)/h$ during the time $T$ is therefore

$$\Delta\varphi \sim 2\pi \left( \frac{\partial E_m}{\partial x} - \frac{\partial E_n}{\partial x} \right) \frac{Td}{h} = \frac{pd}{h} 2\pi\alpha.$$

From equation (35) it follows at once that

$$\Delta\varphi \geq 1. \tag{36}$$

This means complete indeterminateness in the phases.

The calculation can be carried through more concretely if it is restricted to apply only to magnetic fields. Neglecting the electron spin, it is known that the atom precesses like a rigid body when under the influence of a magnetic field $H$; the velocity of this precession is

$$\omega = \frac{e}{2\mu c} H,$$

and its axis coincides with the direction of the field. This velocity is different for various atoms because of the width of the beam and the inhomogeneity of the field. This difference in the precession of different atoms tends to destroy any phase relation which may initially be present. For the uncertainty in $\omega$, we readily obtain

$$\Delta\omega = \frac{ed}{2\mu c} \frac{\partial H}{\partial x},$$

and the angular separation of the two beams is

$$\alpha = \frac{e}{2\mu c} \frac{\partial H}{\partial x} \frac{hT}{2\pi p};$$

as $\alpha$ must be greater than $h/pd$,

$$T\Delta\omega \geq 2\pi.$$

All trace of the original phase has thus been destroyed by the experiment. Some atoms will have executed one rotation more than others, and all intermediate angles are possible. This does not follow if the apparatus is incapable of resolving the two beams, as then $\alpha$ may be less than $h/pd$.

Bohr* has shown that the foregoing consideration resolves one of the paradoxes introduced by the assumption of stationary states. If a beam of atoms, all initially in the normal state, be excited to fluorescence by illumination with light of a resonance frequency, we are compelled to assume that they will radiate coherently. That is, each atom will scatter a spherical wave, whose phase is determined by that of the incident plane wave at the atom. The elementary spherical waves will then be so related that their superposition results in a refracted plane wave. From the observation of this wave it is impossible to determine the quantum state of the emitter—or even its atomic character. But if the beam leaves the illuminated region and is analyzed by means of an inhomogeneous field, only the beam of atoms in the excited state will be luminous. This beam will contain relatively few atoms, widely spaced compared to the probable length of the train of waves emitted. Their radiation must therefore be practically identical with that from independent point sources. This action of the magnetic field was quite incomprehensible as long as the assumption was retained that the resolving power of the apparatus could be increased indefinitely by decreasing the width of the beam of atoms.

* *Loc. cit.*

# CRITIQUE OF THE PHYSICAL CONCEPTS OF THE WAVE THEORY

In the foregoing chapter the simplest concepts of the wave theory, which are well established by experiment, were assumed without question to be "correct." They were taken as the basis of a critique of the corpuscular picture, and it appeared that this picture is only applicable within certain limits, which were determined. The wave theory, as well, is only applicable with certain limitations, which will now be determined. Just as in the case of particles the limitations of a wave representation were not originally taken into account, so that historically we first encounter attempts to develop *three-dimensional* wave theories that could be readily visualized (Maxwell and de Broglie waves). For these theories the term "classical wave theories" will be used; they are related to the quantum theory of waves in the same way as classical mechanics to quantum mechanics. The mathematical scheme of the classical and quantum theories of waves will be found in the Appendix. (The reader must be warned against an unwarrantable confusion of classical wave theory with the Schrödinger theory of waves in a phase space.) After a critique of the wave concept has been added to that of the particle concept all contradictions between the two disappear—provided only that due regard is paid to the limits of applicability of the two pictures.

## § 1. THE UNCERTAINTY RELATIONS FOR WAVES

The concepts of wave amplitude, electric and magnetic field strengths, energy density, etc., were originally derived from primitive experiences of daily life, such as the observation of water waves or the vibrations of elastic bodies. These concepts are also widely applicable to light and even, as we now know, to matter waves. But since we also know that the concepts of the corpuscular theory are applicable to radiation and matter, it follows that the wave picture also has its limitations, which may be derived from the particle representation. These will now be considered, first for the case of radiation.

Before proceeding to the subject proper, however, we must first discuss briefly what is meant by an exact knowledge of a wave amplitude—for instance, that of an electric or magnetic field strength. Such an exact knowledge of the amplitude at every point of a region of space (in the strict mathematical sense) is obviously an abstraction that can never be realized. For every measurement can yield only an average value of the amplitude in a very small region of space and during a very short interval of time. Although it is perhaps possible in principle to diminish these space and time intervals without limit by refinement of the measuring instruments, nevertheless for the physical discussion of the concepts of the wave theory it is advantageous to introduce finite values for the space and time intervals involved in the measurements and only pass to the limit zero for these intervals at the end of the calculations. This is, in fact, exactly the procedure adopted in treating the mathematical theory of wave fields (cf. A, § 9). It is possible that future developments of the quantum theory will show that the limit zero for such intervals is an abstraction without physical meaning; for the present, however, there seems no reason for imposing any limitations.

For precision of thought we therefore assume that our measurements always give average values over a very small space region of volume $\partial v = (\partial l)^3$, which depends on the method of measurement. Since it is a question of the measurement of the field strengths, light of wave-length $\lambda$ much less than $\partial l$ will not be detected by the experiment. The measurement gives, say, the values $E$ and $H$ for the field strengths (averaged over $\delta v$). If these values $E$ and $H$ were exactly known there would be a contradiction to the particle theory, since the energy and momentum of the small volume $\delta v$ are

$$E = \delta v \frac{1}{8\pi} \left( \boldsymbol{E}^2 + \boldsymbol{H}^2 \right), \qquad G = \delta v \frac{1}{4\pi c} \boldsymbol{E} \times \boldsymbol{H}, \qquad (37)$$

and the right-hand members could be made as small as desired by taking $\delta v$ sufficiently small. This is inconsistent with the particle theory, according to which the energy and momentum content of the small volume is made up of discrete and finite amounts $h\nu$ and $h\nu/c$, respectively. For the highest frequency detectable $h\nu \leq (hc/\delta l)$ so that it is clear that the right-hand members of equation (37) must be uncertain by just the magnitudes of these quanta ($h\nu$ and $h\nu/c$) in order that there be no contradiction to the particle theory. Accordingly there must be uncertainty relations between the components of $E$ and $H$ which give rise to an uncertainty in the value of $E$ of the order of magnitude $hc/\delta l$ and in $G$ of the order of magnitude $h/\delta l$ when $E$ and $G$ are calculated by equations (37). Let $\Delta E$ and $\Delta H$ be the uncertainties in $E$ and $H$; then the uncertainties in $E$ and $G$ are

$$\Delta E = \frac{\delta v}{8\pi}\left\{2\,|E \cdot \Delta E| + 2\,|H \cdot \Delta H| + (\Delta E)^2 + (\Delta H)^2\right\},$$

$$\Delta G_x = \frac{\delta v}{4\pi c}\left\{|(E \times \Delta H)_x| + |(\Delta E \times H)_x| + |(\Delta E \times \Delta H)_x|\right\},$$

with cyclic permutation for the $y$- and $z$-directions.

Since the most probable values of $E$ and $H$ may possibly be zero the terms on the right which contain only $\Delta E$ and $\Delta H$ must alone be sufficient to give the necessary uncertainty to $E$ and $G$. This is attained if

$$\Delta E_x \Delta H_y \geq \frac{hc}{\delta v \delta l} = \frac{hc}{(\delta l)^4}, \tag{38}$$

with cyclic permutation for the other components. These uncertainty relations refer to a simultaneous knowledge of $E_x$ and $H_y$ in the same volume element; in different volume elements $E_x$ and $H_y$ can be known to any degree of accuracy.

The relations (38), as in the case of the particle theory, can also be derived directly from the exchange relations for $E$ and $H$ (cf. A, §§ 9, 12). If a division of space into finite cells of magnitude $\delta v$ is used, the integration with respect to $dv$ in the Lagrangian of A (97) becomes a sum over all the cells $\delta v$. The momentum conjugate to $\psi_\alpha(r)$ in the *r*th cell is then [cf. A(104)]

$$\delta v \frac{\partial L}{\partial \dot{\psi}_\alpha (r)} = \delta v \Pi_\alpha (r),$$ (39)

and in place of A(111),

$$\Pi_\alpha (r) \psi_\beta (s) - \psi_\beta (s) \Pi_\alpha (r) = \delta_{\alpha\beta} \delta_{rs} \frac{h}{2\pi i} \frac{1}{\delta v},$$ (40)

where $\delta_{rs}$, is now the usual δ-function,

$$\delta_{rs} = \begin{cases} 1 \text{ for } r = s, \\ 0 \text{ for } r \neq s. \end{cases}$$

In the limit $\delta v \to 0$ (40) becomes A(111).

From (40) and A(134) applied to the case of electric and magnetic fields it follows that

$$E_i (r) \Phi_\alpha (s) - \Phi_\alpha (s) E_i (r) = -2hci\delta_{rs}\delta_{\alpha i}\frac{1}{\delta v}.$$ (41)

When it is remembered that an uncertainty $\Delta\Phi_k$ gives an uncertainty of order of magnitude $\Delta\Phi_k/\delta l$ for the field strengths resulting from $\Phi_k$, it will be seen that (41) leads immediately to the uncertainty relations (38).

Matter waves may be treated in an entirely similar way. It must be noted, however, that no experiment can ever measure the amplitude directly, as is evident from the fact that the de Broglie waves are complex. If exchange relations for the wave amplitudes are derived formally from those for $\psi$ and $\psi^*$, the result is, to be sure, a physically reasonable one in the case of the Bose-Einstein statistics. However, use of the experimentally correct Fermi-Dirac statistics gives the meaningless result that $\psi$ and $\psi^*$ cannot be exactly measured simultaneously at different points of space. It is thus highly satisfactory that there is no experiment which will measure $\psi$ at a given point at a given time. The mathematical reason for this is that even for the interaction of radiation and matter the part of the Lagrangian referring to matter contains only terms of the form $\psi\psi^*$. From the considerations just given it can also be seen that the Bose-Einstein statistics is a physical necessity for light-quanta if one makes the apparently very natural assumption that measurements of the electric and magnetic fields at different points of space must be independent of each other.

## § 2. DISCUSSION OF AN ACTUAL MEASUREMENT OF THE ELECTROMAGNETIC FIELD

As in the case of the corpuscular picture, it must be possible to trace the origin of the uncertainty in a measurement of the electromagnetic field to its experimental source. We therefore discuss an experiment which is capable of simultaneously measuring $E_x$ and $H_z$ in the same element of volume $\delta v$. This can be accomplished by the observation of the deflection in the direction of $x$ of two beams of cathode rays which traverse the volume in opposite directions along the $y$-axis (cf. Fig. 13). It may be assumed that the width of both beams in the $z$-direction is $\delta l$, i.e., the whole width of the volume element, but their widths in the perpendicular direction must be less than this, say $d$, so that they may traverse $\delta v$ without mutual disturbance. If the distance between the two rays is of order of magnitude $\delta l$, the small inhomogeneities of the field in this direction are also averaged out; it would also be possible to vary the distance between them for this purpose. This experimental arrangement will enable the measurement of $E_x$ and $H_z$ in $\delta l$ provided only that the fields are not too inhomogeneous; should this condition not be fulfilled, the method is incapable of giving a definite result, for the field must not vary appreciably across the width of the rays, or else these will become diffuse and no simple method of determining the deflections is then available.

The angular deflection, $\alpha$, of the rays in the distance $\delta l$ is to be observed, and the field can be calculated from the formulas

$$\alpha_\pm = \frac{e}{p_y}\left(E_x \pm \frac{p_y}{\mu c}H_z\right)\frac{\mu \delta l}{p_y}.$$

Because of the natural spreading of the matter rays, the accuracy of the measurements is given by

$$\Delta E_x \geq \frac{h}{ed}\frac{p_y}{\mu \delta l}, \qquad \Delta H_z \geq \frac{h}{ed}\frac{p_y}{\mu \delta l}\frac{\mu c}{p_y}. \tag{42}$$

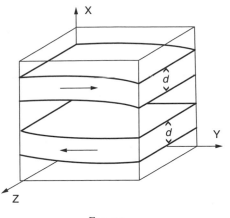

FIG. 13

One essential factor remains to be considered, however. Each of the two electrons which pass through $\delta v$ simultaneously modifies the field, and hence the path of the other electron. The amount of this modification is uncertain to some extent, since it is not known at which point in the cathode ray the electron is to be found. The uncertainty as to the actual fields which arises from this fact is thus

$$\Delta E_x \geq \frac{ed}{(\delta l)^3}, \quad \Delta H_z \geq \frac{ed}{(\delta l)^3} \frac{p_y}{\mu c}, \tag{43}$$

whence

$$\Delta E_x \Delta H_z \geq \frac{hc}{(\delta l)^4},$$

which was to be shown. It is to be noted that the simultaneous consideration of both the corpuscular and wave picture of the process taking place is again fundamental. If the corpuscular picture of the cathode rays had not been invoked, and a continuous distribution of charge assumed as the picture of the rays, then the uncertainty (43) would have disappeared.

# THE STATISTICAL INTERPRETATION OF QUANTUM THEORY

## § 1. MATHEMATICAL CONSIDERATIONS

It is instructive to compare the mathematical apparatus of quantum theory with that of the theory of relativity. In both cases there is an application of the theory of linear algebras. One can therefore compare the matrices of quantum theory with the symmetric tensors of the special theory of relativity. The greatest difference is the fact that the tensors of quantum theory are in a space of infinitely many dimensions, and that this space is not real but imaginary. The orthogonal transformations are replaced by the so-called "unitary" transformations. In order to obtain a picture of this space, we abstract from such differences, fundamental though they be. Then every quantum theoretical "quantity" is characterized by a tensor whose principal directions may be drawn in this space (cf. Fig. 14). In order to obtain a clear picture, one may recall the tensor of the moments of inertia of a rigid body. The principal directions are, in general, different for each quantity; only matrices which commute with one another have coincident principal directions. The exact knowledge of the numerical value of any dynamical variable corresponds to the determination of a definite direction in this space, in the same manner as the exact knowledge of the moment of inertia of a solid body determines the principal direction to which this moment belongs (it is assumed that there is no degeneracy). This direction is thus parallel to the $k$th principal axis of the tensor $T$, along which the component $T_{kk}$ has the value measured. The exact knowledge of the direction (except for a factor of absolute magnitude unity) in unitary space is the maximum information regarding the quantum dynamical variable which can be obtained. Weyl* has called this degree of knowledge a determinate case (*reiner*

---

*H. Weyl, *Zeitschrift für Physik*, **46**, 1, 1927.

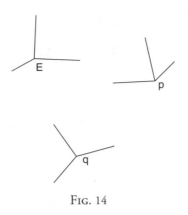

Fig. 14

*Fall*). An atom in a (non-degenerate) stationary state presents such a determinate case: The direction characterizing it is that of the $k$th principal axis of the tensor $E$, which belongs to the energy value $E_{kk}$. There is obviously no significance to be attached to the terms "value of the coordinate $q$," etc., in this direction, just as the specification of the moment of inertia about an axis not coinciding with one of the principal directions is insufficient to determine any type of motion of the rigid body, no matter how simple. Only tensors whose principal axes coincide with those of $E$ have a value in this direction. The total angular momentum of the atom, for example, can be determined simultaneously with its energy. If a measurement of the value of $q$ is to be made, then the exact knowledge of the direction must be replaced by inexact information, which can be considered as a "mixture" of the original directions $E_{kk}$, each with a certain probability coefficient. For example, the indeterminate recoil of the electron when its position is measured by a microscope converts the determinate case $E_{kk}$ into such a mixture (cf. chap. ii, § 2*a*). This mixture must be of such a kind that it may also be considered as a mixture of the principal directions of $q$, though with other probability coefficients. The measurement singles a particular value $q'$ out of this as being the actual result. It follows from this discussion that the value of $q'$ cannot be uniquely predicted from the result of the experiment determining $E$, for a disturbance of the system, which is necessarily indeterminate to a certain degree, must occur between the two experiments involved.

This disturbance is qualitatively determined, however, as soon as one knows that the result is to be an exact value of $q$. In this case, the probability of finding a value $q'$ after $E$ has been measured is given by the square of the cosine of the angle between the original direction $E_k$ and the direction $q'$. More exactly one should say by the analogue to the cosine in the unitary space, which is $|S(E_k, q')|$. This assumption is one of the formal postulates of quantum theory and cannot be derived from any other considerations. It follows from this axiom that the values of two dynamical quantities are causally related if, and only if, the tensors corresponding to them have parallel principal axes. In all other cases there is no causal relationship. The statistical relation by means of probability coefficients is determined by the disturbance of the system produced by the measuring apparatus. Unless this disturbance is produced, there is no significance to be given the terms "value" or "probable value" of a variable in a given direction of unitary space which is not parallel to a principal axis of the corresponding tensor. Thus one becomes entangled in contradictions if one speaks of the probable position of the electron without considering the experiment used to determine it (cf. the paradox of negative kinetic energy, chap. ii, § 2$d$). It must also be emphasized that the statistical character of the relation depends on the fact that the influence of the measuring device is treated in a different manner than the interaction of the various parts of the system on one another. This last interaction also causes changes in the direction of the vector representing the system in the Hilbert space, but these are completely determined. If one were to treat the measuring device as a part of the system—which would necessitate an extension of the Hilbert space—then the changes considered above as indeterminate would appear determinate. But no use could be made of this determinateness unless our observation of the measuring device were free of indeterminateness. For these observations, however, the same considerations are valid as those given above, and we should be forced, for example, to include our own eyes as part of the system, and so on. The chain of cause and effect could be quantitatively verified only if the whole universe were considered as a single system—but then physics has vanished, and only a mathematical scheme remains. The partition of the world into observing and observed system prevents a sharp formulation of the law of cause and effect. (The observing system need not always be a human being; it may also be an inanimate apparatus, such as a photographic plate.)

As examples of cases in which causal relations do exist the following may be mentioned: The conservation theorems for energy and momentum are contained in the quantum theory, for the energies and momenta of different parts of the same system are commutative quantities. Furthermore, the principal axes of $q$ at time $t$ are only infinitesimally different from the principal axes of $q$ at time $t + dt$. Hence, if two position measurements are carried out in rapid succession, it is practically certain that the electron will be in almost the same place both times.

## § 2. INTERFERENCE OF PROBABILITIES

Many paradoxical conclusions may be deduced from the foregoing principles if the perturbation introduced by measuring instruments is not adequately considered. The following idealized experiment furnishes a typical example of such a paradox.

A beam of atoms, all of which are initially in the state $n$, is directed through a field $F_1$ (Fig. 15). This field will cause transitions to other states if it is inhomogeneous in the direction of the beam, but will not separate atoms of one state from those in another. Let $S'_{nm}$ be the transformation function for the transitions in the field $F_1$ so that $|S'_{nm}|^2$

FIG. 15

is the probability of finding an atom in the state $m$ after it has emerged from the field $F_1$. Farther on the atoms encounter a second field $F_2$, similar in properties to $F_1$ for which the corresponding transformation function is $S''_{ml}$. This field is again incapable of separating the atoms in different states, but beyond $F_2$ a determination of the stationary state is made by means of a third field of force. Now, for those atoms that are in the state $m$ after passing through $F_1$ the probability of a transition to state $l$ on passing $F_2$ is given by $[S_{ml}]^2$. Hence the probable fraction of the atoms in the state $l$ beyond $F_2$ should be given by

$$\sum_m |S'_{nm}|^2 |S''_{ml}|^2 . \tag{44}$$

On the other hand, according to equation A(69), the transformation function for the combined fields $F_1$ and $F_2$ is $S'''_{nl} = \sum_m S'_{nm} S''_{ml}$, which results in the value

$$|S'''_{nl}|^2 = \left| \sum_m S'_{nm} S''_{ml} \right|^2 \tag{45}$$

for the same probability as represented by equation (44).

The contradiction disappears when it is remarked that the formulas (44) and (45) really refer to two different experiments. The reasoning leading to (44) is correct only when an experiment permitting the determination of the stationary state of the atom is performed between $F_1$ and $F_2$. The performance of such an experiment will necessarily alter the phase of the de Broglie wave of the atom in state $m$ by an unknown amount of order of magnitude one, as has been shown in chapter ii, § 2d. In applying (45) to this experiment each member $S'_{nm} S''_{ml}$ in the summation must thus be multiplied by the arbitrary factor $exp(i\varphi_m)$ and then averaged over all values of $\varphi_m$. This phase average agrees with (44), which thus applies to this experiment. The rules of the calculus of probabilities can be applied to $|S_{nm}|^2$ only when the causal chain has actually been broken by an observation in the manner explained in the foregoing section. If no break of this sort has occurred it is not reasonable to speak of the atom as having been in a stationary state between $F_1$ and $F_2$, and the rules of quantum mechanics apply.

Three general cases may be illustrated by this experiment, and they must be carefully distinguished in any application of the general principles. They are:

CASE I: The atoms remain undisturbed between $F_1$ and $F_2$. The probability of observing the state $l$ beyond $F_2$ is then

$$\left| \sum_m S'_{nm} S''_{ml} \right|^2 .$$

CASE II: The atoms are disturbed between $F_1$ and $F_2$ by the performance of an experiment which would have made possible the determination of the stationary state. The result of the experiment is not observed, however. The probability of the state $l$ is then

$$\sum_m \left| S'_{nm} \right|^2 \left| S''_{ml} \right|^2 .$$

CASE III: The additional experiment of Case II is performed and its result is observed. The atom is known to have been in state $m$ while passing from $F_1$ to $F_2$. The probability of the state $l$ is then given by

$$\left| S''_{ml} \right|^2 .$$

The difference between Cases II and III is recognized in all treatments of the theory of probability, but the difference between I and II does not exist in classical theories which assume the possibility of observation without perturbation. When stated in a sufficiently generalized form, this distinction is the center of the whole quantum theory.

§ 3. BOHR'S CONCEPT OF COMPLEMENTARITY*

With the advent of Einstein's relativity theory it was necessary for the first time to recognize that the physical world differed from the ideal world conceived in terms of everyday experience. It became apparent that ordinary concepts could only be applied to processes in which

---

*Nature, **121**, 580, 1928.

the velocity of light could be regarded as practically infinite. The experimental material resulting from modern refinements in experimental technique necessitated the revision of old ideas and the acquirement of new ones, but as the mind is always slow to adjust itself to an extended range of experience and concepts, the relativity theory seemed at first repellantly abstract. None the less, the simplicity of its solution for a vexatious problem has gained it universal acceptance. As is clear from what has been said, the resolution of the paradoxes of atomic physics can be accomplished only by further renunciation of old and cherished ideas. Most important of these is the idea that natural phenomena obey exact laws—the principle of causality. In fact, our ordinary description of nature, and the idea of exact laws, rests on the assumption that it is possible to observe the phenomena without appreciably influencing them. To co-ordinate a definite cause to a definite effect has sense only when both can be observed without introducing a foreign element disturbing their interrelation. The law of causality, because of its very nature, can only be defined for isolated systems, and in atomic physics even approximately isolated systems cannot be observed. This might have been foreseen, for in atomic physics we are dealing with entities that are (so far as we know) ultimate and indivisible. There exist no infinitesimals by the aid of which an observation might be made without appreciable perturbation.

Second among the requirements traditionally imposed on a physical theory is that it must explain all phenomena as relations between objects existing in space and time. This requirement has suffered gradual relaxation in the course of the development of physics. Thus Faraday and Maxwell explained electromagnetic phenomena as the stresses and strains of an ether, but with the advent of the relativity theory, this ether was dematerialized; the electromagnetic field could still be represented as a set of vectors in space-time, however. Thermodynamics is an even better example of a theory whose variables cannot be given a simple geometric interpretation. Now, as a geometric or kinematic description of a process implies observation, it follows that such a description of atomic processes necessarily precludes the exact validity of the law of causality—and conversely. Bohr* has pointed out that it is therefore impossible to demand that both requirements be fulfilled by the quantum theory. They represent complementary and mutually exclusive aspects of atomic phenomena. This situation is clearly reflected in the theory which has been developed. There exists a body of exact mathematical laws, but these cannot be interpreted as expressing simple relationships between objects existing in space and time. The observable predictions of this theory can be approximately described in such terms, but not uniquely—the wave and the corpuscular pictures both possess the same approximate validity. This indeterminateness of the picture of the process is a direct result of the interdeterminateness of the concept "observation"—it is not possible to decide, other than arbitrarily, what objects are to be considered as part of the observed system and what as part of the observer's apparatus. In the formulas of the theory this arbitrariness often makes it possible to use quite different analytical methods for the treatment of a single physical experiment. Some examples of this will be given later. Even when this arbitrariness is taken into account the concept "observation" belongs, strictly speaking, to the class of ideas borrowed from the experiences of everyday life.[†] It can only be carried over to atomic phenomena when due regard is paid to the limitations placed on all space-time descriptions by the uncertainty principle.

The general relationships discussed here may be summarized in the following[‡] diagrammatic form:

It is only after attempting to fit this fundamental complementarity of space-time description and causality into one's conceptual scheme that one is in a position to judge the degree of consistency of the methods of quantum theory (particularly of the transformation theory). To mold our thoughts and language to agree with the observed facts of atomic physics is a very difficult task, as it was in the case of the relativity theory. In the case of the latter, it proved advantageous to

---

* *Ibid.*

[†] It need scarcely be remarked that the term "observation" as here used does not refer to the observation of lines on photographic plates, etc., but rather to the observation of "the electrons in a single atom," etc. Cf. p. 1.

[‡] N. Bohr, *loc. cit.*

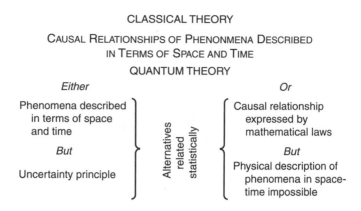

return to the older philosophical discussions of the problems of space and time. In the same way it is now profitable to review the fundamental discussions, so important for epistemology, of the difficulty of separating the subjective and objective aspects of the world. Many of the abstractions that are characteristic of modern theoretical physics are to be found discussed in the philosophy of past centuries. At that time these abstractions could be disregarded as mere mental exercises by those scientists whose only concern was with reality, but today we are compelled by the refinements of experimental art to consider them seriously.

# DISCUSSION OF IMPORTANT EXPERIMENTS

In the preceding chapters the principles of the quantum theory have all been discussed, but a real understanding of them is obtainable only through their relation to the body of experimental facts which the theory must explain. This is particularly true of the general principle of complementarity. A discussion of further experiments of a less idealized type than those previously used to illustrate the separate principles is therefore necessary at this point.

## § 1. THE C. T. R. WILSON EXPERIMENTS

The essential features of the C. T. R. Wilson photographs may be most easily explained with the help of the classical corpuscular picture. This explanation is also completely justified from the standpoint of the quantum theory. The uncertainty relations are not essential to the explanation of the primary fact of the rectilinearity of the tracks of $\alpha$-particles. It is always correct to apply the classical theory to such semi-macroscopic phenomena, and the quantum theory is necessary only for the explanation of the finer features.

Nevertheless it will be profitable to discuss the quantum theory of the Wilson photograph. We encounter at once the arbitrariness in the concept of observation already mentioned, and it appears purely as a matter of expediency whether the molecules to be ionized are regarded as belonging to the observed system or to the observing apparatus. Consider first the latter alternative. The system to be observed then consists of one $\alpha$-particle only, and the position measurement resulting from the ionization will be represented in the mathematical scheme of the theory by a probability packet $|\psi(q')|^2$ in the coordinate space $q = x, y, z$, of the $\alpha$-particle. The calculation will be carried out only for one of the three degrees of freedom.

If the time of this determination be taken as $t = 0$, and if a previous determination at a known time is also available, the momentum of the particle at time $t = 0$ may be determined: let $\bar{p}$ and $\bar{q}$ denote the most probable values of the momentum and co-ordinate at this time, and $\Delta p$, $\Delta q$ the probable errors. The value of the uncertainty product will be considerably greater than $h$ in any actual case, but we may assume that $\Delta p \Delta q = h/2\pi$ (cf. the remarks concerning scintillation measurements, chap. ii, § 2a). This is a determinate case; it is then known [eq. (15)] that

$$\psi\left(q_0'\right) = e^{-(q_0'-\bar{q})^2/2(\Delta q)^2 - \frac{2\pi i}{h}\bar{p}(q'-\bar{q})}.$$

(The index 0 indicates that $q_0'$ is the value of the coordinate at $t = 0$.) The quantum theoretical equations of motion are then

$$p = p_0 = \text{Const.},$$

$$\dot{q} = \frac{1}{\mu}p.$$

Although $p$ and $q$ do not commute, the latter equation may nevertheless be integrated[*] to

$$q = \frac{1}{\mu}pt + q_0.$$

To obtain the probability amplitude $\psi\left(q'\right)$ at time $t$ the transformation function must be calculated from A(41) and A(42):

$$\left(\frac{t}{\mu}\frac{h}{2\pi i}\frac{\partial}{\partial q_0'} + q_0'\right) S\left(q_0'q'\right) = q' S\left(q_0'q'\right).$$

The solution of this equation is

$$S\left(q_0'q'\right) = a e^{\frac{2\pi i\mu}{ht}\left(q'q_0' - q_0'^2/2\right)}; \tag{46}$$

by A(69) the distribution at time $t$ is then to be found from

$$\psi\left(q'\right) = \int_{-\infty}^{+\infty} \psi\left(q_0'\right) S\left(q_0'q'\right) dq_0',$$

which becomes, on evaluation of the integral,

$$\psi\left(q'\right) = b e^{\left[\bar{q}+i(q'-\bar{p}t/\mu)\right]^2/\left[2(\Delta q)^2+(1+i/\beta)^2\right]}, \tag{47}$$

[*]Kennard, *Zeitschrift für Physik*, **44**, 326, 1927.

where

$$\beta = \frac{h}{2\pi} \frac{t}{\mu} \frac{1}{(\Delta q)^2} = \Delta p \frac{t}{\mu \Delta q}.$$

It follows that

$$|\psi(q')|^2 = b' e^{-(q' - p't/\mu - \bar{q})^2 / [(\Delta q)^2 + (t \Delta p/\mu)^2]}. \tag{48}$$

The most probable value for $q'$ is thus $(t/\mu)\bar{p} + \bar{q}$, which is the result to be expected on the classical theory. The mean square error $(\Delta q)^2 + (t \Delta p/\mu)^2$ for $q'$ is made up of two terms corresponding to the uncertainties in $q_0'$ and $p_0'$; its value again agrees with that which would be calculated classically.

If these methods are applied to all three degrees of freedom, $x$, $y$, $z$, it is seen at once that the path of the center of the probability packet is a straight line. It is to be noted, however, that this result applies only while the $\alpha$-particle is undisturbed in its motion. Each successive ionization of a water molecule transforms the packet (48) into an aggregate of such packets (Case II, p. 61). If the ionization is accompanied by an observation of the position, a smaller probability packet of the same form as (48) but with new parameters is separated out of the aggregate (Case III, p. 61). This forms the starting-point of a new orbit—and so on. The angular deviations between successive orbital segments are determined by the relative momenta of the particle and the atomic electron with which it interacts, which accounts for the differences between the paths of $\alpha$- and $\beta$-particles.

As regards the formal aspect of the foregoing calculations, it may be noted that the transformation from $q_0'$ to $q'$ can also be carried out by way of the energy. By equation A(70):

$$S(q_0' q') = \int S(q_0' E) S(E q') dE,$$

and therefore

$$\psi(q') = \int S(E q') dE \int \psi(q_0') S(q_0' E) dq_0'.$$

The functions $S(q' E)$, $S(E q_0')$ are the normalized Schrödinger wave functions for the free electron; the function $\psi(q')$ can thus be built up by superposition of such Schrödinger functions. This method has been used by Darwin in an investigation of the motion of probability packets.

To complete this discussion we shall finally carry through a mathematical treatment of the Wilson photographs under the assumption that the molecules to be ionized are regarded as part of the system. This procedure is more complicated than the preceding method, but has the advantage that the discontinuous change of the probability function recedes one step and seems less in conflict with intuitive ideas. In order to avoid complication we consider only two molecules and one $\alpha$-particle, and suppose the centers of mass of the former to be fixed at the points $x_1, y_1, z_1; x_2, y_2, z_2$. The $\alpha$-particle is in motion with the momenta $p_x, p_y, p_z$, and its co-ordinates are $x, y, z$. The co-ordinates of the electrons in the molecules may be denoted by the single symbols $q_1$ and $q_2$, respectively; the configuration space will thus involve only $x, y, z, q_1$, and $q_2$. We inquire for the probability that both molecules will be ionized and show that it is negligibly small unless the line joining them has nearly the same direction as the vector $(p_x p_y p_z)$. All interaction between the two molecules will be neglected, and their interaction with the $\alpha$-particle will be treated as a perturbation;[*] the energy of this interaction may be written

$$
\left. \begin{aligned}
H^{(1)}(1) + H^{(1)}(2) = H^{(1)}(x - x_1, \ y - y_1, \ z - z_1, \ q_1) \\
+ H^{(1)}(x - x_2, \ y - y_2, \ z - z_2, \ q_2),
\end{aligned} \right\} \quad (49)
$$

regarded as operators acting on the Schrödinger function. The wave equation is then

$$
\left. \begin{aligned}
-\frac{h^2}{8\pi^2 \mu} \underbrace{\nabla^2 \psi}_{\alpha\text{-Particle}} + \underbrace{H^0(q_1)\psi + H^0(q_2)\psi}_{\text{Molecules}} + \underbrace{\epsilon[H^{(1)}(1) + H^{(1)}(2)]\psi}_{\text{Interaction}} \\
+ \frac{h}{2\pi i} \frac{\partial \psi}{\partial t} = 0,
\end{aligned} \right\}
$$

$$(50)$$

*M. Born, *Zeitschrift für Physik*, **38**, 803, 1926.

in which $\nabla^2 = \partial^2/\partial x^2 + \partial^2/\partial y^2 + \partial^2/\partial z^2$, $H^0(q_i)$ is the energy operator of the molecule $i$, and $\epsilon$ is the perturbation parameter in powers of which the wave function is to be expanded: $\psi = \psi^{(0)} + \epsilon \psi^{(1)} + \epsilon^2 \psi^{(2)} \ldots$. Substituting this series into the wave equation and equating each power of $\epsilon$ to zero, we obtain

$$
\left.
\begin{aligned}
&-\frac{h^2}{8\pi^2\mu}\nabla^2\psi^{(0)} + H^{(0)}(1)\psi^{(0)} + H^{(0)}(2)\psi^{(0)} + \frac{h}{2\pi i}\frac{\partial\psi^{(0)}}{\partial t} \\
&\qquad = 0, \\
&-\frac{h^2}{8\pi^2\mu}\nabla^2\psi^{(1)} + H^{(0)}(1)\psi^{(1)} + H^{(0)}(2)\psi^{(1)} + \frac{h}{2\pi i}\frac{\partial\psi^{(1)}}{\partial t} \\
&\qquad = -\left[H^{(1)}(1) + H^{(1)}(2)\right]\psi^{(0)}, \\
&-\frac{h^2}{8\pi^2\mu}\nabla^2\psi^{(2)} + H^{(0)}(1)\psi^{(2)} + H^{(0)}(2)\psi^{(2)} + \frac{h}{2\pi i}\frac{\partial\psi^{(2)}}{\partial t} \\
&\qquad = -\left[H^{(1)}(1) + H^{(1)}(2)\right]\psi^{(1)},
\end{aligned}
\right\}
\quad (51)
$$

$\cdots\cdots\cdots\cdots\cdots\cdots\cdots\cdots\cdots\cdots\cdots\cdots\cdots\cdots\cdots\cdots$

The characteristic solutions of the first equation are

$$
\psi^{(0)} = e^{\frac{2\pi i}{h}\boldsymbol{p}\cdot\boldsymbol{x}}\varphi_{n_1}(q_1)\,\varphi_{n_2}(q_2)\,e^{-\frac{2\pi i}{h}E^{(0)}t}, \quad (52)
$$

where

$$
H^{(0)}(q)\,\varphi_n(q) = E_n\varphi_n(q), \quad (53)
$$

and

$$
E^0 = \frac{1}{2\mu}p^2 + E_{n_1} + E_{n_2}. \quad (54)
$$

These solutions correspond to the case in which the momentum of the $\alpha$-particle is known to be exactly $\boldsymbol{p}$, its position therefore entirely unknown, while the molecules are known to be in the states $n_1$, $n_2$, respectively. All interaction is neglected, and the problem is to determine how the interaction modifies this state of affairs.

This may be solved by determining $\psi^{(1)}$, $\psi^{(2)}$ according to the method of Born. These quantities are first expanded in terms of the orthogonal functions $\varphi_{m_1}(q_1)\,\varphi_{m_2}(q_2)$,

$$
\psi^{(i)} = \sum_{m_1}\sum_{m_2} v^{(i)}_{m_1 m_2}\varphi_{m_1}(q_1)\,\varphi_{m_2}(q_2), \quad (55)
$$

in which the $v^{(i)}_{m_1 m_2}$ are of course functions of $x, y, z$, and $t$. The significance of these quantities is that

$$\left| \sum_i \varepsilon^i v^{(i)}_{m_1 m_2} \right|^2 \tag{56}$$

is the probability of observing the molecule 1 in the state $m_1$, molecule 2 in the state $m_2$, and the electron at $x, y, z$.

Substituting equation (55) for $i = 1$ into the first of equations (51), we obtain

$$\left( -\frac{h^2}{8\pi^2 \mu} \nabla^2 + E_{n_1} + E_{n_2} + \frac{h}{2\pi i} \frac{\partial}{\partial t} \right) v^{(1)}_{n_1 m_2}$$
$$= -\left[ h_{n_1 m_2}(1) \delta_{n_2 m_2} + h_{n_2 m_2}(2) \delta_{n_1 m_1} \right] e^{\frac{2\pi i}{h}[p \cdot x - E \cdot t]},$$

in which the abbreviations

$$\left. \begin{array}{l} h_{n_1 m_1}(1) = \int \varphi^*_{m_1}(q_1) H^{(1)}(1) \varphi_{n_1}(q_1) dq_1 \\ h_{n_2 m_2}(2) = \int \varphi^*_{m_2}(q_2) H^{(1)}(2) \varphi_{n_2}(q_2) dq_2 \end{array} \right\} \tag{57}$$

have been used. The co-ordinates $q_1$ and $q_2$ have thus been eliminated from further consideration; the functions $h(1)$, $h(2)$ are functions of $x, y, z$, and of $x_1, y_1, z_1$ or $x_2, y_2, z_2$, respectively. These equations may be further simplified by writing

$$v^{(1)}_{m_1 m_2}(xyzt) = w^{(1)}_{m_1 m_2}(xyz) e^{-\frac{2\pi i}{h} E \cdot t},$$

whence

$$(\nabla^2 + k^2_{m_1 m_2}) w^{(1)}_{m_1 m_2} = \frac{8\pi^2 \mu}{h^2} (h_{n_1 m_1}(1)\delta_{n_2 m_2} + h_{n_2 m_2}(2)\delta_{n_1 m_1}) e^{\frac{2\pi i}{h} p \cdot x} \tag{58}$$

where

$$\frac{h^2}{8\pi^2 \mu} k^2_{m_1 m_2} = \left[ E_{n_1} + E_{n_2} + \frac{1}{2\mu} p^2 - E_{m_1} - E_{m_2} \right]. \tag{59}$$

In this expression the kinetic energy of the $\alpha$-particle is so much greater than the other terms that, to a sufficient approximation, we may take

$$k^2_{m_1 m_2} = k^2 = \frac{4\pi^2 p^2}{h^2} = \frac{4\pi^2}{\lambda_0^2}. \tag{60}$$

Equations (58) are then all of the form

$$\left(\nabla^2 + k^2\right) w^1_{m_1 m_2} = \rho_{m_1 m_2}(xyz), \tag{61}$$

which is the ordinary equation of wave-motion; $\rho_{m_1 m_2}(xyz)$ is the density of the oscillators producing the wave, and, as it is complex, also determines their phase. The solution of equation (61) is given by Huyghen's principle:

$$w^1_{m_1 m_2} = \int \int \int \rho_{m_1 m_2}(x'y'z')\frac{e^{-ikR}}{R}dx'dy'dz',$$

where $R$ is the distance from $x'$, $y'$, $z'$ to $x$, $y$, $z$.

Since, according to (58), $\rho_{m_1 m_2}$ is zero unless $m_1 = n_1$ or $m_2 = n_2$, all the $w^{(1)}_{m_1 m_2}$ will be zero except $w^{(1)}_{m_1 n_2}$ and $w^{(1)}_{n_1 m_2}$; to the first approximation, only one of the two molecules will be excited. This is in agreement with the classical theory, which says that the probability of two collisions is of second order. The character of the functions $w^{(1)}_{n_1 m_2}$ and $w^{(1)}_{m_1 n_2}$ is readily determined qualitatively; by equation (57)

$$\rho_{m_1 n_2} = \frac{8\pi^2 \mu}{h^2} h_{n_2 m_1}(x - x_1, y - y_1, z - z_1)\, e^{\frac{2\pi i}{h} p \cdot x}.$$

The (fictitious) oscillators producing the wave are thus all located in the region $\Gamma_1$ about $x_1, y_1, z_1$ (cf. Fig. 16) in which $h_{n_2 m_1}$ is appreciably different from zero. They vibrate coherently, their phase being determined essentially by the factor $e^{\frac{2\pi i}{h} p \cdot x}$; in the figure the lines of equal phase are drawn perpendicular to $p$. They are spaced

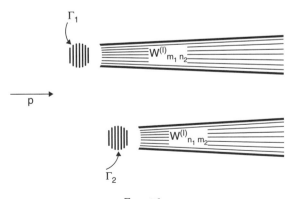

FIG. 16

at distances $\lambda_0$. According to equation (61) the wave-length emitted by the oscillators is also $\lambda_0$, and a simple application of Huyghen's principle shows that the wave disturbance will have an appreciable amplitude only in the conical region which is shaded and whose axis is in the direction of $p$. The cross-section of this region near $x_1, y_1, z_1$ is determined by the cross-section of the molecule: $\Gamma_1$. Its angular opening also depends on $\Gamma_1$, being greater when $\Gamma_1$ is small—i.e., the uncertainty relation $\Delta p_x \Delta_x \sim h/2\pi$ is fulfilled. Similar considerations apply to $w^{(1)}_{n_1 m_2}$; it is different from zero only in a beam originating in $\Gamma_2$ and also having the direction $p$.

We now pass to the second approximation: $v^{(2)}_{m_1 m_2}$ may also be written $w^{(2)}_{m_1 m_2}\, exp(-2\pi\, i/h)\, E^0 t$ and equation (51) reduces to

$$
\left.
\begin{aligned}
(\nabla^2 + k^2)w^{(2)}_{m_1 m_2} &= \frac{8\pi^2\mu}{h^2}\left\{ \sum_l w^{(1)}_{lm_2}h_{lm_1}(1) + \sum_l w^{(1)}_{m_2 l}h_{m_2 l}(2)\right\}, \\
&= \frac{8\pi^2\mu}{h^2}\left\{ w^{(1)}_{n_1 m_2}h_{n_1 m_1}(1) + w^{(1)}_{m_1 n_2}h_{m_2 n_2}(2)\right\}.
\end{aligned}
\right\}
$$

$$(62)$$

The right-hand side of this equation will always be practically zero unless one of the two molecules lies in the beam originating at the other, for $w^{(1)}_{n_1 m_2}$ is different from zero only in the beam originating in $\Gamma_2$ and $h_{n_1 m_1}(1)$ only in $\Gamma_1$. Unless these two regions intersect, the first term will be zero; similarly the second term. Thus the probability of simultaneous ionization or excitation of the two atoms will vanish even in the second approximation unless the line joining their centers of gravity is practically parallel to the direction of motion of the $\alpha$-particle. These considerations may be extended to the case of any number of molecules without essential modification. For each additional molecule the approximation must be carried one step farther, but the principles and results will be the same. It has thus been proved that the ionized molecules will lie practically on straight lines, and that the deviations from rectilinearity satisfy the uncertainty relations. In thus including the molecules in the observed system, it has not been necessary to introduce the discontinuously changing probability packet, but if we wish to consider the methods by which the excitation of the molecule can actually be observed, these discontinuous changes (now of a probability packet in the configuration space $x$, $y$, $z$, $q_1$, $q_2$) will again play a role.

## § 2. DIFFRACTION EXPERIMENTS

The diffraction of light or matter (Davisson-Germer, Thomson, Rupp, Kikuchi) by gratings may be explained most simply by the aid of the classical wave theories. The application of space-time wave theories to these experiments is justified from the point of view of the quantum theory, since the uncertainty relations do not in any way affect the purely geometrical aspects of the waves, but only their amplitude (cf. chap. iii, § 1). The quantum theory need only be invoked when discussing the dynamical relations involving the energy and momentum content of the waves.

The quantum theory of the waves being thus certainly in agreement with the classical theory in so far as the geometric diffraction pattern is concerned, it seems useless to prove it by detailed calculation. On the other hand, Duane has given an interesting treatment of diffraction phenomena from the quantum theory of the corpuscular picture. We imagine for simplicity that the corpuscle is reflected from a plane ruled grating, whose constant is $d$.

Let the grating itself be movable. Its translation in the $x$-direction may be looked upon as a periodic motion, in so far as only the interaction of the incident particles with the grating is considered; for the displacement of the whole grating by an amount $d$ will not change this interaction. Thus we may conclude that the motion of the grating in this direction is quantized and that its momentum $p_x$ may assume only the values $nh/d$ (as follows at once from the earlier form of the theory: $\int p\,dq = nh$). Since the total momentum of grating and particle must remain unchanged, the momentum of the particle can be changed only by an amount $mh/d$ ($m$ an integer):

$$p'_x = p_x + \frac{mh}{d}.$$

Furthermore, because of its large mass, the grating cannot take up any appreciable amount of energy, so that

$$p_x'^2 + p_y'^2 = p_x^2 + p_y^2 = p^2.$$

If $\theta$ is the angle of incidence, $\theta'$ that of reflection, we have

$$\cos\theta = \frac{p_y}{p}, \quad \cos\theta' = \frac{p_y'}{p},$$

whence

$$\sin\theta' - \sin\theta = \frac{mh}{pd}.$$

From equation A(83) for the wave-length of the wave associated with a particle it then follows that

$$d\left(\sin\theta' - \sin\theta\right) = m\lambda,$$

in agreement with the ordinary wave theory.

The dual characters of both matter and light gave rise to many difficulties before the physical principles involved were clearly comprehended, and the following paradox was often discussed. The forces between a part of the grating and the particle certainly diminish very rapidly with the distance between the two. The direction of reflection should therefore be determined only by those parts of the grating which are in the immediate neighborhood of the incident particle, but none the less it is found that the most widely separated portions of the grating are the important factors in determining the sharpness of the diffraction maxima. The source of this contradiction is the confusion of two different experiments (Cases I and II, p. 61). If no experiment is performed which would permit the determination of the position of the particle before its reflection, there is no contradiction with observation if the whole of the grating does act on it. If, on the other hand, an experiment is performed which determines that the particle will strike on a section of length $\Delta x$ of the grating, it must render the knowledge of the particle's momentum essentially uncertain by an amount $\Delta p \sim h/\Delta x$. The direction of its reflection will therefore become correspondingly uncertain. The numerical value of this uncertainty in direction is precisely that which would be calculated from the resolving power of a grating of $\Delta x/d$ lines. If $\Delta x \ll d$ the interference maxima, disappear entirely; not until this case is reached can the path of the particle properly be compared with that expected on the classical particle theory, for not until then can it be determined whether the particle will impinge on a ruling or on one of the plane parts of the surface, etc.

## § 3. THE EXPERIMENT OF EINSTEIN AND RUPP[*]

Another paradox was thought to be presented by the following experiment: An atom (canal ray) is made to pass a slit $S$ of width $d$ with the velocity $v$, and emits light while doing so. This light is analyzed by a spectroscope behind $S$. Since the light can reach the spectroscope only during the time $t = d/v$, the train of waves to be analyzed has a finite length, and the spectroscope will show it as a line whose width corresponds to a frequency range

$$\Delta v = \frac{1}{t} = \frac{v}{d}.$$

On the other hand, the corpuscular theory seems to prohibit such a broadening. The atom emits monochromatic radiation, the energy of each particle of which is $hv$, and the diaphragm (because of its great mass) will not be able to change the energy of the particles.

The fallacy lies in neglecting the Doppler effect and the diffraction of the light at the slit. Those photons which reach $P$ from the atom are not all emitted perpendicularly to the canal ray; the angular aperture of the beam of photons is $\sin \alpha \sim \lambda/d$ because of the diffraction. The Doppler change of frequency due to this is

$$\Delta v = \sin \alpha \frac{v}{c} v,$$

or

$$\Delta v = \frac{\lambda v}{cd} v = \frac{v}{d},$$

---

[*]A. Einstein, *Berliner Berichte*, p. 334, 1926; A. Rupp, *ibid.*, p. 341, 1926.

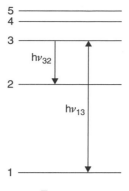

FIG. 17

in agreement with the previous result. In this experiment the exact validity of the energy law for corpuscles is thus in conformity with the requirements of classical optics.

## § 4. EMISSION, ABSORPTION, AND DISPERSION OF RADIATION

*(a) Application of the conservation laws.* The postulate of the existence of stationary states, combined with the theory of photons, is sufficient to give a qualitative explanation of the interaction of atoms and radiation. This was the first decisive success of the Bohr theory. The most important results of this theory may be briefly summarized here. Let the stationary states of the atom be numbered 1, 2, 3 .... $n$ .... (Fig. 17), counting from the normal state. An atom in state 3, for example, can spontaneously perform a transition to state 2, and emit a photon of energy $h\nu_{32} = E_3 - E_2$. In the same way, an atom in state 1 may absorb a photon of energy $h\nu_{31} = E_3 - E_1$ and thus be excited to the state 3. It must be emphasized that these statements are to be taken quite literally, and not as having only a symbolic significance, for it is possible (e.g., by a Stern-Gerlach experiment) to determine the stationary state of the atoms both before and after the emission. It therefore follows that the intensity of an emission line is proportional to the number of atoms in the upper of the two states associated with it, while the intensity of an absorption line is proportional to the number of atoms in the lower state. These results, which have certainly

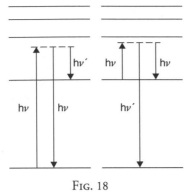

FIG. 18

been amply confirmed by experiment, are entirely characteristic of the quantum theory and can be deduced from no classical theory, either of the wave or particle representation, since even the existence of discrete energy values can never be explained by the classical theory.

An exactly similar situation is met with in the case of scattering. If an atom in state 1 is excited by a photon $h\nu$ it can re-emit the same light quantum without change of state (the mass of the nucleus being assumed infinite), or it can send out the light quantum of energy $h\nu' = h\nu - E_2 + E_1$ by transition to state 2 (Smekal* transition; see Fig. 18). The intensity of both kinds of scattered light is proportional to the number of atoms in state 1. If an atom in state 2 is irradiated with light of frequency $\nu$ it can emit a photon of energy $h\nu' = h\nu + E_2 - E_1$ of shorter wave-length by transition to state 1, and again the intensity of this "anti-Stokes" scattered light is proportional to the number of atoms in state 2. This has been confirmed by Raman's[†] experiments.

*(b) Correspondence principle and the method of virtual charges.*
The postulate of stationary states and the theory of photons, because of their very nature, cannot yield any information either regarding the interference of the emitted light or even regarding the a priori probability of the transitions involved. The interference properties can be completely accounted for by the classical wave theory, but it is

---

* *Naturwissenschaften*, **11**, 873, 1923.
[†] *Nature*, **121**, 501; **122**, 12, 1928.

in turn unable to account for the transitions. To treat these successfully a self-consistent quantum theory of radiation is necessary. It is true that an ingenious combination of arguments based on the correspondence principle can make the quantum theory of matter together with a classical theory of radiation furnish quantitative values for the transition probabilities, i.e., either by the use of Schrödinger's virtual charge density or its equivalent, the element of the matrix representing the electric dipole moment of the atom. Such a formulation of the radiation problem is far from satisfactory, however, and easily leads to false conclusions. These methods may only be applied with the greatest caution, as the following examples may illustrate.

Consider first the case of an atom containing a single electron, and whose nucleus has an infinite mass. If $x \equiv (x, y, z)$ be the co-ordinate of the electron, and $\psi_0(x)$ the Schrödinger function, then

$$-e x_{nm} = -e \int x \psi_n \psi_m^* d\tau \qquad (63)$$

is the element of the matrix representing the dipole moment of the atom. This matrix can enter, strictly speaking, only into calculations based on the principles of the quantum theory of the electron, which in no way involve radiation. It may none the less be interpreted as the dipole moment of the virtual oscillator producing the radiation which is emitted during the transition $n \to m$. This may be deduced from the correspondence principle by remembering that it has been shown that $x_{nm} \to x_n(n - m)$ in the limit of large quantum numbers, where $x_n(n - m)$ is a Fourier coefficient of the classical motion. It may thus be presumed that $x_{nm}$ will enter into the formulas determining the intensity of the radiation in the same way as $x_n(n - m)$, i.e., that $|x_{nm}|^2$ will be the a priori probability of the transition $n \to m$. It must be emphasized that this is a purely formal result; it does not follow from any of the physical principles of quantum theory.

It may be made plausible by another consideration which brings out its unsatisfactory character more clearly. It has been pointed out that the solutions $\psi_n$ of the Schrödinger equation are first approximations to the solutions of the classical matter-wave equations [cf. A(8)].

Denoting by $\psi^c$ a true solution of the latter, the radiation from the charge distribution thus represented will be determined by its dipole moment

$$-e \int \psi^c \psi^{c*} x d\tau$$

provided the extension of this distribution is small compared to the wave-length of the radiation emitted. Now

$$\psi^c \sim \sum_n a_n \psi_n e^{-\frac{2\pi i}{h} E_n t},$$

whence the radiation, calculated by means of this classical distribution, should be determined by

$$-e \sum_{nm} a_n a_m^* x_{nm} e^{\frac{2\pi i}{h}(E_n - E_m)t}. \tag{64}$$

This formula is certainly wrong since it is derived from a purely classical theory; the intensity of the radiation of frequency $(E_n - E_m)/n$ depends on the coefficient $a_m$ of the final state, as well as on $a_n$ of the initial state. This is in direct contradiction to Bohr's fundamental postulate. The contradiction may be eliminated by arbitrarily dissecting the sum into its separate terms, omitting the offending factors and relating each term to the upper level. The formula (63) for the moment of the virtual dipole associated with the transition then appears once more.

*(c) The complete treatment of radiation and matter.* The consistent treatment of radiation phenomena requires the simultaneous application of the quantum theory to radiation and matter, in which case it is naturally immaterial whether the particle or wave representation is used. Dirac,* in his radiation theory, employs the language of the particle representation, but makes use of conclusions drawn from the wave theory of radiation in his derivation of the Hamiltonian function. The fundamental ideas of this theory are briefly outlined here.

The atom will be represented by a single electron moving in an electrostatic force field $\phi_0$. The relativistically invariant equation of

---

* *Proceedings of the Royal Society*, A, **114**, 243, 710, 1927.

the one electron problem is, according to Dirac* ($\phi_0$ scalar potential, $\phi_i$ [$i = 1, 2, 3$], electromagnetic potentials),

$$p_0 + \frac{e}{c}\phi_0 + \alpha_i \left( p_i + \frac{e}{c}\phi_i \right) + \alpha_4 mc = 0, \qquad (65)$$

or

$$H = -e\phi_0 - \alpha_i c \left( p_i + \frac{e}{c}\phi_i \right) - \alpha_4 mc^2. \qquad (66)$$

(The usual summation convention is adopted.) Here, as before, the $p_i$'s are the momenta canonically conjugate to the $q_i$, and the $\alpha$'s are operators which satisfy the equations

$$\alpha_i \alpha_k + \alpha_k \alpha_i = 2\delta_{ik}; \quad \alpha_i \alpha_4 + \alpha_4 \alpha_i = 0; \quad \alpha_4^2 = 1. \qquad (67)$$

From the equations of motion it follows that

$$\dot{p}_i = -\frac{\partial H}{\partial q_i} = \alpha_k c \frac{\partial \phi_k}{\partial x_i}; \quad \dot{q}_i = \frac{\partial H}{\partial p_i} = -\alpha_i c. \qquad (68)$$

Except for a factor ($-c$) the $\alpha_i$'s are thus identical with the velocity matrices. From (66) it follows that the interaction energy of atoms and radiation field can be written in the simple form

$$-\alpha_i e \phi_i = \frac{e}{c}\dot{q}_i \phi_i. \qquad (69)$$

The Hamiltonian function of the complete system atom plus radiation field is thus

$$H_{total\ system} = H_{atom} + \frac{e}{c}\dot{q}_i \phi_i + H_{radiation\ field} \qquad (70)$$

The problem is brought into a simple mathematical form by assuming the radiation field to be in an inclosure, thus providing an orthogonal system of functions on solution of the Maxwell equations subject to the appropriate boundary conditions. The $\phi_i$ may be developed in

---

* *Ibid.*, **117**, 610, 1928.

this system, and the coefficients [cf. A(123) and (124)] may be written in the form

$$a_r = e^{-\frac{2\pi i}{h}\Theta_r} N_r^{1/2},$$

where $N_r$ is the number of light quanta belonging to the $r$th characteristic vibration. The total energy of the radiation field before considering its interaction with the atom is simply

$$H_{radiation\ field} = \sum_r N_r h v_r. \tag{71}$$

In the development of the $\phi_i$ in the orthogonal system the individual terms still depend on the position of the atom in the inclosure. Since the dependence averages out in the final result when the inclosure is sufficiently large, it is convenient to introduce a mean-square amplitude obtained by averaging the square of the true amplitude over all possible positions of the atom. This yields the following expression for $\phi_i$:

$$\phi_i = \left(\frac{h}{2\pi c}\right)^{1/2} \sum_r \cos \alpha_{ir} \left(\frac{v_r}{\sigma_r}\right)^{1/2} \left[N_r^{1/2} e^{\frac{2\pi i}{h}\Theta_r} + e^{-\frac{2\pi i}{h}\Theta_r} N_r^{1/2}\right]. \tag{72}$$

Here $\alpha_{ir}$ is the angle between the electric vector of the $r$th characteristic vibration and the $q_i$-axis, and $\sigma_r$ is the number of characteristic vibrations in the frequency interval $\Delta v_r$ and solid angle $\Delta \omega_r$ divided by $\Delta v_r \Delta \omega_r$. Thus the Hamiltonian function for the complete system is

$$\left.\begin{aligned}
H = H_{atom} + \sum_r N_r h v_r \\
+ \frac{e}{c}\left(\frac{h}{2\pi c}\right)^{1/2} \sum_r \dot{q}_r \left(\frac{v_r}{\sigma_r}\right)^{1/2} \left[N_r^{1/2} e^{\frac{2\pi i}{h}\Theta_r} + e^{-\frac{2\pi i}{h}\Theta_r} N_r^{1/2}\right],
\end{aligned}\right\} \tag{73}$$

where $\dot{q}_r$ is the component of the vector $\dot{q}$ in the direction of the electric vector of the $r$th characteristic vibration.

From equation (73) all the results obtained above by the use of the conservation laws may immediately be deduced. Thus the constancy of $H$ may be proved as in the Appendix (§ 1, p. 121), and it further follows that for the emission or absorption of a light quantum $h\nu_r$ the essential factor is the matrix element of $\dot{q}_r$ corresponding to the transition concerned. Except for certain numerical factors which will not be calculated here the transition probability is given directly by the square of this matrix element. If the calculation is carried out (the interaction terms being regarded as perturbations), emission and absorption processes appear as first-order effects and dispersion phenomena as second order. For the details of the calculation the reader is referred to the papers of Dirac.*

The formulation of the Hamiltonian of the radiation problem in equation (73) has the disadvantage that it does not appear to involve the interference and coherence properties of the radiation. This is only the case, however, when mean amplitudes are used, as in the foregoing. If the correct amplitudes resulting from the development of the $\Phi_i$ in the orthogonal functions are retained, then the fact that these functions are solutions of the Maxwell equations assures interference and coherence properties for the radiation that correspond to the Maxwell equations. For example, solutions of the Maxwell equations appear as factors of the quantities $a_r$ in A (113) and these factors disappear at the position occupied by the atom when the vector potential vanishes there because of interference. Thus there will be no absorption of light in regions where there would be none according to the classical interference theory. From these considerations it follows at once that the classical wave theory is sufficient for the discussion of all questions of coherence and interference.

## § 5. INTERFERENCE AND THE CONSERVATION LAWS

It is very difficult for us to conceive the fact that the theory of photons does not conflict with the requirements of the Maxwell equations. There have been attempts to avoid the contradiction by finding solutions of the latter which represent "needle" radiation (unidirectional beams), but the results could not be satisfactorily interpreted until the

*Dirac (*loc. cit.*) uses the original Schrödinger form in place of the Hamiltonian function (73). With the use of (73) the calculation is somewhat simpler, since the quadratic terms in $\phi i$ drop out of the interaction energy. The results are the same as those of Dirac.

principles of the quantum theory had been elucidated. These show us that whenever an experiment is capable of furnishing information regarding the direction of emission of a photon, its results are precisely those which would be predicted from a solution of the Maxwell equations of the needle type (cf. the reduction of wave-packets, II, § 2c).

As an example, the recoil produced by the emission of a photon will be discussed. Let an atom go from stationary state $n$ to $m$ with the emission of a photon, and an appropriate change of its total momentum. As we are only concerned with the coherence properties of the emitted radiation, we use the correspondence-principle method, in which the radiation is calculated classically. As source of the radiation we take a charge distribution which is modeled after the expression which would be given by the classical theory of matter waves. The atom will be supposed to consist of one electron (of mass $\mu$, charge $-e$, co-ordinates $r_e$) and a nucleus (of mass M, charge $+e$, co-ordinates $r_n$). The Schrödinger function of the $n$th state, in which the atom has the total momentum $P$, is

$$e^{\frac{2\pi i}{h}P \cdot r_c} \psi_n(r_e - r_n) e^{\frac{2\pi i}{h}Et},$$

where $r_c = (\mu r_e + M r_n)/(\mu + M)$ is the vector to the center of gravity of the atom. If the matrix element of the probability density associated to the transition $n \to m$, $P \to P'$, $E \to E'$, be calculated, one obtains

$$e^{\frac{2\pi i}{h}(P-P') \cdot r_c} \psi_n(r_e - r_n)\psi_m^*(r_e - r_n) e^{-\frac{2\pi i}{h}(E-E')t}.$$

By averaging over the co-ordinates of the nucleus, one obtains the charge density due to the electron, by averaging over the co-ordinates of the electron, that due to the nucleus; the total charge density is their sum. This density is to be considered as the virtual source of the emitted radiation, at least in so far as its coherence properties are concerned. The two component densities are [the common factor $e$ is omitted, $r = r_e - r_n$ is the variable of integration, $dv$ the volume element, and $\gamma = M/(\mu + M)$]

$$\rho_e = e^{\frac{2\pi i}{h}(P-P')\cdot r_e} \int e^{\frac{2\pi i}{h}\gamma(P-P')\cdot r}\psi_n\psi_m^* dv \cdot e^{\frac{2\pi i}{h}(E-E')t},$$

$$\rho_n = e^{\frac{2\pi i}{h}(P-P')\cdot r_n} \int e^{\frac{2\pi i}{h}\gamma(P-P')\cdot r}\psi_n\psi_m^* dv \cdot e^{\frac{2\pi i}{h}(E-E')t}.$$

The total density is thus

$$\rho = \text{Const.}\, e^{\frac{2\pi i}{h}[(P-P')\cdot r-(E-E')t]}.$$

in which the value of the constant does not interest us. The current densities are given by analogous expressions. The radiation emitted by these charges is to be calculated from the retarded potentials:

$$\Phi_0 = \int \rho\left(t - R'/c\right)/R' \cdot dv$$

is the scalar potential and analogous expressions may be obtained for the vector potentials $\Phi_i$ ($R'$ is the distance from the point of integration, $r$, to the point of observation $R$). The result is therefore

$$\Phi_0 = \text{Const.} \int \frac{\exp\frac{2\pi i}{h}\left[(P-P')\cdot r - (E-E')(t-R'/c)\right]}{R'}dv.$$

If one supposes that an experiment has determined the position of the atom with a given accuracy (the value of the momentum $P$ must then be correspondingly uncertain), then this means that the density $\rho$ is given by the foregoing expression only in a finite volume $\Delta v$, and is zero elsewhere. If the radiation at a great distance from $\Delta v$ is required, $R'$ may be expanded in terms of $R$ (the co-ordinates of the point of observation) and $r$ (the co-ordinates of the point of integration):

$$R' = R - R_1 \cdot r,$$

where $R_1 = R/R$. The scalar potential is then given by

$$\Phi_0 = \text{Const.}\, e^{\frac{2\pi i}{h}(t-R/c)} \int (1/R)\, e^{\frac{2\pi i}{h}(P-P'-h\nu R_1/c)\cdot r} dv,$$

in which $h\nu = E - E'$.

The integral is appreciably different from zero only in that regions for which the factor of $r$ in the exponential is less in absolute magnitude than the reciprocal of $\Delta l$, the linear dimension of $\Delta v$. In all other regions, the radiation from different portions of $\Delta v$ is destroyed by interference. Hence

$$P - P' = h v R_1 / c \pm h / \Delta l,$$

and the atom recoils with the momentum $h v R_1 / c$ (except for the natural uncertainty $h/\Delta l$). If the direction of recoil is determined by some experimental procedure, the emitted radiation thus behaves like a unidirectional beam. This is only a special case, however, which is realized only when $P$ and $P'$ are determined with sufficient accuracy, and the co-ordinates of the center of gravity are correspondingly unknown. The other extreme is realized when the experiment fixes the position of the atom more precisely than $\Delta l = h / |P - P'| = c/v$, i.e., more precisely than one wave-length of the emitted radiation. The expression for $\Phi_0$ then represents a regular spherical wave and no conclusions can be drawn concerning the recoil, since its uncertainty is greater than its probable value.

This example illustrates very clearly how the quantum theory strips even the light waves of the primitive reality which is ascribed to them by the classical theory. The particular solution of the Maxwell equation which represents the emitted radiation depends on the accuracy with which the co-ordinates of the center of mass of the atom are known.

## § 6. THE COMPTON EFFECT AND THE EXPERIMENT OF COMPTON AND SIMON

There are analogous relations in the theory of the Compton effect, but even though the calculations are the same as those of the preceding paragraph, a summary of the essential results will be given here. It is more interesting to consider bound electrons than free electrons, for then (if one assumes the position of the stationary atomic nucleus as given) there is a certain a priori knowledge concerning the position of the scattering electron. The laws of conservation result in the equations

$$\left.\begin{array}{c} h v + E = h v' + E', \\ \dfrac{h v}{c} \, e \pm \sim \Delta p = \dfrac{h v'}{c} \, e' + p', \end{array}\right\} \tag{74}$$

The unprimed letters refer to variables before the collision, and the primed ones to variables after the collision; $p$ is the linear momentum of the electron, and $e$ and $e'$ signify unit vectors in the direction of motion of the light quantum; $\Delta p$ gives the range of momentum of the electron in the atom. If $\sim \Delta p$ is small compared with $p$ and $h\nu/c$, then (74) enables correspondingly exact conclusions regarding the relation between the directions $e'$ and $p'$ to be drawn. If, for example, $p'$ be measured in a Wilson chamber, then the radiation will have all the properties of needle radiation, since the direction of emission of the light quantum is determined. If $p' \gg \Delta p$, then the translational wave function may be regarded as that of a plane wave, namely, $exp$ $2\pi i/h \cdot (p' \cdot r - E't)$, where $r$ is the vector specifying the position of the electron. Let the wave function of the unperturbed state $E$, which will be assumed to be the normal state, be $\psi_E(r)\, exp\, 2\pi i/h \cdot Et$, where $\psi_E$ is different from zero in an interval $\Delta l[\Delta l \cdot \Delta p \sim h]$.

These wave functions are perturbed by the incident wave of frequency $\nu$, and the perturbation function is a periodic space function of wave-length $\lambda = c/\nu$. Therefore, as the final result for the perturbed charge distribution, one obtains an expression of the form

$$\left. \begin{aligned} \rho &= c f_E\,(r)\, e^{-\frac{2\pi i}{h} Et} e^{\frac{2\pi i}{h}\left(\frac{r \cdot e}{\lambda} - \nu t\right)} e^{-\frac{2\pi i}{h}(p \cdot r - E't)} \\ &= c f_E\,(r)\, e^{\frac{2\pi i}{h}\left[\left(\frac{h\nu}{c} e - p'\right) \cdot r - (E - E' + h\nu)t\right]}, \end{aligned} \right\} \tag{75}$$

Where $f_E$ is different from zero only in the interval $\Delta l$. If one writes the retarded potentials for points at a great distance from the atom, then*

$$\Phi_0\,(R) = c e^{-2\pi i \nu'\left(t - \frac{R}{c}\right)} \int_{\text{atom}} \frac{d\nu'}{R'} f_E\,(r')\, e^{\frac{2\pi i}{h}\left(\frac{h\nu}{c} e - p' - \frac{h\nu'}{c} e'\right) \cdot r'}. \tag{76}$$

In this equation $h\nu' = E - E' + h\nu$, $r'$ is the vector to the point of integration, $R$ to the point of observation, and $R' = R - r'$. The time factor in equation (76) shows that the frequency of the scattered radiation is $\nu'$ and corresponds to that of equation (74). Furthermore, the integral on the right-hand side of equation (76) vanishes because of

---

*G. Breit, *Journal of the Optical Society of America*, **14**, 324, 1927.

interference, if the factor of $r'$ is materially greater than the reciprocal atomic diameter. Accordingly, since $\Delta l \Delta p \sim h$,

$$\frac{h\nu}{c}\boldsymbol{e} = \frac{h\nu'}{c}\boldsymbol{e}' + \boldsymbol{p}' \pm \sim \Delta \boldsymbol{p}, \qquad (77)$$

in agreement with the second equation of (74). The scattered radiation behaves, therefore, in so far as its coherence properties are concerned, like needle radiation. However, the direction of the light quantum is not exactly prescribed, which may be regarded as a consequence of the indeterminateness of the momentum in the original stationary state. This indeterminateness can be diminished if one experiments with more loosely bound electrons, but then the atomic cross-section will be correspondingly greater. If one applies the considerations to an excited state, then $\Delta l \Delta p \sim nh$ appears in place of $\Delta l \Delta p \sim h$ and in the evaluation of the retarded potentials one must take the number of nodes of $\psi(r')$ into account. Since this involves only nonessential complications, we have confined ourselves to the normal state.

If one wishes to explain the Geiger-Bothe experiment on the simultaneity of emission of recoil electron and scattered photon, then if the correspondence principle methods sketched here are used, one must deal with charge distributions which radiate only during a definite time interval. The initial state of the electron will be given, by a wave-packet at rest, whose size depends on the experimental arrangement. The final state will be represented by a morning wave-packet, and the charge density, given by the product of the two wave functions, will then be different from zero only during the time the two packets overlap. The radiation produced will then be a finite wave train moving in a definite direction. A more consequent explanation of the Geiger-Bothe experiment, even though it is equivalent in all its essential points, can only be obtained from the quantum theory of radiation. Moreover, as already shown, in this theory the laws of conservation applied to light quanta and electrons hold, so that one can, without any misgivings, use the customary corpuscular theory of this experiment.

## § 7. Radiation Fluctuation Phenomena

The large mean-square fluctuations, which belong to a corpuscular theory, are contained in the mathematical framework of the quantum theory, as shown in the Appendix. It is especially instructive, however, to study the relations between the various physical pictures with which the quantum theory operates by calculating the fluctuation of a radiation field. Let there be given a black cavity, of volume $V$, containing radiation in temperature equilibrium. The mean energy $\overline{\mathfrak{E}}$ contained in a small volume element $\Delta V$ in the frequency range between $\nu$ and $\nu + \Delta \nu$ is, according to Planck's formula,

$$\overline{\mathfrak{E}} = \frac{8\pi^2 h\nu}{c^3} \frac{\Delta\nu\Delta V}{e^{h\nu/kT} - 1}; \tag{78}$$

$k$ is the Boltzmann constant and $T$ the temperature. According to general thermodynamic laws,* the following relation holds for the mean-square fluctuation of $\mathfrak{E}$:

$$\overline{\Delta\mathfrak{E}^2} = kT^2 \frac{d\overline{\mathfrak{E}}}{dT}.$$

Substituting into equation (78), it was shown by Einstein that

$$\overline{\Delta\mathfrak{E}^2} = \underbrace{h\nu\overline{\mathfrak{E}}}_{\text{corpuscle}} + \underbrace{\frac{c^3}{8\pi^2\nu^2\Delta\nu\Delta V}}_{\text{wave}} \overline{\mathfrak{E}}^2. \tag{79}$$

This value for the mean-square fluctuation can only be derived partially with the help of the classical theory. The corpuscular viewpoint yields

$$\overline{\mathfrak{E}} = h\nu\bar{n}. \tag{80}$$

The classical particle theory thus results only in the first part of formula (79). The classical wave theory of radiation, on the other hand, leads exactly to the second part of (79). The calculations for this will be given later in connection with the quantum theory. Thus, the quantum theory proper is necessary for the derivation of formula (79), in which it is naturally immaterial whether one uses the wave or the corpuscular picture.

---

*J. W. Gibbs, *Elementary Principles in Statistical Mechanics*, pp. 70–72, 1902.

If, in particular, one treats the problem by means of the configuration space of the particles (although it is true that this has not been done in a detailed manner for light quanta), then one must note that the whole term system of the problem can be subdivided into non-combining partial systems, from which a definite one can be chosen as a solution. Because of the exchange relations (84), which become apparent from the corresponding uncertainty relations, that term system must be taken whose characteristic functions are symmetric in the co-ordinates of the light quanta. This choice leads to the Bose statistics for the light quanta and also, as Bose* has shown, to equation (78).

If the wave picture be used, then one obtains the number of light quanta corresponding to the vibration concerned from the amplitudes of the characteristic vibrations, and therefore the same mathematical scheme. In order to avoid unnecessary complications in the calculations, let us treat a vibrating string of length $l$ instead of the black radiation cavity. Let $\varphi(x, t)$ be its lateral displacement, and $c$ the velocity of sound in the string. The Lagrangian function becomes

$$L = \frac{1}{2}\left[\frac{1}{c^2}\left(\frac{\partial \varphi}{\partial t}\right)^2 - \left(\frac{\partial \varphi}{\partial x}\right)^2\right], \tag{81}$$

whence (A § 9)

$$\Pi = \frac{1}{c^2}\frac{\partial \varphi}{\partial t}, \tag{82}$$

and

$$\bar{H} = \frac{1}{2}\int_0^l \left\{c^2\Pi^2 + \left(\frac{\partial \varphi}{\partial x}\right)^2\right\} = \frac{1}{2}\int_0^l \left\{\frac{1}{c^2}\left(\frac{\partial \varphi}{\partial t}\right)^2 + \left(\frac{\partial \varphi}{\partial x}\right)^2\right\} dx. \tag{83}$$

The following exchange relations are to be used:

$$\Pi(x)\,\varphi\left(x'\right) - \varphi\left(x'\right)\Pi(x) = \delta(x - x')\frac{h}{2\pi i}. \tag{84}$$

---

*Zeitschrift für Physik, **26**, 178, 1924.

With the introduction of

$$\varphi(x, t) = \sqrt{\frac{2}{l}} \sum_k q_k(t) \sin \frac{k\pi x}{l},$$

$\bar{H}$ goes over into

$$\bar{H} = \frac{1}{2} \sum_k \left\{ \frac{1}{c^2} \dot{q}_k^2 + \left(\frac{k\pi}{l}\right)^2 q_k^2 \right\}. \tag{85}$$

On introducing the momenta associated to $q_k$,

$$p_k = \frac{1}{c^2} \dot{q}_k, \tag{86}$$

equation (84) becomes

$$p_k q_l - q_l p_k = \delta_{kl} \frac{h}{2\pi i} \tag{87}$$

or

$$\left.\begin{array}{l} p_k = \sqrt{\frac{k\pi}{l}} \sqrt{\frac{h}{2\pi}} \left\{ N_k^{\frac{1}{2}} e^{\frac{2\pi i}{h}\Theta k} + e^{-\frac{2\pi i}{h}\Theta_k} N_k^{\frac{1}{2}} \right\} \\[2ex] q_k = \sqrt{\frac{k\pi}{l}} \sqrt{\frac{h}{2\pi}} \left\{ N_k^{\frac{1}{2}} e^{\frac{2\pi i}{h}\Theta k} - e^{-\frac{2\pi i}{h}\Theta_k} N_k^{\frac{1}{2}} \right\} \frac{1}{i}. \end{array}\right\} \tag{88}$$

The characteristic frequencies of the string are $\nu_k = k(c/2l)$, and therefore

$$\bar{H} = \sum_k h\nu_k \left( N_k + \frac{1}{2} \right). \tag{89}$$

For the energy in a small section $(0, a)$ of the string, one obtains, however,

$$\mathfrak{E} = \frac{1}{l} \int_0^a \sum_{j,k} \left\{ \frac{1}{c^2} \dot{q}_i \dot{q}_k \sin \frac{j\pi x}{l} \sin \frac{k\pi x}{l} \right.$$
$$\left. + q_i q_k jk \left(\frac{\pi}{l}\right)^2 \cos \frac{j\pi x}{l} \cos \frac{k\pi x}{l} \right\} dx. \tag{90}$$

If the terms of this sum with $j = k$ be singled out, then under the explicit hypothesis that the wave-lengths to be considered are all small with respect to $a$, one obtains the value

$$\overline{\mathfrak{E}} = \frac{a}{l} \bar{H}.$$

One thus finds the fluctuation $\Delta \mathfrak{E} = \mathfrak{E} - \bar{E}$ by neglecting the terms with $j = k$ in (90). The integration results in

$$\Delta \mathfrak{E} = \frac{1}{2l} \sum_{j \neq k} \left\{ \frac{1}{c^2} \dot{q}_j \dot{q}_k K_{jk} + jk \left(\frac{\pi}{l}\right)^2 q_j q_k K'_{jk} \right\}, \tag{91}$$

where

$$\left. \begin{aligned} K_{jk} &= c \frac{\sin\left(v_j - v_k\right) a/c}{v_j - v_k} - c \frac{\sin\left(v_j + v_k\right) a/c}{v_j + v_k}, \\ K'_{jk} &= c \frac{\sin\left(v_j - v_k\right) a/c}{v_j - v_k} + c \frac{\sin\left(v_j + v_k\right) a/c}{v_j + v_k}. \end{aligned} \right\} \tag{92}$$

Accordingly, the mean-square fluctuation is given by

$$\overline{\Delta \mathfrak{E}^2} = \frac{1}{2l^2} \sum_{j \neq k} \left\{ \frac{1}{c^4} \overline{\dot{q}_j^2}\, \overline{\dot{q}_k^2} K_{jk}^2 + j^2 k^2 \left(\frac{\pi}{l}\right)^4 \overline{q_j^2}\, \overline{q_k^2} K'_{jk} \right.$$
$$\left. + \left(\frac{\pi}{l}\right)^2 \frac{jk}{c^2} \left(\overline{q_j \dot{q}_j}\, \overline{q_k \dot{q}_k} + \overline{\dot{q}_j q_j}\, \overline{\dot{q}_k q_k}\right) K_{jk} K'_{jk} \right\}.$$

The sums over $j$ and $k$ may be replaced by an integral over the frequencies $v_j$ and $v_k$, respectively, if it be assumed that the string $l$ is very long, so that its characteristic frequencies are close together. In addition, one finally assumes that $a$ is large and uses the relation

$$\lim_{a \to \infty} \frac{1}{a} \int_{-v_1}^{v_2} \frac{\sin^2 va}{v^2} f(v) dv = \pi f(0) \tag{93}$$

if $v_1 > 0$, $v_2 > 0$. The double integral then becomes a simple integral and one finds that

$$\overline{\Delta \mathfrak{E}^2} = \frac{a}{c} \int dv \left\{ \frac{1}{c^4} \left(\overline{\dot{q}_v^2}\right)^2 + \left[\left(\frac{2\pi v}{c}\right)^2 \overline{q_v^2}\right]^2 \right.$$
$$\left. + \frac{1}{c^2} \left(\frac{2\pi v}{c}\right)^2 \left[\left(\overline{q_v \dot{q}_v}\right)^2 + \left(\overline{\dot{q}_v q_v}\right)^2\right] \right\}. \tag{94}$$

Because of the exchange relations (84),

$$\overline{q_v \dot{q}_v} = -\overline{\dot{q}_v q_v} = c^2 \frac{h}{4\pi i}, \tag{95}$$

so that

$$\overline{\mathfrak{E}} = \frac{a}{l} \int dv\, Z_v h v \left( N_v + \frac{1}{2} \right), \tag{96}$$

where $Z_v dv$ denotes the number of characteristic frequencies in the interval $dv$, or, in this case, $Z_v = 2l/c$. If the integral be taken over the frequency interval $\Delta v$, one obtains

$$\overline{\mathfrak{E}} = \frac{a}{l} Z_v \Delta v h v \left( N_v + \frac{1}{2} \right), \tag{97}$$

$$\overline{\Delta \mathfrak{E}^2} = \frac{a}{c} \Delta v \left[ \frac{1}{2} \left( \frac{\overline{\mathfrak{E}} c}{a \Delta v} \right)^2 - \frac{1}{2} (hv)^2 \right]. \tag{98}$$

One then subdivides $\overline{\mathfrak{E}}$ into the thermal energy $\overline{\mathfrak{E}}^*$ and the zero point energy:

$$\overline{\mathfrak{E}} = \overline{\mathfrak{E}}^* + \frac{a}{l} Z_v \Delta v \frac{hv}{2} = \overline{\mathfrak{E}}^* + a \Delta v h v,$$

and finds

$$\overline{\Delta \mathfrak{E}^2} = \frac{a}{2c} \Delta v \left[ \left( \frac{\overline{\mathfrak{E}}^* c}{a \Delta v} \right)^2 + 2 \frac{\overline{\mathfrak{E}}^* c}{a \Delta v} h v \right]$$

$$= h v \overline{\mathfrak{E}}^* + \frac{\overline{\mathfrak{E}}^{*2}}{\Delta v} Z_v \frac{a}{l}. \tag{99}$$

This value corresponds exactly to formula (79). The corresponding relation in the classical wave theory may be obtained by passing to the limit $h = 0$ in (99). The classical wave theory thus leads only to the second term of equation (99). The quantum theory, which one can interpret as a particle theory or as a wave theory as one sees fit, leads to the complete fluctuation formula.

## § 8. Relativistic Formulation of the Quantum Theory

The conditions imposed on all physical theories by the principle of relativity have been neglected in most of the foregoing discussions, and consequently the results obtained are applicable only under those conditions in which the velocity of light may be regarded as infinite.

The reason for this neglect is that all relativistic effects belong to the *terra incognita* of quantum theory; the physical principles which have been elucidated in this book must be valid in this region also and thus it seemed proper not to obscure them with questions that cannot be answered definitely at the present time. None the less, this book would be incomplete without a brief discussion of the attempts to construct theories which shall embody both sets of principles, and the difficulties which have arisen in these attempts.

Dirac* has set up a wave equation which is valid for one electron and is invariant under the Lorentz transformation. It fulfils all requirements of the quantum theory, and is able to give a good account of the phenomena of the "spinning" electron, which could previously only be treated by *ad hoc* assumptions. The essential difficulty which arises with all relativistic quantum theories is not eliminated however. This arises from the relation

$$\frac{1}{c^2} E^2 = \mu^2 c^2 + p_x^2 + p_y^2 + p_z^2 \qquad (100)$$

between the energy and momentum of a free electron. According to this equation there are two values of $E$ which differ in sign associated with each set of values of $p_x$, $p_y$, $p_z$. The classical theory could eliminate this by arbitrarily excluding the one sign, but this is not possible according to the principles of quantum theory. Here spontaneous transitions may occur to the states of negative energy; as these have never been observed, the theory is certainly wrong. Under these conditions it is very remarkable that the positive energy-levels (at least in the case of one electron) coincide with those actually observed.

The difficulty inherent in formula (100) is also shown by a calculation of O. Klein,[†] who proves that if the electron is governed by any equation based on this relation it will be able to pass unhindered through regions in which its potential energy is greater than $2mc^2$.

---

[*] P. A. M. Dirac, *Proceedings of the Royal Society*, A, **117**, 610, 1928.
[†] *Zeitschrift für Physik*, **53**, 157, 1929.

If only motion in the $x$-direction be considered the formulas $(31a)$ $(31c)$ become

$$\frac{E^2}{c^2} = \mu^2 c^2 + p_x^2,$$

$$\frac{(E - V)^2}{c^2} = \mu^2 c^2 + p_x'^2,$$

whence

$$p_x'^2 = p_x^2 + \frac{(E - v)^2 - E^2}{c^2},$$

while the wave function has the form

$$e^{\frac{2\pi i}{h}(p_x' x - Et)}.$$

For very small values of $V$, $p_x'$ is real and there are transmitted waves, just as in chapter ii, § 2f. For larger values, $p_x'$ becomes a pure imaginary, so that the wave is totally reflected at the discontinuity and decreases exponentially in region II. But for very large values of $V$, $p_x'$ again becomes real, i.e., the electron wave again penetrates into the region II with constant amplitude. A more exact calculation verifies this result.

A difficulty of a somewhat different character arises in the calculation of the energy of the field of the electron according to the relativistic theory. For a point electron (one of zero radius) even the classical theory yields an infinite value of the energy, as is well known, so that it becomes necessary to introduce a universal constant of the dimension of a length—the "radius of the electron." It is remarkable that in the non-relativistic theory this difficulty can be avoided in another way—by a suitable choice of the order of non-commutative factors in the Hamiltonian function. This has hitherto not been possible in the relativistic quantum theory.

The hope is often expressed that after these problems have been solved the quantum theory will be seen to be based, in a large measure at least, on classical concepts. But even a superficial survey of the trend of the evolution of physics in the past thirty years shows that it is far more likely that the solution will result in further limitations on the applicability of classical concepts than that it will result in a removal of those already discovered. The list of modifications and limitations of our ideal world—which now contains those required by the relativity theory (for which $c$ is characteristic) and the uncertainty relations (symbolized by Planck's constant $h$)—will be extended by others which correspond to $e$, $\mu$, $M$. But the character of these is as yet not to be anticipated.

# The Development of Quantum Mechanics

By

## Werner Heisenberg

*Nobel lecture, December 11, 1933*

Quantum mechanics, on which I am to speak here, arose, in its formal content, from the endeavour to expand Bohr's principle of correspondence to a complete mathematical scheme by refining his assertions. The physically new viewpoints that distinguish quantum mechanics from classical physics were prepared by the researches of various investigators engaged in analysing the difficulties posed in Bohr's theory of atomic structure and in the radiation theory of light.

In 1900, through studying the law of black-body radiation which he had discovered, Planck had detected in optical phenomena a discontinuous phenomenon totally unknown to classical physics which, a few years later, was most precisely expressed in Einstein's hypothesis of light quanta. The impossibility of harmonizing the Maxwellian theory with the pronouncedly visual concepts expressed in the hypothesis of light quanta subsequently compelled research workers to the conclusion that radiation phenomena can only be understood by largely renouncing their immediate visualization. The fact, already found by Planck and used by Einstein, Debye, and others, that the element of discontinuity detected in radiation phenomena also plays an important part in material processes, was expressed systematically in Bohr's basic postulates of the quantum theory which, together with the Bohr-Sommerfeld quantum conditions of atomic structure, led to a qualitative interpretation of the chemical and optical properties of atoms. The acceptance of these basic postulates of the quantum theory contrasted uncompromisingly with the application of classical mechanics to atomic systems, which, however, at least in its qualitative affirmations, appeared indispensable for understanding the properties of atoms. This circumstance was a fresh argument in support of the assumption that the natural phenomena in which Planck's constant plays an important part can be understood only by largely foregoing a visual description of them. Classical physics seemed the limiting case of visualization of a fundamentally unvisualizable microphysics, the more accurately realizable the more Planck's constant vanishes relative to the parameters of the system. This view of classical mechanics as a limiting case of quantum mechanics also gave rise to Bohr's principle of correspondence which, at least in qualitative terms, transferred a number of conclusions formulated in classical mechanics to quantum mechanics. In connection with the principle of correspondence there was also discussion whether the quantum-mechanical laws could in principle be of a statistical nature; the possibility became particularly apparent in Einstein's derivation of Planck's law of radiation. Finally, the analysis of the relation between radiation theory and atomic theory by Bohr, Kramers, and Slater resulted in the following scientific situation:

According to the basic postulates of the quantum theory, an atomic system is capable of assuming discrete, stationary states, and therefore discrete energy values; in terms of the energy of the atom the emission and absorption of light by such a system occurs abruptly, in the form of impulses. On the other hand, the visualizable properties of the emitted radiation are described by a wave field, the frequency of which is associated with the difference in energy between the initial and final states of the atom by the relation

$$E_1 - E_2 = h\nu$$

To each stationary state of an atom corresponds a whole complex of parameters which specify the probability of transition from this state to another. There is no direct relation between the radiation classically emitted by an orbiting electron and those parameters defining the probability of emission; nevertheless Bohr's principle of correspondence enables a specific term of the Fourier expansion of the classical path to be assigned to each transition of the atom, and the probability for the particular transition follows qualitatively similar laws as the intensity of those Fourier components. Although therefore in the researches carried out by Rutherford, Bohr, Sommerfeld and others, the comparison of the atom with a planetary system of electrons leads to a qualitative interpretation of the optical and chemical properties of atoms, nevertheless the fundamental dissimilarity between the atomic spectrum and the classical spectrum of an electron system imposes the need to relinquish the concept of an electron path and to forego a visual description of the atom.

The experiments necessary to define the electron-path concept also furnish an important aid in revising it. The most obvious answer to the question how the orbit of an electron in its path within the atom could be observed namely, will perhaps be to use a microscope of extreme resolving power. But since the specimen in this microscope would have to be illuminated with light having an extremely short wavelength, the first light quantum from the light source to reach the electron and pass into the observer's eye would eject the electron completely from its path in accordance with the laws of the Compton effect. Consequently only one point of the path would be observable experimentally at any one time.

In this situation, therefore, the obvious policy was to relinquish at first the concept of electron paths altogether, despite its substantiation by Wilson's experiments, and, as it were, to attempt subsequently how much of the electron-path concept can be carried over into quantum mechanics.

In the classical theory the specification of frequency, amplitude, and phase of all the light waves emitted by the atom would be fully equivalent to specifying its electron path. Since from the amplitude and phase of an emitted wave the coefficients of the appropriate term in the Fourier expansion of the electron path can be derived without ambiguity, the complete electron path therefore can be derived from a knowledge of all amplitudes and phases. Similarly, in quantum mechanics, too, the whole complex of amplitudes and phases of the radiation emitted by the atom can be regarded as a complete description of the atomic system, although its interpretation in the sense of an electron path inducing the radiation is impossible. In quantum mechanics, therefore, the place of the electron coordinates is taken by a complex of parameters corresponding to the Fourier coefficients of classical motion along a path. These, however, are no longer classified by the energy of state and the number of the corresponding harmonic vibration, but are in each case associated with two stationary states of the atom, and are a measure for the transition probability of the atom from one stationary state to another. A complex of coefficients of this type is comparable with a matrix such as occurs in linear algebra. In exactly the same way each parameter of classical mechanics, e.g. the momentum or the energy of the electrons, can then be assigned a corresponding matrix in quantum mechanics. To proceed from here beyond a mere description of the empirical state of affairs it was necessary to associate systematically the matrices assigned to the various parameters in the same way as the corresponding parameters in classical mechanics are associated by equations of motions. When, in the interest of achieving the closest possible correspondence between classical and quantum mechanics, the addition and multiplication of Fourier series were tentatively taken as the example for the addition and multiplication of the quantum-theory complexes, the product of two parameters represented by matrices appeared to be most naturally represented by the product matrix in the sense of linear algebra - an assumption already suggested by the formalism of the Kramers-Ladenburg dispersion theory.

It thus seemed consistent simply to adopt in quantum mechanics the equations of motion of classical physics, regarding them as a relation between the matrices representing the classical variables. The Bohr-Sommerfeld quantum conditions could also be re-interpreted in a relation between the matrices, and together with the equations of motion they were sufficient to define all matrices and hence the experimentally observable properties of the atom.

Born, Jordan, and Dirac deserve the credit for expanding the mathematical scheme outlined above into a consistent and practically usable theory. These investigators observed in the first place that the quantum conditions can be written as commutation relations between the matrices representing the momenta and the coordinates of the electrons, to yield the equations ($p_r$, momentum matrices; $q_r$, coordinate matrices):

$$p_r q_s - q_s p_r = \frac{h}{2\pi i} \delta_{rs} \quad q_r q_s - q_s q_r = 0 \quad p_r p_s - p_s p_r = 0$$

$$\partial_{rs} = \begin{cases} 1 \text{ for } r = s \\ 0 \text{ for } r \neq s \end{cases}$$

By means of these commutation relations they were able to detect in quantum mechanics as well the laws which were fundamental to classical mechanics: the invariability in time of energy, momentum, and angular momentum.

The mathematical scheme so derived thus ultimately bears an extensive formal similarity to that of the classical theory, from which it differs outwardly by the commutation relations which, moreover, enabled the equations of motion to be derived from the Hamiltonian function.

In the physical consequences, however, there are very profound differences between quantum mechanics and classical mechanics which impose the need for a thorough discussion of the physical interpretation of quantum mechanics. As hitherto defined, quantum mechanics enables the radiation emitted by the atom, the energy values of the stationary states, and other parameters characteristic for the stationary states to be treated. The theory hence complies with the experimental data contained in atomic spectra. In all those cases, however, where a visual description is required of a transient event, e.g. when interpreting Wilson photographs, the formalism of the theory does not seem to allow an adequate representation of the experimental state of affairs. At this point Schrödinger's wave mechanics, meanwhile developed on the basis of de Broglie's theses, came to the assistance of quantum mechanics.

In the course of the studies which Mr. Schrödinger will report here himself he converted the determination of the energy values of an atom into an eigenvalue problem defined by a boundary-value problem in the coordinate space of the particular atomic system. After Schrödinger had shown the mathematical equivalence of wave mechanics, which he had discovered, with quantum mechanics, the fruitful combination of these two different areas of physical ideas resulted in an extraordinary broadening and enrichment of the formalism of the quantum theory. Firstly it was only wave mechanics which made possible the mathematical treatment of complex atomic systems, secondly analysis of the connection between the two theories led to what is known as the transformation theory developed by Dirac and Jordan. As it is impossible within the limits of the present lecture to give a detailed discussion of the mathematical structure of this theory, I should just like to point out its fundamental physical significance. Through the adoption of the physical principles of quantum mechanics into its expanded formalism, the transformation theory made it possible in completely general terms to calculate for atomic systems the probability for the occurrence of a particular, experimentally ascertainable, phenomenon under given experimental conditions. The hypothesis conjectured in the studies on the radiation theory and enunciated in precise terms in Born's collision theory, namely that the wave function governs the probability for the presence of a corpuscle, appeared to be a special case of a more general pattern of laws and to be a natural consequence of the fundamental assumptions of quantum mechanics. Schrödinger, and in later studies Jordan, Klein, and Wigner as well, had succeeded in developing as far as permitted by the principles of the quantum theory de Broglie's original concept of visualizable matter waves occurring in space and time, a concept formulated even before the development of quantum mechanics. But for that the connection between Schrödinger's concepts and de Broglie's original thesis would certainly have seemed a looser one by this statistical interpretation of wave mechanics and by the greater emphasis on the fact that Schrödinger's theory is concerned with waves in multidimensional space. Before proceeding to discuss the explicit significance of quantum mechanics it is perhaps right for me to deal briefly with this question as to the existence of matter waves in three-dimensional space, since the solution to this problem was only achieved by combining wave and quantum mechanics.

A long time before quantum mechanics was developed Pauli had inferred from the laws in the Periodic System of the elements the well-known principle that a particular quantum state can at all times be occupied by only a single electron. It proved possible to transfer this principle to quantum mechanics on the basis of what at first sight seemed a surprising result: the entire complex of stationary states which an atomic system is capable of adopting breaks down into definite classes such that an atom in a state belonging to one class can never change into a state belonging to another class under the action of whatever perturbations. As finally clarified beyond question by the studies of Wigner and Hund, such a class of states is characterized by a definite symmetry characteristic of the Schrödinger eigenfunction with respect to the transposition of the coordinates of two electrons. Owing to the fundamental identity of electrons, any external perturbation of the atom remains unchanged when two electrons are exchanged and hence causes no transitions between states of various classes. The Pauli principle and the Fermi-Dirac statistics derived from it are equivalent with the assumption that only that class of stationary states is achieved in nature in which the eigenfunction changes its sign when two electrons are exchanged. According to Dirac, selecting the symmetrical system of terms would lead not to the Pauli principle, but to Bose-Einstein electron statistics.

Between the classes of stationary states belonging to the Pauli principle or to Bose-Einstein statistics, and de Broglie's concept of matter waves there is a peculiar relation. A spatial wave phenomenon can be treated according to the principles of the quantum theory by analysing it using the Fourier theorem and then applying to the individual Fourier component of the wave motion, as a system having one degree of freedom, the normal laws of quantum mechanics. Applying this procedure for treating wave phenomena by the quantum theory, a procedure that has also proved fruitful in Dirac's studies of the theory of radiation, to de Broglie's matter waves, exactly the same results are obtained as in treating a whole complex of material particles according to quantum mechanics and selecting the symmetrical system of terms. Jordan and Klein hold that the two methods are mathematically equivalent even if allowance is also made for the interaction of the electrons, i.e. if the field energy originating from the continuous space charge is included in the calculation in de Broglie's wave theory. Schrödinger's considerations of the energy-momentum tensor assigned to the matter waves can then also be adopted in this theory as consistent components of the formalism. The studies of Jordan and Wigner show that modifying the commutation relations underlying this quantum theory of waves results in a formalism equivalent to that of quantum mechanics based on the assumption of Pauli's exclusion principle.

These studies have established that the comparison of an atom with a planetary system composed of nucleus and electrons is not the only visual picture of how we can imagine the atom. On the contrary, it is apparently no less correct to compare the atom with a charge cloud and use the correspondence to the formalism of the quantum theory borne by this concept to derive qualitative conclusions about the behaviour of the atom. However, it is the concern of wave mechanics to follow these consequences.

Reverting therefore to the formalism of quantum mechanics; its application to physical problems is justified partly by the original basic assumptions of the theory, partly by its expansion in the transformation theory on the basis of wave mechanics, and the question is now to expose the explicit significance of the theory by comparing it with classical physics.

In classical physics the aim of research was to investigate objective processes occurring in space and time, and to discover the laws governing their progress from the initial conditions. In classical physics a problem was considered solved when a particular phenomenon had been proved to occur objectively in space and time, and it had been shown to obey the general rules of classical physics as formulated by differential equations. The manner in which the knowledge of each process had been acquired, what observations may possibly have led to its experimental determination, was completely immaterial, and it was also immaterial for the consequences of the classical theory, which possible observations were to verify the predictions of the theory. In the quantum theory, however, the situation is completely different. The very fact that the formalism of quantum mechanics cannot be interpreted as visual description of a phenomenon occurring in space and time shows that quantum mechanics is in no way concerned with the objective determination of space-time phenomena. On the contrary, the formalism of quantum mechanics should be used in such a way that the probability for the outcome of a further experiment may be concluded from the determination of an experimental situation in an atomic system, providing that the system is subject to no perturbations other than those necessitated by performing the two experiments. The fact that the only definite known result to be ascertained after the fullest possible experimental investigation of the system is the probability for a certain outcome of a second experiment shows, however, that each observation must entail a discontinuous change in the formalism describing the atomic process and therefore also a discontinuous change in the physical phenomenon itself. Whereas in the classical theory the kind of observation has no bearing on the event, in the quantum theory the disturbance associated with each observation of the atomic phenomenon has a decisive role. Since, furthermore, the result of an observation as a rule leads only to assertions about the probability of certain results of subsequent observations, the fundamentally unverifiable part of each perturbation must, as shown by Bohr, be decisive for the non-contradictory operation of quantum mechanics. This difference between classical and atomic physics is understandable, of course, since for heavy bodies such as the planets moving around the sun the pressure of the sunlight which is reflected at their surface and which is necessary for them to be observed is negligible; for the smallest building units of matter, however, owing to their low mass, every observation has a decisive effect on their physical behaviour.

The perturbation of the system to be observed caused by the observation is also an important factor in determining the limits within which a visual description of atomic phenomena is possible. If there were experiments which permitted accurate measurement of all the characteristics of an atomic system necessary to calculate classical motion, and which, for example, supplied accurate values for the location and velocity of each electron in the system at a particular time, the result of these experiments could not be utilized at all in the formalism, but rather it would directly contradict the formalism. Again, therefore, it is clearly that fundamentally unverifiable part of the perturbation of the system caused by the measurement itself which hampers accurate ascertainment of the classical characteristics and thus permits quantum mechanics to be applied. Closer examination of the formalism shows that between the accuracy with which the location of a particle can be ascertained and the accuracy with which its momentum can simultaneously be known, there is a relation according to which the product of the probable errors in the measurement of the location and momentum is invariably at least as large as Planck's constant divided by $4\pi$. In a very general form, therefore, we should have

$$\Delta p \Delta q \geq \frac{h}{4\pi}$$

where $p$ and $q$ are canonically conjugated variables. These uncertainty relations for the results of the measurement of classical variables form the necessary conditions for enabling the result of a measurement to be expressed in the formalism of the quantum theory. Bohr has shown in a series of examples how the perturbation necessarily associated with each observation indeed ensures that one cannot go below the limit set by the uncertainty relations. He contends that in the final analysis an uncertainty introduced by the concept of measurement itself is responsible for part of that perturbation remaining fundamentally unknown. The experimental determination of whatever space-time events invariably necessitates a fixed frame - say the system of coordinates in which the observer is at rest - to which all measurements are referred. The assumption that this frame is "fixed" implies neglecting its momentum from the outset, since "fixed" implies nothing other, of course, than that any transfer of momentum to it will evoke no perceptible effect. The fundamentally necessary uncertainty at this point is then transmitted via the measuring apparatus into the atomic event.

Since in connection with this situation it is tempting to consider the possibility of eliminating all uncertainties by amalgamating the object, the measuring apparatuses, and the observer into one quantum-mechanical system, it is important to emphasize that the act of measurement is necessarily visualizable, since, of course, physics is ultimately only concerned with the systematic description of space-time processes. The behaviour of the observer as well as his measuring apparatus must therefore be discussed according to the laws of classical physics, as otherwise there is no further physical problem whatsoever. Within the measuring apparatus, as emphasized by Bohr, all events in the sense of the classical theory will therefore be regarded as determined, this also being a necessary condition before one can, from a result of measurements, unequivocally conclude what has happened. In quantum theory, too, the scheme of classical physics which objectifies the results of observation by assuming in space and time processes obeying laws is thus carried through up to the point where the fundamental limits are imposed by the unvisualizable character of the atomic events symbolized by Planck's constant. A visual description for the atomic events is possible only within certain limits of accuracy - but within these limits the laws of classical physics also still apply. Owing to these limits of accuracy as defined by the uncertainty relations, moreover, a visual picture of the atom free from ambiguity has not been determined. On the contrary the corpuscular and the wave concepts are equally serviceable as a basis for visual interpretation.

The laws of quantum mechanics are basically statistical. Although the parameters of an atomic system are determined in their entirety by an experiment, the result of a future observation of the system is not generally accurately predictable. But at any later point of time there are observations which yield accurately predictable results. For the other observations only the probability for a particular outcome of the experiment can be given. The degree of certainty which still attaches to the laws of quantum mechanics is, for example, responsible for the fact that the principles of conservation for energy and momentum still hold as strictly as ever. They can be checked with any desired accuracy and will then be valid according to the accuracy with which they are checked. The statistical character of the laws of quantum mechanics, however, becomes apparent in that an accurate study of the energetic conditions renders it impossible to pursue at the same time a particular event in space and time.

For the clearest analysis of the conceptual principles of quantum mechanics we are indebted to Bohr who, in particular, applied the concept of complementarity to interpret the validity of the quantum-mechanical laws. The uncertainty relations alone afford an instance of how in quantum mechanics the exact knowledge of one variable can exclude the exact knowledge of another. This complementary relationship between different aspects of one and the same physical process is indeed characteristic for the whole structure of quantum mechanics. I had just mentioned that, for example, the determination of energetic relations excludes the detailed description of space-time processes. Similarly, the study of the chemical properties of a molecule is complementary to the study of the motions of the individual electrons in the molecule, or the observation of interference phenomena complementary to the observation of individual light quanta. Finally, the areas of validity of classical and quantum mechanics can be marked off one from the other as follows: Classical physics represents that striving to learn about Nature in which essentially we seek to draw conclusions about objective processes from observations and so ignore the consideration of the influences which every observation has on the object to be observed; classical physics, therefore, has its limits at the point from which the influence of the observation on the event can no longer be ignored. Conversely, quantum mechanics makes possible the treatment of atomic processes by partially foregoing their space-time description and objectification.

So as not to dwell on assertions in excessively abstract terms about the interpretation of quantum mechanics, I would like briefly to explain with a well-known example how far it is possible through the atomic theory to achieve an understanding of the visual processes with which we are concerned in daily life. The interest of research workers has frequently been focused on the phenomenon of regularly shaped crystals suddenly forming from a liquid, e.g. a supersaturated salt solution. According to the atomic theory the forming force in this process is to a certain extent the symmetry characteristic of the solution to Schrödinger's wave equation, and to that extent crystallization is explained by the atomic theory. Nevertheless this process retains a statistical and - one might almost say - historical element which cannot be further reduced: even when the state of the liquid is completely known before crystallization, the shape of the crystal is not determined by the laws of quantum mechanics. The formation of regular shapes is just far more probable than that of a shapeless lump. But the ultimate shape owes its genesis partly to an element of chance which in principle cannot be analysed further.

Before closing this report on quantum mechanics, I may perhaps be allowed to discuss very briefly the hopes that may be attached to the further development of this branch of research. It would be superfluous to mention that the development must be continued, based equally on the studies of de Broglie, Schrödinger, Born, Jordan, and Dirac. Here the attention of the research workers is primarily directed to the problem of reconciling the claims of the special relativity theory with those of the quantum theory. The extraordinary advances made in this field by Dirac about which Mr. Dirac will speak here, meanwhile leave open the question whether it will be possible to satisfy the claims of the two theories without at the same time determining the

Sommerfeld fine-structure constant. The attempts made hitherto to achieve a relativistic formulation of the quantum theory are all based on visual concepts so close to those of classical physics that it seems impossible to determine the fine-structure constant within this system of concepts. The expansion of the conceptual system under discussion here should, furthermore, be closely associated with the further development of the quantum theory of wave fields, and it appears to me as if this formalism, notwithstanding its thorough study by a number of workers (Dirac, Pauli, Jordan, Klein, Wigner, Fermi) has still not been completely exhausted. Important pointers for the further development of quantum mechanics also emerge from the experiments involving the structure of the atomic nuclei. From their analysis by means of the Gamow theory, it would appear that between the elementary particles of the atomic nucleus forces are at work which differ somewhat in type from the forces determining the structure of the atomic shell; Stern's experiments seem, furthermore, to indicate that the behaviour of the heavy elementary particles cannot be represented by the formalism of Dirac's theory of the electron. Future research will thus have to be prepared for surprises which may otherwise come both from the field of experience of nuclear physics as well as from that of cosmic radiation. But however the development proceeds in detail, the path so far traced by the quantum theory indicates that an understanding of those still unclarified features of atomic physics can only be acquired by foregoing visualization and objectification to an extent greater than that customary hitherto. We have probably no reason to regret this, because the thought of the great epistemological difficulties with which the visual atom concept of earlier physics had to contend gives us the hope that the abstracter atomic physics developing at present will one day fit more harmoniously into the great edifice of Science.

# Quantisation as a Problem of Proper Values; Parts I-IV

By

Erwin Schrodinger

This translation is part of a work originally published as a separate work, *Four Lectures on Wave Mechanics*, which was originally published at Glasgow in 1928.[*]

## Quantisation as a Problem of Proper Values (Part I)

§ 1. In this paper I wish to consider, first, the simple case of the hydrogen atom (non-relativistic and unperturbed), and show that the customary quantum conditions can be replaced by another postulate, in which the notion of "whole numbers", merely as such, is not introduced. Rather when integralness does appear, it arises in the same natural way as it does in the case of the *node-numbers* of a vibrating string. The new conception is capable of generalisation, and strikes, I believe, very deeply at the true nature of the quantum rules.

The usual form of the latter is connected with the Hamilton-Jacobi differential equation,

$$H\left(q, \frac{\partial S}{\partial q}\right) = E. \tag{1}$$

A solution of this equation is sought such as can be represented as the *sum* of functions, each being a function of one only of the independent variables $q$.

Here we now put for $S$ a new unknown $\psi$ such that it will appear as a *product* of related functions of the single co-ordinates, *i.e.* we put

$$S = K \log \psi. \tag{2}$$

*Reprinted courtesy of the American Mathematical Society.

The constant $K$ must be introduced from considerations of dimensions; it has those of *action*. Hence we get

$$H\left(q, \frac{K}{\psi}\frac{\partial\psi}{\partial q}\right) = E. \qquad (1')$$

*Now* we do *not* look for a solution of equation $(1')$, but proceed as follows. If we neglect the relativistic variation of mass, equation $(1')$ can always be transformed so as to become a quadratic form (of $\psi$ and its first derivatives) equated to zero. (For the *one*-electron problem is holds even when mass-variation is not neglected.) We now seek junction $\psi$, such that for any arbitrary variation of it the integral the said quadratic form, taken over the whole co-ordinate space,[1] stationary, $\psi$ being everywhere real, single-valued, finite, and continuously differentiable up to the second order. *The quantum conditions are replaced by this variation problem.*

First, we will take for $H$ the Hamilton function for Keplerian motion, and show that $\psi$ can be so chosen for *all positive*, but only for *discrete set of negative values of E*. That is, the above variation problem has a discrete and a continuous spectrum of proper values.

The discrete spectrum corresponds to the Balmer terms and the ntinuous to the energies of the hyperbolic orbits. For numerical agreement $K$ must have the value $h2\pi$.

The choice of co-ordinates in the formation of the variational equations being arbitrary, let us take rectangular Cartesians. Then $(1')$ becomes in our case

$$\left(\frac{\partial\psi}{\partial x}\right)^2 + \left(\frac{\partial\psi}{\partial y}\right)^2 + \left(\frac{\partial\psi}{\partial z}\right)^2 - \frac{2m}{K^2}\left(E + \frac{e^2}{r}\right)\psi^2 = 0; \quad (1'')$$

$e$ = charge, $m$ = mass of an electron, $r^2 = x^2 + y^2 + z^2$. Our variation problem then reads

$$\delta J = \delta \int\int\int dxdydz \left[\left(\frac{\partial\psi}{\partial x}\right)^2 + \left(\frac{\partial\psi}{\partial y}\right)^2 + \left(\frac{\partial\psi}{\partial z}\right)^2 \right.$$

$$\left. -\frac{2m}{K^2}\left(E + \frac{e^2}{r}\right)\psi^2\right] = 0, \qquad (3)$$

[1] I am aware this formulation is not entirely unambiguous.

252

the integral being taken over all space. From this we find in the usual way

$$\frac{1}{2}\delta J = \int df\, \delta\psi \frac{\partial\psi}{\partial n} - \int\int\int dx\, dy\, dz\, \delta\psi \left[ \nabla^2\psi + \frac{2m}{K^2}\left(E + \frac{e^2}{r}\right)\psi \right] = 0.$$

(4)

Therefore we must have, firstly,

$$\nabla^2\psi + \frac{2m}{K^2}\left(E + \frac{e^2}{r}\right)\psi = 0,$$

(5)

And secondly,

$$\int df\, \delta\psi \frac{\partial\psi}{\partial n} = 0.$$

(6)

*df* is an element of the infinite closed surface over which the integral is taken.

(It will turn out later that this last condition requires us to supplement our problem by a postulate as to the behaviour of $\delta\psi$ at infinity, in order to ensure the existence of the above-mentioned *continuous* spectrum of proper values. See later.)

The solution of (5) can be effected, *for example*, in polar coordinates, *r*, $\Theta$, $\phi$, if $\psi$ be written as the *product* of three functions, each only of *r*, of $\Theta$, or of $\phi$. The method is sufficiently well known. The function of the angles turns out to be a *surface harmonic*, and if that of *r* be called $\chi$, we get easily the differential equation,

$$\frac{d^2\chi}{dr^2} + \frac{2}{r}\frac{d\chi}{dr} + \left( \frac{2mE}{K^2} + \frac{2me^2}{K^2 r} - \frac{n(n+1)}{r^2} \right)\chi = 0. \quad (7)$$
$$n = 0, 1, 2, 3\ldots$$

The limitation of *n* to integral values is *necessary* so that the surface harmonic may be *single-valued*. We require solutions of (7) that will remain finite for all non-negative real values of *r*. Now[1] equation (7) has *two* singularities in the complex *r*-plane, at $r = 0$ and $r = \infty$, of which the second is an "indefinite point" (essential singularity) of *all* integrals, but the first on the contrary is not (for any integral). These two singularities form exactly the *bounding points of our real*

---

[1] For guidance in the treatment of (7) I owe thanks to Hermann Weyl.

*interval.* In such a case it is known now that the postulation of the *finiteness* of $\chi$ at the bounding points is equivalent to a *boundary condition.* The equation has *in general* no integral which remains finite at *both* end points; such an integral exists only for certain special values of the constants in the equation. It is now a question of defining these special values. This is the *jumping-off* point of the whole investigation.[2]

Let us examine first the singularity at $r = 0$. The so-called *indicial* equation which defines the behaviour of the integral at this point, is

$$\rho (\rho - 1) + 2\rho - n (n + 1) = 0, \tag{8}$$

with roots

$$\rho_1 = n, \qquad \rho_2 = - (n + 1). \tag{8'}$$

The two canonical integrals at this point have therefore the exponents $n$ and $-(n + 1)$. Since $n$ is not negative, only the first of these is of use to us. Since it belongs to the greater exponent, it can be represented by an ordinary power series, which begins with $r^n$. (The other integral, which does not interest us, can contain a logarithm, since the difference between the indices is an integer.) The next singularity is at infinity, so the above power series is always convergent and represents a *transcendental integral function.* We therefore have established that:

*The required solution is (except for a constant factor) a single-valued definite transcendental integral function, which at $r = 0$ belongs to the exponent n.*

We must now investigate the behaviour of this function at infinity on the positive real axis. To that end we simplify equation (7) by the substitution

$$\chi = r^{\alpha} U, \tag{9}$$

---

[2] For unproved propositions in what follows, see L. Schlesinger's *Differential Equations* (Collection Schubert, No. 13, Göschen, 1900, especially chapters 3 and 5).

where $\alpha$ is so chosen that the term with $1/r^2$ drops out. It is easy to verify that then $\alpha$ must have one of the two values $n$, $-(n+1)$. Equation (7) then takes the form,

$$\frac{d^2 U}{dr^2} + \frac{2(\alpha+1)}{r}\frac{dU}{dr} + \frac{2m}{K^2}\left(E + \frac{e^2}{r}\right) U = 0. \qquad (7')$$

Its integrals belong at $r = 0$ to the exponents 0 and $-2\alpha -$. For the $\alpha$-value, $\alpha = n$, the *first* of these integrals, and for the second $\alpha$ value, $\alpha = -(n+1)$, the *second* of these integrals is an integral function and leads, according to (9), to the desired solution, which is single-valued. We therefore lose nothing if we confine ourselves to *one* of the two $\alpha$-values. Take, then,

$$\alpha = n. \qquad (10)$$

Our solution $U$ then, at $r = 0$, belongs to the exponent 0. Equation (7') is called Laplace's equation. The general type is

$$U'' + \left(\delta_0 + \frac{\delta_1}{r}\right) U' + \left(\epsilon_0 + \frac{\epsilon_1}{r}\right) U = 0. \qquad (7'')$$

Here the constants have the values

$$\delta_0 = 0, \quad \delta_1 = 2(\alpha+1), \quad \epsilon_0 = \frac{2mE}{K^2}, \quad \epsilon_1 = \frac{2me^2}{K^2}. \qquad (11)$$

This type of equation is comparatively simple to handle for this reason: The so-called Laplace's transformation, which in general leads *again* to an equation of the *second* order, *here* gives one of the *first*. This allows the solutions of (7'') to be represented by complex integrals. The result[1] only is given here. The integral

$$U = \int_L e^{zr} (z - c_1)^{\alpha_1 - 1} (z - c_2)^{\alpha_2 - 1}\, dz \qquad (12)$$

is a solution of (7'') for a path of integration $L$, for which

$$\int_L \frac{d}{dz}\left[e^{zr} (z - c_1)^{\alpha_1} (z - c_2)^{\alpha_2}\right] dz = 0. \qquad (13)$$

[1] Cf. Schlesinger. The theory is due to H. Poincaré eA and J. Horn.

The constants $c_1, c_2, \alpha_1, \alpha_2$ have the following values. $c_1$ and $c_2$ are the roots of the quadratic equation

$$z^2 + \delta_0 z + \epsilon_0 = 0. \tag{14}$$

and

$$\alpha_1 = \frac{\epsilon_1 + \delta_1 c_1}{c_1 - c_2}, \quad \alpha_2 = -\frac{\epsilon_1 + \delta_1 c_2}{c_1 - c_2}. \tag{14'}$$

In the case of equation (7') these become, using (11) and (10),

$$c_1 = +\sqrt{\frac{-2mE}{K^2}}, \quad c_2 = -\sqrt{\frac{-2mE}{K^2}}; \tag{14''}$$

$$\alpha_1 = \frac{me^2}{K\sqrt{-2mE}} + n + 1, \quad \alpha_2 = -\frac{me^2}{K\sqrt{-2mE}} + n + 1.$$

The representation by the integral (12) allows us, not only to survey the asymptotic behaviour of the totality of solutions when $r$ tends to infinity in a definite way, but also to give an account of this behaviour for one *definite* solution, which is always a much more difficult task.

We shall at first *exclude* the case where $\alpha_1$ and $\alpha_2$ are real integers. When this occurs, it occurs for both quantities simultaneously, and when, and only when,

$$\frac{me^2}{K\sqrt{-2mE}} = \text{a real integer.} \tag{15}$$

Therefore we assume that (15) is not fulfilled.

The behaviour of the totality of solutions when $r$ tends to infinity in a definite manner—we think always of $r$ becoming infinite through real positive values—is characterised[1] by the behaviour of the two linearly independent solutions, which we will call $U_1$ and $U_2$, and which are obtained by the following *specialisations* of the path of integration $L$. In *each* case let $z$ come from infinity and return there along the same path, in such a direction that

$$\lim_{z \to \infty} e^{zr} = 0, \tag{16}$$

---

[1] If (15) is satisfied, at least one of the two paths of integration described in the text cannot be used, as it yields a vanishing result.

*i.e.* the real part of $zr$ is to become negative and infinite. In this way condition (13) is satisfied. In the *one* case let $z$ make a circuit once round the point $c_1$ (solution $U_1$), and in the *other*, round $c_2$ (solution $U_2$).

Now for very large real positive values of $r$, these two solutions are represented *asymptotically* (in the sense used by Poincaré) by

$$\begin{cases} U_1 \sim e^{c_1 r} r^{-\alpha_1} (-1)^{\alpha_1} \left(e^{2\pi i \alpha_1} - 1\right) \Gamma(\alpha_1) (c_1 - c_2)^{\alpha_2 - 1}, \\ U_2 \sim e^{c_2 r} r^{-\alpha_2} (-1)^{\alpha_2} \left(e^{2\pi i \alpha_2} - 1\right) \Gamma(\alpha_2) (c_2 - c_1)^{\alpha_1 - 1}, \end{cases} \quad (17)$$

in which we are content to take the first term of the asymptotic series of integral negative powers of $r$.

We have now to distinguish between the two cases.

(1) $\underline{E > 0}$. This guarantees the non-fulfilment of (15), as it makes the left hand a pure imaginary. Further, by (14''), $c_1$ and $c_2$ also become pure imaginaries. The exponential functions in (17), since $r$ is real, are therefore periodic functions which remain finite. The values of $\alpha_1$ and $\alpha_2$ from (14'') show that *both* $U_1$ and $U_2$ tend to zero like $r^{-n-1}$. *This must therefore be valid for our transcendental integral solution $U$, whose behaviour we are investigating, however it may be linearly compounded from $U_1$ and $U_2$.* Further, (9) and (10) show that the function $\chi$, *i.e.* the transcendental integral solution of the *original* equation (7), always tends to zero like $1/r$, as it arises from $U$ through multiplication by $r^n$. We can thus state:

*The Eulerian differential equation (5) of our variation problem has, for every positive $E$, solutions, which are everywhere single-valued, finite, and continuous; and which tend to zero with $1/r$ at infinity, under continual oscillations.* The surface condition (6) has yet to be discussed.

(2) $\underline{E < 0}$. In this case the possibility (15) is not *eo ipso* excluded, yet we will maintain that exclusion provisionally. Then by (14'') and (17), for $r \to \infty$, $U_1$ grows beyond all limits, but $U_2$ vanishes exponentially. Our integral function $U$ (and the same is true for $\chi$) will then remain finite if, and only if, $U$ is identical with $U_2$, save perhaps for a numerical factor. *This, however, can never*

*be*, as is proved thus: If a *closed* circuit round *both* points $c_1$ and $c_2$ be chosen for the path $L$, thereby satisfying condition (13) since the circuit is *really closed* on the Riemann surface of the integrand, on account of $\alpha_1 + \alpha_2$ being an integer, then it is easy to show that the integral (12) represents *our integral function U.* (12) can be developed in a series of positive powers of $r$, which converges, at all events, for $r$ sufficiently small, and since it satisfies equation (7'), it must coincide with the series for $U$. Therefore $U$ is represented by (12) if $L$ be a closed circuit round both points $c_1$ and $c_2$. This closed circuit can be so distorted, however, as to make it appear additively combined from the two paths, considered above, which belonged to $U_1$ and $U_2$; and the factors are non-vanishing, 1 and $e^{2\pi i \alpha_1}$. Therefore $U$ cannot coincide with $U_2$, but must contain also $U_1$. Q.E.D.

Our integral function $U$, which alone of the solutions of (7') is considered for our problem, is therefore not finite for $r$ large, on the above hypothesis. Reserving meanwhile the question of *completeness,* *i.e.* the proving that our treatment allows us to find all the linearly independent solutions of the problem, then we may state:

*For negative values of E which do not satisfy condition (15) our variation problem has no solution.*

We have now only to investigate that discrete set of negative $E$-values which satisfy condition (15). $\alpha_1$ and $\alpha_2$ are then both integers. The first of the integration paths, which previously gave us the fundamental values $U_1$ and $U_2$, must now undoubtedly be modified so as to give a non-vanishing result. For, since $\alpha_1 - 1$ is certainly positive, the point $c_1$ is neither a branch point nor a pole of the integrand, but an ordinary zero. The point $c_2$ can also become regular if $\alpha_2 - 1$ is also not negative. In *every* case, however, two suitable paths are readily found and the integration effected completely in terms of known functions, so that the behaviour of the solutions can be fully investigated.

Let

$$\frac{me^2}{K\sqrt{-2mE}} = l; \quad l = 1, 2, 3, 4 \ldots \tag{15'}$$

Then from (14″) we have

$$\alpha_1 - 1 = l + n, \quad \alpha_2 - 1 = -l + n. \tag{14‴}$$

Two cases have to be distinguished: $l \leq n$ and $l > n$.

(a) $l \leq n$. Then $c_2$ *and* $c_1$ lose every singular character, but instead become starting-points or end-points of the path of integration, in order to fulfil condition (13). A third characteristic point here is at infinity (negative and real). Every path between two of these three points yields a solution, and of these three solutions there are two linearly independent, as is easily confirmed if the integrals are calculated out. In particular, the *transcendental integral solution* is given by the path from $c_1$ to $c_2$. That *this* integral remains regular at $r = 0$ can be seen at once without calculating it. I emphasize this point, as the actual calculation is apt to obscure it. However, the calculation does show that the integral becomes indefinitely great for positive, infinitely great values of $r$. One of the *other* two integrals remains *finite* for $r$ large, but it becomes infinite for $r = 0$.

Therefore when $l \leq n$ we get *no* solution of the problem.

(b) $l > n$. Then from (14‴), $c_1$ is a zero and $c_2$ a pole of the first order at least of the integrand. Two independent integrals are then obtained: one from the path which leads from $z = -\infty$ to the zero, intentionally avoiding the pole; and the other from the *residue* at the pole. The *latter* is the integral function. We will give its calculated value, but multiplied by $r^n$, so that we obtain, according to (9) and (10), the solution $\chi$ of the original equation (7). (The multiplying constant is arbitrary.) We find

$$\chi = f\left(r\frac{\sqrt{-2mE}}{K}\right); \quad f(x) = x^n e^{-x} \sum_{k=0}^{l-n-1} \frac{(-2x)^k}{k!} \begin{pmatrix} l+n \\ l-n-1-k \end{pmatrix}. \tag{18}$$

It is seen that this is a solution that can be utilised, since it remains finite for all real non-negative values of $r$. In addition, it satisfies the surface condition (6) because of its vanishing exponentially at infinity. Collecting then the results for $E$ negative:

*For E negative, our variation problem has solutions if, and only if, E satisfies condition (15). Only values smaller than l (and there is always at least one such at our disposal) can be given to the integer n, which denotes the order of the surface harmonic appearing in the equation. The part of the solution depending on r is given by (18).*

Taking into account the constants in the surface harmonic (known to be $2n + 1$ in number), it is further found that:

*The discovered solution has exactly $2n + 1$ arbitrary constants for any permissible (n, l) combination; and therefore for a prescribed value of l has $l^2$ arbitrary constants.*

We have thus confirmed the main points of the statements originally made about the proper-value spectrum of our variation problem, but there are still deficiencies.

Firstly, we require information as to the completeness of the *collected* system of proper functions indicated above, but I will not concern myself with that in this paper. From experience of similar cases, it may be supposed that no proper value has escaped us.

Secondly, it must be remembered that the proper functions, ascertained for $E$ positive, do not solve the variation problem as originally postulated, because they only tend to zero at infinity as $1/r$, and therefore $\partial \psi / \partial r$ only tends to zero on an infinite sphere as $1/r^2$. Hence the surface integral (6) is still of the same order as $\delta \psi$ at infinity. If it is desired therefore to obtain the continuous spectrum, another condition must be added to the *problem*, viz. that $\delta \psi$ is to vanish at infinity, or at least, that it tends to a constant value independent of the direction of proceeding to infinity; in the latter case the surface harmonics cause the surface integral to vanish.

§ 2. Condition (15) yields

$$-E_l = \frac{me^4}{2K^2 l^2}. \tag{19}$$

Therefore the well-known Bohr energy-levels, corresponding to the Balmer terms, are obtained, if to the constant $K$, introduced into (2) for reasons of dimensions, we give the value

$$K = \frac{h}{2\pi}, \tag{20}$$

from which comes

$$-E_l = \frac{2\pi^2 m e^4}{h^2 l^2}. \tag{19'}$$

Our $l$ is the principal quantum number. $n + 1$ is analogous to the azimuthal quantum number. The splitting up of this number through a closer definition of the surface harmonic can be compared with the resolution of the azimuthal quantum into an "equatorial" and a "polar" quantum. These numbers *here* define the system of node-lines on the sphere. Also the "radial quantum number" $l - n - 1$ gives exactly the number of the "node-spheres", for it is easily established that the function $f(x)$ in (18) has exactly $l - n - 1$ positive real roots. The positive $E$-values correspond to the continuum of the hyperbolic orbits, to which one may ascribe, in a certain sense, the radial quantum number $\infty$. The fact corresponding to this is the proceeding to infinity, under *continual* oscillations, of the functions in question.

It is interesting to note that the range, inside which the functions of (18) differ sensibly from zero, and outside which their oscillations die away, is of the *general order of magnitude* of the major axis of the ellipse in each case. The factor, multiplied by which the radius vector enters as the argument of the constant-free function $f$, is—naturally—the reciprocal of a length, and this length is

$$\frac{K}{\sqrt{-2mE}} = \frac{K^2 l}{me^2} = \frac{h^2 l}{4\pi^2 me^2} = \frac{a_l}{l}, \tag{21}$$

where $a_l =$ the semi-axis of the $l$th elliptic orbit. (The equations follow from (19) plus the known relation $E_l = \frac{-e^2}{2a_l}$)

The quantity (21) gives the order of magnitude of the range of the roots when $l$ and $n$ are small; for then it may be assumed that the roots of $f(x)$ are of the order of unity. That is naturally no longer the case if the coefficients of the polynomial are large numbers. At present

I will not enter into a more exact evaluation of the roots, though I believe it would confirm the above assertion pretty thoroughly.

§ 3. It is, of course, strongly suggested that we should try to connect the function $\psi$ with some *vibration process* in the atom, which would more nearly approach reality than the electronic orbits, the real existence of which is being very much questioned to-day. I originally intended to found the new quantum conditions in this more intuitive manner, but finally gave them the above neutral mathematical form, because it brings more clearly to light what is really essential. The essential thing seems to me to be, that the postulation of "whole numbers" no longer enters into the quantum rules mysteriously, but that we have traced the matter a step further back, and found the "integralness" to have its origin in the finiteness and single-valuedness of a certain space function.

I do not wish to discuss further the possible representations of the vibration process, before more complicated cases have been calculated successfully from the new stand-point. It is not decided that the results will merely re-echo those of the usual quantum theory. For example, if the relativistic Kepler problem be worked out, it is found to lead in a remarkable manner to *half-integral partial* quanta (radial and azimuthal).

Still, a few remarks on the representation of the vibration may be permitted. Above all, I wish to mention that I was led to these deliberations in the first place by the suggestive papers of M. Louis de Broglie,[1] and by reflecting over the space distribution of those "phase waves", of which he has shown that there is always a *whole number*, measured along the path, present on each period or quasi-period of the electron. The main difference is that de Broglie thinks of progressive waves, while we are led to stationary proper vibrations if we interpret our formulae as representing vibrations. I have lately shown[2] that the Einstein gas theory can be based on the consideration of such stationary proper vibrations, to which the dispersion law of

[1] L. de Broglie, *Ann. de Physique* (10) 3, p. 22, 1925. (Thèses, Paris, 1924.)
[2] *Physik. Ztschr.* 27, p. 95, 1926.

de Broglie's phase waves has been applied. The above reflections on the atom could have been represented as a generalisation from those on the gas model.

If we take the separate functions (18), multiplied by a surface harmonic of order $n$, as the description of proper vibration processes, then the quantity. $E$ must have something to do with the related *frequency*. Now in vibration problems we are accustomed to the "parameter" (usually called $\lambda$) being proportional to the *square* of the frequency. However, in the first place, such a statement in our case would lead to *imaginary* frequencies for the *negative $E$-values*, and, secondly, instinct leads us to believe that the energy must be proportional to the frequency itself and not to its square.

The contradiction is explained thus. There has been *no natural zero level* laid down for the "parameter" $E$ of the variation equation (5), especially as the unknown function $\psi$ appears multiplied by a function of $r$, which can be changed by a constant to meet a corresponding change in the zero level of $E$. Consequently, we have to correct our anticipations, in that not $E$ itself—continuing to use the same terminology—but $E$ increased by a certain constant is to be expected to be proportional to the square of the frequency. Let this constant be now *very great* compared with all the admissible negative $E$-values (which are already limited by (15)). Then firstly, the frequencies will become *real*, and secondly, since our $E$-values correspond to only relatively small frequency *differences*, they will actually be very approximately proportional to these frequency differences. This, again, is all that our "quantum-instinct" can require, as long as the zero level of *energy* is not fixed.

The view that the frequency of the vibration process is given by

$$\nu = C'\sqrt{C+E} = C'\sqrt{C} + \frac{C'}{2\sqrt{C}}E + \dots, \qquad (22)$$

where $C$ is a constant very great compared with all the $E$'s, has still another very appreciable advantage. *It permits an understanding of the Bohr frequency condition.* According to the latter the *emission frequencies* are proportional to the *E-differences*, and therefore from (22) also to the differences of the proper frequencies $v$ of those hypothetical vibration processes. But these proper frequencies are all very great compared with the emission frequencies, and they agree very closely among themselves. The emission frequencies appear therefore as deep "difference tones" of the proper vibrations themselves. It is quite conceivable that on the transition of energy from one to another of the normal vibrations, *something*—I mean the light wave—with a *frequency* allied to each frequency *difference*, should make its appearance. One only needs to imagine that the light wave is causally related to the *beats*, which necessarily arise at each point of space during the transition; and that the frequency of the light is defined by the number of times per second the intensity maximum of the beat-process repeats itself.

It may be objected that these conclusions are based on the relation (22), in its *approximate* form (after expansion of the square root), from which the Bohr frequency condition itself seems to obtain the nature of an approximation. This, however, is merely apparently so, and it is wholly avoided when the *relativistic* theory is developed and makes a profounder insight possible. The large constant $C$ is naturally very intimately connected with the rest-energy of the electron ($mc^2$). Also the seemingly *new* and *independent* introduction of the constant $h$ (already brought in by (20)), into the frequency condition, is cleared up, or rather avoided, by the relativistic theory. But unfortunately the correct establishment of the latter meets right away with certain difficulties, which have been already alluded to.

It is hardly necessary to emphasize how much more congenial it would be to imagine that at a quantum transition the energy changes over from one form of vibration to another, than to think of a jumping electron. The changing of the vibration form can take place continuously in space and time, and it can readily last as long as the emission process lasts empirically (experiments on canal rays by W. Wien); nevertheless, if during this transition the atom is placed for a comparatively short time in an electric field which alters the proper frequencies, then the beat frequencies are immediately changed sympathetically, and for just as long as the field operates. It is known that this experimentally established fact has hitherto presented the greatest difficulties. See the well-known attempt at a solution by Bohr, Kramers, and Slater.

Let us not forget, however, in our gratification over our progress in these matters, that the idea of only *one* proper vibration being excited whenever the atom does not radiate—if we must hold fast to this idea—is very far removed from the *natural* picture of a vibrating system. We know that a macroscopic system does not behave like that, but yields in general a *pot-pourri* of its proper vibrations. But we should not make up our minds too quickly on this point. A *pot-pourri* of proper vibrations would also be permissible for a single atom, since thereby no beat frequencies could arise other than those which, according to experience, the atom is capable of emitting *occasionally*. The actual sending out of many of these spectral lines simultaneously by the same atom does not contradict experience. It is thus conceivable that only in the normal state (and approximately in certain "metastable" states) the atom vibrates with *one* proper frequency and just for this reason does *not* radiate, namely, because no beats arise. The *stimulation* may consist of a simultaneous excitation of one or of several other proper frequencies, whereby beats originate and evoke emission of light.

Under all circumstances, I believe, the proper functions, which belong to the *same* frequency, are in general all simultaneously stimulated. Multipleness of the proper values corresponds, namely, in the language of the previous theory to *degeneration*. To the reduction of the quantisation of degenerate systems probably corresponds the arbitrary partition of the energy among the functions belonging to *one* proper value.

## ADDITION AT THE PROOF CORRECTION ON 28. 2. 1926.

In the case of conservative systems in classical mechanics, the variation problem can be formulated in a neater way than was previously shown, and without express reference to the Hamilton-Jacobi differential equation. Thus, let $T(q, p)$ be the kinetic energy, expressees as a function of the co-ordinates and momenta, $V$ the potential energy and $d\tau$ the volume element of the space, "measured rationally", *i.e.* it is not simply the product $dq_1\ dq_2\ dq_3 \ldots dq_n$, but this divided by the square root of the discriminant of the quadratic form $T(q, p)$ (Cf. Gibbs' *Statistical Mechanics*.) Then let $\psi$ be such as to make the "Hamilton integral"

$$\int d\tau \left\{ K^2 T\left(q, \frac{\partial\psi}{\partial q}\right) + \psi^2 V \right\} \tag{23}$$

*stationary*, while fulfilling the *normalising, accessory condition*

$$\int \psi^2 d\tau = 1. \tag{24}$$

The proper values of this variation problem are then *the stationary values* of integral (23) and yield, according to our thesis, *the quantum-levels of the energy.*

It is to be remarked that in the quantity $\alpha_2$ of (14″) we have essentially the well-known Sommerfeld expression $-\frac{B}{\sqrt{A}} + \sqrt{C}$. (Cf. *Atombau*, 4th (German) ed., p. 775.)

Physical Institute of the University of Zürich.
(Received January 27, 1926.)

# Quantisation as a Problem of Proper Values (Part II)

*(Annalen der Physik* (4), vol. 79, 1926)*

## § 1. The Hamiltonian Analogy between Mechanics and Optics

BEFORE we go on to consider the problem of proper values for further special systems, let us throw more light on the *general* correspondence which exists between the Hamilton-Jacobi differential equation of a mechanical problem and the "allied" *wave equation*, i.e. equation (5) of Part I in the case of the Kepler problem. So far we have only briefly described this correspondence on its external analytical side by the transformation (2), which is in itself unintelligible, and by the equally incomprehensible transition from the *equating to zero* of a certain expression to the postulation that the *space integral* of the said expression shall be *stationary.*[1]

The *inner* connection between Hamilton's theory and the process of wave propagation is anything but a new idea. It was not only well known to Hamilton, but it also served him as the starting-point for his theory of mechanics, which grew[2] out of his *Optics of Non-homogeneous Media.* Hamilton's variation principle can be shown to correspond to Fermat's *Principle* for a wave propagation in configuration space (*q*-space), and the Hamilton-Jacobi equation expresses Huygens' *Principle* for this wave propagation. Unfortunately this powerful and momentous conception of Hamilton is deprived, in most modern reproductions, of its beautiful raiment as a superfluous accessory, in favour of a more colourless representation of the analytical correspondence.[3]

---

[1] This procedure will *not be pursued further* in the present paper. It was only intended to give a provisional, quick survey of the external connection between the wave equation and the Hamilton-Jacobi equation. $\psi$ is not actually the action function of a definite motion in the relation stated in (2) of Part I. On the other hand the connection between the wave equation and the variation problem is of course very real; the integrand of the stationary integral is the Lagrange function for the wave process.

[2] Cf. *e.g.* E. T. Whittaker's *Anal. Dynamics*, chap. xi.

[3] Felix Klein has since 1891 repeatedly developed the theory of Jacobi from quasi-optical considerations in non-Euclidean higher space in his lectures on mechanics. Cf. F. Klein, *Jahresber. d. Deutsch. Math. Ver.* 1, 1891, and *Zeits. f. Math. u. Phys.* 46, 1901 (*Ges.-Abh.* ii. pp. 601 and 603). In the second note, Klein remarks reproachfully that his discourse at Halle ten years previously, in which he had discussed

Let us consider the general problem of conservative systems in classical mechanics. The Hamilton-Jacobi equation runs

$$\frac{\partial W}{\partial t} + T\left(q_k, \frac{\partial W}{\partial q_k}\right) + V(q_k) = 0. \tag{1}$$

$W$ is the action function, *i.e.* the time integral of the Lagrange function $T - V$ along a path of the system as a function of the end points and the time. $q_k$ is a representative position co-ordinate; $T$ is the kinetic energy as function of the $q$'s and momenta, being a quadratic form of the latter, for which, as prescribed, the partial derivatives of $W$ with respect to the $q$'s are written. $V$ is the potential energy. To solve the equation put

$$W = -Et + S(q_k), \tag{2}$$

and obtain

$$2T\left(q_k, \frac{\partial W}{\partial q_k}\right) = 2(E - V). \tag{1'}$$

$E$ is an arbitrary integration constant and signifies, as is known, the energy of the system. Contrary to the usual practice, we have let the function $W$ remain itself in (1'), instead of introducing the time-free function of the co-ordinates, $S$. That is a mere superficiality.

Equation (1') can now be very simply expressed if we make use of the method of Heinrich Hertz. It becomes, like all geometrical assertions in configuration space (space of the variables $q_k$), especially simple and clear if we introduce into this space a non-Euclidean metric by means of the kinetic energy of the system.

Let $\bar{T}$ be the kinetic energy as function of the velocities $\dot{q}_k$, not of the *momenta* as above, and let us put for the line element

$$ds^2 = 2\bar{T}(q_k, \dot{q}_k)\, dt^2. \tag{3}$$

The right-hand side now contains $dt$ only externally and represents (since $\dot{q}_k dt = dq_k$) a quadratic form of the $dq_k$'s.

---

this correspondence and emphasized the great significance of Hamilton's optical works, had "not obtained the general attention, which he had expected". For this allusion to F. Klein, I am indebted to a friendly communication from Prof. Sommerfeld. See also *Atombau*, 4th ed., p. 803.

After this stipulation, conceptions such as angle between two line elements, perpendicularity, divergence and curl of a vector, gradient of a scalar, Laplacian operation (= div grad) of a scalar, and others, may be used in the same simple way as in three-dimensional Euclidean space, and we may use in our thinking the Euclidean three-dimensional representation with impunity, except that the analytical expressions for these ideas become a very little more complicated, as the line element (3) must everywhere replace the Euclidean line element. *We, stipulate, that in what follows, all geometrical statements in q-space are to be taken in this non-Euclidean sense.*

One of the most important modifications for the calculation is that we must distinguish carefully between covariant and contravariant components of a vector or tensor. But this complication is not any greater than that which occurs in the case of an oblique set of Cartesian axes.

The $dq_k$'s are the prototype of a contravariant vector. The coefficients of the form $2\bar{T}$, which depend on the $q_k$'s, are therefore of a covariant character and form the covariant fundamental tensor. $2T$ is the contravariant form belonging to $2\bar{T}$, because the momenta are known to form the covariant vector belonging to the speed vector $\dot{q}_k$, the momentum being the velocity vector in covariant form. The left side of (1′) is now simply the contravariant fundamental form, in which the $\frac{\partial W}{\partial q_k}$'s are brought in as variables. The latter form the components of the vector,—according to its nature covariant,

$$\operatorname{grad} W.$$

(The expressing of the kinetic energy in terms of momenta instead of speeds has then *this* significance, that covariant vector components can only be introduced in a contravariant form if something intelligible, *i.e.* invariant, is to result.)

Equation (1′) is equivalent thus to the simple statement

$$(\operatorname{grad} W)^2 = 2(E - V), \tag{1″}$$

or

$$|\operatorname{grad} W| = \sqrt{2\,(E - V)}. \tag{1‴}$$

This requirement is easily analysed. Suppose that a function $W$, of the form (2), has been found, which satisfies it. Then this function can be clearly represented for every definite $t$, if the family of surfaces $W = \text{const}$, be described in $q$-space and to each member a value of $W$ be ascribed.

Now, on the one hand, as will be shown immediately, equation $(1''')$ gives an exact rule for constructing all the other surfaces of the family and obtaining their $W$-values from any single member, *if the latter and its W-value is known*. On the other hand, if the sole necessary data for the construction, viz. *one* surface and its $W$-value be *given quite arbitrarily*, then from the rule, which presents just *two* alternatives, there may be completed one of the functions $W$ fulfilling the given requirement. Provisionally, the time is regarded as constant.—The construction rule therefore *exhausts* the contents of the differential equation; *each* of its solutions can be obtained from a suitably chosen surface and $W$-value.

Let us consider the construction rule. Let the value $W_0$ be given in Fig. 1 to an arbitrary surface. In order to find the surface $W_0 + dW_0$, take *either* side of the given surface as the positive one, erect the normal at each point of it and cut off (with due regard to the sign of $dW_0$) the *step*

$$ds = \frac{dW_0}{\sqrt{2(E - V)}}. \tag{4}$$

The locus of the end points of the steps is the surface $W_0 + dW_0$. Similarly, the family of surfaces may be constructed successively on both sides.

The construction has a *double* interpretation, as the *other* side of the given surface might have been taken as positive for the first step. This ambiguity does not hold for later steps, *i.e.* at any later stage of the process we cannot change arbitrarily the sign of the sides of the surface, at which we have arrived, as this would involve in general a discontinuity in the first differential coefficient of $W$. Moreover,

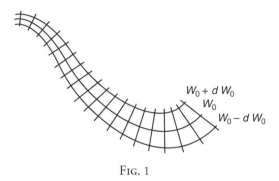

$W_0 + d W_0$
$W_0$
$W_0 - d W_0$

FIG. 1

the two families obtained in the two cases are clearly identical; the $W$-values merely run in the opposite direction.

Let us consider now the very simple dependence on the *time*. For this, (2) shows that at any later (or earlier) instant $t + t'$, the *same* group of surfaces illustrates the $W$-distribution, though different $W$-value are associated with the individual members, namely, from each $W$-value ascribed at time $t$ there must be subtracted $Et'$. The $W$-values wander, as it were, from surface to surface according to a definite, simple law, and for positive $E$ in the direction of $W$ increasing. Instead of this, however, we may imagine that the *surfaces* wander in such a way that each of them continually takes the place and exact form of the following one, and always carries its $W$-value *with* it. The rule for this wandering is given by the fact that the surface $W_0$ at time $t + dt$ must have reached *that* place, which at $t$ was occupied by the surface $W_0 + Edt$. This will be attained according to (4), if each point of the surface $W_0$ is allowed to move in the direction of the positive normal through a distance

$$ds = \frac{E\,dt}{\sqrt{2(E - V)}}. \tag{5}$$

That is, the surfaces move with a *normal velocity*

$$u = \frac{ds}{dt} = \frac{E}{\sqrt{2(E - V)}}, \tag{6}$$

which, when the constant $E$ is given, is a pure function of position.

Now it is seen that our system of surfaces $W=$ const. can be conceived as the system of wave surfaces of a progressive but stationary wave motion in $q$-space, for which the value of the phase velocity at every point in the space is given by (6). For the normal construction can clearly be replaced by the construction of elementary Huygens waves (with radius (5)), and then of their envelope. The "index of refraction" is proportional to the reciprocal of (6), and is dependent on the position but not on the direction. The $q$-space is thus optically non-homogeneous but is isotropic. The elementary waves are "spheres", though of course—let me repeat it expressly once more—in the sense of the line-element (3).

The function of action $W$ plays the part of the *phase* of our wave system. The Hamilton-Jacobi equation is the expression of Huygens' principle. If, now, Fermat's principle be formulated thus,

$$0 = \delta \int_{P_1}^{P_2} \frac{ds}{u} = \delta \int_{P_1}^{P_2} \frac{ds\,\sqrt{2\,(E-V)}}{E} = \delta \int_{t_1}^{t_2} \frac{2T}{E}\,dt = \frac{1}{E}\delta \int_{t_1}^{t_2} 2T dt,$$

(7)

we are led directly to Hamilton's principle in the form given by Maupertuis (where the time integral is to be taken with the usual grain of salt, *i.e.* $T+V=E=$ constant, even during the variation). The "rays", *i.e.* the orthogonal trajectories of the wave surfaces, are therefore the *paths* of the system for the value $E$ of the energy, in agreement with the well-known system of equations

$$p_k = \frac{\partial W}{\partial q_k},$$

(8)

which states, that a set of system paths can be derived from each special function of action, just like a fluid motion from its velocity potential.[1] (The momenta $p_k$ form the covariant velocity vector, which equations (8) assert to be equal to the gradient of the function of action.)

Although in these deliberations on wave surfaces we speak of velocity of propagation and Huygens' principle, we must regard the analogy as one between mechanics and *geometrical* optics, and not

---

[1] See especially A. Einstein, *Verh. d. D. Physik. Ges.* 19, pp. 77, 82, 1917. The framing of the quantum conditions here is the most akin, out of all the older attempts, to the present one. De Broglie has returned to it.

physical or *undulatory* optics. For the idea of "rays", which is the essential feature in the mechanical analogy, belongs to *geometrical* optics; it is only clearly defined in the latter. Also Fermat's principle can be applied in geometrical optics without going beyond the idea of index of refraction. And the system of *W*-surfaces, regarded as wave surfaces, stands in a somewhat looser relationship to mechanical motion, inasmuch as the image point of the mechanical system in no wise moves along the ray with the wave velocity *u*, but, on the contrary, its velocity (for constant *E*) is proportional to $\frac{1}{u}$. It is given directly from (3) as

$$v = \frac{ds}{dt} = \sqrt{2T} = \sqrt{2(E - V)}. \tag{9}$$

This non-agreement is obvious. Firstly, according to (8), the system's point velocity is *great* when grad *W* is great, *i.e.* where the *W*-surfaces are closely crowded together, *i.e.* where *u* is small. Secondly, from the definition of *W* as the time integral of the Lagrange function, *W alters* during the motion (by $(T - V)dt$ in the time *dt*), and so the image point *cannot* remain continuously in contact with the same *W*-surface.

And important ideas in wave theory, such as amplitude, wave length, and frequency—or, speaking more generally, the wave *form*— do not enter into the analogy at all, as there exists no mechanical parallel; even of the wave function itself there is no mention beyond that *W* has the meaning of the *phase* of the waves (and this is somewhat hazy owing to the wave *form* being undefined).

If we find in the whole parallel merely a satisfactory means of contemplation, then this defect is not disturbing, and we would regard any attempt to supply it as idle trifling, believing the analogy to be precisely with *geometrical*, or at furthest, with a very primitive form of wave optics, and not with the fully developed undulatory optics. That geometrical optics is only a rough approximation for *Light* makes no difference. To *preserve* the analogy on the further development of the optics of *q*-space on the lines of wave theory, we must take good care not to depart markedly from the limiting case of geometrical optics, *i.e.* must choose[1] the *wave length* sufficiently small, *i.e.* small compared with all the path dimensions. Then the additions do not teach anything new; the picture is only draped with superfluous ornaments.

So we might think to begin with. But even the first attempt at the development of the analogy to the wave theory leads to such striking results, that a quite different suspicion arises: *we know to-day, in fact, that our classical mechanics fails for very small dimensions of the path and for very great curvatures.* Perhaps this failure is in strict analogy with the failure of geometrical optics, *i.e.* "the optics of infinitely small wave lengths", that becomes evident as soon as the obstacles or apertures are no longer great compared with the real, finite, wave length. Perhaps our classical mechanics is the *complete* analogy of geometrical optics and as such is wrong and not in agreement with reality; it fails whenever the radii of curvature and dimensions of the path are no longer great compared with a certain wave length, to which, in $q$-space, a real meaning is attached. Then it becomes a question of searching[2] for an undulatory mechanics, and the most obvious way is the working out of the Hamiltonian analogy on the lines of undulatory optics.

## § 2. "Geometrical" and "Undulatory" Mechanics

We will at first assume that it is fair, in extending the analogy, to imagine the above-mentioned wave system as consisting of *sine* waves. This is the simplest and most obvious case, yet the *arbitrariness*, which arises from the *fundamental significance* of this assumption, must be emphasized. The wave function has thus only to contain the time in the form of a factor, sin ( . . . ), where the argument is a linear function of *W*. The coefficient of *W* must have the dimensions of the reciprocal of action, since *W* has those of action and the phase of a sine has zero dimensions. We assume that it is quite universal, *i.e.* that it is not only

---

[1] Cf. for the optical case, A. Sommerfeld and Iris Runge, *Ann. d. Phys.* 35, p. 290, 1911. There (in the working out of an oral remark of P. Debye), it is shown, how the equation of *first* order and *second* degree for the *phase* ("Hamiltonian equation") may be accurately derived from the equation of the *second* order and *first* degree for the *wave function* ("wave equation"), in the limiting case of vanishing wave length.
[2] Cf. A. Einstein, *Berl. Ber.* p. 9 *et seq.*, 1925.

independent of $E$, but also of the nature of the mechanical system. We may then at once denote it by $\frac{2\pi}{h}$. The time factor then is

$$\sin\left(\frac{2\pi\,W}{h} + \text{const.}\right) = \sin\left(-\frac{2\pi\,Et}{h} + \frac{2\pi\,S\,(q_k)}{h} + \text{const.}\right).$$

$$(10)$$

Hence the *frequency* $\nu$ of the waves is given by

$$\nu = \frac{E}{h},$$

$$(11)$$

Thus we get the frequency of the $q$-space waves to be proportional to the energy of the system, in a manner which is not markedly artificial.[1] This is only true of course if $E$ is absolute and not, as in classical mechanics, indefinite to the extent of an additive constant. By (6) and (11) the *wave length* is *independent* of this additive constant, being

$$\lambda = \frac{u}{\nu} = \frac{h}{\sqrt{2\,(E - V)}},$$

$$(12)$$

and we know the term under the root to be double the kinetic energy. Let us make a preliminary rough comparison of this wave length with the dimensions of the orbit of a hydrogen electron as given by classical mechanics, taking care to notice that a "step" in $q$-space has not the dimensions of length, but length multiplied by the square root of mass, in consequence of (3). $\lambda$ has similar dimensions. We have therefore to divide $\lambda$ by the dimension of the orbit, $a$.cm., say, and by the square root of $m$, the mass of the electron. The quotient is of the order of magnitude of

$$\frac{h}{mva},$$

where $v$ represents for the moment the electron's velocity (cm./sec.). The denominator $mva$ is of the order of the mechanical moment of momentum, and this is at least of the order of $10^{-27}$ for Kepler orbits, as can be calculated from the values of electronic charge and mass independently of all quantum theories. We thus obtain the correct order for the *limit of the approximate region of validity of classical*

[1] In Part I. this appeared merely as an approximate equation, derived from a pure speculation.

*mechanics*, if we identify our constant $h$ with Planck's quantum of action—and this is only a preliminary attempt.

If in (6), $E$ is expressed by means of (11) in terms of $\nu$, then we obtain

$$u = \frac{h\nu}{\sqrt{2(h\nu - V)}}. \tag{6'}$$

The dependence of the wave velocity on the energy thus becomes a particular kind of dependence on the *frequency, i.e.* it becomes a *law of dispersion* for the waves. This law is of great interest. We have shown in § 1 that the wandering wave surfaces are only loosely connected with the motion of the system point, since their velocities are not equal and cannot be equal. According to (9), (11), and (6') the system's velocity $v$ has thus also a concrete significance for the wave. We verify at once that

$$v = \frac{d\nu}{d\left(\dfrac{\nu}{u}\right)}, \tag{13}$$

*i.e.* the velocity of the system point is that of a *group of waves*, included within a small range of frequencies (signal-velocity). We find here again a theorem for the "phase waves" of the electron, which M. de Broglie had derived, with essential reference to the relativity theory, in those fine researches,[1] to which I owe the inspiration for this work. We see that the theorem in question is of wide generality, and does not arise solely from relativity theory, but is valid for every conservative system of ordinary mechanics.

We can utilise this fact to institute a much more innate connection between wave propagation and the movement of the representative point than was possible before. We can attempt to build up a wave group which will have relatively small dimensions in every direction. Such a wave group will then presumably obey the same laws of motion as a single image point of the mechanical system. It will then give, so to speak, an *equivalent* of the image point, so long as we can look on it as being approximately confined to a point, *i.e.* so long as we can neglect

[1] L. de Broglie, *Ann. de Physique* (10) 3, p. 22, 1925. (Thèses, Paris, 1924.)

any spreading out in comparison with the dimensions of the path of the system. This will only be the case when the path dimensions, and especially the radius of curvature of the path, are very great compared with the wave length. For, in analogy with ordinary optics, it is obvious from what ha. been said that not only must the dimensions of the wave group not be reduced below the order of magnitude of the wave length, but, on the contrary, the group must extend in all directions over a large number of wave lengths, if it is to be *approximately monochromatic*. This, however, must be postulated, since the wave group must move about as a whole with a definite group velocity and correspond to a mechanical system *of definite energy* (cf. equation 11).

So far as I see, such groups of waves can be constructed on exactly the same principle as that used by Debye[1] and von Laue[2] to solve the problem in ordinary optics of giving an exact analytical representation of a cone of rays or of a sheaf of rays. From this there comes a very interesting relation to that part of the Hamilton-Jacobi theory not described in § 1, viz. the well-known derivation of the equations of motion in integrated form, by the differentiation of a complete integral of the Hamilton-Jacobi equation with respect to the constants of integration. As we will see immediately, the system of equations called after Jacobi is equivalent to the statement: the image point of the mechanical system continuously corresponds to *that* point, where a certain continuum of wave trains coalesces in *equal phase*.

In optics, the representation (strictly on the wave theory) of a "sheaf of rays" with a sharply defined finite cross-section, which proceeds to a focus and then diverges again, is thus carried out by Debye. A *continuum* of *plane* wave trains, each of which alone would fill the *whole* space, is superposed. The *continuum* is produced by letting the wave normal vary throughout the given solid angle. The waves then destroy one another almost completely by interference outside a certain double cone; they represent exactly, on the wave theory, the desired limited sheaf of rays and also the diffraction phenomena,

[1] P. Debye, *Ann. d. Phys.* 30, p. 755, 1909.
[2] M. v. Laue, *idem* 44, p. 1197 (§ 2), 1914.

necessarily occasioned by the limitation. We can represent in this manner an *infinitesimal* cone of rays just as well as a finite one, if we allow the wave normal of the group to vary only inside an infinitesimal solid angle. This has been utilised by von Laue in his famous paper on the degrees of freedom of a sheaf of rays.[1] Finally, instead of working with waves, hitherto tacitly accepted as purely monochromatic, we can also allow the *frequency* to vary within an infinitesimal interval, and by a suitable distribution of the amplitudes and phases can confine the disturbance to a region which is relatively small in the longitudinal direction also. So we succeed in representing analytically a "parcel of energy" of relatively small dimensions, which travels with the speed of light, or when dispersion occurs, with the group velocity. Thereby is given the instantaneous *position* of the parcel of energy—if the detailed structure is not in question—in a very plausible way as that point of space where *all* the superposed plane waves meet in *exactly* agreeing phase.

We will now apply these considerations to the $q$-space waves. We select, at a definite time $t$, a definite point $P$ of $q$-space, through which the parcel of waves passes in a given direction $R$, at that time. In addition let the mean frequency $\nu$ or the mean $E$-value for the packet be also given. These conditions correspond exactly to postulating that at a given time the mechanical system is starting from a given configuration with given velocity components. (Energy *plus* direction is equivalent to velocity components.)

In order to carry over the optical construction, we require firstly *one* set of wave surfaces with the desired frequency, *i.e. one* solution of the Hamilton-Jacobi equation (1′) for the given $E$-value. This solution, $W$, say, is to have the following property: the surface of the set which passes through $P$ at time $t$, which we may denote by

$$W = W_0, \tag{14}$$

must have its normal at $P$ in the prescribed direction $R$. But this is still not enough. We must be able to vary to an infinitely small extent

[1] *Loc. cit.*

this set of waves $W$ in an $n$-fold manner ($n$ = number of degrees of freedom), so that the wave normal will sweep out an infinitely small $(n-1)$ dimensional space angle at the point $P$, and so that the frequency $\frac{E}{h}$ will vary in an infinitely small *one*-dimensional region, whereby care is taken that all members of the infinitely small $n$-dimensional continuum of sets of waves meet together at time $t$ in the point $P$ in exactly agreeing phase. Then it is a question of finding at any other time *where* that point lies at which this agreement of phases occurs.

To do this, it will be sufficient if we have at our disposal a solution $W$ of the Hamilton-Jacobi equation, which is dependent not only on the constant $E$, here denoted by $\alpha_1$, but also on $(n-1)$ additional constants $\alpha_2, \alpha_3 \ldots \alpha_n$, in such a way that it cannot be written as a function of less than $n$ combinations of these $n$ constants. For then we can, firstly, bestow on $\alpha_1$ the value prescribed for $E$, and, secondly, define $\alpha_2, \alpha_3 \ldots \alpha_n$, so that the surface of the set passing through the point $P$ has at $P$ the prescribed normal direction. Henceforth we understand by $\alpha_1, \alpha_2 \ldots \alpha_n$, these values, and take (14) as the surface of *this* set, which passes through the point $P$ at time $t$. Then we consider the *continuum of sets* which belongs to the $\alpha_k$-values of an adjacent infinitesimal $\alpha_k$-region. A member of this continuum, *i.e.* therefore *a set*, will be given by

$$W + \frac{\partial W}{\partial \alpha_1} d\alpha_1 + \frac{\partial W}{\partial \alpha_2} d\alpha_2 + \cdots + \frac{\partial W}{\partial \alpha_n} d\alpha_n = \text{const.} \qquad (15)$$

for a *fixed* set of values of $d\alpha_1, d\alpha_2 \ldots d\alpha_n$, and varying constant. That member of *this set, i.e.* therefore that single surface, which goes through $P$ at time $t$ will be defined by the following choice of the const.,

$$W + \frac{\partial W}{\partial \alpha_1} d\alpha_1 + \cdots + \frac{\partial W}{\partial \alpha_n} d\alpha_n = W_0 + \left(\frac{\partial W}{\partial \alpha_1}\right)_0$$

$$d\alpha_1 + \cdots + \left(\frac{\partial W}{\partial \alpha_n}\right)_0 = d\alpha_n, \qquad (15')$$

where $\left(\frac{\partial W}{\partial \alpha_1}\right)_0$, etc , are the *constants* obtained by substituting in the differential coefficients the co-ordinates of the point $P$ and the value $t$ of the time (which latter really only occurs in $\frac{\partial W}{\partial \alpha_1}$).

The surfaces (15′) for all possible sets of values of $d\alpha_1$, $d\alpha_2 \ldots d\alpha_n$, form on their part *a set*. They all go through the point $P$ at time $t$, their wave normals continuously sweep out a little $(n - 1)$ dimensional solid angle and, moreover, their $E$-parameter also varies within a small region. The set of surfaces (15′) is so formed that each of the sets (15) supplies *one* representative to (15′), namely, that member which passes through $P$ at time $t$.

We will now assume that the phase angles of the wave functions which belong to the sets (15) happen to agree precisely for those representatives which enter the set (15′). They agree therefore at time $t$ at the point $P$.

We now ask: Is there, at *any arbitrary time*, a point where all surfaces of the set (15′) cut one another, *and in which, therefore*, all the wave functions which belong to the sets (15) agree in phase? The answer is: *There exists* a point of agreeing phase, but it is *not* the common intersection of the surfaces of set (15′), for such does *not* exist at any subsequent arbitrary time. Moreover, the point of phase agreement arises in *such a way* that the sets (15) *continuously exchange their representatives given to* (15′).

That is shown thus. There must hold

$$W = W_0, \quad \frac{\partial W}{\partial \alpha_1} = \left(\frac{\partial W}{\partial \alpha_1}\right)_0, \quad \frac{\partial W}{\partial \alpha_2} = \left(\frac{\partial W}{\partial \alpha_2}\right)_0 \cdots \frac{\partial W}{\partial \alpha_n} = \left(\frac{\partial W}{\partial \alpha_n}\right)_0,$$

(16)

simultaneously for the common meeting point of all members of (15′) at any time, because the $d\alpha_1$'s are arbitrary within a small region. In these $n + 1$ equations, the right-hand sides are constants, and the left are functions of the $n + 1$ quantities $q_1, q_2, \ldots q_n, t$. The equations are satisfied by the initial system of values, *i.e.* by the co-ordinates of $P$ and the initial time $t$. For another arbitrary value of $t$, they will have *no* solutions in $q_1 \ldots q_n$, but will *more than define* the system of these $n$ quantities.

We may proceed, however, as follows. Let us leave the first equation, $W = W_0$, aside at first, and define the $q_k$'s as functions of the time and the constants according to the remaining $n$ equations. Let this point be called $Q$. By it, naturally, the *first* equation will *not* be satisfied, but the left-hand side will differ from the right by a certain value. If we go back to the derivation of system (16) from (15'), what we have just said means that though $Q$ is not a common point for the set of surfaces (15'), it is so, however, for a set which results from (15'), if we alter the right-hand side of equation (15') by an amount which is constant for all the surfaces. Let this new set be (15''). For it, therefore, $Q$ is a common point. The new set results from (15'), as stated above, by an exchange of the representatives in (15'). This exchange is occasioned by the alteration of the constant in (15), *by the same amount*, for all representatives. Hence the *phase angle* is altered by the same amount for all representatives. The new representatives, *i.e.* the members of the set we have called (15''), which meet in the point $Q$, agree in phase angle just as the old ones did. This amounts therefore to saying:

The point $Q$ which is defined as a function of the time by the $n$ equations

$$\frac{\partial W}{\partial \alpha_1} = \left(\frac{\partial W}{\partial \alpha_1}\right)_0 , \ldots, \frac{\partial W}{\partial \alpha_n} = \left(\frac{\partial W}{\partial \alpha_n}\right)_0 , \tag{17}$$

continues to be a point of agreeing phase for the whole aggregate of wave sets (15).

Of all the $n$-surfaces, of which $Q$ is shown by (17) to be the common point, only the first is variable; the others remain fixed (only the first of equations (17) contains the time). The $n - 1$ fixed surfaces determine the *path* of the point $Q$ as their line of intersection. It is easily shown that this line is the orthogonal trajectory of the set $W = $ const. For, by hypothesis, $W$ satisfies the Hamilton-Jacobi equation (1') identically in $\alpha_1, \alpha_2 \ldots \alpha_n$. If we now differentiate the Hamilton-Jacobi equation with respect to $\alpha_k (k = 2, 3, \ldots n)$, we get the statement that the normal to a surface, $\frac{\partial W}{\partial \alpha_k} = $ const., is *perpendicular*, at every point on it, to the normal of the surface, $W = $ const., which passes through that point, *i.e.* that each of the two surfaces *contains* the normal to the other. If the line of intersection of the $n - 1$ fixed surfaces (17) has no branches, as is generally the case, then must each line element of the intersection, as the *sole common* line element of the $n - 1$ surfaces, coincide with the normal of the *W*-surface, passing through the same point, *i.e.* the line of intersection is the orthogonal trajectory of the W-surfaces. Q.E.D.

We may sum up the somewhat detailed discussion, which has led us to equations (17), in a much shorter or (so to speak) shorthand fashion, as follows: *W* denotes, apart from a universal constant $\left(\frac{1}{h}\right)$, the phase angle of the wave function. If we now deal not merely with *one*, but with a continuous manifold of wave systems, and if these are continuously arranged by means of any continuous parameters $\alpha_i$, then the equations $\frac{\partial W}{\partial \alpha_i} = $ const. express the fact that all infinitely adjacent individuals (wave systems) of this manifold agree in phase. These equations therefore define the geometrical locus of the points of agreeing phase. If the equations are sufficient, this locus shrinks to one point; the equations then define *the point* of phase agreement as a function of the time.

Since the system of equations (17) agrees with the known second system of equations of Jacobi, we have thus shown:

*The point of phase agreement for certain infinitesimal manifolds of wave systems, containing* n *parameters, moves according to the same laws as the image point of the mechanical system.*

I consider it a very difficult task to give an exact proof that the superposition of these wave systems really produces a noticeable disturbance in only a relatively small region surrounding the point of phase agreement, and that everywhere else they practically destroy one another through interference, or that the above statement turns out to be true at least for a suitable choice of the amplitudes, and possibly for a special choice of the *form* of the wave surfaces. I will advance the physical hypothesis, which I wish to attach to what is to be proved, without attempting the proof. The latter will only be worth while if the hypothesis stands the test of trial *and* if its application should *require* the exact proof.

On the other hand, we may be sure that the region to which the disturbance may be confined still contains in all directions a great number of wave lengths. This is directly evident, firstly, because so long as we are only a *few* wave lengths distant from the point of phase agreement, then the agreement of phase is hardly disturbed, as the interference is still almost as favourable as it is at the point itself. Secondly, a glance at the three-dimensional Euclidean case of ordinary optics is sufficient to assure us of this general behaviour.

What I now categorically conjecture is the following:

The true mechanical process is realised or represented in a fitting way by the *wave processes* in *q*-space, and not by the motion of *image points* in this space. The study of the motion of image points, which is the object of classical mechanics, is only an approximate treatment, and has, as such, just as much justification as geometrical or "ray" optics has, compared with the true optical process. A macroscopic mechanical process will be portrayed as a wave signal of the kind described above, which can approximately enough be regarded as confined to a point compared with the geometrical structure of the path. We have seen that the same laws of motion hold exactly for such a signal or group of waves as are advanced by classical mechanics for the motion of the image point. This manner of treatment, however, loses all meaning where the structure of the path is no longer very large compared with the wave length or indeed is comparable with it. Then we *must* treat the matter strictly on the wave theory, *i.e.* we must proceed from the *wave equation* and not from the fundamental equations of mechanics, in order to form a picture of the manifold of the possible processes. These latter equations are just as useless for the elucidation of the micro-structure of mechanical processes as geometrical optics is for explaining the *phenomena of diffraction.*

Now that a certain interpretation of this micro-structure has been successfully obtained as an addition to classical mechanics, although admittedly under new and very artificial assumptions, an interpretation bringing with it practical successes of the highest importance, it seems to me very significant that these theories—I refer to the forms of quantum theory favoured by Sommerfeld, Schwarzschild, Epstein, and others—bear a very close relation to the Hamilton-Jacobi equation and the theory of its solution, *i.e.* to that form of classical mechanics which already points out most clearly the true undulatory character of mechanical processes. The Hamilton-Jacobi equation corresponds to Huygens' Principle (in its old simple form, not in the form due to Kirchhoff). And just as this, supplemented by some rules which, are not intelligible in geometrical optics (Fresnel's construction of zones), can explain to a great extent the phenomena of diffraction, so light can be thrown on the processes in the atom by the theory of the action-function. But we inevitably became involved in irremovable contradictions if we tried, as was very natural, to maintain also the idea of *paths of systems* in these processes; just as we find the tracing of the course of a *light ray* to be meaningless, in the neighbourhood of a diffraction phenomenon.

We can argue as follows. I will, however, not yet give a conclusive picture of the actual process, which positively cannot be arrived at from this starting-point but only from an investigation of the wave equation; I will merely illustrate the matter qualitatively. Let us think of a wave group of the nature described above, which in some way gets into a small closed "path", whose dimensions are of the order of the wave length, and therefore *small* compared with the dimensions of the wave group itself. It is clear that then the "system path" in the sense of classical mechanics, *i.e.* the path of the point of exact phase agreement, will completely lose its prerogative, because there exists a whole continuum of points before, behind, and near the particular point, in which there is almost as complete phase agreement, and which describe totally different "paths". In other words, the wave group not only fills the whole path domain all at once but also stretches far beyond it in all directions.

In *this* sense do I interpret the "phase waves" which, according to de Broglie, accompany the path of the electron; in the sense, therefore, that no special meaning is to be attached to the electronic path itself (at any rate, in the interior of the atom), and still less to the position of the electron on its path. And in this sense I explain the conviction, increasingly evident to-day, *firstly*, that real meaning has to be denied to the *phase* of electronic motions in the atom; *secondly*, that we can never assert that the electron at a definite instant is to be found on *any definite one* of the quantum paths, specialised by the quantum conditions; and *thirdly*, that the true laws of quantum mechanics do not consist of definite rules for the *single path*, but that in these laws the elements of the whole manifold of paths of a system are bound together by equations, so that apparently a certain reciprocal action exists between the different paths.[1]

It is not incomprehensible that a careful analysis of the experimentally known quantities should lead to assertions of this kind, if the experimentally known facts are the outcome of such a structure of the real process as is here represented. All these assertions systematically contribute to the relinquishing of the ideas of "place of the electron" and "path of the electron". If these are not given up, contradictions remain. This contradiction has been so strongly felt that it has even been doubted whether what goes on in the atom could ever be described within the scheme of space and time. From the philosophical standpoint, I would consider a conclusive decision in this sense as equivalent to a complete surrender. For we cannot really alter our manner of thinking in space and time, and what we cannot comprehend within it we cannot understand at all. There *are* such things—but I do not believe that atomic structure is one of them. From our standpoint, however, there is no reason for such doubt, although or rather *because* its appearance is extraordinarily comprehensible. So might a person versed in geometrical optics, after many attempts to explain diffraction phenomena by means of the idea of the ray (trustworthy for his macroscopic optics), which always came to nothing, at last think that the *Laws of Geometry* are not applicable to diffraction, since he continually finds that light rays, which he imagines as *rectilinear* and *independent* of each other, now suddenly

---

[1] Cf. especially the papers of Heisenberg, Born, Jordan, and Dirac quoted later, and further N. Bohr, *Die Naturwissenschaften*, January 1926.

show, even in homogeneous media, the most remarkable *curvatures*, and obviously *mutually influence* one another. I consider this analogy as *very* strict. Even for the unexplained *curvatures*, the analogy in the atom is not lacking—think of the "non-mechanical force", devised for the explanation of anomalous Zeeman effects.

In what way now shall we have to proceed to the undulatory representation of mechanics for those cases where it is necessary? We must start, not from the fundamental equations of mechanics, but from a wave equation for $q$-space and consider the manifold of processes possible *according to it*. The wave equation has not been explicitly used or even put forward in this communication. The only datum for its construction is the *wave velocity*, which is given by (6) or (6') as a function of the mechanical energy parameter or frequency respectively, and by this datum the wave equation is evidently not uniquely defined. It is not even decided that it must be definitely of the second order. Only the striving for simplicity leads us to try this to begin with. We will then say that for the wave function $\psi$ we have

$$\text{div grad } \psi - \frac{1}{u^2}\ddot{\psi} = 0, \tag{18}$$

valid for all processes which only depend on the time through a factor $e^{2\pi i \nu t}$. Therefore, considering (6), (6'), and (11), we get, respectively,

$$\text{div grad } \psi + \frac{8\pi^2}{h^2}\,(h\nu - V)\,\psi = 0, \tag{18'}$$

and

$$\text{div grad } \psi + \frac{8\pi^2}{h^2}\,(E - V)\,\psi = 0. \tag{18''}$$

The differential operations are to be understood with regard to the line element (3). But even under the postulation of second order, the above is not the only equation consistent with (6). For it is possible to generalize by replacing div grad $\psi$ by

$$f(q_k)\,\text{div}\left(\frac{1}{f(q_k)}\text{grad } \psi\right), \tag{19}$$

where $f$ may be an arbitrary function of the $q$'s, which must depend in some plausible way on $E$, $V(q_k)$, and the coefficients of the line element (3). (Think, *e.g.*, of $f = u$.) Our postulation is again dictated by the striving for simplicity, yet I consider in this case that a wrong deduction is not out of the question.[1]

The substitution of a *partial* differential equation for the equations of dynamics in atomic problems appears at first sight a very doubtful procedure, on account of the multitude of solutions that such an equation possesses. Already classical dynamics had led not just to one solution but to a much too extensive manifold of solutions, viz. to a continuous set, while all experience seems to show that only a discrete number of these solutions is realised. The problem of the quantum theory, according to prevailing conceptions, is to select by means of the "quantum conditions" that discrete set of actual paths out of the continuous set of paths possible according to classical mechanics. It seems to be a bad beginning for a new attempt in this direction if the number of possible solutions has been *increased* rather than diminished.

It is true that the problem of classical dynamics also allows itself to be presented in the form of a *partial* equation, namely, the Hamilton-Jacobi equation. But the manifold of solutions of the problem does not correspond to the manifold of solutions of that equation. An arbitrary "complete" solution of the equation solves the mechanical problem *completely*; any *other* complete solution yields the same paths —they are only contained in another way in the manifold of paths.

Whatever the fear expressed about taking equation (18) as the foundation of atomic dynamics comes to, I will not positively assert that no further additional definitions will be required with it. But these will probably no longer be of such a completely strange and incomprehensible nature as the previous "quantum conditions", but will be of the type that we are accustomed to find in physics with a partial differential equation as initial or boundary conditions. They will be, in no way, *analogous* to the quantum conditions—because in all cases of classical dynamics, which I have investigated up till

---

[1] The introduction of $f(q_k)$ means that not only the "density" but also the "elasticity" varies with the position.

now, it turns out that equation (18) *carries within itself the quantum conditions.* It distinguishes in certain cases, and indeed in those where experience demands it, *of itself,* certain frequencies or energy levels as those which alone are possible for stationary processes, without any further assumption, other than the almost obvious demand that, as a physical quantity, the function $\psi$ must be single-valued, finite, and continuous throughout configuration space.

Thus the fear expressed is transformed into its contrary, in any case in what concerns the energy levels, or let us say more prudently, the frequencies. (For the question of the "vibrational energy" stands by itself; we must not forget that it is only in the one electron problem that the interpretation as a vibration in real three-dimensional space is immediately suggested.) The definition of the quantum levels *no longer takes place* in two separated stages: (1) Definition of all paths dynamically possible. (2) *Discarding* of the greater part of those solutions and the selection of a few by special postulations; on the contrary, the quantum levels are *at once* defined as the *proper values* of equation (18), which *carries in itself its natural boundary conditions.*

As to how far an analytical simplification will be effected in this way in more complicated cases, I have not yet been able to decide. I should, however, expect so. Most of the analytical investigators have the feeling that in the two-stage process, described above, there must be yielded in (1) the solution of a more complicated problem than is really necessary for the final result: energy as a (usually) very simple rational function of the quantum numbers. Already, as is known, the application of the Hamilton-Jacobi method creates a great simplification, as the actual calculation of the mechanical solution is avoided. It is sufficient to evaluate the integrals, which represent the momenta, merely for a closed complex path of integration instead of for a variable upper limit, and this gives much less trouble. Still the complete solution of the Hamilton-Jacobi equation must really be known, *i.e.* given by quadratures, so that the integration of the mechanical problem must in principle be effected for arbitrary initial values. In seeking for the proper values of a differential equation, we must usually, in practice, proceed thus. We seek the solution, firstly, without regard to boundary or continuity conditions, and from the form of the solution then pick out those values of the parameters, for which the solution satisfies the given conditions. Part I. supplies an example of this. We see by this example also, however—what is typical of proper value problems—that the solution was only given *generally* in an extremely inaccessible analytical form [equation (12) *loc. cit.*], but that it is extraordinarily simplified for those proper values belonging to the "natural boundary condition". I am not well enough informed to say whether *direct* methods have now been worked out for the calculation of the proper values. This is known to be so for the distribution of proper values of *high order*. But this limiting case is *not* of interest here; it corresponds to the classical, macroscopic mechanics. For spectroscopy and atomic physics, in general just the *first* 5 or 10 proper values will be of interest; even the *first alone* would be a great result—it defines the *ionisation potential*. From the idea, definitely outlined, that every problem of proper values allows itself to be treated as one of maxima and minima without direct reference to the differential equation, it appears to me very probable that direct methods will be found for the calculation, at least approximately, of the proper values, as soon as *urgent* need arises. At least it should be possible to test in individual cases whether the proper values, *known* numerically to all desired accuracy through spectroscopy, *satisfy* the problem or not.

I would not like to proceed without mentioning here that at the present time a research is being prosecuted by Heisenberg, Born, Jordan, and other distinguished workers,[1] to remove the quantum difficulties, which has already yielded such noteworthy success that it cannot be doubted that it contains at least a part of the truth. In its *tendency*, Heisenberg's attempt stands very near the present one, as we have already mentioned. In its method, it is so totally different that I have not yet succeeded in finding the connecting link.

---

[1] W. Heisenberg, *Ztschr. f. Phys.* 33, p. 879, 1925; M. Born and P. Jordan, *ibid.* 34, p. 858, 1925; M. Born, W. Heisenberg, and P. Jordan, *ibid.* 35, p. 557 1926; P. Dirac, *Proc. Roy. Soc.*, London, 109, p. 642, 1925.

I am distinctly hopeful that these two advances will not fight against one another, but on the contrary, just because of the extraordinary difference between the starting-points and between the methods, that they will supplement one another and that the one will make progress where the other fails. The strength of Heisenberg's programme lies in the fact that it promises to give the *line-intensities*, a question that we have not approached as yet. The strength of the present attempt—if I may be permitted to pronounce thereon—lies in the guiding, physical point of view, which creates a bridge between the macroscopic and microscopic mechanical processes, and which makes intelligible the outwardly different modes of treatment which they demand. For me, personally, there is a special charm in the conception, mentioned at the end of the previous part, of the emitted frequencies as "beats", which I believe will lead to an intuitive understanding of the intensity formulae.

## § 3. APPLICATION TO EXAMPLES

We will now add a few more examples to the Kepler problem treated in Part I., but they will only be of the very simplest nature, since we have provisionally confined ourselves to *classical* mechanics, with no magnetic field.[1]

### 1. *The Planck Oscillator. The Question of Degeneracy*

Firstly we will consider the one-dimensional oscillator. Let the coordinate $q$ be the displacement multiplied by the square root of the mass. The two forms of the kinetic energy then are

$$\bar{T} = \tfrac{1}{2}\dot{q}^2, \; T = \tfrac{1}{2}p^2. \tag{20}$$

The potential energy will be

$$V(q) = 2\pi^2 v_0^2 q^2, \tag{21}$$

---

[1] In relativity mechanics and taking a magnetic field into account the statement of the Hamilton-Jacobi equation becomes more complicated. In the case of a single electron, it asserts that the *four-dimensional* gradient of the action function, *diminished* by a given vector (the four-potential), has a constant value. The translation of this statement into the language of the wave theory presents a good many difficulties.

where $\nu_0$ is the proper frequency in the mechanical sense. Then equation (18) reads in this case

$$\frac{d^2\psi}{dq^2} + \frac{8\pi^2}{h^2}(E - 2\pi^2\nu_0^2q^2)\psi = 0. \qquad (22)$$

For brevity write

$$a = \frac{8\pi^2 E}{h^2}, b = \frac{16\pi^4\nu_0^2}{h^2}. \qquad (23)$$

Therefore

$$\frac{d^2\psi}{dq^2} + (a - bq^2)\psi = 0. \qquad (22')$$

Introduce as independent variable

$$x = q\sqrt[4]{b}, \qquad (24)$$

and obtain

$$\frac{d^2\psi}{dx^2} + \left(\frac{a}{\sqrt{b}} - x^2\right)\psi = 0. \qquad (22'')$$

The proper values and functions of this equation are *known*.[1] The proper values are, with the notation used here,

$$\frac{a}{\sqrt{b}} = 1, 3, 5 \ldots (2n + 1) \ldots \qquad (25)$$

The functions are the *orthogonal functions of Hermite*,

$$e^{-\frac{x^2}{2}} H_n(x). \qquad (26)$$

$H_n(x)$ means the $n$th Hermite polynomial, which can be defined as

$$H_n(x) = (-1)^n e^{x^2} \frac{d^n e^{-x^2}}{dx^n}, \qquad (27)$$

or explicitly by

$$H_n(x) = (2x)^n - \frac{n(n-1)}{1!}(2x)^{n-2}$$

$$+ \frac{n(n-1)(n-2)(n-3)}{2!}(2x)^{n-4} - + \cdots \qquad (27')$$

---

[1] Cf. Conrant-Hilbert, *Methods of Mathematical Physics*, i. (Berlin, Springer, 1924), v. § 9, p. 261, eqn. 43, and further ii. § 10, 4, p. 76.

The first of these polynomials are

$$H_0(x) = 1 \qquad\qquad H_1(x) = 2x$$
$$H_2(x) = 4x^2 - 2 \qquad\qquad H_3(x) = 8x^3 - 12x \qquad (27'')$$
$$H_4(x) = 16x^4 - 48x^2 + 12 \ldots$$

Considering next the proper values, we get from (25) and (23)

$$E_n = \frac{2n+1}{2} h v_0; \qquad n = 0, 1, 2, 3, \ldots \qquad (25')$$

Thus as quantum levels appear so-called "half-integral" multiples of the "quantum of energy" peculiar to the oscillator, *i.e.* the *odd* multiples of $\frac{h v_0}{2}$. The intervals between the levels, which alone are important for the radiation, are the same as in the former theory. It is remarkable that our quantum levels are *exactly* those of Heisenberg's theory. In the theory of *specific heat* this deviation from the previous theory is not without significance. It becomes important first when the proper frequency $v_0$ *varies* owing to the dissipation of heat. Formally it has to do with the old question of the "zero-point energy", which was raised in connection with the choice between the first and second forms of Planck's Theory. By the way, the additional term $\frac{h v_0}{2}$ also influences the law of the *band-edges*.

The *proper functions* (26) become, if we reintroduce the original $q$ from (24) and (23),

$$\psi_n(q) = e^{-\frac{2\pi^2 v_0 q^2}{h}} H_n\left(2\pi q \sqrt{\frac{v_0}{h}}\right). \qquad (26')$$

Consideration of (27'') shows that the first function is a *Gaussian Error-curve*; the second vanishes at the origin and for $x$ positive corresponds to a "Maxwell distribution of velocities" in two dimensions, and is continued in the manner of an odd function for $x$ negative. The third function is even, is negative at the origin, and has two symmetrical zeros at $\pm\frac{1}{\sqrt{2}}$, etc. The curves can easily be sketched roughly and it is seen that the roots of consecutive polynomials *separate* one another. From (26') it is also seen that the characteristic points of the proper functions, such as half-breadth (for $n = 0$), zeros, and maxima, are, as regards order of magnitude, within the range of the classical vibration of the oscillator. For the classical *amplitude* of the *n*th vibration is readily found to be given by

$$q_n = \frac{\sqrt{E_n}}{2\pi \nu_0} = \frac{1}{2\pi} \sqrt{\frac{h}{\nu_0}} \sqrt{\frac{2n+1}{2}}. \tag{28}$$

Yet there is in general, as far as I see, no definite meaning that can be attached to the *exact* abscissa of the classical *turning points* in the graph of the proper function. It may, however, be conjectured, because the turning points have *this* significance for the phase space wave, that, at them, the square of the velocity of propagation becomes *infinite* and at greater distances becomes *negative*. In the differential equation (22), however, this only means the *vanishing* of the coefficient of $\psi$ and gives rise to no singularities.

I would not like to suppress the remark here (and it is valid quite generally, not merely for the oscillator), that nevertheless this vanishing and becoming imaginary of the velocity of propagation is something which is very characteristic. It is the analytical reason for the selection of definite proper values, merely through the condition that the function should remain finite. I would like to illustrate this further. A wave equation with a *real* velocity of propagation means just this: there is an *accelerated* increase in the value of the function at all those points where its value is *lower* than the average of the values at neighbouring points, and vice versa. Such an equation, if not immediately and lastingly as in case of the *equation for the conduction of heat*, yet in the course of time, causes a *levelling* of extreme values and does not permit at any point an excessive growth of the function. A wave equation with an *imaginary* velocity of propagation means the exact opposite: values of the function above the average of surrounding values experience an *accelerated increase* (or retarded decrease), and vice versa. We see, therefore, that a function represented by such an equation is in the greatest danger of growing beyond all bounds, and we must order matters skilfully to preserve it from this danger. The sharply defined proper values are just what makes this possible.

Indeed, we can see in the example treated in Part I. that the demand for sharply defined proper values immediately ceases as soon as we choose the quantity $E$ to be *positive*, as this makes the wave velocity real throughout all space.

After this digression, let us return to the oscillator and ask ourselves if anything is altered when we allow it two or more degrees of freedom (space oscillator, rigid body). If *different* mechanical proper frequencies ($\nu_0$-values) belong to the separate co-ordinates, then nothing is changed. $\psi$ is taken as the *product* of functions, each of a single co-ordinate, and the problem splits up into just as many separate problems of the type treated above as there are co-ordinates present. The proper functions are products of Hermite orthogonal functions, and the proper values of the whole problem appear as sums of those of the separate problems, taken in every possible combination. No proper value (for the whole system) is multiple, if we presume that there is no rational relation between the $\nu_0$-values.

If, however, there is such a relation, then the same manner of treatment is still *possible*, but it is certainly not *unique*. Multiple proper values appear and the "separation" can certainly be effected in other co-ordinates, *e.g.* in the case of the isotropic space oscillator in spherical polars.[1]

The proper values that we get, however, are certainly in each case exactly the same, at least in so far as we are able to prove the "completeness" of a system of proper functions, obtained in *one* way. We recognise here a complete parallel to the well-known relations which the method of the previous quantisation meets with in the case of *degeneracy*. Only in one point there is a not unwelcome formal difference. If we applied the Sommerfeld-Epstein quantum conditions *without* regard to a possible degeneracy then we always got the same energy levels, but reached different conclusions as to the paths permitted, according to the choice of co-ordinates.

[1] We are led thus to an equation in $r$, which may be treated by the method shown in the Kepler problem of Part I. Moreover, the *one*-dimensional oscillator leads to the same equation if $q^2$ be taken as variable. I originally solved the problem directly in *that* way. For the hint that it was a question of Hermite polynomials, I have to thank Herr E. Fues. The polynomial appearing in the Kepler problem (eqn. 18 of Part I.) is the $(2n+1)$th differential coefficient of the $(n+l)$th polynomial of Laguerre, as I subsequently found.

Now that is *not* the case here. Indeed we come to a completely different system of proper functions, if we, for example, treat the vibration problem corresponding to unperturbed Kepler motion in *parabolic* co-ordinates instead of the polars used in Part I. However, it is not just the *single proper vibration* that furnishes a *possible state of vibration*, but an arbitrary, finite or infinite, *linear aggregate* of such vibrations. And as such the proper functions found in any second way may always be represented; namely, they may be represented as linear aggregates of the proper functions found in an arbitrary way, provided the latter form a *complete* system.

The question of how the energy is really distributed among the proper vibrations, which has not been taken into account here up till now, will, of course, have to be faced some time. Relying on the former quantum theory, we will be disposed to assume that in the degenerate case only the energy of the set of vibrations belonging to one definite proper value must have a certain prescribed value, which in the non-degenerate case belongs to one single proper vibration. I would like to leave this question still *quite* open—and also the question whether the discovered "energy levels" are really energy steps of the *vibration process* or whether they *merely* have the significance of its frequency. If we accept the beat theory, then the meaning of energy levels is no longer necessary for the explanation of sharp emission frequencies.

## 2. *Rotator with Fixed Axis*

On account of the lack of potential energy and because of the *Euclidean* line element, this is the simplest conceivable example of vibration theory. Let $A$ be the moment of inertia and $\phi$ the angle of rotation, then we clearly obtain as the vibration equation

$$\frac{1}{A}\frac{d^2\psi}{d\phi^2} + \frac{8\pi^2 E}{h^2}\psi = 0, \tag{29}$$

which has the solution

$$\psi = \frac{\sin}{\cos}\left[\sqrt{\frac{8\pi^2 E A}{h^2}} \cdot \phi\right]. \tag{30}$$

Here the argument must be an *integral* multiple of $\phi$, simply because otherwise $\psi$ would neither be single-valued nor continuous throughout the range of the co-ordinate $\phi$, as we know $\phi + 2\pi$ has the same significance as $\phi$. This condition gives the well-known result

$$E_n = \frac{n^2 h^2}{8\pi^2 A} \tag{31}$$

in *complete* agreement with the former quantisation.

*No* meaning, however, can be attached to the result of the application to band spectra. For, as we shall learn in a moment, it is a peculiar fact that our theory gives *another* result for the rotator with *free* axis. *And this is true in general.* It is not allowable in the applications of wave mechanics, to think of the freedom of movement of the system as being more strictly limited, in order to simplify calculation, than it *actually is*, even when we know from the integrals of the mechanical equations that in a single movement certain definite freedoms are not made use of. For micro-mechanics, the fundamental system of mechanical equations is absolutely incompetent; the single paths with which it deals have now no separate existence. A wave process fills the *whole* of the phase space. It is well known that even the *number* of the dimensions in which a wave process takes place is very significant.

### 3. *Rigid Rotator with Free Axis*

If we introduce as co-ordinates the polar angles $\theta$, $\phi$ of the radius from the nucleus, then for the kinetic energy as a function of the momenta we get

$$T = \frac{1}{2A} \left( p_{\theta^2} + \frac{p_{\phi^2}}{\sin^2 \theta} \right). \tag{32}$$

According to its form this is the kinetic energy of a particle constrained to move on a spherical surface. The Laplacian operator is thus simply that part of the spatial Laplacian operator which depends on the polar angles, and the vibration equation (18″) takes the following form,

$$\frac{1}{\sin \theta} \frac{\partial}{\partial \theta} \left( \sin \theta \frac{\partial \psi}{\partial \theta} \right) + \frac{1}{\sin^2 \theta} \frac{\partial^2 \psi}{\partial \phi^2} + \frac{8\pi^2 AE}{h^2} \psi = 0. \tag{33}$$

The postulation that $\psi$ should be single-valued and continuous on the spherical surface leads to the proper value condition

$$\frac{8\pi^2 A}{h^2} E = n(n+1); \quad n = 0, 1, 2, 3, \ldots \tag{34}$$

The proper functions are known to be spherical surface harmonics. The energy levels are, therefore,

$$E_n = \frac{n(n+1)h^2}{8\pi^2 A}; \quad n = 0, 1, 2, 3, \ldots \tag{34'}$$

This definition is different from all previous statements (except perhaps that of Heisenberg?). Yet, from various arguments from experiment we were led to put "half-integral" values for $n$ in formula (31). It is easily seen that (34') gives practically the same as (31) with half-integral values of $n$. For

$$n(n+1) = \left(n + \frac{1}{2}\right)^2 - \frac{1}{4}.$$

The discrepancy consists only of a small additive constant; the level *differences* in (34') are the same as are got from "half-integral quantisation". This is true also for the application to short-wave bands, where the moment of inertia is not the same in the initial and final states, on account of the "electronic jump". For at most a small constant additional part comes in for *all* lines of a band, which is swamped in the large "electronic term" or in the "nuclear vibration term". Moreover, our previous analysis does not permit us to speak of this small part in any more definite way than as, say,

$$\frac{1}{4} \frac{h^2}{8\pi^2} \left(\frac{1}{A} - \frac{1}{A'}\right).$$

The notion of the moment of inertia being fixed by "quantun conditions" for electronic motions and nuclear vibrations follows naturally from the whole line of thought developed here. We will show in the next section how we can treat, approximately at least, the nuclear vibrations and the rotations of the diatomic molecule simultaneously by a synthesis of the cases[1] considered in 1 and 3.

---

[1] Cf. A: Sommerfeld, *Atombau und Spektrallinien*, 4th edit., p. 833. We do not consider here the additional non-harmonic terms in the potential energy.

I should like to mention also that the value $n = 0$ corresponds not to the *vanishing* of the wave function $\psi$ but to a *constant* value for it, and accordingly to a vibration with amplitude constant over the whole sphere.

## 4. *Non-rigid Rotator (Diatomic Molecule)*

According to the observation at the end of section 2, we must state the problem initially with all the six degrees of freedom that the rotator really possesses. Choose Cartesian co-ordinates for the two molecules, viz. $x_1, y_1, z_1; x_2, y_2, z_2$, and let the masses be $m_1$ and $m_2$, and $r$ be their distance apart. The potential energy is

$$V = 2\pi^2 v_0^2 \mu \, (r - r_0)^2 \,, \tag{35}$$

where

$$r^2 = (x_1 - x_2)^2 + (y_1 - y_2)^2 + (z_1 - z_2)^2 \,.$$

Here

$$\mu = \frac{m_1 m_2}{m_1 + m_2} \tag{36}$$

may be called the "resultant mass". Then $v_0$ is the mechanical proper frequency of the nuclear vibration, regarding the line joining the nuclei as fixed, and $r_0$ is the distance apart for which the potential energy is a minimum. These definitions are all in the sense of the usual mechanics.

For the vibration equation $(18'')$ we get the following:

$$\begin{cases} \dfrac{1}{m_1} \left( \dfrac{\partial^2 \psi}{\partial x_1^2} + \dfrac{\partial^2 \psi}{\partial y_1^2} + \dfrac{\partial^2 \psi}{\partial z_1^2} \right) + \dfrac{1}{m_2} \left( \dfrac{\partial^2 \psi}{\partial x_2^2} + \dfrac{\partial^2 \psi}{\partial y_2^2} + \dfrac{\partial^2 \psi}{\partial z_2^2} \right) \\[2mm] + \dfrac{8\pi^2}{h^2} \left[ E - 2\pi^2 v_0^2 \mu \, (r - r_0)^2 \right] \psi = 0. \end{cases} \tag{37}$$

Introduce new independent variables $x, y, z, \xi, \eta, \zeta$, where

$$\begin{aligned} x &= x_1 - x_2; (m_1 + m_2)\, \xi = m_1 x_1 + m_1 x_2 \\ y &= y_1 - y_2; (m_1 + m_2)\, \eta = m_1 y_1 + m_2 y_2 \\ z &= z_1 - z_2; (m_1 + m_2)\, \zeta = m_1 z_1 + m_2 z_2. \end{aligned} \tag{38}$$

The substitution gives

$$\left\{ \frac{1}{\mu} \left( \frac{\partial^2 \psi}{\partial x^2} + \frac{\partial^2 \psi}{\partial y^2} + \frac{\partial^2 \psi}{\partial z^2} \right) + \frac{1}{m_1 + m_2} \left( \frac{\partial^2 \psi}{\partial \xi^2} + \frac{\partial^2 \psi}{\partial \eta^2} + \frac{\partial^2 \psi}{\partial \zeta^2} \right) \right.$$
$$\left. + \left[ \alpha'' - b' (r - r_0)^2 \right] \psi = 0, \right.$$

$$(37')$$

where for brevity

$$a'' = \frac{8\pi^2 E}{h^2}, \quad b' = \frac{16\pi^4 v_0^2 \mu}{h^2}. \tag{39}$$

Now we can put for $\psi$ the product of a function of the relative co-ordinates $x$, $y$, $z$, and a function of the co-ordinates of the centre of mass $\xi$, $\eta$, $\zeta$:

$$\psi = f(x, y, z,) g(\xi, \eta, \zeta) \tag{40}$$

For $g$ we get the defining equation

$$\frac{1}{m_1 + m_2} \left( \frac{\partial^2 g}{\partial \xi^2} + \frac{\partial^2 g}{\partial \eta^2} + \frac{\partial^2 g}{\partial \zeta^2} \right) + \text{const.} \ g = 0. \tag{41}$$

This is of the same form as the equation for the motion, under no forces, of a particle of mass $m_1 + m_2$. The constant would in this case have the meaning

$$\text{const.} = \frac{8\pi^2 E_t}{h^2}, \tag{42}$$

where $E_t$ is the energy of translation of the said particle. Imagine this value inserted in (41). The question as to the values of $E_t$ admissible as proper values depends now on this, whether the whole infinite space is available for the original co-ordinates and hence for those of the centre of gravity without new potential energies coming in, or not. In the first case every non-negative value is permissible and every negative value not permissible. For when $E_t$ is not negative and *only* then, (41) possesses solutions which do not vanish identically and yet remain finite in all space. If, however, the molecule is situated in a "vessel", then the latter must supply boundary conditions for the function $g$, or in other words, equation (41), on account of the introduction of further potential energies, will alter its form very abruptly at the walls of the vessel, and thus a discrete set of $E_t$-values will be selected as proper values. It is a question of the "Quantisation of the motion of translation", the main points of which I have lately discussed, showing that it leads to Einstein's Gas Theory.[1]

For the factor $f$ of the vibration function $\psi$, depending on the relative co-ordinates $x$, $y$, $z$, we get the denning equation

$$\frac{1}{\mu}\left(\frac{\partial^2 f}{\partial x^2} + \frac{\partial^2 f}{\partial y^2} + \frac{\partial^2 f}{\partial z^2}\right) + \left[a' - b'\,(r - r_0)^2\right]f = 0, \quad (43)$$

where for brevity we put

$$a' = \frac{8\pi^2\,(E - E_t)}{h^2}. \quad (39')$$

We now introduce instead of $x$, $y$, $z$, the spherical polars $r$, $\theta$, $\phi$ (which is in agreement with the previous use of $r$). After multiplying by $\mu$ we get

$$\frac{1}{r^2}\frac{\partial}{\partial r}\left(r^2\frac{\partial f}{\partial r}\right) + \frac{1}{r^2}\left\{\frac{1}{\sin\theta}\frac{\partial}{\partial\theta}\left(\sin\theta\frac{\partial f}{\partial\theta}\right) + \frac{1}{\sin^2\theta}\frac{\partial^2 f}{\partial\phi^2}\right\}$$
$$+ \left[\mu a' - \mu b'(r - r_0)^2\right]f = 0. \quad (43')$$

Now break up $f$. The factor depending on the angles is a surface harmonic. Let the order be $n$. The curled bracket is $-n(n+1)f$. Imagine this inserted and for simplicity let $f$ now stand for the factor depending on $r$. Then introduce as new *dependent* variable

$$\chi = rf, \quad (44)$$

and as new *independent* variable

$$\rho = r + r_0. \quad (45)$$

The substitution gives

$$\frac{\partial^2\chi}{\partial\rho^2} + \left[\mu a' - \mu b'\rho^2 - \frac{n(n+1)}{(r_0 + \rho)^2}\right]\chi = 0. \quad (46)$$

To this point the analysis has been exact. Now we will make an approximation, which I well know requires a stricter justification than I will give here. Compare (46) with equation (22') treated earlier. They agree in form and only differ in the coefficient of the unknown

[1] *Physik. Ztschr.* 27, p. 95, 1926.

function by terms of the relative order of magnitude of $\frac{\rho}{r_0}$. This is seen, if we develop thus:

$$\frac{n(n+1)}{(r_0+\rho)^2} = \frac{n(n+1)}{r_0^2}\left(1 - \frac{2\rho}{r_0} + \frac{3\rho^2}{r_0^2} - + \cdots\right), \qquad (47)$$

substitute in (46), and arrange in powers of $\rho/r_0$. If we introduce for $\rho$ a new variable differing only by a small constant, viz.

$$\rho' = \rho - \frac{n(n+1)}{r_0^3\left(\mu b' + \dfrac{3n(n+1)}{r_0^4}\right)}, \qquad (48)$$

then equation (46) takes the form

$$\frac{\partial_2 \chi}{\partial \rho'^2} + \left(a - b\rho'^2 + \left[\frac{\rho'}{r_0}\right]\right)\chi = 0, \qquad (46')$$

where we have put

$$\begin{cases} a = \mu a' - \dfrac{n(n+1)}{r_0^2}\left(1 - \dfrac{n(n+1)}{r_0^4\mu b' + 3n(n+1)}\right) \\ b = \mu b' + \dfrac{3n(n+1)}{r_0^4}. \end{cases} \qquad (49)$$

The symbol $\left[\frac{\rho'}{r_0}\right]$ in (46') represents terms which are small compared with the retained term of the order of $\frac{\rho'}{r_0}$.

Now we know that the *first* proper functions of equation (22'), to which we now compare (46'), only differ markedly from zero in a small range on both sides of the origin. Only those of higher order stretch gradually further out. For moderate orders, the domain for equation (46'), if we *neglect* the term $\left[\frac{\rho'}{r_0}\right]$ and bear in mind the order of magnitude of molecular constants, is indeed small compared with $r_0$. We thus conclude (without rigorous proof, I repeat), that we can in this way obtain a useful approximation for the first proper functions, within the region where they differ at all markedly from zero, and also for the first *proper values*. From the proper value condition (25) and omitting the abbreviations (49), (39'), and (39), though introducing the small quantity

$$\epsilon = \frac{n(n+1)h^2}{16\pi^4 v_0^2 \mu^2 r_0^4} = \frac{n(n+1)h^2}{16\pi^4 v_0^2 A^2} \tag{50}$$

instead, we can easily derive the following *energy steps*,

$$\begin{cases} E = E_t + \dfrac{(n(n+1)h^2}{8\pi^2 A}\left(1 - \dfrac{\epsilon}{1+3\epsilon}\right) + \dfrac{2l+1}{2}hv_0\sqrt{1+3\epsilon} \\ (n = 0, 1, 2\ldots; \quad l = 0, 1, 2\ldots), \end{cases}$$

$$\tag{51}$$

where

$$A = \mu r_0^2 \tag{52}$$

is still written for the *moment of inertia*.

In the language of classical mechanics, $\epsilon$ is the square of the ratio of the frequency of rotation to the vibration frequency $v_0$; it is therefore rlally a small quantity in the application to the molecule, and formula (51) has the usual structure, apart from this small correction and the other differences already mentioned. It is the synthesis of (25') and (34') to which $E_t$ is added as representing the energy of translation. It must be emphasized that the value of the approximation is to be judged not only by the smallness of $\epsilon$ but also by $l$ not being too large. *Practically*, however, only small numbers have to be considered for $l$.

The $\epsilon$-corrections in (51) do *not yet* take account of deviations of the nuclear vibrations from the pure harmonic type. Thus a comparison with Kratzer's formula (*vide* Sommerfeld, *loc. cit.*) and with experience is impossible. I only desired to mention the case provisionally, as an example showing that the intuitive idea of the *equilibrium configuration* of the nuclear system retains its meaning in undulatory mechanics also, and showing the manner in which it does so, provided that the wave amplitude $\psi$ is different from zero practically only in a small neighbourhood of the equilibrium configuration. The direct interpretation of this wave function of *six* variables in *three*-dimensional space meets, at any rate initially, with difficulties of an abstract nature.

The rotation-vibration-problem of the diatomic molecule will have to be re-attacked presently, the non-harmonic terms in the energy of binding *being taken into account*. The method, selected skilfully by Kratzer for the classical mechanical treatment, is also suitable for undulatory mechanics. If, however, we are going to push the calculation as far as is necessary for the fineness of band structure, then we must make use of the theory of the *perturbation of proper values and functions*, that is, of the alteration experienced by a definit proper value and the appertaining proper functions of a differential equation, when there is added to the coefficient of the unknown function in the equation a small "disturbing term". This "perturbation theory" is the complete counterpart of that of classical mechanics, except that it is simpler because in undulatory mechanics we are always in the domain of *linear* relations. As a first approximation we have the statement that the perturbation of the proper value is equal to the perturbing term averaged "over the undistrubed motion".

The perturbation theory broadens the analytical range of the new theory extraordinarily. As an important practical success, let me say here that the *Stark effect* of the first order will be found to be really completely in accord with Epstein's formula, which has become unimpeachable through the confirmation of experience.

Zürich, Physical Institute of the University.
(Received February 23, 1926.)

# QUANTISATION AS A PROBLEM OF PROPER VALUES (PART III)

## PERTURBATION THEORY, WITH APPLICATION TO THE STARK EFFECT OF THE BALMER LINES

(*Annalen der Physik* (4), vol. 80, 1926)

## INTRODUCTION. ABSTRACT

As has already been mentioned at the end of the preceding paper,[1] the available range of application of the proper value theory can by comparatively elementary methods be considerably increased beyond the "directly soluble problems"; for proper values and functions can readily be approximately determined for *such* boundary value problems as are sufficiently closely related to a directly soluble problem. In analogy with ordinary mechanics, let us call the method in question the *perturbation* method. It is based upon the important *property of continuity* possessed by proper values and functions,[2] principally, for our purpose, upon their *continuous* dependence on the *coefficients* of the differential equation, and less upon the extent of the domain and on the boundary conditions, since in our case the domain ("entire *q*-space") and the boundary conditions ("remaining finite") are generally the same for the unperturbed and perturbed problems.

The method is essentially the same as that used by Lord Rayleigh in investigating[3] the vibrations of a string with *small inhomogeneities* in his *Theory of Sound* (2nd edit., vol. i., pp. 115–118, London, 1894). This was a particularly simple case, as the differential equation of the unperturbed problem had *constant* coefficients, and only the perturbing terms were arbitrary functions along the string. A complete generalisation is possible not merely with regard to these points, but also for the specially important case of *several* independent

---

[1] Last two paragraphs of Part II.
[2] Courant-Hilbert, chap. vi. §§ 2, 4, p. 337.
[3] Courant-Hilbert, chap. v. § 5, 2, p. 241.

variables, *i.e.* for *partial* differential equations, in which *multiple proper values* appear in the unperturbed problem, and where the addition of a perturbing term causes the *splitting up* of such values and is of the greatest interest in well-known spectroscopic questions (Zeeman effect, Stark effect, Multiplicities). In the development of the perturbation theory in the following Section I., which really yields nothing new to the mathematician, I put less value on generalising to the *widest possible extent* than on bringing forward the very simple rudiments in the clearest possible manner. From the latter, any desired generalisation arises almost automatically when needed. In Section II., as an example, the Stark effect is discussed and, indeed, by *two* methods, of which the *first* is analogous to Epstein's method, by which he first solved[1] the problem on the basis of classical mechanics, supplemented by quantum conditions, while the *second*, which is much more general, is analogous to the method of secular perturbations.[2] The *first* method will be utilised to show that in wave mechanics also the perturbed problem can be "separated" in *parabolic* co-ordinates, and the perturbation theory will first be applied to the ordinary differential equations into which the original vibration equation is split up. The theory thus merely takes over the task which on the old theory devolved on Sommerfeld's elegant complex integration for the calculation of the quantum integrals.[3] In the *second* method, it is found that in the case of the Stark effect an exact separation co-ordinate system exists, quite by accident, for the perturbed problem also, and the perturbation theory is applied directly to the *partial* differential equation. This latter proceeding proves to be more troublesome in wave mechanics, although it is theoretically superior, being more capable of generalisation.

Also the problem of the intensity of the components in the Stark effect will be shortly discussed in Section II. Tables will be calculated, which, as a whole, agree even better with experiment than the

---

[1] P. S. Epstein, *Ann. d. Phys.* 50, p. 489, 1916.
[2] N. Bohr, *Kopenhagener Akademie* (8), IV., 1, 2, p. 69 *et seq.*, 1918.
[3] A. Sommerfeld, *Atombau*, 4th ed., p. 772.

well-known ones calculated by Kramers with the help of the correspondence principle.[1]

The application (not yet completed) to the *Zeeman effect* will
naturally be of much greater interest. It seems to be indissolubly linked
with a correct formulation in the language of wave mechanics of the
*relativistic* problem, because in the four-dimensional formulation the
vector-potential automatically ranks equally with the scalar. It was
already mentioned in Part I. that the relativistic hydrogen atom may
indeed be treated without further discussion, but that it leads to "half-
integral" azimuthal quanta, and thus contradicts experience. Therefore
"something must still be missing". Since then I have learnt *what* is
lacking from the most important publications of G. E. Uhlenbeck
and S. Goudsmit,[2] and then from oral and written communications
from Paris (P. Langevin) and Copenhagen (W. Pauli), viz., in the
language of the theory of electronic orbits, the *angular momentum* of
the electron round its axis, which gives it a *magnetic moment*. The
utterances of these investigators, together with two highly significant
papers by Slater[3] and by Sommerfeld and Unsöld[4] dealing with the
Balmer spectrum, leave no doubt that, by the introduction of the
paradoxical yet happy conception of the spinning electron, the orbital
theory will be able to master the disquieting difficulties which have
latterly begun to accumulate (anomalous Zeeman effect; Paschen-
Back effect of the Balmer lines; irregular and regular Röntgen doublets;
analogy of the latter with the alkali doublets, etc.). We shall be obliged
to attempt to take over the idea of Uhlenbeck and Goudsmit into
wave mechanics. I believe that the latter is a very fertile soil for this
idea, since in it the electron is not considered as a point charge,
but as continuously flowing through space,[5] and so the unpleasing
conception of a "rotating point-charge" is avoided. In the present
paper, however, the taking over of the idea is not yet attempted.

[1] H. A. Kramers, *Kopenhagener Akademie* (8), III., 3, p. 287, 1919.
[2] G. E. Uhlenbeck and S. Goudsmit, *Physica*, 1925: *Die Naturwissenschaften*, 1926; *Nature*, 20th Feb., 1926; cf. also L. H. Thomas, *Nature*, 10th April, 1926.
[3] J. C. Slater, *Proc. Amer. Nat. Acad.* 11, p. 732, 1925.
[4] A. Sommerfeld and A. Unsöld, *Ztschr. f. Phys.* 36, p. 259, 1926.
[5] Cf. last two pages of previous paper.

To the *third section*, as "mathematical appendix", have been relegated numerous uninteresting calculations—mainly quadratures of products of proper functions, required in the second section. *The formulae of the appendix are numbered (101), (102), etc.*

# I. PERTURBATION THEORY

## § 1. A Single Independent Variable

Let us consider a linear, homogeneous, differential expression of the second order, which we may assume to be in self-adjoint form without loss of generality, viz.

$$L\,[y] = py'' + p'y' - qy. \tag{1}$$

$y$ is the dependent function; $p$, $p'$ and $q$ are continuous functions of the independent variable $x$ and $p \geq 0$. A dash denotes differentiation with respect to $x$ ($p'$ is therefore the derivative of $p$, which is the condition for self-adjointness).

Now let $p(x)$ be another continuous function of $x$, which never becomes negative, and also in general does not vanish. We consider the proper value problem of Sturm and Liouville,[1]

$$L[y] + E\rho y = 0. \tag{2}$$

It is a question, first, of finding all *those* values of the constant $E$ ("proper values") for which the equation (2) possesses solutions $y(x)$, which are continuous and not identically vanishing within a certain domain, and which satisfy certain "boundary conditions" at the bounding points; and secondly of finding these solutions ("proper functions") themselves. In the cases treated in atomic mechanics, domain and boundary conditions are always "natural". The domain, for example, reaches from 0 to $\infty$, when $x$ signifies the value of the radius vector or of an intrinsically positive parabolic co-ordinate, and the boundary conditions are in these cases: *remaining finite*. Or, when $x$ signifies an azimuth, then the domain is the interval from 0 to $2\pi$

---

[1] Cf. Courant-Hilbert, chap. v. § 5, 1, p. 238 *et seq.*

and the condition is: Repetition of the initial values of $y$ and $y'$ at the end of the interval ("periodicity").

It is only in the case of the periodic condition that *multiple*, viz. *double-valued*, proper values appear for *one* independent variable. By this we understand that to the same proper value belong *several* (in the particular case, two) linearly independent proper functions. We will now exclude this case for the sake of simplicity, as it attaches itself easily to the developments of the following paragraph. Moreover, to lighten the formulae, we will not expressly take into account in the notation the possibility that a "band spectrum" (*i.e.* a *continuum* of proper values) may be present when the domain extends to infinity.

Let now $y = u_i(x)$, $i = 1, 2, 3, \ldots$, be the series of Sturm-Liouville proper functions; then the series of functions $u_i(x)\sqrt{\rho(x)}$, $i = 1, 2, 3, \ldots$, forms a *complete orthogonal system* for the domain; *i.e.* in the first place, if $u_i(x)$ and $u_k(x)$ are the proper functions belonging to the values $E_i$ and $E_k$, then

$$\int \rho(x)u_i(x)u_k(x)dx = 0 \quad \text{for } i \neq k. \qquad (3)$$

(Integrals without limits are to be taken over the domain, throughout this paper.) The expression "complete" signifies that an originally arbitrary continuous function is condemned to vanish identically, by the mere postulation that it must be orthogonal with respect to *all* the functions $u_i(x)\sqrt{\rho(x)}$. (More shortly: "There exists no further orthogonal function for the system.") We can and will always regard the proper functions $u_i(x)$ in all general discussions as "normalised", *i.e.* we imagine the constant factor, which is still arbitrary in each of them on account of the homogeneity of (2), to be defined in *such a way* that the integral (3) takes the value unity for $i = k$. Finally we again remind the reader that the proper values of (2) are certainly all *real*.

Let now the proper values $E_i$ and functions $u_i(x)$ be *known*. Let us, from now on, direct our attention specially to a *definite* proper value, $E_k$ say, and the corresponding function $u_k(x)$, and ask how these alter, when we do not alter the problem in any way other than by adding to the left-hand side of (2) a small "perturbing term", which we will initially write in the form

$$-\lambda r(x) y. \tag{4}$$

In this $\lambda$ is a small quantity (the perturbation parameter), and $r(x)$ is an arbitrary continuous function of $x$. It is therefore simply a matter of a slight alteration of the coefficient $q$ in the differential expression (1). From the continuity properties of the proper quantities, mentioned in the introduction, we now know that the altered Sturm-Liouville problem

$$L[y] - \lambda r y + E\rho y = 0 \tag{2'}$$

must have, in any case for a sufficiently small $\lambda$, proper quantities in the near neighbourhood of $E_k$ and $u_k$, which we may write, by way of trial, as

$$E_k^* = E_k + \lambda \epsilon_k; \; u_k^* = u_k(x) + \lambda v_k(x). \tag{5}$$

On substituting in equation (2'), remembering that $u_k$ satisfies (2), neglecting $\lambda^2$ and cutting away a factor $\lambda$ we get

$$L[v_k] + E_k \rho v_k = (r - \epsilon_k \rho) u_k. \tag{6}$$

For the defining of the perturbation $v_k$ of the proper *function*, we thus obtain, as a comparison of (2) and (6) shows, a *non homogeneous* equation, which belongs precisely to *that* homogeneous equation which is satisfied by our unperturbed proper function $u_k$ (for in (6) the special proper value $E_k$ stands in place of $E$). On the right-hand side of this non-homogeneous equation occurs, in addition to known quantities, the still unknown perturbation $\epsilon_k$ of the proper *value*.

This occurrence of $\epsilon_k$ serves for the calculation of this quantity *before* the calculation of $v_k$. It is known that the non-homogeneous equation—and this is *the starting-point of the whole perturbation theory*—for a proper *value* of the homogeneous equation possesses a solution *when*, and *only when*, its right-hand side is *orthogonal*[1] to the allied proper function (to all the allied functions, in the case of

---

[1] Cf. Courant-Hilbert, chap. v. § 10, 2, p. 277.

multiple proper values). (The physical interpretation of this mathematical theorem, for the vibrations of a string, is that if the force is in resonance with a proper vibration it must be distributed in a very special way over the string, namely, so that it does no work in the vibration in question; otherwise the amplitude grows beyond all limits and a stationary condition is impossible.)

The right-hand side of (6) must therefore be orthogonal to $u_k$, *i.e.*

$$\int (r - \epsilon_k \rho) u_k^2 dx = 0, \tag{7}$$

or

$$\epsilon_k = \frac{\int r u_k^2 dx}{\int \rho u_k^2 dx}, \tag{7'}$$

or, if we imagine $u_i$ already normalised, then, more simply,

$$\epsilon_k = \int r u_k^2 dx. \tag{7''}$$

This simple formula expresses the perturbation of the proper value (of first order) in terms of the perturbing function $r(x)$ and the unperturbed proper function $u_k(x)$. If we consider that the proper value of our problem signifies mechanical energy or is analogous to it, and that the proper function $u_k$ is comparable to "motion with energy $E_k$", then we see in (7'') the complete parallel to the well-known theorem in the perturbation theory of classical mechanics, viz. the perturbation of the energy, to a first approximation, is equal to the perturbing function, averaged over the unperturbed motion. (It may be remarked in passing that it is as a rule sensible, or at least aesthetic, to throw into bold relief the factor $\rho(x)$ in the integrands of *all* integrals taken over the entire domain. If we do this, then, in integral (7''), we must speak of $\frac{r(x)}{\rho(x)}$ and not $r(x)$ as the perturbing function, and make a corresponding change in the expression (4). Since the point is quite unimportant, however, we will stick to the notation already chosen.)

We have yet to define $v_k(x)$, the perturbation of the proper *function*, from (6). We solve[1] the non-homogeneous equation by putting

---

[1] Cf. Courant-Hilbert, chap. v. § 5, 1, p. 240, and § 10, p. 279.

for $v_k$ a series of proper functions, viz.

$$v_k(x) = \sum_{i=1}^{\infty} \gamma_{ki} u_i(x), \qquad (8)$$

and by developing the right-hand side, divided by $\rho(x)$, likewise in a series of proper functions, thus

$$\left(\frac{r(x)}{\rho(x)} - \epsilon_k\right) u_k(x) = \sum_{i=1}^{\infty} c_{ki} u_i(x), \qquad (9)$$

where

$$\begin{cases} c_{ki} = \int (r - \epsilon_k \rho) u_k u_i dx \\ \quad = \int r u_k u_i dx \text{ for } i \neq k \\ \quad = 0 \qquad\qquad \text{for } i = k. \end{cases} \qquad (10)$$

The last equality follows from (7). If we substitute from (8) and (9) in (6) we get

$$\sum_{i=1}^{\infty} \gamma_{ki}(L[u_i] + E_k \rho u_i) = \sum_{i=1}^{\infty} c_{ki} \rho u_i. \qquad (11)$$

Since now $u_i$ satisfies equation (2) with $E = E_i$, it follows that

$$\sum_{i=1}^{\infty} \gamma_{ki} \rho (E_k - E_i) u_i = \sum_{i=1}^{\infty} c_{ki} \rho u_i. \qquad (12)$$

By equating coefficients on left and right, all the $\gamma_{ki}$'s, except $\gamma_{kk}$, are defined. Thus

$$\gamma_{ki} = \frac{c_{ki}}{E_k - E_i} = \frac{\int r u_k u_i dx}{E_k - E_i} \text{ for } i \neq k, \qquad (13)$$

while $\gamma_{kk}$, as may be understood, remains completely undefined. This indefiniteness corresponds to the fact that the postulation of normalisation is still available for us for the perturbed proper function. If we make use of (8) in (5) and claim for $u_k^*(x)$ the same normalisation as for $u_k(x)$ (quantities of the order of $\lambda^2$ being neglected), then it is evident that $\gamma_{kk} = 0$. Using (13) we now obtain for the *perturbed proper function*

$$u_k^*(x) = u_k(x) + \lambda \sum_{i=1}^{\infty}{}' \frac{u_i(x) \int r u_k u_i dx}{E_k - E_i}. \qquad (14)$$

(The dash on the sigma denotes that the term $i = k$ has not to be taken.) And the allied perturbed proper value is, from the above,

$$E_k^* = E_k + \lambda \int r u_k^2 dx. \qquad (15)$$

By substituting in (2′) we may convince ourselves that (14) and (15) do really satisfy the proper value problem to the proposed degree of approximation. This verification is necessary since the development, assumed in (5), in *integral* powers of the perturbation parameter is no necessary consequence of continuity.

The procedure, here explained in fair detail for the simplest case, is capable of generalisation in many ways. In the first place, we can of course consider the perturbation in a quite similar manner for the second, and then the third order in $\lambda$, etc., in each case obtaining first the next approximation to the proper value, and then the corresponding approximation for the proper function. In certain circumstances it may be advisable—just as in the perturbation theory of mechanics—to regard the perturbation function itself as a power series in $\lambda$, whose terms come into play one by one in the separate stages. These questions are discussed exhaustively by Herr E. Fues in work which is now appearing in connection with the application to the theory of *band spectra*.

In the second place, in quite similar fashion, we can consider also a perturbation of the term in $y'$ of the differential operator (1) just as we have considered above the term $-qy$. The case is important, for the Zeeman effect leads without doubt to a perturbation of this kind—though admittedly in an equation with several independent variables. Thus the equation loses its self-adjoint form by the perturbation—not an essential matter in the case of a single variable. In a partial differential equation, however, this loss may result in the perturbed proper values no longer being real, though the perturbing term is real; and naturally also conversely, an imaginary perturbing term may have a real, physically intelligible perturbation as its consequence.

We may also go further and consider a perturbation of the term in $y''$. Indeed it is quite possible, in general, to add an arbitrary "infinitely small" linear[1] and homogeneous differential operator, even of higher order than the second, as the perturbing term and to calculate the perturbations in the same manner as above. In these cases, however, we would use with advantage the fact that the second and higher derivatives of the proper functions may be expressed by means of the differential equation itself, in terms of the zero and first derivatives, so that this general case may be reduced, in a certain sense, to the two special cases, first considered—perturbation of the terms in $y$ and $y'$.

Finally, it is obvious that the extension to equations of order higher than the second is possible.

Undoubtedly, however, the most important generalisation is that to several independent variables, *i.e.* to partial differential equations. For *this* really is the problem in the general case, and only in exceptional cases will it be possible to split up the disturbed partial differential equation, by the introduction of suitable variables, into separate differential equations, each only with one variable.

## § 2. Several Independent Variables (Partial Differential Equation)

We will represent the several independent variables in the formulae symbolically by the *one* sign $x$, and briefly write $\int dx$ (instead of $\int \cdots \int dx_1 dx_2 \ldots$) for an integral extending over the multiply-dimensioned *domain*. A notation of this type is already in use in the theory of integral equations, and has the advantage, here as there, that the structure of the formulae is not altered by the increased number of variables as such, but only by *essentially* new occurrences, which *may* be related to it.

Let therefore $L[y]$ now signify a self-adjoint *partial* linear differential expression of the second order, whose explicit form we do not require to specify; and further let $\rho(x)$ again be a positive function of the independent variables, which does not vanish in general. The postulation "self-adjoint" is *now* no longer unimportant, as the

---

[1] Even the limitation "linear" is not absolutely necessary.

property cannot now be generally gained by multiplication by a suitably chosen $f(x)$, as was the case with *one* variable. In the particular differential expression of wave mechanics, however, this is still the case, as it arises from a variation principle.

According to these definitions or conventions, we can regard equation (2) of § 1,

$$L[y] + E\rho y = 0, \tag{2}$$

as the formulation of the Sturm-Liouville proper value problem in the case of several variables also. Everything said there about the proper values and functions, their orthogonality, normalisation, etc., as also *the whole perturbation theory there developed*—in short, the whole of § 1—remains *valid without change, when* all the proper values are *simple*, if we use the abbreviated symbolism just agreed upon above. And only *one* thing does *not* remain valid, namely, that they *must* be simple.

Nevertheless, from the pure mathematical standpoint, the case when the roots are all distinct is to be regarded as the *general* case for several variables also, and multiplicity regarded as a special occurrence, which, it is admitted, *is the rule in applications*, on account of the specially simple and symmetrical structure of the differential expressions $L[y]$ (and the "boundary conditions") which appear. Multiplicity of the proper values corresponds to *degeneracy* in the theory of conditioned periodic systems and is therefore especially interesting for quantum theory.

A proper value $E_k$ is called $\alpha$-fold, when equation (2), for $E = E_k$, possesses not *one* but exactly $\alpha$ linearly independent solutions which satisfy the boundary conditions. We will denote these by

$$u_{k_1}, u_{k_2}, \ldots u_{k_\alpha}. \tag{16}$$

Then it is true that each of these $\alpha$ proper functions is *orthogonal* to each of the *other* proper functions belonging to *another* proper value (the factor $\rho(x)$ being included; cf. (3)). On the contrary, these $\alpha$ functions are *not* in general orthogonal *to one another*, if we merely postulate that they are $\alpha$ linearly independent proper functions for the proper value $E_k$, and nothing more. For then we can equally well replace them by $\alpha$ arbitrary, linearly independent, linear aggregates (with constant coefficients) of themselves. We may express this otherwise, thus. The series of functions (16) is initially *indefinite* to the extent of a linear transformation (with constant coefficients), involving a non-vanishing determinant, and such a transformation *destroys*, in general, the mutual orthogonality.

But through such a transformation this mutual orthogonality can always be *brought about*, and indeed in an infinite number of ways; the latter property arising because *orthogonal* transformation does *not* destroy the mutual orthogonality. We are now accustomed to include this simply in *normalisation*, that orthogonality is secured for *all* proper functions, even for those which belong to the *same* proper value. We will assume that our $u_{ki}$'s are already *normalised* in this way, and of course for *each* proper value. Then we must have

$$\left\{ \int\!\!\int \rho(x)\, u_{ki}(x)\, u_{k'i'}(x)\, dx = 0 \text{ when } (k, i) \neq (k', i') \right.$$
$$= 1 \text{ when } k' = k, \text{ as well as } i' = i.$$

$$(17)$$

Each of the finite series of proper functions $u_{ki}$, obtained for *constant* $k$ and *varying* $i$, is then only still indefinite to this extent, that it is subject to an *orthogonal* transformation.

We will now discuss, first in words, without using formulae, the consequences which follow when a perturbing term is added to the differential equation (2). The addition of the perturbing term will, in general, remove the above-mentioned symmetry of the differential equation, to which the multiplicity of the proper values (or of certain of them) is due. Since, however, the proper values and functions are *continuously* dependent on the coefficients of the differential equation, a small perturbation causes a group of $\alpha$ proper values, which lie close to one another and to $E_k$, to enter in place of the $\alpha$-fold proper value $E_k$. The latter is *split up*. Of course, if the symmetry is not wholly destroyed by the perturbation, it may happen that the splitting up is not complete and that several proper values (still partly multiple) of, *in summa*, equal multiplicity merely appear in the place of $E_k$ ("*partial removal of degeneracy*").

As for the perturbed proper *functions*, those a members which belong to the $\alpha$ values arising from $E_k$ must evidently also on account of continuity lie infinitely near the unperturbed functions belonging to $E_k$, viz. $u_{ki}$; $i = 1, 2, 3 \ldots \alpha$. Yet we must remember that the last-named series of functions, as we have established above, is indefinite to the extent of an *arbitrary orthogonal transformation. One* of the infinitely numerous definitions, which may be applied to the series of functions, $u_{ki}$; $i = 1, 2, 3 \ldots \alpha$, will lie infinitely near the series of perturbed functions; and if the value $E_k$ is completely split up, it will be a *quite definite one*! For to the separate simple proper values, into which the value is split up, there belong proper functions which are quite uniquely defined.

This unique particular specification of the *unperturbed* proper functions (which may fittingly be designated as the "approximations of zero order" for the *perturbed* functions), which is defined by the nature of the perturbation, will naturally *not* generally coincide with that definition of the unperturbed functions which we chanced to adopt to begin with. Each group of the latter, belonging to a definite $\alpha$-fold proper value $E_k$, will have first to be submitted to an orthogonal substitution, defined by the kind of perturbation, before it can serve as the starting-point, the "zero approximation", for a more exact definition of the perturbed proper functions. *The defining of these orthogonal substitutions*—one for each multiple proper value—*is the only essentially new point* that arises because of the increased number of variables, or from the appearance of multiple proper values. The defining of these substitutions forms the exact counterpart to the finding of an approximate separation system for the perturbed motion in the theory of conditioned periodic systems. As we will see immediately, the definition of the substitutions can always be given in a theoretically simple way. It requires, for each $\alpha$-fold proper value, merely the principal axes transformation of a quadratic form of $\alpha$ (and thus of a finite number of) variables.

When the substitution has once been accomplished, the calculation of the approximations of the *first* order runs almost word for word as in § 1. The sole difference is that the dash on the sigma in equation (14) must mean that in the summation *all* the proper functions belonging to the value $E_k$, *i.e. all* the terms whose denominators would vanish, must be left out. It may be remarked in passing that it is not at all necessary, in the calculation of *first* approximations, to have completed the orthogonal substitutions referred to for *all* multiple proper values, but it is sufficient to have done so for the value $E_k$, in whose splitting up we are interested. For the approximations of higher order, we admittedly require them all. In all other respects, however, these higher approximations are from the beginning carried out exactly as for simple proper values.

Of course it may happen, as was mentioned above, that the value $E_k$, either generally or at the initial stages of the approximation, is not completely split up, and that multiplicities ("degeneracies") still remain. This is expressed by the fact that to the substitutions already frequently mentioned there still clings a certain indefiniteness, which either always remains, or is removed step by step in the later approximations.

Let us now represent these ideas by formulae, and consider as before the perturbation caused by (4), § 1,

$$-\lambda r\,(x)\,y, \tag{4}$$

*i.e.* we imagine the proper value problem belonging to (2) *solved*, and now consider the exactly corresponding problem (2′),

$$L\,[y] - \lambda ry + E\rho y = 0. \tag{2′}$$

We again fix our attention on a definite proper value $E_k$. Let (16) be a system of proper functions belonging to it, which we assume to be normalised and orthogonal to one another in the sense described above, but *not yet* fitted to the particular perturbation in the sense explained, because to find the substitution that leads *to this fitting* is precisely our chief task! In place of (5), § 1, we must now put for the perturbed quantities the following,

$$E_{kl}^* = E_k + \lambda \epsilon_l; \quad u_{kl}^*(x) = \sum_{i=1}^{\alpha} \kappa_{li} u_{ki}(x) + \lambda v_l(x)$$

$$(l = 1, 2, 3 \cdots \alpha),$$ \hfill (18)

wherein the $v_l(x)$'s are functions, and the $\epsilon_l$'s and the $k_{li}$'s are systems of constants, which are still to be defined, but which we initially do not limit in any way, although we know that the system of coefficients $k_{li}$ must[1] form an orthogonal substitution. The index $k$ should still be attached to the three types of quantity named, in order to indicate that the whole discussion refers to the $k$th proper value of the unperturbed problem. We refrain from carrying this out, in order to avoid the confusing accumulation of indices. The index $k$ is to be assumed *fixed* in the whole of the following discussion, until the contrary is stated.

Let us select *one* of the perturbed proper functions and values by giving a definite value to the index $l$ in (18), and let us substitute from (18) in the differential equation (2′) and arrange in powers of $\lambda$. Then the terms independent of $\lambda$ disappear exactly as in § 1, because the unperturbed proper quantities satisfy equation (2), by hypothesis. Only terms containing the *first* power of $\lambda$ remain, as we can strike out the others. Omitting a factor $\lambda$, we get

$$L[v_1] + E_k \rho v_l = \sum_{i=1}^{\alpha} \kappa_{li} (r - \epsilon_l \rho) u_{ki},$$ \hfill (19)

and thus obtain again for the definition of the perturbation $v_l$ of the *functions* a *non-homogeneous* equation, to which corresponds as homogeneous equation the equation (2), with the particular value $E = E_k$, *i.e.* the equation satisfied by the set of functions $u_{ki}$; $i = 1, 2, \ldots \alpha$. The form of the left side of equation (19) is independent of the index $l$.

On the right side occur $\epsilon_l$ and $k_{li}$, the constants to be defined, and we are thus enabled to evaluate them, even *before* calculating $v_l$. For, in order that (19) should have a solution at all, it is necessary

---

[1] It follows from the general theory that the perturbed system of functions $u^*_{kl}(x)$ *must* be orthogonal if the perturbation completely removes the degeneracy, and *may* be assumed orthogonal although that is not the case.

and sufficient that its right-hand side should be orthogonal to *all* the proper functions of the homogeneous equation (2) belonging to $E_k$. Therefore, we must have

$$
\left\{
\begin{aligned}
&\sum_{i=1}^{\alpha} \kappa_{li} \int (r - \epsilon_l \rho)\, u_{ki} u_{km} dx = 0 \\
&(m = 1, 2, 3 \ldots \alpha),
\end{aligned}
\right.
\tag{20}
$$

*i.e.* on account of the normalisation (17),

$$
\left\{
\begin{aligned}
&\kappa_{lm}\epsilon_l = \sum_{i=1}^{\alpha} \kappa_{li} \int r\, u_{ki} u_{km} dx \\
&(m = 1, 2, 3 \cdots \alpha).
\end{aligned}
\right.
\tag{21}
$$

If we write, briefly, for the *symmetrical* matrix of constants, which can be evaluated by quadrature,

$$
\left\{
\begin{aligned}
&\int r\, u_{ki} u_{km} dx = \epsilon_{im} \\
&(i, m = 1, 2, 3 \ldots \alpha),
\end{aligned}
\right.
\tag{22}
$$

then we recognise in

$$
\left\{
\begin{aligned}
&\kappa_{lm}\epsilon_l = \sum_{i=1}^{\alpha} \kappa_{li} \epsilon_{mi} \\
&(m = 1, 2, 3 \ldots \alpha)
\end{aligned}
\right.
\tag{21$'$}
$$

a system of $\alpha$ linear homogeneous equations for the calculation of the $\alpha$ constants $k_{lm}$; $m = 1, 2 \ldots \alpha$, where the perturbation $\epsilon_l$ of the proper value still occurs in the coefficients, and is itself unknown. However, this serves for the calculation of $\epsilon_l$ before that of the $k_{lm}$'s. For it is known that the linear homogeneous system (21$'$) of equations has solutions if, and only if, its determinant vanishes. This yields the following algebraic equation of degree $\alpha$ for $\epsilon_l$:

$$
\begin{vmatrix}
\epsilon_{11} - \epsilon_l, & \epsilon_{12} & , \ldots \epsilon_{1\alpha} \\
\epsilon_{21} & , \epsilon_{22} - \epsilon_l & \ldots \epsilon_{21} \\
\multicolumn{3}{c}{\cdots\cdots\cdots\cdots\cdots\cdots} \\
\multicolumn{3}{c}{\cdots\cdots\cdots\cdots\cdots\cdots} \\
\epsilon_{\alpha 1} & , \epsilon_{\alpha 2} & , \ldots \epsilon_{\alpha\alpha} - \epsilon_l
\end{vmatrix} = 0.
\tag{23}
$$

We see that the problem is completely identical with the transformation of the quadratic form in $\alpha$ variables, with coefficients $\epsilon_{mi}$, to its principal axes. The "secular equation" (23) yields $\alpha$ roots for $\epsilon_l$, the "reciprocal of the squares of the principal axes", which in general are different, and on account of the symmetry of the $\epsilon_{mi}$'s *always real.* We thus get all the $\alpha$ perturbations of the proper values ($l = 1, 2 \ldots \alpha$) at the same time, and would have *inferred* the splitting up of an $\alpha$-fold proper value into exactly $\alpha$ simple values, generally different, even had we not assumed it already, as fairly obvious. For *each* of these $\epsilon_l$-values, equations (21') give a system of quantities $k_{li}i = 1, 2, \ldots \alpha$, and, as is known, *only one* (apart from a general constant factor), provided all the $\epsilon_l$'s are really different. Further, it is known that the whole system of $\alpha^2$ quantities $k_{li}$ forms an *orthogonal* system of coefficients, defining as usual, in the principal axes problem, the *directions* of the new co-ordinate axes with reference to the old ones. We may, and will, employ the undefined factors just mentioned to normalise the $k_{li}$'s completely as "direction cosines", and this, as is easily seen, makes the perturbed proper functions $u_{ki}^*(x)$ turn out *normalised* again, according to (18), at least in the "zero approximation" (*i.e.* apart from the $\lambda$-terms).

If the equation (23) has multiple roots, then we have the case previously mentioned, when the perturbation does not completely remove the degeneration. The perturbed equation has then multiple proper values also and the definition of the constants $k_{li}$ becomes partially arbitrary. This has no consequence other than that (as is *always* the case with multiple proper values) we *must* and *may* acquiesce, even *after* the perturbation is applied, in a system of proper functions which in many respects is still arbitrary.

The main task is accomplished with this transformation to principal axes, and we will often find it sufficient in the applications in quantum theory to define the proper values to a first and the functions to zero approximation. The evaluation of the constants $k_{li}$ and $\epsilon_{li}$ cannot be carried out always, since it depends on the solution of an algebraic equation of degree $\alpha$. At the worst there are methods[1] which give the evaluation to any desired approximation by a rational process. We may thus regard these constants as known, and will now

---

[1] Courant-Hilbert, chap. i. § 3. 3, p. 14.

give the calculation of the functions to the *first* approximation, for the sake of completeness. The procedure is exactly as in § 1.

We have to solve equation (19) and to that end we write $v_l$ as a series of the *whole set* of proper functions of (2),

$$v_l(x) = \sum_{(k'i')} \gamma_{l,k'i'} u_{k'i'}(x). \tag{24}$$

The summation is to extend with respect to $k'$ from 0 to $\infty$, and, for each fixed value of $k'$, for $i'$ varying over the finite number of proper functions which belong to $E_{k'}$. (Now, for the first time, we take account of proper functions which do *not* belong to the $\alpha$-fold value $E_k$ we are fixing our attention on.) Secondly, we develop the right-hand side of (19), divided by $\rho(x)$, in a series of the entire set of proper functions,

$$\sum_{i=1}^{\alpha} \kappa_{li} \left( \frac{r}{\rho} - \epsilon_l \right) u_{ki} = \sum_{(k'i')} c_{l,k'i'} u_{k'i'}, \tag{25}$$

wherein

$$\begin{cases} c_{l,k'i'} = \sum_{i=1}^{\alpha} \kappa_{li} \int (r - \epsilon_l \rho) u_{ki} u_{k'i'} dx \\ \quad\quad = \sum_{i=1}^{\alpha} \kappa_{li} \int r u_{ki} u_{k'i'} dx \quad \text{for } k' \neq k \\ \quad\quad = 0 \quad\quad\quad\quad\quad\quad \text{for } k' = k \end{cases} \tag{26}$$

(the last two equalities follow from (17) and (20) respectively). On substituting from (24) and (25) in (19), we get

$$\sum_{(k'i')} \gamma_{l,k'i'} (L[u_{k'i'}] + E_k \rho u_{k'i'}) = \sum_{(k'i')} c_{l,k'i'} \rho u_{k'i'} \tag{27}$$

Since $u_{k'i'}$ satisfies equation (2) with $E = E_k$, this gives

$$\sum_{(k'i')} \gamma_{l,k'i'} \rho (E_k - E_{k'}) u_{k'i'} = \sum_{(k'i')} c_{l,k'i'} \rho u_{k'i'}. \tag{28}$$

By equating coefficients on right and left, all the $\gamma_{l,k'i'}$'s are defined, with the exception of those in which $k' = k$. Thus

$$\gamma_{l,k'i'} = \frac{c_{l,k'i'}}{E_k - E_{k'}} = \frac{1}{E_k - E_{k'}} \sum_{i-1}^{\alpha} \kappa_{li} \int r u_{ki} u_{k'i'} dx \ (\text{for } k' \neq k),$$

$$\tag{29}$$

while those $\gamma$'s for which $k' = k$ are of course not fixed by equation (19). This again corresponds to the fact that we have provisionally normalised the perturbed functions $u_{kl}^*$, of (18), only in the zero approximation (through the normalisation of the $\kappa_{li}$'s), and it is easily recognised again that we have to put the whole of the $\gamma$-quantities in question equal to zero, in order to bring about the normalisation of the $u_{kl}^*$'s even in the first approximation. By substituting from (29) in (24), and then from (24) in (18), we finally obtain for the *perturbed proper functions to a first approximation*

$$u_{kl}^*(x) = \sum_{i=1}^{\alpha} \kappa_{li} \left( u_{ki}(x) + \lambda \sum_{(k'i')}' \frac{u_{k'i'}(x)}{E_k - E_{k'}} \int r u_{ki'} u_{k'i'} dx \right)$$
$$(l = 1, 2, \ldots, \alpha).$$

(30)

The dash on the second sigma indicates that *all* the terms with $k' = k$ are to be omitted. In the application of the formula for an arbitrary $k$, it is to be observed that the $\kappa_{li}$'s, as obviously also the multiplicity $\alpha$ of the proper value $E_k$, to which we have specially directed our attention, still depend on the index $k$, though this is not expressed in the symbols. Let us repeat here that the $\kappa_{li}$'s are to be calculated as a system of solutions of equations (21'), normalised so that the sum of the squares is unity, where the coefficients of the equations are given by (22), while for the quantity $\epsilon_l$ in (21'), *one* of the roots of (23) is to be taken. *This* root then gives the allied perturbed proper *value*, from

$$E_{kl}^* = E_k + \lambda \epsilon_l. \tag{31}$$

Formulae (30) and (31) are the generalisations of (14) and (15) of § 1.

It need scarcely be said that the extensions and generalisations mentioned at the end of § 1 can of course take effect here also. It is hardly worth the trouble to carry out these developments generally. We succeed best in any special case if we do not use ready-made formulae, but go directly by the simple fundamental principles, which have been explained, perhaps too minutely, in the present paper. I would only like to consider briefly the possibility, already mentioned at the end of § 1, that the equation (2) perhaps may lose (and indeed in the case of several variables irreparably lose), its self-adjoint character if the perturbing terms also contain derivatives of the unknown function. From general theorems we know that then the proper values of the perturbed equation no longer need to be real. We can illustrate this further. We can easily see, by carrying out the developments of this paragraph, that the elements of determinant (23) are *no longer symmetrical*, when the perturbing term contains derivatives. It is known that in this case the roots of equation (23) no longer require to be real.

The necessity for the expansion of certain functions in a series of proper functions, in order to arrive at the first or zero approximation of the proper values or functions, can become very inconvenient, and can at least complicate the calculation considerably in cases where an extended spectrum co-exists with the point spectrum and where the point spectrum has a limiting point (point of accumulation) at a finite distance. This is just the case in the problems appearing in the quantum theory. Fortunately it is often—perhaps always—possible, for the purpose of the perturbation theory, to free oneself from the generally very troublesome extended spectrum, and to develop the perturbation theory from an equation which does *not* possess such a spectrum, and whose proper values do *not* accumulate near a finite value, but grow beyond all limits with increasing index. We will become acquainted with an example in the next paragraph. Of course, this simplification is only possible when we are not interested in a proper value of the extended spectrum.

# II. APPLICATION TO THE STARK EFFECT

### § 3. Calculation of Frequencies by the Method Which Corresponds to That of Epstein

If we add a potential energy $+eFz$ to the wave equation (5), Part I. (p. 2), of the Kepler problem, corresponding to the influence of an electric field of strength $F$ in the positive $z$-direction, on a negative electron of charge $e$, then we obtain the following wave equation for the Stark effect of the hydrogen atom,

$$\nabla^2\psi + \frac{8\pi^2 m}{h^2}\left(E + \frac{e^2}{r} - eFz\right)\psi = 0, \tag{32}$$

which forms the basis of the remainder of this paper. In § 5 we will apply the general perturbation theory of § 2 directly to this partial differential equation. Now, however, we will lighten our task by introducing space parabolic co-ordinates $\lambda_1$ $\lambda_2$, $\phi$, by the following equations,

$$\begin{cases} x = \dfrac{\sqrt{\lambda_1\lambda_2}\cos\phi}{+} \\[2mm] y = \dfrac{\sqrt{\lambda_1\lambda_2}\sin\phi}{+} \\[2mm] z = \tfrac{1}{2}\left(\lambda_1 - \lambda_2\right). \end{cases} \tag{33}$$

$\lambda_1$ and $\lambda_2$ run from 0 to infinity; the corresponding co-ordinate surfaces are the two sets of confocal paraboloids of revolution, which have the origin as focus and the positive ($\lambda_2$) or negative ($\lambda_1$) $z$-axis respectively as axes. $\phi$ runs from 0 to $2\pi$, and the co-ordinate surfaces belonging to it are the set of half planes limited by the $z$-axis. The relation of the co-ordinates is *unique*. For the functional determinant we get

$$\frac{\partial\,(x, y, z)}{\partial\,(\lambda_1, \lambda_2, \phi)} = \tfrac{1}{4}\left(\lambda_1 + \lambda_2\right). \tag{34}$$

The *space element* is thus

$$dx\,dy\,dz = \tfrac{1}{4}\left(\lambda_1 + \lambda_2\right)d\lambda_1 d\lambda_2 d\phi. \tag{35}$$

We notice, as consequences of (33),

$$x^2 + y^2 = \lambda_1\lambda_2; \quad r^2 = x^2 + y^2 + z^2 = \left\{\tfrac{1}{2}\left(\lambda_1 + \lambda_2\right)\right\}^2. \tag{36}$$

The expression of (32) in the chosen co-ordinates gives, if we multiply by (34)[1] (to restore the self-adjoint form),

$$
\left\{
\begin{aligned}
&\frac{\partial}{\partial\lambda_1}\left(\lambda_1\frac{\partial\psi}{\partial\lambda_1}\right) + \frac{\partial}{\partial\lambda_2}\left(\lambda_2\frac{\partial\psi}{\partial\lambda_2}\right) + \tfrac{1}{4}\left(\frac{1}{\lambda_1}+\frac{1}{\lambda_2}\right)\frac{\partial^2\psi}{\partial\phi^2} \\
&+ \frac{2\pi^2 m}{h^2}\left[E\left(\lambda_1+\lambda_2\right)+2e^2-\tfrac{1}{2}eF\left(\lambda_1^2-\lambda_2^2\right)\right]\psi = 0.
\end{aligned}
\right.
$$

(32′)

Here we can again take—and this is the why and wherefore of all "methods" of solving linear partial differential equations—the function $\psi$ as the product of three functions, thus,

$$\psi = \Lambda_1\Lambda_2\Phi,\tag{37}$$

each of which depends on only *one* co-ordinate. For these functions we get the ordinary differential equations

$$
\left\{
\begin{aligned}
&\frac{\partial^2\Phi}{\partial\phi^2} = -n^2\Phi \\
&\frac{\partial}{\partial\lambda_1}\left(\lambda_1\frac{\partial\Lambda_1}{\partial\lambda_1}\right) + \frac{2\pi^2 m}{h^2}\left(-\frac{1}{2}eF\lambda_1^2 + E\lambda_1 + e^2 - \beta - \frac{n^2 h^2}{8\pi^2 m}\frac{1}{\lambda_1}\right)\Lambda_1 = 0, \\
&\frac{\partial}{\partial\lambda_2}\left(\lambda_2\frac{\partial\Lambda_2}{\partial\lambda_2}\right) + \frac{2\pi^2 m}{h^2}\left(\frac{1}{2}eF\lambda_2^2 + E\lambda_2 + e^2 + \beta - \frac{n^2 h^2}{8\pi^2 m}\frac{1}{\lambda_2}\right)\Lambda_2 = 0,
\end{aligned}
\right.
$$

(38)

wherein $n$ and $\beta$ are two further "proper value-like" constants of integration (in addition to $E$), still to be defined. By the choice of symbol for the first of these, we have taken into account the fact that the first of equations (38) makes it take integral values, if $\Phi$ and $\frac{\partial\Phi}{\partial\phi}$ are to be continuous and single-valued functions of the azimuth $\phi$. We then have

$$\Phi = \frac{\sin}{\cos}n\phi\tag{39}$$

and it is evidently sufficient if we do not consider negative values of $n$. Thus

$$n = 0, 1, 2, 3\ldots.\tag{40}$$

---

[1] So far as the actual details of the analysis are concerned, the simplest way to get (32′), or, in general, to get the wave equation for any special co-ordinates, is to transform not the wave equation itself, but the corresponding variation problem (cf. Part I. p. 12), and thus to obtain the wave equation afresh as an Eulerian variation problem. We are thus spared the troublesome evaluation of the *second* derivatives. Cf. Courant-Hilbert, chap. iv. § 7, p. 193.

In the symbol used for the second constant $\beta$, we follow Sommerfeld (*Atombau*, 4th edit., p. 821) in order to make comparison easier. (Similarly, below, with $A$, $B$, $C$, $D$.) We treat the last two equations of (38) together, in the form

$$\frac{\partial}{\partial \xi}\left(\xi \frac{\partial \Lambda}{\partial \xi}\right) + \left(D\xi^2 + A\xi + 2B + \frac{C}{\xi}\right)\Lambda = 0, \qquad (41)$$

where

$$\left.\begin{array}{c}D_1\\D_2\end{array}\right\} = \mp\frac{\pi^2 m e\,F}{h^2},\; A = \frac{2\pi^2 m E}{h^2},\; \left.\begin{array}{c}B_1\\B_2\end{array}\right\} = \frac{\pi^2 m}{h^2}\left(e^2 \mp \beta\right),\; C = -\frac{n^2}{4}, \qquad (42)$$

and the upper sign is valid for $\Lambda = \Lambda_1$, $\xi = \lambda_1$, and the lower one for $\Lambda = \Lambda_2$, $\xi = \lambda_2$. (Unfortunately, we have to write $\xi$ instead of the more appropriate $\lambda$, to avoid confusion with the perturbation parameter $\lambda$ of the general theory, §§ 1 and 2.)

If we omit initially in (41) the Stark effect term $D\xi^2$, which we conceive as a perturbing term (limiting case for vanishing field), then this equation has the same general structure as equation (7) of Part I., and the domain is also the same, from 0 to $\infty$. The discussion is almost the same, word for word, and shows that non-vanishing solutions, which, with their derivatives, are continuous and remain finite within the domain, only exist if *either* $A > 0$ (extended spectrum, corresponding to hyperbolic orbits) *or*

$$\frac{B}{\sqrt[+]{-A}} - \sqrt[+]{-C} = k + \tfrac{1}{2}; \quad k = 0, 1, 2, \ldots \qquad (43)$$

If we apply this to the last two equations of (38) and distinguish the two $k$-values by suffixes 1 and 2, we obtain

$$\left\{\begin{array}{l}\sqrt[+]{-A}\left(k + \tfrac{1}{2} + \sqrt[+]{-C}\right) = B_1 \\[2ex] \sqrt[+]{-A}\left(k_2 + \tfrac{1}{2} + \sqrt[+]{-C}\right) = B_2.\end{array}\right. \qquad (44)$$

By addition, squaring and use of (42) we find

$$A = -\frac{4\pi^4 m^2 e^4}{h^4 l^2} \quad \text{and} \quad E = -\frac{2\pi^2 m e^4}{h^2 l^2} \tag{45}$$

These are the well-known Balmer-Bohr elliptic levels, where as *principal quantum number* enters

$$l = k_1 + k_2 + n + 1. \tag{46}$$

We get the *discrete* term spectrum and the allied proper functions in a way *simpler* than that indicated, if we apply results already known in mathematical literature as follows. We transform first the dependent variable $\lambda$ in (41) by putting

$$\Lambda = \xi^{\frac{n}{2}} u \tag{47}$$

and then the independent $\xi$ by putting

$$2\xi \sqrt{-A} = \eta. \tag{48}$$

We find for $u$ as a function of $\eta$ the equation

$$\frac{d^2 u}{d\eta^2} + \frac{n+1}{\eta}\frac{du}{d\eta} + \left( \frac{D}{(2\sqrt{-A})^3}\eta - \frac{1}{4} + \frac{B}{\sqrt{-A}}\frac{1}{\eta} \right) u = 0. \tag{41'}$$

This equation is very intimately connected with the polynomials named after Laguerre. In the mathematical appendix, it will be shown that the product of $e^{-\frac{x}{2}}$ and the *nth* derivative of the $(n+k)$th Laguerre polynomial satisfies the differential equation

$$y'' + \frac{n+1}{x}y' + \left( -\frac{1}{4} + \left( k + \frac{n+1}{2} \right)\frac{1}{x} \right) y = 0, \tag{103}$$

and that, for a fixed $n$, the functions named form the complete system of proper functions of the equation just written, when $k$ runs through all non-negative integral values. Thus it follows that, for vanishing $D$, equation (41') possesses the proper functions

$$u_k(\eta) = e^{-\frac{\eta}{2}} L_{n+k}^n(\eta) \tag{49}$$

and the proper values

$$\frac{B}{\sqrt{-A}_+} = \frac{n+1}{2} + k \quad (k = 0, 1, 2 \ldots) \tag{50}$$

—and no others! (See the mathematical appendix concerning the remarkable loss of the extended spectrum caused by the apparently inoffensive transformation (48); by this loss the development of the perturbation theory is made much easier.)

We have now to calculate the perturbation of the proper values (50) from the general theory of § 1, caused by including the $D$-term in (41'). The equation becomes self-adjoint if we multiply by $\eta^{n+1}$. The density function $\rho(x)$ of the general theory thus becomes $\eta^n$. As perturbation function $r(x)$ appears

$$-\frac{D}{2(\sqrt{-A}_+)^3} \eta^{n+2} . \tag{51}$$

(We formally put the perturbation parameter $\lambda = 1$; if we desired, we could identify $D$ or $F$ with it.) Now formula (7') gives, for the perturbation of the $k$th proper value,

$$\epsilon_k = -\frac{D}{(2\sqrt{-A}_+)^3} \frac{\int_0^\infty \eta^{n+2} e^{-\eta} \left[ L_{n+k}^n (\eta) \right]^2 d\eta}{\int_0^\infty \eta^n e^{-\eta} \left[ L_{n+k}^n (\eta) \right]^2 d\eta} . \tag{52}$$

For the integral in the denominator, which merely provides for the normalisation, formula (115) of the appendix gives the value

$$\frac{[(n+k)!]^3}{k!} , \tag{53}$$

while the integral in the numerator is evaluated in the same place, as

$$\frac{[(n+k)!]^3}{k!} \left( n^2 + 6nk + 6k^2 + 6k + 3n + 2 \right) . \tag{54}$$

Consequently

$$\epsilon_k = -\frac{D}{(2\sqrt{-A}_+)^3} \left( n^2 + 6nk + 6k^2 + 6k + 3n + 2 \right) . \tag{55}$$

The condition for the $k$th perturbed proper value of equation (41′) and therefore, naturally, also for the $k$th discrete proper value of the original equation (41) runs therefore

$$\frac{B}{\underset{+}{\sqrt{-A}}} = \frac{n+1}{2} + k + \epsilon_k \qquad (56)$$

($\epsilon_k$ is retained meantime for brevity).

This result is applied twice, namely, to the last two equations of (38) by substituting the two systems (42) of values of the constants $A$, $B$, $C$, $D$; and it is to be observed that $n$ is the *same* number in the two cases, while the two $k$-values are to be distinguished by the suffixes 1 and 2, as above. First we have

$$\begin{cases} \dfrac{B_1}{\underset{+}{\sqrt{-A}}} = \dfrac{n+1}{2} + k_1 + \epsilon_{k_1} \\[4mm] \dfrac{B_2}{\underset{+}{\sqrt{-A}}} = \dfrac{n+1}{2} + k_2 + \epsilon_{k_2}, \end{cases} \qquad (57)$$

whence comes

$$A = -\frac{(B_1 + B_2)^2}{\left(l + \epsilon_{k_1} + \epsilon_{k_2}\right)^2} \qquad (58)$$

(applying abbreviation (46) for the principal quantum number). In the approximation we are aiming at we may expand with respect to the small quantities $\epsilon_k$ and get

$$A = -\frac{(B_1 + B_2)^2}{l^2}\left[1 - \frac{2}{l}\left(\epsilon_{k_1} + \epsilon_{k_2}\right)\right]. \qquad (59)$$

Further, in the calculation of these small quantities, we may use the approximate value (45) for $A$ in (55). We thus obtain, noticing the two $D$ values, by (42),

$$\begin{cases} \epsilon_{k_1} = +\dfrac{Fh^4l^3}{64\pi^4 m^2 e^5}\left(n^2 + 6nk_1 + 6k_1^2 + 6k_1 + 3n + 2\right) \\[4mm] \epsilon_{k_2} = -\dfrac{Fh^4l^3}{64\pi^4 m^2 e^5}\left(n^2 + 6nk_2 + 6k_2^2 + 6k_2 + 3n + 2\right). \end{cases} \qquad (60)$$

Addition gives, after an easy reduction,

$$\epsilon_{k_1} + \epsilon_{k_2} = \frac{3Fh^4l^4(k_1 - k_2)}{32\pi^4m^2e^5}. \tag{61}$$

If we substitute this, and the values of $A$, $B_1$, and $B_2$ from (42) in (59), we get, after reduction,

$$E = -\frac{2\pi^2me^4}{h^2l^2} - \frac{3}{8}\frac{h^2Fl(k_2 - k_1)}{\pi^2me}. \tag{62}$$

This is our provisional *conclusion*; it is the well-known formula of Epstein for the term values in the Stark effect of the hydrogen spectrum.

$k_1$ and $k_2$ correspond fully to the parabolic quantum numbers; they are capable of taking the value zero. Also the integer $n$, which has evidently to do with the *equatorial* quantum number, may from (40) take the value zero. However, from (46) the sum of these three numbers must still be increased by unity in order to yield the principal quantum number. Thus $(n + 1)$ and not $n$ corresponds to the equatorial quantum number. The value zero for the *latter* is thus *automatically* excluded by wave mechanics, just as by Heisenberg's mechanics.[1] *There is simply no proper function, i.e.* no state of vibration, which corresponds to such a meridional orbit. This important and gratifying circumstance was already brought to light in Part I. in counting the constants, and also afterwards in § 2 of Part I. in connection with the azimuthal quantum number, through the non-existence of states of vibration corresponding to *pendulum orbits*; its full meaning, however, only fully dawned on me through the remarks of the two authors just quoted.

For later application, let us note the system of proper functions of equation (32) or (32′) in "zero approximation", which belongs to the proper values (62). It is obtained from statement (37), from conclusions (39) and (49), and from consideration of transformations (47) and (48) and of the approximate value (45) of $A$. For brevity, let

---

[1] W. Pauli, jun., *Ztschr. f. Phys.* 36, p. 336, 1926; N. Bohr, *Die Naturw.* 1, 1926.

us call $a_0$ the "radius of the first hydrogen orbit". Then we get

$$\frac{1}{2l\sqrt{-A}} = \frac{h^2}{4\pi^2 m e^2} = a_0. \tag{63}$$

The proper functions (not yet normalised!) then read

$$\psi n k_1 k_2 = \lambda_1^{\frac{n}{2}} \lambda_2^{\frac{n}{2}} e^{-\frac{\lambda_1 + \lambda_2}{2l a_0}} L_{n+k_1}^n \left(\frac{\lambda_1}{l a_0}\right) L_{n+k_2}^n \left(\frac{\lambda_2}{l a_0}\right) \frac{\sin}{\cos} n\phi. \tag{64}$$

They belong to the proper values (62), where $l$ has the meaning (46). To each non-negative integral trio of values $n$, $k_1$, $k_2$ belong (on account of the double symbol $\frac{\sin}{\cos}$) *two* proper functions or *one*, according as $n > 0$ or $n = 0$.

## § 4. ATTEMPT TO CALCULATE THE INTENSITIES AND POLARISATIONS OF THE STARK EFFECT PATTERNS

I have lately shown[1] that from the proper functions we can calculate by differentiation and quadrature the elements of the *matrices*, which are allied in Heisenberg's mechanics to functions of the generalised position- and momentum-co-ordinates. For example, for the $(rr')$th element of the matrix, which according to Heisenberg belongs to the generalised co-ordinate $q$ itself, we find

$$\begin{cases} q^{rr'} = \int q\rho(x)\,\psi_r(x)\,\psi_{r'}(x)\,dx \\ \quad \cdot \left\{ \int \rho(x)\,[\psi_r(x)]^2\,dx \cdot \int \rho(x)\,[\psi_{r'}(x)]^2\,dx \right\}^{-\frac{1}{2}}. \end{cases} \tag{65}$$

Here, for our case, the separate indices *each* deputise for a *trio* of indices $n$, $k_1$, $k_2$, and further, $x$ represents the three co-ordinates $r$, $\theta$, $\phi$. $\rho(x)$ is the density function; in our case the quantity (34). (We may compare the self-adjoint equation (32′) with the general form (2)). The "denominator" $(\ldots)^{-\frac{1}{2}}$ in (65) must be put in because our system (64) of functions is not yet normalised.

According to Heisenberg,[2] now, if $q$ means a rectangular Cartesian co-ordinate, then the square of the matrix element (65) is

---

[1] Preceding paper of this collection.
[2] W. Heisenberg, *Ztschr. f. Phys.* 33, p. 879, 1925; M. Born and P. Jordan, *Ztschr. f. Phys.* 34, pp. 867, 886, 1925.

to be a measure of the "probability of transition" from the $r$th state to the $r'$th, or, more accurately, a measure of the intensity of that part of the radiation, bound up with this transition, which is polarised in the $q$-direction. Starting from this, I have shown in the above paper that if we make certain simple assumptions as to the electrodynamical meaning of $\psi$, the "mechanical field scalar", then the matrix element in question is susceptible of a very simple physical interpretation in wave mechanics, namely, *actually:* component of the amplitude of the periodically oscillating electric moment of the atom. The word *component* is to be taken in a double sense: (1) component in the $q$-direction, *i.e.* in the spatial direction in question, and (2) only the part of this spatial component which changes in a time-sinusoidal manner with exactly the frequency of the emitted light, $|E_r - E_{r'}|/h$. (It is a question then of a kind of Fourier analysis: not in harmonic frequencies, but in the actual frequencies of emission.) However, the idea of wave mechanics is not that of a sudden transition from one state of vibration to another, but according to it, the partial moment concerned—as I will briefly name it—arises from the *simultaneous existence* of the two proper vibrations, and lasts just as long as both are excited together.

Moreover, the above assertion that the $q^{rr'}$'s are proportional to the partial moments is more accurately phrased thus. The ratio of, *e.g.*, $q^{rr'}$ to $q^{rr''}$ is equal to the ratio of the partial moments which arise when the proper function $\psi_r$ and the proper functions $\psi_{r'}$ and $\psi_{r''}$ are stimulated, the first *with any strength whatever* and the last two with strengths *equal to one another*—*i.e.* corresponding to normalisation. To calculate the ratio of the *intensities*, the $q$-quotient must first be squared and then multiplied by the ratio of the fourth powers of the emission frequencies. The latter, however, has no part in the intensity ratio of the Stark effect components, for there we only compare intensities of lines which have practically the same frequency.

The known *selection* and *polarisation rules* for Stark effect components can be obtained, almost without calculation, from the integrals in the numerator of (65) and from the form of the proper functions in (64). They follow from the vanishing or non-vanishing of the integral with respect to $\phi$. We obtain the components whose electric vector vibrates *parallel* to the field, *i.e.* to the $z$-direction, by replacing the $q$ in (65) by $z$ from (33). The expression for $z$, *i.e.* $(\frac{1}{2}\lambda_1 - \lambda_2)$, does *not* contain the azimuth $\phi$. Thus we see at once from (64) that a non-vanishing result after integration with respect to $\phi$ can only arise if we combine proper functions whose $n$'s are *equal*, and thus whose equatorial quantum numbers are equal, being in fact equal to $n + 1$. For the components which vibrate *perpendicular* to the field, we must put $q$ equal to $x$ or equal to $y$ (cf. equation (33)). Here $\cos \phi$ or $\sin \phi$ enters, and we see almost as easily as before, that the $n$-values of the two combined proper functions must differ exactly by unity, if the integration with respect to $\phi$ is to yield a non-vanishing result. Hence the known selection and polarisation rules are proved. Further, it should be recalled again that we do not require to exclude any $n$-value after additional reflection, as was necessary in the older theory in order to agree with experience. Our $n$ is smaller by 1 than the equatorial quantum number, and right from the beginning cannot take negative values (quite the same state of affairs exists, we know, in Heisenberg's theory).[1]

The numerical evaluation of the integrals with respect to $\lambda_1$ and $\lambda_2$ which appear in (65) is exceptionally tedious, especially for those of the numerator. The same apparatus for calculating comes into play as served already in the evaluation of (52), only the matter is somewhat more detailed because the two (generalised) Laguerre polynomials, whose product is to be integrated, have not the same argument. By good luck, in the *Balmer lines*, which interest us principally, one of the two polynomials $L_{n+k}^n$, namely that relating to the doubly quantised state, is either a constant or is a linear function of its argument. The method of calculation is described more fully in the mathematical appendix. The following tables and diagrams give the results for the first four Balmer lines, in comparison with the known measurements and estimates of intensity, made by Stark[2] for a field strength of about 100,000 volts per centimetre. The first column

[1] W. Pauli, jun., *Ztschr. f. Physik*, 36, p. 336, 1926.
[2] J. Stark, *Ann. d. Phys.* 48, p. 193, 1915.

indicates the state of polarisation, the second gives the combination of the terms in the usual manner of description, *i.e.* in *our* symbols: of the two trios of numbers $(k_1, k_2, n + 1)$ the *first* trio refers to the higher quantised state and the *second* to the doubly quantised state. The third column, with the heading $\Delta$, gives the term decomposition in multiples of $3h^2 F / 8\pi^2 me$, (see equation (62)). The next column gives the intensities observed by Stark, and 0 there signifies not observed. The question mark was put by Stark at such lines as clash either with irrelevant lines or with possible "ghosts" and thus cannot be guaranteed. On account of the unequal weakening of the two states of polarisation in the spectrograph, according to Stark his results for the || and for the $\perp$ components of vibration are not directly comparable with one another. Finally, the last column gives the results of our calculation in *relative numbers*, which are comparable for the collective components (|| and $\perp$) of *one* line, *e.g.* of $H_\alpha$, but not for those of $H_\alpha$ with $H_\beta$, etc. These relative numbers are reduced to their *smallest integral* values, *i.e.* the numbers in each of the four tables are *prime* to each other.

# INTENSITIES IN THE STARK EFFECT OF THE BALMER LINES

## TABLE 1
### $H_\alpha$

| Polarisation. | Combination. | Δ | Observed Intensity. | Calculated Intensity. |
|---|---|---|---|---|
| ‖ | (111) (011) | 2 | 1 | 729 |
| | (102) (002) | 3 | 1.1 | 2304 |
| | (201) (101) | 4 | 1.2 | 1681 |
| | (201) (011) | 8 | 0 | 1 |
| | | | | Sum: 4715 |
| ⊥ | (003) (002) | 0 | } 2.6 | 4608 |
| | (111) (002) | 0 | | 882 |
| | (102) (101) | 1 | 1 | 1936 |
| | (102) (011) | 5 | 0 | 16 |
| | (201) (002) | 6 | 0 | 18 |
| | | | | Sum*: 4715 |

*Undisplaced components halved.

## TABLE 2
### $H_\beta$

| Polarisation. | Combination. | Δ | Observed Intensity. | Calculated Intensity. |
|---|---|---|---|---|
| ‖ | (112) (002) | 0 | 1.4 | 0 |
| | (211) (101) | 2 | 1.2 | 9 |
| | — | (4) | 1 | 0 |
| | (211) (011) | 6 | 4.8 | 81 |
| | (202) (002) | 8 | 9.1 | 384 |
| | (301) (101) | 10 | 11.5 | 361 |
| | — | (12) | 1 | 0 |
| | (301) (011) | 14 | 0 | 1 |
| | | | | Sum: 836 |
| ⊥ | — | (0) | 1.4 | 0 |
| | (112) (011) | 2 | 3.3 | 72 |
| | (103) (002) | 4 | } 12.6 | 384 |
| | (211) (002) | 4 | | 72 |
| | (202) (101) | 6 | 9.7 | 294 |
| | — | (8) | 1.3 | 0 |
| | (202) (011) | 10 | 1.1? | 6 |
| | (301) (002) | 12 | 1 ? | 8 |
| | | | | Sum: 836 |

TABLE 3

$H_\gamma$

| Polarisation. | Combination. | Δ | Observed Intensity. | Calculated Intensity. |
|---|---|---|---|---|
| ‖ | (221) (011) | 2 | 1.6 | 15 625 |
| | (212) (002) | 5 | 1.5 | 19 200 |
| | (311) (101) | 8 | 1 | 1 521 |
| | (311) (011) | 12 | 2.0 | 16 641 |
| | (302) (002) | 15 | 7.2 | 115 200 |
| | (401) (101) | 18 | 10.8 | 131 769 |
| | (401) (011) | 22 | 1 ? | 729 |
| | | | | Sum: 300 685 |
| ⊥ | (113) (002) | 0 | } 7.2 | 115 200 |
| | (221) (002) | 0 | | 26 450 |
| | (212) (101) | 3 | 3.2 | 46 128 |
| | (212) (011) | 7 | 1.2 | 5 808 |
| | (203) (002) | 10 | } 4.3 | 76 800 |
| | (311) (002) | 10 | | 11 250 |
| | (302) (101) | 13 | 6.1 | 83 232 |
| | (302) (011) | 17 | 1.1 | 2 592 |
| | (401) (002) | 20 | 1 | 4 050 |
| | | | | Sum*: 300 685 |

*Undisplaced components halved.

TABLE 4

$H_\delta$

| Polarisation. | Combination. | Δ | Observed Intensity. | Calculated Intensity. |
|---|---|---|---|---|
| ‖ | (222) (002) | 0 | 0 | 0 |
| | (321) (101) | 4 | 1 | 8 |
| | (321) (011) | 8 | 1.2 | 32 |
| | (312) (002) | 12 | 1.5 | 72 |
| | (411) (101) | 16 | 1.2 | 18 |
| | (411) (011) | 20 | 1.1 | 18 |
| | (402) (002) | 24 | 2.8 | 180 |
| | (501) (101) | 28 | 7.2 | 242 |
| | (501) (011) | 32 | 1 ? | 2 |
| | | | | Sum: 572 |
| ⊥ | (222) (011) | 2 | 1.3 | 36 |
| | (213) (002) | 6 | } 3.2 | 162 |
| | (321) (002) | 6 | | 36 |
| | (312) (101) | 10 | 2.1 | 98 |
| | (312) (011) | 14 | 1 | 2 |
| | (303) (002) | 18 | } 2.0 | 90 |
| | (411) (002) | 18 | | 9 |
| | (402) (101) | 22 | 2.4 | 125 |
| | (402) (011) | 26 | 1.3 | 5 |
| | (501) (002) | 30 | 1 ? | 9 |
| | | | | Sum: 572 |

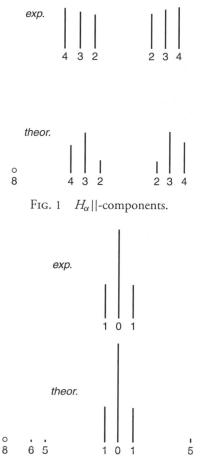

FIG. 1   $H_\alpha \|$-components.

FIG. 2   $H_\alpha \perp$-components.

In the *diagrams* it is to be noticed that, on account of the huge differences in the theoretical intensities, some theoretical intensities cannot be truly represented to scale, as they are much too small. These are indicated by *small circles*.

A consideration of the diagrams shows that the agreement is tolerably good for almost all the strong components, and taken all over it is somewhat better than for the values deduced from correspondence considerations.[1] Thus, for example, is removed one of the most serious contradictions which arose, in that the correspondence principle

[1] H. A. Kramers, *Dänische Akademie* (8), iii. 3, p. 333 *et seq.*, 1919.

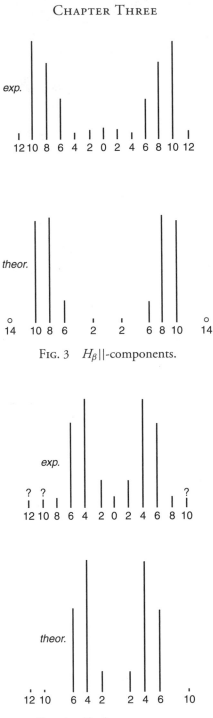

FIG. 3    $H_\beta \| $-components.

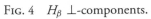

FIG. 4    $H_\beta \perp$-components.

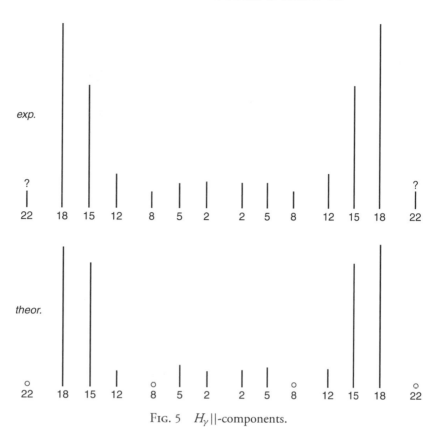

FIG. 5   $H_\gamma\|$-components.

gave the ratio of the intensities of the two strong $\perp$-components of $H_\beta$, for $\Delta = 4$ and 6, inversely and indeed very much out, in fact as almost $1:2$, while experiment requires about $5:4$. A similar thing occurs with the mean ($\Delta = 0$) $\perp$-components of $H_\gamma$, which decidedly preponderate experimentally, but are given as *far* too weak by the correspondence principle. In *our* diagrams also, it is admitted that such "reciprocities" between the intensity ratios of intense components demanded by theory and by experiment are not entirely wanting. The theoretically most intense $\|$-component ($\Delta = 3$) of $H_\alpha$ is furthest out; by experiment, it should lie *between* its neighbours in intensity. And the two strongest $\|$-components of $H_\beta$ and two $\perp$-components ($\Delta = 10, 13$) of $H_\gamma$ are given "reciprocally" by the theory. Of course, in both cases the intensity ratios, both experimentally and theoretically, are pretty near unity.

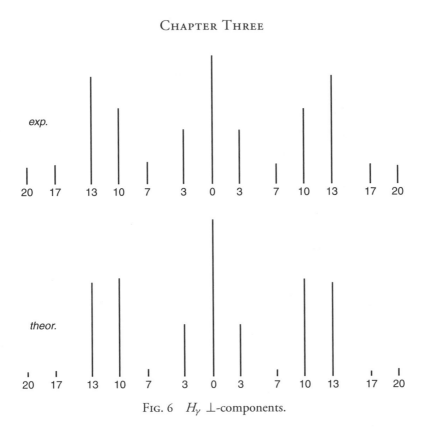

FIG. 6   $H_\gamma$ ⊥-components.

Passing now to the weaker components, we notice first that the contradiction which exists for some weak observed components of $H_\beta$ to the selection and polarisation rules, of course still remains in the new theory, since the latter gives these rules in conformity with the older theory. However, components which are extremely weak theoretically are for the most part *un*observed, or the observations are *questionable*. The strength *ratios* of weaker components to one another or to stronger ones are *almost never* given even approximately correctly; cf. especially $H_\gamma$ and $H_\delta$. Such serious mistakes in the experimental determination of the blackening are of course out of the question.

Considering all this, we might feel inclined to be very sceptical of the thesis that the integrals (65) or their squares are measures of intensity. I am far from wishing to represent this thesis as irrefutable. There are still many alterations conceivable, and these may, perhaps, be necessitated by internal reasons when the theory is further extended.

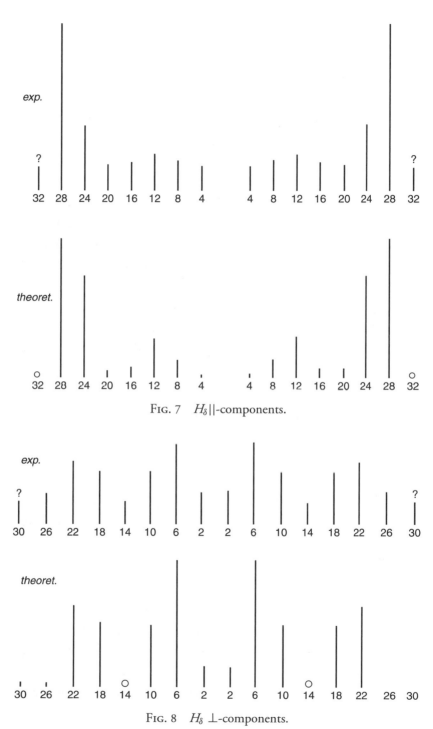

FIG. 7  $H_\delta ||$-components.

FIG. 8  $H_\delta \perp$-components.

Yet the following should be remembered. The whole calculation has been performed with the *unperturbed* proper functions, or more precisely, with the *zero* approximation to the perturbed ones (cf. above § 2). It, therefore, represents an approximation for a *vanishing* field strength! However, just for the weak or almost vanishing components we should expect theoretically a fairly powerful growth with increasing field strength, for the following reason. According to the view of wave mechanics, as explained at the beginning of this section, the integrals (65) represent the amplitudes of the electrical partial moments, which are produced by the distribution of charges which flow round about the nucleus within the atom's domain. When for a line component we get as a zero approximation very weak or even vanishing intensity, this is not caused in any way by the fact that to the simultaneous existence of the two proper vibrations corresponds only an insignificant motion of electricity, or even none at all. The vibrating mass of electricity—if this vague expression is allowed—may be represented as the same in all components, on the ground of *normalisation*. Rather is the reason for the low line intensity to be found in a high degree of *symmetry* in the motion of the electricity, through which only a small, or even no, dipole moment arises (on the contrary, *e.g.*, only a four-pole moment). Therefore it is to be expected that the *vanishing* of a line component in presence of perturbations of any kind is a relatively *unstable* condition, since the symmetry is probably destroyed by the perturbation. And thus it may be expected that weak or vanishing components gain quickly in intensity with increasing field strength.

This has now actually been observed, and the intensity ratios, indeed, alter quite considerably with field strength, for strengths of about 10,000 gauss and upwards; and, if I understand aright, in the way[1] shown by the present general discussion. Certain information on the question whether this really explains these discrepancies could of course only be got from a continuation of the calculation to the next approximation, but this is *very* troublesome and complicated.

---

[1] J. Stark, *Ann. d. Phys.* 43, p. 1001 *et seq.*, 1914.

The present considerations are of course nothing but the "translation" into the language of the new theory of very well-known considerations which Bohr[1] has brought forward in connection with calculation of line intensities by means of the principle of correspondence.

The theoretical intensities given in the tables satisfy a fundamental requirement, which is set up not only by intuition but also by experiment,[2] viz., the sum of the intensities of the ||-components is equal to that of the ⊥-components. (Before adding, *undisplaced* components must be *halved*—as a compensation for the *duplication* of all the others, which occur on both sides.) This makes a very welcome "control" for the arithmetic.

It is also of interest to compare the *total intensities* of the four lines by using the four "sums" given in the tables. For this purpose I take back from my numerical calculations the four factors, which were omitted in order to represent the intensity ratios within each of the four line groups by the smallest integers possible, and multiply by them. Further, I multiply each of these four products by the *fourth* power of the appropriate emission frequency. Thus I obtain the following four numbers:

$$\text{for } H_\alpha \ldots \quad \frac{2^6 \cdot 23 \cdot 41}{3^2 \cdot 5^9} = 0.003433\ldots$$

$$\text{for } H_\beta \ldots \quad \frac{4 \cdot 11 \cdot 19}{3^{12}} = 0.001573\ldots$$

$$\text{for } H_\gamma \ldots \quad \frac{2^6 \cdot 3^6 \cdot 11^2 \cdot 71}{5 \cdot 7^{13}} = 0.0008312\ldots$$

$$\text{for } H_\delta \ldots \quad \frac{11 \cdot 13}{2^{15} \cdot 3^2} = 0.0004849\ldots$$

I give these numbers *with still greater reserve* than the former ones because I am not sure, theoretically, about the *fourth* power of the frequency. Investigations[3] which I have lately published seem to call,

---

[1] N. Bohr, *Dänische Akademie* (8), iv. 1. 1, p. 35, 1918.

[2] J. Stark, *Ann. d. Phys.* 43, p. 1004, 1914.

[3] Equation (38) at end of previous paper of this collection. The *fourth* allows for the fact that for the radiation it is a question of the square of the *acceleration* and not of the electric moment itself. In this equation (38) occurs explicitly another factor $(E_k - E_m)/h$. This is occasioned by the appearance of $\partial/\partial t$ in statement (36). *Addition at proof correction:* Now I recognise this $\partial/\partial t$ to be incorrect, though I

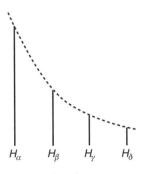

FIG. 9    Total Intensities.

perhaps, for the *sixth*. The above method of calculation corresponds exactly to the assumptions of Born, Jordan, and Heisenberg.[1] Fig. 9 represents the results diagrammatically.

Actual measured intensities of emission lines, which are known to depend greatly on the conditions of excitation, naturally cannot here be used in a comparison with experience. From his researches[2] on dispersion and magneto-rotation in the neighbourhood of $H_\alpha$ and $H_\beta$, R. Ladenburg has, with F. Reiche,[3] calculated the value 4.5 (limits 3 and 6) for the ratio of the so-called "electronic numbers" of these two lines. If I assume that the above numbers may be taken as proportional to Ladenburg's[4] expression,

$$\sum \frac{g_k}{g_i} a_{ki} \nu_0,$$

then they may be reduced to (relative) "electronic numbers" by division by $\nu_0^3$, *i.e.* by

$$\left(\frac{5}{36}\right)^3, \quad \left(\frac{3}{16}\right)^3, \quad \left(\frac{21}{100}\right)^3, \quad \text{and} \quad \left(\frac{2}{9}\right)^3 \quad \text{respectively.}$$

Hence we obtain the four numbers,

$$1.281, \quad 0.2386, \quad 0.08975, \quad 0.04418.$$

hoped it would make the later relativistic generalisation easier. Statement (36), *loc. cit.*, is to be replaced by $\psi\bar\psi$. The above doubts about the *fourth* power are therefore dissolved.

[1] Cf. M. Born and P. Jordan, *Ztschr. f. Phys.* 34, p. 887, 1925.
[2] R. Ladenburg, *Ann. d. Phys.* (4), 38, p. 249, 1912.
[3] R. Ladenburg and F. Reiche, *Die Naturwissenschaften*, 1923, p. 584.
[4] Cf. Ladenburg-Reiche, *loc. cit.*, the first formula in the second column, p. 584. The factor $\nu_0$ in the *above* expression comes from the fact that the "transition probability" $a_{ki}$ is still to be multiplied by the "energy quantum" to give the intensity of the radiation.

The ratio of the first to the second is 5.37, which agrees sufficiently with Ladenburg's value.

## § 5. Treatment of the Stark Effect By the Method Which Corresponds to that of Bohr

Mainly to give an *example* of the general theory of § 2, I wish to outline *that* treatment of the proper value problem of equation (32), which must have been adopted, if we had *not* noticed that the perturbed equation is also exactly "separable" in parabolic co-ordinates. We therefore now keep to the polar co-ordinates $r$, $\theta$, $\phi$, and thus replace $z$ by $r \cos \theta$. We also introduce a new variable $\eta$ for $r$ by the transformation

$$2r\sqrt{-\frac{8\pi^2 m E}{h^2}} = \eta, \qquad (66)$$

(which is closely akin to transformation (48) for the parabolic co-ordinate $\xi$). For one of the unperturbed proper values (45), we get from (66)

$$\eta = \frac{2r}{la_0}, \qquad (66')$$

where $a_0$ is the same constant as in (63). ("Radius of the innermost hydrogen orbit.") If we introduce this and the unperturbed value (45) into the equation (32), which is to be treated, then we obtain

$$\nabla'^2 \psi + \left( -\frac{1}{4} - g\eta \cos\theta + \frac{l}{\eta} \right) \psi = 0, \qquad (67)$$

where for brevity

$$g = \frac{a_0^2 F l^3}{4e}. \qquad (68)$$

The dash on the Laplacian operator is merely to signify that in it the letter $\eta$ is to be written for the radius vector.

In equation (67) we *conceive l to be the proper value*, and the term in *g* to be the perturbing term. The fact that the perturbing term *contains* the proper value need not trouble us in the first approximation. If we neglect the perturbing term, the equation has as proper values the natural numbers

$$l = 1, 2, 3, 4 \ldots \tag{69}$$

*and no others.* (The extended spectrum is again cut out by the artifice (66), which would be valuable for closer approximations.) The allied *proper functions* (not yet normalised) are

$$\psi l_{nm} = P_n^m (\cos \theta) \, \frac{\cos}{\sin} (m\phi) \cdot \eta^n e^{-\frac{\eta}{2}} L_{n+l}^{2n+1} (\eta) . \tag{70}$$

Here $P_n^m$ signifies the $m$th "associated" Legendre function of the $n$th order, and $L_{n+l}^{2n+1}$ is the $(2n + 1)$th derivative of the $(n + l)$th Laguerre polynomial.[1] So we must have

$$n < l,$$

otherwise $L_{n+l}^{2n+1}$ would vanish, because the number of differentiations would be greater than the degree. With reference to this, the numbering of the spherical surface harmonics shows that $l$ is an $l^2$-fold proper value of the unperturbed equation. We now investigate the *splitting up* of a *definite* value of $l$, supposed fixed in what follows, due to the addition of the perturbing term.

To do this we have, *in the first place*, to normalise our proper functions (70), according to § 2. From an uninteresting calculation, which is easily performed with the aid of the formulae in the appendix,[2] we get as the normalising factor

$$\frac{1}{\sqrt{\pi}} \sqrt{\frac{2n + 1}{2}} \sqrt{\frac{(n - m)!}{(n + m)!}} \sqrt{\frac{(l - n - 1)!}{[(n + l)!]^3}} , \tag{71}$$

if $m \neq 0$, but, for $m = 0$, $\frac{1}{\sqrt{2}}$ times this value. *Secondly,* we have to calculate the symmetrical matrix of constants $\epsilon_{im}$, according to (22). The $r$ there is to be identified[2] with our perturbing function $-g\eta^3 \cos \theta \sin \theta$, and the proper functions, there called $u_{ki}$, are to be identified with our functions (70). The fixed suffix $k$, which characterises the proper value, corresponds to the *first* suffix $l$ of $\psi_{lnm}$,

---

[1] I lately gave the proper functions (70) (see Part I.), but without noticing their connection with the Laguerre polynomials. For the proof of the above representation, see the Mathematical Appendix, section 1.

[2] It is to be noticed that the *density function*, generally denoted by $\rho(x)$, reads as $\eta \sin \theta$ in equation (67), because the equation must be multiplied by $\eta^2 \sin \theta$, in order to acquire self-adjoint form.

and the other suffix $i$ of $u_{ki}$ corresponds now to the *pair* of suffixes $n$, $m$ in $\psi_{lnm}$. The matrix (22) of constants forms in our case a square of $l^2$ rows and $l^2$ columns. The quadratures are easily carried out by the formulae of the appendix and yield the following results. Only those elements of the matrix are different from zero, for which the two proper functions $\psi_{lnm}$, $\psi_{ln'm'}$, to be combined, satisfy the following conditions simultaneously:

1. The *upper indices* of the "associated Legendre functions" must agree, *i.e.* $m = m'$.

2. The *orders* of the two Legendre functions must differ exactly by unity, *i.e.* $|n - n'| = 1$.

3. To each trio of indices $lnm$, if $m \neq 0$, there belong, according to (70), *two* Legendre functions, and thus also *two* proper functions $\psi_{lnm}$, which only differ from each other in that one contains a factor cos $m\phi$ and the other sin $m\phi$. The third condition reads: we may only combine sine with sine, or cosine with cosine, and not sine with cosine.

The remaining non-vanishing elements of the desired matrix would have to be characterised from the beginning by *two* index-pairs $(n, m)$ and $(n + 1, m)$. (We renounce any idea of showing the fixed index $l$ explicitly.) Since the matrix is symmetrical, *one* index pair $(n, m)$ is sufficient, if we stipulate that the first index, *i.e.* $n$, shall mean the *greater* of the two orders $n$, $n'$, in every case.

Then the calculation gives

$$\epsilon_{nm} = -6lg\sqrt{\frac{(l^2 - n^2)(n^2 - m^2)}{4n^2 - 1}}. \tag{72}$$

We have now to form the determinant (22) out of these elements. It is advantageous to *arrange* its rows *as well as* its columns on the following principle. (To fix our ideas, let us speak of the columns, and therefore of the index-pair characterising the *first* of the two Legendre functions.) Thus: first come all terms with $m = 0$, then all with $m = 1$, then all with $m = 2$, etc., and finally, all terms with $m = l - 1$, which last is the greatest value that $m$ (like $n$) can take.

*Inside* each of these groups, let us arrange the terms thus: first, all terms with cos $m\phi$, and then all with sin $m\phi$. Within these "half groups" let us arrange them in order of increasing $n$, which runs through the values $m$, $m + 1$, $m + 2 \ldots l - 1$, *i.e.* $(l - m)$ values in all.

If we carry this out, we find that the non-vanishing elements (72) are exclusively confined to the two secondary diagonals, which lie immediately alongside the principal diagonal. On the latter are the proper value perturbations which are to be found, but taken negatively, while everywhere else are zeros. Further, the two secondary diagonals are interrupted by zeros at *those* places, where they break through the *boundaries* between the so-called "half-groups", in very convenient fashion. Hence the whole determinant *breaks up* into a product of just so many smaller determinants as there are "half-groups" present, viz. $(2l - 1)$. It will be sufficient if we consider one of them. We write it here, denoting the desired perturbation of the proper value by $\epsilon$ (without suffix):

$$\begin{vmatrix} -\epsilon & \epsilon_{m+1,m} & 0 & 0 & \cdots & 0 \\ \epsilon_{m+1,m} & -\epsilon & \epsilon_{m+2,m} & 0 & \cdots & 0 \\ 0 & \epsilon_{m+2,m} & -\epsilon & \epsilon_{m+3,m} & \cdots & 0 \\ 0 & 0 & \epsilon_{m+3,m} & -\epsilon & \cdots & 0 \\ \cdots & \cdots & \cdots & \cdots & \cdots & \cdots \\ \cdots & \cdots & \cdots & \cdots & \cdots & \cdots \\ 0 & 0 & 0 & 0 & \epsilon_{l-1,m} & -\epsilon \end{vmatrix} \quad (73)$$

If we divide each term here by the common factor $6lg$ of the $\epsilon_{nm}$'s (cf. (72)), and for the moment regard as the unknown

$$k^* = -\frac{\epsilon}{6lg}, \quad (74)$$

the above equation of the $(l - m)$th degree has the roots

$$k^* = \pm (l - m - 1), \pm (l - m - 3), \pm (l - m - 5) \ldots \quad (75)$$

where the series stops with $\pm 1$ or $0$ (inclusive) according as the *degree* $l - m$ is *even* or *odd*. The proof of this is unfortunately *not* to be found in the appendix, as I have not been successful in obtaining it.

If we form the series (75) for each of the values $m = 0, 1, 2\ldots$ $(l - 1)$, then we have in the numbers

$$\epsilon = -6lgk^*  \tag{76}$$

the complete set of *perturbations of the principal quantum number l.* In order to find the perturbed *proper* values $E$ (term-levels) of the equation (32), we have only to substitute (76) in

$$E = -\frac{2\pi^2 me^4}{h^2(l + \epsilon)^2}  \tag{77}$$

taking into account the signification of the abbreviations $g$ (see (68)) and $a_0$ (see (63)).

After reducing this gives

$$E = -\frac{2\pi^2 me^4}{h^2 l^2} - \frac{3}{8}\frac{h^2 Flk^*}{\pi^2 me}.  \tag{78}$$

Comparison with (62) shows that $k^*$ is the *difference* $k_2 - k_1$ of the parabolic quantum numbers. From (75), bearing in mind the range of values of $m$ referred to above, we see that $k^*$ may also take the same values as the difference just mentioned, viz. $0, 1, 2\ldots(l - 1)$. Also, if we take the trouble to work it out, we will find for the *multiplicity*, in which $k^*$ and the difference $k_2 - k_1$ appear, the same value, viz. $l - |k^*|$.

We have thus obtained the proper value perturbations of the first order also from the general theory. The next step would be the solution of the system (21′) of linear equations of the general theory for the $\kappa$-quantities. These would then yield, according to (18) (provisionally putting $\lambda = 0$), the perturbed proper functions of zero order; this is nothing more than a representation of the proper functions (64) as linear forms of the proper functions (70). In our case the solution of (21′) would naturally be anything but unique, on account of the considerable multiplicity of the roots $\epsilon$. The solution is made much simpler if we notice that the equations break up into *just as many* groups, viz. $(2l - 1)$, or, retaining the former expression, *half-groups*, with completely separated variables, as the determinant investigated above contains factors like (73); and if we further notice that it is allowable, after we have chosen a definite $\epsilon$-value, to regard only the variables $\kappa$ of a *single* half-group as different from zero, of that half-group, in fact, for which the determinant (73) vanishes for the chosen $\epsilon$-value. The definition of this half-group of variables is then *unique*.

But our object, viz. to illustrate the general method of § 2 by an example, has been sufficiently attained. Since the continuation of the calculation is of no special physical interest, I have not troubled to bring the determinantal quotients, which we immediately obtain for the coefficients $\kappa$, into a clearer form, or to work out the transformation to principal axes in any other way.

On the whole, we must admit that in the present case the method of secular perturbations (§ 5) is considerably more troublesome than the direct application of a system of separation (§ 3). I believe that this may ako be true in other cases. In ordinary mechanics it is, as we know, usually quite the reverse.

# III. MATHEMATICAL APPENDIX

*Prefatory Note:*—It is not intended to supply in uninterrupted detail all the calculations omitted from the text. Without that, the present paper has already become too long. In general, only those methods of calculation will be briefly described which another might utilise with advantage in similar work, if something better does not occur to him—as it may easily do.

## § 1. THE GENERALISED LAGUERRE POLYNOMIALS AND ORTHOGONAL FUNCTIONS

The $k$th Laguerre polynomial $L_k(x)$ satisfies the differential equation[1]

$$xy'' + (1 - x)y' + ky = 0. \tag{101}$$

If we first replace $k$ by $n + k$, and then differentiate $n$ times, we find that the $n$th derivative of the $(n + k)$th Laguerre polynomial, which

[1] Courant-Hilbert, chap. ii. § 11, 5, p. 78, equation (72).

we will always denote by $L_{n+k}^n$, satisfies the equation

$$xy'' + (n + 1 - x)y' + ky = 0. \tag{102}$$

Moreover, by an easy transformation, we find that for $e^{-\frac{x}{2}} L_{n+1}^n(x)$ the following equation holds,

$$y'' + \frac{n+1}{x} y' + \left( -\frac{1}{4} + \left( k + \frac{n+1}{2} \right) \frac{1}{x} \right) y = 0. \tag{103}$$

This found an application in equation (41') of § 3. The allied generalised Laguerre orthogonal functions are

$$x^{\frac{n}{2}} e^{-\frac{x}{2}} L_{n+k}^n(x). \tag{104}$$

*Their* equation, it may be remarked in passing, is

$$y'' + \frac{1}{x} y' + \left( -\frac{1}{4} + \left( k + \frac{n+1}{2} \right) \frac{1}{x} - \frac{n^2}{4x^2} \right) y = 0. \tag{105}$$

Let us turn to equation (103), and consider there that $n$ is a fixed (real) integer, and $k$ is the proper value parameter. Then, according to what has been said, in the domain $x \geq 0$, at any rate, the equation has the proper functions,

$$e^{-\frac{x}{2}} L_{n+k}^n(x), \tag{106}$$

belonging to the proper values,

$$k = 0, 1, 2, 3, \ldots \tag{107}$$

In the text it is maintained that it has no further values, and, above all, that it possesses no continuous spectrum. This seems paradoxical, for the equation

$$\frac{d^2y}{d\xi^2} + \frac{n+1}{\xi} \frac{dy}{d\xi} + \left( -\frac{1}{(2k + n + 1)^2} + \frac{1}{\xi} \right) y = 0 \tag{108}$$

into which (103) is transformed by the substitution

$$\xi = \left( k + \frac{n+1}{2} \right) x \tag{109}$$

*does possess* a continuous spectrum, if in it we regard

$$E = -\frac{1}{(2k + n + 1)^2} \tag{110}$$

as proper value parameter, viz. all positive values of $E$ are proper values (cf. Part I., analysis of equation (7)). The reason why *no* proper values $k$ of (103) can correspond to these positive $E$-values is that by (110) the $k$-values in question would be complex, and this is impossible, according to general theorems.[1] Each *real* proper value of (103), by (110), gives rise to a *negative* proper value of (108). Moreover, we know (cf. Part I.) that (108) possesses absolutely no negative proper values other than those that arise, as in (110), from the series (107). There thus remains only the one possibility, that in the series (107) certain negative $k$-values are lacking, which appear on solving (110) for $k$, on account of the double-valuedness when extracting the root. But this also is impossible, because the $k$-values in question turn out to be algebraically less than $-n + \frac{1}{2}$ and thus, from general theorems,[2] cannot be proper values of equation (103). The series of values (107) is thus complete. Q.E.D.

The above supplements the proof that the functions (70) are the proper functions of (67) (with the perturbing term suppressed), allied to the proper values (69). We have only to write the solutions of (67) as a product of a function of $\theta$, $\phi$ and a function of $\eta$. The equation in $\eta$ can readily be brought to the form of (105), the only difference being that out present $n$ is there always an odd number, namely, the $(2n + 1)$ which is to be found there.

## § 2. DEFINITE INTEGRALS OF PRODUCTS OF TWO LAGUERRE ORTHOGONAL FUNCTIONS

The Laguerre polynomials can all be obtained, in the following manner, as coefficients of the powers of the auxiliary variable $t$, in the expansion in a series of a so-called "generating function"[3]

$$\sum_{k=0}^{\infty} L_k(x) \frac{t^k}{k!} = \frac{e^{-\frac{xt}{1-t}}}{1-t}. \tag{111}$$

---

[1] Courant-Hilbert, chap. iii. § 4, 2, p. 115.
[2] Courant-Hilbert, chap. v. § 5, 1, p. 240.
[3] Courant-Hilbert, chap. ii. § 11, 5, p. 78, equation (68).

If we replace $k$ by $n + k$ and then differentiate $n$ times with respect to $x$, we obtain the generating function of our generalised polynomials,

$$\sum_{k=0}^{\infty} L^n_{n+k}(x) \frac{t^k}{(n + k)!} = (-1)^n \frac{e^{-\frac{xt}{1-t}}}{(1 - t)^{n+1}}. \qquad (112)$$

In order to evaluate with its help integrals such as appeared for the first time in the text in expression (52), or, more generally, such as were necessary in § 4 for the calculation of (65), and also in § 5, we proceed as follows. We write (112) over again, providing both the fixed index $n$ and the varying index $k$ with a dash, and replacing the undefined $t$ by $s$. These two equations are then multiplied together, *i.e.* left side by left side, and right side by right. Then we multiply further by

$$x^p e^{-x} \qquad (113)$$

and integrate with respect to $x$ from $0$ to $\infty$. $p$ is to be a positive integer—this being sufficient for our purpose. The integration is practicable by elementary methods on the right-hand side, and we get

$$\sum_{k=0}^{\infty} \sum_{k'=0}^{\infty} \frac{t^k s^{k'}}{(n + k)! \, (n' + k')!} \int_0^{\infty} x^p e^{-x} L^n_{n+k}(x) L^{n'}_{n'+k'}(x) dx$$

$$= (-1)^{n+n'} p! \frac{(1 - t)^{p-n}(1 - s)^{p-n'}}{(1 - ts)^{p+1}}. \qquad (114)$$

We have now, on the left, the desired integrals like pearls on a string, and we merely detach the one we happen to need by searching on the right for the coefficient of $t^k s^{k'}$. This coefficient is always a simple sum, and, in fact, in the cases occurring in the text, always a finite sum with very few terms (up to three). In general, we have

$$\left\{ \begin{array}{l} \displaystyle \int_0^{\infty} x^p e^{-x} L^n_{n+k}(x) L^{n'}_{n'+k'}(x) dx = p! \, (n + k)! \, (n' + k')! \\[2mm] \displaystyle \cdot \sum_{\tau=0}^{\leq k, k'} (-1)^{n+n'+k+k'+\tau} \binom{p-n}{k-\tau} \binom{p-n'}{k'-\tau} \binom{-p-1}{\tau}. \end{array} \right. \qquad (115)$$

The sum stops after the smaller of the two numbers $k, k'$. It often, in actual fact, begins at a positive value of $\tau$, as binomial coefficients, whose lower number is greater than the upper, vanish. For example, in the integral in the denominator of (52), we put $p = n = n'$, and $k' = k$. Then $\tau$ can take only the *one* value $k$, and we can establish statement (53) of the text. In the integral of the numerator in (52), only $p$ has another value, namely $p = n + 2$. $\tau$ now takes the values $k - 2, k - 1$, and $k$, and after an easy reduction we get formula (54) of the text. In the very same way the integrals appearing in § 5 are evaluated by Laguerre polynomials.

We can now, therefore, regard integrals of the type of (115) as known, and we have only to concern ourselves with those occurring in § 4 in the calculation of intensities (cf. expression (65) and functions (64) which have, to be substituted there). In this type, the two Laguerre orthogonal functions, whose product is to be integrated, *have not the same argument*, but, for example, in our case, have the arguments $\lambda_1/la_0$ and $\lambda_1/l'a_0$, where $l$ and $l'$ are the principal quantum numbers of the two levels that we have combined. Let us consider, as typical, the integral

$$J = \int_0^\infty x^p e^{-\frac{\alpha+\beta}{2}x} L^n_{n+k}(\alpha x) L^{n'}_{n'+k'}(\beta x) dx. \qquad (116)$$

Now we can proceed in a superficially different way. At first, the former procedure still goes on smoothly; only on the right-hand side of (114) a somewhat more complicated expression appears. In the denominator occurs the power of a quadrinomial instead of that of a binomial, as before. And this makes the matter somewhat confusing, for the right-hand side of (114) becomes five-fold instead of three-fold, and thus the right side of (115) becomes a three-fold instead of a simple sum. I found that the following substitution made things clearer:

$$\frac{\alpha + \beta}{2} x = y. \qquad (117)$$

Hence

$$\begin{cases} \alpha x = \left(1 + \dfrac{\alpha - \beta}{\alpha + \beta}\right) y \\[2mm] \beta x = \left(1 - \dfrac{\alpha - \beta}{\alpha + \beta}\right) y. \end{cases} \qquad (118)$$

After expanding the two polynomials in their Taylor series, which are finite and have similar polynomials as coefficients, we get, using the abbreviations

$$\sigma = \frac{2}{\alpha + \beta}, \quad \gamma = \frac{\alpha - \beta}{\alpha + \beta}, \tag{119}$$

the following,

$$J = \alpha^{p+1} \sum_{\lambda=0}^{k} \sum_{\mu=0}^{k'} (-1)^{\mu} \frac{\gamma^{\lambda+\mu}}{\lambda! \, \mu!} \int_{0}^{\infty} y^{p+\lambda+\mu} L_{n+k}^{n+\lambda}(y) L_{n'+k'}^{n'+\mu}(y) dy. \tag{120}$$

Thus the calculation of $J$ is reduced to the simpler type of integral (115). In the case of the Balmer lines, the double sum in (120) is comparatively tractable, for one of the two $k$-values, namely, the one referring to the two-quantum level, never exceeds unity, and thus $\lambda$ may have two values at most, and, as it turns out, $\mu$ four values at most. The circumstance that out of the polynomials referring to the two-quantum level, none but

$$L_0 = 1, \quad L_1 = -x + 1, \quad L_1^1 = -1,$$

appear, permits further simplifications. Nevertheless we must calculate out a number of tables, and it is much to be regretted that the figures given in the tables of the text for the intensities do not allow their general construction to be seen. By good fortune the additive relations between the ||- and the ⊥ components hold good, so that we may, with some *probability*, feel ourselves safe from arithmetical blunders at least.

## § 3. INTEGRALS WITH LEGENDRE FUNCTIONS

There are three simple integral relations between associated Legendre functions, which are necessary for the calculations in § 5. For the convenience of others, I will state them here, because I was not able to discover them in any of the places I searched. We use the customary definition,

$$P_n^m(\cos \theta) = \sin^m \theta \frac{d^m P_n(\cos \theta)}{(d \cos \theta)^m}. \tag{121}$$

Then the following holds,

$$\int_0^\pi [P_n^m(\cos\theta)]^2 \sin\theta\, d\theta = \frac{2}{2n+1}\frac{(n+m)!}{(n-m)!} \qquad (122)$$
$$\text{(the normalising relation).}$$

Moreover,

$$\left\{\int_0^\pi P_n^m(\cos\theta)\, P_{n'}^m(\cos\theta)\cos\theta\sin\theta\, d\theta = 0 \atop \text{for } |n-n'| \neq 1. \right. \qquad (123)$$

On the other hand,

$$\int_0^\pi P_n^m(\cos\theta) P_{n-1}^m(\cos\theta)\cos\theta\sin\theta\, d\theta$$
$$= \frac{n+m}{2n+1}\int_0^\pi [P_{n-1}^m(\cos\theta)]^2 \sin\theta\, d\theta = \frac{2(n+m)!}{(4n^2-1)(n-m-1)!}.$$
$$(124)$$

The last two relations decide the "selection" of the determinantal terms on page 408 of the text. They are, moreover, of fundamental importance for the theory of spectra, for it is obvious that the selection principle for the azimuthal quantum number depends on them (and on two others which have $\sin^2\theta$ in place of $\cos\theta\sin\theta$).

### Addition at Proof Correction

Hr. W. Pauli, jun., informs me that he has arrived at the following closed formulae for the total intensity of the lines in the Lyman and Balmer series, through a modification of the method given in section 2 of the Appendix. For the Lyman series these are

$$\nu_{l,1} = R\left(\frac{1}{1^2} - \frac{1}{l^2}\right); \quad J_{l,1} = \frac{2^7\cdot(l-1)^{2l-1}}{l\cdot(l+1)^{2l+1}};$$

and for the Balmer series

$$\nu_{l,2} = R\left(\frac{1}{2^2} - \frac{1}{l^2}\right); \quad J_{l,2} = \frac{4^3\cdot(l-2)^{2l-3}}{l\cdot(l+2)^{2l+3}}(3l^2-4)(5l^2-4).$$

The total emission intensities (square of amplitudes into fourth power of the frequency) are proportional to these expressions, within the series in question. The numbers obtained from the formula for the Balmer series are in complete agreement with those given on pp. 404, 405.

Zürich, Physical Institute of the University.
(Received May 10, 1926.)

# Quantisation as a Problem of Proper Values (Part IV[1])

(*Annalen der Physik* (4), vol. 81, 1926)

ABSTRACT: § 1. Elimination of the energy-parameter from the vibration equation. The real wave equation. Non-conservative systems. § 2. Extension of the perturbation theory to perturbations which explicitly contain the time. Theory of dispersion. § 3. Supplementing § 2. Excited atoms, degenerate systems, continuous spectrum. § 4. Discussion of the resonance case. § 5. Generalisation for an arbitrary perturbation. § 6. Relativistic-magnetic generalisation of the fundamental equations. § 7. On the physical significance of the field scalar.

## § 1. Elimination of the Energy-parameter from the Vibration Equation. The Real Wave Equation. Non-conservative Systems

The wave equation (18) or (18″) of Part II., viz.

$$\nabla^2 \psi - \frac{2(E - V)}{E^2} \frac{\partial^2 \psi}{\partial t^2} = 0 \tag{1}$$

or

$$\nabla^2 \psi + \frac{8\pi^2}{h^2}(E - V)\psi = 0, \tag{1'}$$

which forms the *basis* for the re-establishment of mechanics attempted in this series of papers, suffers from the disadvantage that it expresses the law of variation of the "mechanical field scalar" $\psi$, neither *uniformly* nor *generally*. Equation (1) contains the energy- or frequency-parameter $E$, and is valid, as is expressly emphasized in Part II., with a *definite* $E$-value inserted, for processes which depend on the time exclusively through a *definite* periodic factor:

$$\psi \sim \text{real part of } \left( e^{\pm \frac{2\pi i E t}{h}} \right). \tag{2}$$

---

[1] Cf. *Ann. d. Phys.* **79**, pp. 361, 489; **80**, p. 437, 1926 (Parts I., II., III.); further, on the connection with Heisenberg's theory, *ibid.* 79, p. 734 (p. 45).

Equation (1) is thus not really any more general than equation (1′), which takes account of the circumstance just mentioned and does not contain the time at all.

Thus, when we designated equation (1) or (1′), on various occasions, as "the wave equation", we were really wrong and would have been more correct if we had called it a "vibration-" or an "amplitude-" equation. However, we found it sufficient, because to *it* is linked the Sturm-Liouville proper value problem—just as in the mathematically strictly analogous problem of the free vibrations of strings and membranes—and not to the *real* wave equation.

As to this, we have always postulated up till now that the potential energy $V$ is a pure function of the co-ordinates and does *not* depend explicitly on the time. There arises, however, an urgent need for the extension of the theory to *non-conservative* systems, because it is only in that way that we can study the behaviour of a system under the influence of prescribed external forces, *e.g.* a light wave, or a strange atom flying past. Whenever $V$ contains the time explicitly, it is manifestly *impossible* that equation (1) or (1′) should be satisfied by a function $\psi$, the method of dependence of which on the time is as given by (2). We then find that the amplitude equation is no longer sufficient and that we must search for the real wave equation.

For conservative systems, the latter is easily obtained. (2) is equivalent to

$$\frac{\partial^2 \psi}{\partial t^2} = -\frac{4\pi^2 E^2}{h^2}\psi. \tag{3}$$

We can eliminate $E$ from (1′) and (3) by differentiation, and obtain the following equation, which is written in a symbolic manner, easy to understand:

$$\left(\nabla^2 - \frac{8\pi^2}{h^2}V\right)^2 \psi + \frac{16\pi^2}{h^2}\frac{\partial^2 \psi}{\partial t^2} = 0. \tag{4}$$

This equation must be satisfied by every $\psi$ which depends on the time as in (2), though *with E arbitrary*, and consequently also by every $\psi$ which can be expanded in a Fourier series with respect to the time (naturally with functions of the co-ordinates as coefficients).

Equation (4) is thus evidently *the uniform and general wave equation for the field scalar $\psi$*.

It is evidently no longer of the simple type arising for vibrating membranes, but is of the *fourth* order, and of a type similar to that occurring in many problems in the theory of elasticity.[1] However, we need not fear any excessive complication of the theory, or any necessity to revise the previous methods, associated with equation (1'). If $V$ does *not* contain the time, we can, proceeding from (4), apply (2), and then split up the operator as follows:

$$\left(\nabla^2 - \frac{8\pi^2}{h^2}V + \frac{8\pi^2}{h^2}E\right)\left(\nabla^2 - \frac{8\pi^2}{h^2}V - \frac{8\pi^2}{h^2}E\right)\psi = 0. \quad (4')$$

*By way of trial*, we can resolve this equation into two "alternative" equations, namely, into equation (1') and into another, which only differs from (1') in that its proper value parameter will be *called* minus $E$, instead of plus $E$. According to (2) this does not lead to new solutions. The decomposition of (4') is not absolutely cogent, for the theorem that "a product can only vanish when at least *one* factor vanishes" is not valid for operators. This lack of cogency, however, is a feature common to all the methods of solution of partial differential equations. The procedure finds its subsequent justification in the fact that we can prove the *completeness* of the discovered proper functions, as functions of the co-ordinates. This completeness, coupled with the fact that the imaginary part as well as the real part of (2) satisfies equation (4), allows arbitrary initial conditions to be fulfilled by $\psi$ and $\partial\psi/\partial t$.

Thus we see that the wave equation (4), which contains in itself the law of dispersion, can really stand as the basis of the theory previously developed for conservative systems. The generalisation for the case of a time-varying potential function nevertheless demands caution, because terms with time derivatives of $V$ may then appear, about which no information can be given to us by equation (4), owing to the way we obtained it. In actual fact, if we attempt to apply

---

[1] *E.g.*, for a vibrating *plate*, $\nabla^2\nabla^2 u + \frac{\partial^2 u}{\partial t^2} = 0$. Cf. Courant-Hilbert, chap. v. § 8, p. 256.

equation (4) as it stands to non-conservative systems, we meet with complications, which seem to arise from the term in $\partial V/\partial t$. Therefore, in the following discussions, I have taken a somewhat different route, which is much easier for calculations, and which I consider is justified in principle.

We *need* not raise the order of the wave equation to four, in order to get rid of the energy-parameter. The dependence of $\psi$ on the time, which must exist if (1') is to hold, can be expressed by

$$\frac{\partial \psi}{\partial t} = \pm \frac{2\pi i}{h} E \psi \qquad (3')$$

as well as by (3). We thus arrive at one of the two equations

$$\nabla^2 \psi - \frac{8\pi^2}{h^2} V\psi \mp \frac{4\pi i}{h} \frac{\partial \psi}{\partial t} = 0. \qquad (4'')$$

*We will require the complex wave function $\psi$ to satisfy one of these two equations.* Since the conjugate complex function $\psi$ will then satisfy the *other* equation, we may take the real part of $\psi$ as the real wave function (if we require it). In the case of a conservative system (4'') is essentially equivalent to (4), as the real operator may be split up into the product of the two conjugate complex operators if $V$ does not contain the time.

## § 2. EXTENSION OF THE PERTURBATION THEORY TO PERTURBATIONS CONTAINING THE TIME EXPLICITLY. THEORY OF DISPERSION

Our main interest is not in systems for which the time and spatial variations of the potential energy $V$ are of the same order of magnitude, but in systems, conservative in themselves, which are *perturbed* by the addition of small given functions of the time (and of the co-ordinates) to the potential energy. Let us, therefore, write

$$V = V_0(x) + r(x, t), \qquad (5)$$

where, as often before, $x$ represents the whole of the configuration coordinates. We regard the unperturbed proper value problem ($r = 0$) as *solved*. Then the perturbation problem can be solved by *quadratures*.

However, we will not treat the general problem immediately, but will select the problem of the *dispersion theory* out of the vast number of weighty applications which fall under this heading, on account of its striking importance, which really justifies a separate treatment in any case. Here the perturbing forces originate in an alternating electric field, homogeneous and vibrating synchronously in the domain of the atom; and thus, if we have to do with a linearly polarised monochromatic light of frequency $v$, we write

$$r(x, t) = A(x) \cos 2\pi vt, \qquad (6)$$

and hence

$$V = V_0(x) + A(x) \cos 2\pi vt. \qquad (5')$$

Here $A(x)$ is the negative product of the light-amplitude and the co-ordinate function which, *according to ordinary mechanics*, signifies the component of the electric moment of the atom in the direction of the electric light-vector (say $-F\Sigma_{e_i z_i}$, if $F$ is the light-amplitude, $e_i$, $z_i$ the charges and $z$-co-ordinates of the particles, and the light is polarised in the $z$-direction). We borrow the time-*variable* part of the potential function from ordinary mechanics with just as much or as little right as previously, *e.g.* in the Kepler problem, we borrowed the *constant* part. Using (5′), equation (4″) becomes

$$\nabla^2 \psi - \frac{8\pi^2}{h^2}(V_0 + A \cos 2\pi vt)\psi \mp \frac{4\pi i}{h} \frac{\partial \psi}{\partial t} = 0. \qquad (7)$$

For $A = 0$, these equations are changed by the substitution

$$\psi = u(x)e^{\pm \frac{2\pi iEt}{h}} \qquad (8)$$

(which is now to be taken in the literal sense, and does *not* imply *pars realis*) into the amplitude equation (1′) of the unperturbed problem, and we know (cf. § 1) that the totality of the solutions of the unperturbed problem is found in this way. Let

$$E_k \quad \text{and} \quad u_k(x); \quad k = 1, 2, 3, \ldots$$

be the proper values and normalised proper functions of the unperturbed problem, which we regard as *known*, and which we will assume to be *discrete* and *different* from one another (non-degenerate system with no continuous spectrum), so that we may not become involved in secondary questions, requiring special consideration.

Just as in the case of a perturbing potential independent of the time, we will have to seek solutions of the perturbed problem in the neighbourhood of *each* possible solution of the unperturbed problem, and thus in the neighbourhood of an arbitrary linear combination of the $u_k$'s, which has constant co-efficients [from (8), the $u_k$'s to be combined with the appropriate time factors $e^{\pm\frac{2\pi i E_k t}{h}}$]. The solution of the perturbed problem, lying in the neighbourhood of a *definite* linear combination, will have the following physical meaning. It will be *this solution* which first appears, if, when the light wave arrived, precisely that definite linear combination of free proper vibrations was present (perhaps with trifling changes during the "excitation").

Since, however, the equation of the perturbed problem is also *homogeneous*—let this want of analogy with the "forced vibrations" of acoustics be expressly emphasized—it is evidently sufficient to seek the perturbed solution in the neighbourhood of each *separate*

$$u_k(x)e^{\pm\frac{2\pi i E_k t}{h}},\qquad(9)$$

as we may then linearly combine these *ad lib.*, just as for unperturbed solutions.

To solve the first of equations (7) we therefore now put

$$\psi = u_k(x)e^{\frac{2\pi i E_k t}{h}} + w(x, t).\qquad(10)$$

[The lower symbol, *i.e.* the second of equations (7), is henceforth left on one side, as it would not yield anything new.] The additional term $w(x, t)$ can be regarded as small, and its product with the perturbing potential neglected. Bearing this in mind while substituting from (10) in (7), and remembering that $u_k(x)$ and $E_k$ are proper functions and values of the unperturbed problem, we get

$$\begin{cases}\nabla^2 w - \dfrac{8\pi^2}{h^2}V_0 w - \dfrac{4\pi i}{h}\dfrac{\partial w}{\partial t} = \dfrac{8\pi^2}{h^2}A\cos 2\pi vt \cdot u_k e^{\frac{2\pi i E_k t}{h}},\\[2mm] \qquad\qquad = \dfrac{4\pi^2}{h^2}Au_k \cdot \left(e^{\frac{2\pi i t}{h}(E_k+hv)} + e^{\frac{2\pi i t}{h}(E_k-hv)}\right).\end{cases}$$

$$(11)$$

This equation is readily, and really *only*, satisfied by the substitution

$$w = w_+(x)e^{\frac{2\pi it}{h}(E_k+h\nu)} + w_-(x)e^{\frac{2\pi it}{h}(E_k-h\nu)}, \tag{12}$$

where the two functions $w\pm$ respectively obey the two equations

$$\nabla^2 w_\pm + \frac{8\pi^2}{h^2}(E_k \pm h\nu - V_0)w_\pm = \frac{4\pi^2}{h^2}Au_k. \tag{13}$$

This step is essentially *unique*. At first sight, we apparently can add to (12) an arbitrary aggregate of unperturbed proper vibrations. But this aggregate would necessarily be assumed small, of the first order (since this has been assumed for $w$), and thus does not interest us at present, as it could only produce perturbations of the second order at most.

In equations (13) we have at last those *non-homogeneous* equations we might have expected to encounter—in spite of the lack of analogy with real forced vibrations, as emphasized above. This lack of analogy is extraordinarily important and manifests itself in equations (13) in the two following particulars. *Firstly*, as the "second member" ("exciting force"), the perturbation function $A(x)$ does not appear *alone*, but *multiplied* by the amplitude of the free vibration already present. This is indispensable if the physical facts are to be properly taken into account, for the reaction of an atom to an incident light wave depends almost entirely on the *state* of the atom at that time, whereas the forced vibrations of a membrane, plate, etc., are known to be quite independent of the proper vibrations which may be superimposed on them, and thus would yield an obviously wrong representation of our case. *Secondly*, in place of the proper value on the left-hand side of (13), *i.e.* as "exciting frequency", we do not find the frequency $\nu$ of the perturbing force *alone*, but rather in one case added to, and in the other subtracted from, that of the free vibration already present. This is equally indispensable. Otherwise the proper frequencies *themselves*, which correspond to the *term*-frequencies, would function as *resonance-points*, and not the *differences* of the proper frequencies, as is demanded, and is really given by equation (13). Moreover, we see with satisfaction that the latter gives *only* the differences between a proper frequency *which is actually excited* and all the others, and *not* the differences between pairs of proper frequencies, of which *no member* is excited.

In order to investigate this more closely, let us complete the solution. By well-known methods[1] we find, as *simple* solutions of (13),

$$w_{\pm}(x) = \frac{1}{2} \sum_{n=1}^{\infty} \frac{a'_{kn} u_n(x)}{E_k - E_n \pm h\nu},$$ (14)

where

$$a'_{kn} = \int A(x) u_k(x) u_n(x) \rho(x) dx.$$ (15)

$\rho(x)$ is the "density function", *i.e.* that function of the position-coordinates with which equation (1') must be multiplied to make it self-adjoint. The $u_n(x)$'s are assumed to be normalised. It is further postulated that *h$\nu$ does not agree exactly with any of the differences $E_k - E_n$ of the proper values*. This "resonance case" will be dealt with later (cf. § 4).

If we now form from (14), using (12) and (10), the entire perturbed vibration, we get

$$\left\{ \psi = u_k(x) e^{\frac{2\pi i E_k t}{h}} + \frac{1}{2} \sum_{n=1}^{\infty} a'_{kn} u_n(x) \right.$$

$$\left. \cdot \left( \frac{e^{\frac{2\pi i t}{h}(E_k + h\nu)}}{E_k - E_n + h\nu} + \frac{e^{\frac{2\pi i t}{h}(E_k - h\nu)}}{E_k - E_n - h\nu} \right). \right.$$ (16)

Thus in the perturbed case, along with each *free* vibration $u_k(x)$ occur in small amplitude all those vibrations $u_n(x)$, for which $a'_{kn} \neq 0$. The latter are exactly those, which, if they exist as free vibrations along with $u_k$, give rise to a radiation, which is (wholly or partially) polarised in the direction of polarisation of the incident wave. For apart from a factor, $a'_{kn}$ is just the component amplitude, in this direction of polarisation, of the atom's *electric moment*, which is oscillating with frequency $(E_k - E_n)/h$, *according to wave mechanics*, and which appears when $u_k$ and $u_n$ exist together.[2] The simultaneous oscillation,

---

[1] Cf. Part III. §§ 1 and 2, text beside equations (8) and (24).
[2] Cf. what follows, and § 7.

however, takes place with neither the proper frequency $E_n/h$, peculiar to these vibrations, nor the frequency $v$ of the light wave, but rather with the sum and difference of $v$ and $E_k/h$ (*i.e.* the frequency of the *one* existing *free* vibration).

The real or the imaginary part of (16) can be considered as the *real* solution. In the following, however, we will operate with the complex solution itself.

To see the significance that our result has in the theory of dispersion, we must examine the radiation arising from the simultaneous existence of the excited forced vibrations and the free vibration, already present. For this purpose, we form, following the method we[1] have always adopted above—a criticism follows in § 7—the product of the complex wave function (16) and its conjugate, *i.e.* the norm of the complex wave function $\psi$. We notice that the perturbing terms are small, so that squares and products may be neglected. After a simple reduction[2] we obtain

$$\psi\bar{\psi} = u_k(x)^2 + 2\cos 2\pi v t \sum_{n=1}^{\infty} \frac{(E_k - E_n)a'_{kn}u_k(x)u_n(x)}{(E_k - E_n)^2 - h^2v^2}. \quad (17)$$

According to the *heuristic hypothesis* on the electrodynamical significance of the field scalar $\psi$, the present quantity—apart from a multiplicative constant—represents the electrical density as a function of the space co-ordinates and the time, *if x stands for only three space co-ordinates*, *i.e.* if we are dealing with the problem of *one* electron. We remember that the same hypothesis led us to correct selection and polarisation rules and to a very satisfactory representation of intensity relationships in our discussion of the hydrogen Stark effect. By a natural generalisation of this hypothesis—of which more in § 7— we regard the following as representing in the general case the density of the electricity, which is "associated" with one of the particles of classical mechanics, or which "originates in it", or which "corresponds

---

[1] Cf. end of paper on Quantum Mechanics of Heisenberg, etc., and also the Calculation of Intensities in the Stark Effect in Part III. At the first quoted place, the real part of $\psi\bar{\psi}$ was proposed instead of $\psi\bar{\psi}$. This was a mistake, which was corrected in Part III.

[2] We assume as previously, for the sake of simplicity, the proper functions $u_n(x)$ to be *real*, but notice that it may sometimes be much more convenient or even imperative to work with complex aggregates of the real proper functions, *e.g.* in the proper functions of the Kepler problem to work with $e^{\pm m\phi i}$ instead of $\genfrac{}{}{0pt}{}{\cos}{\sin} m\phi$.

to it in wave mechanics": the *integral* of $\psi\bar{\psi}$ taken over all those co-ordinates of the system, which in classical mechanics fix the position of the *rest of the* particles, multiplied by a certain constant, the classical "charge" of the first particle. The resultant density of charge at any point of space is then represented by the *sum* of such integrals taken over all the particles.

Thus in order to find any space component whatever of the total wave-mechanical dipole moment as a function of the time, we must, on this hypothesis, multiply expression (17) by that function of the co-ordinates which gives that particular dipole - component *in classical mechanics* as a function of the configuration of the point system, *e.g.* by

$$M_y = \sum e_i y_i, \tag{18}$$

if we are dealing with the dipole moment in the *y*-direction. Then we have to integrate over *all* the configuration co-ordinates.

Let us work this out, using the abbreviation

$$b_{kn} = \int M_y(x) u_k(x) u_n(x) \rho(x) dx. \tag{19}$$

Let us elucidate further the definition (15) of the $a'_{kn}$'s by recalling that if the incident electric light-vector is given by

$$\mathfrak{E}_z = F \cos 2\pi \nu t, \tag{20}$$

then

$$\left\{ \begin{array}{l} A(x) = -F \cdot M_z(x), \\ \text{where } M_z(x) = \sum e_i z_i. \end{array} \right. \tag{21}$$

If we put, in analogy with (19),

$$a_{kn} = \int M_z(x) u_k(x) u_n(x) \rho(x) dx, \tag{22}$$

then $a'_{kn} = -F a_{kn}$, and by carrying out the proposed integration we find,

$$\int M_y \psi\bar{\psi} \rho dx = a_{kk} + 2F \cos 2\pi \nu t \sum_{n=1}^{\infty} \frac{(E_n - E_k) a_{kn} b_{kn}}{(E_k - E_n)^2 - h^2 \nu^2} \tag{23}$$

for *the resulting electric moment, to which the secondary radiation*, caused by the incident wave (20), *is to be attributed.*

The radiation depends of course only upon the second (time-variable) part, while the first part represents the time-constant dipole moment, which is possibly connected with the originally existing free vibration. This variable part seems fairly promising and may meet all the demands we are accustomed to make on a "dispersion formula". Above all, let us note the appearance of those so-called "negative" terms, which—in the usual phraseology—correspond to the probability of transition to a lower level ($E_n < E_k$), and to which Kramers[1] was the first to direct attention, from a correspondence standpoint. Generally, our formula—despite very different ways of thought and expression—may be characterised as really identical in form with Kramer's formula for secondary radiation. The important connection between $a_{kn}$, $b_{kn}$, the coefficients of the secondary and of the spontaneous radiation, is brought out, and indeed the secondary radiation is also described accurately with respect to its condition of polarisation.[2]

I would like to believe that the absolute value of the scattered radiation or of the induced dipole moment is also given correctly by formula (23), although it is obviously within the bounds of possibility that an error in the numerical factor may have occurred in applying the heuristic hypothesis introduced above. At any rate the physical dimensions are right, for from (18), (19), (21), and (22) $a_{kn}$ and $b_{kn}$ are electric moments, since the squared integrals of the proper functions were normalised to unity. If $v$ is far removed from the emission frequency in question, the ratio of the induced to the spontaneous dipole moment is of the same order of magnitude as the ratio of the additional potential energy $Fa_{kn}$ to the "energy step" $E_k - E_n$.

---

[1] H. A. Kramers, *Nature*, May 10, 1924; *ibid.* August 30, 1924; Kramers and W. Heisenberg. *Ztschr. f. Phys.* 31, p. 681, 1925. The description given in the latter paper of the polarisation of the scattered light (equation 27) from correspondence principles, is almost identical *formally* with ours.

[2] It is hardly necessary to say that the two directions which, for simplicity, we have designated as "z-direction" and "y-direction" do not require to be exactly perpendicular to one another. The one is the direction of polarisation of the incident wave; the other is that polarisation component of the secondary wave, in which we are specially interested.

## § 3. SUPPLEMENTS TO § 2. EXCITED ATOMS, DEGENERATE SYSTEMS, CONTINUOUS SPECTRUM

For the sake of clearness, we have made some special assumptions, and put many questions aside, in the preceding paragraph. These have now to be discussed by way of supplement.

First: what happens when the light wave meets the atom, when the latter is in a state in which not merely *one* free vibration, $u_k$, is excited as hitherto assumed, but several, say two, $u_k$ and $u_l$? As remarked above, we have in the perturbed case simply to combine additively the two perturbed solutions (16) corresponding to the suffix $k$ and the suffix $l$, after we have provided them with constant (possibly complex) coefficients, which correspond to the *strength* presumed for the free vibrations, and to the phase relationship of their stimulation. Without actually performing the calculation, we see that in the expression for $\psi\bar{\psi}$ and also in the expression (23) for the resulting electric moment, there then occurs *not merely* the corresponding linear aggregate of the terms previously obtained, *i.e.* of the expressions (17) or (23) written with $k$, and then with $l$. We have in addition "combination terms", namely, considering *first* the greatest order of magnitude, a term in

$$u_k(x)u_l(x)e^{\frac{2\pi i}{h}(E_k-E_l)t},\qquad(24)$$

which gives again the *spontaneous* radiation, bound up with the co-existence of the two *free* vibrations; and *secondly* perturbing terms of the first order, which are proportional to the perturbing field amplitude, and which correspond to the interaction of the forced vibrations belonging to $u_k$ with the free vibration $u_l$—and of the forced vibrations belonging to $u_l$ with $u_k$. The *frequency* of these new terms appearing in (17) or (23) is *not* $\nu$ but

$$|\nu \pm (E_k - E_l)/h|,\qquad(25)$$

as can easily be seen, still without carrying out the calculation. (New "resonance denominators", however, do *not* occur in these terms.) Thus we have to do here with a secondary radiation, whose frequency neither coincides with the exciting light-frequency nor with a spontaneous frequency of the system, but is a combination frequency of both.

The existence of this remarkable kind of secondary radiation was first postulated by Kramers and Heisenberg (*loc. cit.*), from correspondence considerations, and then by Born, Heisenberg, and Jordan from consideration of Heisenberg's quantum mechanics.[1] As far as I know, it has not yet been demonstrated experimentally. The *present* theory also shows distinctly that the occurrence of this scattered radiation is dependent on special conditions, which demand researches expressly arranged for the purpose. Firstly, *two* proper vibrations $u_k$ and $u_l$ must be *strongly* excited, so that all experiments made on atoms in their normal state—as happens in the vast majority of cases—are to be rejected. Secondly, at least *one* third state of proper vibration must *exist* (*i.e.* must be possible—it need not be *excited*), which leads to powerful spontaneous emission, when combined with $u_k$ as well as with $u_l$. For the extraordinary scattered radiation, which is to be discovered, is proportional to the product of the spontaneous emission coefficients in question ($a_{kn}b_{ln}$ and $a_{ln}b_{kn}$). The combination ($u_k$, $u_l$) need not, in itself, cause a strong emission. It would not matter if—to use the language of the older theory—this was a "forbidden transition". Yet in practice we must also demand that the line ($u_k$, $u_l$) should actually be emitted strongly during the experiment, for this is the only means of assuring ourselves that *both* proper vibrations are strongly excited in the same individual atoms and in a sufficiently great number of them. If we reflect now that in the powerful term-series mostly examined, *i.e.* in the ordinary $s$-, $p$-, $d$-, $f$-series, the relations are generally such that two terms, which combine strongly with a third, do not do so with one another, then a special choice of the object and conditions of the research seems really necessary, if we are to expect the desired scattered radiation with any certainty, especially as its frequency is not that of the exciting light and *thus* it does not produce dispersion or

---

[1] Born, Heisenberg, and Jordan, *Ztschr. f. Phys.* 35, p. 572, 1926.

rotation of the plane of polarisation, but can only be observed as light scattered on all sides.

As far as I see, the above-mentioned dispersion theory of Heisenberg, Born, and Jordan does *not* allow of such reflections as we have just made, in spite of its great formal similarity to the present one. For it only considers *one* way in which the atom reacts to incident radiation. It conceives the atom as a timeless entity, and up till now is not able to express in its language the undoubted fact that the atom can be in *different* states at different times, and thus, as has been proved, reacts in different ways to incident radiation.[1]

Let us turn now to another question. In § 2 the collective proper values were postulated to be *discrete* and *different* from one another. We now drop the second hypothesis and ask: what is altered when *multiple* proper values occur, *i.e.* when *degeneracy* is present? Perhaps we expect that complications then arise, similar to those we met in the case of a time-constant perturbation (Part III. § 2), *i.e.* that a system of proper functions of the unperturbed atom, suited to the particular perturbation, must be defined by the solution of a "secular equation", and applied to carry out the perturbation calculation. This is indeed so in the case of an *arbitrary* perturbation, represented by $r(x, t)$ as in equation (5), but *not* so in the case of a perturbation by a light wave (equation (6))—at any rate, for our usual first approximation, and as long as we suppose that the light frequency $\nu$ does not coincide with any of the spontaneous emission frequencies considered. Then the parameter value in the double equation (13), for the amplitudes of the perturbed vibrations, is *not* a proper value, and the pair of equations has always the unambiguous pair of solutions (14), in which no vanishing denominators occur even when $E_k$ is a multiple value. Thus the terms in the sum for which $E_n = E_k$ are *not, as might be thought,* to be omitted, any more than the term for $n = k$ itself. It is worth noticing that through these terms—if one of them occurs really, *i.e.* with non-vanishing $a_{kn}$—the frequency $\nu = 0$ also appears

---

[1] Cf. especially the concluding words of Heisenberg's latest exposition of his theory, *Math. Ann.* 95, p. 683, 1926, in connection with this difficulty of comprehending *the course of an event in time.*

among the resonance frequencies. These terms do *not*, of course, contribute to the "ordinary" scattered radiation, as we see from (23), since $E_k - E_n = 0$.

The simplification, that we do not require to consider specially any possible degeneracy present, at least in a first approximation, is always available[1] when the time-averaged value of the perturbation function vanishes, or what is the same thing, when the latter's Fourier expansion in terms of the time contains no constant, *i.e.* time-independent, term. This is the case for a light wave.

While our *first* postulation about the proper values—that they should be *simple*—has thus shown itself to be really a superfluous precaution, a dropping of the *second*—that they should be absolutely discrete—while leading to no alterations *in principle*, brings about, however, very considerable alterations in the external appearance of the calculation, inasmuch as *integrals* taken over the continuous spectrum of equation (1′) are to be added to the discrete sums in (14), (16), (17), and (23). The theory of such representations by integrals has been developed by H. Weyl,[2] and though only for ordinary differential equations, the extension to partials is permissible. In all brevity, the state of the case is this.[3] If the homogeneous equation belonging to the non-homogeneous equations (13), *i.e.* the vibration equation (1′) of the unperturbed system, possesses in addition to a point-spectrum a continuous one, which stretches, say, from $E = a$ to $E = b$, then an arbitrary function $f(x)$ naturally cannot be developed thus,

$$f(x) = \sum_{n=1}^{\infty} \phi_n \cdot u_n(x), \quad \text{where} \quad \phi_n = \int f(x)u_n(x)\rho(x)dx$$

(26)

in terms of the normalised discrete proper functions $u_n(x)$ alone, but there must be added an integral expansion in terms of the proper solutions $u(x, E)$, which belong to the proper values $a \leq E \leq b$, and

---

[1] Further discussed in § 5.

[2] H. Weyl, *Math. Ann.* 68, p. 220, 1910; *Gött. Nachr.* 1910. Cf. also E. Hilb, *Sitz.-Ber. d. Physik. Mediz. Soc. Erlangen*, 43, p. 68, 1911; *Math. Ann.* 71, p. 76, 1911. I have to thank Herr Weyl not only for these references but also for very valuable oral instruction in these not very simple matters.

[3] I have to thank Herr Fues for this exposition.

so we have

$$f(x) = \sum_{n=1}^{\infty} \phi_n \cdot u_n(x) + \int_a^b u(x, E)\phi(E)dE, \qquad (27)$$

where to emphasize the analogy we have intentionally chosen the same letter for the "coefficient function" $\phi(E)$ as for the discrete coefficients $\phi_n$. If now we have *normalised*, once for all, the proper solution $u(x, E)$ by associating with it a suitable function of $E$, in such a way that

$$\int dx\rho(x) \int_{E'}^{E'+\Delta} u(x, E)u(x, E')dE' = 1 \quad \text{or} \quad = 0 \qquad (28)$$

according to whether $E$ *belongs* to the interval $E'$, $E' + \Delta$ or not, then in (27) under the integral sign we substitute from

$$\phi(E) = \lim_{\Delta=0} \frac{1}{\Delta} \int \rho(\xi) f(\xi) \cdot \int_E^{E+\Delta} u(\xi, E')dE' \cdot d\xi, \qquad (29)$$

wherein the *first* integral sign refers as always to the domain of the group of variables $x$.[1] Assuming (28) to be fulfilled and expansion (27) to exist—which statements are proved by Weyl for ordinary differential equations—the definition of the "coefficient functions" from (29) is almost as obvious as the well-known definition of the Fourier coefficients.

The most important and difficult task in any concrete case is the carrying out of the normalisation of $u(x, E)$, *i.e.* the finding of that function of $E$ by which we have to multiply the (as yet not normalised) proper solution of the continuous spectrum, in order that condition (28) may be satisfied. The above-quoted works of Herr Weyl contain very valuable guidance for this practical task, and also some worked-out examples. An example from atomic dynamics on the intensities of band spectra is worked out by Herr Fues in a paper appearing in the present issue of *Annalen der Physik*.

Let us apply this to our problem, *i.e.* to the solution of the pair of equations (13) for the amplitudes. $w_\pm$ of the perturbed vibrations, where we postulate as usual that the *one* excited *free* vibration, $u_k$,

---

[1] As Herr E. Fues informs me, we can very often omit the limiting process in practice and write $u(\xi, E)$ for the inner integral, viz. always, when $\int \rho(\xi) f(\xi) u(\xi, E) d\xi$ exists.

belongs to the discrete point-spectrum. We develop the right-hand side of (13) according to the scheme (27) thus,

$$\frac{4\pi^2}{h^2} A(x)u_k(x) = \frac{4\pi^2}{h^2} \sum_{n=1}^{\infty} a'_{kn}u_n(x) + \frac{4\pi^2}{h^2} \int_a^b u(x, E)a'_k(E)dE,$$

(30)

in which $a'_{kn}$ is given by (15), and $a'_k(E)$ from (29) by

$$a'_k(E) = \lim_{\Delta=0} \frac{1}{\Delta} \int \rho(\xi)A(\xi)u_k(\xi) \cdot \int_E^{E+\Delta} u(\xi, E')dE' \cdot d\xi. \quad (15')$$

If we imagine expansion (30) put into (13), and then expand also the desired solution $w_\pm(x)$ similarly in terms of the proper solutions $u_n(x)$ and $u(x, E)$, and notice that for the last-named functions the left side of (13) takes the value

$$\frac{8\pi^2}{h^2} (E_k \pm h\nu - E_n) u_n (x)$$

or

$$\frac{8\pi^2}{h^2} (E_k \pm h\nu - E) u(x, E),$$

then by "comparison of coefficients" we obtain as the generalisation of (14)

$$w_\pm (x) = \frac{1}{2} \sum_{n=1}^{\infty} \frac{a'_{kn}u_n (x)}{E_k - E_n \pm h\nu} + \frac{1}{2} \int_a^b \frac{\alpha'_k (E) u (x, E)}{E_k - E \pm h\nu} dE.$$

(14')

The further procedure is completely analogous to that of § 2.

Finally, we get as *additional term* for (23)

$$+2 \cos 2\pi \nu t \int d\xi \rho(\xi) M_y(\xi)u_k\xi \int_a^b \frac{(E_k - E) a'_k (E) u (\xi, E)}{(E_k - E)^2 - h^2\nu^2} dE.$$

(23')

Here, perhaps, we may not always change the order of integration without further examination, because the integral with respect to $\xi$ may possibly not converge. However, we can—as an intuitive makeshift for a strict passage to the limit, which may be dispensed with here—decompose the integral $\int_a^b$ into many small parts, each having a range $\Delta$, which is sufficiently small to allow us to regard all the functions of $E$ in question as constant in each part, with the exception of $u(x, E)$, for we know from the general theory that its integral cannot be obtained through such a fixed partition, which is independent of $\xi$. We can then take the remaining functions out of the partial integrals, and as *additional term for the dipole moment* (23) *of the secondary radiation*, obtain finally exactly the following,

$$2F \cos 2\pi \nu t \int_a^b \frac{(E - E_k)\,\alpha_k\,(E)\,\beta_k\,(E)}{(E_k - E)^2 - h^2\nu^2}, \tag{23''}$$

where

$$\alpha_k\,(E) = \lim_{\Delta=0} \frac{1}{\Delta} \int \rho(\xi)M_z(\xi)u_k(\xi) \cdot \int_E^{E+\Delta} u\left(\xi, E'\right) dE' \cdot d\xi, \tag{22'}$$

$$\beta_k\,(E) = \lim_{\Delta=0} \frac{1}{\Delta} \int \rho(\xi)M_y(\xi)u_k(\xi) \cdot \int_E^{E+\Delta} u\left(\xi, E'\right) dE' \cdot d\xi \tag{19'}$$

(please note the complete analogy with the formulae with the same numbers but without the dashes in § 2).

The preceding sketch of the calculation is of course only a general outline, given merely to show that the much-discussed influence of the continuous spectrum on dispersion, which experiment[1] appears to indicate as existing, is required by the present theory exactly in the form expected, and to outline the way in which the calculation of the problem is to be tackled.

## § 4. DISCUSSION OF THE RESONANCE CASE

Up till now we have always assumed that the frequency $\nu$ of the light wave does not agree with any of the emission frequencies that have to be considered. We now assume that, say,

$$h\nu = E_n - E_k > 0, \tag{31}$$

and we revert, moreover, to the limiting conditions of § 2 for the sake of simplicity (simple, discrete proper values, one single free vibration

---

[1] K. F. Herzfeld and K. L. Wolf, *Ann. d. Phys.* 76, p. 71, 567, 1925; H. Kollmann and H. Mark, *Die Nw.* 14, p. 648, 1926.

$u_k$ excited). In the pair of equations (13), the proper value parameter then takes the values

$$E_k \pm E_n \mp E_k = \begin{cases} E_n \\ 2E_k - E_n \end{cases}, \tag{32}$$

*i.e.* for the upper sign there appears a *proper value*, namely, $E_n$. The two cases are possible. Firstly, the right side of equation (13) multiplied by $\rho(x)$, may be *orthogonal* to the proper function $u_n(x)$ corresponding to $E_n$, *i.e.* we have

$$\int A(x) u_k(x) u_n(x) \rho(x) dx = a'_{kn} = 0, \tag{33}$$

which means, physically, that if $u_k$ and $u_n$ exist together as free vibrations they will give rise to no spontaneous emission or to one which is polarised perpendicularly to the direction of polarisation of the incident light. In this case the critical equation (13) also again possesses a solution, which now, as before, is given by (14), in which the catastrophic term vanishes. This means physically—in the old phraseology—that a "forbidden transition" cannot be stimulated through resonance, or that a "transition", even if not forbidden, cannot be caused by light which is vibrating perpendicularly to the direction of polarisation of that light which would be emitted by the "spontaneous transition".

Otherwise, secondly, (33) *is not* fulfilled. Then the critical equation possesses *no* solution. Statement (10), which assumes a vibration which differs very *little*—by quantities of the order of the light amplitude $F$—from the originally existing free vibration, and is the *most general* possible *under this assumption, thus does not then lead to the goal.* No solution, therefore, exists which only differs by quantities of the order of $F$ from the original free vibration. The incident light has thus *a varying influence* on the state of the system, *which bears no relation to the magnitude of the light amplitude.* What influence? We can judge this, still without further calculation, if we start out from the case where the resonance condition (31) is not exactly but only approximately fulfilled. Then we see from (16) that $u_n(x)$ is excited in unusually strong forced vibrations, on account of the small denominator, and that—not less important—the frequency of these forced vibrations approaches the natural proper frequency $E_n/h$ of the proper vibration $u_n$. (All this is, indeed, very *similar* to, yet in a way of its own *different* from, the resonance phenomena encountered elsewhere; otherwise I would not discuss it so minutely.)

In a gradual approach to the critical frequency, the proper vibration $u_n$, formerly not excited, whose possible existence is responsible for the crisis, is stimulated to a stronger and stronger degree, and with a frequency more and more closely approaching its own proper frequency. In contradistinction to ordinary resonance phenomena there comes a point, and that even before the critical frequency is reached, where our solution does not represent the circumstances correctly any longer, even under the assumption that our obviously "undamped" wave postulation is strictly correct. For we have in fact regarded the forced vibration $w$ as small compared with the existing free vibration and neglected a squared term (in equation (11)).

I believe that the present discussion has already shown, with sufficient clearness, that in the resonance case the theory will actually give the result it ought to give, in order to agree with Wood's resonance phenomenon: an increase of the proper vibration $u_n$, which causes the crisis, to a finite magnitude comparable with that of the originally existing $u_k$, from which, of course, "spontaneous emission" of the spectral line $(u_k\ u_n)$ results. I do not wish, however, to attempt to work out the calculation of the resonance case fully here, because the result would be of little value, so long as the *reaction* of the emitted radiation on the emitting system is not taken into account. Such a reaction must exist, not only because there is no ground at all for differentiating on principle between the light wave which is incident from outside, and that which is emitted by the system itself, but also because otherwise, if several proper vibrations were simultaneously excited in a system left to itself, the spontaneous emission would continue indefinitely. This required back-coupling must act so that in this case, along with the light emission, the higher proper vibrations gradually die down, and, finally, the fundamental vibration, corresponding to the normal state of the system, alone remains. The back-coupling is evidently exactly analogous to the reaction of radiation $\left(\frac{2e^2}{3mc^3}\ddot{v}\right)$ in the classical electron theory. This analogy also allays the increasing apprehension caused by the previous neglect of this back-coupling. The influence of the relevant term (probably no longer linear) in the wave equation will generally be small, just as in the electron the back pressure of radiation is generally very small compared with the force of inertia and the external field strength. In the resonance case, however—just as in the electron theory—the coupling with the proper light wave will be of the same order as that with the incident wave, and must be taken into account, if the "equilibrium" between the different proper vibrations, which sets in for the given irradiation, is to be correctly computed.

Let it be expressly remarked, however, that the back-coupling term is *not necessary for averting a resonance catastrophe!* Such can never occur in any circumstances, because according to the theorem of the *persistence of normalisation*, proved below in § 7, the configuration space integral of $\psi\bar{\psi}$ always remains normalised to the same value, even under the influence of arbitrary external forces—and indeed quite automatically, as a consequence of the wave equation (4″). The amplitudes of the $\psi$-vibrations, therefore, cannot grow indefinitely; they have, "on the average", always the same value. If *one* proper vibration waxes, then another must, therefore, wane.

## § 5. GENERALISATION FOR AN ARBITRARY PERTURBATION

If an *arbitrary* perturbation is in question as was assumed in equation (5) at the beginning of § 2, then we shall expand the perturbation energy *r(x, t)* as a Fourier series or Fourier integral in terms of the time. The terms of this expansion have, then, the form (6) of the perturbation potential of a light wave. We see immediately that on the right-hand side of equation (11) we then simply get two *series* (or, possibly, integrals) of imaginary powers of *e*, instead of merely two terms. If none of the exciting frequencies coincide with a critical frequency, we get the solution in exactly the same way as described in § 2, but, naturally, as Fourier series (or possibly Fourier integrals) of the time. It serves no purpose to write down the formal expansions here, and a more exact working out of separate problems lies outside the scope of the present paper. Yet an important point, already touched upon in § 3, must be mentioned.

Among the critical frequencies of equation (13), the frequency $\nu = 0$, from $E_k - E_k = 0$, also generally figures. For in this case also one proper value, namely, $E_k$, appears on the left side as proper value parameter. Thus, if the frequency 0, $i.e.$ a term independent of the time, occurs in the Fourier expansion of the perturbation function $r(x, t)$, we cannot reach our goal by exactly the earlier method. We easily see, however, how it must be modified, for the case of a time-constant perturbation is known from previous work (cf. Part III.). We have then to consider, at the same time, a small alteration and possibly a splitting up of the proper value or values of the excited free vibrations, $i.e.$ in the indices of the powers of $e$ in the first term on the right hand of equation (10) we have to replace $E_k$ by $E_k$ plus a small constant, the perturbation of the proper value. Exactly as described in Part III., § 1 and § 2, this perturbation is defined by the postulation that the right side of the critical Fourier component of our equation (13) is to be orthogonal to $u_k$ (or possibly to *all* the proper functions belonging to $E_k$).

The number of special problems, which fall under the question formulated in the present paragraph, is extraordinarily great. By superposing the perturbations due to a constant electric or magnetic field and a light wave, we obtain magnetic and electric double refraction, and magnetic rotation of the plane of polarisation. Resonance radiation in a magnetic field also comes under this heading, but for this purpose we must first obtain an exact solution for the resonance case discussed in § 4. Further, we can treat the action of an $\alpha$-particle or electron flying past the atom[1] in this way, if the encounter is not too

---

[1] A very interesting and successful attempt to compare the action of flying charged particles with the action of light waves, through a Fourier decomposition of their field, is to be found in a paper by E. Fermi, *Ztschr. f. Phys.* 29, p. 315, 1924.

close for the perturbation of each of the two systems to be calculable from the undisturbed motion of the other. All these questions are mere matters of calculation as soon as the proper values and functions of the unperturbed systems are known. It is, therefore, to be hoped that we will succeed in defining these functions, at least approximately, for heavier atoms also, in analogy with the approximate definition of the Bohr electronic orbits which belong to different types of terms.

## § 6. RELATIVISTIC-MAGNETIC GENERALISATION OF THE FUNDAMENTAL EQUATIONS

As an appendix to the physical problems just mentioned, in which the magnetic field, which has hitherto been completely ignored in this series of papers, plays an important part, I would like to give, briefly, the probable relativistic-magnetic generalisation of the basic equations (4″), although I can only do this meantime for the one electron problem, and only with the greatest possible reserve—the latter for two reasons. *Firstly*, the generalisation is provisionally based on a purely formal analogy. *Secondly*, as was mentioned in Part I., though it does formally lead in the Kepler problem to Sommerfeld's fine-structure formula with, in fact, the "half-integral" azimuthal and radial quantum, which is generally regarded as correct to-day, nevertheless there is *still lacking* the *supplement*, which is necessary to secure numerically correct diagrams of the splitting up of the hydrogen lines, and which is given in Bohr's theory by Goudsmit and Uhlenbeck's electronic spin.

The Hamilton-Jacobi partial differential equation for the Lorentzian electron can readily be written:

$$\left\{ \begin{aligned} &\left( \frac{1}{c}\frac{\partial W}{\partial t} + \frac{e}{c}V \right)^2 - \left( \frac{\partial W}{\partial x} - \frac{e}{c}\mathfrak{A}_x \right)^2 - \left( \frac{\partial W}{\partial y} - \frac{e}{c}\mathfrak{A}_y \right)^2 \\ &\qquad - \left( \frac{\partial W}{\partial z} - \frac{e}{c}\mathfrak{A}_z \right)^2 - m^2 c^2 = 0. \end{aligned} \right. \tag{34}$$

Here $e$, $m$, $c$ are the charge and mass of the electron, and the velocity of light; $V$, $\mathfrak{A}$ are the electro-magnetic potentials of the external electro-magnetic field at the position of the electron, and $W$ is the action function.

From the classical (relativistic) equation (34) I am now attempting to derive the *wave equation* for the electron, by the following *purely formal* procedure, which, we can verify easily, will lead to equations (4″), if it is applied to the Hamiltonian equation of a particle moving in an arbitrary field of force in ordinary (non-relativistic) mechanics. *After* the squaring, in equation (34), I replace the *quantities*

$$\left\{ \begin{array}{l} \dfrac{\partial W}{\partial t}, \quad \dfrac{\partial W}{x\,\partial x}, \quad \dfrac{\partial W}{\partial y}, \quad \dfrac{\partial W}{\partial z}, \\[2ex] \text{by the respective } operators \\[2ex] \pm\dfrac{h}{2\pi i}\dfrac{\partial}{\partial t}, \quad \pm\dfrac{h}{2\pi i}\dfrac{\partial}{\partial x}, \quad \pm\dfrac{h}{2\pi i}\dfrac{\partial}{\partial y}, \quad \pm\dfrac{h}{2\pi i}\dfrac{\partial}{\partial z}. \end{array} \right. \tag{35}$$

The double linear operator, so obtained, is applied to a wave function $\psi$ and the result put equal to zero, thus:

$$\nabla^2\psi - \frac{1}{c^2}\frac{\partial^2\psi}{\partial t^2} \mp \frac{4\pi i e}{hc}\left(\frac{V}{c}\frac{\partial\psi}{\partial t} + \mathfrak{A}\,\mathrm{grad}\,\psi\right)$$
$$+ \frac{4\pi^2 e^2}{h^2 c^2}\left(V^2 - \mathfrak{A}^2 - \frac{m^2 c^4}{e^2}\right)\psi = 0. \tag{36}$$

(The symbols $\nabla^2$ and grad have here their elementary three-dimensional Euclidean meaning.) The pair of equations (36) would be the possible relativistic-magnetic generalisation of (4″) for the case of a single electron, and should likewise be understood to mean that the complex wave function has to satisfy either the one or the other equation.

From (36) the fine structure formula of Sommerfeld for the hydrogen atom may be obtained by exactly the same method as is described in Part I., and also we may derive (neglecting the term in $\mathfrak{A}^2$) the normal Zeeman effect as well as the well-known selection and polarisation rules and intensity formulae. They follow from the integral relations between Legendre functions introduced at the end of Part III.

For the reasons given in the first section of this paragraph, I withhold the detailed reproduction of these calculations meantime, and also in the following final paragraph refer to the "classical", and not to the still incomplete relativistic-magnetic version of the theory.

## § 7. On the Physical Significance of the Field Scalar

The heuristic hypothesis of the electro-dynamical meaning of the field scalar $\psi$, previously employed in the *one*-electron problem, was extended off-hand to an arbitrary system of charged particles in § 2, and there a more exhaustive description of the procedure was promised. We had calculated the density of electricity at an arbitrary point in space as follows. We selected *one* particle, kept the trio of co-ordinates that describes *its* position in ordinary mechanics fixed; integrated $\psi\bar{\psi}$ over all the rest of the co-ordinates of the system and multiplied the result by a certain constant, the "charge" of the selected particle; we did a similar thing for each particle (trio of co-ordinates), in each case giving the selected particle the same position, namely, the position of that point of *space* at which we desired to know the electric density. The latter is equal to the algebraic sum of the partial results.

This rule is now equivalent to the following conception, which allows the true meaning of $\psi$ to stand out more clearly. $\psi\bar{\psi}$ is a kind of *weight-function* in the system's configuration space. The *wave-mechanical* configuration of the system is a *superposition of many*, strictly speaking of *all*, point-mechanical configurations kinematically possible. Thus, each point-mechanical configuration contributes to the true wave-mechanical configuration with a certain *weight*, which is given precisely by $\psi\bar{\psi}$. If we like paradoxes, we may say that the system exists, as it were, simultaneously in all the positions kinematically imaginable, but not "equally strongly" in all. In macroscopic motions, the weight-function is practically concentrated in a small region of positions, which are practically indistinguishable. The centre of gravity of this region in configuration space travels over distances which are macroscopically perceptible. In problems of microscopic motions, we are in any case interested *also*, and in certain cases even *mainly*, in the varying *distribution* over the region.

This new interpretation may shock us at first glance, since we have often previously spoken in such an intuitive concrete way of the "$\psi$-vibrations" as though of something quite real. But there is something tangibly real behind the present conception also, namely, the very real electrodynamically effective fluctuations of the electric spacedensity. The $\psi$-function is to do no more and no less than permit of the totality of these fluctuations being mastered and surveyed mathematically by a single partial differential equation. We have repeatedly called attention[1] to the fact that the $\psi$-function itself cannot and may not be interpreted directly in terms of three-dimensional space—however much the one-electron problem tends to mislead us on this point—because it is in general a function in configuration space, not real space.

Concerning such a weight-function in the above sense, we would wish its integral over the whole configuration space to remain constantly normalised to the same unchanging value, preferably to unity. We can easily verify that this is necessary if the total charge of the system is to remain constant on the above definitions. Even for non-conservative systems, this condition must obviously be postulated. For, naturally, the charge of a system is not to be altered when, *e.g.*, a light wave falls on it, continues for a certain length of time, and then ceases. (*N.B.*—This is also valid for ionisation processes. A disrupted particle is still to be included in the system, until the separation is also *logically*—by decomposition of configuration space—completed.)

The question now arises as to whether the postulated *persistence of normalisation* is actually guaranteed by equations (4″), to which $\psi$ is subject. If this were not the case, our whole conception would practically break down. Fortunately, it is the case. Let us form

$$\frac{d}{dt} \int \psi \bar{\psi} \rho dx = \int \left( \psi \frac{\partial \bar{\psi}}{\partial t} + \bar{\psi} \frac{\partial \psi}{\partial t} \right) \rho dx. \qquad (37)$$

[1] End of Part II. (p. 39); paper on Heisenberg's quantum mechanics (p. 60).

Now, $\psi$ satisfies one of the two equations (4″), and $\bar{\psi}$ the other. Therefore, apart from a multiplicative constant, this integral becomes

$$\int \left(\psi \nabla^2 \bar{\psi} - \bar{\psi} \nabla^2 \psi\right) \rho \, dx = 2i \int \left(J \nabla^2 R - R \nabla^2 J\right) \rho \, dx, \quad (38)$$

where for the moment we put

$$\psi = R + iJ.$$

According to Green's theorem, integral (38) vanishes identically; the sole necessary condition that functions $R$ and $J$ must satisfy for this—vanishing in sufficient degree at infinity—means physically nothing more than that the system under consideration should practically be confined to a *finite* region.

We can put this in a somewhat different way, by not immediately integrating over the whole configuration space, but by merely changing the time-derivative of the weight-function into a divergence by Green's transformation. Through this we get an insight into the question of the flow of the weight-function, and thus of electricity. The two equations

$$\begin{aligned}
\frac{\partial \psi}{\partial t} &= \frac{h}{4\pi i} \left(\nabla^2 - \frac{8\pi^2}{h^2} V\right) \psi \\
\frac{\partial \bar{\psi}}{\partial t} &= -\frac{h}{4\pi i} \left(\nabla^2 - \frac{8\pi^2}{h^2} V\right) \bar{\psi}
\end{aligned} \qquad (4'')$$

are multiplied by $\rho \bar{\psi}$ and $\rho \psi$ respectively, and added. Hence

$$\frac{\partial}{\delta t} \left(\rho \psi \bar{\psi}\right) = \frac{h}{4\pi i} \rho \cdot \left(\bar{\psi} \nabla^2 \psi - \psi \nabla^2 \bar{\psi}\right). \qquad (39)$$

To carry out *in extenso* the transformation of the right-hand side, we must remember the explicit form of our many-dimensional, non-Euclidean, Laplacian operator:[1]

$$\rho \nabla^2 = \sum_k \frac{\partial}{\partial q_k} \left[\rho \, T_{p_k} \left(q_l, \frac{\partial \psi}{\partial q_l}\right)\right]. \qquad (40)$$

---

[1] Cf. paper on Heisenberg's theory, equation (31). The quantity there denoted by $\Delta_D - \frac{1}{2}$ is our "density function" $\rho(x)$ (e.g. $r^2 \sin \theta$ in spherical polars). $T$ is the kinetic energy as function of the position co-ordinates and *momenta*, the suffix at $T$ denoting differentiation with respect to a momentum. In equations (31) and (32), *loc. cit.*, unfortunately by error the suffix $k$ is used twice, once for the summation and then also as a representative suffix in the argument of the functions.

By a small transformation we readily obtain

$$\frac{\partial}{\partial t}\left(\rho \psi \bar{\psi}\right) = \frac{h}{4\pi i} \sum_{k} \frac{\partial}{\partial q_k}\left[\rho \bar{\psi}\, T_{p_k}\left(q_l, \frac{\partial \psi}{\partial q_l}\right) - \rho \psi\, T_{p_k}\left(q_l, \frac{\partial \bar{\psi}}{\partial q_l}\right)\right].$$

(41)

The right-hand side appears as the divergence of a many-dimensional real vector, which is evidently to be interpreted as the *current density of the weight-function* in configuration space. Equation (41) is the *continuity equation* of the weight-function.

From it we can obtain the *equation of continuity of electricity*, and, indeed, a separate equation of this sort is valid for the charge density "originating from each separate particle". Let us fix on the $\alpha$th particle, say. Let its "charge" be $e_\alpha$, its mass $m_\alpha$, and let its co-ordinate space be described by Cartesians $x_\alpha, y_\alpha, z_\alpha$, for the sake of simplicity. We denote the product of the differentials of the *remaining* co-ordinates shortly by $dx'$. Over the latter, we integrate equation (41), keeping $x_\alpha, y_\alpha, z_\alpha$, *fixed*. As the result, all terms except three disappear from the right-hand side, and we obtain

$$\begin{aligned}
\frac{\partial}{\partial t}\left[e_\alpha \int \psi \bar{\psi}\, dx'\right] &= \frac{h e_\alpha}{4\pi i m_\alpha}\left\{\frac{\partial}{\partial x_\alpha}\left[\int\left(\bar{\psi}\frac{\partial \psi}{\partial x_\alpha} - \psi\frac{\partial \bar{\psi}}{\partial x_\alpha}\right)dx'\right]\right.\\
&\qquad\left. + \frac{\partial}{\partial y_\alpha}\left[\int\left(\bar{\psi}\frac{\partial \psi}{\partial y_\alpha} - \psi\frac{\partial \bar{\psi}}{\partial y_\alpha}\right)dx'\right] + \cdots\right\}\\
&= \frac{h e_\alpha}{4\pi i m_\alpha}\mathrm{div}_\alpha\left[\int\left(\bar{\psi}\,\mathrm{grad}_\alpha \psi - \psi\,\mathrm{grad}_\alpha \bar{\psi}\right)dx'\right].
\end{aligned}$$

(42)

In this equation, div and grad have the usual three-dimensional Euclidean meaning, and $x_\alpha, y_\alpha, z_\alpha$ are to be interpreted as Cartesian co-ordinates of real space. The equation is the continuity equation of *that* charge density which "originates from the $\alpha$th particle". If we form all the others in an analogous fashion, and add them together, we obtain the total equation of continuity. Of course, we must emphasize that the interpretation of the integrals on the right-hand side as *components of the current density*, is, as in all such, cases, not absolutely compulsory, because a divergence-free vector could be added thereto.

To give an example, in the conservative *one*-electron problem, if $\psi$ is given by

$$\psi = \sum_k c_k u_k e^{2\pi i v_k t + i\theta_k} \quad (c_k, \theta_k \text{ real constants}), \qquad (43)$$

we get for the *current density J*

$$J = \frac{he_1}{2\pi m_1} \sum_{(k,l)} c_k c_l \left( u_1 \operatorname{grad} u_k - u_k \operatorname{grad} u_l \right)$$

$$\times \sin \left[ 2\pi \left( v_k - v_l \right) t + \theta_k - \theta_l \right]. \qquad (44)$$

We see, and this is valid for conservative systems generally, that, if only a single proper vibration is excited, the current components disappear and the distribution of electricity is constant in time. The latter is also immediately evident from the fact that $\psi \bar{\psi}$ becomes constant with respect to the time. This is still the case even when several proper vibrations are excited, if they all belong to the same proper value. On the other hand, the current density then no longer needs to vanish, but there may be present, and generally is, a *stationary* current distribution. Since the one or the other occurs in the unperturbed normal state at any rate, we may in a certain sense speak of a *return to electrostatic and magnetostatic atomic models*. In this way the lack of radiation in the normal state would, indeed, find a startlingly simple explanation.

I hope and believe that the present statements will prove useful in the elucidation of the magnetic properties of atoms and molecules, and further for explaining the flow of electricity in solid bodies.

Meantime, there is no doubt a certain crudeness in the use of a *complex* wave function. If it were unavoidable *in principle*, and not merely a facilitation of the calculation, this would mean that there are in principle *two* wave functions, which must be used *together* in order to obtain information on the state of the system. This somewhat unacceptable inference admits, I believe, of the very much more congenial interpretation that the state of the system is given by a real function and its time-derivative. Our inability to give more accurate information about this is intimately connected with the fact that, in the pair of equations (4″), we have before us only the *substitute*—extraordinarily convenient for the calculation, to be sure—for a real wave equation of probably the fourth order, which, however, I have not succeeded in forming for the non-conservative case.

Zürich, Physical Institute of the University.
(Received June 23, 1926.)

# Chapter Four

**B**oth Werner Heisenberg's and Erwin Schrodinger's formulations of quantum mechanics were *non-relativistic*, meaning that they did not include Einstein's special theory of relativity. In order to describe very fast moving quantum particles, a relativistic quantum theory would need to be developed. In 1928, a brilliant paper by Paul Dirac entitled "The Quantum Theory of the Electron" presented a relativistic quantum theory and a relativistic replacement to the Schrodinger Equation which is now known as the Dirac Equation. Remarkably, Dirac's theory required the existence of anti-particles even though no such thing had yet been supposed. Thus he was able to predict the existence of the positron (which is the electron's anti-particle) before it had even been experimentally detected! Equally astounding, Dirac was also able to show that by including relativity in quantum mechanics he could explain the intrinsic angular momentum or "spin" of electrons which had been an unsolved problem since its discovery.

It had been found that spin was quantized in units of $\frac{1}{2}\hbar$—that is, all particles have spin with an integer times $\frac{1}{2}\hbar$ (where $\hbar$ is Planck's constant divided by $2\pi$). In his paper "The Connection Between Spin and Statistics," Wolfgang Pauli further developed the theory of spin by showing that particles with half-integer spin must obey Fermi-Dirac statistics. Those particles are now known as *fermions*. Pauli also showed that particles with integral spin obey Bose-Einstein statistics, so are known as *bosons*.

Fermi-Dirac statistics is the statistical system which applies to particles which obey Pauli's famous exclusion principle. The exclusion principle states that no two identical fermions in the same system can be in identical states. Pauli exclusion is the explanation for many physical phenomena, including much of the structure of the periodic table of elements. We have seen that atoms are composed of a positively charged nucleus surrounded by a "cloud" of electrons. Electrons are fermions, so only one can be in the ground (the lowest energy) state. The others must occupy higher energy states. That all the electrons in an atom must occupy increasingly energetic shells explains among other things why metals are shiny and excellent electric conductors. It explains why some elements are gases at room temperature and others are solids. In fact, the variety of interactions between the electrons in various states of the different elements gives rise to much of the field of chemistry. Without the exclusion principle, chemistry would be a very different field!

Another application of the exclusion principle is the stability of white dwarf stars. White dwarfs are the leftover remains of a once active star. They no longer have an internal energy source to stabilize them against the crushing force of their own gravity. Why then, do they not collapse on themselves? The answer is the Pauli exclusion principle. One electron in a white dwarf cannot get too near the state of a nearby electron. This provides a pressure force that stabilizes the star against its own gravity.

Bosons are essentially the opposite of fermions, and they do *not* obey Pauli exclusion. In fact, bosons "like" to be in the same state. Photons are bosons, and the fact that they like to be in the same state is exploited in lasers and explains why lasers produce such coherent monochromatic light. One of the most interesting things about bosons is that many of their properties can be viewed, not just at a very small scale which is typical of quantum theory, but on a macroscopic scale. This is a direct result of their ability to occupy the same quantum state. If a gas of identical bosons is sufficiently cool, all the particles in the gas will tend to occupy their lowest energy state and will begin to act coherently. This gas is known as a Bose-Einstein condensate. If there are a large number of particles in the gas, then the entire gas will exhibit uniquely quantum mechanical properties on a macroscopic scale. In the last the two decades techniques to ultra-cool gases have been developed, allowing the production of Bose-Einstein condensates for the first time. As these techniques are refined, it will be interesting to study the many peculiar quantum phenomena that will be observable in the macroscopic world.

# THE QUANTUM THEORY
# OF THE ELECTRON

By

PAUL A.M. DIRAC

*Proceedings of the Royal Society of London. Series A, Containing papers of a Mathematical and Physical Character*, Vol. 117, No. 778. (Feb. 1, 1929) pp. 610–624.

The new quantum mechanics, when applied to the problem of the structure of the atom with point-charge electrons, does not give results in agreement with experiment. The discrepancies consist of "duplexity" phenomena, the observed number of stationary states for an electron in an atom being twice the number given by the theory. To meet the difficulty, Goudsmit and Uhlenbeck have introduced the idea of an electron with a spin angular momentum of half a quantum and a magnetic moment of one Bohr magneton. This model for the electron has been fitted into the new mechanics by Pauli,* and Darwin,† working with an equivalent theory, has shown that it gives results in agreement with experiment for hydrogen-like spectra to the first order of accuracy.

The question remains as to why Nature should have chosen this particular model for the electron instead of being satisfied with the point-charge. One would like to find some incompleteness in the previous methods of applying quantum mechanics to the point-charge electron such that, when removed, the whole of the duplexity phenomena follow without arbitrary assumptions. In the present paper it is shown that this is the case, the incompleteness of the previous theories lying in their disagreement with relativity, or, alternatetively, with the general transformation theory of quantum mechanics. It appears that the simplest Hamiltonian for a point-charge electron satisfying

---

*Pauli, 'Z. f. Physik,' vol. 43, p. 601 (1927).
† Darwin,' Roy. Soc. Proc.,' A, vol. 116, p. 227 (1927).

the requirements of both relativity and the general transformation theory leads to an explanation of all duplexity phenomena without further assumption. All the same there is a great deal of truth in the spinning electron model, at least as a first approximation. The most important failure of the model seems to be that the magnitude of the resultant orbital angular momentum of an electron moving in an orbit in a central field of force is not a constant, as the model leads one to expect.

## § 1. Previous Relativity Treatments

The relativity Hamiltonian according to the classical theory for a point electron moving in an arbitrary electro-magnetic field with scalar potential $A_0$ and vector potential A is

$$F \equiv \left(\frac{W}{c} + \frac{e}{c}A_0\right)^2 + \left(p + \frac{e}{c}A\right)^2 + m^2c^2,$$

where p is the momentum vector. It has been suggested by Gordon[*] that the operator of the wave equation of the quantum theory should be obtained from this F by the same procedure as in non-relativity theory, namely, by putting

$$W = ih\frac{\partial}{\partial t},$$

$$p_r = -ih\frac{\partial}{\partial x_r}, \quad r = 1, 2, 3,$$

in it. This gives the wave equation

$$F\psi \equiv \left[\left(ih\frac{\partial}{c\partial t} + \frac{e}{c}A_0\right)^2 + \Sigma_r\left(-ih\frac{\partial}{\partial x_r} + \frac{e}{c}A_r\right)^2 + m^2c^2\right]\psi = 0,$$

$$(1)$$

the wave function $\psi$ being a function of $x_1$, $x_2$, $x_3$, $t$. This gives rise to two difficulties.

The first is in connection with the physical interpretation of $\psi$. Gordon, and also independently Klein,[†] from considerations of the conservation theorems, make the assumption that if $\psi_m$, $\psi_n$ are two

[*]Gordon, 'Z. f. Physik,' vol. 40, p. 117 (1926).
[†]Klein, 'Z. f. Physik,' vol. 41, p. 407 (1927).

solutions

$$\rho_{mn} = -\frac{e}{2mc^2}\left\{ih\left(\psi_m\frac{\partial\psi_n}{\partial t} - \overline{\psi}_n\frac{\partial\psi_m}{\partial t}\right) + 2e\,A_0\psi_m\overline{\psi}_n\right\}$$

and

$$I_{mn} = -\frac{e}{2m}\left\{-ih\left(\psi_m \text{ grad } \overline{\psi}_n - \overline{\psi}_n \text{ grad } \psi_m\right) + 2\frac{e}{c}A_m\psi_m\overline{\psi}_n\right\}$$

are to be interpreted as the charge and current associated with the transition $m \to n$. This appears to be satisfactory so far as emission and absorption of radiation are concerned, but is not so general as the interpretation of the non-relativity quantum mechanics, which has been developed* sufficiently to enable one to answer the question: What is the probability of any dynamical variable at any specified time having a value lying between any specified limits, when the system is represented by a given wave function $\psi_n$? The Gordon-Klein interpretation can answer such questions if they refer to the position of the electron (by the use of $\rho_{nn}$), but not if they refer to its momentum, or angular momentum or any other dynamical variable. We should expect the interpretation of the relativity theory to be just as general as that of the non-relativity theory.

The general interpretation of non-relativity quantum mechanics is based on the transformation theory, and is made possible by the wave equation being of the form

$$(H - W)\psi = 0, \tag{2}$$

*i.e.*, being linear in W or $\partial/\partial t$, so that the wave function at any time determines the wave function at any later time. The wave equation of the relativity theory must also be linear in W if the general interpretation is to be possible.

The second difficulty in Gordon's interpretation arises from the fact that if one takes the conjugate imaginary of equation (1), one gets

$$\left[\left(-\frac{W}{c} + \frac{e}{c}A_0\right)^2 + \left(-p + \frac{e}{c}A\right)^2 + m^2c^2\right]\psi = 0,$$

*Jordan, 'Z. f. Physik,' vol. 40, p. 809 (1927); Dirac, 'Roy. Soc. Proc.,' A, vol. **113**, p. 621 (1927).

which is the same as one would get if one put $-e$ for $e$. The wave equation (1) thus refers equally well to an electron with charge $e$ as to one with charge $-e$. If one considers for definiteness the limiting case of large quantum numbers one would find that some of the solutions of the wave equation are wave packets moving in the way a particle of charge $-e$ would move on the classical theory, while others are wave packets moving in the way a particle of charge $e$ would move classically. For this second class of solutions W has a negative value. One gets over the difficulty on the classical theory by arbitrarily excluding those solutions that have a negative W. One cannot do this on the quantum theory, since in general a perturbation will cause transitions from states with W positive to states with W negative. Such a transition would appear experimentally as the electron suddenly changing its charge from $-e$ to $e$, a phenomenon which has not been observed. The true relativity wave equation should thus be such that its solutions split up into two non-combining sets, referring respectively to the charge $-e$ and the charge $e$.

In the present paper we shall be concerned only with the removal of the first of these two difficulties. The resulting theory is therefore still only an approximation, but it appears to be good enough to account for all the duplexity phenomena without arbitrary assumptions.

## § 2. The Hamiltonian for No Field

Our problem is to obtain a wave equation of the form (2) which shall be invariant under a Lorentz transformation and shall be equivalent to (1) in the limit of large quantum numbers. We shall consider first the case of no field, when equation (1) reduces to

$$\left(-p_0^2 + p^2 + m^2 c^2\right) \psi = 0 \tag{3}$$

if one puts

$$p_0 = \frac{W}{c} = ih \frac{\partial}{c \, \partial t}.$$

The symmetry between $p_0$ and $p_1$, $p_2$, $p_3$ required by relativity shows that, since the Hamiltonian we want is linear in $p_0$, it must also be linear in $p_1, p_2$ and $p_3$. Our wave equation is therefore of the form

$$(p_0 + \alpha_1 p_1 + \alpha_2 p_2 + \alpha_3 p_3 + \beta)\, \psi = 0, \tag{4}$$

where for the present all that is known about the dynamical variables or operators $\alpha_1, \alpha_2, \alpha_3, \beta$ is that they are independent of $p_0, p_1, p_2, p_3$, i.e., that they commute with $t, x_1, x_2, x_3$. Since we are considering the case of a particle moving in empty space, so that all points in space are equivalent, we should expect the Hamiltonian not to involve $t, x_1, x_2, x_3$. This means that $\alpha_1, \alpha_2, \alpha_3, \beta$ are independent of $t, x_1, x_2, x_3$, i.e., that they commute with $p_0, p_1, p_2, p_3$. We are therefore obliged to have other dynamical variables besides the co-ordinates and momenta of the electron, in order that $\alpha_1, \alpha_2, \alpha_3, \beta$ may be functions of them. The wave function $\psi$ must then involve more variables than merely $x_1, x_2, x_3, t$.

Equation (4) leads to

$$\begin{aligned}
0 &= (-p_0 + \alpha_1 p_1 + \alpha_2 p_2 + \alpha_3 p_3 + \beta)\,(p_0 + \alpha_1 p_1 + \alpha_2 p_2 + \alpha_3 p_3 + \beta)\psi \\
&= [-p_0^2 + \Sigma \alpha_1^2 p_1^2 + \Sigma(\alpha_1 \alpha_2 + \alpha_2 \alpha_1) p_1 p_2 + \beta^2 + \Sigma(\alpha_1 \beta + \beta \alpha_1) p_1]\psi,
\end{aligned} \tag{5}$$

where the $\Sigma$ refers to cyclic permutation of the suffixes 1, 2, 3. This agrees with (3) if

$$\left.\begin{aligned}
\alpha_r^2 &= 1, & \alpha_r \alpha_s + \alpha_s \alpha_r &= 0 \quad (r \neq s) \\
\beta^2 &= m^2 c^2, & \alpha_r \beta + \beta \alpha_r &= 0
\end{aligned}\right\}\, r, s = 1, 2, 3.$$

If we put $\beta = \alpha_4 mc$, these conditions become

$$\alpha_\mu^2 = 1 \quad \alpha_\mu \alpha_\nu + \alpha_\nu \alpha_\mu = 0\ (\mu \neq \nu) \quad \mu, \nu = 1, 2, 3, 4. \tag{6}$$

We can suppose the $\alpha_\mu$'s to be expressed as matrices in some matrix scheme, the matrix elements of $\alpha_\mu$ being, say, $\alpha_\mu (\zeta'\, \zeta'')$. The wave function $\psi$ must now be a function of $\zeta$ as well as $x_1, x_2, x_3, t$. The result of $\alpha_\mu$ multiplied into $\psi$ will be a function $(\alpha_\mu \psi)$ of $x_1, x_2, x_3, t, \zeta$ denned by

$$(\alpha_\mu \psi)(x,\ t,\ \zeta) = \Sigma'_\zeta \alpha_\mu(\zeta\zeta')\, \psi\,(x,\ t,\ \zeta').$$

We must now find four matrices $\alpha_\mu$ to satisfy the conditions (6). We make use of the matrices

$$\sigma_1 = \begin{pmatrix} 0 & 1 \\ 1 & 0 \end{pmatrix} \quad \sigma_2 = \begin{pmatrix} 0 & -i \\ i & 0 \end{pmatrix} \quad \sigma_3 = \begin{pmatrix} 1 & 0 \\ 0 & -1 \end{pmatrix}$$

which Pauli introduced* to describe the three components of spin angular momentum. These matrices have just the properties

$$\sigma_r^2 = 1 \quad \sigma_r\sigma_s + \sigma_s\sigma_r = 0, \quad (r \neq s), \tag{7}$$

that we require for our $\alpha$'s. We cannot, however, just take the $\sigma$'s to be three of our $\alpha$'s, because then it would not be possible to find the fourth. We must extend the $\sigma$'s in a diagonal manner to bring in two more rows and columns, so that we can introduce three more matrices $\rho_1$, $\rho_2$, $\rho_3$ of the same form as $\sigma_1$, $\sigma_2$, $\sigma_3$, but referring to different rows and columns, thus:—

$$\sigma_1 = \begin{Bmatrix} 0 & 1 & 0 & 0 \\ 1 & 0 & 0 & 0 \\ 0 & 0 & 0 & 1 \\ 0 & 0 & 1 & 0 \end{Bmatrix} \quad \sigma_2 = \begin{Bmatrix} 0 & -i & 0 & 0 \\ i & 0 & 0 & 0 \\ 0 & 0 & 0 & -i \\ 0 & 0 & i & 0 \end{Bmatrix} \quad \sigma_3 = \begin{Bmatrix} 1 & 0 & 0 & 0 \\ 0 & -1 & 0 & 0 \\ 0 & 0 & 1 & 0 \\ 0 & 0 & 0 & -1 \end{Bmatrix},$$

$$\rho_1 = \begin{Bmatrix} 0 & 0 & 1 & 0 \\ 0 & 0 & 0 & 1 \\ 1 & 0 & 0 & 0 \\ 0 & 1 & 0 & 0 \end{Bmatrix} \quad \rho_2 = \begin{Bmatrix} 0 & 0 & -i & 0 \\ 0 & 0 & 0 & -i \\ i & 0 & 0 & 0 \\ 0 & i & 0 & 0 \end{Bmatrix} \quad \rho_3 = \begin{Bmatrix} 1 & 0 & 0 & 0 \\ 0 & 1 & 0 & 0 \\ 0 & 0 & -1 & 0 \\ 0 & 0 & 0 & -1 \end{Bmatrix}.$$

The $\rho$'s are obtained from the $\sigma$'s by interchanging the second and third rows, and the second and third columns. We now have, in addition to equations (7)

$$\left. \begin{array}{c} \rho_r^2 = 1 \quad \rho_r\rho_s + \rho_s\rho_r = 0 \quad (r \neq s), \\ \\ \rho_r\sigma_t = \sigma_t\rho_r. \end{array} \right\} \tag{7'}$$

and also

If we now take

$$\alpha_1 = \rho_1\sigma_1, \quad \alpha_2 = \rho_1\sigma_2, \quad \alpha_3 = \rho_1\sigma_3, \quad \alpha_4 = \rho_3,$$

all the conditions (6) are satisfied, *e.g.*,

$$\alpha_1^2 = \rho_1\sigma_1\rho_1\sigma_1 = \rho_1^2\sigma_1^2 = 1$$

$$\alpha_1\alpha_2 = \rho_1\sigma_1\rho_1\sigma_2 = \rho_1^2\sigma_1\sigma_2 = -\rho_1^2\sigma_2\sigma_1 = -\alpha_2\alpha_1.$$

*Pauli, *loc. cit.*

The following equations are to be noted for later reference

$$\left. \begin{array}{l} \rho_1\rho_2 = i\rho_3 = -\rho_2\rho_1 \\ \sigma_1\sigma_2 = i\sigma_3 = -\sigma_2\sigma_1 \end{array} \right\}, \tag{8}$$

together with the equations obtained by cyclic permutation of the suffixes.

The wave equation (4) now takes the form

$$\left[ p_0 + \rho_1\left(\sigma, \mathbf{p}\right) + \rho_3 mc \right] \psi = 0, \tag{9}$$

where $\sigma$ denotes the vector $(\sigma_1, \sigma_2, \sigma_3)$.

## § 3. PROOF OF INVARIANCE UNDER A LORENTZ TRANSFORMATION

Multiply equation (9) by $\rho_3$ on the left-hand side. It becomes, with the help of (8),

$$\left[ \rho_3 p_0 + i\rho_2 \left( \sigma_1\rho_1 + \sigma_2\rho_2 + \sigma_3\rho_3 \right) + mc \right] \psi = 0.$$

Putting

$$p_0 + ip_4,$$

$$\rho_3 - \gamma_4, \qquad \rho_2\sigma_r = \gamma_r, \qquad r = 1, 2, 3, \tag{10}$$

we have

$$\left[ i\Sigma_{\gamma_\mu p_\mu} + mc \right] \psi = 0, \qquad \mu = 1, 2, 3, 4. \tag{11}$$

The $p_\mu$ transform under a Lorentz transformation according to the law

$$p'_\mu = \Sigma_v a_{\mu v} p_v,$$

where the coefficients $a_{\mu v}$ are c-numbers satisfying

$$\Sigma_\mu a_{\mu v} a_{\mu r} = \delta_{vr}, \qquad \Sigma_r a_{\mu r} a_{vr} = \delta_{\mu v}.$$

The wave equation therefore transforms into

$$\left[ i\Sigma\gamma'_\mu p'_\mu + mc \right] \psi = 0. \tag{12}$$

where

$$\gamma'_\mu = \Sigma_\nu a_{\mu\nu} \gamma_\nu.$$

Now the $\gamma_\mu$, like the $\alpha_\mu$, satisfy

$$\gamma^2_\mu = 1, \quad \gamma_\mu \gamma_\nu + \gamma_\nu \gamma_\mu = 0, \quad (\mu \neq \nu).$$

These relations can be summed up in the single equation

$$\gamma_\mu \gamma_\nu + \gamma_\nu \gamma_\mu = 2\delta_{\mu\nu}.$$

We have

$$\begin{aligned}
\gamma'_\mu \gamma'_\nu + \gamma'_\nu \gamma'_\mu &= \Sigma_{\tau\lambda} a_{\mu\tau} a_{\nu\lambda} (\gamma_\tau \gamma_\lambda + \gamma_\lambda \gamma_\tau) \\
&= 2\Sigma_{\tau\lambda} a_{\mu\tau} a_{\nu\lambda} \delta_{\tau\lambda} \\
&= 2\Sigma_\tau a_{\mu\tau} a_{\nu\tau} = 2\delta_{\mu\nu}.
\end{aligned}$$

Thus the $\gamma'_\mu$ satisfy the same relations as the $\gamma_\mu$. Thus we can put, analogously to (10)

$$\gamma'_4 = \rho'_3 \qquad \gamma'_r = \rho'_2 \sigma'_r$$

where the $\rho''$s and $\sigma''$s are easily verified to satisfy the relations corresponding to (7), (7') and (8), if $\rho'_2$ and $\rho'_1$ are defined by $\rho'_2 = i\gamma'_1 \gamma'_2 \gamma'_3$, $\rho'_1 = -i\rho'_2 \rho'_3$.

We shall now show that, by a canonical transformation, the $\rho''$s and $\sigma''$s may be brought into the form of the $\rho$'s and $\sigma$'s. From the equation $\rho'^2_3 = 1$, it follows that the only possible characteristic values for $\rho'_3$ are $\pm 1$. If one applies to $\rho'_3$ a canonical transformation with the transformation function $\rho'_1$, the result is

$$\rho'_1 \rho'_3 \left( \rho'_1 \right)^{-1} = -\rho'_3 \rho'_1 \left( \rho'_1 \right)^{-1} = -\rho'_3.$$

Since characteristic values are not changed by a canonical transformation, $\rho'_3$ must have the same characteristic values as $-\rho'_3$. Hence the characteristic values of $\rho'_3$ are $+1$ twice and $-1$ twice. The same argument applies to each of the other $\rho''$s, and to each of the $\sigma''$s.

Since $\rho'_3$ and $\sigma'_3$ commute, they can be brought simultaneously to the diagonal form by a canonical transformation. They will then have for their diagonal elements each $+1$ twice and $-1$ twice. Thus, by suitably rearranging the rows and columns, they can be brought into the form $\rho_3$ and $\sigma_3$ respectively. (The possibility $\rho'_3 = \pm\sigma'_3$ is excluded by the existence of matrices that commute with one but not with the other.)

Any matrix containing four rows and columns can be expressed as

$$c + \Sigma_r c_r \sigma_r + \Sigma_r c_r' \rho_r + \Sigma_{rs} c_{rs} \rho_r \sigma_s \tag{13}$$

where the sixteen coefficients $c, c_r, c_r', c_{rs}$ are c-numbers. By expressing $\sigma_1'$ in this way, we see, from the fact that it commutes with $\rho_3' = \rho_3$ and anticommutes* with $\sigma_3' = \sigma_3$, that it must be of the form

$$\sigma_1' = c_1 \sigma_1 + c_2 \sigma_2 + c_{31} \rho_3 \sigma_1 + c_{32} \rho_3 \sigma_2,$$

*i.e.*, of the form

$$\sigma_1' = \begin{Bmatrix} 0 & a_{12} & 0 & 0 \\ a_{21} & 0 & 0 & 0 \\ 0 & 0 & 0 & a_{34} \\ 0 & 0 & a_{43} & 0 \end{Bmatrix}$$

The condition $\sigma_1'^2 = 1$ shows that $a_{12}a_{21} = 1, a_{34}a_{43} = 1$. If we now apply the canonical transformation: first row to be multiplied by $(a_{21}/a_{12})^{\frac{1}{2}}$ and third row to be multiplied by $(a_{43}/a_{34})^{\frac{1}{2}}$, and first and third columns to be divided by the same expressions, $\sigma_1'$ will be brought into the form of $\sigma_1$, and the diagonal matrices $\sigma_3'$ and $\rho_3'$ will not be changed.

If we now express $\rho_1'$ in the form (13) and use the conditions that it commutes with $\sigma_1' = \sigma_1$ and $\sigma_3' = \sigma_3$ and anticommutes with $\rho_3' = \rho_3$, we see that it must be of the form

$$\rho_1' = c_1' \rho_1 + c_2' \rho_2.$$

The condition $\rho_1'^2 = 1$ shows that $c_1'^2 + c_2'^2 = 1$, or $c_1' = \cos \theta, c_2' = \sin \theta$. Hence $\rho_1'$ is of the form

$$\rho_1' = \begin{Bmatrix} 0 & 0 & e^{-i\theta} & 0 \\ 0 & 0 & 0 & e^{-i\theta} \\ e^{i\theta} & 0 & 0 & 0 \\ 0 & e^{i\theta} & 0 & 0 \end{Bmatrix}$$

---

*We say that $a$ anticommutes with $b$ when $ab = -ba$.

If we now apply the canonical transformation: first and second rows to be multiplied by $e^{i\theta}$ and first and second columns to be divided by the same expression, $\rho_1'$ will be brought into the form $\rho_1$, and $\sigma_1, \sigma_3$, $\rho_3$ will not be altered. $\rho_2'$ and $\sigma_2'$ must now be of the form $\rho_2$ and $\sigma_2$, on account of the relations $i\rho_2' = \rho_3'\rho_1'$, $i\sigma_2' = \sigma_3'\sigma_1'$.

Thus by a succession of canonical transformations, which can be combined to form a single canonical transformation, the $\rho'$'s and $\sigma'$'s can be brought into the form of the $\rho$'s and $\sigma$'s. The new wave equation (12) can in this way be brought back into the form of the original wave equation (11) or (9), so that the results that follow from this original wave equation must be independent of the frame of reference used.

## § 4. THE HAMILTONIAN FOR AN ARBITRARY FIELD

To obtain the Hamiltonian for an electron in an electromagnetic field with scalar potential $A_0$ and vector potential $A$, we adopt the usual procedure of substituting $p_0 + e/c$. $A_0$ for $p_0$ and $\mathbf{p} + e/c$. $\mathbf{A}$ for $\mathbf{p}$ in the Hamiltonian for no field. From equation (9) we thus obtain

$$\left[ p_0 + \frac{e}{c}\mathbf{A}_0 + \rho_1\left( \boldsymbol{\sigma}, \mathbf{p} + \frac{e}{c}\mathbf{A} \right) + \rho_3 mc \right] \psi = 0. \qquad (14)$$

This wave equation appears to be sufficient to account for all the duplexity phenomena. On account of the matrices $\rho$ and $\sigma$ containing four rows and columns, it will have four times as many solutions as the non-relativity wave equation, and twice as many as the previous relativity wave equation (1). Since half the solutions must be rejected as referring to the charge $+ e$ on the electron, the correct number will be left to account for duplexity phenomena. The proof given in the preceding section of invariance under a Lorentz transformation applies equally well to the more general wave equation (14).

We can obtain a rough idea of how (14) differs from the previous relativity wave equation (1) by multiplying it up analogously to (5).

This gives, if we write $e'$ for $e/c$

$$0 = \left[ -(p_0 + e'A_0) + \rho_1 (\boldsymbol{\sigma}, \mathbf{p} + e'\mathbf{A}) + \rho_3 mc \right]$$
$$\times \left[ (p_0 + e'A_0) + \rho_1 (\boldsymbol{\sigma}, \mathbf{p} + e'\mathbf{A}) + \rho_3 mc \right] \psi$$
$$= \left[ -(p_0 + e'A_0)^2 + (\boldsymbol{\sigma}, \mathbf{p} + e'\mathbf{A})^2 + m^2 c^2 \right.$$
$$\left. + \rho_1 \{ (\boldsymbol{\sigma}, \mathbf{p} + e'\mathbf{A})(p_0 + e'A_0) - (p_0 + e'A_0)(\boldsymbol{\sigma}, \mathbf{p} + e'\mathbf{A}) \} \right] \psi.$$

$$(15)$$

We now use the general formula, that if $\mathbf{B}$ and $\mathbf{C}$ are any two vectors that commute with $\boldsymbol{\sigma}$

$$(\boldsymbol{\sigma}, \mathbf{B})(\boldsymbol{\sigma}, \mathbf{C}) = \Sigma \sigma_1^2 B_1 C_1 + \Sigma (\sigma_1 \sigma_2 B_1 C_2 + \sigma_2 \sigma_1 B_2 C_1)$$
$$= (\mathbf{B}, \mathbf{C}) + i \Sigma \sigma_3 (B_1 C_2 - B_2 C_1) \qquad (16)$$
$$= (\mathbf{B}, \mathbf{C}) + i(\boldsymbol{\sigma}, \mathbf{B} \times \mathbf{C}).$$

Taking $\mathbf{B} = \mathbf{C} = \mathbf{p} + e'\mathbf{A}$, we find

$$(\boldsymbol{\sigma}, \mathbf{p} + e'\mathbf{A})^2 = (\mathbf{p} + e'\mathbf{A})^2 + i \Sigma \sigma_3$$
$$[(p_1 + e'A_1)(p_2 + e'A_2) - (p_2 + e'A_2)(p_1 + e'A_1)]$$
$$= (\mathbf{p} + e'\mathbf{A})^2 + he'(\boldsymbol{\sigma}, \operatorname{curl} \mathbf{A}).$$

Thus (15) becomes

$$0 = \left[ -(p_0 + e'A_0)^2 + (\mathbf{p} + e'\mathbf{A})^2 + m^2 c^2 + e'h(\boldsymbol{\sigma}, \operatorname{curl} \mathbf{A}) \right.$$
$$\left. - ie'h\rho_1 \left( \boldsymbol{\sigma}, \operatorname{grad} A_0 + \frac{1}{c} \frac{\partial \mathbf{A}}{\partial t} \right) \right] \psi$$
$$= \left[ -(p_0 + e'A_0)^2 + (\mathbf{p} + e'\mathbf{A})^2 + m^2 c^2 + e'h(\boldsymbol{\upsilon}, \mathbf{H}) \right.$$
$$\left. + ie'h\rho_1 (\boldsymbol{\sigma}, \mathbf{E}) \right] \psi,$$

where $\mathbf{E}$ and $\mathbf{H}$ are the electric and magnetic vectors of the field.

This differs from (1) by the two extra terms

$$\frac{eh}{c}(\boldsymbol{\sigma}, \mathbf{H}) + \frac{ieh}{c} \rho_1 (\boldsymbol{\sigma}, \mathbf{E})$$

in F. These two terms, when divided by the factor $2m$, can be regarded as the additional potential energy of the electron due to its new degree of freedom. The electron will therefore behave as though it has a magnetic moment $eh/2mc$. $\boldsymbol{\sigma}$ and an electric moment $ieh/2mc$. $\rho_1 \boldsymbol{\sigma}$. This magnetic moment is just that assumed in the spinning electron model. The electric moment, being a pure imaginary, we should not expect to appear in the model. It is doubtful whether the electric moment has any physical meaning, since the Hamiltonian in (14) that we started from is real, and the imaginary part only appeared when we multiplied it up in an artificial way in order to make it resemble the Hamiltonian of previous theories.

## § 5. THE ANGULAR MOMENTUM INTEGRALS FOR MOTION IN A CENTRAL FIELD

We shall consider in greater detail the motion of an electron in a central field of force. We put $A = 0$ and $e'A_0 = V(r)$, an arbitrary function of the radius $r$, so that the Hamiltonian in (14) becomes

$$F \equiv p_0 + V + \rho_1(\boldsymbol{\sigma}, \mathbf{p}) + \rho_3 mc.$$

We shall determine the periodic solutions of the wave equation $F \psi = 0$, which means that $p_0$ is to be counted as a parameter instead of an operator; it is, in fact, just $1/c$ times the energy level.

We shall first find the angular momentum integrals of the motion. The orbital angular momentum $\mathbf{m}$ is defined by

$$\mathbf{m} = \mathbf{x} \times \mathbf{p},$$

and satisfies the following "Vertauschungs" relations

$$\left. \begin{array}{ll} m_1 x_1 - x_1 m_1 = 0, & m_1 x_2 - x_2 m_1 = i h x_3 \\ m_1 p_1 - p_1 m_1 = 0, & m_1 p_2 - p_2 m_1 = i h p_3 \\ \mathbf{m} \times \mathbf{m} = i h \mathbf{m}, & \mathbf{m}^2 m_1 - m_1 \mathbf{m}^2 = 0, \end{array} \right\}, \qquad (17)$$

together with similar relations obtained by permuting the suffixes. Also $\mathbf{m}$ commutes with $r$, and with $p_r$, the momentum canonically conjugate to $r$.

We have

$$\begin{aligned} m_1 F - F m_1 &= \rho_1 \left\{ m_1(\boldsymbol{\sigma}, \mathbf{p}) - (\boldsymbol{\sigma}, \mathbf{p}) m_1 \right\} \\ &= \rho_1 (\boldsymbol{\sigma}, m_1 \mathbf{p} - \mathbf{p} m_1) \\ &= i h_{\rho_1} (\sigma_2 p_3 - \sigma_3 p_2), \end{aligned}$$

and so

$$\mathbf{m} F - F \mathbf{m} = i h_{\rho 1} \quad \boldsymbol{\sigma} \times \mathbf{p}. \qquad (18)$$

Thus $\mathbf{m}$ is not a constant of the motion. We have further

$$\sigma_1 F - F\sigma_1 = \rho_1 \left\{ \sigma_1 \left( \sigma, \, \mathbf{p} \right) - \left( \sigma, \, \mathbf{p} \right) \sigma_1 \right\}$$
$$= \rho_1 \left( \sigma_1 \sigma - \sigma \sigma_1, \, \mathbf{p} \right)$$
$$= 2i\rho_1 \left( \sigma_3 p_2 - \sigma_2 p_3 \right),$$

with the help of (8), and so

$$\sigma F - F\sigma = -2i\rho_1 \quad \sigma \times \mathrm{p}.$$

Hence

$$\left( \mathbf{m} + \tfrac{1}{2} h \sigma \right) F - F \left( \mathbf{m} + \tfrac{1}{2} h \sigma \right) = 0.$$

Thus $\mathbf{m} + \tfrac{1}{2} h \, \sigma \, (= \mathbf{M}$ say) is a constant of the motion. We can interpret this result by saying that the electron has a spin angular momentum of $\tfrac{1}{2} \, h \, \sigma$, which, added to the orbital angular momentum $\mathbf{m}$, gives the total angular momentum $\mathbf{M}$, which is a constant of the motion.

The Vertauschungs relations (17) all hold when M's are written for the $m$'s. In particular

$$\mathbf{M} \times \mathbf{M} = i h \mathbf{M} \quad \text{and} \quad \mathbf{M}^2 M_3 = M_3 \mathbf{M}^2.$$

$M_3$ will be an action variable of the system. Since the characteristic values of $m_3$ must be integral multiples of $h$ in order that the wave function may be single-valued, the characteristic values of $M_3$ must be half odd integral multiples of $h$. If we put

$$\mathbf{M}^2 = \left( j^2 - \tfrac{1}{4} \right) h^2, \tag{19}$$

$j$ will be another quantum number, and the characteristic values of $M_3$ will extend from $\left( j - \tfrac{1}{2} \right) h$ to $\left( -j + \tfrac{1}{2} \right) h$.[*] Thus $j$ takes integral values.

One easily verifies from (18) that $m^2$ does not commute with F, and is thus not a constant of the motion. This makes a difference between the present theory and the previous spinning electron theory, in which $\mathbf{m}^2$ is constant, and defines the azimuthal quantum number $k$ by a relation similar to (19). We shall find that our $j$ plays the same part as the $k$ of the previous theory.

[*] See 'Roy. Soc. Proc.,' A, vol. 111, p. 281 (1926).

## § 6. THE ENERGY LEVELS FOR MOTION IN A CENTRAL FIELD

We shall now obtain the wave equation as a differential equation in $r$, with the variables that specify the orientation of the whole system removed. We can do this by the use only of elementary non-commutative algebra in the following way.

In formula (16) take $\mathbf{B} = \mathbf{C} = \mathbf{m}$. This gives

$$
\begin{aligned}
(\boldsymbol{\sigma}, \ \mathbf{m})^2 &= \mathbf{m}^2 + i\,(\boldsymbol{\sigma}, \ \mathbf{m} \times \mathbf{m}) \\
&= \left(\mathbf{m} + \tfrac{1}{2}h\boldsymbol{\sigma}\right)^2 - h\,(\boldsymbol{\sigma}, \ \mathbf{m}) - \tfrac{1}{4}h^2\boldsymbol{\sigma}^2 - h\,(\boldsymbol{\sigma}, \ \mathbf{m}) \quad (20) \\
&= \mathbf{M}^2 - 2h\,(\boldsymbol{\sigma}, \ \mathbf{m}) - \tfrac{3}{4}h^2 .
\end{aligned}
$$

Hence

$$
\{(\boldsymbol{\sigma}, \ \mathbf{m}) + h\}^2 = \mathbf{M}^2 + \tfrac{1}{4}h^2 = j^2 h^2 .
$$

Up to the present we have defined $j$ only through $j^2$, so that we could now, if we liked, take $jh$ equal to $(\boldsymbol{\sigma}, \ \mathbf{m}) + h$. This would not be convenient since we want $j$ to be a constant of the motion while $(\boldsymbol{\sigma}, \ \mathbf{m}) + h$ is not, although its square is. We have, in fact, by another application of (16),

$$
(\boldsymbol{\sigma}, \ \mathbf{m})\,(\boldsymbol{\sigma}, \ \mathbf{p}) = i\,(\boldsymbol{\sigma}, \ \mathbf{m} \times \mathbf{p})
$$

since $(\mathbf{m}, \ \mathbf{p}) = 0$, and similarly

$$
(\boldsymbol{\sigma}, \ \mathbf{p})\,(\boldsymbol{\sigma}, \ \mathbf{m}) = i\,(\boldsymbol{\sigma}, \ \mathbf{p} \times \mathbf{m}) ,
$$

so that

$$
\begin{aligned}
(\boldsymbol{\sigma}, \ \mathbf{m})\,(\boldsymbol{\sigma}, \ \mathbf{p}) + (\boldsymbol{\sigma}, \ \mathbf{p})\,(\boldsymbol{\sigma}, \ \mathbf{m}) &= i\Sigma_{\sigma_1}\,(m_2 p_3 - m_3 p_2 + p_2 m_3 - p_3 m_2) \\
&= i\Sigma_{\sigma_1} \cdot 2ihp_1 = -2h\,(\boldsymbol{\sigma}, \ \mathbf{p})
\end{aligned}
$$

or

$$
\{(\boldsymbol{\sigma}, \ \mathbf{m}) + h\}\,(\boldsymbol{\sigma}, \ \mathbf{p}) + (\boldsymbol{\sigma}, \ \mathbf{p})\,\{(\boldsymbol{\sigma}, \ \mathbf{m})\} = 0 .
$$

Thus $(\boldsymbol{\sigma}, \ \mathbf{m}) + h$ anticommutes with one of the terms in F, namely, $\rho_1\,(\boldsymbol{\sigma}, \ \mathbf{p})$, and commutes with the other three. Hence $\rho_3\,\{(\boldsymbol{\sigma}, \ \mathbf{m}) + h\}$ commutes with all four, and is therefore a constant of the motion. But the square of $\rho_3\,\{(\boldsymbol{\sigma}, \ \mathbf{m}) + h\}$ must also equal $j^2 h^2$. We therefore take

$$
jh = \rho_3\,\{(\boldsymbol{\sigma}, \ \mathbf{m}) + h\} . \quad (21)
$$

We have, by a further application of (16)

$$(\boldsymbol{\sigma},\, \mathbf{x})\,(\boldsymbol{\sigma},\, \mathbf{p}) = (\mathbf{x},\, \mathbf{p}) + i\,(\boldsymbol{\sigma},\, \mathbf{m}).$$

Now a permissible definition of $p_r$ is

$$(\mathbf{x},\, \mathbf{p}) = r p_r + ih,$$

and from (21)

$$(\boldsymbol{\sigma},\, \mathbf{m}) = \rho_3 jh - h.$$

Hence

$$(\boldsymbol{\sigma},\, \mathbf{x})\,(\boldsymbol{\sigma},\, \mathbf{p}) = r\rho_r + i\rho_3 jh. \tag{22}$$

Introduce the quantity $\varepsilon$ defined by

$$r\varepsilon = \rho_1(\boldsymbol{\sigma},\, \mathbf{x}). \tag{23}$$

Since $r$ commutes with $\rho_1$ and with $(\boldsymbol{\sigma},\, \mathbf{x})$, it must commute with $\varepsilon$. We thus have

$$r^2\varepsilon^2 = [\rho_1\,(\boldsymbol{\sigma},\, \mathbf{x})]^2 = (\boldsymbol{\sigma},\, \mathbf{x})^2 = \mathbf{x}^2 = r^2$$

or

$$\varepsilon^2 = 1.$$

Since there is symmetry between $\mathbf{x}$ and $\mathbf{p}$ so far as angular momentum is concerned, $\rho_1\,(\boldsymbol{\sigma},\, \mathbf{x})$, like $\rho_1\,(\boldsymbol{\sigma},\, \mathbf{p})$, must commute with $\mathbf{M}$ and $j$. Hence $\varepsilon$ commutes with $\mathbf{M}$ and $j$. Further, $\varepsilon$ must commute with $p_r$, since we have

$$(\boldsymbol{\sigma},\, \mathbf{x})\,(\mathbf{x},\, \mathbf{p}) - (\mathbf{x},\, \mathbf{p})\,(\boldsymbol{\sigma},\, \mathbf{x}) = ih\,(\boldsymbol{\sigma},\, \mathbf{x}),$$

which gives

$$r\varepsilon\,(r p_r + ih) - (r p_r + ih)\,r\varepsilon = ihr\varepsilon,$$

which reduces to

$$\varepsilon p_r - p_r\varepsilon = 0.$$

From (22) and (23) we now have

$$r\varepsilon\rho_1\,(\boldsymbol{\sigma},\, \mathbf{p}) = r p_r + i\rho_3 jh$$

or

$$\rho_1\,(\boldsymbol{\sigma},\, \mathbf{p}) = \varepsilon p_r + i\varepsilon\rho_3 jh/r.$$

Thus

$$F = p_0 + V + \varepsilon p_r + i\varepsilon\rho_3 jh/r + \rho_3 mc. \qquad (24)$$

Equation (23) shows that $\varepsilon$ anticommutes with $\rho_3$. We can therefore by a canonical transformation (involving perhaps the $x$'s and $p$'s as well as the $\sigma$'s and $\rho$'s) bring $\varepsilon$ into the form of the $\rho_2$ of § 2 without changing $\rho_3$, and without changing any of the other variables occurring on the right-hand side of (24), since these other variables all commute with $\varepsilon$. $i\varepsilon\rho_3$ will now be of the form $i\rho_2\rho_3 = -\rho_1$, so that the wave equation takes the form

$$F\psi \equiv [p_0 + V + \rho_2 p_r - \rho_1 jh/r + \rho_3 mc]\,\psi = 0.$$

If we write this equation out in full, calling the components of $\psi$ referring to the first and third rows (or columns) of the matrices $\psi_\alpha$ and $\psi_\beta$ respectively, we get

$$(F\psi)_\alpha \equiv (p_0 + V)\,\psi_\alpha - h\frac{\partial}{\partial r}\psi_\beta - \frac{jh}{r}\psi_\beta + mc\,\psi_\alpha = 0,$$

$$(F\psi)_\beta \equiv (p_0 + V)\,\psi_\beta + h\frac{\partial}{\partial r}\psi_\alpha - \frac{jh}{r}\psi_\alpha - mc\,\psi_\beta = 0.$$

The second and fourth components give just a repetition of these two equations. We shall now eliminate $\psi_\alpha$. If we write $hB$ for $p_0 + V + mc$, the first equation becomes

$$\left(\frac{\partial}{\partial r} + \frac{j}{r}\right)\psi_\beta = B\psi_\alpha,$$

which gives on differentiating

$$\frac{\partial^2}{\partial r^2}\psi_\beta + \frac{j}{r}\frac{\partial}{\partial r}\psi_\beta - \frac{j}{r^2}\psi_\beta = B\frac{\partial}{\partial r}\psi_\alpha + \frac{\partial B}{\partial r}\psi_\alpha$$

$$= \frac{B}{h}\left[-(p_0 + V - mc)\,\psi_\beta + \frac{jh}{r}\psi_\alpha\right] + \frac{1}{h}\frac{\partial V}{\partial r}\psi_\alpha$$

$$= -\frac{(p_0 + V)^2 - m^2 c^2}{h^2}\psi_\beta + \left(\frac{j}{r} + \frac{1}{Bh}\frac{\partial V}{\partial r}\right)\left(\frac{\partial}{\partial r} + \frac{j}{r}\right)\psi_\beta.$$

This reduces to

$$\frac{\partial^2}{\partial r^2}\psi_\beta + \left[\frac{(p_0 + V)^2 - m^2 c^2}{h^2} - \frac{j(j+1)}{r^2}\right]\psi_\beta - \frac{1}{Bh}\frac{\partial V}{\partial r}\left(\frac{\partial}{\partial r} + \frac{j}{r}\right)\psi_\beta = 0.$$

$$(25)$$

The values of the parameter $p_0$ for which this equation has a solution finite at $r = 0$ and $r = \infty$ are $1/c$ times the energy levels of the system. To compare this equation with those of previous theories, we put $\psi_\beta = r\chi$, so that

$$\frac{\partial^2}{\partial r^2}\chi + \frac{2}{r}\frac{\partial}{\partial r}\chi + \left[\frac{(p_0 + V)^2 - m^2c^2}{h^2} - \frac{j(j+1)}{r^2}\right]\chi$$
$$- \frac{1}{Bh}\frac{\partial V}{\partial r}\left(\frac{\partial}{\partial r} + \frac{j+1}{r}\right)\chi = 0. \tag{26}$$

If one neglects the last term, which is small on account of B being large, this equation becomes the same as the ordinary Schroedinger equation for the system, with relativity correction included. Since $j$ has, from its definition, both positive and negative integral characteristic values, our equation will give twice as many energy levels when the last term is not neglected.

We shall now compare the last term of (26), which is of the same order of magnitude as the relativity correction, with the spin correction given by Darwin and Pauli. To do this we must eliminate the $\partial\chi/\partial r$ term by a further transformation of the wave function. We put

$$\chi = B^{-\frac{1}{2}}\chi_1,$$

which gives

$$\frac{\partial^2}{\partial r^2}\chi_1 + \frac{2}{r}\frac{\partial}{\partial r}\chi_1 + \left[\frac{(p_0 + V)^2 - m^2c^2}{h^2} - \frac{j(j+1)}{r^2}\right]\chi_1$$
$$+ \left[\frac{1}{Bh}\frac{j}{r}\frac{\partial V}{\partial r} - \frac{1}{2}\frac{1}{Bh}\frac{\partial^2 V}{\partial r^2} + \frac{1}{4}\frac{1}{B^2h^2}\left(\frac{\partial V}{\partial r}\right)^2\right]\chi_1 = 0. \tag{27}$$

The correction is now, to the first order of accuracy

$$\frac{1}{Bh}\left(\frac{j}{r}\frac{\partial V}{\partial r} - \frac{1}{2}\frac{\partial^2 V}{\partial r^2}\right),$$

where $Bh = 2mc$ (provided $p_0$ is positive). For the hydrogen atom we must put $V = e^2/cr$. The first order correction now becomes

$$-\frac{e^2}{2mc^2r^3}(j+1). \tag{28}$$

If we write $-j$ for $j + 1$ in (27), we do not alter the terms representing the unperturbed system, so

$$\frac{e^2}{2mc^2r^3}j \qquad (28')$$

will give a second possible correction for the same unperturbed term.

In the theory of Pauli and Darwin, the corresponding correcting term is

$$\frac{e^2}{2mhc^2r^3}(\boldsymbol{\sigma}, \mathbf{m})$$

when the Thomas factor $\frac{1}{2}$ is included. We must remember that in the Pauli-Darwin theory, the resultant orbital angular momentum $k$ plays the part of our $j$. We must define $k$ by

$$\mathbf{m}^2 = k(k + 1)h^2$$

instead of by the exact analogue of (19), in order that it may have integral characteristic values, like $j$. We have from (20)

$$(\boldsymbol{\sigma}, \mathbf{m})^2 = k(k + 1)h^2 - h(\boldsymbol{\sigma}, \mathbf{m})$$

or

$$\{(\boldsymbol{\sigma}, \mathbf{m}) + \tfrac{1}{2} h\}^2 = (k + \tfrac{1}{2})^2 h^2,$$

hence

$$(\boldsymbol{\sigma}, \mathbf{m}) = kh \quad \text{or} \quad -(k + 1)h.$$

The correction thus becomes

$$\frac{e^2}{2mc^2r^3}k \quad \text{or} \quad -\frac{e^2}{2mc^2r^3}(k + 1),$$

which agrees with (28) and (28'). The present theory will thus, in the first approximation, lead to the same energy levels as those obtained by Darwin, which are in agreement with experiment.

# THE CONNECTION BETWEEN SPIN AND STATISTICS

By

WOLFGANG PAULI

## Abstract

In the following paper we conclude for the relativistically invariant wave equation for free particles: From postulate (I), according to which the energy must be positive, the necessity of *Fermi-Dirac* statistics for particles with arbitrary half-integral spin; from postulate (II), according to which observables on different space-time points with a space-like distance are commutable, the necessity of *Einstein-Base* statistics for particles with arbitrary integral spin. It has been found useful to divide the quantities which are irreducible against Lorentz transformations into four symmetry classes which have a commutable multiplication like $+1$, $-1$, $+\epsilon$, $-\epsilon$ with $\epsilon^2 = 1$.

## § 1. UNITS AND NOTATIONS

Since the requirements of the relativity theory and the quantum theory are fundamental for every theory, it is natural to use as units the vacuum velocity of light $c$, and Planck's constant divided by $2\pi$ which we shall simply denote by $\hbar$. This convention means that all quantities are brought to the dimension of the power of a length by multiplication with powers of $\hbar$ and $c$. The reciprocal length corresponding to the rest mass $m$ is denoted by $\kappa = mc/\hbar$.

As time coordinate we use accordingly the length of the light path. In specific cases, however, we do not wish to give up the use of the

---

This paper is part of a report which was prepared by the author for the Solvay Congress 1939 and in which slight improvements have since been made. In view of the unfavorable times, the Congress did not take place, and the publication of the reports has been postponed for an indefinite length of time. The relation between the present discussion of the connection between spin and statistics, and the somewhat less general one of Belinfante, based on the concert of charge invariance, has been cleared up by W. Pauli and J. Belinfante, Physica 7, 177 (1940).

---

imaginary time coordinate. Accordingly, a tensor index denoted by small Latin letters $i$, refers to the imaginary time coordinate and runs from 1 to 4. A special convention for denoting the complex conjugate seems desirable. Whereas for quantities with the index 0 an asterisk signifies the complex-conjugate in the ordinary sense (e.g., for the current vector $S_i$ the quantity $S_0^*$ is the complex conjugate of the charge density $S_0$). in general $U_{i\kappa\ldots}^*$ signifies: the complex-conjugate of $U_{i\kappa\ldots}$ multiplied with $(-1)^n$, where $n$ is the number of occurrences of the digit 4 among the $i, k, \ldots$ (e.g. $S_4 = iS_0$, $S_4^* = iS_4^*$).

Dirac's spinors $u_\rho$, with $\rho = 1, \ldots, 4$ have always a Greek index running from 1 to 4, and $u_\rho^*$ means the complex-conjugate of $u_\rho$, in the ordinary sense.

Wave functions, insofar as they are ordinary vectors or tensors, are denoted in general with capital letters, $U_i$, $U_{i\kappa} \ldots$ The symmetry character of these tensors must in general be added explicitly. As classical fields the electromagnetic and the gravitational fields, as well as fields with rest mass zero, take a special place, and are therefore denoted with the usual letters $\varphi_i, f_{i\kappa} = -f_{\kappa i}$ and $g_{i\kappa} = g_{\kappa i}$ respectively.

The energy-momentum tensor $T_{i\kappa}$, is so defined, that the energy-density $W$ and the momentum density $G_\kappa$ are given in natural units by $W = -T_{44}$ and $G_\kappa = -iT_{\kappa 4}$ with $k = 1, 2, 3$.

## § 2. IRREDUCIBLE TENSORS. DEFINITION OF SPINS

We shall use only a few general properties of those quantities which transform according to irreducible representations of the Lorentz group.[1] The proper Lorentz group is that continuous linear group the transformations of which leave the form

$$\sum_{k=1}^{4} x_k^2 = \mathbf{x}^2 - x_0^2$$

invariant and in addition to that satisfy the condition that they have the determinant $+1$ and do not reverse the time. A tensor or spinor which transforms irreducibly under this group can be characterized

---

[1] See B. L. v. d. Waerden, *Die gruppentheoretische Methode in der Quantentheorie* (Berlin, 1932).

by two integral positive numbers $(p, q)$. (The corresponding "angular momentum quantum numbers" $(j, k)$ are then given by $p = 2j + 1, q = 2k + 1$, with integral or half-integral $j$ and $k$.)[1] The quantity $U(j, k)$ characterized by $(j, k)$ has $p \cdot q = (2j + 1)(2k + 1)$ independent components. Hence to $(0, 0)$ corresponds the scalar, to $(\frac{1}{2}, \frac{1}{2})$ the vector, to $(1,0)$ the self-dual skew-symmetrical tensor, to $(1,1)$ the symmetrical tensor with vanishing spur, etc. Dirac's spinor it, reduces to two irreducible quantities $(\frac{1}{2}, 0)$ and $(0, \frac{1}{2})$ each of which consists of two components. If $U(j, k)$ transforms according to the representation

$$U'_r = \sum_{s=1}^{(2j+1)(2k+1)} \Lambda_{rs} U_s,$$

then $U^*$ $(k, j)$ transforms according to the complex-conjugate representation $\Lambda^*$. Thus for $k = j$, $\Lambda^* = \Lambda$. This is true only if the components of $U(j, k)$ and $U(k, j)$ are suitably ordered. For an arbitrary choice of the components, a similarity transformation of $\Lambda$ and $\Lambda^*$ would have to be added. In view of § 1 we represent generally with $U^*$ the quantity the transformation of which is equivalent to $\Lambda^*$ if the transformation of $U$ is equivalent to $\Lambda$.

The most important operation is the reduction of the product of two quantities

$$U_1\left(j_1, k_1\right) \cdot U_2\left(j_2, k_2\right)$$

which, according to the well-known rule of the composition of angular momenta, decompose into several $U(j, k)$ where, independently of each other $j$, $k$ run through the values

$$j = j_1 + j_2, \quad j_1 + j_2 - 1, \ldots, |j_1 - j_2|$$
$$k = k_1 + k_2, \quad k_1 + k_2 - 1, \ldots, |k_1 - k_2|.$$

By limiting the transformations to the subgroup of space rotations alone, the distinction between the two numbers $j$ and $k$ disappears and $U(j, k)$ behaves under this group just like the product of two irreducible quantities $U(j)U(k)$ which in turn reduces into several

---

[1] In the spinor calculus this is a spinor with $2j$ undotted and $2k$ dotted indices.

irreducible $U(l)$ each having $2l + 1$ components, with

$$l = j + k, \ j + k - 1, \ldots, |j - k|.$$

Under the space rotations the $U(l)$ with integral $l$ transform according to single-valued representation, whereas those with half-integral $l$ transform according to double-valued representations. Thus the unreduced quantities $T(j, k)$ with integral (half-integral) $j + k$ are single-valued (double-valued).

If we now want to determine the spin value of the particles which belong to a given field it seems at first that these are given by $l = j + k$. Such a definition would, however, not correspond to the physical facts, for there then exists no relation of the spin value with the number of independent plane waves, which are possible in the absence of interaction for given values of the components $k$ in the phase factor exp $i(\mathbf{kx})$. In order to define the spin in an appropriate fashion,[1] we want to consider first the case in which the rest mass $m$ of all the particles is different from zero. In this case we make a transformation to the rest system of the particle, where all the space components of $k_i$, are zero, and the wave function depends only on the time. In this system we reduce the field components, which according to the field equations do not necessarily vanish, into parts irreducible against space rotations. To each such part, with $r = 2s +1$ componentsi belong $r$ different eigenfunctions which under space rotations transform among themselves and which belong to a particle with spin $s$. If the field equations describe particles with only one spin value there then exists in the rest system only one such irreducible group of components. From the Lorentz invariance, it follows, for an arbitrary system of reference, that $r$ or $\Sigma \ r$ eigenfunctions always belong to a given arbitrary $k_i$. The number of quantities $U(j, k)$ which enter the theory is, however, in a general coordinate system more complicated, since these quantities together with the vector $k_i$ have to satisfy several conditions.

[1] see M. Fierz, Helv. Phys. acta **12**, 3 (1939); also L. de Broglie, Comptes rendus **208**, 1697 (1939); **209**, 265 (1939).

In the case of zero rest mass there is a special degeneracy because, as has been shown by Fierz, this case permits a gauge transformation of the second kind.[1] If the field now describes only one kind of particle with the rest mass zero and a certain spin value, then there are for a given value of $k_i$. only two states, which cannot be transformed into each other by a gauge transformation. The definition of spin may, in this case, not be determined so far as the physical point of view is concerned because the total angular momentum of the field cannot be divided up into orbital and spin angular momentum by measurements. But it is possible to use the following property for a definition of the spin. If we consider, in the $q$ number theory, states where only one particle is present, then not all the eigenvalues $j(j+1)$ of the square of the angular momentum are possible. But $j$ begins with a certain minimum value $s$ and takes then the values $s$, $s + 1, \ldots$.[2] This is only the case for $m = 0$. For photons, $s = 1$, $j = 0$ is not possible for one single photon.[3] For gravitational quanta $s = l$ and the values $j = 0$ and $j = 1$ do not occur.

In an arbitrary system of reference and for arbitrary rest masses, the quantities $U$ all of which transform according to double-valued (single-valued) representations with half-integral (integral) $j + k$ describe only particles with half-integral (integral) spin. A special investigation is required only when it is necessary to decide whether the theory describes particles with one single spin value or with several spin values.

## § 3. Proof of the Indefinite Character of the Charge in Case of Integral and of the Energy in Case of Half-integral Spin

We consider first a theory which contains only $U$ with integral $j + k$, i.e., which describes particles with integral spins only. It is not assumed

---

[1] By "gauge-transformation of the first kind" we understand a transformation $U \rightarrow U e^{i\alpha}$ $U^* \rightarrow U^* e^{-i\alpha}$ with an arbitrary space and time function $\alpha$. By "gauge-transformation of the second kind" we understand a transformation of the type

$$\varphi_k \rightarrow \varphi_k - \frac{1}{\epsilon} i \frac{\partial \alpha}{\partial x_k}$$

as for those of the electromagnetic potentials.
[2] The general proof for this has been given by M. Fierz, Helv. Phys. Acta 13, 45 (1940).
[3] See for instance W. Pauli in the article "Wellen-mechanik" in the Handbuch der Physik, Vol. **24/2**, p. 260.

that only particles with one single spin value will be described, but all particles shall have integral spin.

We divide the quantities $U$ into two classes: (1) the "+ 1 class" with $j$ integral, $k$ integral; (2) the "− 1 class" with $j$ half-integral, $k$ half-integral.

The notation is justified because, according to the indicated rules about the reduction of a product into the irreducible constituents under the Lorentz group, the product of two quantities of the +1 class or two quantities of the −1 class contains only quantities of the +1 class, whereas the product of a quantity of the +1 class with a quantity of the −1 class contains only quantities of the −1 class. It is important that the complex conjugate $U^*$ for which $j$ and $k$ are interchanged belong to the same class as $U$. As can be seen easily from the multiplication rule, tensors with even (odd) number of indices reduce only to quantities of the +1 class (−1 class). The propagation nector $k_i$ we consider as belonging to the −1 class, since it behaves after multiplication with other quantities like a quantity of the −1 class.

We consider now a homogeneous and linear equation in the quantities $U$ which, however, does not necessarily have to be of the first order. Assuming a plane wave, we may put $k_i$ for $-i\partial/\partial x_l$. Solely on account of the invariance against the *proper* Lorentz group it must be of the typical form

$$\sum k U^+ = \sum U^-, \quad \sum k U^- = \sum U^+. \qquad (1)$$

This typical form shall mean that there may be as many different terms of the same type present, as there are quantities $U^*$ and $U^-$. Furthermore, among the $U^*$ may occur the $U^+$ as well as the $(U^+)^*$, whereas other $U$ may satisfy reality conditions $U = U^*$. Finally we have omitted an even number of $k$ factors. These may be present in arbitrary number in the term of the sum on the left- or right-hand side of these equations. It is now evident that these equations remain invariant under the substitution

$$
\begin{aligned}
&k_i \to -k_i; \quad U^+ \to U^+, \quad [(U^+) \to (U^+)^*]; \\
&U^- \to -U^-, \quad [(U^-)^* \to -(U^-)^* \to -(U^-)^*].
\end{aligned}
\qquad (2)
$$

Let us consider now tensors $T$ of even rank (scalars, skew-symmetrical or symmetrical tensors of the 2nd rank, etc.), which are composed quadratically or bilinearly of the $U's$. They are then composed solely of quantities with even $j$ and even $k$ and thus are of the typical form

$$T \sim \sum U^+ U^+ + \sum U^- U^- + \sum U^+ k U^-, \qquad (3)$$

where again a possible even number of $k$ factors is omitted and no distinction between $U$ and $U^*$ is made. Under the substitution (2) they remain unchanged, $T \to T$.

The situation is different for tensors of odd rank $S$ (vectors, etc.) which consist of quantities with half-integral $j$ and half-integral $k$. These are of the typical form

$$S \sim \sum U^+ k U^+ + \sum U^- k U^- + \sum U^- \qquad (4)$$

and hence change the sign under the substitution (2), $S \to -S$. Particularly is this the case for the current vector $s_i$. To the transformation $k_i \to -k_i$, belongs for arbitrary wave packets the transformation $x_i \to -x_i$, and it is remarkable that from the invariance of Eq. (I) against the proper Lorentz group alone there follows an invariance property for the change of sign of all the coordinates. In particular, the indefinite character of the current density and the total charge for even spin follows, since to every solution of the field equations belongs another solution for which the components of $s_k$, change their sign. The definition of a definite particle density for even spin which transforms like the 4-component of a vector is therefore impossible.

We now proceed to a discussion of the somewhat less simple case of half-integral spins. Here we divide the quantities $U$, which have half-integral $j + k$, in the following fashion: (3) the "$+\epsilon$ class" with $j$ integral $k$ half-integral, (4) the "$-\epsilon$ class" with $j$ half-integral $k$ integral.

The multiplication of the classes (1), ..., (4), follows from the rule $\epsilon^2 = 1$ and the commutability of the multiplication. This law remains unchanged if $\epsilon$ is replaced by $-\epsilon$.

We can summarize the multiplication law between the different classes in the following multiplication table:

|      | 1         | $-1$       | $\epsilon$   | $-\epsilon$  |
|------|-----------|------------|--------------|--------------|
| 1    | 1         | $-1$       | $\epsilon$   | $-\epsilon$  |
| $-1$ | $-1$      | $+1$       | $-\epsilon$  | $+\epsilon$  |
| $\epsilon$ | $-\epsilon$ | $-\epsilon$ | $+1$      | $-1$         |
| $-\epsilon$ | $-\epsilon$ | $\epsilon$ | $-1$     | $+1$         |

We notice that these classes have the multiplication law of Klein's "four-group."

It is important that here the complex-conjugate quantities for which $j$ and $k$ are interchanged do not belong to the same class, so that

$$U^{+\epsilon}, \ \left(U^{-\epsilon}\right)^* \quad \text{belong to the} \quad +\epsilon \text{ class}$$
$$U^{-\epsilon}, \ \left(U^{+\epsilon}\right)^* \quad\quad\quad\quad\quad -\epsilon \text{ class.}$$

We shall therefore cite the complex-conjugate quantities explicitly. (One could even choose the $U^{+\epsilon}$ suitably so that *all* quantities of the $-\epsilon$ class are of the form $(U^{+\epsilon})^*$).

Instead of (1) we obtain now as typical form

$$\sum k U^{+\epsilon} + \sum k(U^{-\epsilon})^* = \sum U^{-\epsilon} + \sum (U^{+\epsilon})^*$$
$$\sum k U^{-\epsilon} + \sum k(U^{+\epsilon})^* = \sum U^{+\epsilon} + \sum (U^{-\epsilon})^*, \tag{5}$$

since a factor $k$ or $-i\partial/\partial x$ always changes the expression from one of the classes $+\epsilon$ or $-\epsilon$ into the other. As above, an even number of $k$ factors have been omitted.

Now we consider instead of (2) the substitution

$$k_i \rightarrow -k_i; \quad U^{+\epsilon} \rightarrow iU^{+\epsilon}; \quad (U^{-\epsilon})^* \rightarrow i(U^{-\epsilon})^*;$$
$$(U^{+\epsilon} \rightarrow -i(U^{+\epsilon})^*; \quad U^{-\epsilon} \rightarrow -iU^{-\epsilon}. \tag{6}$$

This is in accord with the algebraic requirement of the passing over to the complex conjugate, as well as with the requirement that quantities of the same class as $U^{+\epsilon}, (U^{-\epsilon})^*$ transform in the same way. Furthermore, it does not interfere with possible reality conditions of the type

$U^{+\epsilon} = (U^{-\epsilon})^*$. or $U^{-\epsilon} = (U^{+\epsilon})^*$. Equations (5) remain unchanged under the substitution (6).

We consider again tensors of even rank (scalars, tensors of 2nd rank, etc.), which are composed bilinearly or quadratically of the $U$ and their complex-conjugate. For reasons similar to the above they must be of the form

$$T \sim \sum U^{+\epsilon}U^{+\epsilon} + \sum U^{-\epsilon}U^{-\epsilon} + \sum U^{+\epsilon}kU^{-\epsilon} + \sum U^{+\epsilon}(U^{-\epsilon})^*$$
$$+ \sum U^{-\epsilon}(U^{+\epsilon})^* + \sum (U^{-\epsilon})^*kU^{-\epsilon} + \sum (U^{+\epsilon})^*kU^{+\epsilon} + \sum (U^{-\epsilon})^*k(U^{+\epsilon})^*$$
$$\sum (U^{-\epsilon})^*(U^{-\epsilon})^* + \sum (U^{-\epsilon})^*(U^{+\epsilon})^*. \tag{7}$$

Furthermore, the tensors of odd rank (vectors, etc.) must be of the form

$$S \sim \sum U^{+\epsilon}kU^{+\epsilon} + \sum U^{-\epsilon}kU^{-\epsilon} + \sum U^{+\epsilon}U^{-\epsilon} + \sum U^{+\epsilon}(U^{-\epsilon})^*$$
$$+ \sum U^{-\epsilon}k(U^{+\epsilon})^* + \sum U^{-\epsilon}(U^{-\epsilon})^* + \sum U^{+\epsilon}(U^{+\epsilon})^* + \sum (U^{-\epsilon})^*k(U^{-\epsilon})^*$$
$$+ \sum (U^{-\epsilon})^*(U^{-\epsilon})^* + \sum (U^{+\epsilon})^*(U^{+\epsilon})^*. \tag{8}$$

*The result of the substitution (6) is now the opposite of the result of the substitution (2): the tensors of even rank change their sign, the tensors of odd rank remain unchanged:*

$$T \rightarrow -T; \quad S \rightarrow +S. \tag{9}$$

In case of half-integral spin, therefore, a positive definite energy density, as well as a positive definite total energy, is impossible. The latter follows from the fact, that, under the above substitution, the energy density in every space-time point changes its sign as a result of which the total energy changes also its sign.

It may be emphasized that it was not only unnecessary to assume that the wave equation is of the first order,[1] but also that the question is left open whether the theory is also invariant with respect to space reflections x' = − x, x'$_0$ = x$_0$). This scheme covers therefore also Dirac's two component wave equations (with rest mass zero).

These considerations do not prove that for integral spins there always exists a definite energy density and for half-integral spins a

---

[1] But we exclude operation like $(k^2 + k^2)^{1/2}$, which operate at finite distances in the coordinate space.

definite charge density. In fact, it has been shown by Fierz[1] that this is not the case for spin $> 1$ for the densities. There exists, however (in the $c$ number theory), a definite total charge for half-integral spins and a definite total energy for the integral spins. The spin value $\frac{1}{2}$ is discriminated through the possibility of a definite charge density, and the spin values 0 and 1 are discriminated through the possibility of defining a definite energy density. Nevertheless, the present theory permits arbitrary values of the spin quantum numbers of elementary particles as well as arbitrary values of the rest mass, the electric charge, and the magnetic moments of the particles.

## § 4. QUANTIZATION OF THE FIELDS IN THE ABSENCE OF INTERACTIONS. CONNECTION BETWEEN SPIN AND STATISTICS

The impossibility of defining in a physically satisfactory way the particle density in the case of integral spin and the energy density in the case of half-integral spins in the $c$–number theory is an indication that a satisfactory interpretation of the theory within the limits of the one-body problem is not possible.[2] In fact, all relativistically invariant theories lead to particles, which in external fields can be emitted and absorbed in pairs of opposite charge for electrical particles and singly for neutral particles. The fields must, therefore, undergo a second quantization. For this we do not wish to apply here the canonical formalism, in which time is unnecessarily sharply distinguished from space, and which is only suitable if there are no supplementary conditions between the canonical variables.[3] Instead, we shall apply here a generalization of this method which was applied for the first time by Jordan and Pauli to the electromagnetic field.[4] This method is especially convenient in the absence of interaction, where all fields

---

[1] M. Fierz, Helv. Phys. Acta 12, 3 (1939).

[2] The author therefore considers as not conclusive the original argument of Dirac. according to which the field equation must be of the first order.

[3] On account of the existence of such conditions the canonical formalism is not applicable for spin $> 1$ and therefore the discussion about the connection between spin and statistics by J. S. de Wet, Phys. Rev. **57**, 646 (1940), which is based on that formalism is not general enough.

[4] The consistent development of this method leads to the "many-time formalism" of Dirac, which has been given by P. A. M. Dirac, Quantum Mechanics (Oxford, second edition, 1935).

$U^{(r)}$ satisfy the wave equation of the second order

$$\square\, U^{(r)} - \kappa^2 U^{(r)} = 0,$$

where

$$\square \equiv \sum_{k=1}^{4} \frac{\partial^2}{\partial x\kappa^2} = \Delta - \frac{\partial^2}{\partial x_0^2}$$

and $\kappa$ is the rest mass of the particles in units *hbar/c*.

An important tool for the second quantization is the invariant $D$ function, which satisfies the wave equation (9) and is given in a periodicity volume $V$ of the eigenfunctions by

$$D(\mathbf{x}, x_0) = \frac{1}{V} \sum \exp\left[i(\mathbf{kx})\right] \frac{\sin k_0 x_0}{k_0} \tag{10}$$

or in the limit $V \to \infty$

$$D(\mathbf{x}, x_0) = \frac{1}{(2\pi)^3} \int d^3k \exp\left[i(\mathbf{kx})\right] \frac{\sin k_0 x_0}{k_0}. \tag{11}$$

By to we understand the positive root

$$k_0 = +(k^2 + \kappa^2)^{1/2} \tag{12}$$

The $D$ function is uniquely determined by the conditions:

$$\square\, D - \kappa^2 D = 0;\ D(\mathbf{x}, 0) = 0;$$
$$\left(\frac{\partial D}{\partial x_0}\right)_{x0=0} = \delta(\mathbf{x}). \tag{13}$$

For $\kappa = 0$ we have simply

$$D(\mathbf{x}, x_0) = \{\delta(r - x_0) - \delta(r - x_0)\}/4\pi r. \tag{14}$$

This expression also determines the singularity of $D(\mathbf{x}, x_0)$ on the light cone for $\kappa \neq 0$. But in the latter case $D$ is no longer different from zero in the inner part of the cone. One finds for this region[1]

$$D(\mathbf{x},\ x_0) = -\frac{1}{4\pi r}\frac{\partial}{\partial r} F(r,\ x_0)$$

with

$$F(r,\ x_0) = \begin{cases} J_0\left[\kappa(x_0^2 - r^2)^{1/2}\right] & \text{for } x_0 > r \\ 0 & \text{for } r > x_0 > -r \\ -J_0\left[\kappa(x_0^2 - r^2)^{1/2}\right] & \text{for } -r > x_0. \end{cases} \tag{15}$$

---

[1] See P. A. M. Dirac, Proc. Camb. Phil. Soc. **30**, 150 (1934).

The jump from $+$ to $-$ of the function $F$ on the light cone corresponds to the $\delta$ singularity of $D$ on this cone. For the following it will be of decisive importance that $D$ vanish in the exterior of the cone (i.e., for $r > x_0 > -r$).

The form of the factor $d^3k/k_0$, is determined by the fact that $d^3k/k_0$ is invariant on the hyper-boloid $(k)$ of the four-dimensional momentum space $(\kappa, k_0)$. It is for this reason that, apart from $D$, there exists just one more function which is invariant and which satisfies the wave equation (9), namely,

$$D_1(\mathbf{x}, \ x_0) = \frac{1}{(2\pi)^3} \int d^3k \exp\left[i(\mathbf{kx})\right] \frac{\cos k_0 x_0}{k_0}. \qquad (16)$$

For $\kappa = 0$ one finds

$$D_1(\mathbf{x}, \ x_0) = \frac{1}{2\pi^2} \frac{1}{r^2 - x_0^2}. \qquad (17)$$

In general it follows

$$D_1(\mathbf{x}, \ x_0) = \frac{1}{4\pi} \frac{1}{r} \frac{\partial}{\partial r} F_1(r, x_0)$$

$$F_1(r, \ x_0) = \begin{cases} N_0\left[\kappa(x_0^2 - r^2)^{1/2}\right] & \text{for } x_0 > r \\ -iH_0^{(1)}\left[i\kappa(r^2 - x_0^2)^{1/2}\right] & \text{for } r > x_0 > -r \\ N_0\left[\kappa(x_0^2 - r^2)^{1/2}\right] & \text{for } -r > x_0. \end{cases} \qquad (18)$$

Here $N_0$ stands for Neumann's function and $H_0^{(1)}$ for the first Hankel cylinder function. The strongest singularity of $D$, on the surface of the light cone is in general determined by (17).

We shall, however, expressively postulate in the following *that all physical quantities at finite distances exterior to the light cone* $\left(\text{for } x_0' - x_0'' | < |\mathbf{x}' - \mathbf{x}''|\right)$ *are commutable.*[1] It follows from this that the bracket expressions of all quantities which satisfy the force-free wave equation (9) can be expressed by the function $D$ and (a finite number) of derivatives of it without using the function $D_1$. This is also true for brackets with the $+$ sign, since otherwise it would follow that gauge invariant quantities, which are constructed bilinearly from

---

[1] For the canonical quantization formalism this postulate is satisfied implicitly. But this postulate is much more general than the canonical formalism.

the $U^{(r)}$, as for example the charge density, are noncommutable in two points with a space-like distance.[1]

The justification for our postulate lies in the fact that measurements at two space points with a space-like distance can never disturb each other, since no signals can be transmitted with velocities greater than that of light. Theories which would make use of the $D_1$ function in their quantization would be very much different from the known theories in their consequences.

At once we are able to draw further conclusions about the number of derivatives of $D$ function which can occur in the bracket expressions, if we take into account the invariance of the theories under the transformations of the restricted Lorentz group and if we use the results of the preceding section on the class division of the tensors. We assume the quantities $U^{(r)}$ to be ordered in such a way that each field component is composed only of quantities of the same class. We consider especially the bracket expression of a field component $U^{(r)}$ with its own complex conjugate

$$\left[ U^{(r)}(\mathbf{x}', x_0'), U^{*9r)}(\mathbf{x}'', x_0'') \right].$$

We distinguish now the two cases of half-integral and integral spin. In the former case this expression transforms according to (8) under Lorentz transformations as a tensor of odd rank. In the second case, however, it transforms as a tensor of even rank. Hence we have for half-integral spin

$$\left[ U^{(r)}(\mathbf{x}', x_0'), U^{*(r)}(\mathbf{x}'', x_0'') \right]$$

= odd number of derivatives of the function

$$D(\mathbf{x}' - \mathbf{x}'', x_0' - x_0'') \qquad (19a)$$

and similarly for integral spin

$$\left[ U^{(r)}(\mathbf{x}', x_0'), U^{*(r)}(\mathbf{x}'', x_0'') \right]$$

= even number of derivatives of the function

$$D(\mathbf{x}' - \mathbf{x}'', x_0' - x_0''). \qquad (19b)$$

[1] See W. Pauli, Ann. de l'Inst. H. Poincare **6**, 137 (1936), esp. § 3.

This must be understood in such a way that on the right-hand side there may occur a complicated sum of expressions of the type indicated. We consider now the following expression, which is symmetrical in the two points

$$X \equiv \left[ U^{(r)}(\mathbf{x}', x_0'), \; U^{*(r)}(\mathbf{x}'', x_0'') \right] + \left[ U^{(r)}(\mathbf{x}'', x_0''), \; U^{*(r)}(\mathbf{x}', x_0') \right].$$
(19)

Since the $D$ function is even in the space coordinates odd in the time coordinate, which can be seen at once from Eqs. (11) or (15), it follows from the symmetry of $X$ that $X =$ even number of space-like times odd numbers of time-like derivatives of $D(x' - x'', x'_0 - x''_0)$. This is fully consistent with the postulate (19a) for half-integral spin, but in contradiction with (19b) for integral spin unless $X$ vanishes. We have therefore the result for integral spin

$$\left[ U^{(r)}(\mathbf{x}', x_0'), \; U^{*(r)}(\mathbf{x}'', x_0'') \right] + \left[ U^{(r)}(\mathbf{x}'', x_0''), U^{*(r)}(\mathbf{x}', x_0') \right] = 0.$$
(20)

So far we have not distinguished between the two cases of Bose statistics and the exclusion principle. In the former case, one has the ordinary bracket with the $-$ sign, in the latter case, according to Jordan and Wigner, the bracket

$$[A, B]_+ = AB + BA$$

with the $+$ sign. *By inserting the brackets with the $+$ sign into (20) we have an algebraic contradiction,* since the left-hand side is essentially positive for $x' = x''$ and cannot vanish unless both $U^{(r)}$ and $U^{*(r)}$ vanish.[1]

Hence we come to the result: *For integral spin the quantization according to the exclusion principle is not possible. For this result it is*

---

[1] This contradiction may be seen also by resolving $U^{(r)}$ into eigenvibrations according to

$$U^{*(r)}(\mathbf{x}, x_0) = V^{-1/2} \sum_k \left\{ U_+^*(k) \exp\left[ i \{-(\mathbf{k}\mathbf{x}) + k_0 x_0\} \right] + U_-(k) \exp\left[ i \{(\mathbf{k}\mathbf{x}) - k_0 x_0\} \right] \right\}$$

$$U^{(r)}(\mathbf{x}, x_0) = V^{-1/2} \sum_k \left\{ U_+(k) \exp\left[ i \{(\mathbf{k}\mathbf{x}) - k_0 x_0\} \right] + U_-^*(k) \exp\left[ i \{-(\mathbf{k}\mathbf{x}) + k_0 x_0\} \right] \right\}.$$

The equation (21) leads then, among others, to the relation

$$[U_+^*(k), \; U_+(k)] + [U_-(k), \; U_-^*(k)] = 0,$$

a relation, which is not possible for brackets with the $+$ sign unless $U_\pm(k)$ and $U_\pm^*(k)$ vanish.

*essential, that the use of the* $D_1$ *function in place of the D function be, for general reasons, discarded.*

On the other hand, it is formally possible to quantize the theory for half-integral spins according to Einstein-Bose-statistics, *but according to the general result of the preceding section the energy of the system would not be positive.* Since for physical reasons it is necessary to postulate this, we must apply the exclusion principle in connection with Dirac's hole theory.

For the positive proof that a theory with a positive total energy is possible by quantization according to Bose-statistics (exclusion principle) for integral (half-integral) spins, we must refer to the already mentioned paper by Fierz. In another paper by Fierz and Pauli[1] the case of an external electromagnetic field and also the connection between the special case of spin 2 and the gravitational theory of Einstein has been discussed. In conclusion we wish to state, that according to our opinion the connection between spin and statistics is one of the most important applications of the special relativity theory.

---

[1] M. Fierz and W. Pauli, Proc. Roy. Soc. **A173**, 211 (1939).

# Exclusion Principle and Quantum Mechanics

By

## Wolfgang Pauli

*Nobel Lecture, December 13, 1946*

The history of the discovery of the "exclusion principle", for which I have received the honor of the Nobel Prize award in the year 1945, goes back to my students days in Munich. While, in school in Vienna, I had already obtained some knowledge of classical physics and the then new Einstein relativity theory, it was at the University of Munich that I was introduced by Sommerfeld to the structure of the atom - somewhat strange from the point of view of classical physics. I was not spared the shock which every physicist, accustomed to the classical way of thinking, experienced when he came to know of Bohr's "basic postulate of quantum theory" for the first time. At that time there were two approaches to the difficult problems connected with the quantum of action. One was an effort to bring abstract order to the new ideas by looking for a key to translate classical mechanics and electrodynamics into quantum language which would form a logical generalization of these. This was the direction which was taken by Bohr's "correspondence principle". Sommerfeld, however, preferred, in view of the difficulties which blocked the use of the concepts of kinematical models, a direct interpretation, as independent of models as possible, of the laws of spectra in terms of integral numbers, following, as Kepler once did in his investigation of the planetary system, an inner feeling for harmony. Both methods, which did not appear to me irreconcilable, influenced me. The series of whole numbers 2, 8, 18, 32... giving the lengths of the periods in the natural system of chemical elements, was zealously discussed in Munich, including the remark of the Swedish physicist, Rydberg, that these numbers are of the simple form $2 n^2$, if $n$ takes on all integer values. Sommerfeld tried especially to connect the number 8 and the number of corners of a cube.

A new phase of my scientific life began when I met Niels Bohr personally for the first time. This was in 1922, when he gave a series of guest lectures at Göttingen, in which he reported on his theoretical investigations on the Periodic System of Elements. I shall recall only briefly that the essential progress made by Bohr's considerations at that time was in explaining, by means of the spherically symmetric atomic model, the formation of the intermediate shells of the atom and the general properties of the rare earths. The question, as to why all electrons for an atom in its ground state were not bound in the innermost shell, had already been emphasized by Bohr as a fundamental problem in his earlier works. In his Göttingen lectures he treated particularly the closing of this innermost K-shell in the helium atom and its essential connection with the two non-combining spectra of helium, the ortho- and para-helium spectra. However, no convincing explanation for this phenomenon could be given on the basis of classical mechanics. It made a strong impression on me that Bohr at that time and in later discussions was looking for a *general* explanation which should hold for the closing of *every* electron shell and in which the number 2 was considered to be as essential as 8 in contrast to Sommerfeld's approach.

Following Bohr's invitation, I went to Copenhagen in the autumn of 1922, where I made a serious effort to explain the so-called "anomalous Zeeman effect", as the spectroscopists called a type of splitting of the spectral lines in a magnetic field which is different from the normal triplet. On the one hand, the anomalous type of splitting exhibited beautiful and simple laws and Landé had already succeeded to find the simpler splitting of the spectroscopic terms from the observed splitting of the lines. The most fundamental of his results thereby was the use of half-integers as magnetic quantum numbers for the doublet-spectra of the alkali metals. On the other hand, the anomalous splitting was hardly understandable from the standpoint of the mechanical model of the atom, since very general assumptions concerning the electron, using classical theory as well as quantum theory, always led to the same triplet. A closer investigation of this problem left me with the feeling that it was even more unapproachable. We know now that at that time one was confronted with two logically different difficulties simultaneously. One was the absence of a general key to translate a given mechanical model into quantum theory which one tried in vain by using classical mechanics to describe the stationary quantum states themselves. The second difficulty was our ignorance concerning the proper classical model itself which could be suited to derive at all an anomalous splitting of spectral lines emitted by an atom in an external magnetic field. It is therefore not surprising that I could not find a satisfactory solution of the problem at that time. I succeeded, however, in generalizing Landé's term analysis for very strong magnetic fields[2], a case which, as a result of the magneto-optic transformation (Paschen-Back effect), is in many respects simpler. This early work was of decisive importance for the finding of the exclusion principle.

Very soon after my return to the University of Hamburg, in 1923, I gave there my inaugural lecture as *Privatdozent* on the Periodic System of Elements. The contents of this lecture appeared very unsatisfactory to me, since the problem of the closing of the electronic shells had been clarified no further. The only thing that was clear was that a closer relation of this problem to the theory of multiplet structure must exist. I therefore tried to examine again critically the simplest case, the doublet structure of the alkali spectra. According to the point of view then orthodox, which was also taken over by Bohr in his already mentioned lectures in Göttingen, a non-vanishing angular momentum of the atomic core was supposed to be the cause of this doublet structure.

In the autumn of 1924 I published some arguments against this point of view, which I definitely rejected as incorrect and proposed instead of it the assumption of a new quantum theoretic property of the electron, which I called a "two-valuedness not describable classically"[3]. At this time a paper of the English physicist, Stoner, appeared[4] which contained, besides improvements in the classification of electrons in subgroups, the following essential remark: For a given value of the principal quantum number is the number of energy levels of a single electron in the alkali metal spectra in an external magnetic field the same as the number of electrons in the closed shell of the rare gases which corresponds to this principal quantum number.

On the basis of my earlier results on the classification of spectral terms in a strong magnetic field the general formulation of the exclusion principle became clear to me. The fundamental idea can be stated in the following way: The complicated numbers of electrons in closed subgroups are reduced to the simple number *one* if the division of the groups by giving the values of the four quantum numbers of an electron is carried so far that every degeneracy is removed. An entirely non-degenerate energy level is already "closed", if it is occupied by a single electron; states in contradiction with this postulate have to be excluded. The exposition of this general formulation of the exclusion principle was made in Hamburg in the spring of 1925[5], after I was able to verify some additional conclusions concerning the anomalous Zeeman effect of more complicated atoms during a visit to Tübingen with the help of the spectroscopic material assembled there.

With the exception of experts on the classification of spectral terms, the physicists found it difficult to understand the exclusion principle, since no meaning in terms of a model was given to the fourth degree of freedom of the electron. The gap was filled by Uhlenbeck and Goudsmit's idea of electron spin[6], which made it possible to understand the anomalous Zeeman effect simply by assuming that the spin quantum number of one electron is equal to $\frac{1}{2}$ and that the quotient of the magnetic moment to the mechanical angular moment has for the spin a value twice as large as for the ordinary orbit of the electron. Since that time, the exclusion principle has been closely connected with the idea of spin. Although at first I strongly doubted the correctness of this idea because of its classical-mechanical character, I was finally converted to it by Thomas' calculations[7] on the magnitude of doublet splitting. On the other hand, my earlier doubts as well as the cautious expression "classically non-describable two-valuedness" experienced a certain verification during later developments, since Bohr was able to show on the basis of wave mechanics that the electron spin cannot be measured by classically describable experiments (as, for instance, deflection of molecular beams in external electromagnetic fields) and must therefore be considered as an essentially quantum-mechanical property of the electron[8,9].

The subsequent developments were determined by the occurrence of the new quantum mechanics. In 1925, the same year in which I published my paper on the exclusion principle, De Broglie formulated his idea of matter waves and Heisenberg the new matrix-mechanics, after which in the next year Schrödinger's wave mechanics quickly followed. It is at present unnecessary to stress the importance and the fundamental character of these discoveries, all the more as these physicists have themselves explained, here in Stockholm, the meaning of their leading ideas[10]. Nor does time permit me to illustrate in detail the general epistemological significance of the new discipline of quantum mechanics, which has been done, among others, in a number of articles by Bohr, using hereby the idea of "complementarity" as a new central concept[11]. I shall only recall that the statements of quantum mechanics are dealing only with possibilities, not with actualities. They have the form "This is not possible" or "Either this or that is possible", but they can never say "That will actually happen then and there". The actual observation appears as an event outside the range of a description by physical laws and brings forth in general a discontinuous selection out of the several possibilities foreseen by the statistical laws of the new theory. Only this renouncement concerning the old claims for an objective description of the physical phenomena, independent of the way in which they are observed, made it possible to reach again the self-consistency of quantum theory, which actually had been lost since Planck's discovery of the quantum of action. Without discussing further the change of the attitude of modern physics to such concepts as "causality" and "physical reality" in comparison with the older classical physics I shall discuss more particularly in the following the position of the exclusion principle on the new quantum mechanics.

As it was first shown by Heisenberg[12], wave mechanics leads to qualitatively different conclusions for particles of the same kind (for instance for electrons) than for particles of different kinds. As a consequence of the impossibility to distinguish one of several like particles from the other, the wave functions describing an ensemble of a given number of like particles in the configuration space are sharply separated into different classes of symmetry which can never be transformed into each other by external perturbations. In the term "configuration space" we are including here the spin degree of freedom, which is described in the wave function of a single particle by an index with only a finite number of possible values. For electrons this number is equal to two; the configuration space of $N$ electrons has therefore $3\,N$ space dimensions and $N$ indices of "two-valuedness". Among the different classes of symmetry, the most important ones (which moreover for two particles are the only ones) are the symmetrical class, in which the wave function does not change its value when the space and spin coordinates of two particles are permuted, and the antisymmetrical class, in which for such a permutation the wave function changes its sign. At this stage of the theory three different hypotheses turned out to be logically possible concerning the actual ensemble of several like particles in Nature.

    I. This ensemble is a mixture of all symmetry classes.

    II. Only the symmetrical class occurs.

    III. Only the antisymmetrical class occurs.

As we shall see, the first assumption is never realized in Nature. Moreover, it is only the third assumption that is in accordance with the exclusion principle, since an antisymmetrical function containing two particles in the same state is identically zero. The assumption III can therefore be considered as the correct and general wave mechanical formulation of the exclusion principle. It is this possibility which actually holds for electrons.

This situation appeared to me as disappointing in an important respect. Already in my original paper I stressed the circumstance that I was unable to give a logical reason for the exclusion principle or to deduce it from more general assumptions. I had always the feeling and I still have it today, that this is a deficiency. Of course in the beginning I hoped that the new quantum mechanics, with the help of which it was possible to deduce so many half-empirical formal rules in use at that time, will also rigorously deduce the exclusion principle. Instead of it there was for electrons still an exclusion: not of particular states any longer, but of whole classes of states, namely the exclusion of all classes different from the antisymmetrical one. The impression that the shadow of some incompleteness fell here on the bright light of success of the new quantum mechanics seems to me unavoidable. We shall resume this problem when we discuss relativistic quantum mechanics but wish to give first an account of further results of the application of wave mechanics to systems of several like particles.

In the paper of Heisenberg, which we are discussing, he was also able to give a simple explanation of the existence of the two non-combining spectra of helium which I mentioned in the beginning of this lecture. Indeed, besides the rigorous separation of the wave functions into symmetry classes with respect to space-coordinates and spin indices together, there exists an approximate separation into symmetry classes with respect to space coordinates alone. The latter holds only so long as an interaction between the spin and the orbital motion of the electron can be neglected. In this way the para- and ortho-helium spectra could be interpreted as belonging to the class of symmetrical and antisymmetrical wave functions respectively in the space coordinates alone. It became clear that the energy difference between corresponding levels of the two classes has nothing to do with magnetic interactions but is of a new type of much larger order of magnitude, which one called exchange energy.

Of more fundamental significance is the connection of the symmetry classes with general problems of the statistical theory of heat. As is well known, this theory leads to the result that the entropy of a system is (apart from a constant factor) given by the logarithm of the number of quantum states of the whole system on a so-called energy shell. One might first expect that this number should be equal to the corresponding volume of the multidimensional phase space divided by $h^f$, where $h$ is Planck's constant and $f$ the number of degrees of freedom of the whole system. However, it turned out that for a system of $N$ like particles, one had still to divide this quotient by $N!$ in order to get a value for the entropy in accordance with the usual postulate of homogeneity that the entropy has to be proportional to the mass for a given inner state of the substance. In this way a qualitative distinction between like and unlike particles was already preconceived in the general statistical mechanics, a distinction which Gibbs tried to express with his concepts of a generic and a specific phase. In the light of the result of wave mechanics concerning the symmetry classes, this division by $N!$, which had caused already much discussion, can easily be interpreted by accepting one of our assumptions II and III, according to both of which only one class of symmetry occurs in Nature. The density of quantum states of the whole system then really becomes smaller by a factor $N!$ in comparison with the density which had to be expected according to an assumption of the type I admitting all symmetry classes.

Even for an ideal gas, in which the interaction energy between molecules can be neglected, deviations from the ordinary equation of state have to be expected for the reason that only one class of symmetry is possible as soon as the mean De Broglie wavelength of a gas molecule becomes of an order of magnitude comparable with the average distance between two molecules, that is, for small temperatures and large densities. For the antisymmetrical class the statistical consequences have been derived by Fermi and Dirac[13], for the symmetrical class the same had been done already before the discovery of the new quantum mechanics by Einstein and Bose[14]. The former case could be applied to the electrons in a metal and could be used for the interpretation of magnetic and other properties of metals.

As soon as the symmetry classes for electrons were cleared, the question arose which are the symmetry classes for other particles. One example for particles with symmetrical wave functions only (assumption II) was already known long ago, namely the photons. This is not only an immediate consequence of Planck's derivation of the spectral distribution of the radiation energy in the thermodynamical equilibrium, but it is also necessary for the applicability of the classical field concepts to light waves in the limit where a large and not accurately fixed number of photons is present in a single quantum state. We note that the symmetrical class for photons occurs together with the integer value I for their spin, while the antisymmetrical class for the electron occurs together with the half-integer value $\frac{1}{2}$ for the spin.

The important question of the symmetry classes for nuclei, however, had still to be investigated. Of course the symmetry class refers here also to the permutation of both the space coordinates and the spin indices of two like nuclei. The spin index can assume $2I+1$ values if $I$ is the spin-quantum number of the nucleus which can be either an integer or a half-integer. I may include the historical remark that already in 1924, before the electron spin was discovered, I proposed to use the assumption of a nuclear spin to interpret the hyperfine-structure of spectral lines[15]. This proposal met on the one hand strong opposition from many sides but influenced on the other hand Goudsmit and Uhlenbeck in their claim of an electron spin. It was only some years later that my attempt to interpret the hyperfine-structure could be definitely confirmed experimentally by investigations in which also Zeeman himself participated and which showed the existence of a magneto-optic transformation of the hyperfine-structure as I had predicted it. Since that time the hyperfine-structure of spectral lines became a general method of determining the nuclear spin.

In order to determine experimentally also the symmetry class of the nuclei, other methods were necessary. The most convenient, although not the only one, consists in the investigation of band spectra due to a molecule with two like atoms[16]. It could easily be derived that in the ground state of the electron configuration of such a molecule the states with even and odd values of the rotational quantum number are symmetric and antisymmetric respectively for a permutation of the space coordinates of the two nuclei. Further there exist among the $(2\,I + 1)^2$ spin states of the pair of nuclei, $(2\,I + 1)\,(I + 1)$ states symmetrical and $(2\,I + 1)I$ states antisymmetrical in the spins, since the $(2\,I + 1)$ states with two spins in the same direction are necessarily symmetrical. Therefore the conclusion was reached: If the total wave function of space coordinates and spin indices of the nuclei is symmetrical, the ratio of the weight of states with an even rotational quantum number to the weight of states with an odd rotational quantum number is given by $(I + 1)$: $I$. In the reverse case of an antisymmetrical total wave function of the nuclei, the same ratio is $I$: $(I + 1)$. Transitions between one state with an even and another state with an odd rotational quantum number will be extremely rare as they can only be caused by an interaction between the orbital motions and the spins of the nuclei. Therefore the ratio of the weights of the rotational states with different parity will give rise to two different systems of band spectra with different intensities, the lines of which are alternating.

The first application of this method was the result that the protons have the spin $\frac{1}{2}$ and fulfill the exclusion principle just as the electrons. The initial difficulties to understand quantitatively the specific heat of hydrogen molecules at low temperatures were removed by Dennison's hypothesis[17], that at this low temperature the thermal equilibrium between the two modifications of the hydrogen molecule (ortho-$H_2$: odd rotational quantum numbers, parallel proton spins; para-$H_2$: even rotational quantum numbers, antiparallel spins) was not yet reached. As you know, this hypothesis was later, confirmed by the experiments of Bonhoeffer and Harteck and of Eucken, which showed the theoretically predicted slow transformation of one modification into the other.

Among the symmetry classes for other nuclei those with a different parity of their mass number $M$ and their charge number $Z$ are of a particular interest. If we consider a compound system consisting of numbers $A_1$, $A_2$, ... of different constituents, each of which is fulfilling the exclusion principle, and a number $S$ of constituents with symmetrical states, one has to expect symmetrical or antisymmetrical states if the sum $A_1 + A_2 + \ldots$ is even or odd. This holds regardless of the parity of $S$. Earlier one tried the assumption that nuclei consist of protons and electrons, so that $M$ is the number of protons, $M - Z$ the number of electrons in the nucleus. It had to be expected then that the parity of $Z$ determines the symmetry class of the whole nucleus. Already for some time the counter-example of nitrogen has been known to have the spin I and symmetrical states[18]. After the discovery of the neutron, the nuclei have been considered, however, as composed of protons and neutrons in such a way that a nucleus with mass number $M$ and charge number $Z$ should consist of $Z$ protons and $M - Z$ neutrons. In case the neutrons would have symmetrical states, one should again expect that the parity of the charge number $Z$ determines the symmetry class of the nuclei. If, however, the neutrons fulfill the exclusion principle, it has to be expected that the parity of $M$ determines the symmetry class: For an even $M$, one should always have symmetrical states, for an odd $M$, antisymmetrical ones. It was the latter rule that was confirmed by experiment without exception, thus proving that the neutrons fulfill the exclusion principle.

The most important and most simple crucial example for a nucleus with a different parity of $M$ and $Z$ is the heavy hydrogen or deuteron with $M = 2$ and $Z = 1$ which has symmetrical states and the spin $I = 1$, as could be proved by the investigation of the band spectra of a molecule with two deuterons[19]. From the spin value I of the deuteron can be concluded that the neutron must have a half-integer spin. The simplest possible assumption that this spin of the neutron is equal to $\frac{1}{2}$, just as the spin of the proton and of the electron, turned out to be correct.

There is hope, that further experiments with light nuclei, especially with protons, neutrons, and deuterons will give us further information about the nature of the forces between the constituents of the nuclei, which, at present, is not yet sufficiently clear. Already now we can say, however, that these interactions are fundamentally different from electromagnetic interactions. The comparison between neutron-proton scattering and proton-proton scattering even showed that the forces between these particles are in good approximation the same, that means independent of their electric charge. If one had only to take into account the magnitude of the interaction energy, one should therefore expect a stable di-proton or $^2_2\mathrm{He}$ ($M = 2$, $Z = 2$) with nearly the same binding energy as the deuteron. Such a state is, however, forbidden by the exclusion principle in accordance with experience, because this state would acquire a wave function symmetric with respect to the two protons. This is only the simplest example of the application of the exclusion principle to the structure of compound nuclei, for the understanding of which this principle is indispensable, because the constituents of these heavier nuclei, the protons and the neutrons, fullfil it.

In order to prepare for the discussion of more fundamental questions, we want to stress here a law of Nature which is generally valid, namely, the connection between spin and symmetry class. *A half-integer value of the spin quantum number is always connected with antisymmetrical states (exclusion principle), an integer spin with symmetrical states.* This law holds not only for protons and neutrons but also for protons and electrons. Moreover, it can easily be seen that it holds for compound systems, if it holds for all of its constituents. If we search for a theoretical explanation of this law, we must pass to the discussion of relativistic wave mechanics, since we saw that it can certainly not be explained by non-relativistic wave mechanics.

We first consider classical fields[20], which, like scalars, vectors, and tensors transform with respect to rotations in the ordinary space according to a one-valued representation of the rotation group. We may, in the following, call such fields briefly "one-valued" fields. So long as interactions of different kinds of field are not taken into account, we can assume that all field components will satisfy a second-order wave equation, permitting a superposition of plane waves as a general solution. Frequency and wave number of these plane waves are connected by a law which, in accordance with De Broglie's fundamental assumption, can be obtained from the relation between energy and momentum of a particle claimed in relativistic mechanics by division with the constant factor equal to Planck's constant divided by $2\pi$. Therefore, there will appear in the classical field equations, in general, a new constant $\mu$ with the dimension of a reciprocal length, with which the rest-mass $m$ in the particle picture is connected by $m = h \, \mu/c$, where $c$ is the vacuum-velocity of light. From the assumed property of one-valuedness of the field it can be concluded, that the number of possible plane waves for a given frequency, wave number and direction of propagation, is for a non-vanishing $\mu$ always odd. Without going into details of the general definition of spin, we can consider this property of the polarization of plane waves as characteristic for fields which, as a result of their quantization, give rise to integer spin values.

The simplest cases of one-valued fields are the scalar field and a field consisting of a four-vector and an antisymmetric tensor like the potentials and field strengths in Maxwell's theory. While the scalar field is simply fulfilling the usual wave equation of the second order in which the term proportional to $\mu^2$ has to be included, the other field has to fulfill equations due to Proca which are a generalization of Maxwell's equations which become in the particular case $\mu = 0$. It is satisfactory that for these simplest cases of one-valued fields the energy density is a positive definite quadratic form of the field-quantities and their first derivatives at a certain point. For the general case of one-valued fields it can at least be achieved that the total energy after integration over space is always positive.

The field components can be assumed to be either real or complex. For a complex field, in addition to energy and momentum of the field, a four-vector can be defined which satisfies the continuity equation and can be interpreted as the four-vector of the electric current. Its fourth component determines the electric charge density and can assume both positive and negative values. It is possible that the charged mesons observed in cosmic rays have integral spins and thus can be described by such a complex field. In the particular case of real fields this four-vector of current vanishes identically.

Especially in view of the properties of the radiation in the thermodynamical equilibrium in which specific properties of the field sources do not play any role, it seemed to be justified first to disregard in the formal process of field quantization the interaction of the field with the sources. Dealing with this problem, one tried indeed to apply the same mathematical method of passing from a classical system to a corresponding system governed by the laws of quantum mechanics which has been so successful in passing from classical point mechanics to wave mechanics. It should not be forgotten, however, that a field can only be observed with help of its interaction with test bodies which are themselves again sources of the field.

The result of the formal process of field quantization were partly very encouraging. The quantized wave fields can be characterized by a wave function which depends on an infinite sequence of (non-negative) integers as variables. As the total energy and the total momentum of the field and, in case of complex fields, also its total electric charge turn out to be linear functions of these numbers, they can be interpreted as the number of particles present in a specified state of a single particle. By using a sequence of configuration spaces with a different number of dimensions corresponding to the different possible values of the total number of particles present, it could easily be shown that this description of our system by a wave function depending on integers is equivalent to an ensemble of particles with wave functions symmetrical in their configuration spaces.

Moreover Bohr and Rosenfeld[21] proved in the case of the electromagnetic field that the uncertainty relations which result for the average values of the field strengths over finite space-time regions from the formal commutation rules of this theory have a direct physical meaning so long as the sources can be treated classically and their atomistic structure can be disregarded. We emphasize the following property of these commutation rules: All physical quantities in two world points, for which the four-vector of their joining straight line is spacelike commute with each other. This is indeed necessary for physical reasons because any disturbance by measurements in a world point $P_\nu$ can only reach such points $P_2$, for which the vector $P_1P_2$, is timelike, that is, for which $c\,(t_1 - t_2) > r_{12}$. The points $P_2$ with a spacelike vector $P_1P_2$ for which $c(t_1 - t_2) < r_{12}$ cannot be reached by this disturbance and measurements in $P_1$ and $P_2$ can then never influence each other.

This consequence made it possible to investigate the logical possibility of particles with integer spin which would obey the exclusion principle. Such particles could be described by a sequence of configuration spaces with different dimensions and wave functions antisymmetrical in the coordinates of these spaces or also by a wave function depending on integers again to be interpreted as the number of particles present in specified states which now can only assume the values 0 or 1. Wigner and Jordan[22] proved that also in this case operators can be defined which are functions of the ordinary spacetime coordinates and which can be applied to such a wave function. These operators do not fulfil any longer commutation rules: instead of the difference, the sum of the two possible products of two operators, which are distinguished by the different order of its factors, is now fixed by the mathematical conditions the operators have to satisfy. The simple change of the sign in these conditions changes entirely the physical meaning of the formalism. In the case of the exclusion principle there can never exist a limiting case where such operators can be replaced by a classical field. Using this formalism of Wigner and Jordan I could prove under very general assumptions that a relativistic invariant theory describing systems of like particles with integer spin obeying the exclusion principle would always lead to the non-commutability of physical quantities joined by a spacelike vector[23]. This would violate a reasonable physical principle which holds good for particles with symmetrical states. In this way, by combination of the claims of relativistic invariance and the properties of field quantization, one step in the direction of an understanding of the connection of spin and symmetry class could be made.

The quantization of one-valued complex fields with a non-vanishing four-vector of the electric current gives the further result that particles both with positive and negative electric charge should exist and that they can be annihilated and generated in external electromagnetic field[22]. This pair-generation and annihilation claimed by the theory makes it necessary to distinguish clearly the concept of charge density and of particle density. The latter concept does not occur in a relativistic wave theory either for fields carrying an electric charge or for neutral fields. This is satisfactory since the use of the particle picture and the uncertainty relations (for instance by analyzing imaginative experiments of the type of the $\gamma$-ray microscope) gives also the result that a localization of the particle is only possible with limited accuracy[24]. This holds both for the particles with integer and with half-integer spins. In a state with a mean value $E$ of its energy, described by a wave packet with a mean frequency $v = E/h$, a particle can only be localized with an error $\Delta x > hc/E$ or $\Delta x > c/v$. For photons, it follows that the limit for the localization is the wavelength; for a particle with a finite rest mass $m$ and a characteristic length $\mu^{-1} = \hbar/mc$, this limit is in the rest system of the center of the wave packet that describes the state of the particles given by $\Delta x > \hbar/mc$ or $\Delta x > \mu^{-1}$.

Until now I have mentioned only those results of the application of quantum mechanics to classical fields which are satisfactory. We saw that the statements of this theory about averages of field strength over finite spacetime regions have a direct meaning while this is not so for the values of the field strength at a certain point. Unfortunately in the classical expression of the energy of the field there enter averages of the squares of the field strengths over such regions which cannot be expressed by the averages of the field strengths themselves. This has the consequence that the zero-point energy of the vacuum derived from the quantized field becomes infinite, a result which is directly connected with the fact that the system considered has an infinite number of degrees of freedom. It is clear that this zero-point energy has no physical reality, for instance it is not the source of a gravitational field. Formally it is easy to subtract constant infinite terms which are independent of the state considered and never change; nevertheless it seems to me that already this result is an indication that a fundamental change in the concepts underlying the present theory of quantized fields will be necessary.

In order to clarify certain aspects of relativistic quantum theory I have discussed here, different from the historical order of events, the one-valued fields first. Already earlier Dirac[25] had formulated his relativistic wave equations corresponding to material particles with spin $\frac{1}{2}$ using a pair of so-called spinors with two components each. He applied these equations to the problem of one electron in an electromagnetic field. In spite of the great success of this theory in the quantitative explanation of the fine structure of the energy levels of the hydrogen atom and in the computation of the scattering cross section of one photon by a free electron, there was one consequence of this theory which was obviously in contradiction with experience. The energy of the electron can have, according to the theory, both positive and negative values, and, in external electromagnetic fields, transitions should occur from states with one sign of energy to states with the other sign. On the other hand there exists in this theory a four-vector satisfying the continuity equation with a fourth component corresponding to a density which is definitely positive.

It can be shown that there is a similar situation for all fields, which, like the spinors, transform for rotations in ordinary space according to two-valued representations, thus changing their sign for a full rotation. We shall call briefly such quantities "two-valued". From the relativistic wave equations of such quantities one can always derive a four-vector bilinear in the field components which satisfies the continuity equation and for which the fourth component, at least after integration over the space, gives an essentially positive quantity. On the other hand, the expression for the total energy can have both the positive and the negative sign.

Is there any means to shift the minus sign from the energy back to the density of the four-vector? Then the latter could again be interpreted as charge density in contrast to particle density and the energy would become positive as it ought to be. You know that Dirac's answer was that this could actually be achieved by application of the exclusion principle. In his lecture delivered here in Stockholm[10] he himself explained his proposal of a new interpretation of his theory, according to which in the actual vacuum all the states of negative energy should be occupied and only deviations of this state of smallest energy, namely holes in the sea of these occupied states are assumed to be observable. It is the exclusion principle which guarantees the stability of the vacuum, in which all states of negative energy are occupied. Furthermore the holes have all properties of particles with positive energy and positive electric charge, which in external electromagnetic fields can be produced and annihilated in pairs. These predicted positrons, the exact mirror images of the electrons, have been actually discovered experimentally.

The new interpretation of the theory obviously abandons in principle the standpoint of the one-body problem and considers a many-body problem from the beginning. It cannot any longer be claimed that Dirac's relativistic wave equations are the only possible ones but if one wants to have relativistic field equations corresponding to particles, for which the value $\frac{1}{2}$ of their spin is known, one has certainly to assume the Dirac equations. Although it is logically possible to quantize these equations like classical fields, which would give symmetrical states of a system consisting of many such particles, this would be in contradiction with the postulate that the energy of the system has actually to be positive. This postulate is fulfilled on the other hand if we apply the exclusion principle and Dirac's interpretation of the vacuum and the holes, which at the same time substitutes the physical concept of charge density with values of both signs for the mathematical fiction of a positive particle density. A similar conclusion holds for all relativsitic wave equations with two-valued quantities as field components. This is the other step (historically the earlier one) in the direction of an understanding of the connection between spin and symmetry class.

I can only shortly note that Dirac's new interpretation of empty and occupied states of negative energy can be formulated very elegantly with the help of the formalism of Jordan and Wigner mentioned before. The transition from the old to the new interpretation of the theory can indeed be carried through simply by interchanging the meaning of one of the operators with that of its hermitian conjugate if they are applied to states originally of negative energy. The infinite "zero charge" of the occupied states of negative energy is then formally analogous to the infinite zero-point energy of the quantized one-valued fields. The former has no physicial reality either and is not the source of an electromagnetic field.

In spite of the formal analogy between the quantization of the one-valued fields leading to ensembles of like particles with symmetrical states and to particles fulfilling the exclusion principle described by two-valued operator quantities, depending on space and time coordinates, there is of course the fundamental difference that for the latter there is no limiting case, where the mathematical operators can be treated like classical fields. On the other hand we can expect that the possibilities and the limitations for the applications of the concepts of space and time, which find their expression in the different concepts of charge density and particle density, will be the same for charged particles with integer and with half-integer spins.

The difficulties of the present theory become much worse, if the interaction of the electromagnetic field with matter is taken into consideration, since the well-known infinities regarding the energy of an electron in its own field, the so-called self-energy, then occur as a result of the application of the usual perturbation formalism to this problem. The root of this difficulty seems to be the circumstance that the formalism of field quantization has only a direct meaning so long as the sources of the field can be treated as continuously distributed, obeying the laws of classical physics, and so long as only averages of field quantities over finite space-time regions are used. The electrons themselves, however, are essentially non-classical field sources.

At the end of this lecture I may express my critical opinion, that a correct theory should neither lead to infinite zero-point energies nor to infinite zero charges, that it should not use mathematical tricks to subtract infinities or singularities, nor should it invent a "hypothetical world" which is only a mathematical fiction before it is able to formulate the correct interpretation of the actual world of physics.

From the point of view of logic, my report on "Exclusion principle and quantum mechanics" has no conclusion. I believe that it will only be possible to write the conclusion if a theory will be established which will determine the value of the fine-structure constant and will thus explain the atomistic structure of electricity, which is such an essential quality of all atomic sources of electric fields actually occurring in Nature.

## References

1. A. Landé, *Z. Physik*, 5 (1921) 231 and *Z. Physik*, 7 (1921) 398, Physik. Z., 22 (1921) 417.
2. W. Pauli, *Z. Physik*, 16 (1923) 155.
3. W. Pauli, *Z. Physik*, 31 (1925) 373.
4. E. C. Stoner, *Phil. Mag.*, 48 (1924) 719.
5. W. Pauli, *Z. Physik*, 31 (1925) 765.
6. S. Goudsmit and G. Uhlenbeck, *Naturwiss.*, 13 (1925) 953, *Nature*, 117 (1926) 264.
7. L. H. Thomas, *Nature*, 117 (1926) 514, and *Phil. Mag.*, 3 (1927) 1. Compare also J. Frenkel, *Z. Physik*, 37 (1926) 243.
8. Compare *Rapport du Sixième Conseil Solvay de Physique, Paris, 1932*, pp. 217–225.
9. For this earlier stage of the history of the exclusion principle compare also the author's note in *Science*, 103 (1946) 213, which partly coincides with the first part of the present lecture.
10. The Nobel Lectures of W. Heisenberg, E. Schrödinger, and P. A. M. Dirac are collected in *Die moderne Atomtheorie*, Leipzig, 1934.
11. The articles of N. Bohr are collected in *Atomic Theory and the Description of Nature*, Cambridge University Press, 1934. See also his article "Light and Life", *Nature*, 131 (1933) 421, 457.

12. W. Heisenberg, *Z. Physik*, 38 (1926) 411 and 39 (1926) 499.

13. E. Fermi, *Z. Physik*, 36 (1926) 902. P. A. M. Dirac, *Proc. Roy. Soc. London*, A 112 (1926) 661.

14. S. N. Bose, *Z. Physik*, 26 (1924) 178 and 27 (1924) 384. A. Einstein, *Berl. Ber.*, (1924) 261; (1925) 1, 18.

15. W. Pauli, *Naturwiss.*, 12 (1924) 741.

16. W. Heisenberg, *Z. Physik*, 41 (1927) 239, F. Hund, *Z. Physik*, 42 (1927) 39.

17. D. M. Dennison, *Proc. Roy. Soc. London*, A 115 (1927) 483.

18. R. de L. Kronig, *Naturwiss.*, 16 (1928) 335. W. Heitler und G. Herzberg, *Naturwiss.*, 17 (1929) 673.

19. G. N. Lewis and M. F. Ashley, *Phys. Rev.*, 43 (1933) 837. G. M. Murphy and H. Johnston, *Phys. Rev.*, 45 (1934) 550 and 46 (1934) 95.

20. Compare for the following the author's report in *Rev. Mod. Phys.*, 13 (1941) 203, in which older literature is given. See also W. Pauli and V. Weisskopf, *Helv. Phys. Acta*, 7 (1934) 809.

21. N. Bohr and L. Rosenfeld, *Kgl. Danske Videnskab. Selskab. Mat. Fys. Medd.*, 12 [8] (1933).

22. P. Jordan and E. Wigner, *Z. Physik*, 47 (1928) 631. Compare also V. Fock, *Z. Physik*, 75 (1932) 622.

23. W. Pauli, *Ann. Inst. Poincaré*, 6 (1936) 137 and *Phys. Rev.*, 58 (1940) 716.

24. L. Landau and R. Peierls, *Z. Physik*, 69 (1931) 56. Compare also the author's article in *Handbuch der Physik*, 24, Part 1, 1933, Chap. A, § 2.

25. P. A. M. Dirac, *Proc. Roy. Soc. London*, A 117 (1928) 610.

# Chapter Five

In the papers covered in the previous chapters, much of the mathematical basis of quantum physics was worked out. However, the more philosophical questions about what quantum theory says about reality and what the "right way" to interpret quantum physics is remain. The standard interpretation of quantum mechanics arose from Niels Bohr and his collaborators, so it has been named the "Copenhagen interpretation." Two fundamental postulates of the Copenhagen interpretation are that we should only be concerned with what is actually observed and that the quantum wave function or state vector of a system contains all possible information for that system. These two postulates seem very reasonable, but they lead to many strange results. Since the wave function contains all possible information about a system, it is crucially important to understand what it means. How do we interpret the wave function? The mostly widely accepted answer comes from Max Born. He argued that the wave function (or more precisely the square of the amplitude of the wave function) represents the probability that an event will occur. In this sense, quantum mechanics is non-deterministic. In a deterministic theory, when a system starts out with a given initial state, its final state can be calculated from the theory at all times. But in quantum mechanics this is not the case. Identical experiments with identical starting conditions can produce different results. All we can do is calculate the *probability* that a system will end in a certain final state.

The inherently statistical nature of quantum theory troubled many physicists. In fact, some of the theory's greatest contributors balked at the strange notions of reality that quantum mechanics seemed to indicate. Notable among these are Albert Einstein and Erwin Schrodinger. Schrodinger was troubled by the idea that according to the standard interpretation of quantum mechanics, a system actually exists in *all* possible states until a measurement is made and collapses the wave function of that system into a single state. In "The Present Situation in Quantum Mechanics," Schrodinger introduces his famous cat experiment, which was meant to show the absurdity of the Copenhagen interpretation of quantum mechanics. In this thought experiment he imagines a cat which is in a sealed box with a radioactive source and detector. When a radioactive decay is detected cyanide is released killing the cat. According to the Copenhagen interpretation so long as an observation is not made the radioactive source will exist in a superposition of decayed and non-decayed states. Schrodinger maintained this results in an absurd conclusion, because the cyanide will be simultaneously released and not-released and the cat will simultaneously exist in dead and living states, which is, of course, impossible.

Another attack on the standard interpretation of quantum theory came from Albert Einstein, Boris Podolsky, and Nathan Rosen in a paper in which they argued that quantum mechanics cannot be a complete theory of reality. Like Schrodinger, they were also troubled by the statistical nature of quantum theory, and they wondered if perhaps there was a deeper reality hidden beyond that which was represented by the quantum mechanical wave function. They were also very troubled by the idea of what is now called *quantum entanglement.* That is, if two systems are allowed to interact and then are separated, a measurement on one system causing the collapse of its wave function can instantaneously cause the wave function of the other system to collapse as well. Einstein called this "spooky action at a distance," and we can imagine why it was especially troubling to him. At first glance it seems to violate his theory of relativity. It appears that some signal must be traveling instantaneously between the two systems carrying the news that a measurement has been made. It has since been shown that no information is actually carried in the collapse, and so it is not actually a violation of relativity. Nevertheless, the idea of instantaneous collapse is still very troubling to many physicists. Einstein, Podolsky, and Rosen concluded that quantum mechanics as it is formulated cannot be a complete description of reality. In other words, they conclude that there must be elements of reality that are "hidden" from quantum mechanics. When these are taken into consideration, the spooky action at a distance that so troubled Einstein vanishes. It seems that it was Einstein's hope until his death that a more complete theory of reality—a so-called local hidden-variables theory—would someday replace quantum mechanics.

In 1952 David Bohm produced two papers in which he attempted to create a hidden-variables interpretation of quantum physics. But in order to make his theory match the experimental observations he was not able to make his theory completely deterministic, nor was he able to eliminate the "non-local" spooky action at a distance which so troubled Einstein. In fact, unfortunately for Einstein it is not likely that such a theory can exist. In the extremely creative paper, "On the Einstein-Podolsky-Rosen Paradox," John Bell showed that any local hidden-variables theory will make predictions about measurable quantities that differ from those of quantum mechanics. Careful experimentation has continually supported the predictions made by quantum mechanics. Despite the objections of Einstein, Podolsky, and Rosen, quantum mechanics seems to be a more accurate representation of reality than deterministic hidden-variable theories.

# The Statistical Interpretation of Quantum Mechanics

### By

### Max Born

*Nobel Lecture, December 11, 1954*

The work, for which I have had the honour to be awarded the Nobel Prize for 1954, contains no discovery of a fresh natural phenomenon, but rather the basis for a new mode of thought in regard to natural phenomena. This way of thinking has permeated both experimental and theoretical physics to such a degree that it hardly seems possible to say anything more about it that has not been already so often said. However, there are some particular aspects which I should like to discuss on what is, for me, such a festive occasion. The first point is this: the work at the Göttingen school, which I directed at that time (1926–1927), contributed to the solution of an intellectual crisis into which our science had fallen as a result of Planck's discovery of the quantum of action in 1900. Today, physics finds itself in a similar crisis - I do not mean here its entanglement in politics and economics as a result of the mastery of a new and frightful force of Nature, but I am considering more the logical and epistemological problems posed by nuclear physics. Perhaps it is well at such a time to recall what took place earlier in a similar situation, especially as these events are not without a definite dramatic flavour.

The second point I wish to make is that when I say that the physicists had accepted the concepts and mode of thought developed by us at the time, I am not quite correct. There are some very noteworthy exceptions, particularly among the very workers who have contributed most to building up the quantum theory. Planck, himself, belonged to the sceptics until he died. Einstein, De Broglie, and Schrödinger have unceasingly stressed the unsatisfactory features of quantum mechanics and called for a return to the concepts of classical, Newtonian physics while proposing ways in which this could be done without contradicting experimental facts. Such weighty views cannot be ignored. Niels Bohr has gone to a great deal of trouble to refute the objections. I, too, have ruminated upon them and believe I can make some contribution to the clarification of the position. The matter concerns the borderland between physics and philosophy, and so my physics lecture will partake of both history and philosophy, for which I must crave your indulgence.

First of all, I will explain how quantum mechanics and its statistical interpretation arose. At the beginning of the twenties, every physicist, I think, was convinced that Planck's quantum hypothesis was correct. According to this theory *energy* appears in finite quanta of magnitude $hv$ in oscillatory processes having a specific frequency $v$ (e.g. in light waves). Countless experiments could be explained in this way and always gave the same value of Planck's constant $h$. Again, Einstein's assertion that light quanta have *momentum $hv/c$* (where $c$ is the speed of light) was well supported by experiment (e.g. through the Compton effect). This implied a revival of the corpuscular theory of light for a certain complex of phenomena. The wave theory still held good for other processes. Physicists grew accustomed to this *duality* and learned how to cope with it to a certain extent.

In 1913 Niels Bohr had solved the riddle of *line spectra* by means of the quantum theory and had thereby explained broadly the amazing stability of the atoms, the structure of their electronic shells, and the Periodic System of the elements. For what was to come later, the most important assumption of his teaching was this: an atomic system cannot exist in all mechanically possible states, forming a continuum, but in a series of discrete "stationary" states. In a transition from one to another, the difference in energy $E_m - E_n$ is emitted or absorbed as a light quantum $hv_{mn}$ (according to whether $E_m$ is greater or less than $E_n$). This is an interpretation in terms of energy of the fundamental law of spectroscopy discovered some years before by W. Ritz. The situation can be taken in at a glance by writing the energy levels of the stationary states twice over, horizontally and vertically. This produces a square array

|        | $E_1$, | $E_2$, | $E_3$ .... |   |
|--------|--------|--------|------------|---|
| $E_1$  | 11     | 12     | 13         | - |
| $E_2$  | 21     | 22     | 23         | - |
| $E_3$  | 31     | 32     | 33         | - |
|        | -      | -      | -          | - |

in which positions on a diagonal correspond to states, and non-diagonal positions correspond to transitions.

It was completely clear to Bohr that the law thus formulated is in conflict with mechanics, and that therefore the use of the energy concept in this connection is problematical. He based this daring fusion of old and new on his *principle of correspondence.* This consists in the obvious requirement that ordinary classical mechanics must hold to a high degree of approximation in the limiting case where the numbers of the stationary states, the so-called quantum numbers, are very large (that is to say, far to the right and to the lower part in the above array) and the energy changes relatively little from place to place, in fact practically continuously.

Theoretical physics maintained itself on this concept for the next ten years. The problem was this: an harmonic oscillation not only has a frequency, but also an intensity. For each transition in the array there must be a corresponding intensity. The question is how to find this through the considerations of correspondence? It meant guessing the unknown from the available information on a known limiting case. Considerable success was attained by Bohr himself, by Kramers, Sommerfeld, Epstein, and many others. But the decisive step was again taken by Einstein who, by a fresh derivation of Planck's radiation formula, made it transparently clear that the classical concept of intensity of radiation must be replaced by the statistical concept of *transition probability.* To each place in our pattern or array there belongs (together with the frequency $v_{mn} = (E_n - E_m)/h$)

a definite probability for the transition coupled with emission or absorption.

In Göttingen we also took part in efforts to distil the unknown mechanics of the atom from the experimental results. The logical difficulty became ever sharper. Investigations into the scattering and dispersion of light showed that Einstein's conception of transition probability as a measure of the strength of an oscillation did not meet the case, and the idea of an *amplitude* of oscillation associated with each transition was indispensable. In this connection, work by Ladenburg[*], Kramer[†], Heisenberg[‡], Jordan and me[§] should be mentioned. The art of guessing correct formulae, which deviate from the classical formulae, yet contain them as a limiting case according to the correspondence principle, was brought to a high degree of perfection. A paper of mine, which introduced, for the first time I think, the expression *quantum mechanics* in its title, contains a rather involved formula (still valid today) for the reciprocal disturbance of atomic systems.

Heisenberg, who at that time was my assistant, brought this period to a sudden end[‖]. He cut the Gordian knot by means of a philosophical principle and replaced guess-work by a mathematical rule. The principle states that concepts and representations that do not correspond to physically observable facts are not to be used in theoretical description. Einstein used the same principle when, in setting up his theory of relativity, he eliminated the concepts of absolute velocity of a body and of absolute simultaneity of two events at different places. Heisenberg banished the picture of electron orbits with definite radii and periods of rotation because these quantities are not observable, and insisted that the theory be built up by means of the square arrays mentioned above. Instead of describing the motion by giving a coordinate as a function of time, $x(t)$, an array of transition amplitudes

[*] R. Ladenburg, *Z. Physik*, 4 (1921) 451; R. Ladenburg and F. Reiche, *Naturwiss.*, 11 (1923) 584.
[†] H. A. Kramers, *Nature*, 113 (1924) 673.
[‡] H. A. Kramers and W. Heisenberg, *Z. Physik*, 31 (1925) 681.
[§] M. Born, *Z. Physik*, 26 (1924) 379; M. Born and P. Jordan, *Z. Physik*, 33 (1925) 479.
[‖] W. Heisenberg, *Z. Physik*, 33 (1925) 879.

$x_{mn}$ should be determined. To me the decisive part of his work is the demand to determine a rule by which from a given

array $\begin{bmatrix} x_{11} \ x_{12} \dots \\ x_{21} \ x_{22} \dots \\ \text{-------} \end{bmatrix}$ the array for the square $\begin{bmatrix} (x^2)_{11} \ (x^2)_{12} \dots \\ (x^2)_{21} \ (x^2)_{22} \dots \\ \text{----------} \end{bmatrix}$

can be found (or, more general, the *multiplication rule* for such arrays).

By observation of known examples solved by guess-work he found this rule and applied it successfully to simple examples such as the harmonic and anharmonic oscillator.

This was in the summer of 1925. Heisenberg, plagued by hay fever took leave for a course of treatment by the sea and gave me his paper for publication if I thought I could do something with it.

The significance of the idea was at once clear to me and I sent the manuscript to the *Zeitschrift für Physik*. I could not take my mind off Heisenberg's multiplication rule, and after a week of intensive thought and trial I suddenly remembered an algebraic theory which I had learned from my teacher, Professor Rosanes, in Breslau. Such square arrays are well known to mathematicians and, in conjunction with a specific rule for multiplication, are called matrices. I applied this rule to Heisenberg's quantum condition and found that this agreed in the diagonal terms. It was easy to guess what the remaining quantities must be, namely, zero; and at once there stood before me the peculiar formula

$$pq - qp = h/2\pi i$$

This meant that coordinates $q$ and momenta $p$ cannot be represented by figure values but by symbols, the product of which depends upon the order of multiplication - they are said to be "non-commuting".

I was as excited by this result as a sailor would be who, after a long voyage, sees from afar, the longed-for land, and I felt regret that Heisenberg was not there. I was convinced from the start that we had stumbled on the right path. Even so, a great part was only guess-work, in particular, the disappearance of the non-diagonal elements in the above-mentioned expression. For help in this problem I obtained the assistance and collaboration of my pupil Pascual Jordan, and in a few days we were able to demonstrate that I had guessed correctly. The joint paper by Jordan and myself* contains the most important principles of quantum mechanics including its extension to electrodynamics. There followed a hectic period of collaboration among the three of us, complicated by Heisenberg's absence. There was a lively exchange of letters; my contribution to these, unfortunately, have been lost in the political disorders. The result was a three-author paper† which brought the formal side of the investigation to a definite conclusion. Before this paper appeared, came the *first dramatic surprise:* Paul Dirac's paper on the same subject‡. The inspiration afforded by a lecture of Heisenberg's in Cambridge had led him to similar results as we had obtained in Göttingen except that he did not resort to the known matrix theory of the mathematicians, but discovered the tool for himself and worked out the theory of such non-commutating symbols.

The first non-trivial and physically important application of quantum mechanics was made shortly afterwards by W. Pauli§ who calculated the stationary energy values of the *hydrogen atom* by means of the matrix method and found complete agreement with Bohr's formulae. From this moment onwards there could no longer be any doubt about the correctness of the theory.

What this formalism really signified was, however, by no means clear. Mathematics, as often happens, was cleverer than interpretative thought. While we were still discussing this point there came the *second dramatic surprise*, the appearance of Schrödinger's famous papers‖. He took up quite a different line of thought which had originated from Louis de Broglie#.

A few years previously, the latter had made the bold assertion, supported by brilliant theoretical considerations, that wave-corpuscle duality, familiar to physicists in the case of light, must also be valid for

---

*M. Born and P. Jordan, *Z. Physik*, 34 (1925) 858.
†M. Born, W. Heisenberg, and P. Jordan, *Z. Physik*, 35 (1926) 557.
‡P. A. M. Dirac, *Proc. Roy. Soc. (London)*, A 109 (1925) 642.
§W. Pauli, *Z. Physik*, 36 (1926) 336.
‖E. Schrödinger, *Ann. Physik*, [4] 79 (1926) 361, 489, 734; 80 (1926) 437; 81(1926) 109.
#L. de Broglie, *Thesis Paris, 1924; Ann. Phys. (Paris)*, [10] 3 (1925) 22.

electrons. To each electron moving free of force belongs a plane wave of a definite wavelength which is determined by Planck's constant and the mass. This exciting dissertation by De Broglie was well known to us in Göttingen. One day in 1925 I received a letter from C. J. Davisson giving some peculiar results on the reflection of electrons from metallic surfaces. I, and my colleague on the experimental side, James Franck, at once suspected that these curves of Davisson's were crystal-lattice spectra of De Broglie's electron waves, and we made one of our pupils, Elsasser[*], to investigate the matter. His result provided the first preliminary confirmation of the idea of De Broglie's, and this was later proved independently by Davisson and Germer[†] and G. P. Thomson[‡] by systematic experiments.

But this acquaintance with De Broglie's way of thinking did not lead us to an attempt to apply it to the electronic structure in atoms. This was left to Schrödinger. He extended De Broglie's wave equation which referred to force-free motion, to the case where the effect of force is taken into account, and gave an exact formulation of the *subsidiary conditions*, already suggested by De Broglie, to which the wave function $\psi$ must be subjected, namely that it should be single-valued and finite in space and time. And he was successful in deriving the stationary states of the hydrogen atom in the form of those monochromatic solutions of his wave equation which do not extend to infinity.

For a brief period at the beginning of 1926, it looked as though there were, suddenly, two self-contained but quite distinct systems of explanation extant: matrix mechanics and wave mechanics. But Schrödinger himself soon demonstrated their complete equivalence.

Wave mechanics enjoyed a very great deal more popularity than the Göttingen or Cambridge version of quantum mechanics. It operates with a wave function $\psi$, which in the case of *one* particle at least, can be pictured in space, and it uses the mathematical methods of partial differential equations which are in current use by physicists.

[*] W. Elasser, *Naturwiss.*, 13 (1925) 711.
[†] C. J. Davisson and L. H. Germer, *Phys. Rev.*, *30* (1927) 707.
[‡] G. P. Thomson and A. Reid, *Nature*, *119* (1927) 890; G. P. Thomson, *Proc. Roy. Soc.* (London), A 117 (1928) 600.
[§] E. Schrödinger, *Brit. J. Phil. Sci.*, *3* (1952) 109, 233.

Schrödinger thought that his wave theory made it possible to return to deterministic classical physics. He proposed (and he has recently emphasized his proposal anew's), to dispense with the particle representation entirely, and instead of speaking of electrons as particles, to consider them as a continuous density distribution $|\psi|^2$ (or electric density $e|\psi|^2$).

To us in Göttingen this interpretation seemed unacceptable in face of well established experimental facts. At that time it was already possible to count particles by means of scintillations or with a Geiger counter, and to photograph their tracks with the aid of a Wilson cloud chamber.

It appeared to me that it was not possible to obtain a clear interpretation of the $\psi$-function, by considering bound electrons. I had therefore, as early as the end of 1925, made an attempt to extend the matrix method, which obviously only covered oscillatory processes, in such a way as to be applicable to aperiodic processes. I was at that time a guest of the Massachusetts Institute of Technology in the USA, and I found there in Norbert Wiener an excellent collaborator. In our joint paper* we replaced the matrix by the general concept of an operator, and thus made it possible to describe aperiodic processes. Nevertheless we missed the correct approach. This was left to Schrödinger, and I immediately took up his method since it held promise of leading to an interpretation of the $\psi$-function. Again an idea of Einstein's gave me the lead. He had tried to make the duality of particles - light quanta or photons - and waves comprehensible by interpreting the square of the optical wave amplitudes as probability density for the occurrence of photons. This concept could at once be carried over to the $\psi$-function: $|\psi|^2$ ought to represent the probability density for electrons (or other particles). It was easy to assert this, but how could it be proved?

The atomic collision processes suggested themselves at this point. A swarm of electrons coming from infinity, represented by an incident

*M. Born and N. Wiener, Z. Physik, 36 (1926) 174.

wave of known intensity (i.e., $|\psi|^2$), impinges upon an obstacle, say a heavy atom. In the same way that a water wave produced by a steamer causes secondary circular waves in striking a pile, the incident electron wave is partially transformed into a secondary spherical wave whose amplitude of oscillation $\psi$ differs for different directions. The square of the amplitude of this wave at a great distance from the scattering centre determines the relative probability of scattering as a function of direction. Moreover, if the scattering atom itself is capable of existing in different stationary states, then Schrödinger's wave equation gives automatically the probability of excitation of these states, the electron being scattered with loss of energy, that is to say, inelastically, as it is called. In this way it was possible to get a theoretical basis[*] for the assumptions of Bohr's theory which had been experimentally confirmed by Franck and Hertz. Soon Wentzel[†] succeeded in deriving Rutherford's famous formula for the scattering of $\alpha$-particles from my theory.

However, a paper by Heisenberg[‡], containing his celebrated uncertainty relationship, contributed more than the above-mentioned successes to the swift acceptance of the statistical interpretation of the $\psi$-function. It was through this paper that the revolutionary character of the new conception became clear. It showed that not only the determinism of classical physics must be abandonded, but also the naive concept of reality which looked upon the particles of atomic physics as if they were very small grains of sand. At every instant a grain of sand has a definite position and velocity. This is not the case with an electron. If its position is determined with increasing accuracy, the possibility of ascertaining the velocity becomes less and *vice versa*. I shall return shortly to these problems in a more general connection, but would first like to say a few words about the theory of collisions.

The mathematical approximation methods which I used were quite primitive and soon improved upon. From the literature, which

[*] M. Born, *Z. Physik, 37* (1926) 863; 38 (1926) 803; *Göttinger Nachr. Math. Phys. Kl.*, (1926) 146.
[†] G. Wentzel, *Z. Physik, 40* (1926) 590.
[‡] W. Heisenberg, *Z. Physik, 43* (1927) 172.

has grown to a point where I cannot cope with, I would like to mention only a few of the first authors to whom the theory owes great progress: Faxén in Sweden, Holtsmark in Norway*, Bethe in Germany[†], Mott and Massey in England[‡].

Today, collision theory is a special science with its own big, solid textbooks which have grown completely over my head. Of course in the last resort all the modern branches of physics, quantum electrodynamics, the theory of mesons, nuclei, cosmic rays, elementary particles and their transformations, all come within range of these ideas and no bounds could be set to a discussion on them.

I should also like to mention that in 1926 and 1927 I tried another way of supporting the statistical concept of quantum mechanics, partly in collaboration with the Russian physicist Fock[§]. In the above-mentioned three-author paper there is a chapter which anticipates the Schrödinger function, except that it is not thought of as a function $\psi(x)$ in space, but as a function $\psi_n$ of the discrete index $n = 1, 2, \ldots$ which enumerates the stationary states. If the system under consideration is subject to a force which is variable with time, $\psi_n$ becomes also time-dependent, and $|\psi_n(t)|^2$ signifies the probability for the existence of the state $n$ at time $t$. Starting from an initial distribution where there is only one state, transition probabilities are obtained, and their properties can be examined. What interested me in particular at the time, was what occurs in the adiabatic limiting case, that is, for very slowly changing action. It was possible to show that, as could have been expected, the probability of transitions becomes ever smaller. The theory of transition probabilities was developed independently by Dirac with great success. It can be said that the whole of atomic and nuclear physics works with this system of concepts, particularly in the very elegant form given to them by Dirac[‖]. Almost all experiments lead to statements about relative frequencies of events, even when

*H. Faxén and J. Holtsmark, *Z. Physik*, 45 (1927) 307.
[†]H. Bethe, *Ann. Physik*, 5 (1930) 325.
[‡]N. F. Mott, *Proc. Roy. Soc. (London)*, A 124 (1929) 422, 425; *Proc. Cambridge Phil. Soc.*, 25 (1929) 304.
[§]M. Born, *Z. Physik*, 40 (1926) 167; M. Born and V. Fock, *Z. Physik*, 51 (1928) 165.
[‖]P. A. M. Dirac, *Proc. Roy. Soc. (London)*, A 109 (1925) 642; 110 (1926) 561; 111 (1926) 281; 112 (26) 674.

they occur concealed under such names as effective cross section or the like.

How does it come about then, that great scientists such as Einstein, Schrödinger, and De Broglie are nevertheless dissatisfied with the situation? Of course, all these objections are levelled not against the correctness of the formulae, but against their interpretation. Two closely knitted points of view are to be distinguished: the question of determinism and the question of reality.

Newtonian mechanics is deterministic in the following sense:

If the initial state (positions and velocities of all particles) of a system is accurately given, then the state at any other time (earlier or later) can be calculated from the laws of mechanics. All the other branches of classical physics have been built up according to this model. Mechanical determinism gradually became a kind of article of faith: the world as a machine, an automaton. As far as I can see, this idea has no forerunners in ancient and medieval philosophy. The idea is a product of the immense success of Newtonian mechanics, particularly in astronomy. In the 19th century it became a basic philosophical principle for the whole of exact science. I asked myself whether this was really justified. Can absolute predictions really be made for all time on the basis of the classical equations of motion? It can easily be seen, by simple examples, that this is only the case when the possibility of absolutely exact measurement (of position, velocity, or other quantities) is assumed. Let us think of a particle moving without friction on a straight line between two end-points (walls), at which it experiences completely elastic recoil. It moves with constant speed equal to its initial speed $v_0$ backwards and forwards, and it can be stated exactly where it will be at a given time provided that $v_0$ is accurately known. But if a small inaccuracy $\Delta v_0$ is allowed, then the inaccuracy of prediction of the position at time $t$ is $t\Delta v_0$ which increases with $t$. If one waits long enough until time $t_c = l/\Delta v_0$ where $l$ is the distance between the elastic walls, the inaccuracy $\Delta x$ will have become equal to the whole space $l$. Thus it is impossible to forecast anything about the position at a time which is later than $t_c$.

Thus determinism lapses completely into indeterminism as soon as the slightest inaccuracy in the data on velocity is permitted. Is there any sense - and I mean any physical sense, not metaphysical sense - in which one can speak of absolute data? Is one justified in saying that the coordinate $x = \pi$ cm where $\pi = 3.1415$ is the familiar transcendental number that determines the ratio of the circumference of a circle to its diameter? As a mathematical tool the concept of a real number represented by a nonterminating decimal fraction is exceptionally important and fruitful. As the measure of a physical quantity it is nonsense. If $\pi$ is taken to the 20th or the 25th place of decimals, two numbers are obtained which are indistinguishable from each other and the true value of $\pi$ by any measurement. According to the heuristic principle used by Einstein in the theory of relativity, and by Heisenberg in the quantum theory, concepts which correspond to no conceivable observation should be eliminated from physics. This is possible without difficulty in the present case also. It is only necessary to replace statements like $x = \pi$ cm by: the probability of distribution of values of $x$ has a sharp maximum at $x = \pi$ cm; and (if it is desired to be more accurate) to add: of such and such a breadth. In short, ordinary mechanics must also be statistically formulated. I have occupied myself with this problem a little recently, and have realized that it is possible without difficulty. This is not the place to go into the matter more deeply. I should like only to say this: the determinism of classical physics turns out to be an illusion, created by overrating mathematico-logical concepts. It is an idol, not an ideal in scientific research and cannot, therefore, be used as an objection to the essentially indeterministic statistical interpretation of quantum mechanics.

Much more difficult is the objection based on reality. The concept of a particle, e.g. a grain of sand, implicitly contains the idea that it is in a definite position and has definite motion. But according to quantum mechanics it is impossible to determine simultaneously with any desired accuracy both position and velocity (more precisely: momentum, i.e. mass times velocity). Thus two questions arise: what prevents us, in spite of the theoretical assertion, to measure both quantities to any desired degree of accuracy by refined experiments? Secondly, if it really transpires that this is not feasible, are we still justified in applying to the electron the concept of particle and therefore the ideas associated with it?

Referring to the first question, it is clear that if the theory is correct - and we have ample grounds for believing this - the obstacle to simultaneous measurement of position and motion (and of other such pairs of so-called conjugate quantities) must lie in the laws of quantum mechanics themselves. In fact, this is so. But it is not a simple matter to clarify the situation. Niels Bohr himself has gone to great trouble and ingenuity* to develop a theory of measurements to clear the matter up and to meet the most refined and ingenious attacks of Einstein, who repeatedly tried to think out methods of measurement by means of which position and motion could be measured simultaneously and accurately. The following emerges: to measure space coordinates and instants of time, rigid measuring rods and clocks are required. On the other hand, to measure momenta and energies, devices are necessary with movable parts to absorb the impact of the test object and to indicate the size of its momentum. Paying regard to the fact that quantum mechanics is competent for dealing with the interaction of object and apparatus, it is seen that no arrangement is possible that will fulfil both requirements simultaneously. There exist, therefore, mutually exclusive though complementary experiments which only as a whole embrace everything which can be experienced with regard to an object.

This idea of *complementarity* is now regarded by most physicists as the key to the clear understanding of quantum processes. Bohr has generalized the idea to quite different fields of knowledge, e.g. the connection between consciousness and the brain, to the problem of free will, and other basic problems of philosophy. To come now to the last point: can we call something with which the concepts of

---

*N. Bohr, *Naturwiss., 16* (1928) 245; 17 (1929) 483; 21 (1933) 13. "Kausalität und Komplementarität" (Causality and Complementarity), *Die Erkenntnis,* 6 (1936) 293.

position and motion cannot be associated in the usual way, a thing, or a particle? And if not, what is the reality which our theory has been invented to describe?

The answer to this is no longer physics, but philosophy, and to deal with it thoroughly would mean going far beyond the bounds of this lecture. I have given my views on it elsewhere*. Here I will only say that I am emphatically in favour of the retention of the particle idea. Naturally, it is necessary to redefine what is meant. For this, well-developed concepts are available which appear in mathematics under the name of invariants in transformations. Every object that we perceive appears in innumerable aspects. The concept of the object is the invariant of all these aspects. From this point of view, the present universally used system of concepts in which particles and waves appear simultaneously, can be completely justified.

The latest research on nuclei and elementary particles has led us, however, to limits beyond which this system of concepts itself does not appear to suffice. The lesson to be learned from what I have told of the origin of quantum mechanics is that probable refinements of mathematical methods will not suffice to produce a satisfactory theory, but that somewhere in our doctrine is hidden a concept, unjustified by experience, which we must eliminate to open up the road.

---

*M. Born, *Phil. Quart.*, 3 (1953) 134; *Physik. Bl.*, 10 (1954) 49.

# The Present Situation in Quantum Mechanics

By

Erwin Schrodinger

A translation of Schrodinger's "cat paradox" paper

Translator: John D. Trimmer

This translation was originally published in *Proceedings of the American Philosophical Society*, 124, 323–38. [And then appeared as Section I.11 of Part I of *Quantum Theory and Measurement* (J.A. Wheeler and W.H. Zurek, eds., Princeton university Press, New Jersey 1983).]

## 5. Are the Variables Really Blurred?

One can even set up quite ridiculous cases. A cat is penned up in a steel chamber, along with the following device (which must be secured against direct interference by the cat): in a Geiger counter there is a tiny bit of radioactive substance, *so* small, that *perhaps* in the course of the hour one of the atoms decays, but also, with equal probability, perhaps none; if it happens, the counter tube discharges and through a relay releases a hammer which shatters a small flask of hydrocyanic acid. If one has left this entire system to itself for an hour, one would say that the cat still lives *if* meanwhile no atom has decayed. The psi-function of the entire system would express this by having in it the living and dead cat (pardon the expression) mixed or smeared out in equal parts.

It is typical of these cases that an indeterminacy originally restricted to the atomic domain becomes transformed into macroscopic indeterminacy, which can then be *resolved* by direct observation. That prevents us from so naively accepting as valid a "blurred model" for representing reality.

# Can Quantum-Mechanical Description of Physical Reality Be Considered Complete?

## By

## Albert Einstein, Boris Podolsky and Nathan Rosen

In a complete theory there is an element corresponding to each element of reality. A sufficient condition for the reality of a physical quantity is the possibility of predicting it with certainty, without disturbing the system. In quantum mechanics in the case of two physical quantities described by non-commuting operators, the knowledge of one precludes the knowledge of the other. Then either (1) the description of reality given by the wave function in quantum mechanics is not complete or (2) these two quantities cannot have simultaneous reality. Consideration of the problem of making predictions concerning a system on the basis of measurements made on another system that had previously interacted with it leads to the result that if (1) is false then (2) is also false. One is thus led to conclude that the description of reality as given by a wave function is not complete.

## 1.

Any serious consideration of a physical theory must take into account the distinction between the objective reality, which is independent of any theory, and the physical concepts with which the theory operates. These concepts are intended to correspond with the objective reality, and by means of these concepts we picture this reality to ourselves.

In attempting to judge the success of a physical theory, we may ask ourselves two questions: (1) "Is the theory correct?" and (2) "Is the description given by the theory complete?" It is only in the case in which positive answers may be given to both of these questions, that the concepts of the theory may be said to be satisfactory. The correctness of the theory is judged by the degree of agreement between the conclusions of the theory and human experience. This experience, which alone enables us to make inferences about reality, in physics takes the form of experiment and measurement. It is the second question that we wish to consider here, as applied to quantum mechanics.

Whatever the meaning assigned to the term *complete*, the following requirement for a complete theory seems to be a necessary one: *every element of the physical reality must have a counterpart in the physical theory*. We shall call this the condition of completeness. The second question is thus easily answered, as soon as we are able to decide what are the elements of the physical reality.

The elements of the physical reality cannot be determined by *a priori* philosophical considerations, but must be found by an appeal to results of experiments and measurements. A comprehensive definition of reality is, however, unnecessary for our purpose. We shall be satisfied with the following criterion, which we regard as reasonable. *If, without in any way disturbing a system, we can predict with certainty (i.e., with probability equal to unity) the value of a physical quantity, then there exists an element of physical reality corresponding to this physical quantity.* It seems to us that this criterion, while far from exhausting all possible ways of recognizing a physical reality, at least provides us with one such way, whenever the conditions set down in it occur. Regarded not as a necessary, but merely as a sufficient, condition of reality, this criterion is in agreement with classical as well as quantum-mechanical ideas of reality.

To illustrate the ideas involved let us consider the quantum-mechanical description of the behavior of a particle having a single degree of freedom. The fundamental concept of the theory is the concept of *state*, which is supposed to be completely characterized by the wave function $\psi$, which is a function of the variables chosen to describe the particle's behavior. Corresponding to each physically observable quantity $A$ there is an operator, which may be designated by the same letter.

If $\psi$ is an eigenfunction of the operator $A$, that is, if

$$\psi' \equiv A\psi = a\psi, \tag{1}$$

where $a$ is a number, then the physical quantity $A$ has with certainty the value $a$ whenever the particle is in the state given by $\psi$. In accordance with our criterion of reality, for a particle in the state given by $\psi$ for which Eq. (1) holds, there is an element of physical reality corresponding to the physical quantity $A$. Let, for example,

$$\psi = e^{(2\pi i / h)p_0 x}, \tag{2}$$

where $h$ is Planck's constant, $p_0$ is some constant number, and $x$ the independent variable. Since the operator corresponding to the momentum of the particle is

$$p = (h/2\pi i)\partial/\partial x, \tag{3}$$

we obtain

$$\psi' = p\psi = (h/2\pi i)\partial\psi/\partial x = p_0\psi. \tag{4}$$

Thus, in the state given by Eq. (2), the momentum has certainly the value $p_0$. It thus has meaning to say that the momentum of the particle in the state given by Eq. (2) is real.

On the other hand if Eq. (1) does not hold, we can no longer speak of the physical quantity $A$ having a particular value. This is the case, for example, with the coordinate of the particle. The operator corresponding to it, say $q$, is the operator of multiplication by the independent variable. Thus,

$$q\psi = x\psi \neq a\psi. \tag{5}$$

In accordance with quantum mechanics we can only say that the relative probability that a measurement of the coordinate will give a result lying between $a$ and $b$ is

$$P(a, b) = \int_a^b \bar{\psi}\psi \, dx = \int_a^b dx = b - a. \tag{6}$$

Since this probability is independent of $a$, but depends only upon the difference $b - a$, we see that all values of the coordinate are equally probable.

A definite value of the coordinate, for a particle in the state given by Eq. (2), is thus not predictable, but may be obtained only by a direct measurement. Such a measurement however disturbs the particle and thus alters its state. After the coordinate is determined, the particle will no longer be in the state given by Eq. (2). The usual conclusion from this in quantum mechanics is that *when the momentum of a particle is known, its coordinate has no physical reality.*

More generally, it is shown in quantum mechanics that, if the operators corresponding to two physical quantities, say $A$ and $B$, do not commute, that is, if $AB \neq BA$, then the precise knowledge of one of them precludes such a knowledge of the other. Furthermore, any attempt to determine the latter experimentally will alter the state of the system in such a way as to destroy the knowledge of the first.

From this follows that either (1) *the quantum-mechanical description of reality given by the wave function is not complete* or (2) *when the operators corresponding to two physical quantities do not commute the two quantities cannot have simultaneous reality.* For if both of them had simultaneous reality—and thus definite values—these values would enter into the complete description, according to the condition of completeness. If then the wave function provided such a complete description of reality, it would contain these values; these would then be predictable. This not being the case, we are left with the alternatives stated.

In quantum mechanics it is usually assumed that the wave function *does* contain a complete description of the physical reality of the system in the state to which it corresponds. At first sight this assumption is entirely reasonable, for the information obtainable from a wave function seems to correspond exactly to what can be measured without altering the state of the system. We shall show, however, that this assumption, together with the criterion of reality given above, leads to a contradiction.

## 2.

For this purpose let us suppose that we have two systems, I and II, which we permit to interact from the time $t = 0$ to $t = T$, after which time we suppose that there is no longer any interaction between the two parts. We suppose further that the states of the two systems before $t = 0$ were known. We can then calculate with the help of Schrödinger's equation the state of the combined system I + II at any subsequent time; in particular, for any $t > T$. Let us designate the corresponding wave function by $\Psi$. We cannot, however, calculate the state in which either one of the two systems is left after the interaction. This, according to quantum mechanics, can be done only with the help of further measurements, by a process known as the *reduction of the wave packet*. Let us consider the essentials of this process.

Let $a_1, a_2, a_3, \ldots$ be the eigenvalues of some physical quantity $A$ pertaining to system I and $u_1(x_1), u_2(x_1), u_3(x_1), \ldots$ the corresponding eigenfunctions, where $x_1$ stands for the variables used to describe the first system. Then $\Psi$, considered as a function of $x_1$, can be expressed as

$$\Psi(x_1, \ x_2) = \sum_{n=1}^{\infty} \psi_n(x_2) u_n(x_1), \tag{7}$$

where $x_2$ stands for the variables used to describe the second system. Here $\psi_n(x_2)$ are to be regarded merely as the coefficients of the expansion of $\Psi$ into a series of orthogonal functions $u_n(x_1)$. Suppose now that the quantity $A$ is measured and it is found that it has the value $a_k$. It is then concluded that after the measurement the first system is left in the state given by the wave function $u_k(x_1)$, and that the second system is left in the state given by the wave function $\psi_k(x_2)$. This is the process of reduction of the wave packet; the wave packet given by the infinite series (7) is reduced to a single term $\psi_k(x_2) u_k(x_1)$.

The set of functions $u_n(x_1)$ is determined by the choice of the physical quantity $A$. If, instead of this, we had chosen another quantity, say $B$, having the eigenvalues $b_1, b_2, b_3, \ldots$ and eigenfunctions $v_1(x_1), v_2(x_1), v_3(x_1), \ldots$ we should have obtained, instead of Eq. (7), the expansion

$$\Psi(x_1, x_2) = \sum_{n=1}^{\infty} \varphi_s(x_2)v_s(x_1), \tag{8}$$

where $\varphi_s$'s are the new coefficients. If now the quantity $B$ is measured and is found to have the value $b_r$, we conclude that after the measurement the first system is left in the state given by $v_r(x_1)$ and the second system is left in the state given by $\varphi_r(x_2)$.

We see therefore that, as a consequence of two different measurements performed upon the first system, the second system may be left in states with two different wave functions. On the other hand, since at the time of measurement the two systems no longer interact, no real change can take place in the second system in consequence of anything that may be done to the first system. This is, of course, merely a statement of what is meant by the absence of an interaction between the two systems. Thus, *it is possible to assign two different wave functions* (in our example $\psi_k$ and $\varphi_r$) *to the same reality* (the second system after the interaction with the first).

Now, it may happen that the two wave functions, $\psi_k$ and $\varphi_r$, are eigenfunctions of two non-commuting operators corresponding to some physical quantities $P$ and $Q$, respectively. That this may actually be the case can best be shown by an example. Let us suppose that the two systems are two particles, and that

$$\Psi(x_1, x_2) = \int_{-\infty}^{\infty} e^{(2\pi i/h)(x_1-x_2+x_0)p}\,dp, \tag{9}$$

where $x_0$ is some constant. Let $A$ be the momentum of the first particle; then, as we have seen in Eq. (4), its eigenfunctions will be

$$u_p(x_1) = e^{(2\pi i/h)px_1} \tag{10}$$

corresponding to the eigenvalue $p$. Since we have here the case of a continuous spectrum, Eq. (7) will now be written

$$\Psi(x_1, x_2) = \int_{-\infty}^{\infty} \psi_p(x_2)u_p(x_1)dp, \tag{11}$$

where

$$\psi_p(x_2) = e^{-(2\pi i/h)(x_2-x_0)p}.$$ (12)

This $\psi_p$ however is the eigenfunction of the operator

$$P = (h/2\pi i)\partial/\partial x_2,$$ (13)

corresponding to the eigenvalue $-p$ of the momentum of the second particle. On the other hand, if $B$ is the coordinate of the first particle, it has for eigenfunctions

$$v_x(x_1) = \delta(x_1 - x),$$ (14)

corresponding to the eigenvalue $x$, where $\delta(x_1 - x)$ is the well-known Dirac delta-function. Eq. (8) in this case becomes

$$\Psi(x_1, x_2) = \int_{-\infty}^{\infty} \varphi_x(x_2)v_x(x_1)dx,$$ (15)

where

$$\varphi_z(x_2) = \int_{-\infty}^{\infty} e^{(2\pi i/h)(x-x_2+x_0)p} dp$$
$$= h\delta(x - x_2 + x_0).$$ (16)

This $\varphi_x$, however, is the eigenfunction of the operator

$$Q = x_2$$ (17)

corresponding to the eigenvalue $x + x_0$ of the coordinate of the second particle. Since

$$PQ - QP = h/2\pi i,$$ (18)

we have shown that it is in general possible for $\psi_k$ and $\varphi_r$ to be eigenfunctions of two noncommuting operators, corresponding to physical quantities.

Returning now to the general case contemplated in Eqs. (7) and (8), we assume that $\psi_k$ and $\varphi_r$ are indeed eigenfunctions of some non-commuting operators $P$ and $Q$, corresponding to the eigenvalues $p_k$ and $q_r$, respectively. Thus, by measuring either $A$ or $B$ we are in a position to predict with certainty, and without in any way disturbing the second system, either the value of the quantity $P$ (that is $p_k$) or the value of the quantity $Q$ (that is $q_r$). In accordance with our criterion of reality, in the first case we must consider the quantity $P$ as being an element of reality, in the second case the quantity $Q$ is an element of reality. But, as we have seen, both wave functions $\psi_k$ and $\varphi_r$ belong to the same reality.

Previously we proved that either (1) the quantum-mechanical description of reality given by the wave function is not complete or (2) when the operators corresponding to two physical quantities do not commute the two quantities cannot have simultaneous reality. Starting then with the assumption that the wave function does give a complete description of the physical reality, we arrived at the conclusion that two physical quantities, with noncommuting operators, can have simultaneous reality. Thus the negation of (1) leads to the negation of the only other alternative (2). We are thus forced to conclude that the quantum-mechanical description of physical reality given by wave functions is not complete.

One could object to this conclusion on the grounds that our criterion of reality is not sufficiently restrictive. Indeed, one would not arrive at our conclusion if one insisted that two or more physical quantities can be regarded as simultaneous elements of reality *only when they can be simultaneously measured or predicted*. On this point of view, since either one or the other, but not both simultaneously, of the quantities $P$ and $Q$ can be predicted, they are not simultaneously real. This makes the reality of $P$ and $Q$ depend upon the process of measurement carried out on the first system, which does not disturb the second system in any way. No reasonable definition of reality could be expected to permit this.

While we have thus shown that the wave function does not provide a complete description of the physical reality, we left open the question of whether or not such a description exists. We believe, however, that such a theory is possible.

# Can Quantum-Mechanical Description of Physical Reality Be Considered Complete?

## By

## Niels Bohr

It is shown that a certain "criterion of physical reality" formulated in a recent article with the above title by A. Einstein, B. Podolsky and N. Rosen contains an essential ambiguity when it is applied to quantum phenomena. In this connection a viewpoint termed "complementarity" is explained from which quantum-mechanical description of physical phenomena would seem to fulfill, within its scope, all rational demands of completeness.

In a recent article* under the above title A. Einstein, B. Podolsky and N. Rosen have presented arguments which lead them to answer the question at issue in the negative. The trend of their argumentation, however, does not seem to me adequately to meet the actual situation with which we are faced in atomic physics. I shall therefore be glad to use this opportunity to explain in somewhat greater detail a general viewpoint, conveniently termed "complementarity," which I have indicated on various previous occasions,† and from which quantum mechanics within its scope would appear as a completely rational description of physical phenomena, such as we meet in atomic processes.

The extent to which an unambiguous meaning can be attributed to such an expression as "physical reality" cannot of course be deduced from *a priori* philosophical conceptions, but—as the authors of the article cited themselves emphasize—must be founded on a direct

---

*A. Einstein, B. Podolsky and N. Rosen, Phys. Rev. 47, 777 (1935).
†Cf. N. Bohr, *Atomic Theory and Description of Nature*, I (Cambridge, 1934).

Reprinted with permission from the American Physical Society: N. Bohr, *Physical Review*, Volume 48, 1935.

appeal to experiments and measurements. For this purpose they propose a "criterion of reality" formulated as follows: "If, without in any way disturbing a system, we can predict with certainty the value of a physical quantity, then there exists an element of physical reality corresponding to this physical quantity." By means of an interesting example, to which we shall return below, they next proceed to show that in quantum mechanics, just as in classical mechanics, it is possible under suitable conditions to predict the value of any given variable pertaining to the description of a mechanical system from measurements performed entirely on other systems which previously have been in interaction with the system under investigation. According to their criterion the authors therefore want to ascribe an element of reality to each of the quantities represented by such variables. Since, moreover, it is a well-known feature of the present formalism of quantum mechanics that it is never possible, in the description of the state of a mechanical system, to attach definite values to both of two canonically conjugate variables, they consequently deem this formalism to be incomplete, and express the belief that a more satisfactory theory can be developed.

Such an argumentation, however, would hardly seem suited to affect the soundness of quantum-mechanical description, which is based on a coherent mathematical formalism covering automatically any procedure of measurement like that indicated.* The apparent

---

*The deductions contained in the article cited may in this respect be considered as an immediate consequence of the transformation theorems of quantum mechanics, which perhaps more than any other feature of the formalism contribute to secure its mathematical completeness and its rational correspondence with classical mechanics. In fact, it is always possible in the description of a mechanical system, consisting of two partial systems (1) and (2), interacting or not, to replace any two pairs of canonically conjugate variables $(q_1 p_1)$, $(q_2 p_2)$ pertaining to systems (1) and (2), respectively, and satisfying the usual commutation rules

$$[q_1 p_1] = [q_2 p_2] = ih/2\pi,$$
$$[q_1 q_2] = [p_1 p_2] = [q_1 p_2] = [q_2 p_1] = 0,$$

by two pairs of new conjugate variables $(Q_1 P_1)$, $(Q_2 P_2)$ related to the first variables by a simple orthogonal transformation, corresponding to a rotation of angle $\theta$ in the planes $(q_1 q_2)$, $(p_1 p_2)$

$$q_1 = Q_1 \cos \theta - Q_2 \sin \theta \quad p_1 = P_1 \cos \theta - P_2 \sin \theta$$
$$q_2 = Q_1 \sin \theta + Q_2 \cos \theta \quad p_2 = P_1 \sin \theta + P_2 \cos \theta.$$

Since these variables will satisfy analogous commutation rules, in particular

$$[Q_1 P_1] = ih/2\pi, \qquad [Q_1 P_2] = 0,$$

it follows that in the description of the state of the combined system definite numerical values may not be assigned to both $Q_1$ and $P_1$, but that we may clearly assign such values to both $Q_1$ and $P_2$. In that case it further results from the expressions of these variables in terms of $(q_1 p_1)$ and $(q_2 p_2)$, namely

$$Q_1 = q_1 \cos \theta + q_2 \sin \theta, \quad P_2 = -p_1 \sin \theta + p_2 \cos \theta,$$

that a subsequent measurement of either $q_2$ or $p_2$ will allow us to predict the value of $q_1$ or $p_1$ respectively.

contradiction in fact discloses only an essential inadequacy of the customary viewpoint of natural philosophy for a rational account of physical phenomena of the type with which we are concerned in quantum mechanics. Indeed the *finite interaction between object and measuring agencies* conditioned by the very existence of the quantum of action entails—because of the impossibility of controlling the reaction of the object on the measuring instruments if these are to serve their purpose—the necessity of a final renunciation of the classical ideal of causality and a radical revision of our attitude towards the problem of physical reality. In fact, as we shall see, a criterion of reality like that proposed by the named authors contains—however cautious its formulation may appear—an essential ambiguity when it is applied to the actual problems with which we are here concerned. In order to make the argument to this end as clear as possible, I shall first consider in some detail a few simple examples of measuring arrangements.

Let us begin with the simple case of a particle passing through a slit in a diaphragm, which may form part of some more or less complicated experimental arrangement. Even if the momentum of this particle is completely known before it impinges on the diaphragm, the diffraction by the slit of the plane wave giving the symbolic representation of its state will imply an uncertainty in the momentum of the particle, after it has passed the diaphragm, which is the greater the narrower the slit. Now the width of the slit, at any rate if it is still large compared with the wave-length, may be taken as the uncertainty $\Delta q$ of the position of the particle relative to the diaphragm, in a direction perpendicular to the slit. Moreover, it is simply seen from de Broglie's relation between momentum and wavelength that the uncertainty $\Delta p$ of the momentum of the particle in this direction is correlated to $\Delta q$ by means of Heisenberg's general principle

$$\Delta p \, \Delta q \sim h,$$

which in the quantum-mechanical formalism is a direct consequence of the commutation relation for any pair of conjugate variables.

Obviously the uncertainty $\Delta p$ is inseparably connected with the possibility of an exchange of momentum between the particle and the diaphragm; and the question of principal interest for our discussion is now to what extent the momentum thus exchanged can be taken into account in the description of the phenomenon to be studied by the experimental arrangement concerned, of which the passing of the particle through the slit may be considered as the initial stage.

Let us first assume that, corresponding to usual experiments on the remarkable phenomena of electron diffraction, the diaphragm, like the other parts of the apparatus,—say a second diaphragm with several slits parallel to the first and a photographic plate,—is rigidly fixed to a support which defines the space frame of reference. Then the momentum exchanged between the particle and the diaphragm will, together with the reaction of the particle on the other bodies, pass into this common support, and we have thus voluntarily cut ourselves off from any possibility of taking these reactions separately into account in predictions regarding the final result of the experiment,—say the position of the spot produced by the particle on the photographic plate. The impossibility of a closer analysis of the reactions between the particle and the measuring instrument is indeed no peculiarity of the experimental procedure described, but is rather an essential property of any arrangement suited to the study of the phenomena of the type concerned, where we have to do with a feature of *individuality* completely foreign to classical physics. In fact, any possibility of taking into account the momentum exchanged between the particle and the separate parts of the apparatus would at once permit us to draw conclusions regarding the "course" of such phenomena,—say through what particular slit of the second diaphragm the particle passes on its way to the photographic plate—which would be quite incompatible with the fact that the probability of the particle reaching a given element of area on this plate is determined not by the presence of any particular slit, but by the positions of all the slits of the second diaphragm within reach of the associated wave diffracted from the slit of the first diaphragm.

By another experimental arrangement, where the first diaphragm is not rigidly connected with the other parts of the apparatus, it would at least in principle* be possible to measure its momentum with any desired accuracy before and after the passage of the particle, and thus to predict the momentum of the latter after it has passed through the slit. In fact, such measurements of momentum require only an unambiguous application of the classical law of conservation of momentum, applied for instance to a collision process between the diaphragm and some test body, the momentum of which is suitably controlled before and after the collision. It is true that such a control will essentially depend on an examination of the space-time course of some process to which the ideas of classical mechanics can be applied; if, however, all spatial dimensions and time intervals are taken sufficiently large, this involves clearly no limitation as regards the accurate control of the momentum of the test bodies, but only a renunciation as regards the accuracy of the control of their space-time coordination. This last circumstance is in fact quite analogous to the renunciation of the control of the momentum of the fixed diaphragm in the experimental arrangement discussed above, and depends in the last resort on the claim of a purely classical account of the measuring apparatus, which implies the necessity of allowing a latitude corresponding to the quantum-mechanical uncertainty relations in our description of their behavior.

The principal difference between the two experimental arrangements under consideration is, however, that in the arrangement suited for the control of the momentum of the first diaphragm, this body can no longer be used as a measuring instrument for the same purpose as in the previous case, but must, as regards its position relative to the rest of the apparatus, be treated, like the particle traversing the slit, as an object of investigation, in the sense that the quantum-mechanical uncertainty relations regarding its position and momentum must be

---

*The obvious impossibility of actually carrying out, with the experimental technique at our disposal, such measuring procedures as are discussed here and in the following does clearly not affect the theoretical argument, since the procedures in question are essentially equivalent with atomic processes, like the Compton effect, where a corresponding application of the conservation theorem of momentum is well established.

taken explicitly into account. In fact, even if we knew the position of the diaphragm relative to the space frame before the first measurement of its momentum, and even though its position after the last measurement can be accurately fixed, we lose, on account of the uncontrollable displacement of the diaphragm during each collision process with the test bodies, the knowledge of its position when the particle passed through the slit. The whole arrangement is therefore obviously unsuited to study the same kind of phenomena as in the previous case. In particular it may be shown that, if the momentum of the diaphragm is measured with an accuracy sufficient for allowing definite conclusions regarding the passage of the particle through some selected slit of the second diaphragm, then even the minimum uncertainty of the position of the first diaphragm compatible with such a knowledge will imply the total wiping out of any interference effect—regarding the zones of permitted impact of the particle on the photographic plate—to which the presence of more than one slit in the second diaphragm would give rise in case the positions of all apparatus are fixed relative to each other.

In an arrangement suited for measurements of the momentum of the first diaphragm, it is further clear that even if we have measured this momentum before the passage of the particle through the slit, we are after this passage still left with a *free choice* whether we wish to know the momentum of the particle or its initial position relative to the rest of the apparatus. In the first eventuality we need only to make a second determination of the momentum of the diaphragm, leaving unknown forever its exact position when the particle passed. In the second eventuality we need only to determine its position relative to the space frame with the inevitable loss of the knowledge of the momentum exchanged between the diaphragm and the particle. If the diaphragm is sufficiently massive in comparison with the particle, we may even arrange the procedure of measurements in such a way that the diaphragm after the first determination of its momentum will remain at rest in some unknown position relative to the other parts of the apparatus, and the subsequent fixation of this position may therefore simply consist in establishing a rigid connection between the diaphragm and the common support.

My main purpose in repeating these simple, and in substance well-known considerations, is to emphasize that in the phenomena concerned we are not dealing with an incomplete description characterized by the arbitrary picking out of different elements of physical reality at the cost of sacrificing other such elements, but with a rational discrimination between essentially different experimental arrangements and procedures which are suited either for an unambiguous use of the idea of space location, or for a legitimate application of the conservation theorem of momentum. Any remaining appearance of arbitrariness concerns merely our freedom of handling the measuring instruments, characteristic of the very idea of experiment. In fact, the renunciation in each experimental arrangement of the one or the other of two aspects of the description of physical phenomena,—the combination of which characterizes the method of classical physics, and which therefore in this sense may be considered as *complementary* to one another,—depends essentially on the impossibility, in the field of quantum theory, of accurately controlling the reaction of the object on the measuring instruments, i.e., the transfer of momentum in case of position measurements, and the displacement in case of momentum measurements. Just in this last respect any comparison between quantum mechanics and ordinary statistical mechanics,—however useful it may be for the formal presentation of the theory,—is essentially irrelevant. Indeed we have in each experimental arrangement suited for the study of proper quantum phenomena not merely to do with an ignorance of the value of certain physical quantities, but with the impossibility of defining these quantities in an unambiguous way.

The last remarks apply equally well to the special problem treated by Einstein, Podolsky and Rosen, which has been referred to above, and which does not actually involve any greater intricacies than the simple examples discussed above. The particular quantum-mechanical state of two free particles, for which they give an explicit mathematical expression, may be reproduced, at least in principle, by a simple experimental arrangement, comprising a rigid diaphragm with two parallel slits, which are very narrow compared with their separation, and through each of which one particle with given initial momentum passes independently of the other. If the momentum of this diaphragm is measured accurately before as well as after the passing of the particles, we shall in fact know the sum of the components perpendicular to the slits of the momenta of the two escaping particles, as well as the difference of their initial positional coordinates in the same direction; while of course the conjugate quantities, i.e., the difference of the components of their momenta, and the sum of their positional coordinates, are entirely unknown.* In this arrangement, it is therefore clear that a subsequent single measurement either of the position or of the momentum of one of the particles will automatically determine the position or momentum, respectively, of the other particle with any desired accuracy; at least if the wave-length corresponding to the free motion of each particle is sufficiently short compared with the width of the slits. As pointed out by the named authors, we are therefore faced at this stage with a completely free choice whether we want to determine the one or the other of the latter quantities by a process which does not directly interfere with the particle concerned.

Like the above simple case of the choice between the experimental procedures suited for the prediction of the position or the momentum of a single particle which has passed through a slit in a diaphragm, we are, in the "freedom of choice" offered by the last arrangement, just concerned with a *discrimination between different experimental procedures which allow of the unambiguous use of complementary classical concepts.* In fact to measure the position of one of the particles can mean nothing else than to establish a correlation between its behavior and some instrument rigidly fixed to the support which defines the space frame of reference. Under the experimental conditions described such a measurement will therefore also provide us with the knowledge

---

*As will be seen, this description, a part from a trivial normalizing factor, corresponds exactly to the transformation of variables described in the preceding footnote if $(q_1 p_1)$, $(q_2 p_2)$ represent the positional coordinates and components of momenta of the two particles and if $\theta = -\pi/4$. It may also be remarked that the wave function given by formula (9) of the article cited corresponds to the special choice of $P_2 = 0$ and the limiting case of two infinitely narrow slits.

of the location, otherwise completely unknown, of the diaphragm with respect to this space frame when the particles passed through the slits. Indeed, only in this way we obtain a basis for conclusions about the initial position of the other particle relative to the rest of the apparatus. By allowing an essentially uncontrollable momentum to pass from the first particle into the mentioned support, however, we have by this procedure cut ourselves off from any future possibility of applying the law of conservation of momentum to the system consisting of the diaphragm and the two particles and therefore have lost our only basis for an unambiguous application of the idea of momentum in predictions regarding the behavior of the second particle. Conversely, if we choose to measure the momentum of one of the particles, we lose through the uncontrollable displacement inevitable in such a measurement any possibility of deducing from the behavior of this particle the position of the diaphragm relative to the rest of the apparatus, and have thus no basis whatever for predictions regarding the location of the other particle.

From our point of view we now see that the wording of the above-mentioned criterion of physical reality proposed by Einstein, Podolsky and Rosen contains an ambiguity as regards the meaning of the expression "without in any way disturbing a system." Of course there is in a case like that just considered no question of a mechanical disturbance of the system under investigation during the last critical stage of the measuring procedure. But even at this stage there is essentially the question of *an influence on the very conditions which define the possible types of predictions regarding the future behavior of the system.* Since these conditions constitute an inherent element of the description of any phenomenon to which the term "physical reality" can be properly attached, we see that the argumentation of the mentioned authors does not justify their conclusion that quantum-mechanical description is essentially incomplete. On the contrary this description, as appears from the preceding discussion, may be characterized as a rational utilization of all possibilities of unambiguous interpretation of measurements, compatible with the finite and uncontrollable interaction between the objects and the measuring instruments in the field of quantum theory. In fact, it is only the mutual exclusion of any two experimental procedures, permitting the unambiguous definition of complementary physical quantities, which provides room for new physical laws, the coexistence of which might at first sight appear irreconcilable with the basic principles of science. It is just this entirely new situation as regards the description of physical phenomena, that the notion of *complementarity* aims at characterizing.

The experimental arrangements hitherto discussed present a special simplicity on account of the secondary role which the idea of time plays in the description of the phenomena in question. It is true that we have freely made use of such words as "before" and "after" implying time-relationships; but in each case allowance must be made for a certain inaccuracy, which is of no importance, however, so long as the time intervals concerned are sufficiently large compared with the proper periods entering in the closer analysis of the phenomenon under investigation. As soon as we attempt a more accurate time description of quantum phenomena, we meet with well-known new paradoxes, for the elucidation of which further features of the interaction between the objects and the measuring instruments must be taken into account. In fact, in such phenomena we have no longer to do with experimental arrangements consisting of apparatus essentially at rest relative to one another, but with arrangements containing moving parts,—like shutters before the slits of the diaphragms,—controlled by mechanisms serving as clocks. Besides the transfer of momentum, discussed above, between the object and the bodies defining the space frame, we shall therefore, in such arrangements, have to consider an eventual exchange of energy between the object and these clock-like mechanisms.

The decisive point as regards time measurements in quantum theory is now completely analogous to the argument concerning measurements of positions outlined above. Just as the transfer of momentum to the separate parts of the apparatus,—the knowledge of the relative positions of which is required for the description of the phenomenon,—has been seen to be entirely uncontrollable, so the exchange of energy between the object and the various bodies, whose relative motion must be known for the intended use of the apparatus, will defy any closer analysis. Indeed, it is *excluded in principle to control the energy which goes into the clocks without interfering essentially with their use as time indicators.* This use in fact entirely relies on the assumed possibility of accounting for the functioning of each clock as well as for its eventual comparison with other clocks on the basis of the methods of classical physics. In this account we must therefore obviously allow for a latitude in the energy balance, corresponding to the quantum-mechanical uncertainty relation for the conjugate time and energy variables. Just as in the question discussed above of the mutually exclusive character of any unambiguous use in quantum theory of the concepts of position and momentum, it is in the last resort this circumstance which entails the complementary relationship between any detailed time account of atomic phenomena on the one hand and the unclassical features of intrinsic stability of atoms, disclosed by the study of energy transfers in atomic reactions on the other hand.

This necessity of discriminating in each experimental arrangement between those parts of the physical system considered which are to be treated as measuring instruments and those which constitute the objects under investigation may indeed be said to form a *principal distinction between classical and quantum-mechanical description of physical phenomena.* It is true that the place within each measuring procedure where this discrimination is made is in both cases largely a matter of convenience. While, however, in classical physics the distinction between object and measuring agencies does not entail any difference in the character of the description of the phenomena concerned, its fundamental importance in quantum theory, as we have seen, has its root in the indispensable use of classical concepts in the interpretation of all proper measurements, even though the classical theories do not suffice in accounting for the new types of regularities with which we are concerned in atomic physics. In accordance with this situation there can be no question of any unambiguous interpretation of the symbols of quantum mechanics other than that embodied in the well-known rules which allow to predict the results to be obtained by a given experimental arrangement described in a totally classical way, and which have found their general expression through the transformation theorems, already referred to. By securing its proper correspondence with the classical theory, these theorems exclude in particular any imaginable inconsistency in the quantum-mechanical description, connected with a change of the place where the discrimination is made between object and measuring agencies. In fact it is an obvious consequence of the above argumentation that in each experimental arrangement and measuring procedure we have only a free choice of this place within a region where the quantum-mechanical description of the process concerned is effectively equivalent with the classical description.

Before concluding I should still like to emphasize the bearing of the great lesson derived from general relativity theory upon the question of physical reality in the field of quantum theory. In fact, notwithstanding all characteristic differences, the situations we are concerned with in these generalizations of classical theory present striking analogies which have often been noted. Especially, the singular position of measuring instruments in the account of quantum phenomena, just discussed, appears closely analogous to the well-known necessity in relativity theory of upholding an ordinary description of all measuring processes, including a sharp distinction between space and time coordinates, although the very essence of this theory is the establishment of new physical laws, in the comprehension of which we must renounce the customary separation of space and time ideas.* The dependence on the reference system, in relativity theory, of all readings of scales and clocks may even be compared with the essentially uncontrollable exchange of momentum or energy between the objects of

---

*Just this circumstance, together with the relativistic invariance of the uncertainty relations of quantum mechanics, ensures the compatibility between the argumentation outlined in the present article and all exigencies of relativity theory. This question will be treated in greater detail in a paper under preparation, where the writer will in particular discuss a very interesting paradox suggested by Einstein concerning the application of gravitation theory to energy measurements, and the solution of which offers an especially instructive illustration of the generality of the argument of complementarity. On the same occasion a more thorough discussion of space-time measurements in quantum theory will be given with all necessary mathematical developments and diagrams of experimental arrangements, which had to be left out of this article, where the main stress is laid on the dialectic aspect of the question at issue.

measurements and all instruments defining the space-time system of reference, which in quantum theory confronts us with the situation characterized by the notion of complementarity. In fact this new feature of natural philosophy means a radical revision of our attitude as regards physical reality, which may be paralleled with the fundamental modification of all ideas regarding the absolute character of physical phenomena, brought about by the general theory of relativity.

# A Suggested Interpretation of the Quantum Theory in Terms of "Hidden" Variables. I

## By

## David Bohm

The usual interpretation of the quantum theory is self-consistent, but it involves an assumption that cannot be tested experimentally, *viz.*, that the most complete possible specification of an individual system is in terms of a wave function that determines only probable results of actual measurement processes. The only way of investigating the truth of this assumption is by trying to find some other interpretation of the quantum theory in terms of at present "hidden" variables, which in principle determine the precise behavior of an individual system, but which are in practice averaged over in measurements of the types that can now be carried out. In this paper and in a subsequent paper, an interpretation of the quantum theory in terms of just such "hidden" variables is suggested. It is shown that as long as the mathematical theory retains its present general form, this suggested interpretation leàds to precisely the same results for all physical processes as does the usual interpretation. Nevertheless, the suggested interpretation provides a broader conceptual framework than the usual interpretation, because it makes possible a precise and continuous description of all processes, even at the quantum level. This broader conceptual framework allows more general mathematical formulations of the theory than those allowed by the usual interpretation. Now, the usual mathematical formulation seems to lead to insoluble difficulties when it is extrapolated into the domain of distances of the order of $10^{-13}$ cm or less. It is therefore entirely possible that the interpretation suggested here may be needed for the resolution of these difficulties. In any case, the mere

*Now at Universidade de São Paulo, Faculdade de Filosofia, Ciencias, e Letras, São Paulo, Brasil.

Reprinted with permission from the American Physical Society: D. Bohm, *Physical Review*, Volume 85, Number 2, 1952. © 1952 by the American Physical Society.

possibility of such an interpretation proves that it is not necessary for us to give up a precise, rational, and objective description of individual systems at a quantum level of accuracy.

# 1. INTRODUCTION

The usual interpretation of the quantum theory is based on an assumption having very far-reaching implications, *viz.*, that the physical state of an individual system is completely specified by a wave function that determines only the probabilities of actual results that can be obtained in a statistical ensemble of similar experiments. This assumption has been the object of severe criticisms, notably on the part of Einstein, who has always believed that, even at the quantum level, there must exist precisely definable elements or dynamical variables determining (as in classical physics) the actual behavior of each individual system, and not merely its probable behavior. Since these elements or variables are not now included in the quantum theory and have not yet been detected experimentally, Einstein has always regarded the present form of the quantum theory as incomplete, although he admits its internal consistency.[1-5]

Most physicists have felt that objections such as those raised by Einstein are not relevant, first, because the present form of the quantum theory with its usual probability interpretation is in excellent agreement with an extremely wide range of experiments, at least in the domain of distances[6] larger than $10^{-13}$ cm, and, secondly, because no consistent alternative interpretations have as yet been suggested. The purpose of this paper (and of a subsequent paper hereafter denoted by II) is, however, to suggest just such an alternative interpretation.

[1] Einstein, Podolsky, and Rosen, Phys. Rev. **47**, 777 (1933).

[2] D. Bohm, *Quantum Theory* (Prentice-Hall, Inc., New York, 1951), see p. 611.

[3] N. Bohr, Phys. Rev. **48**, 696 (1935).

[4] W. Furry, Phys. Rev. **49**, 393, 476 (1936).

[5] Paul Arthur Schilp, editor, *Albert Einstein, Philosopher-Scientist* (Library of Living Philosophers, Evanston, Ilinois, 1949). This book contains a thorough summary of the entire controversy.

[6] At distances of the order of $10^{-13}$ cm or smaller and for times of the order of this distance divided by the velocity of light or smaller, present theories becomes so inadequate that it is generally believed that they are probably not applicable, except perhaps in a very crude sense. Thus, it is generally expected that in connection with phenomena associated with this so-called "fundamental length," a totally new theory will probably be needed. It is hoped that this theory could not only deal precisely with such processes as meson production and scattering of elementary particles, but that it would also systematically predict the masses, charges, spins, etc., of the large number of so-called "elementary" particles that have already been found, as well as those of new particles which might be found in the future.

In contrast to the usual interpretation, this alternative interpretation permits us to conceive of each individual system as being in a precisely definable state, whose changes with time are determined by definite laws, analogous to (but not identical with) the classical equations of motion. Quantum-mechanical probabilities are regarded (like their counterparts in classical statistical mechanics) as only a practical necessity and not as a manifestation of an inherent lack of complete determination in the properties of matter at the quantum level. As long as the present general form of Schroedinger's equation is retained, the physical results obtained with our suggested alternative interpretation are precisely the same as those obtained with the usual interpretation. We shall see, however, that our alternative interpretation permits modifications of the mathematical formulation which could not even be described in terms of the usual interpretation. Moreover, the modifications can quite easily be formulated in such a way that their effects are insignificant in the atomic domain, where the present quantum theory is in such good agreement with experiment, but of crucial importance in the domain of dimensions of the order of $10^{-13}$ cm, where, as we have seen, the present theory is totally inadequate. It is thus entirely possible that some of the modifications describable in terms of our suggested alternative interpretation, but not in terms of the usual interpretation, may be needed for a more thorough understanding of phenomena associated with very small distances. We shall not, however, actually develop such modifications in any detail in these papers.

After this article was completed, the author's attention was called to similar proposals for an alternative interpretation of the quantum theory made by de Broglie[7] in 1926, but later given up by him partly as a result of certain criticisms made by Pauli[8] and partly because of additional objections raised by de Broglie[7] himself.[†] As we shall

---

[7] L. de Broglie, *An Introduction to the Study of Wave Mechanics* (E. P. Dutton and Company, Inc., New York, 1930), see Chapters 6, 9, and 10. See also Compt. rend. **183**, 447 (1926); **184**, 273 (1927); **185**, 380 (1927).

[8] *Reports on the Solvay Congress* (Gauthiers-Villars et Cie., Paris, 1928), see p. 280.

[†] *Note added in proof.*—Madelung has also proposed a similar interpretation of the quantum theory, but like de Broglie he did not carry this interpretation to a logical conclusion. See E. Madelung, Z. 1. Physik **40**, 332 (1926), also G. Temple, *Introduction to Quantum Theory* (London, 1931).

show in Appendix B of Paper II, however, all of the objections of de Broglie and Pauli could have been met if only de Broglie had carried his ideas to their logical conclusion. The essential new step in doing this is to apply our interpretation in the theory of the measurement process itself as well as in the description of the observed system. Such a development of the theory of measurements is given in Paper II,[9] where it will be shown in detail that our interpretation leads to precisely the same results for all experiments as are obtained with the usual interpretation. The foundation for doing this is laid in Paper I, where we develop the basis of our interpretation, contrast it with the usual interpretation, and apply it to a few simple examples, in order to illustrate the principles involved.

## 2. THE USUAL PHYSICAL INTERPRETATION OF THE QUANTUM THEORY

The usual physical interpretation of the quantum theory centers around the uncertainty principle. Now, the uncertainty principle can be derived in two different ways. First, we may start with the assumption already criticized by Einstein,[1] namely, that a wave function that determines only probabilities of actual experimental results nevertheless provides the most complete possible specification of the so-called "quantum state" of an individual system. With the aid of this assumption and with the aid of the de Broglie relation, $\mathbf{p} = h\mathbf{k}$, where $\mathbf{k}$ is the wave number associated with a particular fourier component of the wave function, the uncertainty principle is readily deduced.[10] From this derivation, we are led to interpret the uncertainty principle as an inherent and irreducible limitation on the precision with which it is correct for us even to conceive of momentum and position as simultaneously defined quantities. For if, as is done in the usual

---

[9]In Paper II, Sec. 9, we also discuss von Neumann's proof [see J. von Neumann, *Mathematische Grundlagen der Quantenmechanik* (Verlag, Julius Springer, Berlin, 1932)] that quantum theory cannot be understood In terms of a statistical distribution of "hidden" causal parameters. We shall show this his conclusions do not apply to our interpretation, because he implicitly assumes that the hidden parameters must be associated only with the observed system, whereas, as will become evident in these papers, our interpretation requires that the hidden parameters shall also be associated with the measuring apparatus.

[10]See reference 2, Chapter 5.

interpretation of the quantum theory, the wave intensity is assumed to determine only the probability of a given position, and if the **k**th Fourier component of the wave function is assumed to determine only the probability of a corresponding momentum, $\mathbf{p} = h\mathbf{k}$, then it becomes a contradiction in terms to ask for a state in which momentum and position are simultaneously and precisely defined.

A second possible derivation of the uncertainty principle is based on a theoretical analysis of the processes with the aid of which physically significant quantities such as momentum and position can be measured. In such an analysis, one finds that because the measuring apparatus interacts with the observed system by means of indivisible quanta, there will always be an irreducible disturbance of some observed property of the system. If the precise effects of this disturbance could be predicted or controlled, then one could correct for these effects, and thus one could still in principle obtain simultaneous measurements of momentum and position, having unlimited precision. But if one could do this, then the uncertainty principle would be violated. The uncertainty principle is, as we have seen, however, a necessary consequence of the assumption that the wave function and its probability interpretation provide the most complete possible specification of the state of an individual system. In order to avoid the possibility of a contradiction with this assumption, Bohr[3,5,10,11] and others have suggested an additional assumption, namely, that the process of transfer of a single quantum from observed system to measuring apparatus is inherently unpredictable, uncontrollable, and not subject to a detailed rational analysis or description. With the aid of this assumption, one can show[10] that the same uncertainty principle that is deduced from the wave function and its probability interpretation is also obtained as an inherent and unavoidable limitation on the precision of all possible measurements. Thus, one is able to obtain a set of assumptions, which permit a self-consistent formulation of the usual interpretation of the quantum theory.

[11] N. Bohr, *Atomic Theory and the Description of Nature* (Cambridge University Press, London, 1934).

The above point of view has been given its most consistent and systematic expression by Bohr,[3,5,10] in terms of the "principle of complementarity." In formulating this principle, Bohr suggests that at the atomic level we must renounce our hitherto successful practice of conceiving of an individual system as a unified and precisely definable whole, all of whose aspects are, in a manner of speaking, simultaneously and unambiguously accessible to our conceptual gaze. Such a system of concepts, which is sometimes called a "model," need not be restricted to pictures, but may also include, for example, mathematical concepts, as long as these are supposed to be in a precise (i.e., one-to-one) correspondence with the objects that are being described. The principle of complementarity requires us, however, to renounce even mathematical models. Thus, in Bohr's point of view, the wave function is in no sense a conceptual model of an individual system, since it is not in a precise (one-to-one) correspondence with the behavior of this system, but only in a statistical correspondence.

In place of a precisely defined conceptual model, the principle of complementarity states that we are restricted to complementarity pairs of inherently imprecisely defined concepts, such as position and momentum, particle and wave, etc. The maximum degree of precision of definition of either member of such a pair is reciprocally related to that of the opposite member. This need for an inherent lack of complete precision can be understood in two ways. First, it can be regarded as a consequence of the fact that the experimental apparatus needed for a precise measurement of one member of a complementary pair of variables must always be such as to preclude the possibility of a simultaneous and precise measurement of the other member. Secondly, the assumption that an individual system is completely specified by the wave function and its probability interpretation implies a corresponding unavoidable lack of precision in the very conceptual structure, with the aid of which we can think about and describe the behavior of the system.

It is only at the classical level that we can correctly neglect the inherent lack of precision in all of our conceptual models; for here, the incomplete determination of physical properties implied by the uncertainty principle produces effects that are too small to be of practical significance. Our ability to describe classical systems in terms of precisely definable models is, however, an integral part of the usual interpretation of the theory. For without such models, we would have no way to describe, or even to think of, the result of an observation, which is of course always finally carried out at a classical level of accuracy. If the relationships of a given set of classically describable phenomena depend significantly on the essentially quantum-mechanical properties of matter, however, then the principle of complementarity states that no single model is possible which could provide a precise and rational analysis of the connections between these phenomena. In such a case, we are not supposed, for example, to attempt to describe in detail how future phenomena arise out of past phenomena. Instead, we should simply accept without further analysis the fact that future phenomena do in fact somehow manage to be produced, in a way that is, however, necessarily beyond the possibility of a detailed description. The only aim of a mathematical theory is then to predict the statistical relations, if any, connecting these phenomena.

## 3. CRITICISM OF THE USUAL INTERPRETATION OF THE QUANTUM THEORY

The usual interpretation of the quantum theory can be criticized on many grounds.[5] In this paper, however, we shall stress only the fact that it requires us to give up the possibility of even conceiving precisely what might determine the behavior of an individual system at the quantum level, without providing adequate proof that such a renunciation is necessary.[9] The usual interpretation is admittedly consistent; but the mere demonstration of such consistency does not exclude the possibility of other equally consistent interpretations, which would involve additional elements or parameters permitting a detailed causal and continuous description of all processes, and not requiring us to forego the possibility of conceiving the quantum level in precise terms. From the point of view of the usual interpretation, these additional elements or parameters could be called "hidden" variables. As a matter of fact, whenever we have previously had recourse to statistical theories, we have always ultimately found that the laws governing the individual members of a statistical ensemble could be expressed in terms of just such hidden variables. For example, from the point of view of macroscopic physics, the coordinates and momenta of individual atoms are hidden variables, which in a large scale system manifest themselves only as statistical averages. Perhaps then, our present quantum-mechanical averages are similarly a manifestation of hidden variables, which have not, however, yet been detected directly.

Now it may be asked why these hidden variables should have so long remained undetected. To answer this question, it is helpful to consider as an analogy the early forms of the atomic theory, in which the existence of atoms was postulated in order to explain certain large-scale effects, such as the laws of chemical combination, the gas laws, etc. On the other hand, these same effects could also be described directly in terms of existing macrophysical concepts (such as pressure, volume, temperature, mass, etc.); and a correct description in these terms did not require any reference to atoms. Ultimately, however, effects were found which contradicted the predictions obtained by extrapolating certain purely macrophysical theories to the domain of the very small, and which could be understood correctly in terms of the assumption that matter is composed of atoms. Similarly, we suggest that if there are hidden variables underlying the present quantum theory, it is quite likely that in the atomic domain, they will lead to effects that can also be described adequately in the terms of the usual quantum-mechanical concepts; while in a domain associated with much smaller dimensions, such as the level associated with the "fundamental length" of the order of $10^{-13}$ cm, the hidden variables may lead to completely new effects not consistent with the extrapolation of the present quantum theory down to this level.

If, as is certainly entirely possible, these hidden variables are actually needed for a correct description at small distances, we could easily be kept on the wrong track for a long time by restricting ourselves to the usual interpretation of the quantum theory, which excludes such hidden variables as a matter of principle. It is therefore very important for us to investigate our reasons for supposing that the usual physical interpretation is likely to be the correct one. To this end, we shall begin by repeating the two mutually consistent assumptions on which the usual interpretation is based (see Sec. 2):

(1) The wave function with its probability interpretation determines the most complete possible specification of the state of an individual system.

(2) The process of transfer of a single quantum from observed system to measuring apparatus is inherently unpredictable, uncontrollable, and unanalyzable.

Let us now inquire into the question of whether there are any experiments that could conceivably provide a test for these assumptions. It is often stated in connection with this problem that the mathematical apparatus of the quantum theory and its physical interpretation form a consistent whole and that this combined system of mathematical apparatus and physical interpretation is tested adequately by the extremely wide range of experiments that are in agreement with predictions obtained by using this system. If assumptions (1) and (2) implied a unique mathematical formulation, then such a conclusion would be valid, because experimental predictions could then be found which, if contradicted, would clearly indicate that these assumptions were wrong. Although assumptions (1) and (2) do limit the possible forms of the mathematical theory, they do not limit these forms sufficiently to make possible a unique set of predictions that could in principle permit such an experimental test. Thus, one can contemplate practically arbitrary changes in the Hamiltonian operator, including, for example, the postulation of an unlimited range of new kinds of meson fields each having almost any conceivable rest mass, charge, spin, magnetic moment, etc. And if such postulates should prove to be inadequate, it is conceivable that we may have to introduce nonlocal operators, nonlinear fields, $S$-matrices, etc. This means that when the theory is found to be inadequate (as now happens, for example, at distances of the order of $10^{-13}$ cm), it is always possible, and, in fact, usually quite natural, to assume that the theory can be made to agree with experiment by some as yet unknown change in the mathematical formulation alone, not requiring any fundamental changes in the physical interpretation. This means that as long as we accept the usual physical interpretation of the quantum theory, we cannot be led by any conceivable experiment to give up this interpretation, even if it should happen to be wrong. The usual physical interpretation therefore presents us with a considerable danger of falling into a trap, consisting of a self-closing chain of circular hypotheses, which are in principle unverifiable if true. The only way of avoiding the possibility of such a trap is to study the consequences of postulates that contradict assumptions (1) and (2) at the outset. Thus, we could, for example, postulate that the precise outcome of each individual measurement process is in principle determined by some at present "hidden" elements or variables; and we could then try to find experiments that depended in a unique and reproducible way on the assumed state of these hidden elements or variables. If such predictions are verified, we should then obtain experimental evidence favoring the hypothesis that hidden variables exist. If they are not verified, however, the correctness of the usual interpretation of the quantum theory is not necessarily proved, since it may be necessary instead to alter the specific character of the theory that is supposed to describe the behavior of the assumed hidden variables.

We conclude then that a choice of the present interpretation of the quantum theory involves a real physical limitation on the kinds of theories that we wish to take into consideration. From the arguments given here, however, it would seem that there are no secure experimental or theoretical grounds on which we can base such a choice because this choice follows from hypotheses that cannot conceivably be subjected to an experimental test and because we now have an alternative interpretation.

## 4. NEW PHYSICAL INTERPRETATION OF SCHROEDINGER'S EQUATION

We shall now give a general description of our suggested physical interpretation of the present mathematical formulation of the quantum theory. We shall carry out a more detailed description in subsequent sections of this paper.

We begin with the one-particle Schroedinger equation, and shall later generalize to an arbitrary number of particles. This wave equation is

$$i h \partial \psi / \partial t = -(h^2 / 2m) \nabla^2 \psi + V(\mathbf{x}) \psi. \tag{1}$$

Now $\psi$ is a complex function, which can be expressed as

$$\psi = R \exp(i S / h), \tag{2}$$

where $R$ and $S$ are real. We readily verify that the equations for $R$ and $S$ are

$$\frac{\partial R}{\partial t} = -\frac{1}{2m} [R \nabla^2 S + 2 \nabla R \cdot \nabla S], \tag{3}$$

$$\frac{\partial S}{\partial t} = -\left[ \frac{(\nabla S)^2}{2m} + V(\mathbf{x}) - \frac{h^2}{2m} \frac{\nabla^2 R}{R} \right]. \tag{4}$$

It is convenient to write $P(\mathbf{x}) = R^2(\mathbf{x})$, or $R = P^{\frac{1}{2}}$ where $P(\mathbf{x})$ is the probability density. We then obtain

$$\frac{\partial P}{\partial t} + \nabla \cdot \left( P \frac{\nabla S}{m} \right) = 0, \tag{5}$$

$$\frac{\partial S}{\partial t} + \frac{(\nabla S)^2}{2m} + V(\mathbf{x}) - \frac{h^2}{4m} \left[ \frac{\nabla^2 P}{P} - \frac{1}{2} \frac{(\nabla P)^2}{P^2} \right] = 0. \tag{6}$$

Now, in the classical limit ($h \to 0$) the above equations are subject to a very simple interpretation. The function $S(\mathbf{x})$ is a solution of the Hamilton-Jacobi equation. If we consider an ensemble of particle trajectories which are solutions of the equations of motion, then it is a well-known theorem of mechanics that if all of these trajectories are normal to any given surface of constant $S$, then they are normal to all surfaces of constant $S$, and $\nabla S(\mathbf{x}) / m$ will be equal to the velocity vector, $\mathbf{v}(\mathbf{x})$, for any particle passing the point $\mathbf{x}$. Equation (5) can therefore be re-expressed as

$$\partial P/\partial t + \nabla \cdot (P\mathbf{v}) = 0. \tag{7}$$

This equation indicates that it is consistent to regard $P(\mathbf{x})$ as the probability density for particles in our ensemble. For in that case, we can regard $P\mathbf{v}$ as the mean current of particles in this ensemble, and Eq. (7) then simply expresses the conservation of probability.

Let us now see to what extent this interpretation can be given a meaning even when $h \neq 0$. To do this, let us assume that each particle is acted on, not only by a "classical" potential, $V(\mathbf{x})$ but also by a "quantum-mechanical" potential,

$$U(\mathbf{x}) = \frac{-h^2}{4m} \left[ \frac{\nabla^2 P}{P} - \frac{1}{2} \frac{(\nabla P)^2}{P^2} \right] = \frac{-h^2}{2m} \frac{\nabla^2 R}{R}. \tag{8}$$

Then Eq. (6) can still be regarded as the Hamilton-Jacobi equation for our ensemble of particles, $\nabla S(\mathbf{x})/m$ can still be regarded as the particle velocity, and Eq. (5) can still be regarded as describing conservation of probability in our ensemble. Thus, it would seem that we have here the nucleus of an alternative interpretation for Schroedinger's equation.

The first step in developing this interpretation in a more explicit way is to associate with each electron a particle having precisely definable and continuously varying values of position and momentum. The solution of the modified Hamilton-Jacobi equation (4) defines an ensemble of possible trajectories for this particle, which can be obtained from the Hamilton-Jacobi function, $S(\mathbf{x})$, by integrating the velocity, $\mathbf{v}(\mathbf{x}) = \nabla S(\mathbf{x})/m$. The equation for $S$ implies, however, that the particles moves under the action of a force which is not entirely derivable from the classical potential, $V(\mathbf{x})$, but which also obtains a contribution from the "quantum-mechanical" potential, $U(\mathbf{x}) = (-h^2/2m) \times \nabla^2 R/R$. The function, $R(\mathbf{x})$, is not completely arbitrary, but is partially determined in terms of $S(\mathbf{x})$ by the differential Eq. (3). Thus $R$ and $S$ can be said to codetermine each other. The most convenient way of obtaining $R$ and $S$ is, in fact, usually to solve Eq. (1) for the Schroedinger wave function, $\psi$, and then to use the relations,

$$\psi = U + iW = R[\cos(S/h) + i\,\sin(S/h)],$$
$$R^2 = U^2 + V^2; \quad S = h\,\tan^{-1}(W/U).$$

Since the force on a particle now depends on a function of the absolute value, $R(\mathbf{x})$, of the wave function, $\psi(\mathbf{x})$, evaluated at the actual location of the particle, we have effectively been led to regard the wave function of an individual electron as a mathematical representation of an objectively real field. This field exerts a force on the particle in a way that is analogous to, but not identical with, the way in which an electromagnetic field exerts a force on a charge, and a meson field exerts a force on a nucleon. In the last analysis, there is, of course, no reason why a particle should not be acted on by a $\psi$-field, as well as by an electromagnetic field, a gravitational field, a set of meson fields, and perhaps by still other fields that have not yet been discovered.

The analogy with the electromagnetic (and other) field goes quite far. For just as the electromagnetic field obeys Maxwell's equations, the $\psi$-field obeys Schroedinger's equation. In both cases, a complete specification of the fields at a given instant over every point in space determines the values of the fields for all times. In both cases, once we know the field functions, we can calculate force on a particle, so that, if we also know the initial position and momentum of the particle, we can calculate its entire trajectory.

In this connection, it is worth while to recall that the use of the Hamilton-Jacobi equation in solving for the motion of a particle is only a matter of convenience and that, in principle, we can always solve directly by using Newton's laws of motion and the correct boundary conditions. The equation of motion of a particle acted on by the classical potential, $V(\mathbf{x})$, and the "quantum-mechanical" potential, Eq. (8), is

$$md^2\mathbf{x}/dt^2 = -\nabla\{V(\mathbf{x}) - (h^2/2m)\nabla^2 R/R\}. \qquad (8a)$$

It is in connection with the boundary conditions appearing in the equations of motion that we find the only fundamental difference between the $\psi$-field and other fields, such as the electromagnetic field. For in order to obtain results that are equivalent to those of the usual interpretation of the quantum theory, we are required to restrict the value of the initial particle momentum to $\mathbf{p} = \nabla S(\mathbf{x})$. From the application of Hamilton-Jacobi theory to Eq. (6), it follows that this restriction is consistent, in the sense that if it holds initially, it will hold for all time. Our suggested new interpretation of the quantum theory implies, however, that this restriction is not inherent in the conceptual structure. We shall see in Sec. 9, for example, that it is quite consistent in our interpretation to contemplate modifications in the theory, which permit an arbitrary relation between $\mathbf{p}$ and $\nabla S(\mathbf{x})$. The law of force on the particle can, however, be so chosen that in the atomic domain, $\mathbf{p}$ turns out to be very nearly equal to $\nabla S(\mathbf{x})/m$, while in processes involving very small distances, these two quantities may be very different. In this way, we can improve the analogy between the $\psi$-field and the electromagnetic field (as well as between quantum mechanics and classical mechanics).

Another important difference between the $\psi$-field and the electromagnetic field is that, whereas Schroedinger's equation is homogeneous in $\psi$, Maxwell's equations are inhomogeneous in the electric and magnetic fields. Since inhomogeneities are needed to give rise to radiation, this means that our present equations imply that the $\psi$-field is not radiated or absorbed, but simply changes its form while its integrated intensity remains constant. This restriction to a homogeneous equation is, however, like the restriction to a homogeneous equation is, however, like the restriction to $\mathbf{p} = \nabla S(\mathbf{x})$, not inherent in the conceptual structure of our new interpretation. Thus, in Sec. 9, we shall show that one can consistently postulate inhomogeneities in the equation governing $\psi$, which produce important effects only at very small distances, and negligible effects in the atomic domain. If such inhomogeneities are actually present, then the $\psi$-field will be subject to being emitted and absorbed, but only in connection with processes associated with very small distances. Once the $\psi$-field has been emitted, however, it will in all atomic processes simply obey

Schroedinger's equation as a very good approximation. Nevertheless, at very small distances, the value of the $\psi$-field would, as in the case of the electromagnetic field, depend to some extent on the actual location of the particle.

Let us now consider the meaning of the assumption of a statistical ensemble of particles with a probability density equal to $P(\mathbf{x}) = R^2(\mathbf{x}) = |\psi(\mathbf{x})|^2$. From Eq. (5), it follows that this assumption is consistent, provided that $\psi$ satisfies Schroedinger's equation, and $\mathbf{v} = \nabla S(\mathbf{x})/m$. This probability density is numerically equal to the probability density of particles obtained in the usual interpretation. In the usual interpretation, however, the need for a probability description is regarded as inherent in the very structure of matter (see Sec. 2), whereas in our interpretation, it arises, as we shall see in Paper II, because from one measurement to the next, we cannot in practice predict or control the precise location of a particle, as a result of corresponding unpredictable and uncontrollable disturbances introduced by the measuring apparatus. Thus, in our interpretation, the use of a statistical ensemble is (as in the case of classical statistical mechanics) only a practical necessity, and not a reflection of an inherent limitation on the precision with which it is correct for us to conceive of the variables defining the state of the system. Moreover, it is clear that if in connection with very small distances we are ultimately required to give up the special assumptions that $\psi$ satisfies Schroedinger's equation and that $\mathbf{v} = \nabla S(\mathbf{x})/m$, then $|\psi|^2$ will cease to satisfy a conservation equation and will therefore also cease to be able to represent the probability density of particles. Nevertheless, there would still be a true probability density of particles which is conserved. Thus, it would become possible in principle to find experiments in which $|\psi|^2$ could be distinguished from the probability density, and therefore to prove that the usual interpretation, which gives $|\psi|^2$ only a probability interpretation must be inadequate. Moreover, we shall see in Paper II that with the aid of such modifications in the theory, we could in principle measure the particle positions and momenta precisely, and thus violate the uncertainty principle.

As long as we restrict ourselves to conditions in which Schroedinger's equation is satisfied, and in which $\mathbf{v} = \nabla S(\mathbf{x})/m$, however, the uncertainty principle will remain an effective practical limitation on the possible precision of measurements. This means that at present, the particle positions and momenta should be regarded as "hidden" variables, since as we shall see in Paper II, we are not now able to obtain experiments that localize them to a region smaller than that in which the intensity of the $\psi$-field is appreciable. Thus, we cannot yet find clear-cut experimental proof that the assumption of these variables is necessary, although it is entirely possible that, in the domain of very small distances, new modifications in the theory may have to be introduced, which would permit a proof of the existence of the definite particle position and momentum to be obtained.

We conclude that our suggested interpretation of the quantum theory provides a much broader conceptual framework than that provided by the usual interpretation, for all of the results of the usual interpretation are obtained from our interpretation if we make the following three special assumptions which are mutually consistent:

(1) That the $\psi$-field satisfies Schroedinger's equation.

(2) That the particle momentum is restricted to $\mathbf{p} = \nabla S(\mathbf{x})$.

(3) That we do not predict or control the precise location of the particle, but have, in practice, a statistical ensemble with probability density $P(\mathbf{x}) = |\psi(\mathbf{x})|^2$. The use of statistics is, however, not inherent in the conceptual structure, but merely a consequence of our ignorance of the precise initial conditions of the particle.

As we shall see in Sec. 9, it is entirely possible that a better theory of phenomena involving distances of the order of $10^{-13}$ cm or less would require us to go beyond the limitations of these special assumptions. Our principal purpose in this paper (and in Paper II) is to show, however, that if one makes these special assumptions, our interpretation leads in all possible experiments to the same predictions as are obtained from the usual interpretation.[9]

It is now easy to understand why the adoption of the usual interpretation of the quantum theory would tend to lead us away from the direction of our suggested alternative interpretation. For in a theory involving hidden variables, one would normally expect that the behavior of an individual system should not depend on the statistical ensemble of which it is a member, because this ensemble refers to a series of similar but disconnected experiments carried out under equivalent initial conditions. In our interpretation, however, the "quantum-mechanical" potential, $U(\mathbf{x})$, acting on an individual particle depends on a wave intensity, $P(\mathbf{x})$, that is also numerically equal to a probability density in our ensemble. In the terminology of the usual interpretation of the quantum theory, in which one tacitly assumes that the wave function has only one interpretation; namely, in terms of a probability, our suggested new interpretation would look like a mysterious dependence of the individual on the statistical ensemble of which it is a member. In our interpretation, such a dependence is perfectly rational, because the wave function can consistently be interpreted both as a force and as a probability density.[12]

It is instructive to carry our analogy between the Schroedinger field and other kinds of fields a bit further. To do this, we can derive the wave Eqs. (5) and (6) from a Hamiltonian functional. We begin by writing down the expression for the mean energy as it is expressed in the usual quantum theory:

$$\bar{H} = \int \psi^* \left( -\frac{h^2}{2m}\nabla^2 + V(\mathbf{x}) \right) \psi \, d\mathbf{x}$$

$$= \int \left\{ \frac{h^2}{2m}|\nabla\psi|^2 + V(\mathbf{x})|\psi|^2 \right\} d\mathbf{x}.$$

Writing $\psi = P^{\frac{1}{2}}\exp(iS/h)$, we obtain

$$\bar{H} = \int P(\mathbf{x}) \left\{ \frac{(\nabla S)^2}{2m} + V(\mathbf{x}) + \frac{h^2}{8m}\frac{(\nabla P)^2}{P^2} \right\} d\mathbf{x}. \qquad (9)$$

We shall now reinterpret $P(\mathbf{x})$ as a field coordinate, defined at each point, $\mathbf{x}$, and we shall tentatively assume that $S(\mathbf{x})$ is the momentum,

---

[12] This consistency is guaranteed by the conservation Eq. (7). The questions of why an arbitrary statistical ensemble tends to decay into an ensemble with a probability density equal to $\psi^*\psi$ will be discussed in Paper II, Sec. 7.

canonically conjugate to $P(\mathbf{x})$. That such an assumption is appropriate can be verified by finding the Hamiltonian equations of motion for $P(\mathbf{x})$ and $S(\mathbf{x})$, under the assumption that the Hamiltonian functional is equal to $\bar{H}$ (See Eq. (9)). These equations of motion are

$$\dot{P} = \frac{\delta \bar{H}}{\delta S} = -\frac{1}{m} \nabla \cdot (P \nabla S),$$

$$\dot{S} = -\frac{\partial \bar{H}}{\partial P} = -\left[ \frac{(\nabla S)^2}{2m} + V(\mathbf{x}) - \frac{\hbar^2}{4m} \left( \frac{\nabla^2 P}{P} - \frac{1}{2} \frac{(\nabla P)^2}{P^2} \right) \right].$$

These are, however, the same as the correct wave Eqs. (5) and (6).

We can now show that the mean particle energy averaged over our ensemble is equal to the usual quantum mechanical mean value of the Hamiltonian, $\bar{H}$. To do this, we note that according to Eqs. (3) and (6), the energy of a particle is

$$E(\mathbf{x}) = -\frac{\partial S(\mathbf{x})}{\partial t} = \left[ \frac{(\nabla S)^2}{2m} + V(\mathbf{x}) - \frac{\hbar^2}{2m} \frac{\nabla^2 R}{R} \right]. \quad (10)$$

The mean particle energy is found by averaging $E(\mathbf{x})$ with the weighting function, $P(\mathbf{x})$. We obtain

$$\langle E \rangle_{\substack{\text{ensemble} \\ \text{average}}} = \int P(\mathbf{x}) E(\mathbf{x}) d\mathbf{x}$$

$$= \int P(\mathbf{x}) \left[ \frac{(\nabla S)^2}{2m} + V(\mathbf{x}) \right] d\mathbf{x} - \frac{\hbar^2}{2m} \int R \nabla^2 R d\mathbf{x}.$$

A little integration by parts yields

$$\langle E \rangle_{\substack{\text{ensemble} \\ \text{average}}} = \int P(\mathbf{x}) \left[ \frac{(\nabla S)^2}{2m} + V(\mathbf{x}) \right.$$

$$\left. + \frac{\hbar^2}{8m} \frac{(\nabla P)^2}{P^2} \right] d\mathbf{x} = \bar{H}. \quad (11)$$

## 5. THE STATIONARY STATE

We shall now show how the problem of stationary states is to be treated in our interpretation of the quantum theory.

The following seem to be reasonable requirements in our interpretation for a stationary state:

(1) The particle energy should be a constant of the motion.

(2) The quantum-mechanical potential should be independent of time.

(3) The probability density in our statistical ensemble should be independent of time.

It is easily verified that these requirements can be satisfied with the assumption that

$$\psi(\mathbf{x}, t) = \psi_0(\mathbf{x}) \exp(-iEt/h)$$
$$= R_0(\mathbf{x}) \exp[i(\Phi(\mathbf{x}) - Et)/h]. \tag{12}$$

From the above, we obtain $S = \Phi(\mathbf{x}) - Et$. According to the generalized Hamilton-Jacobi Eq. (4), the particle energy is given by

$$\partial S/\partial t = -E.$$

Thus, we verify that the particle energy is a constant of the motion. Moreover, since $P = R^2 = |\psi|^2$, it follows that $P$ (and $R$) are independent of time. This means that both the probability density in our ensemble and the quantum-mechanical potential are also time independent.

The reader will readily verify that no other form of solution of Schroedinger's equation will satisfy all three of our criteria for a stationary state.

Since $\psi$ is now being regarded as a mathematical representation of an objectively real force field, it follows that (like the electromagnetic field) it should be everywhere finite, continuous, and single valued. These requirements will guarantee in all cases that occur in practice that the allowed values of the energy in a stationary state, and the corresponding eigenfunctions are the same as are obtained from the usual interpretation of the theory.

In order to show in more detail what a stationary state means in our interpretation, we shall now consider three examples of stationary states.

## CASE 1: "s" STATE

The first case that we shall consider is an "s" state. In an "s" state, the wave function is

$$\psi = f(r) \exp[i(\alpha - Et)/h], \tag{13}$$

where $\alpha$ is an arbitrary constant and $r$ is the radius taken from the center of the atom. We conclude that the Hamilton-Jacobi function is

$$S = \alpha - Et.$$

The particle velocity is

$$\mathbf{v} = \nabla S = 0.$$

The particle is therefore simply standing still, wherever it may happen to be. How can it do this? The absence of motion is possible because the applied force, $-\nabla V(\mathbf{x})$, is balanced by the "quantum-mechanical" force, $(h^2/2m)\nabla(\nabla^2 R/R)$, produced by the Schroedinger $\psi$-field acting on its own particle. There is, however, a statistical ensemble of possible positions of the particle, with a probability density, $P(\mathbf{x}) = (f(r))^2$.

## CASE 2: STATE WITH NONZERO ANGULAR MOMENTUM

In a typical state of nonzero angular momentum, we have

$$\psi = f_n^{l}(r) P_l^{m}(\cos\theta) \exp[i(\beta - Et + hm\phi)/h], \tag{14}$$

where $\theta$ and $\phi$ are the colatitude and azimuthal polar angles, respectively, $P_l^{m}$ is the associated Legendre polynomial, and $\beta$ is a constant. The Hamilton-Jacobi function is $S = \beta - Et + hm\phi$. From this result it follows that the $z$ component of the angular momentum is equal to $hm$. To prove this, we write

$$L_s = xp_y - yp_x = x\partial S/\partial y - y\partial S/\partial x = \partial S/\partial \phi = hm. \tag{15}$$

Thus, we obtain a statistical ensemble of trajectories which can have different forms, but all have the same "quantized" value of the $z$ component of the angular momentum.

## CASE 3: A SCATTERING PROBLEM

Let us now consider a scattering problem. Because it is comparatively easy to analyze, we shall discuss a hypothetical experiment, in which an electron is incident in the $z$ direction with an initial momentum, $p_0$, on a system consisting of two slits.[13] After the electron passes through the slit system, its position is measured and recorded, for example, on a photographic plate.

Now, in the usual interpretation of the quantum theory, the electron is described by a wave function. The incident part of the wave function is $\psi_0 \sim \exp(ip_0z/h)$; but when the wave passes through the slit system, it is modified by interference and diffraction effects, so that it will develop a characteristic intensity pattern by the time it reaches the position measuring instrument. The probability that the electron will be detected between $\mathbf{x}$ and $\mathbf{x} + d\mathbf{x}$ is $|\psi(\mathbf{x})|^2 d\mathbf{x}$. If the experiment is repeated many times under equivalent initial conditions, one eventually obtains a pattern of hits on the photographic plate that is very reminiscent of the interference patterns of optics.

In the usual interpretation of the quantum theory, the origin of this interference pattern is very difficult to understand. For there may be certain points where the wave function is zero when both slits are open, but not zero when only one slit is open. How can the opening of a second slit prevent the electron from reaching certain points that it could reach if this slit were closed? If the electron acted completely like a classical particle, this phenomenon could not be explained at all. Clearly, then the wave aspects of the electron must have something to do with the production of the interference pattern. Yet, the electron cannot be identical with its associated wave, because the latter spreads out over a wide region. On the other hand, when the electron's position is measured, it always appears at the detector as if it were a localized particle.

The usual interpretation of the quantum theory not only makes no attempt to provide a single precisely defined conceptual model

---

[13]This experiment is discussed in some detail in reference 2, Chapter 6, Sec. 2.

for the production of the phenomena described above, but it asserts that no such model is even conceivable. Instead of a single precisely defined conceptual model, it provides, as pointed out in Sec. 2, a pair of complementary models, *viz.*, particle and wave, each of which can be made more precise only under conditions which necessitate a reciprocal decrease in the degree of precision of the other. Thus, while the electron goes through the slit system, its position is said to be inherently ambiguous, so that if we wish to obtain an interference pattern, it is meaningless to ask through which slit an individual electron actually passed. Within the domain of space within which the position of the electron has no meaning we can use the wave model and thus describe the subsequent production of interference. If, however, we tried to define the position of the electron as it passed the slit system more accurately by means of a measurement, the resulting disturbance of its motion produced by the measuring apparatus would destroy the interference pattern. Thus, conditions would be created in which the particle model becomes more precisely defined at the expense of a corresponding decrease in the degree of definition of the wave model. When the position of the electron is measured at the photographic plate, a similar sharpening of the degree of definition of the particle model occurs at the expense of that of the wave model.

In our interpretation of the quantum theory, this experiment is described causally and continuously in terms of a single precisely definable conceptual model. As we have already shown, we must use the same wave function as is used in the usual interpretation; but instead we regard it as a mathematical representation of an objectively real field that determines part of the force acting on the particle. The initial momentum of the particle is obtained from the incident wave function, $\exp(ip_0z/h)$, as $p = \partial s/\partial z = p_0$. We do not in practice, however, control the initial location of the particle, so that although it goes through a definite slit, we cannot predict which slit this will be. The particle is at all times acted on by the "quantum-mechanical" potential, $U = (-h^2/2m)\nabla^2 R/R$. While the particle is incident, this potential vanishes because $R$ is then a constant; but after it passes through the slit system, the particle encounters a quantum-mechanical potential that changes rapidly with position. The subsequent motion of the particle may therefore become quite complicated. Nevertheless, the probability that a particle shall enter a given region, $d\mathbf{x}$, is as in the usual interpretation, equal to $|\psi(\mathbf{x})|^2 d\mathbf{x}$. We therefore deduce that the particle can never reach a point where the wave function vanishes. The reason is that the "quantum-mechanical" potential, $U$, becomes infinite when $R$ becomes zero. If the approach to infinity happens to be through positive values of $U$, there will be an infinite force repelling the particle away from the origin. If the approach is through negative values of $U$, the particle will go through this point with infinite speed, and thus spend no time there. In either case, we obtain a simple and precisely definable conceptual model explaining why particles can never be found at points where the wave function vanishes.

If one of the slits is closed, the "quantum-mechanical" potential is correspondingly altered, because the $\psi$-field is changed, and the particle may then be able to reach certain points which it was unable to reach when both slits were open. The slit is therefore able to affect the motion of the particle only indirectly, through its effect on the Schroedinger $\psi$-field. Moreover, as we shall see in Paper II, if the position of the electron is measured while it is passing through the slit system, the measuring apparatus will, as in the usual interpretation, create a disturbance that destroys the interference pattern. In our interpretation, however, the necessity for this destruction is not inherent in the conceptual structure; and as we shall see, the destruction of the interference pattern could in principle be avoided by means of other ways of making measurements, ways which are conceivable but not now actually possible.

## 6. THE MANY-BODY PROBLEM

We shall now extend our interpretation of the quantum theory to the problem of many bodies. We begin with the Schroedinger equation for two particles. (For simplicity, we assume that they have equal masses, but the extension of our treatment to arbitrary masses will be obvious.)

$$ih\frac{\partial\psi}{\partial t} = -\frac{h^2}{2m}(\nabla_1{}^2\psi + \nabla_2{}^2\psi) + V(\mathbf{x}_1, \mathbf{x}_2)\psi.$$

Writing $\psi = R(\mathbf{x}_1, \mathbf{x}_2)\exp[iS(\mathbf{x}_1, \mathbf{x}_2)/h]$ and $R^2 = P$, we obtain

$$\frac{\partial P}{\partial t} + \frac{1}{m}[\nabla_1 \cdot P\nabla_1 S + \nabla_2 \cdot P\nabla_2 S] = 0, \qquad (16)$$

$$\frac{\partial S}{\partial t} + \frac{(\nabla_1 S)^2 + (\nabla_2 S)^2}{2m} + V(\mathbf{x}_1, \mathbf{x}_2)$$
$$- \frac{h^2}{2mR}[\nabla_1{}^2 R + \nabla_2{}^2 R] = 0. \qquad (17)$$

The above equations are simply a six-dimensional generalization of the similar three-dimensional Eqs. (5) and (6) associated with the one-body problem. In the two-body problem, the system is described therefore by a six-dimensional Schroedinger wave and by a six-dimensional trajectory, specifying the actual location of each of the two particles. The velocity of this trajectory has components, $\nabla_1 S/m$ and $\nabla_2 S/m$, respectively, in each of the three-dimensional surfaces associated with a given particle. $P(\mathbf{x}_1, \mathbf{x}_2)$ then has a dual interpretation. First, it defines a "quantum-mechanical" potential, acting on each particle

$$U(\mathbf{x}_1, \mathbf{x}_2) = -(h^2/2mR)[\nabla_1{}^2 R + \nabla_2{}^2 R].$$

This potential introduces an additional effective interaction between particles over and above that due to the classically inferrable potential $V(\mathbf{x})$. Secondly, the function $P(\mathbf{x}_1, \mathbf{x}_2)$ can consistently be regarded as the probability density of representative points $(\mathbf{x}_1, \mathbf{x}_2)$ in our six-dimensional ensemble.

The extension to an arbitrary number of particles is straightforward, and we shall quote only the results here. We introduce the wave function, $\psi = R(\mathbf{x}_1, \mathbf{x}_2, \dots \mathbf{x}_n)\exp[iS(\mathbf{x}_1, \mathbf{x}_2 \dots \mathbf{x}_n)/h]$ and define a $3n$-dimensional trajectory, where $n$ is the number of particles, which describes the behavior of every particle in the system. The velocity of the $i$th particle is $\mathbf{v}_i = \nabla_i S(\mathbf{x}_1, \mathbf{x}_2 \ldots \mathbf{x}_n)/m$. The function $P(\mathbf{x}_1, \mathbf{x}_2 \ldots \mathbf{x}_n = R^2$ has two interpretations. First, it defines a "quantum-mechanical" potential

$$U(\mathbf{x}_1, \mathbf{x}_2 \ldots \mathbf{x}_n) = -\frac{h^2}{2mR} \sum_{s=1}^{n} \nabla_s^2 R(\mathbf{x}_1, \mathbf{x}_2 \ldots \mathbf{x}_n). \qquad (18)$$

Secondly, $P(\mathbf{x}_1, \mathbf{x}_2 \ldots \mathbf{x}_n)$ is equal to the density of representative points $(\mathbf{x}_1, \mathbf{x}_2 \ldots \mathbf{x}_n)$ in our $3n$-dimensional ensemble.

We see here that the "effective potential," $U(\mathbf{x}_1, \mathbf{x}_2, \ldots \mathbf{x}_n)$, acting on a particle is equivalent to that produced by a "many-body" force, since the force between any two particles may depend significantly on the location of every other particle in the system. An example of the effects of such a force is given by the exclusion principle. Thus, if the wave function is antisymmetric, we deduce that the "quantum-mechanical" forces will be such as to prevent two particles from ever reaching the same point in space, for in this case, we must have $P = 0$.

# 7. TRANSITIONS BETWEEN STATIONARY STATES—THE FRANCK-HERTZ EXPERIMENT

Our interpretation of the quantum theory describes all processes as basically causal and continuous. How then can it lead to a correct description of processes such as the Franck-Hertz experiment, the photoelectric effect, and the Compton effect, which seem to call most strikingly for an interpretation in terms of discontinuous and incompletely determined transfers of energy and momentum? In this section, we shall answer this question by applying our suggested interpretation of the quantum theory in the analysis of the Franck-Hertz experiment. Here, we shall see that the apparently discontinuous nature of the process of transfer of energy from the bombarding particle to the atomic electron is brought about by the "quantum-mechanical" potential, $U = (-h^2/2m)\nabla^2 R/R$, which does not necessarily become small when the wave intensity becomes small. Thus, even if the force of interaction between two particles is very weak, so that a correspondingly small disturbance of the Schroedinger wave function is produced by the interaction of these particles, this disturbance is capable of bringing about very large transfers of energy and momentum between the particles in a very short time. This means that if we view only the end results, this process presents the aspect of being discontinuous. Moveover, we shall see that the precise value of the energy transfer is in principle determined by the initial position of each particle and by the initial form of the wave function. Since we cannot in practice predict or control the initial particle positions with complete precision, we are also unable to predict or control the final outcome of such an experiment, and can, in practice, predict only the probability of a given outcome. Because the probability that the particles will enter a region with coordinates, $\mathbf{x}_1$, $\mathbf{x}_2$, is proportional to $R^2(\mathbf{x}_1, \mathbf{x}_2)$, we conclude that although a Schroedinger wave of low intensity can bring about large transfers of energy, such a process is (as in the usual interpretation) highly improbable.

In Appendix A of Paper II, we shall see that similar possibilities arise in connection with the interaction of the electromagnetic field with charged matter, so that electromagnetic waves can very rapidly transfer a full quantum of energy (and momentum) to an electron, even after they have spread out and fallen to a very low intensity. In this way, we shall explain the photoelectric effect and the Compton effect. Thus, we are able in our interpretation to understand by means of a causal and continuous model just those properties of matter and light which seem most convincingly to require the assumption of discontinuity and incomplete determinism.

Before we discuss the process of interaction between two particles, we shall find it convenient to analyze the problem of an isolated single particle that happens to be in a nonstationary state. Because the field function $\psi$ is a solution of Schroedinger's equation, we can linearly suppose stationary-state solutions of this equation and in this way obtain new solutions. As an illustration, let us consider a superposition of two solutions

$$\psi = C_1\psi_1(\mathbf{x})\exp(-iE_1t/h) + C_2\psi_2(\mathbf{x})\exp(-iE_2t/h),$$

where $C_1$, $C_2$, $\psi_1$, and $\psi_2$ are real. Thus we write $\psi_1 = R_1$, $\psi_2 = R_2$, and

$$\psi = \exp[-i(E_1 + E_2)t/2h]\{C_1R_1\exp[-i(E_1 - E_2)t/2h]$$
$$+ C_2R_2\exp[i(E_1 - E_2)t/2h]\}.$$

Writing $\psi = R\exp(iS/h)$, we obtain

$$R^2 = C_1{}^2R_1{}^2(\mathbf{x}) + C_2{}^2R_2{}^2(\mathbf{x})$$
$$+2C_1C_2R_1(\mathbf{x})R_2(\mathbf{x})\cos[(E_1 - E_2)t/2h], \qquad (19)$$

$$\tan\left\{\frac{S + (E_1 - E_2)t/2}{h}\right\}$$
$$= \frac{C_2R_2(\mathbf{x}) - C_1R_1(\mathbf{x})}{C_2R_2(\mathbf{x}) + C_1R_1(\mathbf{x})}\tan\left\{\frac{(E_1 - E_2)t}{2h}\right\}. \qquad (20)$$

We see immediately that the particle experiences a "quantum-mechanical" potential, $U(\mathbf{x}) = (-h/2m)\nabla^2R/R$, which fluctuates with angular frequency, $w = (E_1 - E_2)/h$, and that the energy of this particle, $E = -\partial S/\partial t$, and its momentum $\mathbf{p} = \nabla S$, fluctuate with the same angular frequency. If the particle happens to enter a region of space where $R$ is small, these fluctuations can become quite violent. We see then that, in general, the orbit of a particle in a nonstationary state is very irregular and complicated, resembling Brownian motion more closely than it resembles the smooth track of a planet around the sun.

If the system is isolated, these fluctuations will continue forever. The result is quite reasonable, since as is well known, a system can make a transition from one stationary state to another only if it can exchange energy with some other system. In order to treat the problem of transition between stationary states, we must therefore introduce another system capable of exchanging energy with the system of interest. In this section, we shall discuss the Franck-Hertz experiment, in which this other system consists of a bombarding particle. For the sake of illustration, let us suppose that we have hydrogen atoms of energy $E_0$ and wave function, $\psi_0(\mathbf{x})$, which are bombarded by particles that can be scattered inelastically, leaving the atom with energy $E_n$ and wave function, $\psi_n(\mathbf{x})$.

We begin by writing down the initial wave function, $\Psi_i(\mathbf{x}, \mathbf{y}, t)$. The incident particle, whose coordinates are represented by $\mathbf{y}$ must be associated with a wave packet, which can be written as

$$f_0(\mathbf{y}, t) = \int e^{i\mathbf{k}\cdot\mathbf{y}} f(\mathbf{k} - \mathbf{k}_0) \exp(-i\hbar k^2 t/2m) d\mathbf{k}. \qquad (21)$$

The center of this packet occurs where the phase has an extremum as a function of $\mathbf{k}$, or where $\mathbf{y} = \hbar\mathbf{k}_0 t/m$.

Now, as in the usual interpretation, we begin by writing the incident wave function for the combined system as a product

$$\Psi_i = \psi_0(\mathbf{x}) \exp(-iE_0 t/\hbar) f_0(\mathbf{y}, t). \qquad (22)$$

Let us now see how this wave function is to be understood in our interpretation of the theory. As pointed out in Sec. 6, the wave function is to be regarded as a mathematical representation of a six-dimensional but objectively real field, capable of producing forces that act on the particles. We also assume a six-dimensional representative point, described by the coordinates of the two particles, $\mathbf{x}$ and $\mathbf{y}$. We shall now see that when the combined wave function takes the form (22) involving a product of a function of $\mathbf{x}$ and a function of $\mathbf{y}$, the six-dimensional system can correctly be regarded as being made up of two independent three-dimentional subsystems. To prove this, we write

$$\psi_0(\mathbf{x}) = R_0(\mathbf{x}) \exp[iS_0(\mathbf{x})/\hbar] \quad \text{and}$$
$$f_0(\mathbf{y}, t) = M_0(\mathbf{y}, t) \exp[iN_0(\mathbf{y}, t)/\hbar].$$

We then obtain for the particle velocities

$$d\mathbf{x}/dt = (1/m)\nabla S_0(\mathbf{x}); \quad d\mathbf{y}/dt = (1/m)\nabla N_0(y, t), \qquad (23)$$

and for the "quantum-mechanical" potential

$$U = -\frac{h^2\left\{(\nabla_x{}^2 + \nabla_y{}^2) R(\mathbf{x}, \mathbf{y})\right\}}{2m R(\mathbf{x}, \mathbf{y})}$$

$$= \frac{-h^2}{2m}\left\{\frac{\nabla^2 R_0(\mathbf{x})}{R_0(\mathbf{x})} + \frac{\nabla^2 M_0(\mathbf{y}, t)}{M_0(\mathbf{y}, t)}\right\}. \tag{24}$$

Thus, the particle velocities are independent and the "quantum-mechanical" potential reduces to a sum of terms, one involving only $\mathbf{x}$ and the other involving only $\mathbf{y}$. This means that the particles move independently. Moreover, the probability density, $P = R_0{}^2(\mathbf{x}) \times M_0{}^2(\mathbf{y}, t)$, is a product of a function of $\mathbf{x}$ and a function of $\mathbf{y}$, indicating that the distribution in $\mathbf{x}$ is statistically independent of that in $\mathbf{y}$. We conclude, then, that whenever the wave function can be expressed as a product of two factors, each involving only the coordinates of a single system, then the two systems are completely independent of each other.

As soon as the wave packet in y space reaches the neighborhood of the atom, the two systems begin to interact. If we solve Schroedinger's equation for the combined system, we obtain a wave function that can be expressed in terms of the following series:

$$\Psi = \Psi_i + \sum_n \psi_n(\mathbf{x}) \exp(-i E_n t / h) f_n(\mathbf{y}, t), \tag{25}$$

where the $f_n(\mathbf{y}, t)$ are the expansion coefficients of the complete set of functions, $\psi_n(\mathbf{x})$. The asymptotic form of the wave function is[14]

$$\Psi = \Psi_i(\mathbf{x}, \mathbf{y}) + \sum_n \psi_n(\mathbf{x}) \exp\left(-\frac{i E_n t}{h}\right) \int f(\mathbf{k} - \mathbf{k}_0)$$

$$\times \frac{\exp[i k_n \cdot \mathbf{r} - (h k_n{}^2/2n)t]}{r} g_n(\theta, \phi, \mathbf{k}) d\mathbf{k}, \tag{26}$$

where

$$h^2 k_n{}^2 / 2m = (h^2 k_0{}^2 / 2m) + E_0 - E_n \tag{27}$$

(conservation of energy).

The additional terms in the above equation represent outgoing wave packets, in which the particle speed, $h k_n / m$, is correlated with

[14] N. F. Mott and H. S. W. Massey, *The Theory of Atomic Collisions* (Clarendon Press, Oxford, 1933).

the wave function, $\psi_n(\mathbf{x})$, representing the state in which the hydrogen atom is left. The center of the $n$th packet occurs at

$$r_n = (\hbar k_n / m)t. \tag{28}$$

It is clear that because the speed depends on the hydrogen atom quantum number, $n$, every one of these packets will eventually be separated by distances which are so large that this separation is classically describable.

When the wave function takes the form (25), the two particles system must be described as a single six-dimensional system and not as a sum of two independent three-dimensional subsystems, for at this time, if we try to express the wave function as $\psi(\mathbf{x}, \mathbf{y}) = R(\mathbf{x}, \mathbf{y}) \times \exp[i S(\mathbf{x}, \mathbf{y})/\hbar]$, we find that the resulting expressions for $R$ and $S$ depend on $\mathbf{x}$ and $\mathbf{y}$ in a very complicated way. The particle momenta, $\mathbf{p}_1 = \nabla_2 S(\mathbf{x}, \mathbf{y})$ and $\mathbf{p}_2 = \nabla_y S(\mathbf{x}, \mathbf{y})$, therefore become inextricably interdependent. The "quantum-mechanical" potential,

$$U = -\frac{\hbar^2}{2m R(\mathbf{x}, \mathbf{y})}(\nabla_x^2 R + \nabla_y^2 R)$$

ceases to be expressible as the sum of a term involving $\mathbf{x}$ and a term involving $\mathbf{y}$. The probability density, $R^2(\mathbf{x}, \mathbf{y})$ can no longer be written as a product of a function of $\mathbf{x}$ and a function of $\mathbf{y}$, from which we conclude that the probability distributions of the two particles are no longer statistically independent. Moreover, the motion of the particle is exceedingly complicated, because the expressions for $R$ and $S$ are somewhat analogous to those obtained in the simpler problem of a nonstationary state of a single particle [see Eqs. (19) and (20)]. In the region where the scattered waves $\psi_n(\mathbf{x}) f_n(\mathbf{y}, t)$ have an amplitude comparable with that of the incident wave, $\psi_0(\mathbf{x}) f_0(\mathbf{y}, t)$, the functions $R$ and $S$, and therefore the "quantum-mechanical" potential and the particle momenta, undergo rapid and violent fluctuations, both as functions of position and of time. Because the quantum-mechanical potential has $R(\mathbf{x}, \mathbf{y}, t)$ in the denominator, these fluctuations may become very large in this region where $R$ is small. If the particles happen to enter such a region, they may exchange very large quantities of energy and momentum in a very short time, even if the classical potential, $V(\mathbf{x}, \mathbf{y})$ is very small. A small value of $V(\mathbf{x}, \mathbf{y})$ implies, however, a correspondingly small value of the scattered wave amplitudes, $f_n(\mathbf{y}, t)$. Since the fluctuations become large only in the region where the scattered wave amplitude is comparable with the incident wave amplitude and since the probability that the particles shall enter a given region of $\mathbf{x}, \mathbf{y}$ space is proportional to $R^2(\mathbf{x}, \mathbf{y})$, it is clear that a large transfer of energy is improbable (although still always possible) when $V(\mathbf{x}, \mathbf{y})$ is small.

While interaction between the two particles takes place then, their orbits are subject to wild fluctuations. Eventually, however, the behavior of the system quiets down and becomes simple again. For after the wave function takes its asymptotic form (26), and the packets corresponding to different values of $n$ have obtained classically describable separations, we can deduce that because the probability density is $|\psi|^2$, the outgoing particle must enter one of these packets and stay with that packet thereafter (since it does not enter the space between packets in which the probability density is negligibly different from zero). In the calculation of the particle velocities, $\mathbf{V}_1 = \nabla_x S/m$, $\mathbf{V}_2 = \nabla_y S/m$, and of the quantum-mechanical potential, $U = (-h^2/2mR)(\nabla_x{}^2 R + \nabla_y{}^2 R)$, we can therefore ignore all parts of the wave function other than the one actually containing the outgoing particle. It follows that the system acts as if it had the wave function

$$\Psi_n = \psi_n(\mathbf{x}) \exp\left(\frac{iE_n t}{h}\right) \int f(\mathbf{k} - \mathbf{k}_0)$$
$$\times \frac{\exp\{i[\mathbf{k}_n \cdot \mathbf{r} - (hk_n{}^2 t/2m)t]\}}{r} g_n(\theta, \phi, \mathbf{k}) d\mathbf{k}, \tag{29}$$

where $n$ denotes the packet actually containing the outgoing particle. This means that for all practical purposes the complete wave function (26) of the system may be replaced by Eq. (29), which corresponds to an atomic electron in its $n$th quantum state, and to an outgoing particle with a correlated energy, $E_n{}' = h^2 k_n{}^2/2m$. Because the wave function is a product of a function of $\mathbf{x}$ and a function of $\mathbf{y}$, each system once again acts independently of the other. The wave function can now be renormalized because the multiplication of $\Psi_n$ by a constant changes no physically significant quantity, such as the particle velocity or the "quantum-mechanical" potential. As shown in Sec. 5, when the electronic wave function is $\psi_n(\mathbf{x})\exp(-iE_n t/h)$, its energy must be $E_n$. Thus, we have obtained a description of how it comes about that the energy is always transferred in quanta of size $E_n - E_0$.

It should be noted that while the wave packets are still separating, the electron energy is not quantized, but has a continuous range of values, which fluctuate rapidly. It is only the final value of the energy, appearing after the interaction is over that must be quantized. A similar result is obtained in the usual interpretation if one notes that because of the uncertainty principle, the energy of either system can become definite only after enough time has elapsed to complete the scattering process.[15]

In principle, the actual packet entered by the outgoing particle could be predicted if we knew the initial position of both particles and, of course, the initial form of the wave function of the combined system.[16] In practice, however, the particle orbits are very complicated and very sensitively dependent on the precise values of these initial positions. Since we do not at present know how to measure these initial positions precisely, we cannot actually predict the outcome of such an interaction process. The best that we can do is to predict the probability that an outgoing particle enters the $n$th packet within a given range of solid angle, $d\Omega$, leaving the hydrogen atom in its $n$th quantum state. In doing this, we use the fact that the probability density in $\mathbf{x}$, $\mathbf{y}$ space is $|\psi(\mathbf{x}, \mathbf{y})|^2$ and that as long as we are restricted to the $n$th packet, we can replace the complete wave function (26) by the wave function (29), corresponding to the packet that actually contains the particle. Now, by definition, we have $\int |\psi_n(\mathbf{x})|^2 d\mathbf{x} = 1$.

---

[15] See reference 2, Chapter 18, Sec. 19.

[16] Note that in the usual interpretation one assumes that *nothing* determines the precise outcome of an individual scattering process. Instead, one assumes that all descriptions are inherently and unavoidably statistical (see Sec. 2).

The remaining integration of

$$\left| \int f(\mathbf{k} - \mathbf{k}_0) \frac{\exp\{i[k_n r - (\hbar k_n{}^2/2m)t]\}}{r} g_n(\theta, \phi, k)d\mathbf{k} \right|^2$$

over the region of space corresponding to the $n$th outgoing packet leads, however, to precisely the same probability of scattering as would have been obtained by applying the usual interpretation. We conclude, then, that if $\psi$ satisfies Schroedinger's equation, that if $\mathbf{v} = \nabla S/m$, and that if the probability density of particles is $P(\mathbf{x}, \mathbf{y}) = R^2(\mathbf{x}, \mathbf{y})$, we obtain in every respect exactly the same physical predictions for this problem as are obtained when we use the usual interpretation.

There remains only one more problem; namely, to show that if the outgoing packets are subsequently brought together by some arrangement of matter that does not act on the atomic electron, the atomic electron and the scattered particle will continue to act independently.[17] To show that these two particles will continue to act independently, we note that in all practical applications, the outgoing particle soon interacts with some classically describable system. Such a system might consist, for example, of the host of atoms of the gas with which it collides or of the walls of a container. In any case, if the scattering process is ever to be observed, the outgoing particle must interact with a classically describable measuring apparatus. Now all classically describable systems have the property that they contain an enormous number of internal "thermo-dynamic" degrees of freedom that are inevitably excited when the outgoing particle interacts with the system. The wave function of the outgoing particle is then coupled to that of these internal thermodynamic degrees of freedom, which we represent as $y_1, y_2, \ldots y_s$. To denote this coupling, we write the wave function for the entire system as

$$\Psi = \sum_n \psi_n(\mathbf{x}) \exp(-iE_n t/\hbar) f_n(\mathbf{y}, y_1, y_2 \ldots y_s). \qquad (30)$$

Now, when the wave function takes this form, the overlapping of different packets in $\mathbf{y}$ space is not enough to produce interference between the different $\psi_n(\mathbf{x})$. To obtain such interference, it is necessary

---

[17] See reference 2, Chapter 22, Sec. 11, for a treatment of a similar problem.

that the packets $f_n(\mathbf{y}, y_1, y_2, \ldots y_s)$ overlap in every one of the $S + 3$ dimensions, $\mathbf{y}, y_1, y_2 \ldots y_s$. The reader will readily convince himself, by considering a typical case such as a collision of the outgoing particle with a metal wall, that it is overwhelmingly improbable that two of the packets $f_n(\mathbf{y}_1, y_1, y_2 \ldots y_s)$ will overlap with regard to every one of the internal thermodynamic coordinates, $y_1, y_2, \ldots y_s$, even if they are successfully made to overlap in $\mathbf{y}$ space. This is because each packet corresponds to a different particle velocity and to a different time of collision with the metal wall. Because the myriads of internal thermodynamic degrees of freedom are so chaotically complicated, it is very likely that as each of the $n$ packets interacts with them, it will encounter different conditions, which will make the combined wave packet $f_n(\mathbf{y}, y_1, \ldots y_s)$ enter very different regions of $y_1, y_2 \ldots y_s$ space. Thus, for all practical purposes, we can ignore the possibility that if two of the packets are made to cross in $\mathbf{y}$ space, the motion either of the atomic electron or of the outgoing particle will be affected.[18]

## 8. PENETRATION OF A BARRIER

According to classical physics, a particle can never penetrate a potential barrier having a height greater than the particle kinetic energy. In the usual interpretation of the quantum theory, it is said to be able, with a small probability, to "leak" through the barrier. In our interpretation of the quantum theory, however, the potential provided by the Schroedinger $\psi$-field enables it to "ride" over the barrier, but only a few particles are likely to have trajectories that carry them all the way across without being turned around.

We shall merely sketch in general terms how the above results can be obtained. Since the motion of the particle is strongly affected by its $\psi$-field, we must first solve for this field with the aid of "Schroedinger's

---

[18] It should be noted that exactly the same problem arises in the usual interpretation of the quantum theory for (reference 16), for whenever two packets overlap, then even in the usual interpretation, the system must be regarded as, in some sense, covering the states corresponding to both packets simultaneously. See reference 2, Chapter 6 and Chapter 16, Sec. 25. Once two packets have obtained classically describable separations, then, both in the usual interpretation and in our interpretation the probability that there will be significant interference between them is so overwhelmingly small that it may be compared to the probability that a tea kettle placed on a fire will happen to freeze instead of boil. Thus, we may for all practical purposes neglect the possibility of interference between packets corresponding to the different possible energy states in which the hydrogen atom may be left.

equation." Initially, we have a wave packet incident on the potential barrier; and because the probability density is equal to $|\psi(\mathbf{x})|^2$, the particle is certain to be somewhere within this wave packet. When the wave packet strikes the repulsive barrier, the $\psi$-field undergoes rapid changes which can be calculated[19] if desired, but whose precise form does not interest us here. At this time, the "quantum-mechanical" potential, $U = (-h^2/2m)\nabla^2 R/R$, undergoes rapid and violent fluctuations, analogous to those described in Sec. 7 in connection with Eqs. (19), (20), and (25). The particle orbit then becomes very complicated and, because the potential is time dependent, very sensitive to the precise initial relationship between the particle position and the center of the wave packet. Ultimately, however, the incident wave packet disappears and is replaced by two packets, one of them a reflected packet and the other a transmitted packet having a much smaller intensity. Because the probability density is $|\psi|^2$, the particle must end up in one of these packets. The other packet can, as shown in Sec. 7, subsequently be ignored. Since the reflected packet is usually so much stronger than the transmitted packet, we conclude that during the time when the packet is inside the barrier, most of the particle orbits must be turned around, as a result of the violent fluctuations in the "quantum-mechanical" potential.

## 9. POSSIBLE MODIFICATIONS IN MATHEMATICAL FORMULATION LEADING TO EXPERIMENTAL PROOF THAT NEW INTERPRETATION IS NEEDED

We have already seen in a number of cases and in Paper II we shall prove in general, that as long as we assume that $\psi$ satisfies Schroedinger's equation, that $\mathbf{v} = \nabla S(\mathbf{x})/m$, and that we have a statistical ensemble with a probability density equal to $|\psi(\mathbf{x})|^2$, our interpretation of the quantum theory leads to physical results that are identical with those

---

[19]See, for example, reference 2, Chapter 11, Sec. 17, and Chapter 12, Sec. 18.

obtained from the usual interpretation. Evidence indicating the need for adopting our interpretation instead of the usual one could therefore come only from experiments, such as those involving phenomena associated with distances of the order of $10^{-13}$ cm or less, which are not now adequately understood in terms of the existing theory. In this paper we shall not, however, actually suggest any specific experimental methods of distinguishing between our interpretation and the usual one, but shall confine ourselves to demonstrating that such experiments are conceivable.

Now, there are an infinite number of ways of modifying the mathematical form of the theory that are consistent with our interpretation and not with the usual interpretation. We shall confine ourselves here, however, to suggesting two such modifications, which have already been indicated in Sec. 4, namely, to give up the assumption that $\mathbf{v}$ is necessarily equal to $\nabla S(\mathbf{x})/m$, and to give up the assumption that $\psi$ must necessarily satisfy a homogeneous linear equation of the general type suggested by Schroedinger. As we shall see, giving up either of those first two assumptions will in general also require us to give up the assumption of a statistical ensemble of particles, with a probability density equal to $|\psi(\mathbf{x})|^2$.

We begin by noting that it is consistent with our interpretation to modify the equations of motion of a particle (8a) by adding any conceivable force term to the right-hand side. Let us, for the sake of illustration, consider a force that tends to make the difference, $\mathbf{p} - \nabla S(\mathbf{x})$, decay rapidly with time, with a mean decay time of the order of $\tau = 10^{-13}/c$ seconds, where $c$ is the velocity of light. To achieve this result, we write

$$m\frac{d^2\mathbf{x}}{dt^2} = -\nabla\left\{V(\mathbf{x}) - \frac{\hbar^2}{2m}\frac{\nabla^2 R}{R}\right\} + \mathbf{f}(\mathbf{p} - \nabla S(\mathbf{x})), \qquad (31)$$

where $\mathbf{f}(\mathbf{p} - \nabla S(\mathbf{x}))$ is assumed to be a function which vanishes when $\mathbf{p} = \nabla S(\mathbf{x})$ and more generally takes such a form that it implies a force tending to make $\mathbf{p} - \nabla S(\mathbf{x})$ decrease rapidly with the passage of time. It is clear, moreover, that $f$ can be so chosen that it is large only in processes involving very short distances (where $\nabla S(\mathbf{x})$ should be large).

If the correct equations of motion resembled Eq. (31), then the usual interpretation would be applicable only over times much longer than $\tau$, for only after such times have elapsed will the relation $\mathbf{p} = \nabla S(\mathbf{x})$ be a good approximation. Moreover, it is clear that such modifications of the theory cannot even be described in the usual interpretation, because they involve the precisely definable particle variables which are not postulated in the usual interpretation.

Let us now consider a modification that makes the equation governing $\psi$ inhomogeneous. Such a modification is

$$i h \psi / \partial t = H \psi + \xi(\mathbf{p} - \nabla S(\mathbf{x}_i)). \tag{32}$$

Here, $H$ is the usual Hamiltonian operator, $\mathbf{x}_i$, represents the actual location of the particle, and $\xi$ is a function that vanishes when $\mathbf{p} = \nabla S(\mathbf{x}_i)$. Now, if the particle equations are chosen, as in Eq. (31), to make $\mathbf{p} - \nabla S(\mathbf{x}_i)$ decay rapidly with time, it follows that in atomic processes, the inhomogeneous term in Eq. (32) will become negligibly small, so that Schroedinger's equation is a good approximation. Nevertheless, in processes involving very short distances and very short times, the inhomogeneities would be important, and the $\psi$-field would, as in the case of the electromagnetic field, depend to some extent on the actual location of the particle.

It is clear that Eq. (32) is inconsistent with the usual interpretation of the theory. Moreover, we can contemplate further generalizations of Eq. (32), in the direction of introducing nonlinear terms that are large only for processes involving small distances. Since the usual interpretation is based on the hypothesis of linear superposition of "state vectors" in a Hilbert space, it follows that the usual interpretation could not be made consistent with such a nonlinear equation for a one-particle theory. In a many-particle theory, operators can be introduced, satisfying a nonlinear generalization of Schroedinger's equation; but these must ultimately operate on wave functions that satisfy a linear homogeneous Schroedinger equation.

Finally, we repeat a point already made in Sec. 4, namely, that if the theory is generalized in any of the ways indicated here, the probability density of particles will cease to equal $|\psi(\mathbf{x})|^2$. Thus, experiments would become conceivable that distinguish between $|\psi(\mathbf{x}))|^2$ and this probability; and in this way we could obtain an experimental proof that the usual interpretation, which gives $|\psi(\mathbf{x})|^2$ *only* a probability interpretation, must be inadequate. Moreover, we shall show in Paper II that modifications like those suggested here would permit the particle position and momentum to be measured simultaneously, so that the uncertainty principle could be violated.

## ACKNOWLEDGMENT

The author wishes to thank Dr. Einstein for several interesting and stimulating discussions.

# A Suggested Interpretation of the Quantum Theory in Terms of "Hidden" Variables. II

By

David Bohm

In this paper, we shall show how the theory of measurements is to be understood from the point of view of a physical interpretation of the quantum theory in terms of "hidden" variables, developed in a previous paper. We find that in principle, these "hidden" variables determine the precise results of each individual measurement process. In practice, however, in measurements that we now know how to carry out, the observing apparatus disturbs the observed system in an unpredictable and uncontrollable, way, so that the uncertainty principle is obtained as a practical limitation on the possible precision of measurements. This limitation is not, however, inherent in the conceptual structure of our interpretation. We shall see, for example, that simultaneous measurements of position and momentum having unlimited precision would in principle be possible if, as suggested in the previous paper, the mathematical formulation of the quantum theory needs to be modified at very short distances in certain ways that are consistent with our interpretation but not with the usual interpretation.

We give a simple explanation of the origin of quantum-mechanical correlations of distant objects in the hypothetical experiment of Einstein, Podolsky, and Rosen, which was suggested by these authors as a criticism of the usual interpretation.

Finally, we show that von Neumann's proof that quantum theory is not consistent with hidden variables does not apply to our interpretation, because the hidden variables contemplated here depend both

*Now at Universidade de São Paulo, Faculdade de Filosofia, Ciencias e Letras, São Paulo, Brasil.

Reprinted with permission from the American Physical Society: D. Bohm, *Physical Review*, Volume **85**, Number 2, 1952. © 1952 by the American Physical Society.

on the state of the measuring apparatus and the observed system and therefore go beyond certain of von Neumann's assumptions.

In two appendixes, we treat the problem of the electromagnetic field in our interpretation and answer certain additional objections which have arisen in the attempt to give a precise description for an individual system at the quantum level.

# 1. INTRODUCTION

In a previous paper,[1] to which we shall hereafter refer as I, we have suggested an interpretation of the quantum theory in terms of "hidden" variables. We have shown that although this interpretation provides a conceptual framework that is broader than that of the usual interpretation, it permits of a set of three mutually consistent special assumptions, which lead to the same physical results as are obtained from the usual interpretation of the quantum theory. These three special assumptions are: (1) The $\psi$-field satisfies Schroedinger's equation. (2) If we write $\psi = R \exp(is/h)$, then the particle momentum is restricted to $\mathbf{p} = \nabla S(\mathbf{x})$. (3) We have a statistical ensemble of particle positions, with a probability density, $P = |\psi(\mathbf{x})|^2$. If the above three special assumptions are not made, then one obtains a more general theory that cannot be made consistent with the usual interpretation. It was suggested in Paper I that such generalizations may actually be needed for an understanding of phenomena associated with distances of the order of $10^{-13}$ cm or less, but may produce changes of negligible importance in the atomic domain.

In this paper, we shall apply the interpretation of the quantum theory suggested in Paper I to the development of a theory of measurements in order to show that as long as one makes the special assumptions indicated above, one is led to the same predictions for all measurements as are obtained from the usual interpretation. In our interpretation, however, the uncertainty principle is regarded, not as an inherent limitation on the precision with which we can correctly

[1] D. Bohm, Phys. Rev. **84**, 166 (1951).

conceive of the simultaneous definition of momentum and position, but rather as a practical limitation on the precision with which these quantities can simultaneously be measured, arising from unpredictable and uncontrollable disturbances of the observed system by the measuring apparatus. If the theory needs to be generalized in the ways suggested in Paper I, Secs. 4 and 9, however, then these disturbances could in principle either be eliminated, or else be made subject to prediction and control, so that their effects could be corrected for. Our interpretation therefore demonstrates that measurements violating the uncertainty principle are at least conceivable.

## 2. QUANTUM THEORY OF MEASUREMENTS

We shall now show how the quantum theory of measurements is to be expressed in terms of our suggested interpretation of the quantum theory.[2]

In general, a measurement of any variable must always be carried out by means of an interaction of the system of interest with a suitable piece of measuring apparatus. The apparatus must be so constructed that any given state of the system of interest will lead to a certain range of states of the apparatus. Thus, the interaction introduces correlations between the state of the observed system and the state of the apparatus. The range of indefiniteness in this correlation may be called the uncertainty, or the error, in the measurement.

Let us now consider an observation designed to measure an arbitrary (hermitian) "observable" $Q$, associated with an electron. Let $\mathbf{x}$ represent the position of the electron, $y$ that of the significant apparatus coordinate (or coordinates if there are more than one). Now, one can show[2] that it is enough to consider an impulsive measurement, i.e., a measurement utilizing a very strong interaction between apparatus and system under observation, which lasts for so short a time that the changes of the apparatus and the system under observation that

---

[2] For a treatment of how the theory of measurements can be carried out with the usual interpretation, see D. Bohm, *Quantum Theory* (Prentice-Hall, Inc., New York, 1951), Chapter 22.

would have taken place in the absence of interaction can be neglected. Thus, at least while the interaction is taking place, we can neglect the parts of the Hamiltonian associated with the apparatus alone and with the observed system alone, and we need retain only the part of the Hamiltonian, $H_1$, representing the interaction. Moreover, if the Hamiltonian operator is chosen to be a function only of quantities that commute with $Q$, then the interaction process will produce no uncontrollable changes in the observable, $Q$, but only in observables that do not commute with $Q$. In order that the apparatus and the system under observation shall be coupled, however, it is necessary that $H_1$ shall also depend on operators involving $y$.

For the sake of illustration of the principles involved, we shall consider the following interaction Hamiltonian:

$$H_1 = -a\,Qp_y, \tag{1}$$

where $a$ is a suitable constant and $p_y$ is the momentum conjugate to $y$.

Now, in our interpretation, the system is to be described by a four-dimensional but objectively real wave field that is a function of $\mathbf{x}$ and $y$ and by a corresponding four-dimensional representative point, specified by the coordinates, $\mathbf{x}$, of the electron and the coordinate, $y$, of the apparatus. Since the motion of the representative point is in part determined by forces produced by the $\psi$-field acting on both electron and apparatus variables, our first step in solving this problem is to calculate the $\psi$-field. This is done by solving Schroedinger's equation, with the appropriate boundary conditions on $\psi$.

Now, during interaction, Schroedinger's equation is approximated by

$$i\hbar\,\partial\Psi/\partial t = -a\,Qp_y\Psi = (ia/\hbar)\,Q\partial\Psi/\partial y. \tag{2}$$

It is now convenient to expand $\Psi$ in terms of the complete set $\psi_q(\mathbf{x})$ of eigenfunctions of the operator, $Q$, where $q$ denotes an eigenvalue of $Q$. For the sake of simplicity, we assume that the spectrum of $Q$ is discrete, although the results are easily generalized to a continuous spectrum. Denoting the expansion coefficients by $f_q(y, t)$, we obtain

$$\Psi(\mathbf{x}, y, t) = \sum_q \psi_q(\mathbf{x}) f_q(y, t). \tag{3}$$

Noting that $Q\psi_q(\mathbf{x}) = q\psi_q(\mathbf{x})$, we readily verify that Eq. (2) can now be reduced to the following series of equations for $f_q(y, t)$:

$$i\hbar \partial f_q(y, t)/\partial t = (ia/\hbar)q f_q(y, t). \tag{4}$$

If the initial value of $f_q(y, t)$ and $f_q{}^0(y)$ we obtain as a solution

$$f_q(y, t) = f_q{}^0(y - aqt/\hbar^2), \tag{5}$$

and

$$\Psi(\mathbf{x}, y, t) = \sum_q \psi_q(\mathbf{x}) f_q{}^0(y - aqt/\hbar^2). \tag{6}$$

Now, Initially the apparatus and the electron were independent. As shown in Paper I, Sec. 7, in our interpretation (as in the usual interpretation), independent systems must have wave fields $\Psi(\mathbf{x}, y, t)$ that are equal to a product of a function of $\mathbf{x}$ and a function of $y$. Initially, we therefore have

$$\Psi_0(\mathbf{x}, y) = \psi_0(\mathbf{x})g_0(y) = g_0(y) \sum_q c_q \psi_q(\mathbf{x}), \tag{7}$$

where the $c_q$ are the (unknown) expansion coefficients of $\psi_q(\mathbf{x})$, and $g_0(y)$ is the initial wave function of the apparatus coordinate, $y$. The function $g_0(y)$ will take the form of a packet. For the sake of convenience, we assume that this packet is centered at $y = 0$ and that its width is $\Delta y$. Normally, because the apparatus is classically describable, the definition of this packet is far less precise than that allowed by the limits of precision set by the uncertainty principle.

From Eqs. (7) and (3), we shall readily deduce that $f_q{}^0(y) = c_q g_0(y)$ When this value of $f_q{}^0(y)$ is inserted into Eq. (6), we obtain

$$\Psi(\mathbf{x}, y, t) = \sum_q c_q \psi_q(\mathbf{x})g_0(y - aqt/\hbar^2). \tag{8}$$

Equation (8) indicates already that the interaction has introduced a correlation between $q$ and the apparatus coordinate, $y$. In order to show what this correlation means in our interpretation of the quantum theory, we shall use some arguments that have been developed in more detail in Paper I, Sec. 7, in connection with a similar problem involving the interaction of two particles in a scattering process. First we note that while the electron and the apparatus are interacting, the wave function (8) becomes very complicated, so that if it is expressed as

$$\Psi(\mathbf{x}, y, t) = R(\mathbf{x}, y, t) \exp[i S(\mathbf{x}, y, t)/h],$$

then $R$ and $S$ undergo rapid oscillations both as a function of position and of time. From this we deduce that the "quantum-mechanical" potential,

$$U = (-h^2/2m R)(\nabla_x{}^2 R + \partial^2 R/\partial y^2),$$

undergoes violent fluctuations, especially where $R$ is small, and that the particle momenta, $\mathbf{p} = \nabla_x S(\mathbf{x}, y, t)$ and $p_y = \partial S(\mathbf{x}, y, t)/\partial y$, also undergo corresponding violent and extremely complicated fluctuations. Eventually, however, if the interaction continues long enough, the behavior of the system will become simpler because the packets $g_0(y - aqt/h^2)$, corresponding to different values of $q$, will cease to overlap in $y$ space. To prove this, we note that the center of the $q$th packet in $y$ space is at

$$y = aqt/h^2; \quad \text{or} \quad q = h^2 y/at. \tag{9}$$

If we denote the separation of adjacent values of $q$ by $\delta q$, we then obtain for the separation of the centers of adjacent packets in $y$ space

$$\delta y = at\delta q/h^2. \tag{10}$$

It is clear that if the product of the strength of interaction $a$, and the duration of Interaction, $t$, is large enough, then $\delta y$ can be made much larger than the width $\Delta y$ of the packet. Then packets corresponding to different values of $q$ will cease to overlap in $y$ space and will, in fact, obtain separations large enough to be classically describable.

Because the probability density is equal to $|\Psi|^2$, we deduce that the apparatus variable, $y$, must finally enter one of the packets and remain with that packet thereafter (since it does not enter the intermediate space between packets in which the probability density is practically zero). Now, the packet entered by the apparatus variable $y$ determines the actual result of the measurement, which the observer will obtain when he looks at the apparatus. The other packets can (as shown in Paper I, Sec. 7) be ignored, because they affect neither the quantum-mechanical potential acting on the particle coordinates $\mathbf{x}$ and $y$, nor the particle momenta, $\mathbf{p}_x = \nabla_x S$ and $p_y = \partial S / \partial y$. Moreover, the wave function can also be renormalized without affecting any of the above quantities. Thus, for all practical purposes, we can replace the complete wave function, Eq. (8), by a new renormalized wave function

$$\Psi(\mathbf{x}, y) = \psi_q(\mathbf{x}) g_0(y - aqt/h^2), \tag{11}$$

where $q$ now corresponds to the packet actually containing the apparatus variable, $y$. From this wave function, we can deduce, as shown in Paper I, Sec. 7, that the apparatus and the electron will subsequently behave independently. Moreover, by observing the approximate value of the apparatus coordinate within an error $\Delta y \ll \delta y$, we can deduce with the aid of Eq. (9) that since the electron wave function can for all practical purposes be regarded as $\psi_q(\mathbf{x})$, the observable, $Q$, must have the definite value, $q$. However, if the product, "$at\delta q/h^2$," appearing in Eqs. (8), (9), (10), and (11), had been less than $\Delta y$, then no clear measurement of $Q$ would have been possible, because packets corresponding to different $q$ would have overlapped, and the measurement would not have had the requisite accuracy.[3]

Finally, we note that even if the apparatus packets are subsequently caused to overlap, none of those conclusions will be altered. For the apparatus variable $y$ will inevitably be coupled to a whole host of internal thermodynamic degrees of freedom, $y_1, y_2, \ldots y_s$, as a result of effects such as friction and brownian motion. As shown in Paper I, Sec. 7, interference between packets corresponding to different values of $q$ would be possible only if the packets overlapped in the space of $y_1, y_2, \ldots y_s$, as well as in $y$ space. Such an overlap, however, is so improbable that for all practical purposes, we can ignore the possibility that it will ever occur.

---

[3] A similar requirement is obtained in the usual interpretation. See references 2, Chapter 22, Sec. 8.

# 3. THE ROLE OF PROBABILITY IN MEASUREMENTS—THE UNCERTAINTY PRINCIPLE

In principle, the final result of a measurement is determined by the initial form of the wave function of the combined system, $\Psi^0(\mathbf{x}, y)$, and by the initial position of the electron particle, $\mathbf{x}_0$, and the apparatus variable, $y_0$. In practice, however, as we have seen, the orbit fluctuates violently while interaction takes place, and is very sensitive to the precise initial values of $\mathbf{x}$ and $y$, which we can neither predict nor control. All that we can predict in practice is that in an ensemble of similar experiments performed under equivalent initial conditions, the probability density is $|\Psi(\mathbf{x}, y)|^2$. From this information, however, we are able to calculate only the probability that in an individual experiment, the result of a measurement of $Q$ will be a specific number $q$. To obtain the probability of a given value of $q$, we need only integrate the above probability density over all $\mathbf{x}$ and over all values of $y$ in the neighborhood of the $q$th packet. Because the packets do not overlap, the $\Psi$-field in this region is equal to $c_q \psi_q(\mathbf{x}) g_0(y - aqt/h^2)$ [see Eq. (8)]. Since, by definition, $\psi_q(\mathbf{x})$ and $g_0(y)$ are normalized, the total probability that a particle is in the $q$th packet is

$$P_q = |c_q|^2. \tag{12}$$

The above is, however, just what is obtained from the usual interpretation. We conclude then that our interpretation is capable of leading in all possible experiments to identical predictions with those obtained from the usual interpretation (provided, or course, that we make the special assumptions indicated in the introduction).

Let us now see what a measurement of the observable, $Q$, implies with regard to the state of the electron particle and its $\Psi$-field. First, we note that the process of interaction with an apparatus designed to measure the observable, $Q$, effectively transforms the electron $\psi$-field from whatever it was before the measurement took place into an eigenfunction $\psi_q(\mathbf{x})$ of the operator $Q$. The precise value of $q$ that comes out of this process is as we have seen, not, in general, completely predictable or controllable. If, however, the same measurement is repeated after the $\psi$-field has been transformed into $\psi_q(\mathbf{x})$, we can then predict that (as in the usual interpretation), the same value of $q$, and therefore the same wave function, $\psi_q(\mathbf{x})$, will be obtained again. If, however, we measure an observable "$P$" that does not commute with $Q$, then the results of this measurement are not, in practice, predictable or controllable. For as shown in Eq. (8), the $\Psi$-field after interaction with the measuring apparatus is now transformed into

$$\Psi(\mathbf{x}, z, t) = \sum_p a_{p,q} \phi_p(\mathbf{x}) g_0(z - apt/h^2), \qquad (13)$$

where $\phi_p(\mathbf{x})$ is an eigenfunction of the operator, $P$, belonging to an eigenvalue, $p$, and where $a_{p,q}$ is an expansion coefficient defined by

$$\psi_q(\mathbf{x}) = \sum_p a_{p,q} \phi_p(\mathbf{x}). \qquad (14)$$

Since the packets corresponding to different $p$ ultimately become completely separate in $z$ space, we deduce, as in the case of the measurement of $Q$, that for all practical purposes, this wave function may be replaced by

$$\Psi = a_{p\,q} \phi_p(\mathbf{x}) g_0(z - apt/h^2),$$

where $p$ now represents the packet actually entered by the apparatus coordinate, $y$. As in the case of measurement of $Q$, we readily show that the precise value of $p$ that comes out of this experiment cannot be predicted or controlled and that the probability of a given value of $p$ is equal to $|a_{pq}|^2$. This is, however, just what is obtained in the usual interpretation of this process.

It is clear that if two "observables," $P$ and $Q$, do not commute, one cannot carry out a measurement of both simultaneously on the same system. The reason is that each measurement disturbs the system in a way that is incompatible with carrying out the process necessary for the measurement of the other. Thus, a measurement of $P$ requires that wave field, $\psi$, shall become an eigenfunction of $P$, while a measurement of $Q$ requires that it shall become an eigenfunction of $Q$. If $P$ and $Q$ do not commute, then by definition, no $\psi$-function can be simultaneously an eigenfunction of both. In this way, we understand in our interpretation why measurements, of complementary quantities, must (as in the usual interpretation) necessarily be limited in their precision by the uncertainty principle.

## 4. PARTICLE POSITIONS AND MOMENTA AS "HIDDEN VARIABLES"

We have seen that in measurements that can now be carried out, we cannot make precise inferences about the particle position, but can say only that the particle must be somewhere in the region in which $|\psi|$ is appreciable. Similarly, the momentum of a particle that happens to be at the point, $\mathbf{x}$, is given by $\mathbf{p} = \nabla S(\mathbf{x})$, so that since $\mathbf{x}$ is not known, the precise value of $\mathbf{p}$ is also not, in general, inferrable. Hence, as long as we are restricted to making observations of this kind, the precise values of the particle position and momentum must, in general, be regarded as "hidden," since we cannot at present measure them. They are, however, connected with real and already observable properties of matter because (along with the $\psi$-field) they determine in principle the actual result of each individual measurement. By way of contrast, we recall here that in the usual interpretation of the theory, it is stated that although each measurement admittedly leads to a definite number, nothing determines the actual value of this number. The result of each measurement is assumed to arise somehow in an inherently indescribable way that is not subject to a detailed analysis. Only the statistical results are said to be predictable. In our interpretation, however, we assert that the at present "hidden" precisely definable particle positions and momenta determine the results of each individual measurement process, but in a way whose precise details are so complicated and uncontrollable, and so little known, that one must for all practical purposes restrict oneself to a statistical description of the connection between the values of these variables and the directly observable results of measurements. Thus, we are unable at present to obtain direct experimental evidence for the existence of precisely definable particle positions and momenta.

# 5. "OBSERVABLES" OF USUAL INTERPRETATION ARE NOT A COMPLETE DESCRIPTION OF SYSTEM IN OUR INTERPRETATION

We have seen in Sec. 3 that in the measurement of an "observable," $Q$, we cannot obtain enough information to provide a complete specification of the state of an electron, because we cannot infer the precisely defined values of the particle momentum and position, which are, for example, needed if we wish to make precise predictions about the future behavior of the electron. Moreover, the process of measuring an observable does not provide any unambiguous information about the state that existed before the measurement took place; for in such a measurement, the $\psi$-field is transformed into an in practice unpredictable and uncontrollable eigenfunction, $\psi_q(\mathbf{x})$, of the measured "observable" $Q$. This means that the measurement of an "observable" is not really a measurement of any physical property belonging to the observed system alone. Instead, the value of an "observable" measures only an incompletely predictable and controllable potentiality belonging just as much to the measuring apparatus as to the observed system itself.[4] At best, such a measurement provides unambiguous information only at a classical level of accuracy, where the disturbance of the $\psi$-field by the measuring apparatus can be neglected. The usual "observables" are therefore not what we ought to try to measure at a quantum level of accuracy. In Sec. 6, we shall see that it is conceivable that we may be able to carry out new kinds of measurements, providing information not about "observables" having a very ambiguous significance, but rather about physically significant properties of a system, such as the actual values of the particle position and momentum.

As an example of the rather indirect and ambiguous significance of the "observable," we may consider the problem of measuring the momentum of an electron. Now, in the usual interpretation, it is stated that one can always measure the momentum "observable" without

---

[4]Even in the usual interpretation, an observation must be regarded as yielding a measure of such a potentiality. See reference 2, Chapter 6, Sec. 9.

changing the value of the momentum. The result is said, for example, to be obtainable with the aid of an impulsive interaction involving only operators which commute with the momentum operator, $\mathbf{p}_x$. To represent such a measurement, we could choose $H_I = -a p_x p_y$ in Eq. (1). In our interpretation, however, we cannot in general conclude that such an interaction will enable us to measure the actual particle momentum without changing its value. In fact, in our interpretation, a measurement of particle momentum that does not change the value of this momentum is possible only if the $\psi$-field initially takes the special form, $\exp(i\mathbf{p} \cdot \mathbf{x}/h)$. If, however, $\psi$ initially takes its most general possible form,

$$\psi = \sum_p a_1 \mathbf{p} \exp\left(i\mathbf{p} \cdot \mathbf{x}/h\right), \tag{15}$$

then as we have seen in Secs. 2 and 3, the process of measuring the "observable" $p_x$ will effectively transform the $\psi$-field of the electron into

$$\exp(i p_x / h) \tag{16}$$

with a probability $|a_p|^2$ that a given value of $p_x$ will be obtained. When the $\psi$-field is altered in this way, large quantities of momentum can be transferred to the particle by the changing $\psi$-field, even though the interaction Hamiltonian, $H_1$, commutes with the momentum operator, $\mathbf{p}$.

As an example, we may consider a stationary state of an atom, of zero angular momentum. As shown in Paper I, Sec. 5, the $\psi$-field for such a state is real, so that we obtain

$$\mathbf{p} = \nabla S = 0.$$

Thus, the particle is at rest. Nevertheless, we see from Eqs. (14) and (15) that if the momentum "observable" is measured, a large value of this "observable" may be obtained if the $\psi$-field happens to have a large fourier coefficient, $a_p$, for a high value of $\mathbf{p}$. The reason is that in the process of interaction with the measuring apparatus, the $\psi$-field is altered in such a way that it can give the electron particle a correspondingly large momentum, thus transforming some of the potential energy of interaction of the particle with its $\psi$-field into kinetic energy.

A more striking illustration of the points discussed above is afforded by the problem of a "free" particle contained between two impenetrable and perfectly reflecting walls, separated by a distance $L$. For this case, the spatial part of the $\psi$-field is

$$\psi = \sin(2\pi nx/L),$$

where $n$ is an integer and the energy of the electron is

$$E = (1/2m)\,(nh/L)^2.$$

Because the $\psi$-field is real, we deduce that the particle is at rest.

Now, at first sight, it may seem puzzling that a particle having a high energy should be at rest in the empty space between two walls. Let us recall, however, that the space is not really empty, but contains an objectively real $\psi$-field that can act on the particle. Such an action is analogous to (but of course not identical with) the action of an electromagnetic field, which could create non-uniform motion of the particle in this apparently "empty" enclosure. We observe that in our problem, the $\psi$-field is able to bring the particle to rest and to transform the entire kinetic energy into potential energy of interaction with the $\psi$-field. To prove this, we evaluate the "quantum-mechanical potential" for this $\psi$-field

$$U = \frac{-h^2}{2m}\frac{\nabla^2 R}{R} = \frac{-h^2}{2m}\frac{\nabla^2 \psi}{\psi} = \frac{1}{2m}\left(\frac{nh}{L}\right)^2$$

and note that it is precisely equal to the total energy, $E$.

Now, as we have seen, any measurement of the momentum "observable" must change the $\psi$-field in such a way that in general some (and in our case, all) of this potential energy is transformed into kinetic energy. We may use as an illustration of this general result a very simple specific method of measuring the momentum "observable," namely, to remove the confining walls suddenly and then to measure the distance moved by the particle after a fairly long time. We can compute the momentum by dividing this distance by the time of transit. If (as in the usual interpretation of the quantum theory) we assume that the electron is "free," then we conclude that the process of removing the walls should not appreciably change the momentum if we do it fast enough, for the probability that the particle is near a wall when this happens can then in principle be made arbitrarily small. In our interpretation, however, the removal of the walls alters the particle momentum indirectly, because of its effect on the $\psi$-field, which acts on the particle. Thus, after the walls are removed, two wave packets moving in opposite directions begin to form, and ultimately they become completely separate in space. Because the probability density is $|\psi|^2$, we deduce that the particle must end up in one packet or the other. Moreover, the reader will readily convince himself that the particle momentum will be very close to $\pm nh/L$, the sign depending on which packet the particle actually enters. As in Sec. (2), the packet not containing the particle can subsequently be ignored. In principle, the final particle momentum is determined by the initial form of the $\psi$-field and by the initial particle position. Since we do not in practice know the latter, we can at best predict a probability of $\frac{1}{2}$ that the particle ends up in either packet. We conclude then that this measurement of the momentum "observable" leads to the same result as is predicted in the usual interpretation. However, the actual particle momentum existing before the measurement took place is quite different from the numerical value obtained for the momentum "observable," which, in the usual interpretation, is called the "momentum."

## 6. ON THE POSSIBILITY OF MEASUREMENTS OF UNLIMITED PRECISION

We have seen that the so-called "observables" do not measure any very readily interpretable properties of a system. For example, the momentum "observable" has in general no simple relation to the actual particle momentum. It may therefore be fruitful to consider how we might try to measure properties which, according to our interpretation, are (along with the $\psi$-field) the physically significant properties of an electron, namely, the actual particle position and momentum. In connection with this problem, we shall show that if, as suggested in Paper I, Secs. 4 and 9, we give up the three mutually consistent special assumptions leading to the same results as those of the usual interpretation of the quantum theory, then in our interpretation, the particle position and momentum can in principle be measured simultaneously with unlimited precision.

Now, for our purposes, it will be adequate to show that precise predictions of the future behavior of a system are in principle possible. In our interpretation, a sufficient condition for precise predictions is as we have seen that we shall be able to prepare a system in a state in which the $\psi$-field and the initial particle position and momentum are precisely known. We have shown that it is possible, by measuring the "observable," $Q$, with the aid of methods that are now available, to prepare a state in which the $\psi$-field is effectively transformed into a known form, $\psi_q(\mathbf{x})$; but we cannot in general predict or control the precise position and momentum of the particle. If we could now measure the position and momentum of the particle without altering the $\psi$-field, then precise predictions would be possible. However, the results of Secs. 2, 3, and 4 prove that as long as the three special assumptions indicated above are valid, we cannot measure the particle position more accurately without effectively transforming the $\psi$-function into an incompletely predictable and controllable packet that is much more localized than $\psi_q(\mathbf{x})$. Thus, efforts to obtain more precise definition of the state of the system will be defeated. But it is clear that the difficulty originates in the circumstance that the potential energy of interaction between electron and apparatus, $V(\mathbf{x}, y)$, plays two roles. For it not only introduces a direct interaction between the two particles, proportional in strength to $V(\mathbf{x}, y)$ itself, but it introduces an indirect interaction between these particles, because this potential also appears in the equation governing the $\psi$-field. This indirect interaction may involve rapid and violent fluctuations, even when $V(\mathbf{x}, y)$ is small. Thus, we are led to lose control of the effects of this interaction, because no matter how small $V(\mathbf{x}, y)$ is, very large and chaotically complicated disturbances in the particle motion may occur.

If, however, we give up the three special assumptions mentioned previously, then it is not inherent in our conceptual structure that every interaction between particles must inevitably also produce large and uncontrollable changes in the $\psi$-field. Thus, in Paper I, Eq. (31), we give an example in which we postulate a force acting on a particle that is not necessarily accompanied by a corresponding change in the $\psi$-field. Equation (31), Paper I, is concerned only with a one-particle system, but similar assumptions can be made for systems of two or more particles. In the absence of any specific theory, our interpretation permits an infinite number of kinds of such modifications, which can be chosen to be important at small distances but negligible in the atomic domain. For the sake of illustration, suppose that it should turn out that in certain processes connected with very small distances, the force acting on the apparatus variable is

$$F_y = ax,$$

where $a$ is a constant. Now if "$a$" is made large enough so that the interaction is impulsive, we can neglect all changes in $y$ that are brought about by the forces that would have been present in the absence of this interaction. Moreover, for the sake of illustration of the principles involved, we are permitted to make the assumption, consistent with our interpretation, that the force on the electron is zero. The equation of motion of $y$ is then

$$\ddot{y} = ax/m.$$

The solution is

$$y - y_0 = (axt^2/2m) + \dot{y}_0 t,$$

where $\dot{y}_0$ is the initial velocity of the apparatus variable and $y_0$ its initial position. Now, if the product, $at^2$, is large enough, then $y - y_0$ can be made much larger than the uncertainty in $y$ arising from the uncertainty of $y_0$, and the uncertainty of $\dot{y}_0$. Thus, $y - y_0$ will be determined primarily by the particle position, $x$. In this way, it is conceivable that we could obtain a measurement of $x$ that does not significantly change $x$, $\dot{x}$, or the $\psi$-function. The particle momentum can then be obtained from the relation, $p = \nabla S(\mathbf{x})$, where $S/h$ is the phase of the $\psi$-function. Thus, precise predictions would in principle be possible.

## 7. THE ORIGIN OF THE STATISTICAL ENSEMBLE IN THE QUANTUM THEORY

We shall now see that even if, because of a failure of the three special assumptions mentioned in Secs. 1 and 6, we are able to determine the particle positions and momenta precisely, we shall nevertheless ultimately obtain a statistical ensemble again at the atomic level, with a probability density equal to $|\psi|^2$. The need for such an ensemble arises from the chaotically complicated character of the coupling between the electron and classical systems, such as volumes of gas, walls of containers, pieces of measuring apparatus, etc., with which this particle must inevitably in practice interact. For as we have seen in Sec. 2, and in Paper I, Sec. 7, during the course of such an interaction, the "quantum-mechanical" potential undergoes violent and rapid fluctuations, which tend to make the particle orbit wander over the whole region in which the $\psi$-field is appreciable. Moreover, these fluctuations are further complicated by the effects of molecular chaos in the very large number of internal thermodynamic degrees of freedom of these classically describable systems, which are inevitably excited in any interaction process. Thus, even if the initial particle variables were well defined, we should soon in practice lose all possibility of following the particle motion and would be forced to have recourse to some kind of statistical theory. The only question that remains is to show why the probability density that ultimately comes about should be equal to $|\psi|^2$ and not to some other quantity.

To answer this question, we first note that a statistical ensemble with a probability density $|\psi(\mathbf{x})|^2$ has the property that under the action of forces which prevail at the atomic level, where our three special assumptions are satisfied, it will be preserved by the equations of motion of the particles, once it comes into existence. There remains only the problem of showing that an arbitrary deviation from this ensemble tends, under the action of the chaotically complicated forces described in the previous paragraph, to decay into an ensemble with a probability density of $|\psi(\mathbf{x})|^2$. This problem is very similar to that of proving Boltzmann's $H$ theorem, which shows in connection with a different but analogous problem that an arbitrary ensemble tends as a result of molecular chaos to decay into an equilibrium Gibbs ensemble. We shall not carry out a detailed proof here, but we merely suggest that it seems plausible that one could along similar lines prove that in our problem, an arbitrary ensemble tends to decay into an ensemble with a density of $|\psi(\mathbf{x})|^2$. These arguments indicate that in our interpretation, quantum fluctuations and classical fluctuations (such as the Brownian motion) have basically the same origin; *viz.*, the chaotically complicated character of motion at the microscopic level.

## 8. THE HYPOTHETICAL EXPERIMENT OF EINSTEIN, PODOLSKY, AND ROSEN

The hypothetical experiment of Einstein, Podolsky, and Rosen[5] is based on the fact that if we have two particles, the sum of their momenta, $p = p_1 + p_2$, commutes with the difference of their positions, $\xi = x_1 - x_2$. We can therefore define a wave function in which $p$ is zero, while $\xi$ has a given value, $a$. Such a wave function is

$$\psi = \delta(x_1 - x_2 - a). \tag{17}$$

In the usual interpretation of the quantum theory, $p_1 - p_2$ and $x_1 + x_2$ are completely undetermined in a system having the above wave function.

[5] Einstein, Podolsky, and Rosen, Phys. Rev. **47**, 777 (1933).

The whole experiment centers on the fact that an observer has a choice of measuring either the momentum or the position of any one of the two particles. Whichever of these quantities he measures, he will be able to infer a definite value of the corresponding variable in the other particle, because of the fact that the above wave function implies correlations between variables belonging to each particle. Thus, if he obtains a position $x_1$ for the first particle, he can infer a position of $x_2 = a - x_1$ for the second particle; but he loses all possibility of making any inferences about the momenta of either particle. On the other hand, if he measures the momentum of the first particle and obtains a value of $p_1$, he can infer a value of $p_2 = -p_1$ for the momentum of the second particle; but he loses all possibility of making any inferences about the position of either particle. Now, Einstein, Podolsky, and Rosen believe that this result is itself probably correct, but they do not believe that quantum theory as usually interpreted can give a complete description of how these correlations are propagated. Thus, if these were classical particles, we could easily understand the propagation of correlations because each particle would then simply move with a velocity opposite to that of the other. But in the usual interpretation of quantum theory, there is no similar conceptual model showing in detail how the second particle, which is not in any way supposed to interact with the first particle, is nevertheless able to obtain either an uncontrollable disturbance of its position or an uncontrollable disturbance of its momentum depending on what kind of measurement the observer decided to carry out on the first particle. Bohr's point of view is, however, that no such model should be sought and that we should merely accept the fact that these correlations somehow manage to appear. We must note, of course, that the quantum-mechanical description of these processes will always be consistent, even though it gives us no precisely definable means of describing and analyzing the relationships between the classically describable phenomena appearing in various pieces of measuring apparatus.

In our suggested new interpretation of the quantum theory, however, we can describe this experiment in terms of a single precisely definable conceptual model, for we now describe the system in terms of a combination of a six-dimensional wave field and a precisely definable trajectory in a six-dimensional space (see Paper I, Sec. 6). If the wave function is initially equal to Eq. (17), then since the phase vanishes, the particles are both at rest. Their possible positions are, however, described by an ensemble, in which $x_1 - x_2 = a$. Now, if we measure the position of the first particle, we introduce uncontrollable fluctuations in the wave function for the entire system, which, through the "quantum-mechanical" forces, bring about corresponding uncontrollable fluctuations in the momentum of each particle. Similarly, if we measure the momentum of the first particle, uncontrollable fluctuations in the wave function for the system bring about, through the "quantum-mechanical" forces, corresponding uncontrollable changes in the position of each particle. Thus, the "quantum-mechanical" forces may be said to transmit uncontrollable disturbances instantaneously from one particle to another through the medium of the $\psi$-field.

What does this transmission of forces at an infinite rate mean? In nonrelativistic theory, it certainly causes no difficulties. In a relativistic theory, however, the problem is more complicated. We first note that as long as the three special assumptions mentioned in Sec. 2 are valid, our interpretation can give rise to no inconsistencies with relativity, because it leads to precisely the same predictions for all physical processes as are obtained from the usual interpretation (which is known to be consistent with relativity). The reason why no contradictions with relativity arise in our interpretation despite the instantaneous transmission of momentum between particles is that no signal can be carried in this way. For such a transmission of momentum could constitute a signal only if there were some practical means of determining precisely what the second particle would have done if the first particle had not been observed; and as we have seen, this information cannot be obtained as long as the present form of the quantum theory is valid. To obtain such information, we require conditions (such as might perhaps exist in connection with distances of the order of $10^{-13}$ cm) under which the usual form of the quantum theory breaks down (see Sec. 6), so that the positions and momenta of the particles can be determined simultaneously and precisely. If such conditions should exist, then there are two ways in which contradictions might be avoided. First, the more general physical laws appropriate to the new domains may be such that they do not permit the transmission of controllable aspects of interparticle forces faster than light. In this way, Lorentz covariance could be preserved. Secondly, it is possible that the application of the usual criteria of Lorentz covariance may not be appropriate when the usual interpretation of quantum theory breaks down. Even in connection with gravitational theory, general relativity indicates that the limitation of speeds to the velocity of light does not necessarily hold universally. If we adopt the spirit of general relativity, which is to seek to make the properties of space dependent on the properties of the matter that moves in this space, then it is quite conceivable that the metric, and therefore the limiting velocity, may depend on the $\psi$-field as well as on the gravitational tensor $g^{\mu,\nu}$. In the classical limit, the dependence on the $\psi$-field could be neglected, and we would get the usual form of covariance. In any case, it can hardly be said that we have a solid experimental basis for requiring the same form of covariance at very short distances that we require at ordinary distances.

To sum up, we may assert that wherever the present form of the quantum theory is correct, our interpretation cannot lead to inconsistencies with relativity. In the domains where the present theory breaks down, there are several possible ways in which our interpretation could continue to treat the problem of covariance consistently. The attempt to maintain a consistent treatment of covariance in this problem might perhaps serve as an important heuristic principle in the search for new physical laws.

## 9. ON VON NEUMANN'S DEMONSTRATION THAT QUANTUM THEORY IS INCONSISTENT WITH HIDDEN VARIABLES

Von Neumann[6] has studied the following question: "If the present mathematical formulation of the quantum theory and its usual probability interpretation are assumed to lead to absolutely correct results for every experiment that can ever be done, can quantum-mechanical probabilities be understood in terms of any conceivable distribution over hidden parameters?" Von Neumann answers this question in the negative. His conclusions are subject, however, to the criticism that in his proof he has implicitly restricted himself to an excessively narrow class of hidden parameters and in this way has excluded from consideration precisely those types of hidden parameters which have been proposed in this paper.

To demonstrate the above statements, we summarize Von Neumann's proof briefly. This proof (which begins on p. 167 of his book), shows that the usual quantum-mechanical rules of calculating probabilities imply that there can be no "dispersionless states," i.e., states in which the values of all possible observables are simultaneously determined by physical parameters associated with the observed system. For example, if we consider two noncommuting observables, $p$ and $q$, then Von Neumann shows that it would be inconsistent with the usual rules of calculating quantum-mechanical probabilities to assume that there were in the observed system a set of hidden parameters which simultaneously determined the results of measurements of position and momentum "observables." With this conclusion, we are in agreement. However, in our suggested new interpretation of the theory, the so-called "observables" are, as we have seen in Sec. 5, not properties belonging to the observed system alone, but instead potentialities whose precise development depends just as much on the observing apparatus as on the observed system. In fact, when we measure the momentum "observable," the final result is determined by hidden parameters in

[6]J. von Neumann, *Mathematics Grundlagen der Quanienmechanik* (Verlag. Julius Springer, Berlin, 1932).

the momentum-measuring device as well as by hidden parameters in the observed electron. Similarly, when we measure the position "observable," the final result is determined in part by hidden parameters in the position-measuring device. Thus, the statistical distribution of "hidden" parameters to be used in calculating averages in a momentum measurement is different from the distribution to be used in calculating averages in a position measurement. Von Neumann's proof (see p. 171 in his book) that no single distribution of hidden parameters could be consistent with the results of the quantum theory is therefore irrelevant here, since in our interpretation of measurements of the type that can now be carried out, the distribution of hidden parameters varies in accordance with the different mutually exclusive experimental arrangements of matter that must be used in making different kinds of measurements. In this point, we are in agreement with Bohr, who repeatedly stresses the fundamental role of the measuring apparatus as an inseparable part of the observed system. We differ from Bohr, however, in that we have proposed a method by which the role of the apparatus can be analyzed and described in principle in a precise way, whereas Bohr asserts that a precise conception of the details of the measurement process is as a matter of principle unattainable.

Finally, we wish to stress that the conclusions drawn thus far refer only to the measurement of the so-called "observables" carried out by the methods that are now available. If the quantum theory needs to be modified at small distances, then, as suggested in Sec. 6, precise measurements can in principle be made of the actual position and momentum of a particle. Here, it should be noted that Von Neumann's theorem is likewise irrelevant, this time because we are going beyond the assumption of the unlimited validity of the present general form of quantum theory, which plays an integral part in his proof.

## 10. SUMMARY AND CONCLUSIONS

The usual interpretation of the quantum theory implies that we must renounce the possibility of describing an individual system in terms of a single precisely defined conceptual model. We have, however, proposed an alternative interpretation which does not imply such a renunciation, but which instead leads us to regard a quantum-mechanical system as a synthesis of a precisely definable particle and a precisely definable $\psi$-field which exerts a force on this particle. An experimental choice between these two interpretations cannot be made in a domain in which the present mathematical formulation of the quantum theory is a good approximation; but such a choice is conceivable in domains, such as those associated with dimensions of the order of $10^{-13}$ cm, where the extrapolation of the present theory seems to break down and where our suggested new interpretation can lead to completely different kinds of predictions.

At present, our suggested new interpretation provides a consistent alternative to the usual assumption that no objective and precisely definable description of reality is possible at the quantum level of accuracy. For, in our description, the problem of objective reality at the quantum level is at least in principle not fundamentally different from that at the classical level, although new problems of measurement of the properties of an individual system appear, which can be solved only with the aid of an improvement in the theory, such as the possible modifications in the nuclear domain suggested in Sec 6. In this connection, we wish to point out that what we can measure depends not only on the type of apparatus that is available, but also on the existing theory, which determines the kind of inference that can be used to connect the directly observable state of the apparatus with the state of the system of interest. In other words, our epistemology is determined to a large extent by the existing theory. It is therefore not wise to specify the possible forms of future theories in terms of purely epistemological limitations deduced from existing theories.

The development of the usual interpretation of the quantum theory seems to have been guided to a considerable extent by the principle of not postulating the possible existence of entities which cannot now be observed. This principle, which stems from a general philosophical point of view known during the nineteenth century as "positivism" or "empiricism" represents an extraphysical limitation on the possible kinds of theories that we shall choose to take into consideration.[7] The word "extraphysical" is used here advisedly, since we can in no way deduce, either from the experimental data of physics, or from its mathematical formulation, that it will necessarily remain forever impossible for us to observe entities whose existence cannot now be observed. Now, there is no reason why an extraphysical general principle is necessarily to be avoided, since such principles could conceivably serve as useful working hypotheses. The particular extraphysical principle described above cannot, however, be said to be a good working hypothesis. For the history of scientific research is full of examples in which it was very fruitful indeed to assume that certain objects or elements might be real, long before any procedures were known which would permit them to be observed directly. The atomic theory is just such an example. For the possibility of the actual existence of individual atoms was first postulated in order to explain various macrophysical results which could, however, also be understood directly in terms of macrophysical concepts without the need for assuming the existence of atoms. Certain nineteenth-century positivists (notably Mach) therefore insisted on purely philosophical grounds that it was incorrect to suppose that individual atoms actually existed, because they had never been observed as such. The atomic theory, they thought, should be regarded only as an interesting way of calculating various observable large-scale properties of matter. Nevertheless, evidence for the existence of individual atoms was ultimately discovered by people who took the atomic hypothesis seriously enough to suppose that individual atoms might actually exist, even though no one had yet observed them. We may have here, perhaps, a close analogy to the usual interpretation of the quantum theory, which avoids considering the possibility that the wave function of an individual system may represent objective reality, because we cannot observe it with the aid of existing experiments and theories.

[7] A leading nineteenth-century exponent of the positivist point of view was Mach. Modern positivists appear to have retreated from this extreme position, but its reflection still remains is the philosophical point of view implicitly adopted by a large number of modern theoretical physicists.

Finally, as an alternative to the positivist hypothesis of assigning reality only to that which we can now observe, we wish to prevent here another hypothesis, which we believe corresponds more closely to conclusions that can be drawn from general experience in actual scientific research. This hypothesis is based on the simple assumption that the world as a whole is objectively real and that, as far as we now know, it can correctly be regarded as having a precisely describable and analyzable structure of unlimited complexity. The pattern of this structure seems to be reflected completely but indirectly at every level, so that from experiments done at the level of size of human beings, it is very probably possible ultimately to draw inferences concerning the properties of the whole structure at all levels. We should never expect to obtain a complete theory of this structure, because there are almost certainly more elements in existence than we possibly can be aware of at any particular stage of scientific development. Any specified element, however, can in principle ultimately be discovered, but never all of them. Of course, we must avoid postulating a new element for each new phenomenon. But an equally serious mistake is to admit into the theory only those elements which can now be observed. For the purpose of a theory is not only to correlate the results of observations that we already know how to make, but also to suggest the need for new kinds of observations and to predict their results. In fact, the better a theory is able to suggest the need for new kinds of observations and to predict their results correctly, the more confidence we have that this theory is likely to be good representation of the actual properties of matter and not simply an empirical system especially chosen in such a way as to correlate a group of already known facts.

# APPENDIX A. PHOTOELECTRIC AND COMPTON EFFECTS

In this appendix, we shall show how the electromagnetic field is to be described in our new interpretation, with the purpose of making possible a treatment of the photoelectric and Compton effects. For our purposes, it is adequate to restrict ourselves to a gauge in which div A = 0, and to consider only the transverse part of the electromagnetic field, for in this gauge, the longitudinal part of the field can be expressed through Poisson's equation entirely in terms of the charge density. The Fourier analysis of the vector potential is then

$$A(\mathbf{x}) = (4\pi / V)^{\frac{1}{2}} \sum_{\mathbf{k},\,\mu} \epsilon_{\mathbf{k},\,\mu} q_{\mathbf{k},\,\mu} e^{i\mathbf{k}\cdot\mathbf{x}} \qquad (A1)$$

with

$$q_{\mathbf{k},\,\mu}{}^{*} = q_{-\mathbf{k},\,\mu}.$$

The $q_{\mathbf{k},\mu}$ are coordinates of the electromagnetic field, associated with oscillations of wave number, k, and polarization direction, $\mu$, where $\epsilon_{\mathbf{k},\mu}$ is a unit vector normal to k and $\mu$ runs over two indices, corresponding to a pair of orthogonal directions of polarization. $V$ is the volume of the box, which is assumed to be very large.

We also introduce the momenta $\prod_{\mathbf{k},\mu} = \partial q_{\mathbf{k},\mu}{}^{*}/\partial t$, canonically conjugate[8] to the $q_{\mathbf{k},\mu}$. We have for the transverse part of the electric field

$$\mathfrak{E}(\mathbf{x}) = -\frac{1}{c}\frac{\partial A(\mathbf{x})}{\partial t} = -\left(\frac{4\pi}{Vc^2}\right)^{\frac{1}{2}} \sum_{\mathbf{k},\mu} \epsilon_{\mathbf{k},\mu} \prod_{\mathbf{k},\mu}{}^{*} e^{i\mathbf{k}\cdot\mathbf{x}} \qquad (A2)$$

and for the magnetic field

$$\mathfrak{H}(\mathbf{x}) = \nabla \times A = -(4\pi / V)^{\frac{1}{2}} i \sum_{\mathbf{k},\,\mu} \left(\mathbf{k} \times \epsilon_{\mathbf{k},\mu}\right) q_{\mathbf{k},\,\mu} e^{i\mathbf{k}\cdot\mathbf{x}}. \qquad (A3)$$

The Hamiltonian of the radiation field corresponds to a collection of independent harmonic oscillators, each with angular frequency, $\omega = kc$. This Hamiltonian is

$$H^{(R)} = \sum_{\mathbf{k},\mu} \left(\prod_{\mathbf{k},\mu} \prod_{\mathbf{k},\mu}{}^{*} + k^2 c^2 q_{\mathbf{k},\mu} q_{\mathbf{k},\mu}{}^{*}\right). \qquad (A4)$$

Now, in our interpretation of the quantum theory, the quantity $q_{\mathbf{k},\mu}$ is assumed to refer to the actual value of the $\mathbf{k}$, $\mu$ Fourier component of the vector potential. As in the case of the electron, however, there is present an objectively real superfield that is a function of all

[8] See G. Wentsel, *Quantum Theory of Fields* (Interscience Publishers, Inc., New York, 1948).

the electromagnetic field coordinates $q_{k,\mu}$. Thus, we have

$$\Psi^{(R)} = \Psi^{(R)}(\dots q_{k,\,\mu} \dots). \tag{A5}$$

Writing $\Psi^{(R)} = R \exp(iS/h)$, we obtain (in analogy with Paper I, Sec. 4)

$$\partial q_{k,\,\mu}/\partial t = \prod_{k,\,\mu}{}^{*} = \partial S/\partial q_{k,\,\mu}{}^{*}. \tag{A6}$$

The function $R(\dots q_{k,\mu} \dots)$ has two interpretations. First, it defines an additional quantum-mechanical term appearing in Maxwell's equations. To see the origin of this term, let us write the generalized Hamilton-Jacobi equation of the electromagnetic field, analogous to Paper I, Eq. (4),

$$\frac{\partial s}{\partial t} + \sum_{k,\,\mu} \frac{\partial s}{\partial q_{k,\,\mu}} \frac{\partial s}{\partial q_{k,\,\mu}{}^{*}} + \sum_{k,\,\mu} (kc)^2 q_{k,\,\mu} q_{k,\,\mu}{}^{*}$$

$$- \frac{h^2}{2R} \sum_{k,\,\mu} \frac{\partial^2 R(\dots q_{k,\,\mu} \dots)}{\partial q_{k,\,\mu} \partial q_{k,\,\mu}{}^{*}} = 0. \tag{A7}$$

The equation of motion of $q_{k,\mu}$, derived from the Hamiltonian implied by Eq. (A7) becomes

$$\ddot{q}_{k,\,\mu} + k^2 c^2 q_{k,\,\mu} = \frac{\partial}{\partial q_{k,\,\mu}{}^{*}} \left( \frac{h^2}{2R} \sum_{k',\,\mu'} \frac{\partial^2 R}{\partial q_{k',\,\mu'} \partial q_{k',\,\mu'}{}^{*}} \right). \tag{A8}$$

Since Maxwell's equations for empty space follow when the right-hand side is zero, we see that the "quantum-mechanical" terms can profoundly modify the behavior of the electromagnetic field. In fact, it is this modification which will contribute to the explanation of the ability of an oscillator, $q_{k,\mu}$, to transfer large quantities of energy and momentum rapidly even when $q_{k,\mu}$ is very small, for when $q_{k,\mu}$ is small, the right-hand side of Eq. (A8) may become very large.

The second interpretation of $R$ is that as in Paper I, Eq. (5), it defines a conserved probability density that each of the $q_{k,\mu}$ has a certain specified Value. From this fact, we see that although large transfers of energy and momentum to a radiation oscillator can occur in a short time when $R$ is small, the probability of such a process is (as was also shown in Paper I, Sec. 7) very small.

In the lowest state (when no quanta are present) every oscillator is in the ground state. The super wave fields is then

$$\Psi_0^{(R)} = \exp\left[-\sum_{k,\,\mu}\left(kcq_{k,\,\mu}q_{k,\mu}^* + \tfrac{1}{2}ikct\right)\right]. \qquad (A9)$$

If the $\mathbf{k}'$, $\mu'$ oscillator is excited to the $n$th quantum state, the super wave field is

$$\Psi^{(R)} = h_n(q_{k',\,\mu'})e^{-ink'ct}\Psi_0^{(R)}, \qquad (A10)$$

where $h_n$ is the $n$th hermite polynomial. As shown in Paper I, Sec. 5, the stationary states of such a system correspond to a quantized energy equal to the same value, $E_n = (n + \tfrac{1}{2})hkc$, obtained from the usual Interpretation. In nonstationary states, however, Eqs. (A7) and (A8) imply that the energy of each oscillator may fluctuate violently, as was also true of nonstationary states of the hydrogen atom (see Paper I, Sec. 7).

A nonstationary state of particular interest in the photoelectric and Compton effects is a state corresponding to the presence of an electromagnetic wave packet containing a single quantum. The super wave field for such a state is

$$\Psi_P^{(R)} = \sum_{k,\,\mu} f_\mu(\mathbf{k} - \mathbf{k}_0)q_{k,\,\mu}e^{-ikct}\Psi_0^{(R)}, \qquad (A11)$$

where $f_\mu(\mathbf{k} - \mathbf{k}_0)$ is a function that is large only near $\mathbf{k} = \mathbf{k}_0$ and the first hermite polynomial is represented by $q_{k,\mu}$, to which it is proportional.

To prove that Eq. (A11) represents an electromagnetic wave packet, we can evaluate the difference

$$\langle \Delta W \rangle_{Av} = \langle W \rangle_{Av} - \langle W_0 \rangle_{Av,} \qquad (A12)$$

where $\langle W(\mathbf{x}) \rangle_{Av}$ is the actual mean energy density present (averaged over the ensemble), and $\langle W_0(\mathbf{x}) \rangle_{Av}$ is the mean energy that would be present even in the ground state, because of zero-point fluctuations.

We have

$$\langle W(\mathbf{x})\rangle_{\text{Av}} = \int \int \cdots \int \Psi_P^{*(R)}\left(\ldots q_{\text{k},\mu}\ldots\right)$$
$$\times \frac{[\mathfrak{E}^2(\mathbf{x}) + \mathfrak{H}^2(\mathbf{x})]}{8\pi}\Psi_P^{(R)}\left(\ldots q_{\text{k},\mu}\ldots\right)$$
$$\times \left(\ldots dq_{\text{k},\mu}\ldots\right), \tag{A13}$$

$$\langle W_0(\mathbf{x})\rangle_{\text{Av}} = \int \int \cdots \int \Psi_0^{*(R)}\left(\ldots q_{\text{k},\mu}\ldots\right)$$
$$\times \frac{[\mathfrak{E}^2(\mathbf{x}) + \mathfrak{H}^2(\mathbf{x})]}{8\pi}\Psi_0^{(R)}\left(\ldots q_{\text{k},\mu}\ldots\right)$$
$$\times \left(\ldots dq_{\text{k},\mu}\ldots\right). \tag{A14}$$

Obtaining $\mathfrak{E}(\mathbf{x})$ from Eq. (A2), $\mathfrak{H}(\mathbf{x})$ from Eq. (A3), $\Psi_P^{(R)}$ from Eq. (A10), $\Psi_0^{(R)}$ from Eq. (A9), we readily show that

$$\langle \Delta W(\mathbf{x})\rangle_{\text{Av}} = \sum_{\text{k},\mu}\sum_{\text{k}',\mu'} f_\mu(\mathbf{k}-\mathbf{k}_0) f_{\mu'}(\mathbf{k}'-\mathbf{k}_0)$$
$$\times e^{i(\mathbf{k}+\mathbf{k}')\cdot\mathbf{x}}\epsilon_{\text{k},\mu}\cdot\epsilon_{\text{k}',\mu'}. \tag{A15}$$

This means that the wave packet implies an excess over zero-point energy that is localized within a region in which the packet function, g(x) is appreciable, where

$$g(\mathbf{x}) = \sum_{\text{k},\mu} f_\mu(\mathbf{k}-\mathbf{k}_0)e^{i\mathbf{k}\cdot\mathbf{x}}\epsilon_{\text{k},\mu}. \tag{A16}$$

We are now ready to treat the photoelectric and Compton effects. The entire treatment is to similar to that of the Franck-Hertz experiment (Paper I, Sec. 7) that we need merely sketch it here. We begin by adding to the radiation Hamiltonian, $H^{(R)}$, the particle Hamiltonian,

$$H^{(P)} = (1/2m)[\mathbf{p} - (e/c)\text{A}(\mathbf{x})]^2. \tag{A17}$$

(We restrict ourselves here to nonrelativistic treatment.) The photo-electric effect corresponds to the transition of a radiation oscillator from an excited state to the ground state, while the atomic electron is ejected, with an energy $E = h\nu - I$, where $I$ is the ionization potential of the atom. The initial super wave field, corresponding to an incident packet containing only one quantum, plus an atom in the ground state is (see Eq. (A11))

$$\Psi_i = \psi_0(\mathbf{x}) \exp(-i E_0 t / b) \Psi_0^{(R)}(\ldots q_{k,\mu} \ldots)$$
$$\times \sum_{k,\mu} f_\mu(\mathbf{k} - \mathbf{k}_0) q_{k,\mu} e^{-ikct}. \qquad (A18)$$

By solving Schroedinger's equation for the combined system, we obtain an asymptotic wave field analogous to Paper I, Eq. (26), containing terms corresponding to the photoelectric effect. These terms, which must be added to $\Psi_i$, to yield the complete superfield, are (asymptotically)

$$\delta \Psi_a = \Psi_0^{(R)}(\ldots q_{k,\mu} \ldots) \sum_{k,\mu} f_\mu(\mathbf{k} - \mathbf{k}_0)$$
$$\times \frac{\exp[i\mathbf{k}' \cdot \mathbf{r} - ib(k'^2/2m)t]}{r} g_\mu(\theta, \phi, k'), \qquad (A19)$$

where the energy of the outgoing electron is $E = b^2 k'^2 / 2m = bkc + E_0$. The function $g_\mu(\theta, \phi, k')$ is the amplitude associated with the $\psi$-field of the outgoing electron. This quantity can be calculated from the matrix element of the interaction term, $-(e/c)\mathbf{p} \cdot \mathbf{A}(\mathbf{x})$, by methods that are easily deducible from the usual perturbation theory.[6]

The outgoing electron packet has its center at $r = (bk'/m)t$. Eventually, this packet will become completely separated from the initial electron wave function, $\psi_0(\mathbf{x})$. If the electron happens to enter the outgoing packet, the initial wave function can subsequently be ignored. The system then acts for all practical purposes as if its wave field were given by Eq. (A9), from which we conclude that the radiation field is in the ground state, while the electron has been liberated. It is readily shown that, as in the usual interpretation, the probability that the electron appears in the direction $\theta, \phi$ can be calculated from $|g_\mu(\theta, \phi, k')|^2$ (see Paper I, Sec. 7).

To describe the Compton effect, we need only add to the super wave field the term corresponding to the appearance of an outgoing electromagnetic wave, as well as an outgoing electron. This part is asymptotically

$$\delta\Psi_b = \Psi_0{}^{(R)}(\ldots q_{k,\mu}\ldots)\sum_{k',\mu'} f_\mu(\mathbf{k}-\mathbf{k}_0)$$

$$\times c_{k',\mu'}{}^{k,\mu} q_{k',\mu'} g_{k',\mu'}{}^{k,\mu}(\theta,\phi)\frac{e^{ik''r}}{r}$$

$$\times \exp\left(-ik'ct - \frac{i\hbar k''^2 t}{2m}\right), \tag{A20}$$

where

$$(\hbar^2 k''^2/2m) + \hbar k'c = \hbar kc + E_0.$$

The quantity, $c_{k',\mu'}{}^{k,\mu}$ is proportioned to the matrix element for a transition in which the $\mathbf{k}$, $\mu$-radiation oscillator falls from the first excited state, to the ground state, while the $\mathbf{k}'$, $\mu'$-oscillator rises from the ground state to the first excited state. This matrix element is determined mainly by the term $(e^2/8mc^2)A^2(\mathbf{x})$ in the hamiltonian.

It is easily seen that the outgoing electron packet eventually becomes completely separated both from the initial wave field, $\Psi_i(\mathbf{x}, \ldots q_{k,\mu}\ldots)$, and from the packet for the photoelectric effect, $\delta\Psi_a$ [defined in Eq. (A19)]. If the electron should happen to enter this packet, then the others can be ignored, and the system acts for all practical purposes like an outgoing electron, plus an independent outgoing light quantum. The reader will readily verify that the probability that the light quantum $\mathbf{k}'$, $\mu'$ appears along with an electron with angles $\theta$, $\phi$ is precisely the same as in the usual interpretation.

# APPENDIX B. A DISCUSSION OF INTERPRETATIONS OF THE QUANTUM THEORY PROPOSED BY DE BROGLIE AND ROSEN

After this article had been prepared, the author's attention was called to two papers in which an interpretation of the quantum theory similar to that suggested here was proposed, first by L. de Broglie,[9] and

[9] L. de Broglie, *An Introduction to the Study of Wave Mechanics* (E. P. Dutton and Company, Inc., New York, 1930), see Chapters 6, 9, and 10.

later by N. Rosen.[10] In both of these papers, it was suggested that if one writes $\psi = R \exp(is/h)$, then one can regard $R^2$ as a probability density of particles having a velocity, $\mathbf{v} = \nabla s/m$. De Broglie regarded the $\psi$-field as an agent "guiding" the particle, and therefore referred to $\psi$ as a "pilot wave." Both of these authors came to the conclusion that this interpretation could not consistently be carried through in those cases in which the field contained a linear combination of stationary state wave functions. As we shall see in this appendix, however, the difficulties encountered by the above authors could have been overcome by them, if only they had carried their ideas to a logical conclusion.

De Broglie's suggestions met strong objections on the part of Pauli,[11] in connection with the problem of inelastic scattering of a particle by a rigid rotator. Since this problem is conceptually equivalent to that of inelastic scattering of a particle by a hydrogen atom, which we have already treated in Paper I, Sec. 7, we shall discuss the objections raised by Pauli in terms of the latter example.

Now, according to Pauli's argument, the initial wave function in the scattering problem should be $\Psi = \exp(i\mathbf{p}_0 \cdot \mathbf{y}/h)\psi_0(\mathbf{x})$. This corresponds to a stationary state for the combined system, in which the particle momentum is $\mathbf{p}_0$, while the hydrogen atom is in its ground state, with a wave function, $\psi_0(\mathbf{x})$. After interaction between the incident particle and the hydrogen atom, the combined wave function can be represented as

$$\Psi = \sum_n f_n(\mathbf{y})\psi_n(\mathbf{x}), \tag{B1}$$

where $\psi_n(\mathbf{x})$ is the wave function for the $n$th excited state of the hydrogen atom, and $f_n(\mathbf{y})$ is the associated expansion coefficient. It is easily shown[12] that asymptotically, $f_n(\mathbf{y})$ takes the form of an outgoing wave, $f_n(\mathbf{y}) \sim g_n(\theta, \phi)e^{iknr}/r$, where $(hkn)^2/2m = [(hk_0)^2/2m] + E_n - E_0$. Now, if we write $\psi = R \exp(iS/h)$, we find that the particle momenta, $p_x = \nabla_x S(\mathbf{x}, \mathbf{y})$ and $\mathbf{p}_y = \nabla_y S(\mathbf{x}, \mathbf{y})$, fluctuate violently in a way that depends strongly on the position of each

---

[10] N. Rosen, J. Elisha Mitchel Sci. Soc. 61, Nos. 1 and 2 (August, 1945).

[11] *Reports on the 1927 Solvay Congress* (Gauthiers-Villars et Cie., Paris, 1928), see p. 280.

[12] N. F. Mott and H. S. W. Massey, *The Theory of Atomic Collisions* (Clarendon Press, Oxford, 1933).

particle. Thus, neither atom nor the outgoing particle ever seem to approach a stationary energy. On the other hand, we know from experiment that both the atom and the outgoing particle do eventually obtain definite (but presumably unpredictable) energy values. Pauli therefore concluded that the interpretation proposed by de Broglie was untenable. De Broglie seems to have agreed with the conclusion, since he subsequently gave up his suggested interpretation.[9]

Our answer to Pauli's objection is already contained in Paper I, Sec. 7, as well as in Sec. 2 of this paper. For as is well known, the use of an incident plane wave of infinite extent is an excessive abstraction, not realizable in practice. Actually, both the incident and outgoing parts of the $\psi$-field will always take the form of bounded packets. Moreover, as shown in Paper I, Sec. 7, all packets corresponding to different values of $n$ will ultimately obtain classically describable separations. The outgoing particle must enter one of these packets, and it will remain with that particular packet thereafter, leaving the hydrogen atom in a definite but correlated stationary state. Thus, Pauli's objection is seen to be based on the use of the excessively abstract model of an infinite plane wave.

Although the above constitutes a complete answer to Pauli's specific objections to our suggested interpretation, we wish here to amplify our discussion somewhat, in order to anticipate certain additional objections that might be made along similar lines. For at this point, one might argue that even though the wave packet is bounded, it can nevertheless in principle be made arbitrarily large in extent by means of a suitable adjustment of initial conditions. Our interpretation predicts that in the region in which incident and outgoing $\psi$-waves overlap, the momentum of each particle will fluctuate violently, as a result of corresponding fluctuations in the "quantum-mechanical" potential produced by the $\psi$-field. The question arises, however, as to whether such fluctuations can really be in accord with experimental fact, especially since in principle they could occur when the particles were separated by distances much greater than that over which the "classical" interaction potential, $V(\mathbf{x}, \mathbf{y})$, was appreciable.

To show that these fluctuations are not in disagreement with any experimental facts now available, we first point out that even in the usual interpretation the energy of each particle cannot correctly be regarded as definite under the conditions which are assumed here, namely, that the incident and outgoing wave packets overlap. For as long as interference between two stationary state wave function is possible, the system acts as if it, in some sense, covered both states simultaneously.[13] In such a situation, the usual interpretation implies that a precisely defined value for the energy of either particle is meaningless. From such a wave function, one can predict only the probability that if the energy is measured, a definite value will be obtained. On the other hand, the very experimental conditions needed for measuring the energy play a key role in making a definite value of the energy possible because the effect of the measuring apparatus is to destroy interference between parts of the wave function corresponding to different values of the energy.[14]

In our interpretation, the overlap of incident and outgoing wave packets signifies not that the precise value of the energy of either particle can be given no meaning, but rather that this value fluctuates violently in an, in practice, unpredictable and uncontrollable way. When the energy of either particle is measured, however, then our interpretation predicts, in agreement with the usual interpretation, that the energy of each particle will become definite and constant, as a result of the effects of the energy-measuring apparatus on the observed system. To show how this happens, let us suppose that the energy of the hydrogen atom is measured by means of an interaction in which the "classical" potential, $V$, is a function only of the variables associated with the electron in the hydrogen atom and with the apparatus, but is not a function of variables associated with the outgoing particle. Let $z$ be the coordinate of the measuring apparatus. Then as shown in Sec. 2, interaction with an apparatus that measures the energy of the hydrogen atom will transform the $\Psi$-function (B1),

[13] Reference 2, Chapter 16, Sec. 25.
[14] Reference 2, Chapter 6, Secs. 3 to 8; Chapter 22, Secs. 8 to 10.

into

$$\Psi = \sum_n f_n(\mathbf{y})\psi_n(\mathbf{x})g_0(z - aEnt/h^2). \qquad \text{(B2)}$$

Now, we have seen that if the product $at$ is large enough to make a distinct measurement possible, packets corresponding to different values of $n$ will ultimately obtain classically describable separations in $z$ space. The apparatus variable, $z$, must enter one of these packets; and, thereafter, all other packets can for practical purposes be ignored. The hydrogen atom is then left in a state having a definite and constant energy, while the outgoing particle has a correspondingly definite but correlated constant value for its energy. Thus, we find that as with the usual interpretation, our interpretation predicts that whenever we measure the energy of either particle by methods that are now available, a definite and constant value will always be obtained. Nevertheless, under conditions in which incident and outgoing wave packets overlap, and in which neither particle interacts with an energy-measuring device, our interpretation states unambiguously that real fluctuations in the energy of each particle will occur. These fluctuations are moreover, at least in principle, observable (for example, by methods discussed in Sec. 6). Meanwhile, under conditions in which we are limited by present methods of observation, our interpretation leads to predictions that are precisely the same as those obtained from the usual interpretation, so that no experiments supporting the usual interpretation can possibly contradict our interpretation.

In his book,[9] de Broglie raises objections to his own suggested interpretation of the quantum theory, which are very similar to those raised by Pauli. It is therefore not necessary to answer de Broglie's objections in detail here, since the answer is essentially the same as that which has been given to Pauli. We wish, however, to add one point. De Broglie assumes that not only electrons, but also light quanta, are associated with particles. A consistent application of the interpretation suggested here requires, however, as shown in Appendix A, that light quanta be described as electromagnetic wave packets. The only precisely definable quantities in such a packet are the Fourier components,

$q_{k,\mu}$, of the vector potential and the corresponding canonically conjugate momenta, $\Pi_{k,\mu}$. Such packets have many particle-like properties, including the ability to transfer rapidly a full quantum of energy at great distances. Nevertheless, it would not be consistent to assume the existence of a "photon" particle, associated with each light quantum.

We shall now discuss Rosen's paper briefly.[10] Rosen gave up his suggested interpretation of the quantum theory, because of difficulties arising in connection with the interpretation of standing waves. In the case of the stationary states of a free particle in a box, which we have already discussed in Sec. 8, our interpretation leads to the conclusion that the particle is standing still. Rosen did not wish to accept this conclusion, because it seemed to disagree with the statement of the usual interpretation that in such a state the electron is moving with equal probability that the motion is in either direction. To answer Rosen's objections, we need merely point out again that the usual interpretation can give no meaning to the motion of particles in a stationary state; at best, it can only predict the probability that a given result will be obtained, if the velocity is measured. As we saw in Sec. 8, however, our interpretation leads to precisely the same predictions as are obtained from the usual interpretation, for any process which could actually provide us with a measurement of the velocity of the electron. One must remember, however, that the value of the momentum "observable" as it is now "measured" is not necessarily equal to the particle momentum existing before interaction with the measuring apparatus took place.

We conclude that the objections raised by Pauli, de Broglie, and Rosen, to interpretations of the quantum theory similar to that suggested here, can all be answered by carrying every aspect of our suggested interpretation to its logical conclusion.

# ON THE EINSTEIN PODOLSKY ROSEN PARADOX

## BY
## JOHN S. BELL

Originally published in *Physics*, 1, 195–200 (1964).

## I. Introduction

The paradox of Einstein, Podolsky and Rosen [1] was advanced as an argument that quantum mechanics could not be a complete theory but should be supplemented by additional variables. These additional variables were to restore to the theory causality and locality [2]. In this note that idea will be formulated mathematically and shown to be incompatible with the statistical predictions of quantum mechanics. It is the requirement of locality, or more precisely that the result of a measurement on one system be unaffected by operations on a distant system with which it has interacted in the past, that creates the essential difficulty. There have been attempts [3] to show that even without such a separability or locality requirement no "hidden variable" interpretation of quantum mechanics is possible. These attempts have been examined elsewhere [4] and found wanting. Moreover, a hidden variable interpretation of elementary quantum theory [5] has been explicitly constructed. That particular interpretation has indeed a grossly nonlocal structure. This is characteristic, according to the result to be proved here, of any such theory which reproduces exactly the quantum mechanical predictions.

## II. Formulation

With the example advocated by Bohm and Aharonov [6], the EPR argument is the following. Consider a pair of spin one-half particles

* Work supported in part by the U.S. Atomic Energy Commission
† On leave of absence from SLAC and CERN

formed somehow in the singlet spin state and moving freely in opposite directions. Measurements can be made, say by Stern-Gerlach magnets, on selected components of the spins $\vec{\sigma}_1$ and $\vec{\sigma}_2$. If measurement of the component $\vec{\sigma}_1 \cdot \vec{a}$, where $\vec{a}$ is some unit vector, yields the value $+1$ then, according to quantum mechanics, measurement of $\vec{\sigma}_2 \cdot \vec{a}$ must yield the value $-1$ and vice versa. Now we make the hypothesis [2], and it seems one at least worth considering, that if the two measurements are made at places remote from one another the orientation of one magnet does not influence the result obtained with the other. Since we can predict in advance the result of measuring any chosen component of $\vec{\sigma}_2$, by previously measuring the same component of $\vec{\sigma}_1$, it follows that the result of any such measurement must actually be predetermined. Since the initial quantum mechanical wave function does *not* determine the result of an individual measurement, this predetermination implies the possibility of a more complete specification of the state.

Let this more complete specification be effected by means of parameters $\lambda$. It is a matter of indifference in the following whether $\lambda$ denotes a single variable or a set, or even a set of functions, and whether the variables are discrete or continuous. However, we write as if $\lambda$ were a single continuous parameter. The result $A$ of measuring $\vec{\sigma}_1 \cdot \vec{a}$ is then determined by $\vec{a}$ and $\lambda$, and the result $B$ of measuring $\vec{\sigma}_2 \cdot \vec{b}$ in the same instance is determined by $\vec{b}$ and $\lambda$, and

$$A(\vec{a}, \lambda) = \pm 1, \, B(\vec{b}, \lambda) = \pm 1. \tag{1}$$

The vital assumption [2] is that the result $B$ for particle 2 does not depend on the setting $\vec{a}$, of the magnet for particle 1, nor $A$ on $\vec{b}$.

If $\rho(\lambda)$ is the probability distribution of $\lambda$ then the expectation value of the product of the two components $\vec{\sigma}_1 \cdot \vec{a}$ and $\vec{\sigma}_2 \cdot \vec{b}$ is

$$P(\vec{a}, \vec{b}) = \int d\lambda \rho(\lambda) A(\vec{a}, \lambda) B(\vec{b}, \lambda) \tag{2}$$

This should equal the quantum mechanical expectation value, which for the singlet state is

$$\langle \vec{\sigma}_1 \cdot \vec{a}\ \vec{\sigma}_2 \cdot \vec{b} \rangle = -\vec{a} \cdot \vec{b}. \tag{3}$$

But it will be shown that this is not possible.

Some might prefer a formulation in which the hidden variables fall into two sets, with $A$ dependent on one and $B$ on the other; this possibility is contained in the above, since $\lambda$ stands for any number of variables and the dependences thereon of $A$ and $B$ are unrestricted. In a complete physical theory of the type envisaged by Einstein, the hidden variables would have dynamical significance and laws of motion; our $\lambda$ can then be thought of as initial values of these variables at some suitable instant.

## III. Illustration

The proof of the main result is quite simple. Before giving it, however, a number of illustrations may serve to put it in perspective.

Firstly, there is no difficulty in giving a hidden variable account of spin measurements on a single particle. Suppose we have a spin half particle in a pure spin state with polarization denoted by a unit vector $\vec{p}$. Let the hidden variable be (for example) a unit vector $\vec{\lambda}$ with uniform probability distribution over the hemisphere $\vec{\lambda} \cdot \vec{p} > 0$. Specify that the result of measurement of a component $\vec{\sigma} \cdot \vec{a}$ is

$$\text{sign } \vec{\lambda} \cdot \vec{a}', \tag{4}$$

where $\vec{a}'$ is a unit vector depending on $\vec{a}$ and $\vec{p}$ in a way to be specified, and the sign function is $+1$ or $-1$ according to the sign of its argument. Actually this leaves the result undetermined when $\lambda \cdot d = 0$, but as the probability of this is zero we will not make special prescriptions for it. Averaging over $\vec{\lambda}$ the expectation value is

$$\langle \vec{\sigma} \cdot \vec{a} \rangle = 1 - 2\theta'/\pi, \tag{5}$$

where $\theta'$ is the angle between $\vec{a}'$ and $\vec{p}$. Suppose then that $\vec{a}'$ is obtained from $\vec{a}$ by rotation towards $\vec{p}$ until

$$1 - \frac{2\theta'}{\pi} = \cos\theta \tag{6}$$

where $\theta$ is the angle between $\vec{a}$ and $\vec{p}$. Then we have the desired result

$$\langle \vec{\sigma} \cdot \vec{a} \rangle = \cos\theta \tag{7}$$

So in this simple case there is no difficulty in the view that the result of every measurement is determined by the value of an extra variable, and that the statistical features of quantum mechanics arise because the value of this variable is unknown in individual instances.

Secondly, there is no difficulty in reproducing, in the form (2), the only features of (3) commonly used in verbal discussions of this problem:

$$\left.\begin{array}{l} P(\vec{a}, \vec{a}) = -P(\vec{a}, -\vec{a}) = -1 \\ P(\vec{a}, \vec{b}) = 0 \quad \text{if} \quad \vec{a} \cdot \vec{b} = 0 \end{array}\right\} \tag{8}$$

For example, let $\lambda$ now be unit vector $\vec{\lambda}$, with uniform probability distribution over all directions, and take

$$\left.\begin{array}{l} A(\vec{a}, \vec{\lambda}) = \text{sign } \vec{a} \cdot \vec{\lambda} \\ B(a, b) = -\text{sign } \vec{b} \cdot \vec{\lambda} \end{array}\right\} \tag{9}$$

This gives

$$P(\vec{a}, \vec{b}) = -1 + \frac{2}{\pi}\theta, \tag{10}$$

where $\theta$ is the angle between $a$ and $b$, and (10) has the properties (8). For comparison, consider the result of a modified theory [6] in which the pure singlet state is replaced in the course of time by an isotropic mixture of product states; this gives the correlation function

$$-\frac{1}{3}\vec{a} \cdot \vec{b} \tag{11}$$

It is probably less easy, experimentally, to distinguish (10) from (3), than (11) from (3).

Unlike (3), the function (10) is not stationary at the minimum value $-1$ (at $\theta = 0$). It will be seen that this is characteristic of functions of type (2).

Thirdly, and finally, there is no difficulty in reproducing the quantum mechanical correlation (3) if the results $A$ and $B$ in (2) are allowed to depend on $\vec{b}$ and $\vec{a}$ respectively as well as on $\vec{a}$ and $\vec{b}$. For example, replace $\vec{a}$ in (9) by $\vec{a}'$, obtained from $\vec{a}$ by rotation towards $\vec{b}$ until

$$1 - \frac{2}{\pi}\theta' = \cos\theta,$$

where $\theta'$ is the angle between $\vec{a}'$ and $\vec{b}$. However, for given values of the hidden variables, the results of measurements with one magnet now depend on the setting of the distant magnet, which is just what we would wish to avoid.

## IV. Contradiction

The main result will now be proved. Because $\rho$ is a normalized probability distribution,

$$\int d\lambda \rho(\lambda) = 1, \tag{12}$$

and because of the properties (1), $P$ in (2) cannot be less than $-1$. It can reach $-1$ at $\vec{a} = \vec{b}$ only if

$$A(\vec{a}, \lambda) = -B(\vec{a}, \lambda) \tag{13}$$

except at a set of points $\lambda$ of zero probability. Assuming this, (2) can be rewritten

$$P(\vec{a}, \vec{b}) = -\int d\lambda \rho(\lambda) A(\vec{a}, \lambda) A(\vec{b}, \lambda). \tag{14}$$

It follows that $\vec{c}$ is another unit vector

$$P(\vec{a}, \vec{b}) - P(\vec{a}, \vec{c}) = -\int d\lambda \rho(\lambda)[A(\vec{a}, \lambda)A(\vec{b}, \lambda) - A(\vec{a}, \lambda)A(\vec{c}, \lambda)]$$

$$= \int d\lambda \rho(\lambda) A(\vec{a}, \lambda) A(\vec{b}, \lambda)[A(\vec{b}, \lambda)A(\vec{c}, \lambda) - 1]$$

using (1), whence

$$|P(\vec{a}, \vec{b}) - P(\vec{a}, \vec{c})| \leq \int d\lambda \rho(\lambda)[1 - A(\vec{b}, \lambda)A(\vec{c}, \lambda)]$$

The second term on the right is $P(\vec{b}, \vec{c})$, whence

$$1 + P(\vec{b}, \vec{c}) \geq |P(\vec{a}, \vec{b}) - P(\vec{a}, \vec{c})| \tag{15}$$

Unless $P$ is constant, the right hand side is in general of order $|\vec{b} - \vec{c}|$ for small $|\vec{b} - \vec{c}|$. Thus $P(\vec{b}, \vec{c})$ cannot be stationary at the minimum value ($-1$ at $\vec{b} = \vec{c}$) and cannot equal the quantum mechanical value (3).

Nor can the quantum mechanical correlation (3) be arbitrarily closely approximated by the form (2). The formal proof of this may be set out as follows. We would not worry about failure of the approximation at isolated points, so let us consider instead of (2) and (3) the functions

$$\bar{P}(\vec{a}, \vec{b}) \quad \text{and} \quad \overline{-\vec{a} \cdot \vec{b}}$$

where the bar denotes independent averaging of $P(\vec{a}', \vec{b}')$ and $-\vec{a}' \cdot \vec{b}'$ over vectors $\vec{a}'$ and $\vec{b}'$ within specified small angles of $\vec{a}$ and $\vec{b}$. Suppose that for all $\vec{a}$ and $\vec{b}$ the difference is bounded by $\varepsilon$:

$$|\bar{P}(\vec{a}, \vec{b}) + \vec{a} \cdot \vec{b}| \leq \varepsilon \tag{16}$$

Then it will be shown that $\varepsilon$ cannot be made arbitrarily small.

Suppose that for all $a$ and $b$

$$|\overline{\vec{a} \cdot \vec{b}} - \vec{a} \cdot \vec{b}| \leq \delta \tag{17}$$

Then from (16)

$$|\bar{P}(\vec{a}, \vec{b}) + \vec{a} \cdot \vec{b}| \leq \varepsilon + \delta \tag{18}$$

From (2)

$$\bar{P}(\vec{a}, \vec{b}) = \int d\lambda \rho(\lambda) \bar{A}(\vec{a}, \lambda) \bar{B}(\vec{b}, \lambda) \tag{19}$$

where

$$|\bar{A}(\vec{a}, \lambda)| \leq 1 \quad \text{and} \quad |\bar{B}(\vec{b}, \lambda)| \leq 1 \tag{20}$$

From (18) and (19), with $\vec{a} = \vec{b}$,

$$d\lambda \rho(\lambda)[\bar{A}(\vec{b}, \lambda) \bar{B}(\vec{b}, \lambda) + 1] \leq \varepsilon + \delta \tag{21}$$

From (19)

$$\bar{P}(\vec{a}, \vec{b}) - \bar{P}(\vec{a}, \vec{c}) = \int d\lambda \rho(\lambda)[\bar{A}(\vec{a}, \lambda)\bar{B}(\vec{b}, \lambda) - \bar{A}(\vec{a}, \lambda)\bar{B}(\vec{c}, \lambda)]$$

$$= \int d\lambda \rho(\lambda)\bar{A}(\vec{a}, \lambda)\bar{B}(\vec{b}, \lambda)[1 + \bar{A}(\vec{b}, \lambda)\bar{B}(\vec{c}, \lambda)]$$

$$- \int d\lambda \rho(\lambda)\bar{A}(\vec{a}, \lambda)\bar{B}(\vec{c}, \lambda)[1 + \bar{A}(\vec{b}, \lambda)\bar{B}(\vec{b}, \lambda)]$$

Using (20) then

$$|\bar{P}(\vec{a}, \vec{b}) - \bar{P}(\vec{a}, \vec{c})| \le \int d\lambda \propto (\lambda)[1 + \bar{A}(\vec{b}, \lambda)\bar{B}(\vec{c}, \lambda)]$$

$$+ \int d\lambda \rho(\lambda)[1 + \bar{A}(\vec{b}, \lambda)\bar{B}(\vec{b}, \lambda)]$$

Then using (19) and 21)

$$|\bar{P}(\vec{a}, \vec{b}) - \bar{P}(\vec{a}, \vec{c})| \le 1 + \bar{P}(\vec{b}, \vec{c}) + \varepsilon + \delta$$

Finally, using (18),

$$|\vec{a} \cdot \vec{c} - \vec{a} \cdot \vec{b}| - 2(\varepsilon + \delta) \le 1 - \vec{b} \cdot \vec{c} + 2(\varepsilon + \delta)$$

or

$$4(\varepsilon + \delta) \ge |\vec{a} \cdot \vec{c} - \vec{a} \cdot \vec{b}| + \vec{b} \cdot \vec{c} - 1 \qquad (22)$$

Take for example $\vec{a} \cdot \vec{c} = 0, \vec{a} \cdot \vec{b} = \vec{b} \cdot \vec{c} = 1/\sqrt{2}$   Then

$$4(\varepsilon + \delta) \ge \sqrt{2} - 1$$

Therefore, for small finite $\delta$, $\varepsilon$ cannot be arbitrarily small.

Thus, the quantum mechanical expectation value cannot be represented, either accurately or arbitrarily closely, in the form (2).


# V. Generalization

The example considered above has the advantage that it requires little imagination to envisage the measurements involved actually being made. In a more formal way, assuming [7] that any Hermitian operator with a complete set of eigenstates is an "observable", the result is easily extended to other systems. If the two systems have state spaces of dimensionality greater than 2 we can always consider two dimensional subspaces and define, in their direct product, operators $\vec{\sigma}_1$ and $\vec{\sigma}_2$ formally analogous to those used above and which are zero for states outside the product subspace. Then for at least one quantum mechanical state, the "singlet" state in the combined subspaces, the statistical predictions of quantum mechanics are incompatible with separable predetermination.

# VI. Conclusion

In a theory in which parameters are added to quantum mechanics to determine the results of individual measurements, without changing the statistical predictions, there must be a mechanism whereby the setting of one measuring device can influence the reading of another instrument, however remote. Moreover, the signal involved must propagate instantaneously, so that such a theory could not be Lorentz invariant.

Of course, the situation is different if the quantum mechanical predictions are of limited validity. Conceivably they might apply only to experiments in which the settings of the instruments are made sufficiently in advance to allow them to reach some mutual rapport by exchange of signals with velocity less than or equal to that of light. In that connection, experiments of the type proposed by Bohm and Aharonov [6], in which the settings are changed during the flight of the particles, are crucial.

*I am indebted to Drs. M. Bander and J. K. Perring for very useful discussions of this problem. The first draft of the paper was written during a stay at Brandeis University; I am indebted to colleagues there and at the University of Wisconsin for their interest and hospitality.*

## REFERENCES

1. A. EINSTEIN, N. ROSEN and B. PODOLSKY, *Phys. Rev.* **47**. 777 (1935); see also N. BOHR, *Ibid.* **48**, 696 (1935), W. H. FURRY, *Ibid.* **49**, 393 and 476 (1936), and D. R. INGLIS, *Rev. Mod. Phys.* **33**, 1 (1961).

2. "But on one supposition we should, in my opinion, absolutely hold fast: the real factual situation of the system $S_2$ is independent of what is done with the system $S_1$, which is spatially separated from the former." A. EINSTEIN in *Albert Einstein, Philosopher Scientist*, (Edited by P. A. SCHILP) p. 85, Library of Living Philosophers, Evanston, Illinois (1949).

3. J. VON NEUMANN, *Mathematishe Grundlagen der Quanten-mechanik.* Verlag Julius-Springer, Berlin (1932), [English translation: Princeton University Press (1955)]; J. M. JAUCH and C. PIRON, *Helv. Phys. Acta* **36**, 827 (1963).

4. J. S. BELL, to be published.

5. D. BOHM, *Phys. Rev.* **85**, 166 and 180 (1952).

6. D. BOHM and Y. AHARONOV, *Phys. Rev.* **108**, 1070 (1957).

7. P. A. M. DIRAC, *The Principles of Quantum Mechanics* (3rd Ed.) p. 37. The Clarendon Press, Oxford (1947).

# Chapter Six

The formulations of quantum mechanics by Werner Heisenberg and Erwin Schrodinger and the relativistic extension by Paul Dirac provide a way to calculate the dynamics of physical systems under the influence of any instantaneous force. This is very important but is not the complete picture. Electrons, or any charged particle, when accelerated will produce electromagnetic radiation, and equivalently, when electromagnetic radiation impacts a charged particle, the particle will accelerate. Understanding how quantum particles interact with electromagnetic fields, how they emit and absorb radiation, requires a quantum theory of electrodynamics, and the founders of quantum mechanics immediately began to develop this theory.

Dirac first attempted to tackle this problem in a 1927 paper entitled "The Quantum Theory of the Emission and Absorption of Radiation." In order to develop a complete theory of quantum electrodynamics, Dirac looked for inspiration from the classical theory. There are at least two independent formulations of classical mechanics, known as the Lagrangian and Hamiltonian methods. Quantum mechanics had been built by analogy with the classical Hamiltonian, but it was known that the Lagrangian method was easier to reformulate into a relativistic theory. Since electromagnetics is the theory upon which relativity is built, Dirac's hope was that by finding a Lagrangian method for quantum mechanics it would be easier to develop a theory of quantum electrodynamics. In 1932, Dirac published "The Langrangian in Quantum Mechanics" in which he applied the Lagrangian method to quantum mechanics. His hope proved correct, and Dirac's paper helped form the basis of both Richard Feynman's and Julian Schwinger's approaches to quantum electrodynamics.

Also in 1932, Dirac published, with the collaborators Vladimir Fock and Boris Podolsky, "On Quantum Electrodynamics," in which they provided a formulation of quantum electrodynamics and then proved that it was equivalent to an earlier theory produced in 1930 by Heisenberg and Pauli. It was quickly realized that both the Dirac and Heisenberg-Pauli formulations of quantum electrodynamics suffered from a serious problem. Many of the computations required in these theories led to predictions of infinite energy. For example, when calculating the energy of an electron a term must be included that accounts for the self-energy stored in its electric field. For a point particle, the mathematical integral representing this term is infinite. This is clearly not physical, and it indicates a weakness in the theory. These infinities were deeply troubling, and for the next decade the elite of quantum theory struggled to find a way to eliminate them. The progress of this work is summarized in Robert Oppenheimer's 1947 paper "Electron Theory."

One important method for removing the infinities from the theory was published in 1934 by Max Born and Leopold Infeld. In "Foundations of the New Field Theory," they modified Maxwell's equations, which are the equations that govern the theory of electrodynamics, in such a way as to keep the self-energy of a point particle finite. Just how in Einstein's theory of relativity there is a cutoff velocity—the speed of light—above which nothing can travel, Born and Infeld postulated that there may be a cutoff above which no electric field is possible. They modified Maxwell's equations to incorporate this assumption. In the next chapters we will find that such an assumption was not necessary. In the late 1940s a method was developed for regularizing the infinities without having to explicitly modify the classical form of Maxwell's equations.

# The Quantum Theory of the Emission and Absorption of Radiation

By

Paul A.M. Dirac

From Proceedings of the Royal Society of London, Series A,
Vol. 114, p. 243 (1927)

## § 1. Introduction and Summary.

The new quantum theory, based on the assumption that the dynamical variables do not obey the commutative law of multiplication, has by now been developed sufficiently to form a fairly complete theory of dynamics. One can treat mathematically the problem of any dynamical system composed of a number of particles with instantaneous forces acting between them, provided it is describable by a Hamiltonian function, and one can interpret the mathematics physically by a quite definite general method. On the other hand, hardly anything has been done up to the present on quantum electrodynamics. The questions of the correct treatment of a system in which the forces are propagated with the velocity of light instead of instantaneously, of the production of an electromagnetic field by a moving electron, and of the reaction of this field on the electron have not yet been touched. In addition, there is a serious difficulty in making the theory satisfy all the requirements of the restricted principle of relativity, since a Hamiltonian function can no longer be used. This relativity question is, of course, connected with the previous ones, and it will be impossible to answer any one question completely without at the same time answering them all. However, it appears to be possible to build up a fairly satisfactory theory of the emission of radiation and of the reaction of the radiation field on the emitting system on the basis of a kinematics and dynamics which are not strictly relativistic. This is the main object of the present paper. The theory is non-relativistic only on account of the time being counted throughout as a c-number, instead of being treated symmetrically with the space co-ordinates. The relativity variation of mass with velocity is taken into account without difficulty.

The underlying ideas of the theory are very simple. Consider an atom interacting with a field of radiation, which we may suppose for definiteness to be confined in an enclosure so as to have only a discrete set of degrees of freedom. Resolving the radiation into its Fourier components, we can consider the energy and phase of each of the components to be dynamical variables describing the radiation field. Thus if $E_r$ is the energy of a component labelled $r$ and $\theta_r$ is the corresponding phase (defined as the time since the wave was in a standard phase), we can suppose each $E_r$ and $\theta_r$ to form a pair of canonically conjugate variables. In the absence of any interaction between the field and the atom, the whole system of field plus atom will be describable by the Hamiltonian

$$H = \Sigma_r E_r + H_0 \qquad (1)$$

equal to the total energy, $H_0$ being the Hamiltonian for the atom alone, since the variables $E_r$, $\theta_r$ obviously satisfy their canonical equations of motion

$$\dot{E}_r = -\frac{\partial H}{\partial \theta_r} = 0, \quad \dot{\theta}_r = \frac{\partial H}{\partial E_r} = 1.$$

When there is interaction between the field and the atom, it could be taken into account on the classical theory by the addition of an interaction term to the Hamiltonian (1), which would be a function of the variables of the atom and of the variables $E_r$, $\theta_r$ that describe the field. This interaction term would give the effect of the radiation on the atom, and also the reaction of the atom on the radiation field.

In order that an analogous method may be used on the quantum theory, it is necessary to assume that the variables $E_r$, $\theta_r$ are q-numbers satisfying the standard quantum conditions $\theta_r E_r - E_r \theta_r = ih$, etc., where $h$ is $(2\pi)^{-1}$ times the usual Planck's constant, like the other dynamical variables of the problem. This assumption immediately gives light-quantum properties to the radiation.* For if $\nu_r$ is the frequency of the component $r$, $2\pi \nu_r \theta_r$ is an angle variable, so that its canonical conjugate $E_r/2\pi\nu_r$ can only assume a discrete set of values differing by multiples of $h$, which means that $E_r$ can change only by integral multiples of the quantum $(2\pi h)\nu_r$. If we now add an interaction term (taken over from the clasical theory) to the Hamiltonian (1), the problem can be solved according to the rules of quantum mechanics, and we would expect to obtain the correct results for the action of the radiation and the atom on one another. It will be shown that we actually get the correct laws for the emission and absorption of radiation, and the correct values for Einstein's A's and B's. In the author's previous theory,[†] where the energies and phases of the components of radiation were c-numbers, only the B's could be obtained, and the reaction of the atom on the radiation could not be taken into account.

It will also be shown that the Hamiltonian which describes the interaction of the atom and the electromagnetic waves can be made identical with the Hamiltonian for the problem of the interaction of the atom with an assembly of particles moving with the velocity of light and satisfying the Einstein-Bose statistics, by a suitable choice of the interaction energy for the particles. The number of particles having any specified direction of motion and energy, which can be used as a dynamical variable in the Hamiltonian for the particles, is equal to the number of quanta of energy in the corresponding wave in the Hamiltonian for the waves. There is thus a complete harmony between the wave and light-quantum descriptions of the interaction. We shall actually build up the theory from the light-quantum point of view, and show that the Hamiltonian transforms naturally into a form which resembles that for the waves.

The mathematical development of the theory has been made possible by the author's general transformation theory of the quantum

---

* Similar assumptions have been used by Born and Jordan ['Z. f. Physik,' vol. 34, p. 886 (1925)] for the purpose of taking over the classical formula for the emission of radiation by a dipole into the quantum theory, and by Born, Heisenberg and Jordan ['Z. f. Physik,' vol. 35, p. 606 (1925)] for calculating the energy fluctuations in a field of black-body radiation.
    † 'Roy. Soc. Proc.,' A, vol. 112, p. 661, § 5 (1926). This is quoted later by, *loc. cit.*, I.

matrices.* Owing to the fact that we count the time as a c-number, we are allowed to use the notion of the value of any dynamical variable at any instant of time. This value is a q-number, capable of being represented by a generalised "matrix" according to many different matrix schemes, some of which may have continuous ranges of rows and columns, and may require the matrix elements to involve certain kinds of infinities (of the type given by the $\delta$ functions[†]). A matrix scheme can be found in which any desired set of constants of integration of the dynamical system that commute are represented by diagonal matrices, or in which a set of variables that commute are represented by matrices that are diagonal at a specified time.[‡] The values of the diagonal elements of a diagonal matrix representing any q-number are the characteristic values of that q-number. A Cartesian co-ordinate or momentum will in general have all characteristic values from $-\infty$ to $+\infty$, while an action variable has only a discrete set of characteristic values. (We shall make it a rule to use unprimed letters to denote the dynamical variables or q-numbers, and the same letters primed or multiply primed to denote their characteristic values. Transformation functions or eigenfunctions are functions of the characteristic values and not of the q-numbers themselves, so they should always be written in terms of primed variables.)

If $f(\xi, \eta)$ is any function of the canonical variables $\xi_k$, $\eta_k$, the matrix representing $f$ at any time $t$ in the matrix scheme in which the $\xi_k$ at time $t$ are diagonal matrices may be written down without any trouble, since the matrices representing the $\xi_k$ and $\eta_k$ themselves at time $t$ are known, namely,

$$
\left.
\begin{aligned}
\xi_k(\xi'\xi'') &= \xi'_k\delta(\xi'\xi''), \\
\eta_k(\xi'\xi'') &= -ih\,\delta(\xi'_1 - \xi''_1)\ldots\delta(\xi'_{k-1} - \xi''_{k-1})\delta'(\xi'_k - \xi''_k)\delta(\xi'_{k+1} - \xi''_{k+1})\ldots
\end{aligned}
\right\}.
$$

$$(2)$$

* 'Roy. Soc. Proc.,' A, vol. 113, p. 621 (1927). This is quoted later by *loc. cit.*, II. An essentially equivalent theory has been obtained independently by Jordan ['Z. f. Physik,' vol. 40, p. 809 (1927)]. See also, F. London, 'Z. f. Physik,' vol. 40, p. 193 (1926).

† *Loc. cit.* II, § 2.

‡ One can have a matrix scheme in which a set of variables that commute are at all times represented by diagonal matrices if one will sacrifice the condition that the matrices must satisfy the equations of motion. The transformation function from such a scheme to one in which the equations of motion are satisfied will involve the time explicitly. See p. 628 in *loc. cit.*, II.

Thus if the Hamiltonian H is given as a function of the $\xi_k$ and $\eta_k$, we can at once write down the matrix $H(\xi' \ \xi'')$. We can then obtain the transformation function, $(\xi'/\alpha')$ say, which transforms to a matrix scheme $(\alpha)$ in which the Hamiltonian is a diagonal matrix, as $(\xi'/\alpha')$ must satisfy the integral equation

$$\int H(\xi'\xi'')d\xi''(\xi''/\alpha') = W(\alpha')\cdot(\xi'/\alpha'), \qquad (3)$$

of which the characteristic values $W(\alpha')$ are the energy levels. This equation is just Schrödinger's wave equation for the eigenfunctions $(\xi'/\alpha')$, which becomes an ordinary differential equation when H is a simple algebraic function of the $\xi_k$ and $\eta_k$ on account of the special equations (2) for the matrices representing $\xi_k$ and $\eta_k$. Equation (3) may be written in the more general form

$$\int H(\xi'\xi'')d\xi''(\xi''/\alpha') = ih \ \partial(\xi'/\alpha')/\partial t, \qquad (3')$$

in which it can be applied to systems for which the Hamiltonian involves the time explicitly.

One may have a dynamical system specified by a Hamiltonian H which cannot be expressed as an algebraic function of any set of canonical variables, but which can all the same be represented by a matrix $H(\xi'\xi'')$. Such a problem can still be solved by the present method, since one can still use equation (3) to obtain the energy levels and eigenfunctions. We shall find that the Hamiltonian which describes the interaction of a light-quantum and an atomic system is of this more general type, so that the interaction can be treated mathematically, although one cannot talk about an interaction potential energy in the usual sense.

It should be observed that there is a difference between a light-wave and the de Broglie or Schrodinger wave associated with the light-quanta. Firstly, the light-wave is always real, while the de Broglie wave associated with a light-quantum moving in a definite direction must be taken to involve an imaginary exponential. A more important difference is that their intensities are to be interpreted in different ways. The number of light-quanta per unit volume associated with a monochromatic light-wave equals the energy per unit volume of the wave divided by the energy $(2\pi h)\nu$ of a single light-quantum. On the other hand a monochromatic de Broglie wave of amplitude $a$ (multiplied into the imaginary exponential factor) must be interpreted as representing $a^2$ light-quanta per unit volume for all frequencies. This is a special case of the general rule for interpreting the matrix analysis,* according to which, if $(\xi'/\alpha')$ or $\psi_{a'}(\xi'_k)$ is the eigenfunction in the variables $\xi_k$ of the state $\alpha'$ of an atomic system (or simple particle), $|\psi_{a'}(\xi'_k)|^2$ is the probability of each $\xi_k$ having the value $\xi'_k$, [or $|\psi_{a'}(\xi'_k)|^2 d\xi'_1 d\xi'_2 \ldots$ is the probability of each $\xi_k$ lying between the values $\xi'_k$ and $\xi'_k + d\xi'_k$, when the $\xi_k$ have continuous ranges of characteristic values] on the assumption that all phases of the system are equally probable. The wave whose intensity is to be interpreted in the first of these two ways appears in the theory only when one is dealing with an assembly of the associated particles satisfying the Einstein-Bose statistics. There is thus no such wave associated with electrons.

## § 2. THE PERTURBATION OF AN ASSEMBLY OF INDEPENDENT SYSTEMS.

We shall now consider the transitions produced in an atomic system by an arbitrary perturbation. The method we shall adopt will be that previously given by the author,[†] which leads in a simple way to equations which determine the probability of the system being in any stationary state of the unperturbed system at any time.[‡] This, of course, gives immediately the probable number of systems in that state at that time for an assembly of the systems that are independent of one another and are all perturbed in the same way. The object of the present section is to show that the equations for the rates of change of these probable numbers can be put in the Hamiltonian form in a

---

*Loc. cit.*, II, § § 6, 7.
[†] *Loc. cit.* I.
[‡] The theory has recently been extended by Born ['Z. f. Physik,' vol. 40, p. 167 (1926)] so as to take into account the adiabatic changes in the stationary states that may be produced by the perturbation as well as the transitions. This extension is not used in the present paper.

simple manner, which will enable further developments in the theory to be made.

Let $H_0$ be the Hamiltonian for the unperturbed system and V the perturbing energy, which can be an arbitrary function of the dynamical variables and may or may not involve the time explicitly, so that the Hamiltonian for the perturbed system is $H = H_0 + V$. The eigenfunctions for the perturbed system must satisfy the wave equation

$$ih \, \partial\psi/\partial t = (H_0 + V)\psi,$$

where $(H_0 + V)$ is an operator. If $\psi = \Sigma_r a_r \psi_r$ is the solution of this equation that satisfies the proper initial conditions, where the $\psi_r$'s are the eigenfunctions for the unperturbed system, each associated with one stationary state labelled by the suffix $r$, and the $a_r$'s are functions of the time only, then $|a_r|^2$ is the probability of the system being in the state $r$ at any time. The $a_r$'s must be normalised initially, and will then always remain normalised. The theory will apply directly to an assembly of N similar independent systems if we multiply each of these $a_r$'s by $N^{\frac{1}{2}}$ so as to make $\Sigma_r |a_r|^2 = N$. We shall now have that $|a_r|^2$ is the probable number of systems in the state $r$.

The equation that determines the rate of change of the $a_r$'s is*

$$ih \, \dot{a}_r = \Sigma_s V_{rs} a_s, \tag{4}$$

where the $V_{rs}$'s are the elements of the matrix representing V. The conjugate imaginary equation is

$$-ih \, \dot{a}_r^* = \Sigma_s V_{rs}^* a_s^* = \Sigma_s a_s^* V_{sr}. \tag{4'}$$

If we regard $a_r$ and $ih \, a_r^*$ as canonical conjugates, equations (4) and (4') take the Hamiltonian form with the Hamiltonian function $F_1 = \Sigma_{rs} a_r^* V_{rs} a_s$, namely,

$$\frac{da_r}{dt} = \frac{1}{ih} \frac{\partial F_1}{\partial a_r^*}, \quad ih \frac{da_r^*}{dt} = -\frac{\partial F_1}{\partial a_r}.$$

* *Loc. cit.*, I, equation (25).

We can transform to the canonical variables $N_r$, $\phi_r$ by the contact transformation

$$a_r = N_r^{\frac{1}{2}} e^{-i\phi_r/h}, \qquad a_r^* = N_r^{\frac{1}{2}} e^{i\phi_r/h}.$$

This transformation makes the new variables $N_r$ and $\phi_r$ real, $N_r$ being equal to $a_r a_r^* = |a_r|^2$, the probable number of systems in the state $r$, and $\phi_r/h$ being the phase of the eigenfunction that represents them. The Hamiltonian $F_1$ now becomes

$$F_1 = \Sigma_{rs} V_{rs} N_r^{\frac{1}{2}} N_s^{\frac{1}{2}} e^{i(\phi_r - \phi_s)/h},$$

and the equations that determine the rate at which transitions occur have the canonical form

$$N_r = -\frac{\partial F_1}{\partial \phi_r}, \qquad \dot{\phi}_r = \frac{\partial F_1}{\partial N_r}.$$

A more convenient way of putting the transition equations in the Hamiltonian form may be obtained with the help of the quantities

$$b_r = a_r e^{-iW_r t/h}, \qquad b_r^* = a_r^* e^{iW_r t/h},$$

$W_r$ being the energy of the state $r$. We have $|b_r|^2$ equal to $|a_r|^2$, the probable number of systems in the state $r$. For $b_r$ we find

$$ih\, \dot{b}_r = W_r b_r + ih\, \dot{a}_r e^{-iW_r t/h}$$
$$= W_r b_r + \Sigma_s V_{rs} b_s e^{i(W_s - W_r)t/h}$$

with the help of (4). If we put $V_{rs} = v_{rs} e^{i(W_r - W_s)t/h}$; so that $v_{rs}$, is a constant when V does not involve the time explicitly, this reduces to

$$ih\, \dot{b}_r = W_r b_r + \Sigma_s v_{rs} b_s$$
$$= \Sigma_s H_{rs} b_s, \tag{5}$$

where $H_{rs} = W_r \delta_{rs} + v_{rs}$, which is a matrix element of the total Hamiltonian $H = H_0 + V$ with the time factor $e^{i(W_r - W_s)t/h}$ removed, so that $H_{rs}$ is a constant when H does not involve the time explicitly. Equation (5) is of the same form as equation (4), and may be put in the Hamiltonian form in the same way.

It should be noticed that equation (5) is obtained directly if one writes down the Schrödinger equation in a set of variables that specify the stationary states of the unperturbed system. If these variables are

$\xi_h$, and if $H(\xi'\xi'')$ denotes a matrix element of the total Hamiltonian H in the $(\xi)$ scheme, this Schrödinger equation would be

$$ih\, \partial\psi(\xi')/\partial t = \Sigma_{\xi''} H(\xi'\xi'')\psi(\xi''), \tag{6}$$

like equation (3′). This differs from the previous equation (5) only in the notation, a single suffix $r$ being there used to denote a stationary state instead of a set of numerical values $\xi'_k$ for the variables $\xi_k$, and $b_r$ being used instead of $\psi(\xi')$. Equation (6), and therefore also equation (5), can still be used when the Hamiltonian is of the more general type which cannot be expressed as an algebraic function of a set of canonial variables, but can still be represented by a matrix $H(\xi'\xi'')$ or $H_{rs}$.

We now take $b_r$ and $ih\, b_r^*$ to be canonioally conjugate variables instead of $a_r$ and $ih\, a_r^*$. The equation (5) and its conjugate imaginary equation will now take the Hamiltonian form with the Hamiltonian function

$$F = \Sigma_{rs} b_r^* H_{rs} b_s. \tag{7}$$

Proceeding as before, we make the contact transformation

$$b_r = N_r^{\frac{1}{2}} e^{-i\theta_r/h}, \qquad b_r^* = N_r^{\frac{1}{2}} e^{i\theta_r/h}, \tag{8}$$

to the new canonical variables $N_r$, $\theta_r$, where $N_r$ is, as before, the probable number of systems in the state $r$, and $\theta_r$ is a new phase. The Hamiltonian F will now become

$$F = \Sigma_{rs} H_{rs} N_r^{\frac{1}{2}} N_r^{\frac{1}{2}} e^{i(\theta_r - \theta_s)/h},$$

and the equations for the rates of change of $N_r$ and $\theta_r$ will take the canonical form

$$\dot{N}_r = -\frac{\partial F}{\partial \theta_r}, \qquad \dot{\theta}_r = \frac{\partial F}{\partial N_r}.$$

The Hamiltonian may be written

$$F = \Sigma_r W_r N_r + \Sigma_{rs} v_{rs} N_r^{\frac{1}{2}} N_s^{\frac{1}{2}} e^{i(\theta_r - \theta_s)/h}. \tag{9}$$

The first term $\Sigma_r W_r N_r$ is the total proper energy of the assembly, and the second may be regarded as the additional energy due to the perturbation. If the perturbation is zero, the phases $\theta_r$ would increase linearly with the time, while the previous phases $\phi_r$ would in this case be constants.

## § 3. THE PERTURBATION OF AN ASSEMBLY SATISFYING THE EINSTEIN-BOSE STATISTICS.

According to the preceding section we can describe the effect of a perturbation on an assembly of independent systems by means of canonical variables and Hamiltonian equations of motion. The development of the theory which naturally suggests itself is to make these canonical variables q-numbers satisfying the usual quantum conditions instead of c-numbers, so that their Hamiltonian equations of motion become true quantum equations. The Hamiltonian function will now provide a Schrödinger wave equation, which must be solved and interpreted in the usual manner. The interpretation will give not merely the probable number of systems in any state, but the probability of any given distribution of the systems among the various states, this probability being, in fact, equal to the square of the modulus of the normalised solution of the wave equation that satisfies the appropriate initial conditions. We could, of course, calculate directly from elementary considerations the probability of any given distribution when the systems are independent, as we know the probability of each system being in any particular state. We shall find that the probability calculated directly in this way does not agree with that obtained from the wave equation except in the special case when there is only one system in the assembly. In the general case it will be shown that the wave equation leads to the correct value for the probability of any given distribution when the systems obey the Einstein-Bose statistics instead of being independent.

We assume the variables $b_r$, $ih\, b_r^*$ of § 2 to be canonical q-numbers satisfying the quantum conditions

$$b_r \cdot ih\, b_r^* - ih\, b_r^* \cdot b_r = ih$$

or
$$b_r b_r^* - b_r^* b_r = 1,$$

and
$$b_r b_s - b_s b_r = 0, \qquad b_r^* b_s^* - b_s^* b_r^* = 0,$$
$$b_r b_s^* - b_s^* b_r = 0 \qquad (s \neq r).$$

The transformation equations (8) must now be written in the quantum form

$$\left.\begin{array}{l} b_r = (\mathrm{N}_r + 1)^{\frac{1}{2}} e^{-i\theta_r/h} = e^{-i\theta_r/h} \mathrm{N}_r^{\frac{1}{2}} \\[2mm] b_r^* = \mathrm{N}_r^{\frac{1}{2}} e^{i\theta_r/h} = e^{i\theta_r/h} (\mathrm{N}_r + 1)^{\frac{1}{2}}, \end{array}\right\} \qquad (10)$$

in order that the $\mathrm{N}_r$, $\theta_r$ may also be canonical variables. These equations show that the $\mathrm{N}_r$ can have only integral characteristic values not less than zero,* which provides us with a justification for the assumption that the variables are q-numbers in the way we have chosen. The numbers of systems in the different states are now ordinary quantum numbers.

The Hamiltonian (7) now becomes

$$\begin{aligned} \mathrm{F} &= \Sigma_{rs} b_r^* \mathrm{H}_{rs} b_s = \Sigma_{rs} \mathrm{N}_r^{\frac{1}{2}} e^{i\theta_r/h} \mathrm{H}_{rs} (\mathrm{N}_s + 1)^{\frac{1}{2}} e^{-i\theta_r/h} \\[2mm] &= \Sigma_{rs} \mathrm{H}_{rs} \mathrm{N}_r^{\frac{1}{2}} (\mathrm{N}_s + 1 - \delta_{rs})^{\frac{1}{2}} e^{i(\theta_r - \theta_s)/h} \end{aligned} \qquad (11)$$

in which the $\mathrm{H}_{rs}$ are still c-numbers. We may write this F in the form corresponding to (9)

$$\mathrm{F} = \Sigma_r \mathrm{W}_r \mathrm{N}_r + \Sigma_{rs} v_{rs} \mathrm{N}_r^{\frac{1}{2}} (\mathrm{N}_s + 1 - \delta_{rs})^{\frac{1}{2}} e^{i(\theta_r - \theta_s)/h} \qquad (11')$$

in which it is again composed of a proper energy term $\Sigma_r \mathrm{W}_r \mathrm{N}_r$ and an interaction energy term.

The wave equation written in terms of the variables $\mathrm{N}_r$ is[†]

$$ih \frac{\partial}{\partial t} \psi(\mathrm{N}_1', \mathrm{N}_2', \mathrm{N}_s' \ldots) = \mathrm{F}\psi(\mathrm{N}_1', \mathrm{N}_2', \mathrm{N}_3' \ldots), \qquad (12)$$

where F is an operator, each $\theta_r$ occurring in F being interpreted to mean $ih\, \partial/\partial \mathrm{N}_r'$. If we apply the operator $e^{\pm i\theta_r/h}$ to any function $f(\mathrm{N}_1', \mathrm{N}_2', \ldots \mathrm{N}_r', \ldots)$ of the variables $\mathrm{N}_1', \mathrm{N}_2', \ldots$ the result is

$$\begin{aligned} e^{\pm i\theta_r/h} f(\mathrm{N}_1', \mathrm{N}_2', \ldots \mathrm{N}_r', \ldots) &= e^{\mp\delta/\delta \mathrm{N}_r'} f(\mathrm{N}_1', \mathrm{N}_2', \ldots \mathrm{N}_r' \cdots) \\[2mm] &= f(\mathrm{N}_1', \mathrm{N}_2', \ldots \mathrm{N}_r' \mp 1, \ldots). \end{aligned}$$

---

*See § 8 of the author's paper 'Roy. Soc. Proc.,' A, vol. 111, p. 281 (1926). What are there called the c-number values that a q-number can take are here given the more precise name of the characteristic values of that q-number.
†We are supposing for definiteness that the label $r$ of the stationary states takes the values 1, 2, 3, . . .

If we use this rule in equation (12) and use the expression (11) for F we obtain*

$$ih \frac{\partial}{\partial t} \psi(N_1', N_2', N_3' \ldots)$$
$$= \Sigma_{rs} H_{rs} N_r'^{\frac{1}{2}} (N_s' + 1 - \delta_{rs})^{\frac{1}{2}} \psi(N_1', N_2' \ldots N_r' - 1, \ldots N_s' + 1, \ldots).$$

$$(13)$$

We see from the right-hand side of this equation that in the matrix representing F, the term in F involving $e^{i(\theta_r - \theta_s)/h}$ will contribute only to those matrix elements that refer to transitions in which $N_r$ decreases by unity and $N_s$ increases by unity, *i.e.*, to matrix elements of the type $F(N_1', N_2', \ldots N_r', \ldots N_s'; N_1', N_2' \ldots N_r' - 1 \ldots N_s' + 1 \ldots)$. If we find a solution $\psi(N_1', N_2' \ldots)$ of equation (13) that is normalised [*i.e.*, one for which $\Sigma_{N_1', N_2' \ldots} |\psi(N_1', N_2' \ldots)|^2 = 1$] and that satisfies the proper initial conditions, then $|\psi(N_1', N_2' \ldots)|^2$ will be the probability of that distribution in which $N_1'$ systems are in state 1, $N_2'$ in state 2, ... at any time.

Consider first the case when there is only one system in the assembly. The probability of its being in the state $q$ is determined by the eigenfunction $\psi(N_1', N_2', \ldots)$ in which all the $N'$'s are put equal to zero except $N_q'$, which is put equal to unity. This eigenfunction we shall denote by $\psi\{q\}$. When it is substituted in the left-hand side of (13), all the terms in the summation on the right-hand side vanish except those for which $r = q$, and we are left with

$$ih \frac{\partial}{\partial t} \psi\{q\} = \Sigma_r H_{qs} \psi\{s\},$$

which is the same equation as (5) with $\psi\{q\}$ playing the part of $b_q$. This establishes the fact that the present theory is equivalent to that of the preceding section when there is only one system in the assembly.

Now take the general case of an arbitrary number of systems in the assembly, and assume that they obey the Einstein-Bose statistical mechanics. This requires that, in the ordinary treatment of the problem, only those eigenfunctions that are symmetrical between all

---

*When $s = r$, $\psi(N_1', N_2' \ldots N_r' - 1 \ldots N_s' + 1)$ is to be taken to mean $\psi(N_1' N_2' \ldots N_r' \ldots)$.

the systems must be taken into account, these eigenfunctions being by themselves sufficient to give a complete quantum solution of the problem.* We shall now obtain the equation for the rate of change of one of these symmetrical eigenfunctions, and show that it is identical with equation (13).

If we label each system with a number $n$, then the Hamiltonian for the assembly will be $H_A = \Sigma_n H(n)$, where $H(n)$ is the H of § 2 (equal to $H_0 + V$) expressed in terms of the variables of the $n$th system. A stationary state of the assembly is defined by the numbers $r_1, r_2 \ldots r_n \ldots$ which are the labels of the stationary states in which the separate systems lie. The Schrödinger equation for the assembly in a set of variables that specify the stationary states will be of the form (6) [with $H_A$ instead of H], and we can write it in the notation of equation (5) thus:—

$$ih\, \dot{b}(r_1 r_2 \ldots) = \Sigma_{s_1, s_2 \ldots} H_A(r_1 r_2 \ldots; s_1 s_2 \ldots) b(s_1 s_2 \ldots), \quad (14)$$

where $H_A (r_1 r_2 \ldots; s_1 s_2 \ldots)$ is the general matrix element of $H_A$ [with the time factor removed]. This matrix element vanishes when more than one $s_n$ differs from the corresponding $r_n$; equals $H_{r_m s_m}$ when $s_m$ differs from $r_m$ and every other $s_n$ equals $r_n$; and equals $\Sigma_n H_{r_n r_n}$ when every $s_n$ equals $r_n$. Substituting these values in (14), we obtain

$$ih\, \dot{b}(r_1 r_2 \ldots) = \Sigma_m \Sigma_{s_m \neq r_m} H_{r_m s_m} b(r_1 r_2 \ldots r_{m-1} s_m\, r_{m+1} \ldots)$$
$$+ \Sigma_n H_{r_n r_n} b(r_1 r_2 \ldots). \quad (15)$$

We must now restrict $b(r_1 r_2 \ldots)$ to be a symmetrical function of the variables $r_1, r_2 \ldots$ in order to obtain the Einstein-Bose statistics. This is permissible since if $b(r_1 r_2 \ldots)$ is symmetrical at any time, then equation (15) shows that $\dot{b}(r_1 r_2 \ldots)$ is also symmetrical at that time, so that $b(r_1 r_2 \ldots)$ will remain symmetrical.

Let $N_r$ denote the number of systems in the state $r$. Then a stationary state of the assembly describable by a symmetrical eigenfunction may be specified by the numbers $N_1, N_2 \ldots N_r \ldots$ just as well as by the numbers $r_1, r_2 \ldots r_n \ldots$, and we shall be able to transform

---

*Loc. cit., I, § 3.

equation (15) to the variables $N_1$, $N_2 \ldots$. We cannot actually take the new eigenfunction $b$ $(N_1, N_2 \ldots)$ equal to the previous one $b$ $(r_1 r_2 \ldots)$, but must take one to be a numerical multiple of the other in order that each may be correctly normalised with respect to its respective variables. We must have, in fact,

$$\Sigma_{r_1, r_2 \ldots} |b(r_1, r_2 \ldots)|^2 = 1 = \Sigma_{N_1, N_2 \ldots} |b(N_1, N_2 \ldots)|^2,$$

and hence we must take $|b(N_1, N_2 \ldots)|^2$ equal to the sum of $|b(r_1 r_2 \ldots)|^2$ for all values of the numbers $r_1$, $r_2 \ldots$ such that there are $N_1$ of them equal to 1, $N_2$ equal to 2, etc. There are $N!/N_1!$ $N_2! \ldots$ terms in this sum, where $N = \Sigma_r N_r$ is the total number of systems, and they are all equal, since $b(r_1 r_2 \ldots)$ is a symmetrical function of its variables $r_1$, $r_2 \ldots$. Hence we must have

$$b(N_1, N_2 \ldots) = (N!/N_1!N_2! \ldots)^{\frac{1}{2}} b(r_1 r_2 \ldots).$$

If we make this substitution in equation (15), the left-hand side will become $ih (N_1!N_2! \ldots /N!)^{\frac{1}{2}} b(N_1, N_2 \ldots)$. The term $H_{r_m s_m} b(r_1 r_2 \ldots r_{m-1} s_m r_{m+1} \ldots)$ in the first summation on the right-hand side will become

$$[N_1!N_2! \ldots (N_r - 1)! \ldots (N_s + 1)! \ldots /N!]^{\frac{1}{2}}$$
$$\times H_{rs} b(N_1, N_2 \ldots N_r - 1 \ldots N_s + 1 \ldots), \qquad (16)$$

where we have written $r$ for $r_m$ and $s$ for $s_m$. This term must be summed for all values of $s$ except $r$, and must then be summed for $r$ taking each of the values $r_1$, $r_2 \ldots$. Thus each term (16) gets repeated by the summation process until it occurs a total of $N_r$ times, so that it contributes

$$N_r [N_1! N_2! \ldots (N_r - 1)! \ldots (N_s + 1)! \ldots /N!]^{\frac{1}{2}}$$
$$\times H_{rs} b(N_1, N_2 \ldots N_r - 1 \ldots N_s + 1 \ldots)$$
$$= N_r^{\frac{1}{2}} (N_s + 1)^{\frac{1}{2}} (N_1!N_2! \ldots /N!)^{\frac{1}{2}}$$
$$\times H_{rs} b(N_1, N_2 \ldots N_r - 1 \ldots N_s + 1 \ldots)$$

to the right-hand side of (15). Finally, the term $\Sigma_n H_{r_n r_n} b(r_1, r_2 \ldots)$ becomes

$$\Sigma_r N_r H_{rr} \cdot b(r_1 r_2 \ldots) = \Sigma_r N_r H_{rr} \cdot (N_1! N_2! \ldots /N!)^{\frac{1}{2}} b(N_1, N_2 \ldots).$$

Hence equation (15) becomes, with the removal of the factor $(N_1!\ N_2!\ \ldots/N!)^{\frac{1}{2}}$,

$$i\hbar\,\dot{b}(N_1, N_2\ldots) = \Sigma_r\Sigma_{s\neq r}N_r^{\frac{1}{2}}(N_s+1)^{\frac{1}{2}}$$
$$\times H_{rs}\,b(N_1, N_2\ldots N_r-1\ldots N_s+1\ldots)$$
$$+\Sigma_r N_r H_{rr}\,b\,(N_1, N_2\ldots), \tag{17}$$

which is identical with (13) [except for the fact that in (17) the primes have been omitted from the N's, which is permissible when we do not require to refer the N's as q-numbers]. We have thus established that the Hamiltonian (11) describes the effect of a perturbation on an assembly satisfying the Einstein-Bose statistics.

## § 4. THE REACTION OF THE ASSEMBLY ON THE PERTURBING SYSTEM.

Up to the present we have considered only perturbations that can be represented by a perturbing energy V added to the Hamiltonian of the perturbed system, V being a function only of the dynamical variables of that system and perhaps of the time. The theory may readily be extended to the case when the perturbation consists of interaction with a perturbing dynamical system, the reaction of the perturbed system on the perturbing system being taken into account. (The distinction between the perturbing system and the perturbed system is, of course, not real, but it will be kept up for convenience.)

We now consider a perturbing system, described, say, by the canonical variables $J_k$, $\omega_k$, the J's being its first integrals when it is alone, interacting with an assembly of perturbed systems with no mutual interaction, that satisfy the Einstein-Bose statistics. The total Hamiltonian will be of the form

$$H_T = H_P(J) + \Sigma_n H(n),$$

where $H_p$ is the Hamiltonian of the perturbing system (a function of the J's only) and H($n$) is equal to the proper energy $H_0(n)$ plus the perturbation energy V($n$) of the $n$th system of the assembly. H($n$) is a function only of the variables of the $n$th system of the assembly and of the J's and $w$'s, and does not involve the time explicitly.

The Schrödinger equation corresponding to equation (14) is now

$$ih\, b\left(J', r_1 r_2 \ldots\right) = \Sigma_{J''} \Sigma_{s_1, s_2 \ldots} H_r\left(J', r_1 r_2 \ldots; J'', s_1 s_2 \ldots\right) b\left(J'', s_1 s_2 \ldots\right),$$

in which the eigenfunction $b$ involves the additional variables $J'_k$. The matrix element $H_T(J', r_1 r_2 \ldots; J'', s_1 s_2 \ldots)$ is now always a constant. As before, it vanishes when more than one $s_n$ differs from the corresponding $r_n$. When $s_m$ differs from $r_m$ and every other $s_n$ equals $r_n$, it reduces to $H\left(J'_{r_m}; J''_{s_m}\right)$, which is the $\left(J'_{r_m}; J''_{s_m}\right)$ matrix element (with the time factor removed) of $H = H_0 + V$, the proper energy plus the perturbation energy of a single system of the assembly; while when every $s_n$ equals $r_n$, it has the value $H_P(J')\delta_{J'J''} + \Sigma_n H(J'_{r_n}; J''_{r_n})$. If, as before, we restrict the eigenfunctions to be symmetrical in the variables $r_1, r_2 \ldots$, we can again transform to the variables $N_1, N_2 \ldots$, which will lead, as before, to the result

$$ih\, b\left(J', N'_1, N'_2 \ldots\right) = H_P\left(J'\right) b\left(J', N'_1, N'_2 \ldots\right)$$
$$+\Sigma_{J''} \Sigma_{r,s} N'^{\frac{1}{2}}_r \left(N'_s + 1 - \delta_{rs}\right)^{\frac{1}{2}} H\left(J'r; J''s\right)$$
$$\times b\left(J', N'_1, N'_2 \ldots N'_r - 1 \ldots N'_s + 1 \ldots\right) \quad (18)$$

This is the Schrödinger equation corresponding to the Hamiltonian function

$$F = H_P\left(J\right) + \Sigma_{r,s} H_{rs} N^{\frac{1}{2}}_r \left(N_s + 1 - \delta_{rs}\right)^{\frac{1}{2}} e^{i(\theta_1 - \theta_s)/h}, \quad (19)$$

in which $H_{rs}$ is now a function of the J's and $w$'s, being such that when represented by a matrix in the (J) scheme its $(J'J'')$ element is $H(J'r; J''s)$. (It should be noticed that $H_{rs}$ still commutes with the N's and $\theta$'s.)

Thus the interaction of a perturbing system and an assembly satisfying the Einstein-Bose statistics can be described by a Hamiltonian of the form (19). We can put it in the form corresponding to (11') by observing that the matrix element $H(J'r; J''s)$ is composed of the sum of two parts, a part that comes from the proper energy $H_0$, which equals $W_r$ when $J''_k = J'_k$ and $s = r$ and vanishes otherwise, and a part that comes from the interaction energy V which may be denoted by $v(J'r; J''s)$. Thus we shall have

$$H_{rs} = W_r \, \delta_{rs} + v_{rs},$$

where $v_{rs}$ is that function of the J's and $w$'s which is represented by the matri whose ($J'J''$) element is $v$ ($J'r; J''s$), and so (19) becomes

$$F = H_P (J) + \Sigma_r W_r N_r + \Sigma_{r,s} v_{rs} N_r^{\frac{1}{2}} (N_s + 1 - \delta_{rs})^{\frac{1}{2}} e^{i(\theta_1 - \theta_s)/h}.$$

(20)

The Hamiltonian is thus the sum of the proper energy of the perturbing system $H_p$ (J), the proper energy of the perturbed systems $\Sigma_r W_r N_r$ and the perturbation energy $\Sigma_{r,s} v_{rs} N_r^{\frac{1}{2}} (N_s + 1 - \delta_{rs})^{\frac{1}{2}} e^{i(\theta_r - \theta_s)/h}$.

## § 5. Theory of Transitions in a System from One State to Others of the Same Energ

Before applying the results of the preceding sections to light-quanta, shall consider the solution of the problem presented by a Hamiltonian of t type (19). The essential feature of the problem is that it refers to a dynamic system which can, under the influence of a perturbation energy which do not involve the time explicitly, make transitions from one state to others the same energy. The problem of collisions between an atomic system and electron, which has been treated by Born,[*] is a special case of this type. Born method is to find a *periodic* solution of the wave equation which consists so far as it involves the co-ordinates of the colliding electron, of plane was representing the incident electron, approaching the atomic system, which are scattered or diffracted in all directions. The square of the amplitude of the waves scattered in any direction with any frequency is then assumed by Born to be the probability of the electron being scattered in that direction with the corresponding energy.

This method does not appear to be capable of extension in any simple manner to the general problem of systems that make transitions

---

[*] Born, 'Z. f. Physik,' vol. 38, p. 803 (1926).

from one state to others of the same energy. Also there is at present no very direct and certain way of interpreting a periodic solution of a wave equation to apply to a non-periodic physical phenomenon such as a collision. (The more definite method that will now be given shows that Born's assumption is not quite right, it being necessary to multiply the square of the amplitude by a certain factor.)

An alternative method of solving a collision problem is to find a *non-periodic* solution of the wave equation which consists initially simply of plane waves moving over the whole of space in the necessary direction with the necessary frequency to represent the incident electron. In course of time waves moving in other directions must appear in order that the wave equation may remain satisfied. The probability of the electron being scattered in any direction with any energy will then be determined by the rate of growth of the corresponding harmonic component of these waves. The way the mathematics is to be interpreted is by this method quite definite, being the same as that of the beginning of § 2.

We shall apply this method to the general problem of a system which makes transitions from one state to others of the same energy under the action of a perturbation. Let $H_0$ be the Hamiltonian of the unperturbed system and V the perturbing energy, which must not involve the time explicitly. If we take the case of a continuous range of stationary states, specified by the first integrals, $\alpha_k$ say, of the unperturbed motion, then, following the method of § 2, we obtain

$$i h \, \dot{a} \left( \alpha' \right) = \int \mathrm{V} \left( \alpha' \alpha'' \right) d\alpha'' \cdot a \left( \alpha'' \right), \qquad (21)$$

corresponding to equation (4). The probability of the system being in a state for which each $\alpha_k$ lies between $\alpha'_k$ and $\alpha'_k + d\alpha'_k$ at any time is $|a(\alpha')|^2 d\alpha'_1 \cdot d\alpha'_2 \ldots$ when $a(\alpha')$ is properly normalised and satisfies the proper initial conditions. If initially the system is in the state $\alpha^0$, we must take the initial value of $a\left( \alpha' \right)$ to be of the form $a^0 \cdot \delta \left( \alpha' - \alpha^0 \right)$. We shall keep $a^0$ arbitrary, as it would be inconvenient to normalise $a\left( \alpha' \right)$ in the present case. For a first approximation we may substitute for $a\left( \alpha'' \right)$ in the right-hand side of (21) its initial value.

This gives

$$ih\,\dot{a}\left(\alpha'\right) = a^{0}V\left(\alpha'\alpha^{0}\right) = a^{0}v\left(\alpha'\alpha^{0}\right)e^{i\left[W(\alpha')-W(\alpha^{0})\right]t/h},$$

where $v\left(\alpha'\alpha^{0}\right)$ is a constant and $W(\alpha')$ is the energy of the state $\alpha'$. Hence

$$ih\,a\left(\alpha'\right) = a^{0}\delta\left(\alpha'-\alpha^{0}\right) + a^{0}v\left(\alpha'\alpha^{0}\right)\frac{e^{i\left[W(\alpha')-W(\alpha^{0})\right]t/h}-1}{i\left[W\left(\alpha'\right)-W\left(\alpha^{0}\right)\right]/h}.$$

$$(22)$$

For values of the $\alpha'_{k}$ such that $W\left(\alpha'\right)$ differs appreciably from $W\left(\alpha^{0}\right)$, $a\left(\alpha'\right)$ is a periodic function of the time whose amplitude is small when the perturbing energy $V$ is small, so that the eigenfunctions corresponding to these stationary states are not excited to any appreciable extent. On the other hand, for values of the $\alpha'_{k}$ such that $W(\alpha') = W\left(\alpha^{0}\right)$, and $\alpha'_{k} \neq \alpha^{0}_{k}$ for some $k$, $a\left(\alpha'\right)$ increases uniformly with respect to the time, so that the probability of the system being in the state $\alpha'$ at any time increases proportionally with the square of the time. Physically, the probability of the system being in a state with exactly the same proper energy as the initial proper energy $W\left(\alpha^{0}\right)$ is of no importance, being infinitesimal. We are interested only in the integral of the probability through a small range of proper energy values about the initial proper energy, which, as we shall find, increases linearly with the time, in agreement with the ordinary probability laws.

We transform from the variables $\alpha_{1}, \alpha_{2} \ldots \alpha_{u}$ to a set of variables that are arbitrary independent functions of the $\alpha$'s such that one of them is the proper energy $W$, say, the variables $W, \gamma_{1}, \gamma_{2}, \ldots \gamma_{u-1}$. The probability at any time of the system lying in a stationary state for which each $\gamma_{k}$ lies between $\gamma'_{k}$ and $\gamma'_{k} + d\gamma'_{k}$ is now (apart from the normalising factor) equal to

$$d\gamma'_{1}\cdot d\gamma'_{2}\ldots d\gamma'_{u-1}\int\left|a\left(\alpha'\right)\right|^{2}\frac{\partial\left(\alpha'_{1},\alpha'_{2}\ldots\alpha'_{u}\right)}{\partial\left(W',\gamma'_{1}\ldots\gamma'_{u-1}\right)}d\,W'. \quad (23)$$

For a time that is large compared with the periods of the system we shall find that practically the whole of the integral in (23) is contributed by values of $W'$ very close to $W^0 = W(\alpha^0)$. Put

$$a(\alpha') = a(W', \gamma') \quad \text{and} \quad \partial(\alpha'_1, \alpha'_2 \ldots \alpha'_u) / \partial(W', \gamma'_1 \ldots \gamma'_{u-1}) = J(W', \gamma').$$

Then for the integral in (23) we find, with the help of (22) (provided $\gamma'_k \neq \gamma^0_k$ for some $k$)

$$\int |a(W', \gamma')|^2 J(W', \gamma') \, dW'$$

$$= |a^0|^2 \int |v(W', \gamma'; W^0, \gamma^0)|^2 J(W', \gamma')$$

$$\times \frac{\left[e^{i(W'-W^0)t/h} - 1\right]\left[e^{i(W'-W^0)t/h} - 1\right]}{(W' - W^0)^2} dW'$$

$$= 2|a^0|^2 \int |v(W', \gamma'; W^0, \gamma^0)|^2 J(W', \gamma')$$

$$\times \left[1 - \cos(W' - W^0)t/h\right] / (W' - W^0)^2 \cdot dW'$$

$$= 2|a^0|^2 t/h \cdot \int |v(W^0 + hx/t, \gamma'; W^0, \gamma^0)|^2$$

$$\times J(W^0 + hx/t, \gamma')(1 - \cos x)/x^2 \cdot dx,$$

if one makes the substitution $(W' - W^0)t/h = x$. For large values of $t$ this reduces to

$$2|a^0|^2 t/h \cdot |v(W^0, \gamma'; W^0, \gamma^0)|^2 J(W^0, \gamma') \int_{-\infty}^{\infty} (1 - \cos x)/x^2 \cdot dx$$

$$= 2\pi |a^0|^2 t/h \cdot |v(W^0, \gamma'; W^0, \gamma^0)|^2 J(W^0, \gamma').$$

The probability per unit time of a transition to a state for which each $\gamma_k$ lies between $\gamma'_k$ and $\gamma'_k + d\gamma'_k$ is thus (apart from the normalising factor)

$$2\pi |a^0|^2 /h \cdot |v(W^0, \gamma'; W^0, \gamma^0)|^2 J(W^0, \gamma') \, d\gamma'_1 \cdot d\gamma'_2 \ldots d\gamma'_{u-1},$$

$$(24)$$

which is proportional to the square of the matrix element associated with that transition of the perturbing energy.

To apply this result to a simple collision problem, we take the $\alpha$'s to be the components of momentum $p_x$, $p_y$, $p_z$ of the colliding electron and the $\gamma$'s to be $\theta$ and $\phi$, the angles which determine its direction of motion. If, taking the relativity change of mass with velocity into account, we let P denote the resultant momentum, equal to $(p_x^2 + p_y^2 + p_z^2)^{\frac{1}{2}}$, and E the energy, equal to $(m^2c^4 + P^2c^2)^{\frac{1}{2}}$, of the electron, $m$ being its rest-mass, we find for the Jacobian

$$J = \frac{\partial\,(p_x,\,p_y,\,p_z)}{\partial\,(E,\,\theta,\,\phi)} = \frac{EP}{c^2}\sin\theta.$$

Thus the J $(W^0,\,\gamma')$ of the expression (24) has the value

$$J\,(W^0,\,\gamma') = E'P'\sin\theta'/c^2, \tag{25}$$

where E′ and P′ refer to that value for the energy of the scattered electron which makes the total energy equal the initial energy $W^0$ (*i.e.*, to that value required by the conservation of energy).

We must now interpret the initial value of $a(\alpha')$, namely, $a^0\,\delta\,(\alpha' - \alpha^0)$, which we did not normalise. According to § 2 the wave function in terms of the variables $\alpha_k$ is $b\,(\alpha') = a(\alpha')\,e^{-iW't/h}$, so that its initial value is

$$a^0\delta\,(\alpha' - \alpha^0)\,e^{-iW't/h} = a^0\delta\,(p_x' - p_x^0)\,\delta\,(p_y' - p_y^0)\,\delta\,(p_z' - p_z^0)\,e^{-iW't/h}.$$

If we use the transformation function*

$$(x'/p') = (2\pi h)^{-3/2}\,e^{i\Sigma_{xyz}p_x'x'/h},$$

and the transformation rule

$$\psi\,(x') = \int\,(x'/p')\psi\,(p')\,dp_x'\,dp_y'\,dp_x',$$

we obtain for the initial wave function in the co-ordinates $x$, $y$, $z$ the value

$$a^0\,(2\pi h)^{-3/2}\,e^{i\Sigma_{xyz}p_x^0x'/h}\,e^{-iW't/h}.$$

This corresponds to an initial distribution of $|a^0|^2\,(2\pi h)^{-3}$ electrons per unit volume. Since their velocity is $P^0c^2/E^0$, the number per unit time striking a unit surface at right-angles to their direction of motion is $|a^0|^2P^0c^2/(2\pi h)^3E^0$. Dividing this into the expression (24) we obtain, with the help of (25),

$$4\pi^2(2\pi h)^2\frac{E'E^0}{c^4}|v(p';p^0)|^2\frac{P'}{P^0}\sin\theta'd\theta'd\phi', \tag{26}$$

---

*The symbol $x$ is used for brevity to denote $x, y, z$.

This is the effective area that must be hit by an electron in order that it shall be scattered in the solid angle $\sin \theta' \, d\theta' \, d\phi'$ with the energy E'. This result differs by the factor $(2\pi h)^2/2mE'$. $P'/P^0$ from Born's.[*] The necessity for the factor $P'/P^0$ in (26) could have been predicted from the principle of detailed balancing, as the factor $|v(p'; p^0)|^2$ is symmetrical between the direct and reverse processes.[†]

## § 6. APPLICATION TO LIGHT-QUANTA.

We shall now apply the theory of § 4 to the case when the systems of the assembly are light-quanta, the theory being applicable to this case since lightquanta obey the Einstein-Bose statistics and have no mutual interaction. A light-quantum is in a stationary state when it is moving with constant momentum in a straight line. Thus a stationary state $r$ is fixed by the three components of momentum of the light-quantum and a variable that specifies its state of polarisation. We shall work on the assumption that there are a finite number of these stationary states, lying very close to one another, as it would be inconvenient to use continuous ranges. The interaction of the light-quanta with an atomic system will be described by a Hamiltonian of the form (20), in which $H_P$ ( J) is the Hamiltonian for the atomic system alone, and the coefficients $v_{r,s}$ are for the present unknown. We shall show that this form for the Hamiltonian, with the $v_{rs}$ arbitrary, leads to Einstein's laws for the emission and absorption of radiation.

The light-quantum has the peculiarity that it apparently ceases to exist when it is in one of its stationary states, namely, the zero state, in which its momentum, and therefore also its energy, are zero. When a light-quantum is absorbed it can be considered to jump into this zero state, and when one is emitted it can be considered to jump from the zero state to one in which it is physically in evidence, so that it appears to have been created. Since there is no limit to the number of light-quanta that may be created in this way, we must suppose that

[*] In a more recent paper ('Nachr. Gesell. d. Wiss.,' Gottingen, p. 146 (1926)) Born has obtained a result in agreement with that of the present paper for non-relativity mechanics, by using an interpretation of the analysis based on the conservation theorems. I am indebted to Prof. N. Bohr for seeing an advance copy of this work.
[†] See Klein and Rosseland, 'Z. f. Physik,' vol. 4, p. 46, equation (4) (1921).

there are an infinite number of light-quanta in the zero state, so that the $N_0$ of the Hamiltonian (20) is infinite. We must now have $\theta_0$, the variable canonically conjugate to $N_0$, a constant, since

$$\dot{\theta}_0 = \partial F/\partial N_0 = W_0 + \text{terms involving } N_0^{-\frac{1}{2}} \text{ or } (N_0 + 1)^{-\frac{1}{2}}$$

and $W_0$ is zero. In order that the Hamiltonian (20) may remain finite it is necessary for the coefficients $v_{r0}$, $v_{0r}$ to be infinitely small. We shall suppose that they are infinitely small in such a way as to make $v_{r0}N_0^{\frac{1}{2}}$ and $v_{0r}N_0^{\frac{1}{2}}$ finite, in order that the transition probability coefficients may be finite. Thus we put

$$v_{r0}(N_0 + 1)^{\frac{1}{2}} e^{-i\theta_0/h} = v_r, \qquad v_{0r}N_0^{\frac{1}{2}} e^{i\theta_0/h} = v_r^*,$$

where $v_r$ are and $v_r^*$ finite and conjugate imaginaries. We may consider the $v_r$ and $v_r^*$ to be functions only of the J's and $w$'s of the atomic system, since their factors $(N_0 + 1)^{\frac{1}{2}} e^{-i\theta_0/h}$ and $N_0^{\frac{1}{2}} e^{i\theta_0/h}$ are practically constants, the rate of change of $N_0$ being very small compared with $N_0$. The Hamiltonian (20) now becomes

$$F = H_P(J) + \Sigma_r W_r N_r + \Sigma_{r \neq 0} \left[ v_r N_r^{\frac{1}{2}} e^{i\theta_r/h} + v_r^* (N_r + 1)^{\frac{1}{2}} e^{-i\theta_r/h} \right]$$

$$+ \Sigma_{r \neq 0} \Sigma_{s \neq 0} v_{rs} N_r^{\frac{1}{2}} (N_s + 1 - \delta_{rs})^{\frac{1}{2}} e^{i(\theta_r - \theta_s)/h}. \tag{27}$$

The probability of a transition in which a light-quantum in the state $r$ is absorbed is proportional to the square of the modulus of that matrix element of the Hamiltonian which refers to this transition. This matrix element must come from the term $v_r N_r^{\frac{1}{2}} e^{i\theta_r/h}$ in the Hamiltonian, and must therefore be proportional to $N_r'^{\frac{1}{2}}$ where $N_r'$ is the number of light-quanta in state $r$ before the process. The probability of the absorption process is thus proportional to $N_r'$. In the same way the probability of a light-quantum in state $r$ being emitted is proportional to $(N_r' + 1)$, and the probability of a light-quantum in state $r$ being scattered into state $s$ is proportional to $N_r' (N_s' + 1)$. Radiative processes of the more general type considered by Einstein and Ehrenfest,* in which more than one light-quantum take part simultaneously, are not allowed on the present theory.

*'Z. f. Physik,' vol. 19, p. 301 (1923).

To establish a connection between the number of light-quanta per stationary state and the intensity of the radiation, we consider an enclosure of finite volume, A say, containing the radiation. The number of stationary states for light-quanta of a given type of polarisation whose frequency lies in the range $v_r$ to $v_r + dv_r$ and whose direction of motion lies in the solid angle $d\omega_r$ about the direction of motion for state $r$ will now be $Av_r^2 dv_r d\omega_r / c^3$. The energy of the light-quanta in these stationary states is thus $N_r' \cdot 2\pi h v_r \cdot Av_r^2 dv_r d\omega_r / c^3$. This must equal $Ac^{-1}I_r dv_r d\omega_r$, where $I_r$ is the intensity per unit frequency range of the radiation about the state $r$. Hence

$$I_r = N_r' \, (2\pi h) \, v_r^3 / c^2, \tag{28}$$

so that $N_r'$ is proportional to $I_r$ and $(N_r' + 1)$ is proportional to $I_r + (2\pi h)v_r^3 / c^2$. We thus obtain that the probability of an absorption process is proportional to $I_r$, the incident intensity per unit frequency range, and that of an emission process is proportional to $I_r + (2\pi h)v_r^3 / c^2$, which are just Einstein's laws.* In the same way the probability of a process in which a light-quantum is scattered from a state $r$ to a state $s$ is proportional to $I_r[I_s + (2\pi h)v_r^3 / c^2]$, which is Pauli's law for the scattering of radiation by an electron.[†]

## § 7. THE PROBABILITY COEFFICIENTS FOR EMISSION AND ABSORPTION.

We shall now consider the interaction of an atom and radiation from the wave point of view. We resolve the radiation into its Fourier components, and suppose that their number is very large but finite. Let each component be labelled by a suffix $r$, and suppose there are $\sigma_r$ components associated with the radiation of a definite type of polarisation per unit solid angle per unit frequency range about the component $r$. Each component $r$ can be described by a vector potential $k_r$ chosen so as to make the scalar potential zero. The perturbation term to be added to the Hamiltonian will now be, according to

---

*The ratio of stimulated to spontaneous emission in the present theory is just twice its value in Einstein's. This is because in the present theory either polarised component of the incident radiation can stimulate only radiation polarised in the same way, while in Einstein's the two polarised components are treated together. This remark applies also to the scattering process.

[†] Pauli, 'Z. f. Physik,' vol. 18, p. 272 (1923).

the classical theory with neglect of relativity mechanics, $c^{-1}\Sigma_r \kappa_r \dot{X}_r$, where $X_r$ is the component of the total polarisation of the atom in the direction of $\kappa_r$, which is the direction of the electric vector of the component $r$.

We can, as explained in § 1, suppose the field to be described by the canonical variables $N_r, \theta_r$, of which $N_r$ is the number of quanta of energy of the component $r$, and $\theta_r$ is its canonically conjugate phase, equal to $2\pi h\nu_r$ times the $\theta_r$ of § 1. We shall now have $\kappa_r = a_r \cos \theta_r/h$, where $a_r$ is the amplitude of $\kappa_r$, which can be connected with $N_r$ as follows:—The flow of energy per unit area per unit time for the component $r$ is $\frac{1}{2}\pi c^{-1} a_r^2 \nu_r^2$. Hence the intensity per unit frequency range of the radiation in the neighbourhood of the component $r$ is $I_r = \frac{1}{2}\pi c^{-1} a_r^2 \nu_r^2 \sigma_r$. Comparing this with equation (28), we obtain $a_r = 2(h\nu_r/c\sigma_r)^{\frac{1}{2}} N_r^{\frac{1}{2}}$, and hence

$$\kappa_r = 2\,(h\nu_r/c\sigma_r)^{\frac{1}{2}}\, N_r^{\frac{1}{2}} \cos \theta_r/h.$$

The Hamiltonian for the whole system of atom plus radiation would now be, according to the classical theory,

$$F = H_P\,(J) + \Sigma_r\,(2\pi\, h\nu_r)\, N_r + 2c^{-1}\Sigma_r\,(h\nu_r/c\sigma_r)^{\frac{1}{2}}\, X_r N_r^{\frac{1}{2}} \cos \theta_r/h,$$

$$(29)$$

where $H_P(J)$ is the Hamiltonian for the atom alone. On the quantum theory we must make the variables $N_r$ and $\theta_r$ canonical q-numbers like the variables $J_k$, $w_k$ that describe the atom. We must now replace the $N_r^{\frac{1}{2}} \cos \theta_r/h$ in (29) by the real q-number

$$\tfrac{1}{2}\left\{ N_r^{\frac{1}{2}} e^{i\theta r/h} + e^{-i\theta r/h} N_r^{\frac{1}{2}} \right\} = \tfrac{1}{2}\left\{ N_r^{\frac{1}{2}} e^{i\theta r/h} + (N_r + 1)^{\frac{1}{2}}\, e^{-i\theta r/h} \right\}$$

so that the Hamiltonian (29) becomes

$$F = H_P\,(J) + \Sigma_r\,(2\pi h\nu_r)\, N_r$$
$$+ h^{\frac{1}{2}} c^{-\frac{1}{2}} \Sigma_r\,(\nu_r/\sigma_r)^{\frac{1}{2}}\, X_r \left\{ N_r^{\frac{1}{2}} e^{i\theta r/h} + (N_r + 1)^{\frac{1}{2}}\, e^{-i\theta r/h} \right\}. \quad (30)$$

This is of the form (27), with

$$v_r = v_r^* = h^{\frac{1}{2}} c^{-\frac{1}{2}}\,(\nu_r/\sigma_r)^{\frac{1}{2}}\, \dot{X}_r$$

and
$$v_{rs} = 0 \quad (r, s \neq 0). \quad (31)$$

The wave point of view is thus consistent with the light-quantum point of view and gives values for the unknown interaction coefficient $v_{rs}$ in the light-quantum theory. These values are not such as would enable one to express the interaction energy as an algebraic function of canonical variables. Since the wave theory gives $v_{rs} = 0$ for $r, s \neq 0$, it would seem to show that there are no direct scattering processes, but this may be due to an incompleteness in the present wave theory.

We shall now show that the Hamiltonian (30) leads to the correct expressions for Einstein's A's and B's. We must first modify slightly the analysis of § 5 so as to apply to the case when the system has a large number of discrete stationary states instead of a continuous range. Instead of equation (21) we shall now have

$$ i h \, \dot{a} \left( \alpha' \right) = \Sigma_{a''} V \left( \alpha' \alpha'' \right) a \left( \alpha'' \right) . $$

If the system is initially in the state $\alpha^0$, we must take the initial value of $a \left( \alpha' \right)$ to be $\delta_{\alpha'\alpha^0}$, which is now correctly normalised. This gives for a first approximation

$$ i h \, \dot{a} \left( \alpha' \right) = V \left( \alpha'\alpha^0 \right) = v \left( \alpha'\alpha^0 \right) e^{i \left[ W(\alpha') - W(\alpha^0) \right] t/h}, $$

which leads to

$$ i h \, a \left( \alpha' \right) = \delta_{\alpha'\alpha^0} + v \left( \alpha'\alpha^0 \right) \frac{e^{i \left[ W(\alpha') - W(\alpha^0) \right] t/h} - 1}{i \left[ W \left( \alpha' \right) - W \left( \alpha^0 \right) \right]/h}, $$

corresponding to (22). If, as before, we transform to the variables W, $\gamma_1, \gamma_2 \ldots \gamma_{u-1}$, we obtain (when $\gamma' \neq \gamma^0$)

$$ a \left( W'\gamma' \right) = v \left( W', \gamma'; W^0, \gamma^0 \right) \left[ 1 - e^{i(W' - W^0)t/h} \right] / \left( W' - W^0 \right) . $$

The probability of the system being in a state for which each $\gamma_k$ equals $\gamma'_k$ is $\Sigma_{W'} |a \left( W'\gamma' \right)|^2$. If the stationary states lie close together and if the time $t$ is not too great, we can replace this sum by the integral $(\Delta W)^{-1} \int |a \left( W'\gamma' \right)|^2 dW'$, where $\Delta W$ is the separation between the energy levels. Evaluating this integral as before, we obtain for the probability per unit time of a transition to a state for which each $\gamma_k = \gamma'_k$

$$ 2\pi/h\Delta W \cdot \left| v \left( W^0, \gamma'; W^0, \gamma^0 \right) \right|^2 . \tag{32} $$

In applying this result we can take the $\gamma$'s to be any set of variables that are independent of the total proper energy W and that together with W define a stationary state.

We now return to the problem defined by the Hamiltonian (30) and consider an absorption process in which the atom jumps from the state $J^0$ to the state $J'$ with the absorption of a light-quantum from state $r$. We take the variables $\gamma'$ to be the variables $J'$ of the atom together with variables that define the direction of motion and state of polarisation of the absorbed quantum, but not its energy. The matrix element $v$ $(W^0, \gamma'; W^0, \gamma^0)$ is now

$$h^{1/2}c^{-3/2}\left(v_r/\sigma_r\right)^{1/2}\dot{X}_r\left(J^0J'\right)N_r^0,$$

where $\dot{X}_r(J^0J')$ is the ordinary $(J^0J')$ matrix element of $\dot{X}_r$. Hence from (32) the probability per unit time of the absorption process is

$$\frac{2\pi}{h\Delta W}\cdot\frac{hv_r}{c^3\sigma_r}\left|\dot{X}_r\left(J^0J'\right)\right|^2 N_r^0.$$

To obtain the probability for the process when the light-quantum comes from any direction in a solid angle $d\omega$, we must multiply this expression by the number of possible directions for the light-quantum in the solid angle $d\omega$, which is $d\omega\,\sigma_r\Delta W/2\pi h$. This gives

$$d\omega\frac{v_r}{hc^3}\left|\dot{X}_r\left(J^0J'\right)\right|^2 N_r^0 = d\omega\frac{1}{2\pi h^2 c v_r^2}\left|\dot{X}_r\left(J^0J'\right)\right|^2 I_r$$

with the help of (28). Hence the probability coefficient for the absorption process is $1/2\pi h^2 c v_r^2$. $|\dot{X}_r(J^0J')|^2$, in agreement with the usual value for Einstein's absorption coefficient in the matrix mechanics. The agreement for the emission coefficients may be verified in the same manner.

The present theory, since it gives a proper account of spontaneous emission, must presumably give the effect of radiation reaction on the emitting system, and enable one to calculate the natural breadths of spectral lines, if one can overcome the mathematical difficulties involved in the general solution of the wave problem corresponding to the Hamiltonian (30). Also the theory enables one to understand how it comes about that there is no violation of the law of the conservation of energy when, say, a photo-electron is emitted from an atom under the action of extremely weak incident radiation. The energy of interaction of the atom and the radiation is a q-number that does not commute with the first integrals of the motion of the atom alone or with the intensity of the radiation. Thus one cannot specify this energy by a c-number at the same time that one specifies the stationary state of the atom and the intensity of the radiation by c-numbers. In particular, one cannot say that the interaction energy tends to zero as the intensity of the incident radiation tends to zero. There is thus always an unspecifiable amount of interaction energy which can supply the energy for the photo-electron.

I would like to express my thanks to Prof. Niels Bohr for his interest in this work and for much friendly discussion about it.

## SUMMARY.

The problem is treated of an assembly of similar systems satisfying the Einstein-Bose statistical mechanics, which interact with another different system, a Hamiltonian function being obtained to describe the motion. The theory is applied to the interaction of an assembly of light-quanta with an ordinary atom, and it is shown that it gives Einstein's laws for the emission and absorption of radiation.

The interaction of an atom with electromagnetic waves is then considered, and it is shown that if one takes the energies and phases of the waves to be q-numbers satisfying the proper quantum conditions instead of c-numbers, the Hamiltonian function takes the same form as in the light-quantum treatment. The theory leads to the correct expressions for Einstein's A's and B's.

# The Lagrangian in Quantum Mechanics

## By

## Paul A.M. Dirac

*Physikalische Zeitschrift der Sowjetunion*, Band 3 Heft 1 (1933)

Quantum mechanics was built up on a foundation of analogy with the Hamiltonian theory of classical mechanics. This is because the classical notion of canonical coordinates and momenta was found to be one with a very simple quantum analogue, as a result of which the whole of the classical Hamiltonian theory, which is just a structure built up on this notion, could be taken over in all its details into quantum mechanics.

Now there is an alternative formulation for classical dynamics, provided by the Lagrangian. This requires one to work in terms of coordinates and velocities instead of coordinates and momenta. The two formulations are, of course, closely related, but there are reasons for believing that the Lagrangian one is the more fundamental.

In the first place the Lagrangian method allows one to collect together all the equations of motion and express them as the stationary property of a certain action function. (This action function is just the time-integral of the Lagrangian). There is no corresponding action principle in terms of the coordinates and momenta of the Hamiltonian theory. Secondly the Lagrangian method can easily be expressed relativistically, on account of the action function being a relativistic invariant; while the Hamiltonian method is essentially non-relativistic in form, since it marks out a particular time variable as the canonical conjugate of the Hamiltonian function.

For these reasons it would seem desirable to take up the question of what corresponds in the quantum theory to the Lagrangian method of the classical theory. A little consideration shows, however, that one cannot expect to be able to take over the classical Lagrangian equations in any very direct way. These equations involve partial derivatives of the Lagrangian with respect to the coordinates and velocities and no meaning can be given to such derivatives in quantum mechanics. The only differentiation process that can be carried out with respect to the dynamical variables of quantum mechanics is that of forming Poisson brackets and this process leads to the Hamiltonian theory.*

We must therefore seek our quantum Lagrangian theory in an indirect way. We must try to take over the ideas of the classical Lagrangian theory, not the equations of the classical Lagrangian theory.

## CONTACT TRANSFORMATIONS.

Lagrangian theory is closely connected with the theory of contact transformations. We shall therefore begin with a discussion of the analogy between classical and quantum contact transformations. Let the two sets of variables be $p_r$, $q_r$ and $P_r$, $Q_r$, $(r = 1, 2 \ldots n)$ and suppose the $q$'s and $Q$'s to be all independent, so that any function of the dynamical variables can be expressed in terms of them. It is well known that in the classical theory the transformation equations for this case can be put in the form

$$p_r = \frac{\partial S}{\partial q_r}, \qquad P_r = -\frac{\partial S}{\partial Q_r}, \qquad (1)$$

where $S$ is some function of the $q$'s and $Q$'s.

In the quantum theory we may take a representation in which the $q$'s are diagonal, and a second representation in which the $Q$'s are diagonal. There will be a transformation function $(q' | Q')$ connecting the two representations. We shall now show that this transformation function is the quantum analogue of $e^{\,iS/h}$.

---

*Processes for partial differentiation with respect to matrices have been given by Born, Heisenberg and Jordan (ZS. f. Physik **35**, 561, 1926) but these processes do not give us means of differentiation with respect to dynamical variables, since they are not independent of the representation chosen. As an example of the difficulties involved in differentiation with respect to quantum dynamical variables, consider the three components of an angular momentum, satisfying

$$m_x m_y - m_y m_x = i h\, m_z.$$

We have here $m_z$ expressed explicitly as a function of $m_x$ and $m_y$, but we can give no meaning to its partial derivative with respect to $m_x$ or $m_y$.

If $\alpha$ is any function of the dynamical variables in the quantum theory, it will have a "mixed" representative $(q'|\alpha|Q')$, which may be defined in terms of either of the usual representatives $(q'|\alpha|q'')$, $(Q'|\alpha|Q'')$ by

$$(q'|\alpha|Q') = \int (q'|\alpha|q'')dq''(q''|Q') = \int (q'|Q'')dQ''(Q''|\alpha|Q').$$

From the first of these definitions we obtain

$$(q'|q_r|Q') = q'_r(q'|Q') \tag{2}$$

$$(q'|p_r|Q') = -ih\frac{\partial}{\partial q'_r}(q'|Q') \tag{3}$$

and from the second

$$(q'|Q_r|Q') = Q'_r(q'|Q') \tag{4}$$

$$(q'|P_r|Q') = ih\frac{\partial}{\partial Q'_r}(q'|Q'). \tag{5}$$

Note the difference in sign in (3) and (5).

Equations (2) and (4) may be generalised as follows. Let $f(q)$ be any function of the $q$'s and $g(Q)$ any function of the $Q$'s. Then

$$(q'|f(q)g(Q)|Q') = \iint (q'|f(q)|q'')dq''(q''|Q'')dQ''(Q''|g(Q)|Q')$$
$$= f(q')g(Q')(q'|Q').$$

Further, if $f_k(q)$ and $g_k(Q)$, $(k = 1, 2\dots, m)$ denote two sets of functions of the $q$'s and $Q$'s respectively,

$$(q'|\Sigma_k f_k(q)g_k(Q)|Q') = \Sigma_k f_k(q')g_k(Q')\cdot(q'|Q').$$

Thus if $\alpha$ is any function of the dynamical variables and we suppose it to be expressed as a function $\alpha(qQ)$ of the $q$'s and $Q$'s in a "well-ordered" way, that is, so that it consists of a sum of terms of the form $f(q)g(Q)$, we shall have

$$(q'|\alpha(qQ)|Q') = \alpha(q'Q')(q'|Q'). \tag{6}$$

This is a rather remarkable equation, giving us a connection between $\alpha(qQ)$, which is a function of operators, and $\alpha(q'Q')$, which is a function of numerical variables.

Let us apply this result for $\alpha = p_r$. Putting

$$\left(q' \,|\, Q'\right) = e^{i\,U/h}, \tag{7}$$

where $U$ is a new function of the $q'$'s and $Q'$'s we get from (3)

$$\left(q' \,|\, p_r \,|\, Q'\right) = \frac{\partial U\left(q' Q'\right)}{\partial q'_r}\left(q' \,|\, Q'\right).$$

By comparing this with (6) we obtain

$$p_r = \frac{\partial U\left(q\, Q\right)}{\partial q_r}$$

as an equation between operators or dynamical variables, which holds provided $\partial U/\partial q_r$ is well-ordered. Similarly, by applying the result (6) for $\alpha = P_r$ and using (5), we get

$$P_r = -\frac{\partial U\left(q\, Q\right)}{\partial Q_r},$$

provided $\partial U/\partial Q_r$ is well-ordered. These equations are of the same form as (1) and show that the $U$ defined by (7) is the analogue of the classical function $S$, which is what we had to prove.

Incidentally, we have obtained another theorem at the same time, namely that equations (1) hold also in the quantum theory provided the right-hand sides are suitably interpreted, the variables being treated classically for the purpose of the differentiations and the derivatives being then well-ordered. This theorem has been previously proved by Jordan by a different method.[*]

## THE LAGRANGIAN AND THE ACTION PRINCIPLE.

The equations of motion of the classical theory cause the dynamical variables to vary in such a way that their values $q_t$, $p_t$ at any time $t$ are connected with their values $q_T$, $p_T$. at any other time $T$ by a contact transformation, which may be put into the form (1) with $q$, $p = q_t$, $p_t$; $Q$, $P = q_T$, $p_T$ and $S$ equal to the time integral of the Lagrangian over the range: $T$ to $t$. In the quantum theory the $q_t$, $p_t$

[*]Jordan, ZS. f. Phys, **38**, 513, 1926.

will still be connected with the $q_T$, $p_T$ by a contact transformation and there will be a transformation function $(q_t|q_T)$ connecting the two representations in which the $q_t$ and the $q_T$ are diagonal respectively. The work of the preceding section now shows that

$$(q_t \,|\, q_T) \text{ corresponds to } \exp\left[ i \int_T^t L\,dt/h \right], \tag{8}$$

where $L$ is the Lagrangian. If we take $T$ to differ only infinitely little from $t$, we get the result

$$(q_{t+dt} \,|\, q_t) \text{ corresponds to } \exp\left[ i L\,dt/h \right]. \tag{9}$$

The transformation functions in (8) and (9) are very fundamental things in the quantum theory and it is satisfactory to find that they have their classical analogues, expressible simply in terms of the Lagrangian. We have here the natural extension of the well-known result that the phase of the wave function corresponds to Hamilton's principle function in classical theory. The analogy (9) suggests that we ought to consider the classical Lagrangian, not as a function of the coordinates and velocities, but rather as a function of the coordinates at time $t$ and the coordinates at time $t + dt$.

For simplicity in the further discussion in this section we shall take the case of a single degree of freedom, although the argument applies also to the general case. We shall use the notation

$$\exp\left[ i \int_T^t L\,dt/h \right] = A(tT),$$

so that $A(tT)$ is the classical analogue of $(q_t|q_T)$.

Suppose we divide up the time interval $T \to t$ into a large number of small sections $T \to t_1,\ t_1 \to t_2, \ldots,\ t_{m-1} \to t_m,\ t_m \to t$ by the introduction of a sequence of intermediate times $t_1, t_2, \ldots t_m$. Then

$$A(tT) = A(tt_m)\,A(t_m t_{m-1}) \cdots A(t_2 t_1)\,A(t_1 T). \tag{10}$$

Now in the quantum theory we have

$$(q_t|q_T) = \int (q_t|q_m)dq_m\,(q_m|q_{m-1})\,dq_{m-1}\cdots(q_2|q_1)\,dq_1\,(q_1|q_T)\,,$$

(11)

where $q_k$ denotes $q$ at the intermediate time $t_k$, $(k = 1, 2 \ldots m)$. Equation (11) at first sight does not seem to correspond properly to equation (10), since on the right-hand side of (11) we must integrate after doing the multiplication while on the right-hand side of (10) there is no integration.

Let us examine this discrepancy by seeing what becomes of (11) when we regard $t$ as extremely small. From the results (8) and (9) we see that the integrand in (11) must be of the form $e^{iF/h}$ where $F$ is a function of $q_T, q_1, q_2 \ldots q_m, q_t$ which remains finite as $h$ tends to zero. Let us now picture one of the intermediate $q$'s, say $q_k$, as varying continuously while the others are fixed. Owing to the smallness of $h$, we shall then in general have $F|h$ varying extremely rapidly. This means that $e^{iF/h}$ will vary periodically with a very high frequency about the value zero, as a result of which its integral will be practically zero. The only important part in the domain of integration of $q_k$ is thus that for which a comparatively large variation in $q_k$ produces only a very small variation in $F$. This part is the neighbourhood of a point for which $F$ is stationary with respect to small variations in $q_k$.

We can apply this argument to each of the variables of integration in the right-hand side of (11) and obtain the result that the only important part in the domain of integration is that for which $F$ is stationary for small variations in all the intermediate $q$'s. But, by applying (8) to each of the small time sections, we see that $F$ has for its classical analogue

$$\int_{t_m}^{t} L\,dt + \int_{t_{m-1}}^{t_m} L\,dt + \cdots + \int_{t_1}^{t_2} L\,dt + \int_{T}^{t_1} L\,dt = \int_{T}^{t} L\,dt,$$

which is just the action function which classical mechanics requires to be stationary for small variations in all the intermediate $q$'s. This shows the way in which equation (11) goes over into classical results when $h$ becomes extremely small.

We now return to the general case when $h$ is not small. We see that, for comparison with the quantum theory, equation (10) must be interpreted in the following way. Each of the quantities $A$ must be considered as a function of the $q$'s at the two times to which it refers. The right-hand side is then a function, not only of $q_T$ and $q_t$, but also of $q_1$, $q_2$, ... $q_m$, and in order to get from it a function of $q_T$ and $q_t$ only, which we can equate to the left-hand side, we must substitute for $q_1$, $q_2$ ... $q_m$ their values given by the action principle. This process of substitution for the intermediate $q$'s then corrésponds to the process of integration over all values of these $q$'s in (11).

Equation (11) contains the quantum analogue of the action principle, as may be seen more explicitly from the following argument. From equation (11) we can extract the statement (a rather trivial one) that, if we take specified values for $q_T$ and $q_t$, then the importance of our considering any set of values for the intermediate $q$'s is determined by the importance of this set of values in the integration on the righthand side of (11). If we now make $h$ tend to zero, this statement goes over into the classical statement that, if we take specified values for $q_T$ and $q_t$, then the importance of our considering any set of values for the intermediate $q$'s is zero unless these values make the action function stationary. This statement is one way of formulating the classical action principle.

## APPLICATION TO FIELD DYNAMICS.

We may treat the problem of a vibrating medium in the classical theory by Lagrangian methods which form a natural generalisation of those for particles. We choose as our coordinates suitable field quantities or potentials. Each coordinate is then a function of the four space-time variables $x$, $y$, $z$, $t$, corresponding to the fact that in particle theory it is a function of just the one variable $t$. Thus the one independent variable $t$ of particle theory is to be generalised to four independent variables $x$, $y$, $z$, $t$.*

We introduce at each point of space-time a Lagrangian density, which must be a function of the coordinates and their first derivatives with respect to $x$, $y$, $z$ and $t$, corresponding to the Lagrangian in particle theory being a function of coordinates and velocities. The integral of the Lagrangian density over any (four-dimensional) region of space-time must then be stationary for all small variations of the coordinates inside the region, provided the coordinates on the boundary remain invariant.

It is now easy to see what the quantum analogue of all this must be. If $S$ denotes the integral of the classical Lagrangian density over a particular region of space-time, we should expect there to be a quantum analogue of $e^{iS/h}$ corresponding to the $(q_t|q_T)$ of particle theory. This $(q_t|q_T)$ is a function of the values of the coordinates at the ends of the time interval to which it refers and so we should expect the quantum analogue of $e^{iS/h}$ to be a function (really a functional) of the values of the coordinates on the boundary of the space-time region. This quantum analogue will be a sort of, "generalized transformation function". It cannot in general be interpreted, like $(q_t|q_T)$, as giving a transformation between one set of dynamical variables and another, but it is a four-dimensional generalization of $(q_t|q_T)$ in the following sense.

Corresponding to the composition law for $(q_t|q_T)$

$$(q_t|q_T) = \int (q_t|q_1)\, dq_1\, (q_1|q_T),\qquad (12)$$

the generalized transformation function (g.t.f.) will have the following composition law. Take a given region of space-time and divide it up into two parts. Then the g.t.f. for the whole region will equal the product of the g.t.f.'s for the two parts, integrated over all values for the coordinates on the common boundary of the two parts.

---

*It is customary in field dynamics to regard the values of a field quantity for two different values of $(x, y, z)$ but the same value of $t$ as two different coordinates, instead of as two values of the same coordinate-for two different points in the domain of independent variables, and in this way to keep to the idea of a single independent variable $t$. This point of view is necessary for the Hamiltonian treatment, but for the Lagrangian treatment the point of view adopted in the text seems preferable on account of its greater space-time symmetry.

Repeated application of (12) gives us (11) and repeated application of the corresponding law for g.t.f.'s will enable us in a similar way to connect the g.t.f. for any region with the g.t.f.'s for the very small sub-regions into which that region may be divided. This connection will contain the quantum analogue of the action principle applied to fields.

The square of the modulus of the transformation function $(q_t|q_T)$ can be interpreted as the probability of an observation of the coordinates at the later time $t$ giving the result $q_t$ for a state for which an observation of the coordinates at the earlier time $T$ is certain to give the result $q_T$. A corresponding meaning for the square of the modulus of the g.t.f. will exist only when the g.t.f. refers to a region of space-time bounded by two separate (three-dimensional) surfaces, each extending to infinity in the space directions and lying entirely outside any light-cone having its vertex on the surface. The square of the modulus of the g. t. f. then gives the probability of the coordinates having specified values at all points on the later surface for a state for which they are given to have definite values at all points on the earlier surface. The g.t.f. may in this case be considered as a transformation function connecting the values of the coordinates and momenta on one of the surfaces with their values on the other.

We can alternatively consider $|(q_t|q_T)|^2$ as giving the relative a priori probability of any state yielding the results $q_T$ and $q_t$ when observations of the $q$'s are made at time $T$ and at time $t$ (account being taken of the fact that the earlier observation will alter the state and affect the later observation). Correspondingly we can consider the square of the modulus of the g.t.f. for any space-time region as giving the relative a priori probability of specified results being obtained when observations are made of the coordinates at all points on the boundary. This interpretation is more general than the preceding one, since it does not require a restriction on the shape of the space-time region.

ST JOHN'S COLLEGE, CAMBRIDGE.

# On Quantum Electrodynamics

Paul A.M. Dirac, V.A. Flock, and Boris Podolsky

*Proceedings of the Royal Society of London*, Series A, Vol. 114, p. 243 (1927)

In the first part of this paper the equivalence of the new form of relativistic Quantum Mechanics* to that of Heisenberg and Pauli[†] is proved in a new way which has the advantage of showing their physical relation and serves to suggest further development considered in the second part.

## PART I. EQUIVALENCE OF DIRAC'S AND HEISENBERG PAULI'S THEORIES.

§ 1. Recently Rosenfeld showed[‡] that the new form of relativistic Quantum Mechanics* is equivalent to that of Heisenberg and Pauli.[†] Rosenfeld's proof is, however, obscure and. does not bring out some features of the relation of the two theories. To assist in the further development of the theory we give here a simplified proof of the equivalence.

Consider a system, with a Hamiltonian $H$, consisting of two parts $A$ and $B$ with their respective Hamiltonians $H_a$ and $H_b$ and the interaction $V$. We have

$$H = H_a + H_b + V, \tag{1}$$

where

$$H_a = H_a(p_a q_a T); \qquad H_b = H_b(p_b q_b T);$$
$$V = V(p_a q_a p_b q_b T)$$

*Dirac, Proc, Roy. Soc. A **136**, 453, 1932.
[†] Heisenberg and Pauli, ZS. f. Physik, **56**, 1, 1929 and **59**, 168, 1930.
[‡] Rosenfeld, ZS. f. Physik **76**, 729, 1932.

and $T$ is the time for the entire system. The wave function for the entire system will satisfy the equation*

$$(H - i\hbar \, \partial/\partial T) \, \psi \, (q_a q_b T) = 0 \tag{2}$$

and will be a function of the variables indicated.

Now, upon performing the canonical transformation

$$\psi^* = e^{\frac{i}{\hbar} H_b T} \psi, \tag{3}$$

by which dynamical variables, say $F$, transform as follows

$$F^* = e^{\frac{i}{\hbar} H_b T} F e^{-\frac{i}{\hbar} H_b T}, \tag{4}$$

Eq. (2) takes the form

$$\left(H_a^* + V^* - i\hbar \, \partial/\partial T\right) \psi^* = 0. \tag{5}$$

Since $H_a$ commutes with $H_b$, $H_a^* = H_a$. On the other hand, since the functional relation between variables is not disturbed by the canonical transformation (3), $V^*$ is the same function of the transformed variables $p^*$, $q^*$ as $V$ is of $p$, $q$. But $p_a$ and $q_a$ commute with $H_b$ so that $p_a^* = p_a$, $q_a^* = q_a$.
Therefore

$$V^* = V\left(p_a q_a p_b^* q_b^*\right), \tag{6}$$

where

$$\left. \begin{array}{l} q_b^* = e^{\frac{i}{\hbar} H_b T} q_b e^{-\frac{i}{\hbar} H_b T} \\[2mm] p_b^* = e^{\frac{i}{\hbar} H_b T} p_b e^{-\frac{i}{\hbar} H_b T} \end{array} \right\} \tag{7}$$

It will be shown in § 7, after suitable notation is developed, that Eqs. (7) are equivalent to

$$\left. \begin{array}{l} \partial q_b^*/\partial t = \dfrac{i}{\hbar} \left(H_b q_b^* - q_b^* H_b\right) \\[3mm] \partial p_b^*/\partial t = \dfrac{i}{\hbar} \left(H_b p_b^* - p_b^* H_b\right) \end{array} \right\} \tag{8}$$

where $t$ is the separate time of the part $B$.

These, however, are just the equations of motion for the part $B$ alone, unperturbed by the presence of part $A$.

---

*$\hbar$ is Planck's constant divided by $2\pi$.

§ 2. Now let part $B$ correspond to the field and part $A$ to the particles present. Eqs. (8) must then be equivalent to Maxwell's equations for empty space. Eq. (2) is then the wave equation of Heisenberg-Pauli's theory, while Eq. (5), in which the perturbation is expressed in terms of potentials corresponding to empty space, is the wave equation of the new theory. Thus, this theory corresponds to treating separately a part of the system, which is in some problems more convenient.*

Now, $H_a$ can be represented as a sum of the Hamiltonians for the separate particles. The interaction between the particles is not included in $H_a$ for this is taken to be the result of interaction between the particles and the field. Similarly, $V$ is the sum of interactions between the field and the particles. Thus, we may write

$$H_a = \sum_{s=1}^{n} \left( c\alpha_s \cdot p_s + m_s c^2 \alpha_s^{(4)} \right) = \sum_{s=1}^{n} H_s$$

and

$$V^* = \sum_{s=1}^{n} V_s^* = \sum_{s=1}^{n} \Theta_s \left[ \Phi \left( r_s,\, T \right) - \alpha_s \cdot A \left( r_s,\, T \right) \right]$$

$$\left. \vphantom{\sum_{s=1}^{n}} \right\} \quad (9)$$

where $r_s$ are the coordinates of the $s$-th particle and $n$ is the number of particles.

Eq. (5) takes the form

$$\left[ \sum_{s=1}^{n} \left( H_s + V_s^* \right) - ih\, \partial/\partial T \right] \psi^* \left( r_s;\, J;\, T \right) = 0, \qquad (10)$$

$J$ stands for the variables describing the field. Besides the common time $T$ and the field time $t$ an individual time $t_s = t_1, t_2, \ldots t_n$ is introduced for each particle. Eq. (10) is satisfied by the common solution of the set of equations

$$\left( R_s - ih\, \partial/\partial t_s \right) \psi^* = 0,$$

where

$$R_s = c\alpha_s \cdot p_s + m_s c^2 \alpha_s^4 + \varepsilon_s \left[ \Phi \left( r_s t_s \right) - \alpha_s \cdot A \left( r_s t_s \right) \right]$$

$$\left. \vphantom{\sum} \right\} \quad (11)$$

and $\psi^* = \psi^* \left( r_1\, r_2 \ldots r_n;\, t_1\, t_2 \ldots t_n;\, J \right)$, when all the $t$'s are put equal to the common time $T$.

---

*This is somewhat analogous to Frenkel's method of treating incomplete systems, see Frenkel, Sow. Phys. 1, 99, 1932.

Now, Eqs. (11) are the equations of Dirac's theory. They are obviously relativistically invariant and form a generalization of Eq. (10). This obvious relativistic invariance is achieved by the introduction of separate time for each particle.

§ 3. For further development we shall need some formulas of quantization of electromagnetic fields and shall use amplitudes $F(k)$ and $F^+(k)$ are introduced by the equation

$$F = \left(\frac{1}{2\pi}\right)^{3/2} \int \{F(k)\, e^{-ic|k|t+ik\cdot r} + F^+(k)\, e^{+ic|k|t-ik\cdot r}\}\, dk \quad (17)$$

where $r = (xyz)$ is the position vector, $k = (k_x k_y k_z)$ is the wave vector having the magnitude $|k| = 2\pi/\lambda$, $dk = dk_x dk_y dk_z$, the integration being performed for each component of $k$ from $-\infty$ to $\infty$. In terms of the amplitudes equations of motion can be written

$$\left. \begin{aligned} P(k) &= \frac{i}{c}[k\Phi(k) - |k|A(k)] = -\frac{1}{c}\mathfrak{E}(k) \\ P_0(k) &= \frac{i}{c}[|k|\Phi(k) - k\cdot A(k)] \end{aligned} \right\} \quad (18)$$

the other two equations being algebraic consequences of these.

The commutation rules for the potentials are

$$\left. \begin{aligned} \Phi^+(k)\,\Phi(k') - \Phi(k')\,\Phi^+(k) &= \frac{ch}{2|k|}\delta(k-k') \\ A_i^+(k)\,A_{in}(k') - A_m(k')\,A_i^+(k) &= -\frac{ck}{2|k|}\delta_{lm}\delta(k-k') \end{aligned} \right\} \quad (19)$$

all other combinations of amplitudes commuting.

# PART II. THE MAXWELLIAN CASE.

§ 4. For the Maxwellian case the following additional considerations are necessary. In obtaining the field variables, besides the regular equations of motion of the electromagnetic field one must use the additional condition $P_0 = 0$, or $-cP_0 = \text{div}\, A + \dot{\Phi}/c = 0$. This condition cannot be regarded as aquantum mechanical equation, but rather as a condition on permissible $\psi$ functions. This can be seen, for example, from the fact that, when regarded as a quantum mechanical equation, $\operatorname{div} A + \dot{\Phi}/c = 0$ contradicts the commutation rules. Thus, only those $\psi$'s should be regarded as physically permissible which satisfy the condition

$$-c\,P_0\psi = \left(\operatorname{div} A + \frac{1}{c}\dot{\Phi}\right)\psi = 0. \tag{20}$$

for this purpose some formulas obtained by Fock and Podolsky.* Starting with the Lagrangian function

$$L = \frac{1}{2}\left(\mathfrak{E}^2 - \mathfrak{H}^2\right) - \frac{1}{2}\left(\operatorname{div} A + \frac{1}{c}\dot{\Phi}\right)^2, \tag{12}$$

taking as coordinates $(Q_0\ Q_1\ Q_2\ Q_3)$ the potentials $(\Phi\ A_1\ A_2\ A_3)$, and retaining the usual relations

$$\mathfrak{E} = -\operatorname{grad}\Phi - \frac{1}{c}\dot{A}; \quad \mathfrak{H} = \operatorname{curl} A, \tag{13}$$

one obtains

$$\left.\begin{array}{l}(P_1\,P_2\,P_3) = P = -\dfrac{1}{c}\mathfrak{E}; \\[2mm] P_0 = -\dfrac{1}{c}\left(\operatorname{div} A + \dfrac{1}{c}\dot{\Phi}\right); \end{array}\right\} \tag{14}$$

and the Hamiltonian

$$H = \frac{c^2}{2}\left(P^2 - P_0^2\right) + \frac{1}{2}\sum_{1,2,3}\left(\frac{\delta Q_1}{\delta x_2} - \frac{\delta Q_2}{\delta x_1}\right)^2$$
$$-c\,P_0\sum_{i=1}^{3}\frac{\delta Q_l}{\delta x_l} - c\,P\cdot\operatorname{grad} Q_0. \tag{15}$$

The equations of motion are[†]

$$\left.\begin{array}{l}\dot{A} = c^2 P - c\,\operatorname{grad}\Phi, \\ \dot{\Phi} = -c^2 P_0 - c\,\operatorname{div} A, \\ \dot{P} = \Delta A - \operatorname{grad}\operatorname{div} A - c\,\operatorname{grad} P_0, \\ \dot{P}_0 = -c\,\operatorname{div} P, \end{array}\right\} \tag{16}$$

---

*Fock and Podolsky, Sow. Phys. **1**, 801, 1932, later quoted as 1. c. For other treatments see Jordan and Pauli, ZS. f. Physik, **47**, 151, 1928 or Fermi, Rend. Lincei, **9**, 881, 1929. The Lagrangian (12) differs from that of Fermi only by a four-dimensional divergence.
[†]A dot over a field quantity will be used to designate a derivative with respect to the field time $t$.

On elimination of $P$ and $P_0$, Eqs. (16) give the D'Alembert equations for the potentials $\Phi$ and $A$. To obtain Maxwell's equation for empty space one must set $P_0 = 0$. The quantization rules are expressed in terms of the amplitudes of the Fourier's integral. Thus, for every $F = F(xyst)$,

Condition (20), expressed in terms of amplitudes by the use of Eq. (18), takes the form

$$i[k \cdot A(k) - |k| \Phi(k)] \psi = 0$$

and

$$-i[k \cdot A^+(k) - |k| \Phi^+(k)] \psi = 0. \tag{20'}$$

To these must, of course, be added the wave equation

$$(H_b - ih \, \partial/\partial t) \psi = 0, \tag{21}$$

where $H_b$ is the Hamiltonian for the field

$$H_b = 2 \int \{A^+(k) \cdot A(k) - \Phi^+(k) \, \Phi(k)\} |k|^2 dk, \tag{22}$$

as in 1. c.

If a number of equations $A\psi = 0$, $B\psi = 0$, etc., are simultaneously satisfied, then $AB\psi = 0$, $BA\psi = 0$, etc.; and therefore $(AB - BA)\psi = 0$, etc. All such new equations must be consequences of the old, i.e. must not give any new conditions on $\psi$. This may be regarded as, a test of consistency of the original equations. Applying this to our Eqs. (20') and (21) we have

$$P_0(k) \, P_0^+(k') - P_0^+(k') \, P_0(k)$$
$$= c^2 [k \cdot A(k) \, k' \cdot A^+(k') - k' \cdot A^+(k') \, k \cdot A(k)] \tag{23}$$
$$+ c^2 |k| |k'| [\Phi(k) \, \Phi^+(k') - \Phi^+(k') \, \Phi(k)]$$

since $A$'s commute with $\Phi$'s. Applying now the commutation rules of Eq. (19), we obtain

$$P_0(k) \, P_0^+(k') - P_0^+(k') \, P_0(k)$$
$$= \frac{c^3 h}{2|k|} \left( \sum_{l, m} k_l k_m \delta_{lm} - |k|^2 \right) \delta(k - k') = 0. \tag{24}$$

Eq. (24) is satisfied in consequence of quantum-mechanical equations, hence

$$[P_0(k)\, P_0^+(k') - P_0^+(k')\, P_0(k)]\psi = 0$$

is not a condition on $\psi$. Thus, conditions (20′) are consistent. Since $P_0(k)$ and $P_0^+(k)$ commute with $\partial/\partial t$, to test the consistence of condition (20) with (21) one must test the condition

$$(H_b P_0 - P_0 H_b)\, \psi = 0 \tag{25}$$

However, since $\dot{P}_0 = (i|h)(H_b P_0 - P_0 H_b)$, Eq. (25) takes the form $\dot{P}_0 \psi = 0$, or in Fourier's components

$$\dot{P}_0(k)\, \psi = -ic|k|P_0(k)\, \psi = 0$$

and

$$\dot{P}_0^+(k)\, \psi = ic|k|P_0^+(k)\, \psi = 0.$$

But these are just the conditions (20′). Thus, conditions (20) and (21) are consistent.

§ 5. The extra condition of Eq. (20) is not an equation of motion, but is a "constraint" imposed on the initial coordinates and velocities, which the equations of motion then preserve for all time. The existence of this constraint for the Maxwellian case is the reason for the additional considerations, mentioned at the beginning of § 4. It turns out that we must modify this constraint when particles are present, in order to get something which the equations of motion will preserve for all time.

The conditions (20′) as they stand, when applied to $\psi$, are not consistent with Eqs. (11). It is, however, not difficult to see that they can be replaced by a somewhat different set of conditions*

$$C(k)\, \psi = 0 \quad \text{and} \quad C^+(k)\, \psi = 0, \tag{26′}$$

where

$$C(k) = i[k \cdot A(k) - |k|\Phi(k)]$$

$$+\frac{i}{2\,(2\pi)^{3/2}\,|k|}\sum_{s-1}^{n}\varepsilon_s\, e^{ic|k|t_s - ik\cdot r_s}\,. \tag{27′}$$

---

*We shall drop the asterisk and in the following use $\psi$ instead $\psi^*$.

Terms in $C(k)$ not contained in $— cP_0(k)$ are functions of the coordinates and the time for the particles. They commute with $H_b$ $— ih\, \partial/\partial t$, with $P_0(k)$ and with each other. Therefore Eqs. (26′) are consistent with each other and with Eq. (21). It remains to show that Eqs. (26′) are consistent with Eqs. (11). In fact $C(k)$ and $C^+(k)$ commute with $R_s — ih\, \partial/\partial t_s$. We shall show this for $C(k)$.

Designating, in the usual way, $AB — BA$ as $[A, B]$, we see that it is sufficient to show that

$$\left[ C(k),\, p_s - \frac{\varepsilon_s}{c} A\,(r_s\, t_s) \right]. = 0 \qquad (28)$$

and

$$[C(k),\, ih\, \partial/\partial t_s - \varepsilon_s\, \Phi\,(r_s\, t_s)] = 0. \qquad (29)$$

By considering the form of $C(k)$, these become respectively

$$[k.A\,(k),\, A(r_s\, t_s)] - \frac{c}{2(2\pi)^{\frac{3}{2}}|k|} e^{ic|k|t_s}\, [e^{-ik\cdot r_s},\, p_s] = 0. \qquad (30)$$

and

$$[|k|\Phi\,(k),\, \Phi\,(r_s\, t_s)] + \frac{1}{2(2\pi)^{3/2}|k|} e^{-ik\cdot r_s}\, |e^{ic|k|t_s},\, ih\, \partial/\partial t_s] = 0. \qquad (31)$$

Now

$$[k\cdot A\,(k),\, A\,(r_s\, t_s)] = \left(\frac{1}{2\pi}\right)^{3/2} \int [k\cdot A\,(k),\, A^+\,(k')]\, e^{ic|k'|\, t_s - ik'\cdot r_s}\, dk',$$

by Eq. (17) and because $A(k)$ commutes with $A(k')$. Using the commutation formulas and performing the integration it becomes

$$\frac{chk}{2(2\pi)^{3/2}|k|} e^{ic\,|k|\, t_s - ik\cdot r_s}. \qquad (32)$$

On the other hand

$$[e^{-ik\cdot r_s},\, p_s] = hi\ \mathrm{grad}\ e^{-ik\cdot r_s} = h\, k\, e^{-ik\cdot r_s} \qquad (33)$$

Thus, Eq. (30) is satisfied. Similarly Eq. (31) is satisfied because

$$[|k|\,\Phi\,(k),\, \Phi\,(r_s\, t_s)] = \frac{-ch}{2\,(2\pi)^{3/2}} e^{ic|k|t_s - ik\cdot r_s} \qquad (34)$$

and

$$\left[ e^{ic|k|t_s},\, ih\, \partial/\partial t_s \right] = ch|k|e^{ic|k|t_s}. \qquad (35)$$

Thus, conditions (26′) satisfy all the requirements of consistence. It can be shown that these requirements determine $C(k)$ uniquely up to an additive constant, if it is taken to have the form $i[k \cdot A(k) - |k|\Phi(k)] + f(r_s t_s)$.

§ 6. We shall now show that the introduction of separate time for the field and for each particle allows the use of the entire vacuum electrodynamics of § 3 and 1. c., except for the change discussed in § 5. In fact, we shall show that Maxwell's equations of electrodynamics, in which enter current or charge densities, become conditions on $\psi$ function.

For convenience we collect together our fundamental equations. The equations of vacuum electrodynamics are

$$\mathfrak{E} = -\text{grad } \Phi = \frac{1}{c} \text{ div } A; \quad \mathfrak{H} = \text{curl } A \tag{13}$$

$$\Delta \Phi - \frac{1}{c^2}\ddot{\Phi} = 0; \qquad \Delta A - \frac{1}{c^2}\ddot{A} = 0. \tag{36}$$

The wave equations are

$$\left.\begin{array}{c} (R_s - ih\,\partial/\partial t_s)\,\psi = 0, \\[2mm] R_s = c\alpha_s \cdot p_s + m_s c^2 \alpha_s^{(4)} - \varepsilon_s \alpha_s \cdot A\,(r_s t_s) + \varepsilon_s \Phi\,(r_s t_s)\,. \end{array}\right\} \tag{11}$$

where

The additional conditions on $\psi$ function are

$$C\,(k)\,\psi = 0 \quad \text{and} \quad C^+\,(k)\,\psi = 0, \tag{26′}$$

where

$$C\,(k) = i\,[k \cdot A\,(k) - |k|\Phi\,(k)]$$
$$+ \frac{i}{2\,(2\pi)^{3/2}\,|k|} \sum_{s=1}^{n} \varepsilon_s\, e^{ic|k|t_s - ik \cdot r_s.} \tag{27′}$$

We transform the last two equations by passing from the amplitudes $C(k)$ and $C^+(k)$ to $C(r, t)$ by means of Eq. (17). Thus we obtain

$$C\,(r, t)\,\psi = 0 \tag{26}$$

and

$$C (r, t) = \text{div } A + \frac{1}{c} \frac{\partial \Phi}{\partial t} - \sum_{s=1}^{n} \frac{\varepsilon_s}{4\pi} \Delta (X - X_s), \qquad (27)$$

where $X$ and $X_s$ are four dimensional vectors $X = (xyzt)$, $X_s = (x_s y_s z_s t_s)$ and $\Delta$ is the so-called invariant delta function*

$$\Delta (X) = \frac{1}{|r|} [\delta(|r| + ct) - \delta(|r| - ct)]. \qquad (37)$$

From Eqs. (13) follows immediately

$$\text{div } \mathfrak{H} = 0 \quad \text{and} \quad \text{curl } \mathfrak{E} + \frac{1}{c} \frac{\partial}{\partial t} \mathfrak{H} = 0 \qquad (38)$$

so that these remain as quantum-mechanical equations. Using Eqs. (13) and (36) and condition (26) we obtain by direct calculation

$$\left( \text{curl } \mathfrak{H} - \frac{1}{c} \frac{\partial \mathfrak{E}}{\partial t} \right) \psi = \text{grad} \sum_{S=1}^{n} \frac{\varepsilon_s}{4\pi} \Delta (X - X_s) \psi \qquad (39)$$

and

$$(\text{div } \mathfrak{E}\psi = -\frac{1}{c} \left( \frac{\partial}{\partial t} \sum_{s=1}^{n} \frac{\varepsilon_s}{4\pi} \Delta (X - X_s) \right) \psi. \qquad (40)$$

Now, let us consider what becomes of these equations when we put $t = t_1 = t_2 = \ldots = t_n = T$, which is implied in Maxwell's equations and which we shall write for short as $t_s = T$.

For any quantity $f = f(t t_1 t_2 \ldots t_n)$

$$\frac{\partial f (TTT \ldots T)}{\partial T} = \left[ \left( \frac{\partial f}{\partial t} \right) + \left( \frac{\partial f}{\partial t_1} \right) + \ldots + \left( \frac{\partial f}{\partial t_n} \right) \right]_{t_s = T} \qquad (41)$$

and for each of the $n$ derivatives $\partial / \partial t_s$ we have an equation of motion

$$\frac{\partial f}{\partial t_s} = \frac{i}{h} (R_s f - f R_s). \qquad (42)$$

If we put $f = A(r, t)$ or $f = \Phi (r, t)$, then, since both commute with $R_s$, $\partial f / \partial t_s = 0$ and we get

$$\frac{\partial A}{\partial t} = \frac{\partial A}{\partial T} \quad \text{and} \quad \frac{\partial \Phi}{\partial t} = \frac{\partial \Phi}{\partial T}. \qquad (43)$$

---

*See Jordan and Pauli, ZS. f. Physik **47**, 159, 1928.

It follows that

$$\mathfrak{E} = -\frac{1}{c}\frac{\partial A}{\partial T} - \text{grad }\Phi; \qquad \mathfrak{H} = \text{curl } A, \qquad (44)$$

so that the form of the connection between the field and the potentials is preserved. Remembering that for $t = t_s$ we have $\Delta(X - X_s) = 0$ and hence grad $\Delta(X - X_s) = 0$, and using Eqs. (26), (39) and (40) we obtain

$$\left(\text{div } A + \frac{1}{c}\frac{\partial \Phi}{\partial T}\right)\psi = 0, \qquad (45)$$

$$\left(\text{curl }\mathfrak{H} - \frac{1}{c}\frac{\partial \mathfrak{E}}{\partial t}\right)_{t_s = T}\psi = 0 \qquad (46)$$

and

$$(\text{div }\mathfrak{E}\psi) = -\sum_{s=1}^{n}\frac{\varepsilon_s}{4\pi}\left[\frac{1}{c}\frac{\partial}{\partial t}\Delta(X - X_s)\right]_{t=t_s}\psi. \qquad (47)$$

For further reduction of Eq. (46) we must use Eqs. (41) and (42), from which follows

$$\left(\frac{1}{c}\frac{\partial \mathfrak{E}}{\partial t}\right)_{t_s = T} = \frac{1}{c}\frac{\partial \mathfrak{E}}{\partial T} - \sum_{s=1}^{n}\frac{i}{ch}[R_s, \mathfrak{E}] \qquad (48)$$

and $[R_s, \mathfrak{E}]$ is easily calculated, because the only term in $R_s$ which does not commute with $\mathfrak{E}$ is $-\varepsilon_s\,\alpha_s\cdot A(r_s\ t_s)$, and $-\mathfrak{E}/c$ is the momentum conjugate to $A$. In this way we obtain

$$[R_s, \mathfrak{E}] = ich\varepsilon_s\alpha_s\delta\,(r - r_s). \qquad (49)$$

For the reduction of Eq. (47) we need only remember* that

$$\left|\frac{1}{c}\frac{\partial}{\partial t}\Delta\,(X)\right|_{t=0} = -4\pi\delta\,(r). \qquad (50)$$

Thus, Eqs. (46) and (47) become

$$\left(\text{curl }\mathfrak{H} - \frac{1}{c}\frac{\partial \mathfrak{E}}{\partial T}\right)\psi = \sum_{s=1}^{n}\varepsilon_s\alpha_s\delta\,(r - r_s)\,\psi \qquad (51)$$

*Heisenberg und Pauli, ZS. f. Physik **56**, 34, 1929.

and

$$(\text{div } \mathfrak{E}) \, \psi = \sum_{s=1}^{n} \varepsilon_s \delta \left(r - r_s\right) \psi, \tag{52}$$

which are just the remaining Maxwell's equations appearing as conditions on $\psi$. Eq. (52) is the additional condition of Heisenberg-Pauli's theory.

§ 7. We shall now derive Eq. (8) of § 1. For this we need to recall that the transformation (7) is a canonical transformation which preserves the form of the algebraic relations between the variables, as well as the equations of motion. These will be, in the exact notation now developed,

$$\frac{\partial q_b^*}{\partial T} = \frac{i}{h} \left[ H^*, q_b^* \right]_{t_s = T}; \qquad \frac{\partial p_b^*}{\partial T} = \frac{i}{h} \left[ H^*, p_b^* \right]_{t_s = T}. \tag{53}$$

As we have seen in the discussion following Eq. (5)

$$H^* = H_a + H_b + V^* \tag{54}$$

and since $q_b$ and $p_b$ commute with $H_a$, $q_b^*$ und $p_b^*$ commute with $H_a^*$ and hence with $H_a$. Therefore Eqs. (53) become

$$\left. \begin{aligned} \frac{\partial q_b^*}{\partial T} &= \frac{i}{h} \left\{ \left[ H_b, q_b^* \right] + \left[ V^*, q_b^* \right] \right\}_{t_s = T} \\ \frac{\partial p_b^*}{\partial T} &= \frac{i}{h} \left\{ \left[ H_b, p_b^* \right] + \left[ V^*, p_b^* \right] \right\}_{t_s = T} \end{aligned} \right\} \tag{55}$$

On the other hand, we have from Eqs. (41) and (42)

and

$$\left. \begin{aligned} \frac{\partial q_b^*}{\partial T} &= \left\{ \frac{\partial q_b^*}{\partial t} + \frac{i}{h} \sum_{s=1}^{n} \left[ R_s, q_b^* \right] \right\}_{t_s = T} \\ \frac{\partial p_b^*}{\partial T} &= \left\{ \frac{\partial p_b^*}{\partial t} + \frac{i}{h} \sum_{s=1}^{n} \left[ R_s, p_b^* \right] \right\}_{t_s = T} \end{aligned} \right\} \tag{56}$$

Now the only term in $R_s$ which does not commute with $p_b^*$ and $q_b^*$ is $V_s^*$ so that

$$\left[ R_s, q_b^* \right] = \left[ V_s^*, q_b^* \right] \text{ and } \left[ R_s, p_b^* \right] = \left[ V_s^*, p_b^* \right]. \tag{57}$$

Since $\Sigma V_s^* = V^*$, Eqs. (56) become

$$\left. \begin{array}{l} \dfrac{\partial q_b}{\partial T} = \left\{ \dfrac{\partial q_b^*}{\partial t} + \dfrac{i}{h} \left[ V^*, q_b^* \right] \right\}_{t_s = T} \\[4mm] \dfrac{\partial p_b^*}{\partial T} = \left\{ \dfrac{\partial p_b^*}{\partial t} + \dfrac{i}{h} \left[ V^*, p_b^* \right] \right\}_{t_s = T} \end{array} \right\} \tag{58}$$

Comparison of Eqs. (55) with (58) finally gives

$$\left. \begin{array}{l} \left( \dfrac{\partial q_b^*}{\partial t} \right)_{t=T} = \dfrac{i}{h} \left[ H_b, q_b^* \right]_{t=T} \\[4mm] \left( \dfrac{\partial p_b^*}{\partial t} \right)_{t=T} = \dfrac{i}{h} \left[ H_b, p_b^* \right]_{t=T} \end{array} \right\} \tag{59}$$

which is, in the more exact notation, just Eqs. (8).

CAMBRIDGE, LENINGRAD AND KHARKOV.

# FOUNDATIONS OF THE NEW FIELD THEORY

## MAX BORN AND LEOPOLD INFELD

*Proceedings of the Royal Society of London. Series A, Containing Papers of a Mathematical and Physical Character*, Vol. 144, No. 852. (Mar. 29, 1934), pp. 425–451.

## § 1. INTRODUCTION.

The relation of matter and the electromagnetic field can be interpreted from two opposite standpoints:—

The first which may be called the *unitarian standpoint*[†] assumes only *one* physical entity, the electromagnetic field. The particles of matter are considered as singularities of the field and mass is a derived notion to be expressed by field energy (electromagnetic mass).

The second or *dualistic standpoint* takes field and particle as two essentially different agencies. The particles are the sources of the field, are acted on by the field but are not a part of the field; their characteristic property is inertia measured by a specific constant, the mass.

At the present time nearly all physicists have adopted the dualistic view, which is supported by three facts.

1. *The failure of any attempt to develop a unitarian theory.*—Such attempts have been made with two essentially different tendencies: (*a*) The theories started by Heaviside, Searle and J. J. Thomson, and completed by Abraham, Lorentz, and others, make geometrical assumptions about the "shape" and kinematic behaviour of the electron and distribution of charge density (rigid electron of Abraham, contracting electron of Lorentz); they break down because they are compelled to introduce cohesive forces of non-electromagnetic origin; (*b*)

* Research Fellow of the Rockefeller Foundation. I should like to thank the Rockefeller Foundation for giving me the opportunity to work in Cambridge.

† This expression has nothing to do with "unitary" field theory due to Einstein, Weyl, Eddington, and others where the problem consists of uniting the theories of gravitational and electro-magnetic fields into a kind of non-Riemannian geometry. Specially some of Eddington's formulæ, developed in § 101 of his book "The Mathematical Theory of Relativity" (Cambridge), have a remarkable formal analogy to those of this paper, in spite of the entirely different physical interpretation.

the theory of Mie* formally avoids this difficulty by a generalization of Maxwell's equations making them non-linear; this attempt breaks down because Mie's field equations have the unacceptable property, that their solutions depend on the absolute value of the potentials.

2. *The result of relativity theory*, that the observed dependance of mass on velocity is in no way characteristic of electromagnetic mass, but can be derived from the transformation law.

3. Last, but not least, *the great success of quantum mechanics* which in its present form is essentially based on the dualistic view. It started from the consideration of oscillators and particles moving in a Coulomb field; the methods developed in these cases have then been applied even to the electro-magnetic field, the Fourier coefficients of which behave like harmonic oscillators.

But there are indications that this quantum electrodynamics meets considerable difficulties and is quite insufficient to explain several facts.

The difficulties are chiefly connected with the fact that the self-energy of a point charge is infinite.[†] The facts unexplained concern the existence of elementary particles, the construction of the nuclei, the conversion of these particles into other particles or into photons, etc.

In all these cases there is sufficient evidence that the present theory (formulated by Dirac's wave equation) holds as long as the wavelengths (of the Maxwell or of the de Broglie waves) are long compared with the "radius of the electron" $e^2/mc^2$, but breaks down for a field containing shorter waves. The non-appearance of Planck's constant in this expression for the radius indicates that in the first place the electromagnetic laws are to be modified; the quantum laws may then be adapted to the new field equations.

Considerations of this sort together with the conviction of the great philosophical superiority of the unitarian idea have led to the

---

*'Ann. Physik,' vol. 37, p. 511 (1912); vol. 39, p. 1 (1912); vol. 40, p. 1 (1913). Also Born, 'Göttinger Nachr,' p. 23 (1914).
  [†]The attempt to avoid this difficulty by a new definition of electric force acting on a particle in a given field, made by Wentzel ('Z. Physik,' vol. 86, pp. 479, 635 (1933); vol. 87, p. 726 (1934)), is very ingenious, but rather artificial and leads to new difficulties.

recent attempt[*] to construct a new electrodynamics, based on two rather different lines of thought: a new theory of the electromagnetic field and a new method of quantum mechanical treatment.

It seems desirable to keep these two lines separate in the further development. The purpose of this paper is to give a deeper foundation of the new field equations on classical lines, without touching the question of the quantum theory.

In the papers cited above, the new field theory has been introduced rather dogmatically, by assuming that the Lagrangian underlying Maxwell's theory

$$\mathbf{L} = \tfrac{1}{2}\left(\mathbf{H}^2 - \mathbf{E}^2\right) \qquad (1.1)$$

(**H** and **E** are space-vectors of the electric and magnetic field) has to be replaced by the expression[†]

$$\mathbf{L} = b^2\left(\sqrt{1 + \frac{1}{b^2}(\mathbf{H}^2 - \mathbf{E}^2)} - 1\right). \qquad (1.2)$$

The obvious physical idea of this modification is the following:-

The failure in the present theory may be expressed by the statement that it violates the *principle of finiteness* which postulates that a satisfactory theory should avoid letting physical quantities become infinite. Applying this principle to the velocity one is led to the assumption of an upper limit of velocity $c$ and to replace the Newtonian action function $\tfrac{1}{2} mv^2$ of a free particle by the relativity expression $mc^2\left(1 - \sqrt{1 - v^2/c^2}\right)$. Applying the same condition to the space itself one is lead to the idea of closed space as introduced by Einstein's cosmological theory.[‡] Applying it to the electromagnetic field one is lead immediately to the assumption of an upper limit of the field strength and to the modification of the action function (1.1) into (1.2).

This argument seems to be quite convincing. But we believe that a deeper foundation of such an important law is necessary, just as in

---

[*] Born, 'Nature,' vol. 132, p. 282 (1933); 'Proc. Roy. Soc.,' A, vol. 143, p. 410 (1934), cited here as I.
[†] See Born and Infeld, 'Nature,' vol. 132, p. 1004 (1933).
[‡] See Eddington, "The Expanding Universe," Cambridge, 1933.

Einstein's mechanics the deeper foundation is provided by the postulate of relativity. Assuming that the expression $mc^2\left(1 - \sqrt{1 - v^2/c^2}\right)$ has been found by the idea of a velocity limit it is seen that it can be written in the form

$$mc^2\left(1 - d\tau/dt\right),$$

where

$$c^2 d\tau^2 = c^2 dt^2 - dx^2 - dy^2 - dz^2,$$

and therefore it has the property that the time integral of $mc^2\, d\tau/dt$ is invariant for all transformations for which $d\tau^2$ is invariant. This four-dimensional group of transformations is larger than the three-dimensional group of transformations for which the time integral of the Newtonian function

$$\tfrac{1}{2}mv^2 = \tfrac{1}{2}m\left(ds/dt\right)^2; \quad ds^2 = dx^2 + dy^2 + dz^2,$$

is invariant.

So we believe that we ought to search for a group of transformations for which the new Lagrangian expression has an invariant space-time integral and which is larger than that for the old expression (1.1). This latter group is the known group of special relativity but not the group of general space-time transformations.* Now it is very satisfying that the new Lagrangian belongs to this group of general relativity; we shall show that it can be derived from the postulate of general invariance with a few obvious additional assumptions. Therefore the new field theory seems to be a consequence of this very general principle, and the old one not more than a useful practical approximation, just in the same way as for the mechanics of Newton and Einstein.

In this paper we develop the whole theory from this general standpoint. We shall be obliged to repeat some of the formulæ published in the previous paper. The connection with the problems of gravitation and of quantum theory will be treated later.

*The adaptation of the function L (1.1) to the general relativity by multiplication with $\sqrt{-g}$ is quite formal. Any expression can be made generally invariant in this way.

## § 2. POSTULATE OF INVARIANT ACTION.

We start from the general principle that all laws of nature have to be expressed by equations covariant for all space-time transformations. This, however, should not be taken to mean that the gravitational forces play an essential part in the constitution of the physical world; therefore we neglect the gravitational field so that there exist co-ordinate systems in which the metrical tensor $g_{kl}$ has the value assumed in special relativity even in the centre of an electron. But we postulate that the natural laws are independent of the choice of the space-time co-ordinate system.

We denote space-time co-ordinates by

$$x^2, x^2, x^3, x^4 = x, y, z, ct.$$

The differential $dx^k$ is, as usual, considered to be a contravariant vector. One can pull the indices up and down with help of the metrical tensor which in any cartesian co-ordinate system (as used in special relativity) has the form

$$(g_{kl}) = \begin{bmatrix} -1 & 0 & 0 & 0 \\ 0 & -1 & 0 & 0 \\ 0 & 0 & -1 & 0 \\ 0 & 0 & 0 & 1 \end{bmatrix} = (\delta_{kl}). \qquad (2.1)$$

It is not the unit matrix, because of the different signs in the diagonal. Therefore we have to distinguish between covariant and contravariant tensors even in the co-ordinate systems of special relativity. In this case, however, the rule of pulling up and down of indices is very simple. This operation on the index 4 does not change the value of the tensor component, that on one of the indices 1, 2, 3 changes only the sign.

We use the well-known convention that one has to sum over any index which appears twice.

To obtain the laws of nature we use a variational principle of least action of the form

$$\delta \int \mathscr{L} d\tau = 0, \quad (d\tau = dx^1 dx^2 dx^3 dx^4). \qquad (2.2)$$

*We postulate: the action integral has to be an invariant.* We have to find the form of $\mathscr{L}$ satisfying this condition.

We consider a covariant tensor field $a_{kl}$; we do not assume any symmetry property of $a_{kl}$. The question is to define $\mathscr{L}$ to be such a function of $a_{kl}$ that (2.2) is invariant. The well-known answer is that $\mathscr{L}$ must have the form*

$$\mathscr{L} = \sqrt{|a_{kl}|}; \quad (|a_{kl}| = \text{determinant of } a_{kl}). \qquad (2.3)$$

If the field is determined by several tensors of the second order, $\mathscr{L}$ can be any homogeneous function of the determinants of the covariant tensors of the order $\frac{1}{2}$.

Each arbitrary tensor $a_{kl}$ can be split up into a symmetrical and anti-symmetrical part:

$$a_{kl} = g_{kl} + f_{kl}; \quad g_{kl} = g_{lk}; \quad f_{kl} = -f_{lk}. \qquad (2.4)$$

The simplest simultaneous description of the metrical and electromagnetic field is the introduction of *one* arbitrary (unsymmetrical) tensor $a_{kl}$; we identify its symmetrical part $g_{kl}$ with the metrical field, its antisymmetrical part with the electromagnetic field.[†]

We have then three expressions which multiplied by $d\tau$ are invariant

$$\sqrt{-|a_{kl}|} = \sqrt{-|g_{kl} + f_{kl}|}; \quad \sqrt{-|g_{kl}|}; \quad \sqrt{|f_{kl}|}, \qquad (2.5)$$

where the minus sign is added in order to get real values of the square roots; for (2.1) shows that $|\delta_{kl}| = -1$, therefore always $|g_{kl}| < 0$.

The simplest assumption for $\mathscr{L}$ is any linear function of (2.5):

$$\mathscr{L} = \sqrt{-|g_{kl} + f_{kl}|} + A\sqrt{-|g_{kl}|} + B\sqrt{|f_{kl}|}. \qquad (2.6)$$

---

[*] See Eddington, "The Mathematical Theory of Relativity," Cambridge, § 48 and 101 (1923).

The proof is simple: by a transformation with the Jacobian $I = \frac{\partial(\bar{x}^1 \dots \bar{x}^4)}{\partial(\bar{x}^1 \dots x^4)}$ $d\tau$ is changed into $d\bar{\tau} = I d\tau$ and $|a_{kl}|$ into $|\bar{a}_{kl}| = |a_{kl}| I^{-2}$; for the $dx^k$ are contravariant, $a_{kl}$ convariant.

[†] This assumption has already been considered by Einstein, 'Berl. Ber.,' pp. 75/37 (1923) and p. 414 (1925), from the standpoint of the affine field theory.

But the last term can be omitted. For if $f_{kl}$ is the rotation of a potential vector, as we shall assume, its space-time integral can be changed into a surface integral and has no influence on the variational equation of the field.* Therefore we can take

$$\mathbf{B} = 0. \qquad (2.7)$$

We need another condition for the determination of A. Its choice is obvious. In the limiting case of the cartesian co-ordinate system and of small values of $f_{kl}$, $\mathscr{L}$ has to give the classical expression

$$\mathbf{L} = \tfrac{1}{4} f_{kl} f^{kl}. \qquad (2.8)$$

We now leave the general co-ordinate system which has guided us to the expression (2.6) for $\mathscr{L}$ and calculate $\mathscr{L}$ in cartesian co-ordinates. Then we have with $g_{kl} = \delta_{kl}$ (see (2.1))

$$-|\delta_{kl} + f_{kl}| = - \begin{vmatrix} -1 & f_{12} & f_{13} & f_{14} \\ f_{21} & -1 & f_{23} & f_{24} \\ f_{31} & f_{32} & -1 & f_{34} \\ f_{41} & f_{42} & f_{43} & 1 \end{vmatrix}$$

$$= 1 + \left(f_{23}^2 + f_{31}^2 + f_{12}^2 - f_{14}^2 - f_{24}^2 - f_{34}^2\right)$$

$$\quad - \left(f_{23} f_{14} + f_{31} f_{24} + f_{12} f_{34}\right)^2$$

$$= 1 + \left(f_{23}^2 + f_{31}^2 + f_{12}^2 - f_{14}^2 - f_{24}^2 - f_{34}^2\right) - |f_{kl}|.$$

For small values of $f_{kl}$ the last determinant can be neglected and (2.6) becomes equal to (2.8) only if

$$A = -1. \qquad (2.9)$$

We have therefore the result:—

The action function of the electromagnetic field is in general co-ordinates

$$\mathscr{L} = \sqrt{-|g_{kl} + f_{kl}|} - \sqrt{-|g_{kl}|}, \qquad (2.10)$$

---

*See Eddington, *loc. cit.*, § 1 01.

and in cartesian co-ordinates

$$\mathbf{L} = \sqrt{1 + F - G^2} - 1, \tag{2.11}$$

where

$$F = f_{23}^2 + f_{31}^2 + f_{12}^2 - f_{14}^2 - f_{24}^2 - f_{34}^2 \tag{2.12}$$

$$G = f_{23}f_{14} + f_{31}f_{24} + f_{12}f_{34}. \tag{2.13}$$

Let us go back to the expression for $\mathscr{L}$ in a general co-ordinate system. We denote as usual

$$|g_{kl}| = g,$$

and we develop the determinant $|g_{kl} + f_{kl}|$ into a power series in $f_{kl}$. We have then

$$|g_{kl} + f_{kl}| = g + \Phi\,(g_{kl}, f_{kl}) + |f_{kl}|.$$

The transformation properties of $|g_{kl} + f_{kl}|, g, |f_{kl}|$ and therefore also of $\Phi\,(g_{kl}, f_{kl})$ are the same. They transform in the same way as $g$. If we write

$$g + \Phi + |f_{kl}| = g\left(1 + \frac{\Phi}{g} + \frac{|f_{kl}|}{g}\right), \tag{2.14}$$

we see, that all expressions in the bracket on the right side of (2.14) are invariant. We have calculated their value in a geodetic co-ordinate system and have found:

$$\frac{\Phi}{g} = \tfrac{1}{2} f_{kl} f^{kl} = F = \tfrac{1}{2} f_{kr} f_{ls} g^{lk} g^{sr}.$$

$\Phi/g$ is an invariant. We have therefore in an arbitrary co-ordinate system:

$$\left.\begin{aligned} |g_{kl} + f_{kl}| &= g\left(1 + F - G^2\right) \\ \mathscr{L} &= \sqrt{-g}\left(\sqrt{1 + F - G^2} - 1\right) \end{aligned}\right\}. \tag{2.15}$$

$$F = \frac{1}{2} f_{kl} f^{kl}; \quad G^2 = \frac{|f_{kl}|}{-g} = \frac{(f_{23}f_{14} + f_{31}f_{24} + f_{12}f_{34})^2}{-g}. \tag{2.16}$$

Both F and G are invariant. We shall bring G into such a form, that its invariance will be evident. For this purpose let us define an antisymmetrical tensor $j^{sklm}$ for any pair of indices, that is*

$$
j^{sklm} = \begin{cases} \dfrac{1}{2\sqrt{-g}} \text{if } sklm \text{ is an even permutation of 1, 2, 3, 4} \\[2mm] \dfrac{-1}{2\sqrt{-g}} \text{if } sklm \text{ is an odd permutation of 1, 2, 3, 4} \\[2mm] 0 \text{ in any other case} \end{cases}
$$

(2.17)

We can write now G in the following form:

$$
G = \tfrac{1}{4} j^{sklm} f_{sk} f_{lm}. \tag{2.18}
$$

From the last equation we can deduce the tensor character of $j^{sklm}$. We can also write G in the form

$$
G = \tfrac{1}{4} f_{sk} f^{*sk}, \tag{2.19}
$$

where $f^{*sk}$ is the *dual tensor* denned by

$$
f^{*sk} = j^{sklm} f_{lm}, \tag{2.20}
$$

that is

$$
\left. \begin{array}{lll} f^{*23} = \dfrac{1}{\sqrt{-g}} f_{14}, & f^{*31} = \dfrac{1}{\sqrt{-g}} f_{24}, & f^{*12} = \dfrac{1}{\sqrt{-g}} f_{34} \\[3mm] f^{*14} = \dfrac{1}{\sqrt{-g}} f_{23}, & f^{*24} = \dfrac{1}{\sqrt{-g}} f_{31}, & f^{*34} = \dfrac{1}{\sqrt{-g}} f_{12} \end{array} \right\},
$$

(2.21)

or also

$$
\left. \begin{array}{lll} f_{23}^* = -\sqrt{-g}\, f^{14}, & f_{31}^* = -\sqrt{-g}\, f^{24}. & f_{12}^* = -\sqrt{-g}\, f_{34} \\[3mm] f_{14}^* = -\sqrt{-g}\, f^{23}, & f_{24}^* = -\sqrt{-g}\, f^{31}, & f_{34}^* = -\sqrt{-g}\, f^{12} \end{array} \right\},
$$

(2.22)

*Einstein and Mayer, 'Berl. Ber.,' p. 3 (1932).

because

$$f^*_{sk} = j_{s\,klm}\,f^{lm}; \quad j_{s\,klm} = g\,j^{s\,klm} = g_{as}g_{bk}g_{cl}g_{dm}\,j^{abcd}. \quad (2.23)$$

We shall need later the following formulæ:

$$f^{*kl}\,f^*_{kl} = -f^{kl}\,f_{kl} \qquad (2.24)$$

$$j^{ls\,ab}\,f_{ks}\,f_{ab} = f^{*ls}\,f_{ks} = G\delta^l_k \qquad (2.25)$$

$$f^{**ls} = -f^{ls}. \qquad (2.26)$$

(2.24)–(2.26) follow from the definition of $f^*_{kl}$, $f^{*kl}$ and G given above.

The function $\mathscr{L}$ represented by (2.15) is the simplest Lagrangian satisfying the principle of general invariance. But it differs from that considered in I by the term $G^2$. This is of the fourth order in the $f_{kl}$ and can, therefore, be neglected except in the immediate neighbourhood of singularities (*i.e.*, electrons, see § 6). But the Lagrangian used in I can also be expressed in a general covariant form; for $G^2$ is a determinant, namely, $|f_{kl}|$, therefore

$$\int \left( \sqrt{-|g_{kl} + f_{kl}|} + |f_{kl}| - \sqrt{-|g_{kl}|} \right) d\tau \qquad (2.27)$$

is also invariant; in cartesian co-ordinates it has exactly the form

$$\int \left( \sqrt{1 + F} - 1 \right) d\tau. \qquad (2.28)$$

Which of these action principles is the right one can only be decided by their consequences. We take the expression given by (2.15) and can then easily return to the other (2.27) or (2.28) by putting G = 0. In any case the solution of the *statical* problem is identical for both action functions because one has G = 0 in this special case.

## § 3. ACTION PRINCIPLE, FIELD EQUATION AND CONSERVATION LAW.

We write (2.15) in the general form

$$\mathscr{L} = \sqrt{-g}\, \mathbf{L} = \sqrt{-g}\, \mathbf{L}(g_{kl}, \mathbf{F}, \mathbf{G}).$$

We shall see that all considerations hold if $\mathbf{L}$ is an invariant function of these arguments. As usual we assume the existence of a potential vector $\phi_k$, so that

$$f_{kl} = \frac{\partial \phi_l}{\partial x^k} - \frac{\partial \phi_k}{\partial x^l}. \tag{3.1}$$

Then we have the identity

$$\frac{\partial f_{lm}}{\partial x^k} + \frac{\partial f_{mk}}{\partial x^l} + \frac{\partial f_{kl}}{\partial x^m} = 0, \tag{3.2}$$

which can with the help of (2.20) be written:

$$\frac{\partial \sqrt{-g}\, f^{*kl}}{\partial x^l} = 0. \tag{3.2A}$$

We introduce a second kind of antisymmetrical field tensor $p_{kl}$, which has to $f_{kl}$ a relation similar to that which, in Maxwell's theory of macrospic bodies, the dielectric displacement and magnetic induction have to the field strengths:

$$\begin{aligned}
\sqrt{-g}\, p^{kl} = \frac{\partial \mathscr{L}}{\partial f_{kl}} &= \sqrt{-g} \left( 2 \frac{\partial \mathbf{L}}{\partial \mathbf{F}} f^{kl} + \frac{\partial \mathbf{L}}{\partial \mathbf{G}} f^{*kl} \right) \\
&= \frac{\left( f^{kl} - \mathbf{G} f^{*kl} \right) \sqrt{-g}}{\sqrt{1 + \mathbf{F} - \mathbf{G}^2}}.
\end{aligned} \tag{3.3}$$

The variation principle (2.2) gives the Eulerian equations

$$\frac{\partial \sqrt{-g}\, p^{kl}}{\partial x^l} = 0. \tag{3.4}$$

The equation (3.2) (or (3.2A)) and (3.4) are the complete set of field equations.

We prove the validity of the conservation law as in Maxwell's theory. Assuming a geodetic co-ordinate system, we multiply (3.2) by $p^{lm}$:

$$p^{lm} \left( \frac{\partial f_{lm}}{\partial x^k} + \frac{\partial f_{mk}}{\partial x^l} + \frac{\partial f_{kl}}{\partial x^m} \right) = 0. \tag{3.5}$$

In the second and third term we can take $p^{lm}$ under the differentiation symbol because of (3.4); in the first term we use the definition (3.3) of $p^{lm}$:

$$2\frac{\partial}{\partial x^l}\left(p^{lm} f_{mk}\right) + \frac{\partial \mathbf{L}}{\partial f_{lm}}\frac{\partial f_{lm}}{\partial x^k} = 0,$$

or

$$-2\frac{\partial}{\partial x^l}\left(p^{ml} f_{mk}\right) + 2\frac{\partial \mathbf{L}}{\partial x^k} = 0.$$

If we introduce the tensor

$$T^l_k = \mathbf{L}\delta^l_k - p^{ml} f_{mk}, \tag{3.6}$$

where

$$\delta^l_k = \left\{ \begin{array}{l} 1 \text{ if } k = l \\ 0 \text{ if } k \neq l \end{array} \right\}, \tag{3.7}$$

we have

$$\frac{\partial T^l_k}{\partial x^l} = 0. \tag{3.8}$$

In an arbitrary co-ordinate system we have

$$\frac{\partial \sqrt{-g}\, T^l_k}{\partial x^l} - \frac{1}{2}\sqrt{-g}\, T^{ab}\frac{\partial g_{ab}}{\partial x^k} = 0, \tag{3.9}$$

or, with the usual notation of covariant differentiation

$$T^l_{k/l} = 0. \tag{3.9A}$$

It follows from (3.3) and (2.25) that we can write also $T^l_k$ in the form

$$T^l_k = \mathbf{L}\delta^l_k - \frac{f^{ml} f_{mk} - G^2\delta^l_k}{\sqrt{1 + F - G^2}}. \tag{3.6A}$$

## § 4. LAGRANGIAN AND HAMILTONIAN.

$\mathscr{L}$ can be considered as a function of $g^{kl}$ and $f_{kl}$. We shall show that $\frac{-2}{\sqrt{-g}}\frac{\partial \mathscr{L}}{\partial g^{kl}}$ is the energy-impulse tensor. We find

$$\frac{\partial \sqrt{-g}}{\partial g^{kl}} = -\frac{1}{2}\sqrt{-g}\, g_{kl} \tag{4.1}$$

$$\frac{\partial F}{\partial g^{kl}} = g^{sr} f_{ks} f_{lr} \tag{4.2}$$

$$\frac{\partial G}{\partial g^{kl}} = \frac{1}{2} G g_{kl}.$$ (4.3)

Therefore

$$-2\frac{\partial \mathscr{L}}{\partial g^{kl}} = \sqrt{-g}\left\{\mathbf{L}g_{kl} - 2\left(\frac{\partial \mathbf{L}}{\partial F}\frac{\partial F}{\partial g^{kl}} + \frac{\partial \mathbf{L}}{\partial G}\frac{\partial G}{\partial g^{kl}}\right)\right\}$$

$$= \sqrt{-g}\left\{\mathbf{L}g_{kl} - \frac{f_{ks}f_{lr}g^{sr} - G^2 g_{kl}}{\sqrt{1+F-G^2}}\right\}.$$ (4.4)

It follows from (3.6A) and (3.6)

$$-2\frac{\partial \mathscr{L}}{\partial g^{kl}} = \sqrt{-g}\,T_{kl} = \sqrt{-g}\left(\mathbf{L}g_{kl} - f_{ks}\,p_{lr}g^{sr}\right).$$ (4.5)

Now it is very easy to generalize our action principle in such a way that it contains Einstein's gravitation laws; one has only to add to the action integral the term $\int R\sqrt{-g}\,d\tau$, where R is the scalar of curvature. But we do not discuss problems connected with gravitation in this paper.

$\mathscr{L}$ was regarded as a function of $g^{kl}$ and $f_{kl}$. We can, however, express $\mathscr{L}$ also as a function of $g^{kl}$ and $p_{kl}$. It can be shown that it is possible to solve the equations

$$p^{kl} = \frac{f^{kl} - G f^{*kl}}{\sqrt{1+F-G^2}}$$ (3.3)

with respect to $f^{kl}$. For this purpose we have to calculate

$$\tfrac{1}{2} p^{*kl} p^*_{kl} = P,$$ (4.6)

$$\tfrac{1}{4} p^{kl} p^*_{kl} = Q,$$ (4.7)

*i.e.*, **P** and **Q** corresponding to F and G. Using the formulæ (3.3) and (2.24)–(2.26), we obtain

$$P = \frac{-F + G^2 F + 4G^2}{1 + F - G^2}$$ (4.8)

$$Q = G.$$ (4.9)

The last equations can be written in a more symmetrical form:

$$\frac{1 + F - G^2}{1 + G^2} = \frac{1 + Q^2}{1 + P - Q^2} \qquad (4.8\text{\scriptsize A})$$

$$G = Q. \qquad (4.9\text{\scriptsize A})$$

We are now able to solve the equations (3.3). It follows from (3.3) and (2.26) that

$$p^{*kl} = \frac{f^{*kl} + G f^{kl}}{\sqrt{1 + F - G^2}}. \qquad (3.3\text{\scriptsize A})$$

Solving (3.3) and (3.3A) we obtain (taking into account (4.8A) and (4.9A))

$$f^{kl} = \frac{p^{kl} + Q p^{*kl}}{\sqrt{1 + P - Q^2}}; \qquad f^{*kl} = \frac{p^{*kl} - Q p^{kl}}{\sqrt{1 + P - Q^2}}. \qquad (4.10)$$

The tensors $f_{kl}$ and $p_{kl}$ can now be treated completely symmetrically. Instead of the Lagrangian **L** we can use in the principle of action the Hamiltonian function **H**:

$$\mathbf{H} = \mathbf{L} - \tfrac{1}{2} p^{kl} f_{kl}, \qquad (4.11)$$

where **H** has to be regarded as a function of $g^{kl}$ and $p_{kl}$. From (4.8), (4.9), and (4.10) it follows for **H** as a function of $g^{kl}$ and $p_{kl}$:

$$\mathcal{H} = \mathbf{H}\sqrt{-g} = \sqrt{-g}\left(\sqrt{1 + P - Q^2} - 1\right), \qquad (4.12)$$

and this can be expressed in the form

$$\mathcal{H} = \sqrt{-\left|g_{kl} + p^{*}_{kl}\right|} - \sqrt{-\left|g_{kl}\right|}. \qquad (4.12\text{\scriptsize A})$$

The function **H** leads us to exactly the same equations of the field as the function **L**. We see that the equations

$$p^{*}_{kl} = \frac{\partial \psi^{*}_{l}}{\partial x^{k}} - \frac{\partial \psi^{*}_{k}}{\partial x^{l}} \qquad (\psi^{*}_{k} = \text{anti-potential-vector})$$

$$(4.13)$$

$$\sqrt{-g}\, f^{*kl} = \frac{\partial \mathcal{H}}{\partial p^{*}_{kl}} \qquad (4.14)$$

$$\frac{\partial \sqrt{-g}\, f^{*kl}}{\partial x^{l}} = 0, \qquad (4.15)$$

are entirely equivalent to the equations (3.4), (4.10), (3.2A).

CHAPTER SIX

The energy-impulse tensor (3.6A) can also be expressed with help of **H** instead of **L**. One has

$$T_k^l = \mathbf{H}\delta_k^l - f^{*ml}p_{mk}^* = \left(\mathbf{L} - \tfrac{1}{2}p^{ab}f_{ab}\right)\delta_k^l - f^{*ml}p_{mk}^*. \quad (4.16)$$

The identity of this expression with (3.6) is evident,* if we appeal to the following formula which can be deduced from (2.21), (2.22)

$$f^{*ml}p_{ml}^* = p^{ml}f_{mk} - \tfrac{1}{2}p^{ab}f_{ab}\delta_k^l \quad (4.17)$$

Generally, from each equation containing $\mathscr{L}$, $\phi_k$, $f_{kl}$, $p_{kl}$, one obtains another correct equation changing these quantities correspondingly into $\mathcal{H}$, $\psi_k^*$, $p_{kl}^*$, $f_{kl}^*$.

## § 5. Field Equations in Space-vector Form.

We now introduce the conventional units instead of the natural units. We denote by **B**, **E** and **D**, **E**, the space-vectors which characterize the electromagnetic field in the conventional units. We have in a cartesian co-ordinate system:

$$\left(x^1, x^2, x^3, x^4\right) \rightarrow (x, y, z, ct) \quad (5.1)$$

$$(\phi_1, \phi_2, \phi_3, \phi_4) \rightarrow (\mathbf{A}, \phi) \quad (5.2)$$

$$\left.\begin{array}{l}(f_{23}, f_{31}, f_{12} \rightarrow \mathbf{B} \\ f_{14}, f_{24}, f_{34}) \rightarrow \mathbf{E}\end{array}\right\} \quad (5.3)$$

$$\left.\begin{array}{l}(p_{23}, p_{31}, p_{12} \rightarrow \mathbf{H} \\ p_{14}, p_{24}, p_{34}) \rightarrow \mathbf{D}\end{array}\right\}. \quad (5.4)$$

The quotient of the field strength expressed in the conventional units divided by the field strength in the natural units may be denoted by $b$. This constant of a dimension of a field strength may be called the *absolute field*; later we shall determine the value of $b$, which turns out to be very great, *i.e.*, of the order of magnitude $10^{16}$ e.s.u.

---

*In I it has been stated that the two expressions for $T_k^l$, obtained with help of **L** and **H**, are different; this has turned out to be a mistake.

We have

$$\mathbf{L} = \sqrt{1 + F - G^2} - 1, \qquad (2.11)$$

$$F = \frac{1}{b^2}\left(\mathbf{B}^2 - \mathbf{E}^2\right); \qquad G = \frac{1}{b^2}\,(\mathbf{B.E}) \qquad (2.12\text{A; } 2.13\text{A})$$

$$\left.\begin{array}{l} \mathbf{H} = b^2\dfrac{\partial \mathbf{L}}{\partial \mathbf{B}} = \dfrac{\mathbf{B} - G\mathbf{E}}{\sqrt{1 + F - G^2}}, \\[3mm] \mathbf{D} = b^2\dfrac{\partial \mathbf{L}}{\partial \mathbf{E}} = \dfrac{\mathbf{E} - G\mathbf{B}}{\sqrt{1 + F - G^2}}. \end{array}\right\} \qquad (3.3\text{A})$$

$$\mathbf{B} = \text{rot } \mathbf{A}; \quad \mathbf{E} = -\frac{1}{c}\frac{\partial \mathbf{A}}{\partial t} - \text{grad } \phi \qquad (3.1\text{A})$$

$$\text{rot } \mathbf{E} + \frac{1}{c}\frac{\partial \mathbf{B}}{\partial t} = 0; \quad \text{div } \mathbf{B} = 0 \qquad (3.2\text{B})$$

$$\text{rot } \mathbf{H} - \frac{1}{c}\frac{\partial \mathbf{D}}{\partial t} = 0; \quad \text{div } \mathbf{D} = 0. \qquad (3.4\text{A})$$

Our field equations (3.2B) and (3.4A) are formally identical with Maxwell's equations for a substance which has a dielectric constance and a susceptibility, being certain functions of the field strength, but without a spatial distribution of charge and current.

For the energy-impulse tensor we find:

$$\left(\frac{1}{4\pi}\mathbf{T}^{kl}\right) = \begin{bmatrix} X_x & X_y & X_z & cG_x \\ Y_x & Y_y & Y_z & cG_y \\ Z_x & Z_y & Z_z & cG_z \\ \frac{1}{c}S_x & \frac{1}{c}S_y & \frac{1}{c}S_z & U \end{bmatrix} \qquad (3.6\text{A})$$

$$\left.\begin{array}{l} 4\pi X_x = H_y B_y + H_z B_z - D_x E_x - b^2\mathbf{L} \\[1mm] 4\pi Y_x = 4\pi X_y = -H_y B_x - D_x E_y \\[1mm] \dfrac{4\pi}{c}S_x = 4\pi cG_x = D_y B_z - D_z B_y \\[1mm] 4\pi U = E_x E_x + D_y E_y + D_z E_z + b^2\mathbf{L} \end{array}\right\}. \qquad (3.6\text{B})$$

One gets another set of expressions for these quantities by changing **L, B, E, H, D** into **H, H, D, B, E**.

The conservation laws are:

$$\left.\begin{array}{l} \dfrac{\partial X_x}{\partial x} + \dfrac{\partial X_y}{\partial y} + \dfrac{\partial X_z}{\partial z} = -\dfrac{1}{c^2}\dfrac{\partial S_x}{\partial t} \\[2mm] \dfrac{\partial S_x}{\partial x} + \dfrac{\partial S_y}{\partial y} + \dfrac{\partial S_z}{\partial z} = -\dfrac{\partial U}{\partial t} \end{array}\right\}. \qquad (3.8\text{A})$$

The function **H** is given by

$$\mathbf{H} = \sqrt{1 + P - Q^2} - 1 \qquad (4.12\text{A})$$

$$P = \frac{1}{b^2}\left(\mathbf{D}^2 - \mathbf{H}^2\right); \quad Q = \frac{1}{b^2}\left(\mathbf{D.H}\right). \qquad (4.6\text{A});\ (4.7\text{A})$$

Solving (3.3A) we obtain:

$$\left.\begin{array}{l} \mathbf{B} = b^2\dfrac{\partial \mathbf{H}}{\partial \mathbf{H}} = \dfrac{\mathbf{H} + Q\mathbf{D}}{\sqrt{1 + P - Q^2}} \\[4mm] \mathbf{E} = b^2\dfrac{\partial \mathbf{H}}{\partial \mathbf{D}} = \dfrac{\mathbf{D} + Q\mathbf{H}}{\sqrt{1 + P - Q^2}} \end{array}\right\}. \qquad (3.10\text{A})$$

## § 6. STATIC SOLUTION OF THE FIELD EQUATIONS.

We consider (in the cartesian co-ordinate system) the electrostatic case where $\mathbf{B} = \mathbf{H} = 0$ and all other field components are independent of $t$. Then the field equations reduce to:

$$\text{rot } \mathbf{E} = 0 \qquad (6.1)$$

$$\text{div } \mathbf{D} = 0. \qquad (6.2)$$

We solve this equation for the case of central symmetry. Then (6.2) is simply

$$\frac{d}{dr}\left(r^2 D_r\right) = 0, \qquad (6.3)$$

and (6.3) has the solution

$$D_r = e/r^2. \qquad (6.4)$$

In this case the field **D** is exactly the same as in Maxwell's theory: the sources of **D** are point charges given by the surface integral

$$4\pi e = \int D_r \, d\sigma. \tag{6.5}$$

The equation (6.1) gives

$$E_r = -\frac{d\phi}{dr} = -\phi'(r) \tag{6.6}$$

and from (3.3A)

$$D_r = \frac{E_r}{\sqrt{1 - \frac{1}{b^2} E_r^2}} = -\frac{\phi'(r)}{\sqrt{1 - \frac{1}{b^2} \phi'^2}}. \tag{6.7}$$

The combination of (6.4) and (6.7) gives a differential equation for $\phi(r)$ of the first order, with the solution

$$\phi(r) = \frac{e}{r_0} f\left(\frac{r}{r_0}\right); \quad f(x) = \int_x^\infty \frac{dy}{\sqrt{1 + y^4}}; \quad r_0 = \sqrt{\frac{e}{b}}. \tag{6.8}$$

*This is the elementary potential of a point charge e,* which has to replace Coulomb's law; the latter is an approximation for $x \gg 1$, as is seen immediately, but the new potential is finite everywhere.

With help of the substitution $x = \tan \frac{1}{2}\beta$ one obtains

$$f(x) = \frac{1}{2} \int_{\bar{\beta}(x)}^\pi \frac{d\beta}{\sqrt{1 - \frac{1}{2} \sin^2 \beta}} = f(0) - \frac{1}{2} F\left(\frac{1}{\sqrt{2}}, \bar{\beta}\right), \tag{6.9}$$

where

$$\bar{\beta} = 2 \arctan x, \tag{6.10}$$

and $F(k, \beta)$ is the Jacobian elliptic integral of the first kind for $k = \frac{1}{\sqrt{2}} = \sin \frac{1}{4}\pi$ (tabulated in many books)*

$$F\left(\frac{1}{\sqrt{2}}, \bar{\beta}\right) = \int_0^{\bar{\beta}} \frac{d\beta}{\sqrt{1 - \frac{1}{2} \sin^2 \beta}}. \tag{6.11}$$

---

*E.g., Jahnke-Emde, "Tables of functions" (Teubner 1933), p. 127.

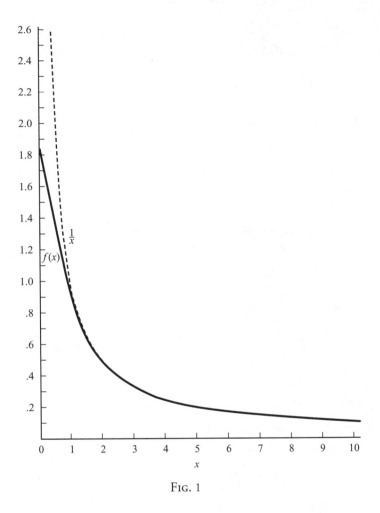

FIG. 1

For $x = 0$ one has

$$f(0) = F\left(\frac{1}{\sqrt{2}}, \frac{1}{2}\pi\right) = 1{,}8541. \qquad (6.12)$$

The potential has its maximum in the centre and its value is

$$\phi(0) = 1{,}8541 \, e/r_0. \qquad (6.13)$$

The function $f(x)$ is plotted in fig. 1. It has very similar properties to the function arc cot $x$. For example, one has

$$\bar{\beta}(1/x) = 2 \arctan 1/x = 2\left(\tfrac{1}{2}\pi - \arctan x\right) = \pi - \bar{\beta}(x);$$

on the other hand

$$F\left(\frac{1}{\sqrt{2}}, \bar{\beta}\right) + F\left(\frac{1}{\sqrt{2}}, \pi - \bar{\beta}\right) = F\left(\frac{1}{\sqrt{2}}, \pi\right).$$

Therefore one has

$$f(x) + f(1/x) = f(0). \tag{6.14}$$

It is sufficient, therefore, to calculate $f(x)$ from $x = 0$ to $x = 1$ or from $\beta = 0$ to $\beta = \frac{1}{4}\pi$.

One sees, that the **D** field is infinite for $r = 0$; **E** and $\phi$, however, are always finite. One has

$$D_r = e/r^2, \tag{6.4}$$

$$E_r = \frac{e}{r_0^2 \sqrt{1 + (r/r_0)^4}}. \tag{6.15}$$

The components $E_x$, $E_y$, $E_z$ are finite at the centre, but have there a discontinuity.

## § 7. Sources of the Field.

In the older theories, which we have called dualistic, because they considered matter and field as essentially different, the ideal would be to assume the particles to be point charges; this was impossible because of the infinite self-energy. Therefore it was necessary to assume the electron having a finite diameter and to make arbitrary assumptions about its inner structure, which lead to the difficulties pointed out in the introduction. In our theory these difficulties do not appear. We have seen that the $p_{kl}$ field (or **D**-field) has a singularity which corresponds to a point charge as the source of the field. **D** and **E** are identical only at large distances ($r \gg r_0$) from the point charge, but differ in its neighbourhood, and one can call their quotient (which is function of **E**) "dielectric constant" of the space. But we shall now show that another interpretation is also possible which corresponds to the old idea of a spatial distribution of charge in the electron. It consists in taking div **E** (instead of div **D** $= 0$) as definition of charge density $\rho$, which we propose to call "free charge density."

Let us now write our set of field equations in the following form:

$$\frac{\partial \sqrt{-g}\, p^{kl}}{\partial x^l} = 0, \tag{3.4}$$

$$\frac{\partial f_{kl}}{\partial x^m} + \frac{\partial f_{lm}}{\partial x^k} + \frac{\partial f_{mk}}{\partial x^l} = 0, \quad \text{or} \quad \frac{\partial \sqrt{-g}\, f^{*kl}}{\partial x^l} = 0. \tag{3.2}$$

$p^{kl}$ is a given function of $f^{kl}$ and if we put in (3.4) for $p^{kl}$ the expression (3.3), in which **L** is not specified, we obtain:

$$\frac{\partial}{\partial x^l}\left\{\left(2\frac{\partial \mathbf{L}}{\partial \mathbf{F}} f^{kl} + \frac{\partial \mathbf{L}}{\partial \mathbf{G}} f^{*kl}\right)\sqrt{-g}\right\} = 0. \tag{7.1}$$

We can now write the equation (7.1) in the form:

$$\frac{\partial \sqrt{-g}\, f^{kl}}{\partial x^l} = 4\pi\rho^k \sqrt{-g}, \tag{7.2}$$

where

$$-4\pi\rho^k = \frac{1}{2\partial \mathbf{L}/\partial \mathbf{F}}\left\{2f^{kl}\frac{\partial}{\partial x^l}\left(\frac{\partial \mathbf{L}}{\partial \mathbf{F}}\right) + f^{*kl}\frac{\partial}{\partial x^l}\left(\frac{\partial \mathbf{L}}{\partial \mathbf{G}}\right)\right\}. \tag{7.3}$$

The equations (7.2) and (3.2) are formally identical with the equations of the Lorentz theory. But the important difference consists in this, that $\rho^k$ is *not* a given function of the space-time co-ordinates, but is a function of the unknown field strength. If we have a solution of our set of equations, we are able to find the density of the "free charge" or the "free current" with help of (7.2) or (7.3).

We see immediately that $\rho^k$ satisfies the conservation law:

$$\frac{\partial \sqrt{-g}\,\rho^k}{\partial x^k} = 0. \tag{7.4}$$

This follows from (7.2), that is from the antisymmetrical character of $f^{kl}$, and can also be checked from (7.3).

In Lorentz's theory there exists the energy-impulse tensor of the electromagnetic field, defined by

$$4\pi S^l_k = \tfrac{1}{2}\delta^l_k \mathbf{F} - f^{ls} f_{ks}, \tag{7.5}$$

but its divergence does not vanish, where the density of charge is not zero. Therefore to preservethe conservation principle in the Lorentz's theory it was necessary to introduce an energy-impulse tensor of matter, $M^{kl}$, the meaning of which is obscure. The tensor $M^l_k$ had to fulfil the condition that the divergence of $S^l_k + M^l_k$ vanishes. This difficulty does not appear in our theory. We do not need to introduce the matter tensor $M^l_k$ because the conservation laws are always satisfied by our energy-impulse tensor $T^l_k$.

We shall, however, show that it is possible by introducing the free charges to bring our conservation law

$$T^k_{s|k} = 0$$

into the form used in the Lorentz theory, namely,

$$S^k_{s|k} = f^{sk} \rho_k. \tag{7.6}$$

The calculations are similar to those used in § 3. The simplest way is to choose a geodetic co-ordinate system. We have then:

$$\frac{\partial f^{kl}}{\partial x^l} = 4\pi \rho^k \tag{7.2A}$$

$$\frac{\partial f_{kl}}{\partial x^m} + \frac{\partial f_{lm}}{\partial x^k} + \frac{\partial f_{mk}}{\partial x^k} = 0. \tag{3.2}$$

Multiplying (3.2) by $f^{kl}$ we find:

$$\frac{1}{2} \frac{\partial}{\partial x^s} \left( f_{kl} f^{kl} \right) - 2 \frac{\partial}{\partial x^l} \left( f^{lk} f_{sk} \right) = 2 f_{sk} \frac{\partial f^{lk}}{\partial x^l} \tag{7.7}$$

and therefore

$$\frac{1}{4} \frac{\partial}{\partial x^s} \left( f_{kl} f^{kl} \right) - \frac{\partial}{\partial x^l} \left( f^{lk} f_{sk} \right) = f_{sk} \rho^k, \tag{7.8}$$

and taking account of (7.5):

$$\frac{\partial S^k_s}{\partial x^k} = f_{sk} \rho^k. \tag{7.9}$$

One can derive the same equation directly from the conservation formula (3.8) writing it, in a geodetic co-ordinate system, in the form

$$\frac{\partial T^k_s}{\partial x^k} = 0, \tag{3.8}$$

and introducing the expression (3.6) for $T^k_s$. The two methods are equivalent.

Let us now specialize our equations for the case in which **L** has the form given in (2.15). We obtain then for $\rho^k$ in (7.3)

$$-4\pi\rho^k = \frac{1}{\sqrt{1+F-G^2}}\left\{ f^{kl}\frac{\partial}{\partial x^l}\left(\frac{1}{\sqrt{1+F-G^2}}\right) \right.$$
$$\left. - f^{*kl}\frac{\partial}{\partial x^l}\left(\frac{1}{\sqrt{1+F-G^2}}\right) \right\}. \qquad (7.10)$$

In the space-vector notation, where

$$(\rho_1, \rho_2, \rho_3, \rho_4) \rightarrow \left(\frac{I}{c}, \rho\right),$$

we have

$$-4\pi\frac{I}{c} = \frac{1}{\sqrt{1+F-G^2}}\left\{ \mathbf{B}\times\mathrm{grad}\left(\frac{1}{\sqrt{1+F-G^2}}\right)\right.$$
$$\left. -\mathbf{E}\times\mathrm{grad}\left(\frac{G}{\sqrt{1+F-G^2}}\right)\right\}$$
$$+\frac{1}{c\sqrt{1+F-G^2}}\left\{ \mathbf{E}\frac{\partial}{\partial t}\left(\frac{1}{\sqrt{1+F-G^2}}\right)\right.$$
$$\left. -\mathbf{B}\frac{\partial}{\partial t}\left(\frac{G}{\sqrt{1+F-G^2}}\right)\right\} \qquad (7.10\text{A})$$
$$4\pi\rho = -\frac{1}{\sqrt{1+F-G^2}}\left\{ \mathbf{E}.\,\mathrm{grad}\left(\frac{1}{\sqrt{1+F-G^2}}\right)\right.$$
$$\left. -\mathbf{B}.\,\mathrm{grad}\left(\frac{1}{\sqrt{1+F-G^2}}\right)\right\}$$

We shall now apply the results here obtained to the case of the statical field. In this case **E** is always finite and has a non-vanishing divergence, which represents the free charge. We can, therefore, regard an electron either as a point charge, *i.e.*, as a source of the **D** ($p_{kl}$) field, or as a continuous distribution of the space charge which is a source of the **E** ($f_{kl}$) field. It can easily be shown that the whole charge is in both cases the same (as is to be expected). Both

$$\int \mathrm{div}\,\mathbf{D}\,dv \quad \text{and} \quad \int \mathrm{div}\,\mathbf{E}\,dv$$

have the same value, *i.e.*, $4\pi e$. For the first integral it has been shown in § 6. For the second we have

$$E_r \sim \frac{e}{r^2} \quad \text{as} \quad r \to \infty;$$

everywhere else $E_r$ is finite. The discontinuity of $E_x$, $E_y$, $E_z$ at the origin is also finite and gives no contribution to the integral. Therefore

$$\int \text{div } \mathbf{E} \, dv = 4\pi \int \rho \, dv = 4\pi e.$$

Let us now calculate the distribution of the free charge in the statical case.

We could calculate it from the equation (7.10A), but it is easier to do it from the equation

$$\text{div } \mathbf{E} = \frac{1}{r^2} \frac{d}{dr} \left( r^2 E_r \right) = 4\pi \rho, \tag{7.11}$$

where

$$E_r = \frac{e}{r_0^2} \frac{1}{\sqrt{1 + (r/r_0)^4}}. \tag{6.15}$$

The result is

$$\rho = \frac{e}{2\pi r_0^3 \frac{r}{r_0} \left( 1 + \left( \frac{r}{r_0} \right)^4 \right)^{3/2}}. \tag{7.12}$$

For $r \gg r_0$, $\rho \propto r^{-7}$, therefore diminishing very rapidly as $r$ increases. For $r < r_0$, $\rho \propto 1/r$, therefore $\rho \to \infty$, but $r^2\rho \to 0$ for $r \to 0$. It is easy to verify that the space integral of $\rho$ is equal to $e$. For one has, putting $r/r_0 = \sqrt{\tan \phi}$

$$\int \rho \, dv = \frac{2e}{r_0^3} \int_0^\infty \frac{r^2 dr}{\frac{r}{r_0} \left( 1 + \frac{r^4}{r_0^4} \right)^{3/2}} = e \int_0^{\pi/2} \cos \phi \, dy = e.$$

Our theory combines the two possible aspects of the field; true point charges and free spatial densities are entirely equivalent. The question whether the one or the other picture of the electron is right has no meaning. This confirms the idea which has proved so fruitful in quantum mechanics, that one has to be careful in applying notions from the macroscopic world to the world of atoms: it may happen that two notions contradictory in macroscopic use are quite compatible in microphysics.

## § 8. LORENTZ'S EQUATIONS OF POINT MOTION AND MASS.

We consider once more the problem of the electron at rest. We intend to calculate the mass and to determine the absolute field constant $b$ in terms of observable quantities. It is convenient to use the space vector notation.

The impulse-energy tensor is according to (3.6B)

$$4\pi X_x = -D_x E_x - b^2 \mathbf{L} = -\frac{E_x^2}{\sqrt{1 - 1/b^2 \mathbf{E}^2}}$$

$$-b^2 \left( \sqrt{1 - \frac{1}{b^2} \mathbf{E}^2} - 1 \right)$$

$$4\pi X_y = -D_x E_y = -\frac{E_x E_y}{\sqrt{1 - 1/b^2 \mathbf{E}^2}}$$

$$S_x = S_y = S_z = 0$$

$$4\pi U = \mathbf{D}.\mathbf{E} + b^2 \mathbf{L} = b^2 \mathbf{H} = b^2 \left( \sqrt{1 - \frac{1}{b^2} \mathbf{E}^2} - 1 \right)$$

$$+ \frac{\mathbf{E}^2}{\sqrt{1 - 1/b^2 \mathbf{E}^2}} = b^2 \left( \frac{1}{\sqrt{1 - 1/b^2 \mathbf{E}^2}} - 1 \right)$$

$$\left. \right\} . \quad (3.6\text{C})$$

We calculate the space integrals of these quantities. Obviously one has with $dv = dx\, dy\, dz$:

$$4\pi \int X_x\, dv = 4\pi \int Y_y\, dv = 4\pi \int Z_s\, dv = -\frac{1}{3} \int$$

$$\times \frac{\mathbf{E}^2}{\sqrt{1 - 1/b^2 \mathbf{E}^2}} dv - b^2 \int \left( \sqrt{1 - \frac{1}{b^2} \mathbf{E}^2} - 1 \right) dv \quad (8.1)$$

$$\int X_y\, dv = \int X_z\, dv = \int Z_y\, dv = 0. \quad (8.2)$$

Using (6.15) and (6.8) we find:

$$\int X_x\, dv = b^2 (I_1 - I_2), \quad (8.3)$$

where

$$I_1 = \int_0^\infty \left(1 - \frac{x^2}{\sqrt{1+x^4}}\right) x^2 dx$$

$$I_2 = \frac{1}{3} \int_0^\infty \frac{dx}{\sqrt{1+x^4}} = \frac{1}{3} f(0)$$

(8.4)

The integral $I_1$ can be transformed by partial integration:

$$I_1 = \frac{1}{3} \int_0^\infty \left(1 - \frac{x^2}{\sqrt{1+x^4}}\right) \frac{dx^3}{dx} dx$$

$$= \left[\frac{1}{3} \left(1 - \frac{x^2}{\sqrt{1+x^4}}\right) x^3\right]_0^\infty + \frac{2}{3} \int_0^\infty \frac{x^4 dx}{\left(1+x^4\right)^{3/2}}.$$

The first term vanishes; the second can be transformed by another partial integration:

$$I_1 = -\frac{1}{3} \int_0^\infty x \frac{d}{dx} \left(\frac{1}{\sqrt{1+x^4}}\right) dx$$

$$= \frac{1}{3} \int_0^\infty \frac{dx}{\sqrt{1+x^4}} = I_2 = \frac{1}{3} f(0).$$

The result is the so-called "theorem of Laue"*

$$\int X_x \, dv = \int Y_y \, dv = \int Z_z \, dv = 0.$$

In the statical case and in a co-ordinate system in which the electron is at rest the integrals of all components of the tensor $T_k^l$ vanish except the total energy

$$E = \int U \, dv = \frac{b^2}{4\pi} \int H \, dv.$$

(8.5)

We find from (3.6c), (6.15), and (8.4)

$$E = m_0 c^2 = \delta r_0^3 (3I_2 - I_1) = \frac{e^2}{r_0} 2I_1 = \frac{2}{3} \frac{e^2}{r_0} f(0) = 1 \cdot 2361 \frac{e^2}{r_0}.$$

(8.6)

We have obtained a *finite value of the energy or the mass of the electron* with a definite numerical factor. This relation enables us to complete

---

*Mie, 'Ann. Physik,' vol. 40, p. 1 (1913).

our theory concerning the value of the absolute field $b$ in the conventional units. For (8.6) gives the "radius" of the electron expressed in terms of its charge and mass:

$$r_0 = 1 \cdot 2361 \frac{e^2}{m_0 c^2} = 2,28 \times 10^{-13} \mathrm{cm}. \qquad (8.7)$$

and

$$b = \frac{e}{r_0^2} = 9,18.10^{15} \text{ e.s.u.} \qquad (8.8)$$

The enormous magnitude of this field justifies the application of the Maxwell's equations in their classical form in all cases, except those where the inner structure of the electron is concerned (field of the order $b$, distance or wavelength of the order $r_0$).

It can be shown that the motion of an elementary charge, on which an external field is acting, satisfies an equation which is an obvious generalization of the classical equation of Lorentz. To find this equation we shall use here a cartesian co-ordinate system.

We assume that the strength of the external field in a region surrounding the electron is very small compared with the proper field of the point charge. We denote the proper field of the electron by

$$p_{kl}^{(0)}, f_{kl}^{(0)}, \qquad (8.9)$$

and the external field by

$$p_{kl}^{(e)} = f_{kl}^{(e)}; \qquad (8.10)$$

we do not take into consideration the sources of the external field. From the assumption, that

$$p_{kl}^{(0)} \gg f_{kl}^{(e)}; \quad f_{kl}^{(0)} \gg f_{kl}^{(e)} \qquad (8.11)$$

inside the sphere surrounding the electron, it follows evidently that the real solution of the field equations cannot be very different from that obtained by adding the unperturbed proper field and the external field. We construct therefore a sphere $S^{(0)}$ with its centre at the singularity of $\mathbf{H}$ and with a radius $r^{(0)}$, which is so small, that inside the sphere (8.11) is always satisfied. But the radius $r^{(0)}$ of the sphere has to be great compared with the radius of the electron, so that we can assume the validity of Maxwell's equations on the surface of the sphere just as outside the sphere.

We make the further assumption that the acceleration (curvature of the world line) is not too large, *i.e.*, one can choose the radius in such a way that the field $p_{kl}^{(0)}$ inside $S^{(0)}$ is essentially identical with that of the charge $e$ in uniform motion and can be derived from the formula of § 7 by a Lorentz transformation. Now we split the integral

$$\int \mathbf{H}\, d\pi \tag{8.12}$$

into a part corresponding to the sphere $S^{(0)}$ and the rest of space R. In $S^{(0)}$ we have

$$\left. \begin{aligned} \mathbf{H} &= \sqrt{1 - \tfrac{1}{2} p_{kl}\, p^{kl}} - 1 \\ &= \sqrt{1 - \tfrac{1}{2} p_{kl}^{(0)}\, p^{(0)kl} - p_{kl}^{(0)}\, f^{(e)kl} - \tfrac{1}{2} f_{kl}^{(e)}\, f^{(e)kl}} - 1 \\ Q &= 0 \end{aligned} \right\}. \tag{8.13}$$

Corresponding to (8.11) one can consider the terms $f_{kl}^{(0)}\, f^{(0)kl}$ as small of the first order (compared with $f_{kl}^{(0)}\, f^{(0)kl}$), the terms $f_{kl}^{(e)}\, f^{(e)kl}$ as small of the second order, and these latter will be neglected. Then we have by developing (8.13) and using (4.14):

$$\mathbf{H} = \sqrt{1 - \tfrac{1}{2} p_{kl}^{(0)}\, p^{(0)kl}} - 1 - \tfrac{1}{2} f_{kl}^{(0)}\, f^{(e)kl}, \tag{8.14}$$

which holds inside the sphere $S^{(0)}$. We can write (8.14) in another form:

$$\left. \begin{aligned} \mathbf{H} &= \mathbf{H}^{(0)} - \tfrac{1}{2} f_{kl}^{0}\, f^{(e)kl} - \tfrac{1}{4} f_{kl}^{(e)}\, f^{(e)kl} \\ \mathbf{H}^{(0)} &= \sqrt{1 - \tfrac{1}{2} p_{kl}^{(0)}\, p^{(0)kl}} - 1 \end{aligned} \right\} \tag{8.15}$$

(8.15) differs from (8.14) only in the terms of the second order. But (8.15) holds not only inside but also outside the sphere. For in R the equation (8.15) takes, according to our assumptions about $r^{(0)}$, the following form:

$$\begin{aligned} \mathbf{H} &= -\tfrac{1}{4} p_{kl}^{(0)}\, p^{(0)kl} - \tfrac{1}{2} f_{kl}^{(0)}\, f^{(e)kl} - \tfrac{1}{4} f_{kl}^{(e)}\, f^{(e)kl} \\ &= -\tfrac{1}{4} f_{kl}^{(0)}\, f^{(0)kl} - \tfrac{1}{2} f_{kl}^{(e)}\, f^{(0)kl} - \tfrac{1}{4} f_{kl}^{(e)}\, f^{(e)kl}. \end{aligned} \tag{8.16}$$

This is, however, the known expression for **H** in Maxwell's theory; (**L** = − **H**). Therefore (8.15) holds as well in the sphere $S^{(0)}$ as in R. One has

$$\int \mathbf{H}d\tau = \int \mathbf{H}^{(0)}d\tau - \frac{1}{2}\int f_{kl}^{(0)} f^{(e)kl}d\tau - \frac{1}{4}\int f_{kl}^{(e)} f^{(e)kl}d\tau.$$

(8.17)

We introduce the notation

$$4\pi\Lambda = \int \mathbf{H}^{(0)}dv - \frac{1}{2}\int f_{kl}^{(0)} f^{(e)kl}dv - \frac{1}{4}\int f_{kl}^{(e)} f^{(e)kl}dv,$$

(8.18)

and have for the action principle

$$\delta \int \Lambda dt = 0.$$

(8.19)

The integral

$$\int f_{kl}^{(e)} f^{(e)kl}d\tau$$

(8.20)

in (8.17) gives zero, because

$$\frac{\partial f^{(e)kl}}{\partial x^l} = 0; \quad \left(\rho_k^{(e)} = 0\right).$$

(8.21)

If we bear in mind that in the co-ordinate system, where the point charge is at rest, $\int \mathbf{H}^{(0)} dv$ is proportional to the mass, we have:

$$\int \mathbf{H}^{(0)}dv = m_0 c^2 \int \sqrt{1 - \mathbf{v}^2/c^2}dv,$$

(8.22)

where **v** is the velocity of the centre of the electron. In the second integral of (8.17) we have

$$f_{kl}^{(e)} = \frac{\partial \phi_l^{(e)}}{\partial x^k} - \frac{\partial \phi_l^{(e)}}{\partial x^l}$$

(8.23)

and by partial integration we find, using $\frac{\partial f^{(0)kl}}{\partial x^l} = 4\pi\rho^k$:

$$\frac{1}{2}\int f^{0kl} f_{kl}^{(e)} dv dt = -4\pi \int \phi_l^{(e)} \rho^l \, dv \, dt.$$

(8.24)

The additional surface integral over the infinitely large surface can be omitted, because it gives no contribution to the variation (8.19).

The result is:

$$\Lambda = m_0 c^2 \sqrt{1 - \mathbf{v}^2/c^2} - \int \phi_l^{(e)} \rho^l \, dv. \qquad (8.25)$$

We can write (8.25) in the space-vector form:

$$\Lambda = m_0 c^2 \sqrt{1 - \mathbf{v}^2/c^2} - \int \phi^{(e)} \rho \, dv + \frac{1}{c} \int \mathbf{A} \mathbf{I} \, dv. \qquad (8.25\text{A})$$

An electron behaves therefore like a mechanical system[*] with the rest mass $m_0$, acted on by the external field $f_{ke}^{(e)}$.[†]

If the external potential is essentially constant in a region surrounding the electron considered, the diameter of which is large compared with $r_0$, one gets instead of (8.25):

$$\int \Lambda \, dt = \int m_0 c^2 \sqrt{1 - \mathbf{v}^2/c^2} + e \left( \phi^{(e)} - \mathbf{v} \mathbf{A}^{(e)}/c \right), \qquad (8.26)$$

and this is entirely equivalent to Lorentz's equations of motion. But our formula (8.25) holds also for fields which are not constant. Any field can be split up in Fourier components or elementary waves; we may consider each of those separately, and choosing the Z-axis parallel to the propagation of the wave, we can assume that $\phi_s^{(e)}$ is proportional to $e^{2\pi i z/\lambda}$. Then we see that this Fourier component gives a contribution to the integral (8.25) of the form (8.26), where $\phi_s^{(e)}$ is now the amplitude of this component and $e$ has to be replaced by an "effective" charge $\bar{e}$, given by

$$\bar{e} = \int \rho \, e^{2\pi i z/\lambda} dv.$$

Using the expression of $\rho$ given by (7.12), and putting $z = r \cos \vartheta$,

$$dv = r^2 \sin \vartheta \, d\vartheta \, d\phi \, dr,$$

[*]Born, 'Ann. Physik,' vol. 28, p. 571 (1909); Pauli, "Relativitätstheorie," p. 642 (Teubner).
[†]The method used in I for deriving the equation of motion is not correct. It started from the action principle in the form

$$\delta \int \mathbf{L} \, d\tau = 0 \quad \left( \text{instead } \delta \int \mathbf{H} \, d\tau = 0 \right);$$

then in the development instead of the coefficients $f_{kl}^{(0)}$ the $p_{kl}^{(0)}$ appear, which become infinite at the centre of the electron. Therefore the transformation of the space integral is not allowed. In the first approximation we have

$$p_{kl} = p_{kl}^{(0)} + p_{kl}^{(e)}$$

and *not*

$$f_{kl} = f_{kl}^{(0)} + f_{kl}^{(e)}.$$

The mistake in the former derivation is also shown by the wrong result for the mass (the numerical factor was half of that given here).

one has

$$e = \frac{e}{r_0^3} \int_0^\infty \int_0^\pi \frac{r^2 dr}{\frac{r}{r_0}\left(1 + \frac{r^4}{r_0^4}\right)^{3/2}} e^{\frac{2\pi i r}{\lambda} \cos\vartheta} \sin\vartheta \; d\vartheta.$$

The $\vartheta$ integration can be performed, and one can write

$$\bar{e} = eg\left(\frac{2\pi r_0}{\lambda}\right); \quad g(x) = \frac{2}{x} \int_0^\infty \frac{\sin xy}{\left(1 + y^4\right)^{\frac{3}{2}}} dy.$$

For waves long compared with $r_0$ one has $\bar{e} = e$, because $g(0) = 1$. But for decreasing wave-lengths the effective charge diminishes, as the little table for $g(x)$ shows:—

Table of $g(x)$.*

| x. | g (x). | x. | g (x). |
|------|--------|------|--------|
| 0 | 1 | 1.25 | 0.796 |
| 0.1 | 0.988 | 1.50 | 0.730 |
| 0.2 | 0.984 | 1.75 | 0.659 |
| 0.3 | 0.968 | 2.00 | 0.588 |
| 0.4 | 0.959 | 2.25 | 0.526 |
| 0.5 | 0.949 | 2.50 | 0.457 |
| 0.6 | 0.929 | 3.00 | 0.347 |
| 0.7 | 0.917 | 3.50 | 0.252 |
| 0.8 | 0.901 | 4.00 | 0.186 |
| 0.9 | 0.880 | 5.00 | 0.094 |
| 1.0 | 0.856 | | |

The decrease begins to become remarkable where $x \sim 1$, or $\lambda \sim 2\pi r_0$. For large $x$ one has $g(x) \cong 2/x^2$.

If we introduce the quantum energy corresponding to the wave-length $\lambda$ by $E = hc/\lambda$, then using (8.6) one has

$$x = \frac{2\pi r_0}{\lambda} = 1.236\frac{2\pi}{hc}\frac{E}{m_0 c^2} = \frac{1.236}{137.1}\frac{E}{m_0 c^2} = \frac{1}{111}\frac{E}{m_0 c^2}.$$

$x = 1$ corresponds to a quantum energy of about $100 \ m_0 c^2 = 5 \cdot 10^7$ e. volt. For energies larger than this the interaction of electrons with other

*Calculated by Mr. Devonshire.

electrons (or light waves excited by those) should become smaller than that calculated by the accepted theories. This consequence seems to be confirmed by the astonishingly high penetrating power of the cosmic rays.[*]

## Summary.

The new field theory can be considered as a revival of the old idea of the electromagnetic origin of mass. The field equations can be derived from the postulate that there exists an "absolute field" $b$ which is the natural unit for all field components and the upper limit of a purely electric field. From the standpoint of relativity transformations the theory can be founded on the assumption that the field is represented by a non-symmetrical tensor $a_{kl}$, and that the Lagrangian is the square root of its determinant; the symmetrical part $g_{kl}$ of $a_{kl}$ represents the metric field, the antisymmetrical part $f_{kl}$ the electromagnetic field. The field equations have the form of Maxwell's equations for a polarizable medium for which the dielectric constant and the magnetic susceptibility are special functions of the field components. The conservation laws of energy and momentum can be derived. The static solution with spherical symmetry corresponds to an electron with finite energy (or mass); the true charge can be considered as concentrated in a point, but it is also possible to introduce a free charge with a spatial distribution law. The motion of the electron in an external field obeys a law of the Lorentz type where the force is the integral of the product of the field and the free charge density. From this follows a decrease of the force for alternating fields of short wavelengths (of the order of the electronic radius), in agreement with the observations of the penetrating power of high frequency (cosmic) rays.

[*]Born, 'Nature,' vol. 133, p. 63 (1934).

# ELECTRON THEORY

## BY

## J. ROBERT OPPENHEIMER

From *Rapports du 8 Conseil de Physique, Solvay*, p. 269 (1950)

In this report I shall try to give an account of the developments of the last year in electrodynamics. It will not be useful to give a complete presentation of the formalism; rather I shall try to pick out the essential logical points of the development, and raise at least some of the questions which may be open, and which bear on an evaluation of the scope of the recent developments, and their place in physical theory. I shall divide the report into three sections: (1) a brief summary of related past work in electrodynamics; (2) an account of the logical and procedural aspects of the recent developments; and (3) a series of remarks and questions on applications of these developments to nuclear problems and on the question of the closure of electrodynamics.

## 1. HISTORY

The problems with which we are concerned go back to the very beginnings of the quantum electrodynamics of Dirac, of Heisenberg and Pauli.[1] This theory, which strove to explore the consequences of complementarity for the electromagnetic field and its interactions with matter, led to great success in the understanding of emission, absorption and scattering processes, and led as well to a harmonious synthesis of the description of static fields and of light quantum phenomena. But it also led, as was almost at once recognized,[2] to paradoxical results, of which the infinite displacement of spectral terms and lines was an example. One recognized an analogy between these results and the infinite electromagnetic inertia of a point electron in classical theory, according to which electrons moving with different mean velocity should have energies infinitely displaced. Yet no attempt at a quantitative interpretation was made, nor was the question raised in a serious way of isolating from the infinite displacements new and typical finite parts clearly separable from the inertial effects. In fact such a program could hardly have been carried through before the discovery of pair production, and an understanding of the far-reaching differences in the actual problem of the singularities of quantum electrodynamics from the classical analogue of a point electron interacting with its field. In the former, the field and charge fluctuations of the vacuum—which clearly have no such classical counterpart—play a decisive part; whereas on the other hand the very phenomena of pair production, which so seriously limit the usefulness of a point model of the electron for distances small compared to its Compton wave length $h/mc$, in some measure ameliorate, though they do not resolve, the problems of the infinite electromagnetic inertia and of the instability of the electron's charge distribution. These last points first were made clear by the self-energy calculations of Weisskopf,[3] and were still further emphasized by the finding, by Pais,[4] and by Sakata,[5] that to the order $e^2$ (and to this limitation we shall have repeatedly to return) the electron's self-energy could be made finite, and indeed small, and its stability insured, by introducing forces of small magnitude and essentially *arbitrarily* small range, corresponding to a new field, and quanta of arbitrarily high rest mass.[6]

On the other hand the decisive, if classically unfamiliar, role of vacuum fluctuations was perhaps first shown—albeit in a highly academic situation—by Rosenfeld's calculation[7] of the (infinite) gravitational energy of the light quantum, and came prominently into view with the discovery of the problem of the self-energy of the photon due to the current fluctuations of the electron-positron field, and the related problems of the (infinite) polarizability of that field. Here for the first time the notion of renormalization was introduced. The infinite polarization of vacuum refers in fact just to situations in which a classical definition of charge should be possible (weak, slowly varying fields); if the polarization were finite, the linear constant term could not be measured directly, nor measured in any classically interpretable experiment; only the sum of "true" and induced charge could be measured. Thus it seemed natural to ignore the infinite linear constant polarizability of vacuum, but to attach significance to the finite deviations from this polarization in rapidly varying and in strong fields.[8] Direct attempts to measure these deviations were not successful; they are in any case intimately related to those which do describe the Lamb-Retherford level shift,[9] but are too small and of wrong sign to account for the bulk of this observation.[10] But the renormalization procedure and philosophy here applied to charge, was to prove, in its obvious extension to the electron's mass, the starting point for new developments.

In their application to level shifts, these developments, which could have been carried out at any time during the last fifteen years, required the impetus of experiment to stimulate and verify. Nevertheless, in other closely related problems, results were obtained essentially identical with those required to understand the Lamb-Retherford shift and the Schwinger corrections to the electron's gyromagnetic ratio.

Thus there is the problem—first studied by Bloch, Nordsieck,[11] Pauli and Fierz,[12] of the radiative corrections to the scattering of a slow electron (of velocity $v$) by a static potential $V$. The contribution of electromagnetic inertia is readily eliminated in non-relativistic calculations, and involves some subtlety in relativistic treatment only in the case of spin $1/2$ (rather than spin zero) charges.[13] It was even pointed out[14] that the new effects of radiation could be summarized by a small supplementary potential

I.
$$\sim \left(\frac{2}{3}\right)\left(\frac{e^2}{\hbar c}\right)\left(\frac{\hbar}{mc}\right)^2 \Delta V \ln\left(\frac{c}{v}\right)$$

(where $e, \hbar, m, c$ have their customary meaning). This of course gives the essential explanation of the Lamb shift.

On the other hand the anomalous $g$-value of the electron was foreshadowed by the remark,[15] that in meson theory, and even for neutral mesons, the coupling of nucleon spin and meson fluctuations would give to the sum of neutron and proton moments a value different from (and in non-relativistic estimates less than) the nuclear magneton.

Yet until the advent of reliable experiments on the electron's interaction, these points hardly attracted serious attention; and interest attached rather to exploring the possibilities of a consistent and reasonable modification of electrodynamics, which should preserve its agreement with experience, and yet, for high fields or short wave lengths, introduce such alterations as to make self-energies finite and the electron stable. In this it has proved decisive that it is *not* sufficient to develop a satisfactory classical analogue; rather one must cope directly with the specific quantum phenomena of fluctuation and pair production.[6] Within the framework of a continuum theory, with the point interactions of what Dirac[16] calls a "localizable" theory—no such satisfactory theory has been found; one may doubt whether, *within* this framework, such a theory can be formed that is expansible in powers of the electron's charge $e$. On the other hand, as mentioned earlier, many families of theories are possible which give satisfactory and consistent results to the order $e^2$.

A further general point which emerged from the study of electrodynamics is that—although the singularities occurring in solutions indicate that it is not a completed consistent theory, the structure of the theory itself gives no indication of a field strength, a maximum frequency of minimum length, beyond which it can no longer consistently be supposed to apply. This last remark holds in particular for the actual electron—for the theory of the Dirac electron-positron field coupled to the Maxwell field. For particles of lower and higher spin, some rough and necessarily ambiguous indications of limiting frequencies and fields do occur.

To these purely theoretical findings, there is a counterpart in experience. No credible evidence, despite much searching, indicates any departure, in the behaviour of electrons and gamma rays, from the expectations of theory. There are, it is true, the extremely weak couplings of $\beta$ decay; there are the weak electromagnetic interactions of gamma rays, and electrons, with the mesons and nuclear matter.

Yet none of these should give appreciable corrections to the present theory in its characteristic domains of application; they serve merely to suggest that for very small (nuclear) distances, and very high energies, electron theory and electrodynamics will no longer be so clearly separable from other atomic phenomena. In the theory of the electron and the electromagnetic field, we have to do with an almost closed, almost complete system, in which however we look precisely to the absence of complete closure to bring us away from the paradoxes that still inhere in it.

## 2. PROCEDURES

The problem then is to see to what extent one can isolate, recognize and postpone the consideration of those quantities, like the electron's mass and charge, for which the present theory gives infinite results— results which, if finite, could hardly be compared with experience in a world in which arbitrary values of the ratio $e^2/\hbar c$ cannot occur. What one can hope to compare with experience is the totality of other consequences of the coupling of charge and field, consequences of which we need to ask: does theory give for them results which are finite, unambiguous and in agreement with experiment?

Judged by these criteria the earliest methods must be characterized as encouraging but inadequate. They rested, as have to date *all* treatments not severely limited throughout by the neglect of relativity, recoil, and pair formation, on an expansion in powers of $e$, going characteristically to the order $e^2$. One carried out the calculation of the problem in question; (for radiative scattering corrections, Lewis[17]; for the Lamb shift, Lamb and Kroll,[18] Weisskopf and French,[19] Bethe[20]; for the electron's $g$-value, Luttinger[21]); one also calculated to the same order the electron's electromagnetic mass, its charge, and the charge induced by external fields, and the light quantum mass; finally one asked for the effect of these changes in charge and mass on the problem in question, and sought to delete the corresponding terms from the direct calculation. Such a procedure would no doubt be satisfactory—if cumbersome—were all quantities involved finite and unambiguous. In fact, since mass and charge corrections are in general represented by logarithmically divergent integrals, the above outlined procedure serves to obtain finite, but not necessarily unique or correct, reactive corrections for the behaviour of an electron in an external field; and a special tact is necessary, such as that implicit in Luttinger's derivation of the electron's anomalous gyromagnetic ratio, if results are to be, not merely plausible, but unambiguous and sound. Since, in more complex problems, and in calculations carried to higher order in $e$, this straightforward procedure becomes more and more ambiguous, and the results more dependent on the choice of Lorentz frame and of gauge, more powerful methods are required. Their development has occurred in two steps, the first largely, the second almost wholly, due to Schwinger.[22]

The first step is to introduce a change in representation, a contact transformation, which seeks, for a single electron not subject to external fields, and in the absence of light quanta, to describe the electron in terms of classically measurable charge $e$ and mass $m$, and eliminate entirely all "virtual" interaction with the fluctuations of electromagnetic and pair fields. In the non-relativistic limit, as was discussed in connection with Kramer's report,[23] and as is more fully described in Bethe's,[24] this transformation can be carried out rigorously to all powers of $e$, without expansion; in fact, the unitary transformation is given by

II.
$$U = \exp \frac{e}{mc} [\mathcal{Z} \cdot \nabla]$$

where $\mathcal{Z}$ is the (transverse) Hertz vector of the electromagnetic field minus the quasi-static field of the electron. When this formalism is applied to the problem of an electron in an external field, it yields reactive corrections which do not converge for frequencies $v > mc^2/\hbar$, thus indicating the need for a fuller consideration of typical relativistic effects.

This generalization is in fact straightforward; yet here it would appear essential that the power series expansion in $e$ is no longer avoidable, not only because no such simple solution as II now exists, but because, owing to the possibilities of pair creation and annihilation, and of interactions of light quanta with each other, the very definition of states of single electrons or single photons depends essentially on the expansion in question.[25] However that may be, the work has so far been carried out only by treating $e^2/\hbar c$ as small, and essentially only to include corrections of the first order in that quantity.

In this form, the contact transformation clearly yields:

(a) an infinite term in the electron's electromagnetic inertia;

(b) an ambiguous light quantum self-energy;

(c) no other effects for a single electron or photon;

(d) interactions of order $e^2$ between electrons, positrons, and photons, which in this order, correspond to the familiar Møller interactions and Compton effect and pair production probabilities;

(e) an infinite vacuum polarizability;

(f) the familiar frequency-dependent finite polarizability for external electromagnetic fields;

(g) emission and absorption probabilities equivalent to those of the Dirac theory for an electron in an external e.m. field;

(h) new reactive corrections of order $e^2$ to the effective charge and current distribution of an electron, which correspond to vanishing total supplementary charge, and to currents of the order $e^3/\hbar c$ distributed over dimensions of the order $\hbar/mc$, and which include the supplementary potential $I$, and the supplementary magnetic moment

$$\left(\frac{e^2}{2\pi\hbar c}\right)\left(\frac{e\hbar}{2mc}\right)(\vec{\sigma})$$

as special (non-relativistic) limiting cases.

Were such calculations to be carried further, to higher order in $e$, they would lead to still further renormalizations of charge and mass, to the successive elimination of all "virtual" interactions, and to reactive corrections, in the form of an expansion in powers of $e^2/\hbar c$, to the probabilities of transitions: pair production, collisions, scattering, etc. Nevertheless, before such a program could be undertaken, or the physically interesting new terms (h) above be taken as correct, a new development is required. The reason for this is the following: the results (h) are not in general independent of gauge and Lorentz frame. Historically this was first discovered by comparison of the supplementary magnetic interaction energy in a uniform magnetostatic field $H$

$$\left(\frac{e^2}{2\pi\hbar c}\right)\left(\frac{e\hbar}{2mc}\right)\left(\vec{\sigma}\cdot\vec{H}\right)$$

with the supplementary (imaginary) electric dipole interaction which appeared with an electron in a homogeneous electric field $E$ derived from a static scalar potential

$$\left(\frac{e^2}{2\pi\hbar c}\right)\left(\frac{e\hbar}{2mc}\right)i\rho_2\left(\vec{\sigma}\cdot\vec{E}\right)$$

a manifestly non-covariant result.

Now it is true that the fundamental equations of quantum-electrodynamics are gauge and Lorentz covariant. But they have in a strict sense no solutions expansible in powers of $e$. If one wishes to explore these solutions, bearing in mind that certain infinite terms will, in a later theory, no longer be infinite, one needs a covariant way of identifying these terms; and for that, not merely the field equations themselves, but the whole method of approximation and solution must at all stages preserve covariance. This means that the familiar Hamiltonian methods, which imply a fixed Lorentz frame $t =$ constant, must be renounced; neither Lorentz frame nor gauge can be specified until after, in a given order in $e$, all terms have been identified, and those bearing on the definition of charge and mass recognized and relegated; then of course, in the actual calculation of transition probabilities and the reactive corrections to them, or in the determination of stationary states in fields which can be treated as static, and in the reactive corrections thereto, the introduction of a definite coordinate system and gauge for these no longer singular and completely well-defined terms can lead to no difficulty.

It is probable that, at least to order $e^2$, more than one covariant formalism can be developed. Thus Stueckelberg's four-dimensional perturbation theory[26] would seem to offer a suitable starting point, as also do the related algorithms of Feynman.[27] But a method originally suggested by Tomonaga,[28] and independently developed and applied by Schwinger,[22] would seem, apart from its practicality, to have the advantage of very great generality and a complete conceptual consistency. It has been shown by Dyson[29] how Feynman's algorithms can be derived from the Tomonaga equations.

The easiest way to come to this is to start with the equations of motion of the coupled Dirac and Maxwell field. These are gauge and Lorentz covariants. The commutation laws, through which the typical quantum features are introduced, can readily be rewritten in covariant form to show: (1) at points outside the light cone from each other, all field quantities commute; and (2) the integral over an *arbitrary space-like* hypersurface yields a simple finite value for the commutator of a field variable at a variable point on the hypersurface, and that of another field variable at a fixed point on the hypersurface.

In this Heisenberg representation, the state vector is of course constant; commutators of field quantities separated by time-like intervals, depending on the solution of the coupled equation of motion, can not be known *a priori*; and no direct progress at either a rigorous or an approximate solution in powers of $e$ has been made.* But a simple change to a mixed representation, that introduced by Tomonaga and called by Schwinger the "interaction representation," makes it possible to carry out the covariant analogue of the power series contact transformation of the Hamiltonian theory.

The change of representation involved is a contact transformation to a system in which the state vector is no longer constant, but in which it would be constant if there were no coupling between the fields, i.e.,

---

*Author's note, 1956. Approximate solutions of the Heisenberg equations of motion were obtained by Yang and Feldman, *Phys. Rev.*, **79**, 972, 1950; and Källén, *Arkiv För Fysik*, **2**, 371, 1950.

if the elementary charge $e = 0$. The basis of this representation is the solution of the uncoupled field equations, which, together with their commutators at all relative positions, are of course well known. This transformation leads directly to the Tomonaga equation for the variation of the state vector $\Psi$:

III.
$$i h \frac{\delta \Psi}{\delta \sigma} = -\frac{1}{c} j^{\mu(P)} A_\mu^{(P)} \Psi$$

Here $\sigma$ is an arbitrary space-like surface through the point $P$. $\delta \Psi$ is the variation in $\Psi$ when a small variation is made in $\sigma$, localized near the point $P$; $\delta \sigma$ is the four-volume between varied and unvaried surfaces; $A_\mu^{(P)}$ is the operator of the four-vector electromagnetic potential at $P$; $j\mu^{(P)}$. is the (charge-symmetrized) operator of electron-positron four-vector current density at the same point.

It may be of interest, in judging the range of applicability of these methods, to note that in the theory of the charged particle of zero spin (the scalar and not Dirac pair field), the Tomonaga equation does not have the simple form III; the operator on $\Psi$ on the right involves explicitly an arbitrary time-like unit vector.[30]

Schwinger's program is then to eliminate the terms of order $e$, $e^2$, and so, in so far as possible, from the right-hand side of III. As before, only the "virtual" transitions can be eliminated by contact transformation; the real transitions of course remain, but with transition amplitudes eventually themselves modified by reactive corrections.

Apart from the obvious resulting covariance of mass and charge corrections, a new point appears for the light quantum self-energy, which now appears in the form of a product of a factor which must be zero on invariance grounds, and an infinite factor. As long as this term is identifiable, it must of course be zero in any gauge and Lorentz invariant formulation; in these calculations for the first time it is possible to make it zero. Yet even here, if one attempts to evaluate directly the product of zero factor and infinite integral, indeterminate, infinite, or even finite[31] values may result. A somewhat similar situation obtains in the problem, so much studied by Pais, of the direct evaluation of the stress in the electron's rest system, where a direct calculation yields the value $(-e^2/2\pi\hbar c)mc^2$, instead of the value zero which follows at once as the limit of the zero value holding uniformly, in this order $e^2$, for the theory rendered convergent by the $f$-quantum hypothesis, even for arbitrarily high $f$-quantum mass. These examples, far from casting doubt on the usefulness of the formalism, may just serve to emphasize the importance of identifying and evaluating such terms without any specialization of coordinate system, and utilizing throughout the covariance of the theory.

To order $e^2$, one again finds the terms (a) to (h) listed above; the covariance of the new reactive terms is now apparent; and they exhibit themselves again but more clearly as supplementary currents, corresponding to charge distribution of order $e^3/\hbar c$ (but vanishing total charge) extended throughout the interior of the light cones about the electron's position, and of spatial dimensions $\sim\hbar/mc$; inversely, they may also be interpreted as corrections of relative order $e^2/\hbar c$ and static range $\hbar/mc$ to the external fields. The supplementary currents immediately make possible simple treatments of the electron in external fields (where neither the electron's velocity, nor the derivatives of the fields need be treated small), and so give corrections for emission, absorption and scattering processes to the extent at least in which the fields may be classically described[32]; the reactive corrections to the Møller interaction and to pair production can probably not be derived without carrying the contact transformation to order $e^4$, since for these typical exchange effects, not included in the classical description of fields, must be expected to appear.

At the moment, to my best present knowledge, the reactive corrections agree with the $S$ level displacements of $H$ to about 1%, the present limit of experimental accuracy. For ionized helium, and for the correction to the electron's $g$-value, the agreement is again within experimental precision, which in this case, however, is not yet so high.

# 3. QUESTIONS

Even this brief summary of developments will lead us to ask a number of questions:

(1) Can the development be carried further, to higher powers of $e$, (a) with finite results, (b) with unique results, (c) with results in agreement with experiment?

(2) Can the procedure be freed of the expansion in $e$, and carried out rigorously?

(3) How general is the circumstance that the only quantities which are not, in this theory, finite, are those like the electromagnetic inertia of electrons, and the polarization effects of charge, which cannot directly be measured within the framework of the theory? Will this hold for charged particles of other spin?

(4) Can these methods be applied to the Yukawa-meson fields of nucleons? Does the resulting power series in the coupling constant converge at all? Do the corrections improve agreement with experience? Can one expect that when the coupling is large there is any valid content to the Maxwell-Yukawa analogy?

(5) In what sense, or to what extent, is electrodynamics—the theory of Dirac pairs and the e.m. field—"closed"?

There is very little experience to draw on for answering this battery of questions. So far there has not yet been a complete treatment of the electron problem in order higher than $e^2$, although preliminary study[33] indicates that here too the physically interesting corrections will be finite.

The experience in the meson fields is still very limited. With the pseudo-scalar theory, Case[34] has indeed shown that the magnetic moment of the neutron is finite (this has nothing to do with the present technical developments), and that the sum of neutron and proton moments, minus the nuclear magneton (which is the analogue of the electron's anomalous $g$-value) is of the same order as the neutron moment, finite, and in disagreement with experience. The proton-neutron mass difference is infinite and of the wrong sign; the reactive corrections to nuclear forces, formally analogous to the corrections to the Moller interaction, have not been evaluated. Despite these discouragements, it would seem premature to evaluate the prospects without further evidence.

Yet it is tempting to suppose that these new successes of electro-dynamics, which extend its range very considerably beyond what had earlier been believed possible, can themselves be traced to a rather simple general feature. As we have noted, both from the formal and from the physical side, electrodynamics is an almost closed subject; changes limited to very small distances, and having little effect even in the typical relativistic domain $E \sim mc^2$, could suffice to make a consistent theory; in fact, only weak and remote interactions appear to carry us out of the domain of electrodynamics, into that of the mesons, the nuclei, and the other elementary particles. Similar successes could perhaps be expected for those mesons (which may well also be described by Dirac-fields), which also show only weak non-electromagnetic interactions. But for mesons and nucleons generally, we are in a quite new world, where the special features of almost complete closure that characterizes electrodynamics are quite absent. That electrodynamics is also not quite closed is indicated, not alone by the fact that for finite $e^2/\hbar c$ the present theory is not after all self-consistent, but equally by the existence of those small interactions with other forms of matter to which we must in the end look for a clue, both for consistency, and for the actual value of the electron's charge.

I hope that even these speculations may suffice as a stimulus and an introduction to further discussion.

## REFERENCES

1. Heisenberg and Pauli, *Zeits. f. Physik.*, **56**, 1, 1929.
2. J. R. Oppenheimer, *Phys. Rev.*, **35**, 461, 1930.
3. V. Weisskopf, *Zeits. f. Physik.*, **90**, 817, 1934.
4. A. Pais, *Verhandelingen Roy. Ac., Amsterdam*, **19**, 1, 1946.
5. Sakata and Hara, *Progr. Theor, Phys.*, **2**, 30, 1947.

6. For a recent summary of the state of theory, see A. Pais, *Developments in the Theory of the Electron*, Princeton University Press, 1948.

7. L. Rosenfeld, *Zeits. f. Physik.*, **65**, 589, 1930.

8. General treatments: R. Serber, *Phys. Rev.*, **48**, 49, 1938, and V. Weisskopf, *Kgl. Dansk. Vidensk. Selskab. Math.-fys. Medd.*, **14**, 6, 1936.

9. Lamb and Retherford, *Phys. Rev.*, **72**, 241, 1947.

10. E. Uehling, *Phys. Rev.*, **48**, 55, 1935.

11. Bloch and Nordsieck, *Phys. Rev.*, **52**, 54, 1937.

12. Pauli and Fierz, *Il Nuovo Cimento*, **15**, 167, 1938.

13. S. Dancoff, *Phys. Rev.*, **55**, 959, 1939; H. Lewis, *Phys. Rev.*, **73**, 173, 1948.

14. Shelter Island Conference, June, 1947.

15. Fröhlich, Heitler and Kemmer, *Proc. Roy. Soc.*, A **166**, 154, 1938

16. P. Dirac, *Phys. Rev.*, **73**, 1092, 1948.

17. H. Lewis, *Phys. Rev.*, **73**, 173, 1948.

18. Lamb and Kroll, *Phys. Rev.*, in press.

19. Weisskopf and French, *Phys. Rev.*, in press.

20. H. Bethe, *Phys. Rev.*, **72**, 339, 1947.

21. P. Luttinger, *Phys. Rev.*, **74**, 893, 1948.

22. J. Schwinger, *Phys. Rev.*, **74**, 1439, 1948, and in press.

23. Report to the 8th Solvay Conference.

24. Report to the 8th Solvay Conference.

25. This may be seen very strikingly in writing down an explicit solution for the Tomonaga equation III below. Formally it is:

$$\Psi (\sigma) = \text{`` exp ''} \left[ \frac{i}{hc} \int_{\sigma_0}^{\sigma} j_\mu A^\mu d_4 x \right] \Psi (\sigma_0)$$

In order to define the "exp", we have at present no other resort than to approximate by a power series, where the ordering of the non-commuting factors for $j_\mu A\mu$ at different points of space-time can be simply prescribed (e.g., the later factor to the left). Cf. especially F. J. Dyson, *Phys. Rev.*, in press.

26. Stueckelberg, *An/n.der Phys.*, **21**, 367, 1934.

27. R. Feynman, *Phys. Rev.*, **74**, 1430, 1948.

28. S. Tomonaga, *Progr. Theor. Phys.*, **1**, 27, 109, 1946.

29. F. Dyson, *Phys. Rev.*, in press.

30. Kanesawa and Tomonaga, *Progr. Theor. Phys.*, **3**, 1, 107, 1948.

31. G. Wentzel, *Phys. Rev.*, **74**, 1070, 1948.

32. See for instance results reported to this conference by Pauli on corrections to the Compton effect for long wave lengths.
33. F. Dyson, *Phys. Rev.*, in press.
    *Author's note, 1956. Questions 1(a) and 1(b) were indeed answered by Dyson, *Phys. Rev.*, **75**, 1736, 1949.
34. K. Case, *Phys. Rev.*, **74**, 1884, 1948.

# Chapter Seven

In 1947 Willis Lamb and Robert Rutherford published a paper entitled "Fine Structure of the Hydrogen Atom by a Microwave Method" in which they described using microwave electromagnetic radiation to demonstrate a small shift in energy among two hydrogen atom states labeled $^2S_{1/2}$ and $^2P_{1/2}$. In Dirac's relativistic quantum theory, electrons in these two states ought to have identical energies. The fact that they do not is a clear indication that the theory is incomplete and demonstrates the need for quantum electrodynamics. Lamb and Rutherford showed that the $^2S_{1/2}$ level has an energy slightly higher than the $^2P_{1/2}$ level. The difference in energy corresponds to a photon frequency of about 1,058 Mhz or a wavelength of 28 cm and is now known as the Lamb shift.

Within days of Lamb and Rutherford's publication, Hans Bethe published an explanation. In the paper "The Electromagnetic Shift of Energy Levels" he showed that the Lamb shift was due to the electron bound in the hydrogen atom interacting with its own electromagnetic field. In today's terms we would say that the electron emits a photon and then quickly reabsorbs it. This has the net effect of slightly altering the electron's position, which perturbs the Coulomb force and slightly shifts the energy state. In the Dirac and Heisenberg-Pauli quantum electrodynamic theories, the calculation of the electron interacting with its own electric field is one of the terms that is infinite, as mentioned in the last chapter. Bethe suggested that it may be possible to remove the infinite result by considering that a free electron already has this infinity included in its measured rest mass. By subtracting the divergent free electron expression, from the divergent bound electron expression the infinity could be removed and a finite answer calculated. Bethe did this and arrived at a predicted value for the Lamb shift that closely matched the observed value. The idea of how to remove infinite energy from calculations has formed the basis of renormalized quantum theory, which we will be presented in the next chapter. Measuring the Lamb shift was the first experiment that required an explanation beyond Dirac's relativistic quantum theory, and it set the stage for the completion of quantum electrodynamics by Sin-Itiro Tomanaga, Julian Schwinger and Richard Feynman in the following years.

# FINE STRUCTURE OF THE HYDROGEN ATOM BY A MICROWAVE METHOD

By

WILLIS E. LAMB JR. AND ROBERT C. RUTHERFORD

THE spectrum of the simplest atom, hydrogen, has a fine structure[1] which according to the Dirac wave equation for an electron moving in a Coulomb field is due to the combined effects of relativistic variation of mass with velocity and spin-orbit coupling. It has been considered one of the great triumphs of Dirac's theory that it gave the "right" fine structure of the energy levels. However, the experimental attempts to obtain a really detailed confirmation through a study of the Balmer lines have been frustrated by the large Doppler effect of the lines in comparison to the small splitting of the lower or $n = 2$ states. The various spectroscopic workers have alternated between finding confirmation[2] of the theory and discrepancies[3] of as much as eight percent. More accurate information would clearly provide a delicate test of the form of the correct relativistic wave equation, as well as information on the possibility of line shifts due to coupling of the atom with the radiation field and clues to the nature of any non-Coulombic interaction between the elementary particles: electron and proton.

The calculated separation between the levels $2^2 P_{\frac{1}{2}}$ and $2^2 P_{3/2}$ is $0.365 \text{ cm}^{-1}$ and corresponds to a wave-length of 2.74 cm. The great

*Publication assisted by the Ernest Kempton Adams Fund for Physical Research of Columbia University, New York.
**Work supported by the Signal Corps under contract number W 36–039 sc-32003.
[1] For a convenient account, see H. E. White, *Introduction to Atomic Spectra* (McGraw-Hill Book Company, New York, 1934), Chap. 8.
[2] J. W. Drinkwater, O. Richardson, and W. E. Williams, Proc. Rov. Soc. **174**, 164 (1940).
[3] W. V. Houston, Phys. Rev. **51**, 446 (1937); R. C. Williams, Phys. Rev. **54**, 558 (1938); S. Pasternack, Phys. Rev. **54**, 1113 (1938) has analyzed these results in terms of an upward shift of the $S$ level by about 0.03 cm$^{-1}$.

wartime advances in microwave techniques in the vicinity of three centimeters wave-length make possible the use of new physical tools for a study of the $n = 2$ fine structure states of the hydrogen atom. A little consideration shows that it would be exceedingly difficult to detect the direct absorption of radiofrequency radiation by excited H atoms in a gas discharge because of their small population and the high background absorption due to electrons. Instead, we have found a method depending on a novel property of the $2^2 S_{\frac{1}{2}}$ level. According to the Dirac theory, this state exactly coincides in energy with the $2^2 P_{\frac{1}{2}}$ state which is the lower of the two $P$ states. The $S$ state in the absence of external electric fields is metastable. The radiative transition to the ground state $1^2 S_{\frac{1}{2}}$ is forbidden by the selection rule $\Delta L = \pm 1$. Calculations of Breit and Teller[4] have shown that the most probable decay mechanism is double quantum emission with a lifetime of 1/7 second. This is to be contrasted with a lifetime of only $1.6 \times 10^{-9}$ second for the nonmetastable $2^2 P$ states. The metastability is very much reduced in the presence of external electric fields[5] owing to Stark effect mixing of the $S$ and $P$ levels with resultant rapid decay of the combined state. If for any reason, the $2^2 S_{\frac{1}{2}}$ level does not exactly coincide with the $2^2 P_{\frac{1}{2}}$ level, the vulnerability of the state to external fields will be reduced. Such a removal of the accidental degeneracy may arise from any defect in the theory or may be brought about by the Zeeman splitting of the levels in an external magnetic field.

In brief, the experimental arrangement used is the following: Molecular hydrogen is thermally dissociated in a tungsten oven, and a jet of atoms emerges from a slit to be cross-bombarded by an electron stream. About one part in a hundred million of the atoms is thereby excited to the metastable $2^2 S_{\frac{1}{2}}$ state. The metastable atoms (with a small recoil deflection) move on out of the bombardment region and are detected by the process of electron ejection from a metal target. The electron current is measured with an FP-54 electrometer tube and a sensitive galvanometer.

[4] H. A. Bethe in *Handbuch der Physik*, Vol. 24/1, §43.
[5] G. Breit and E. Teller, Astrophys. J. 91, 215 (1940).

If the beam of metastable atoms is subjected to any perturbing fields which cause a transition to any of the $2^2P$ states, the atoms will decay while moving through a very small distance. As a result, the beam current will decrease, since the detector does not respond to atoms in the ground state. Such a transition may be induced by the application to the beam of a static electric field somewhere between source and detector. Transitions may also be induced by radiofrequency radiation for which $h\nu$ corresponds to the energy difference between one of the Zeeman components of $2^2S_{\frac{1}{2}}$ and any component of either $2^2P_{\frac{1}{2}}$ or $2^2P_{3/2}$. Such measurements provide a precise method for the location of the $2^2S_{\frac{1}{2}}$ state relative to the $P$ states, as well as the distance between the latter states.

We have observed an electrometer current of the order of $10^{-14}$ ampere which must be ascribed to metastable hydrogen atoms. The strong quenching effect of static electric fields has been observed, and the voltage gradient necessary for this has a reasonable dependence on magnetic field strength.

We have also observed the decrease in the beam of metastable atoms caused by microwaves in the wave-length range 2.4 to 18.5 cm in various magnetic fields. In the measurements, the frequency of the r-f is fixed, and the change in the galvanometer current due to interruption of the r-f is determined as a function of magnetic field strength. A typical curve of quenching *versus* magnetic field is shown in Fig. 1. We have plotted in Fig. 2 the resonance magnetic fields for various frequencies in the vicinity of 10,000 Mc/sec. The theoretically calculated curves for the Zeeman effect are drawn as solid curves, while for comparison with the observed points, the calculated curves have been shifted downward by 1000 Mc/sec. (broken curves). The results indicate clearly that, contrary to theory but in essential agreement with Pasternack's hypothesis,[3] the $2^2S_{\frac{1}{2}}$ state is higher than the $2^2P_{\frac{1}{2}}$ by about 1000 Mc/sec. (0.033 cm$^{-1}$ or about 9 percent of the spin relativity doublet separation. The lower frequency transitions $^2S_{\frac{1}{2}}$ $(m=\frac{1}{2}) \rightarrow$ $^2P_{\frac{1}{2}}$ $(m=\pm\frac{1}{2})$ have also been observed and agree well with such a shift of the $^2S_{\frac{1}{2}}$ level. With the present precision,

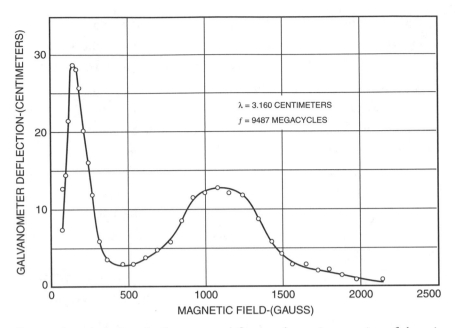

FIG. 1 A typical plot of galvanometer deflection due to interruption of the microwave radiation as a function of magnetic field. The magnetic field was calibrated with a flip coil and may be subject to some error which can be largely eliminated in a more refined apparatus. The width of the curves is probably due to the following causes: (1) the radiative line width of about 100 Mc/sec. of the $^2P$ states, (2) hyperfine splitting of the $^2S$ state which amounts to about 88 Mc/sec., (3) the use of an excessive intensity of radiation which gives increased absorption in the wings of the lines, and (4) inhomogeneity of the magnetic field. No transitions from the state $2^2S_{\frac{1}{2}}$ ($m = -\frac{1}{2}$) have been observed, but atoms in this state may be quenched by stray electric fields because of the more nearly exact degeneracy with the Zeeman pattern of the $^2P$ states.

we have not yet detected any discrepancy between the Dirac theory and the doublet separation of the $P$ levels. (According to most of the imaginable theoretical explanations of the shift, the doublet separation would not be affected as much as the relative location of the $S$ and $P$ states.) With proposed refinements in sensitivity, magnetic field homogeneity, and calibration, it is hoped to locate the $S$ level with respect to each $P$ level to an accuracy of at least ten Mc/sec. By addition of these frequencies and assumption of the theoretical formula $\Delta v = \frac{1}{16}\alpha^2 R$ for the doublet separation, it should be possible to measure

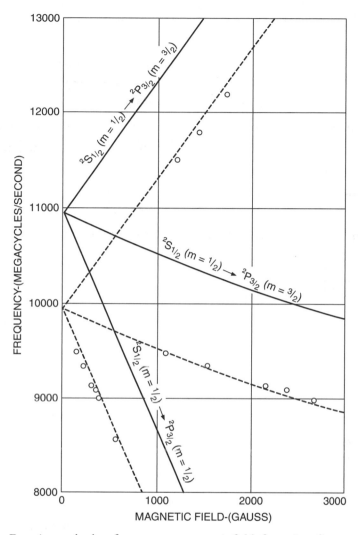

FIG. 2 Experimental values for resonance magnetic fields for various frequencies are shown by circles. The solid curves show three of the theoretically expected variations, and the broken curves are obtained by shifting these down by 1000 Mc/sec. This is done merely for the sake of comparison, and it is not implied that this would represent a "best fit." The plot covers only a small range of the frequency and magnetic field scale covered by our data, but a complete plot would not show up clearly on a small scale, and the shift indicated by the remainder of the data is quite compatible with a shift of 1000 Mc.

the square of the fine structure constant times the Rydberg frequency to an accuracy of 0.1 percent.

By a slight extension of the method, it is hoped to determine the hyperfine structure of the $2\,^2S_{\frac{1}{2}}$ state. All of these measurements will be repeated for deuterium and other hydrogen-like atoms.

A paper giving a fuller account of the experimental and theoretical details of the method is being prepared, and this will contain later and more accurate data.

The experiments described here were discussed at the Conference on the Foundations of Quantum Mechanics held at Shelter Island on June 1–3, 1947 which was sponsored by the National Academy of Sciences.

# The Electromagnetic Shift
# of Energy Levels

By

Hans. A. Bethe

B Y very beautiful experiments, Lamb and Retherford[1] have shown that the fine structure of the second quantum state of hydrogen does not agree with the prediction of the Dirac theory. The $2s$ level, which according to Dirac's theory should coincide with the $2p_{\frac{1}{2}}$ level, is actually higher than the latter by an amount of about 0.033 cm$^{-1}$ or 1000 megacycles. This discrepancy had long been suspected from spectroscopic measurements.[2,3] However, so far no satisfactory theoretical explanation has been given. Kemble and Present, and Pasternack[4] have shown that the shift of the $2s$ level cannot be explained by a nuclear interaction of reasonable magnitude, and Uehling[5] has investigated the effect of the "polarization of the vacuum" in the Dirac hole theory, and has found that this effect also is much too small and has, in addition, the wrong sign.

Schwinger and Weisskopf, and Oppenheimer have suggested that a possible explanation might be the shift of energy levels by the interaction of the electron with the radiation field. This shift comes out infinite in all existing theories, and has therefore always been ignored. However, it is possible to identify the most strongly (linearly) divergent term in the level shift with an electromagnetic *mass* effect which must exist for a bound as well as for a free electron. This effect should properly be regarded as already included in the observed mass

---

[1] Phys. Rev. **72**, 241 (1947).
[2] W. V. Houston, Phys. Rev. **51**, 446 (1937).
[3] R. C. Williams, Phys. Rev. **54**, 558 (1938).
[4] E. C. Kemble and R. D. Present, Phys. Rev. **44**, 1031 (1932); S. Pasternack, Phys. Rev. **54**, 1113 (1938).
[5] E. A. Uehling, Phys. Rev. **48**, 55 (1935).

---

Reprinted with permission from the American Physical Society: Betje, *Physical Review*, Volume 72, p. 339, 1947.

of the electron, and we must therefore subtract from the theoretical expression, the corresponding expression for a free electron of the same average kinetic energy. The result then diverges only logarithmically (instead of linearly) in non-relativistic theory: Accordingly, it may be expected that in the hole theory, in which the *main* term (self-energy of the electron) diverges only logarithmically, the result will be *convergent* after subtraction of the free electron expression.[6] This would set an effective upper limit of the order of $mc^2$ to the frequencies of light which effectively contribute to the shift of the level of a bound electron. I have not carried out the relativistic calculations, but I shall assume that such an effective relativistic limit exists.

The ordinary radiation theory gives the following result for the self-energy of an electron in a quantum state $m$, due to its interaction with transverse electromagnetic waves:

$$W = -(2e^2/3\pi hc^3) \times \int_0^K kdk \sum_n |v_{mn}^2|(E_n - E_m + k), \quad (1)$$

where $k = kw$ is the energy of the quantum and v is the velocity of the electron which, in non-relativistic theory, is given by

$$\mathbf{v} = \mathbf{p}/m = (h/im)\nabla. \quad (2)$$

Relativistically, $\mathbf{v}$ should be replaced by $c\boldsymbol{\alpha}$ where $\boldsymbol{\alpha}$ is the Dirac operator. Retardation has been neglected and can actually be shown to make no substantial difference. The sum in (1) goes over all atomic states $n$, the integral over all quantum energies $k$ up to some maximum $K$ to be discussed later.

For a free electron, $\mathbf{v}$ has only diagonal elements and (1) is replaced by

$$W_0 = -(2e^2/3\pi hc^3)\int kdk\mathbf{v}^2/k. \quad (3)$$

This expression represents the change of the kinetic energy of the electron for fixed momentum, due to the fact that electromagnetic mass is added to the mass of the electron. This electromagnetic mass is

---

[6] It was first suggested by Schwinger and Weisskopf that hole theory must be used to obtain convergence in this problem.

already contained in the experimental electron mass; the contribution (3) to the energy should therefore be disregarded. For a bound electron, $\mathbf{v}^2$ should be replaced by its expectation value, $(\mathbf{v}^2)_{mm}$. But the matrix elements of $\mathbf{v}$ satisfy the sum rule

$$\sum_n |v_{mn}|^2 = (\mathbf{v}^2)_{mm}. \tag{4}$$

Therefore the relevant part of the self-energy becomes

$$W' = W - W_0 = +\frac{2e^2}{3\pi\,hc^3} \times \int_0^K dk \sum_n \frac{|v_{mn}|^2(E_n - E_m)}{E_n - E_m + k}. \tag{5}$$

This we shall consider as a true shift of the levels due to radiation interaction.

It is convenient to integrate (5) first over $k$. Assuming $K$ to be large compared with all energy differences $E_n - E_m$ in the atom,

$$W' = \frac{2e^2}{3\pi\,hc^3} \sum_n |v_{mn}|^2(E_n - E_m) \ln\frac{K}{|E_n - E_m|}. \tag{6}$$

(If $E_n - E_m$ is negative, it is easily seen that the principal value of the integral must be taken, as was done in (6).) Since we expect that relativity theory will provide a natural cut-off for the frequency $k$, we shall assume that in (6)

$$K \approx mc^2. \tag{7}$$

(This does not imply the same limit in Eqs. (2) and (3).) The argument in the logarithm in (6) is therefore very large; accordingly, it seems permissible to consider the logarithm as constant (independent of $n$) in first approximation.

We therefore should calculate

$$A = \sum_n A_{nm} = \sum_n |p_{nm}|^2(E_n - E_m). \tag{8}$$

This sum is well known; it is

$$A = \sum |p_{nm}|^2(E_n - E_m) = -h^2 \int \psi_m * \nabla V \cdot \nabla \psi_m d\tau$$

$$= \frac{1}{2}h^2 \int \nabla^2 V \psi_m{}^2 d\tau = 2\pi h^2 e^2 Z \psi_m{}^2(0), \tag{9}$$

for a nuclear charge $Z$. For any electron with angular momentum $l \neq 0$, the wave function vanishes at the nucleus; therefore, the sum $A = 0$. For example, for the $2p$ level the negative contribution $A_{1s2,p}$ balances the positive contributions from all other transitions. For a state with $l = 0$, however,

$$\psi_m^2(0) = (Z/na)^3/\pi, \tag{10}$$

where $n$ is the principal quantum number and $a$ is the Bohr radius.

Inserting (10) and (9) into (6) and using relations between atomic constants, we get for an $S$ state

$$W'_{ns} = \frac{8}{3\pi}\left(\frac{e^2}{hc}\right)^3 \mathrm{Ry}\frac{Z^4}{n^3}\ln\frac{K}{\langle E_n - E_m\rangle_{\mathrm{Av}}}, \tag{11}$$

where Ry is the ionization energy of the ground state of hydrogen. The shift for the $2p$ state is negligible; the logarithm in (11) is replaced by a value of about $-0.04$. The average excitation energy $(E_n - E_m)_{\mathrm{AV}}$ for the $2s$ state of hydrogen has been calculated numerically[7] and found to be 17.8 Ry, an amazingly high value. Using this figure and $K = \mathrm{mc}^2$, the logarithm has the value 7.63, and we find

$$W'_{ne} = 136\,\ln[K/(E_n - E_m)] = 1040\,\text{megacycles}. \tag{12}$$

This is in excellent agreement with the observed value of 1000 megacycles.

A relativistic calculation to establish the limit $K$ is in progress. Even without exact knowledge of $K$, however, the agreement is sufficiently good to give confidence in the basic theory. This shows

(1) that the level shift due to interaction with radiation is a real effect and is of finite magnitude,

(2) that the effect of the infinite electromagnetic mass of a point electron can be eliminated by proper identification of terms in the Dirac radiation theory,

(3) that an accurate experimental and theoretical investigation of the level shift may establish relativistic effects (e.g., Dirac hole

---

[7] I am indebted to Dr. Stehn and Miss Steward for the numerical calculations.

theory). These effects will be of the order of unity in comparison with the logarithm in Eq. (11).

If the present theory is correct, the level shift should increase roughly as $Z^4$ but not quite so rapidly, because of the variation of $(E_n - E_m)_{AV}$ in the logarithm. For example, for $He^+$, the shift of the $2s$ level should be about 13 times its value for hydrogen, giving 0.43 cm$^{-1}$, and that of the $3s$ level about 0.13 cm$^{-1}$. For the x-ray levels $LI$ and $LII$, this effect should be superposed upon the effect of screening which it partly compensates. An accurate theoretical calculation of the screening is being undertaken to establish this point.

This paper grew out of extensive discussions at the Theoretical Physics Conference on Shelter Island, June 2 to 4, 1947. The author wishes to express his appreciation to the National Academy of Science which sponsored this stimulating conference.

# Chapter Eight

A s we have seen, the problem of infinite energy plagued the development of quantum electrodynamics for much of the 1930s and 40s. In 1947, Hans Bethe figured out a method for removing the infinity from the electron self-energy calculation. His calculation was non-relativistic, but set the stage for the relativistic completion of quantum electrodynamics done independently by Sin-Itiro Tomanaga, Julian Schwinger, and Richard Feynman. The method for removing infinite energy from calculations in quantum electrodynamics is now called *renormalization*. The basic idea of renormalization is to impose a cutoff in the energy and carefully take the limit as the cutoff approaches infinity. A cutoff in energy amounts to imposing a length scale below which variations in the field are ignored. A higher cutoff energy corresponds to a smaller length scale. With a finite length scale the mathematical terms that produced infinite energy become regularized and finite. Most importantly they cancel other terms that also produce infinite energies. With these troubling terms safely canceling each other, the limit that the length scale approaches zero can be taken in such a way that the calculation remains finite.

Schwinger's and Tomonaga's approaches to quantum electrodynamics were similar and are now grouped together. Feynman's, on the other hand, was completely different. Schwinger's approach was more straightforwardly based on previous work, but involved mathematics so complicated that it obscured the basic physical processes. In fact Freeman Dyson once said that it was "something that needed such skills that nobody besides Schwinger could do it."[1] Feynman's approach was more intuitive. He constructed pictorial representations, known as Feynman diagrams, of the process in question and a set of rules describing how to interpret the diagrams. Using these rules it is

---

[1] Jagdish Mehra and Helmut Rechenberg (2001) *The Historical Development of Quantum Theory* Springer pp. 1099

possible to set up mathematical expressions that when evaluated give contributions to the probability for a particular transition to occur. In 1948 Dyson published a proof showing the equivalency of the Tomonaga, Schwinger and Feynman methods. Incidentally, because Feynman was slow to publish his work, Dyson's proof appeared before Feynman had even formally published his method!

In developing his method for solving quantum electrodynamics, Feynman, with the aid of John Wheeler, gave a new interpretation to the positron. They claimed that a positron is an electron moving backward through time. It had been known for some time that when matter and anti-matter interact the result is a total annihilation of both and the production of high-energy photons. In Feynman's method, the annihilation of a positron-electron pair can be understood as a single electron interacting with an electromagnetic field in the form two photons and then reversing direction in time — ie. becoming the incident positron. Although an electron moving backward in time seems strange, it is in fact completely equivalent to a positron moving forward in time.

The renormalization method is not without its detractors. It relies on mathematics that at the time seemed murky. Paul Dirac said about renormalization, "This is just not sensible mathematics. Sensible mathematics involves neglecting a quantity when it is small—not neglecting it just because it is infinitely great and you do not want it!"[2] Nevertheless, renormalized quantum electrodynamics is one of the most successful theories ever produced. Its predictions match observations to an astounding degree of accuracy.

---

[2] Helge Kragh (1990). *Dirac: A Scientific Biography*, Cambridge University Press, pp. 184

# On a Relativistically Invariant Formulation of the Quantum Theory of Wave Fields

By

Sin-Itiro Tomanaga

From *Progress of Theoretical Physics*, Vol. I, pg. 27 (1946)

## 1. THE FORMALISM OF THE ORDINARY QUANTUM THEORY OF WAVE FIELDS

Recently Yukawa[1] has made a comprehensive consideration about the basis of the quantum theory of wave fields. In his article he has pointed out the fact that the existing formalism of the quantum field theory is not yet perfectly relativistic.

Let $v(xyz)$ be the quantity specifying the field, and $\lambda(xyz)$ denote its canonical conjugate. Then the quantum theory requires commutation relations of the form:

$$\begin{cases} \left[v(xyzt), v(x'y'z't)\right] = \left[\lambda(xyzt), \lambda(x'y'z't)\right] = 0 \\ \left[v(xyzt), \lambda(x'y'z't)\right] = i\hbar\delta(x - x')\delta(y - y')\delta(z - z'), \end{cases} \tag{1}†$$

but these have quite non-relativistic forms.

The equations (1) give the commutation relations between the quantities at different points $(xyz)$ and $(x'y'z')$ at the same instant of time $t$. The concept "same instant of time at different points" has, however, a definite meaning only if one specifies some definite Lorentz frame of reference. Thus this is not a relativistically invariant concept.

---

*Translated from the paper, *Bull. I. P. C. R. (Riken-iho)*, **22** (1943), 545, appeared originally in Japanese.

† $[A, B] = AB - BA$. We assume that the field obeys the Bose statistics. Our considerations apply also to the case of Fermi statistics.

Further, the Schrödinger equation for the $\psi$-vector representing the state of the system has the form:

$$\left(\bar{H} + \frac{\hbar}{i}\frac{\partial}{\partial t}\right)\psi = 0, \tag{2}$$

where $\bar{H}$ is the operator representing the total energy of the field which is given by the space integral of a function of $v$ and $\lambda$. As we adopt here the Schrödinger picture, $v$ and $\lambda$ are operators independent of time. The vector representing the state is in this picture a function of the time, and its dependence on $t$ is determined by (2).

Also the differential equation (2) is no less non-relativistic. In this equation the time variable $t$ plays a role quite distinct from the space coordinates $x$, $y$ and $z$. This situation is closely connected with the fact that the notion of probability amplitude does not fit with the relativity theory.

As is well known, the vector $\psi$ has, as the probability amplitude, the following physical meaning: Suppose the representation which makes the field quantity $v(xyz)$ diagonal. Let $\psi[v'(xyz)]$ denote the representative of $\psi$ in this representation.* Then the representative $\psi[v'(xyz)]$ is called probability amplitude, and its absolute square

$$W[v'(xyz)] = |\psi[v'(xyz)]|^2 \tag{3}$$

gives the relative probability of $v(xyz)$ having the specified functional form $v'(xyz)$ at the instant of time $t$. In other words: Suppose a plane† which is parallel to the $xyz$-plane and intercepts the time axis at $t$. Then the probability that the field has the specified functional form $v'(xyz)$ on this plane is given by (3).

As one sees, a plane parallel to the $xyz$-plane plays here a significant role. But such a plane is defined only by referring to a certain frame of reference. Thus the probability amplitude is not a relativistically invariant concept in the space-time world.

---

*We use the square brackets to indicate a functional. Thus $\psi[v'(xyz)]$ means that $\psi$ is a functional of the variable function $v'(xyz)$. When we use ordinary parentheses ( ), as $\psi(v'(xyz))$, we consider $\psi$ as an ordinary function of the function $v'(xyz)$. For example: the energy density is written as $H(v(xyz), \lambda(xyz))$ and this is also a function of $x$, $y$ and $z$, whereas the total energy $H = \int H(v(xyz), \lambda(xyz))dv$ is a functional of $v(xyz)$ and $\lambda(xyz)$ and is written as $\bar{H}[v(xyz), \lambda(xyz)]$.
†We call a three-dimensional manifold in the four-dimensional space-time world simply a "surface."

# 2. FOUR-DIMENSIONAL FORM OF THE COMMUTATION RELATIONS

As stated above, the laws of the quantum theory of wave fields are usually expressed as mathematical relations between quantities having their meanings only in some specified Lorentz frame of reference. But since it is proved that the whole contents of the theory are of course relativistically invariant, it must be certainly possible to build up the theory on the basis of concepts having relativistic space-time meanings. Thus, in his consideration, Yukawa has required with Dirac[2] a generalization of the notion of probability amplitude to fit with the relativity theory. We shall now show below that the generalization of the theory on these lines is in fact possible to the relativistically necessary and sufficient extent. Our results are, however, not so general as expected by Dirac and by Yukawa, but are already sufficiently general in so far as it is required by the relativity theory.

Let us suppose for simplicity that there are only two fields interacting with each other. The case of a greater number of fields can also be treated in the same way. Let $v_1$ and $v_2$ denote the quantities specifying the fields. The canonically conjugate quantities are $\lambda_1$ and $\lambda_2$ respectively. Then between these quantities the commutation relations

$$\begin{cases} \left[v_r(xyzt),\ v_s(x'y'z't)\right] = 0 \\ \left[\lambda_r(xyzt),\ \lambda_s(x'y'z't)\right] = 0 \qquad\qquad r, s = 1, 2 \\ \left[v_r(xyzt),\ \lambda_s(x'y'z't)\right] = i\hbar\delta(x - x')\delta(y - y')\delta(z - z')\delta_{rs} \end{cases}$$

$$(4)$$

must hold. The $\psi$-vector satisfies the Schrödinger equation

$$\left(\bar{H}_1 + \bar{H}_2 + \bar{H}_{12} + \frac{\hbar}{i}\frac{\partial}{\partial t}\right)\psi = 0. \qquad (5)$$

In this equation $\bar{H}_1$ and $\bar{H}_2$ mean respectively the energy of the first and the second field. $\bar{H}_1$ is given by the space integral of a function of $v_1$ and $\lambda_1$, $\bar{H}_2$ by the space integral of a function of $v_2$ and $\lambda_2$. Further, $\bar{H}_{12}$ is the interaction energy of the fields and is given by the space integral of a function of both $v_1$, $\lambda_1$ and $v_2$, $\lambda_2$. We assume (i) that the integrand of $\bar{H}_{12}$, i.e. the interaction-energy density, is a scalar quantity, and (ii) that the energy densities at two different points (but at the same instant of time) commute with each other. In general, these two facts follow from the single assumption: the interaction term in the Lagrangian does not contain the time derivatives of $v_1$ and $v_2$.

If this energy density is denoted by $H_{12}$, then we have

$$\bar{H}_{12} = \int H_{12}\, dx\, dy\, dz. \tag{6}$$

As we adopt here the Schrödinger picture, the quantities $v$ and $\lambda$ in $H_1$, $H_2$ and $H_{12}$ are all operators independent of time.

Thus far we have merely summarized the well-known facts. Now, as the first stage of making the theory relativistic, we suppose the unitary operator

$$U = \exp\left\{ \frac{i}{\hbar}\left( \bar{H}_1 + \bar{H}_2 \right) t \right\} \tag{7}$$

and introduce the following unitary transformations of $v$ and $\lambda$, and the corresponding transformation of $\psi$:

$$\begin{cases} V_r = U_{v,}\, U^{-1}, & \Lambda_r = U\lambda_r U^{-1} \\ \qquad \Psi = U\psi. \end{cases} \quad r = 1, 2 \tag{8}$$

As stated above, $v$ and $\lambda$ in (5) are quantities independent of time. But $V$ and $\Lambda$ obtained from them by means of (8) contain $t$ through $U$. Thus they depend on $t$ by

$$\begin{cases} i\hbar\dot{V}_r = V_r\bar{H}_r - \bar{H}_r V_r \\ i\hbar\dot{\Lambda}_r = \Lambda_r\bar{H}_r - \bar{H}_r\Lambda_r. \end{cases} \quad r = 1, 2 \tag{9}$$

These equations must necessarily have covariant forms against Lorentz transformations, because they are just the field equations for the fields when they are left alone without interacting with each other.

Now, the solutions of these "vacuum equations," the equations which the fields must satisfy, when they are left alone, together with the commutation relations (4), give rise to the relations of the following forms:

$$\begin{cases} \left[V_r(xyzt), V_s(x'y'z't')\right] = A_{rs}(x - x', y - y', z - z', t - t') \\ \left[\Lambda_r(xyzt), \Lambda_s(x'y'z't')\right] = B_{rs}(x - x', y - y', z - z', t - t') \\ \left[V_r(xyzt), \Lambda_s(x'y'z't')\right] = C_{rs}(x - x', y - y', z - z', t - t') \end{cases}$$

(10)

where $A_{rs}$, $B_{rs}$ and $C_{rs}$ are functions which are combinations of the so-called four-dimensional $\delta$-functions and their derivatives.[3] One denotes usually these four-dimensional $\delta$-functions by $D_r(xyzt)$, $r = 1, 2$. They are defined by

$$D_r(xyzt) = \frac{1}{16\pi^3} \int \int \int$$

$$\times \left\{ \frac{e^{i(k_x x + k_y y + k_z z + c k_r t)}}{i k_r} - \frac{e^{i(k_x x + k_y y + k_z z - c k_r t)}}{i k_r} \right\} dk_x \, dk_y \, dk_z$$

(11)

with

$$k_r = \sqrt{k_x^2 + k_y^2 + k_z^2 + \chi_r^2},$$

(12)

$\chi_r$ being the constant characteristic to the field $r$. It can be easily proved that these functions are relativistically invariant.*

Since (10) gives, in contrast with (4), the commutation relations between the fields at two different world points $(xyzt)$ and $(x'y'z't)$, it contains no more the notion of same instant of time. Therefore, (10) is sufficiently relativistic presupposing no special frame of reference. We call (10) four-dimensional form of the commutation relations.

One property of $D(xyzt)$ will be mentioned here: When the world point $(xyzt)$ lies outside the light cone whose vertex is at the origin, then $D(xyzt)$ vanishes identically:

$$D(xyzt) = 0 \quad \text{for} \quad x^2 + y^2 + z^2 - c^2 t^2 > 0.$$

(13)

---

*Suppose that a surface in the $k_x \, k_y \, k_z$ $k$-space is defined by means of the equation $k^2 = k_x^2 + k_y^2 + k_z^2 + \chi^2$. Then this surface has the invariant meaning in this space, since $k_x^2 + k_y^2 + k_z^2 - k^2$ is invariant against Lorentz transformations. The area of the surface element of this surface is given by

$$dS = \sqrt{\left(\frac{\partial k}{\partial k_x}\right)^2 + \left(\frac{\partial k}{\partial k_y}\right)^2 + \left(\frac{\partial k}{\partial k_z}\right)^2 - 1} \, dk_x \, dk_y \, dk_z = x \frac{dk_x \, dk_y \, dk_z}{k}.$$

Now, since $dS$ has the invariant meaning, we can thus conclude that $\frac{dk_x dk_y dk_z}{k}$ is invariant, and this implies that the function defined by (11) is invariant.

It follows directly from (13) that, if the world point $(x'y'z't')$ lies outside the light cone whose vertex is at the world point $(xyzt)$, the right-hand sides of (10) always vanish. In words: Suppose two world points $P$ and $P^r$. When these points lie outside each other's light cones, the field quantities at $P$ and field quantities at $P'$ commute with each other.

## 3. GENERALIZATION OF THE SCHRÖDINGER EQUATION

Next we observe the vector $\Psi$ obtained from $\psi$ by means of the unitary transformation $U$. We see from (5), (7) and (8) that this $\Psi$, considered as a function of $t$, satisfies

$$\left\{ \iint H_{12}(V_1(xyzt), \Lambda_1(xyzt), \right.$$

$$\left. V_2(xyzt), \Lambda_2(xyzt))\, dx\, dy\, dz + \frac{\hbar}{i}\frac{\partial}{\partial t} \right\} \Psi = 0. \quad (14)$$

One sees that $t$ plays also here a role distinct from $x$, $y$ and $z$: also here a plane parallel to the $xyz$-plane has a special significance. So we must in some way remove this unsatisfactory feature of the theory.

This improvement can be attained in the way similar to that in which Dirac[4] has built up the so-called many-time formalism of the quantum mechanics. We will now recall this theory.

The Schrödinger equation for the system containing $\mathcal{N}$ charged particles interacting with the electromagnetic field is given by

$$\left\{ \bar{H}_{el} + \sum_{n=1}^{\mathcal{N}} H_n(q_n, p_n, \mathfrak{a}(q_n)) + \frac{\hbar}{i}\frac{\partial}{\partial t} \right\} \psi = 0. \quad (15)$$

Here $\bar{H}_{el}$ means the energy of the electromagnetic field, $H_n$ the energy of the $n$th particle. $H_n$ contains, besides the kinetic energy of the $n$th particle, the interaction energy between this particle and the field through $\mathfrak{a}(q_n)$, $q_n$ being the coordinates of the particle and $\mathfrak{a}$ the potential of the field. $p_n$ in (15) means as usual the momentum of the $n$th particle.

We consider now the unitary operator

$$u = \exp\left\{\frac{i}{\hbar}\bar{H}_{el}t\right\} \tag{16}$$

and introduce the unitary transformation of $\mathfrak{a}$:

$$\mathfrak{A} = u\mathfrak{a}u^{-1} \tag{17}$$

and the corresponding transformation of $\psi$:

$$\Phi = u\psi. \tag{18}$$

Then we see that $\Phi$ satisfies the equation

$$\left\{\sum_n H_n(q_n, p_n, \mathfrak{A}(q_n, t)) + \frac{\hbar}{i}\frac{\partial}{\partial t}\right\}\Phi = 0. \tag{19}$$

In contrast with $\mathfrak{a}$, which was independent of time (Schrödinger picture), $\mathfrak{A}$ contains $t$ through $u$. To emphasize this, we have written $t$ explicitly as the argument of $\mathfrak{A}$. We can prove that $\mathfrak{A}$ satisfies the Maxwell equations in vacuo (accurately speaking, we need special considerations for the equation div $\mathfrak{E} = 0$).

The equation (19) is the starting point of the many-time theory. In this theory one introduces then the function $\Phi(q_1t_1, q_2t_2, \ldots, q_N, t_N)$ containing as many time variables $t_1, t_2, \ldots t_N$ as the number of the particles in place of the function $\Phi\,(q_1, q_2, \ldots, q_N, t)$ containing only one time variable,* and suppose that this $\Phi(q_1t_1, q_2t_2, \ldots, q_Nt_N)$ satisfies simultaneously the following $N$ equations:

$$\left\{H_n(q_n, p_n, \mathfrak{A}(q_n, t_n)) + \frac{\hbar}{i}\frac{\partial}{\partial t_n}\right\}\Phi\,(q_1t_1, q_2t_2, \ldots, q_Nt_N) = 0$$

$$n = 1, 2, \ldots, N. \tag{20}$$

This $\Phi(t_1, t_2, \ldots, t_N)$, which is a fundamental quantity in the many-time theory, is related to the ordinary probability amplitude $\Phi(t)$ by

$$\Phi\,(t) = \Phi\,(t, t, \ldots, t). \tag{21}$$

---

*Here we suppose the representation which makes the coordinates $q^1, q_2, \ldots, q_N$ diagonal. Thus the vector $\Phi$ is represented by a function of these coordinates.

Now, the simultaneous equations (20) can be solved when and only when the $\mathcal{N}^2$ conditions

$$\left(H_n H'_{n'} - H'_{n'} H_n\right) \Phi\left(q_1 t_1, q_2 t_2, \ldots, q_N t_N\right) = 0 \qquad (22)$$

are satisfied for all pairs of $n$ and $n'$. If the world point $(q_n t_n)$ lies outside the light cone whose vertex is at the point $(q'_{n'} t'_{n'})$, we can prove that $H_n H'_{n'} - H'_{n'} H_n = 0$. As the result, the function satisfying (20) can exist in the region where

$$\left(q_n - q'_{n'}\right)^2 - c^2 \left(t_n - t'_{n'}\right)^2 \geqq 0 \qquad (23)$$

is satisfied simultaneously for all values of $n$ and $n'$.

According to Bloch[5] we can give $\Phi(q_1 t_1, q_2 t_2, \ldots, q_N t_N)$ a physical meaning when its arguments lie in the region given by (23). Namely

$$W\left(q_1 t_1, q_2 t_2, \ldots, q_N t_N\right) = \left|\Phi\left(q_1 t_1, q_2 t_2, \ldots, q_N t_N\right)\right|^2 \qquad (24)$$

gives the relative probability that one finds the value $q_1$ in the measurement of the position of the first particle at the instant of time $t_1$, the value $q_2$ in the measurement of the position of the second particle at the instant of time $t_2$, . . . and the value $q_N$ in the measurement of the position of the $\mathcal{N}$th particle at the instant of time $t_N$.

This is the outline of the many-time formalism of the quantum mechanics. We will now return to our main subject. If we compare our equation (14) with the equation (19) of the many-time theory, we notice a marked similarity between these two equations. In (19) stands the suffix $n$, which designates the particle, while in (14) stand the variables $x$, $y$ and $z$, which designate the position in space. Further, $\Phi$ is a function of the $\mathcal{N}$ independent variables $q_1, q_2, \ldots, q_N, q_n$ giving the position of the $n$th particle, while $\Psi$ is a functional of the infinitely many "independent variables" $v_1(xyz)$ and $v_2(xyz)$, $v_2(xyz)$ and $v_2(xyz)$ giving the fields at the position $(xyz)$. Corresponding to the sum $\sum_n H_n$ in (19) the integral $\int H_{12} dx\, dy\, dz$ stands in (14). In this way, to the suffix $n$ in (19) which takes the values $1, 2, 3, \ldots, \mathcal{N}$ correspond the variables $x$, $y$ and $z$ which take continuously all values from $-\infty$ to $+\infty$.

Such a similarity suggests that we introduce infinitely many time variables $t_{xyz}$, which we may call local time,* each for one position $(xyz)$ in the space, just as we have introduced $\mathcal{N}$ time variables, particle times, $t_1, t_2, \ldots, t_N$, each for one particle. The only difference is that we use in our case infinitely many time variables whereas we have used $\mathcal{N}$ time variables in the ordinary many-time theory.

Corresponding to the transition from the use of the function with one time variable to the use of the function of $\mathcal{N}$ time variables, we must now consider the transition from the use of $\Psi(t)$ to the use of a functional $\Psi[t_{xyz}]$ of infinitely many time variables $t_{xyz}$.

We now regard $t_{xyz}$ as a function of $(xyz)$ and consider its variation $\varepsilon_{xyz}$ which differs from zero only in a small domain $V_0$ in the neighbourhood of the point $(x_0 y_0 z_0)$. We will define the partial differential coefficient of the functional $\Psi[t_{xyz}]$ with respect to the variable $t_{x_0 y_0 z_0}$ in the following manner:

$$\frac{\delta \Psi}{\delta t_{x_0 y_0 z_0}} = \lim_{\substack{\varepsilon \to 0 \\ V_0 \to 0}} \frac{\Psi\left[t_{xyz} + \varepsilon_{xyz}\right] - \Psi\left[t_{xyz}\right]}{\int \int \int \varepsilon_{xyz}\, dx\, dy\, dz} \qquad (25)$$

We then generalize (14), and regard

$$\left\{ H_{12}(x, y, z, t) + \frac{\hbar}{i} \frac{\delta}{\delta t_{xyz}} \right\} \Psi = 0, \qquad (26)$$

the infinitely many simultaneous equations corresponding to the $\mathcal{N}$ equations (20), as the fundamental equations of our theory. In (26) we have written, for simplicity, $H_{12}(x, y, z, t)$ in place of $H_{12}(V_1(xyz,t), V_2(xyz, t), \ldots)$. In general, when we have a function $F(V, \Lambda)$ of $V$ and $\Lambda$, we will write simply $F(x, y, z, t)$ for $F(V(xyz, t_{xyz}), \Lambda\ (xyz, t_{xyz}))$, or still simpler $F(P)$, $P$ denoting the world point with the coordinates $(xyz, t_{xyz})$. Thus $F(P')$ means $F(x', y', z', t')$ or, more precisely, $F(V(x'y'z', t_{x'y'z'}), (x'y'z', t_{x'y'z'}))$.

We will now adopt the equation (26) as the basis of our theory. For $V_1(P)$, $V_2(P)$, $\Lambda_1(P)$ and $\Lambda_2(P)$ in $H_{12}$ the commutation relations (10) hold, where $D(xyzt)$, has the property (13). As the consequence,

---

*The notion of local time of this kind has been occasionally introduced by Stueckelberg.[6]

we have

$$H_{12}(P)H_{12}(P') - H_{12}(P')H_{12}(P) = 0 \qquad (27)$$

when the point $P$ lies a finite distance apart from $P'$ and outside the light cone whose vertex is at $P$. Further, from our assumption (ii) the relation (27) holds also when $P$ and $P'$ are two adjacent points approaching in a space-like direction. Thus our system of equations (26) is integrable when the surface defined by the equations $t = t_{xyz}$, considering $t_{xyz}$ as a function of $x$, $y$ and $z$, is space-like.

In this way, a functional of the variable surface in the space-time world is determined by the functional partial differential equations (26). Corresponding to the relation (21) in case of many-time theory, $\Psi[t_{xyz}]$ reduces to the ordinary $\Psi(t)$ when the surface reduces to a plane parallel to the $xyz$-plane.

The dependent variable surface $t = t_{xyz}$ can be of any (space-like) form in the space-time world, and we need not presuppose any Lorentz frame of reference to define such a surface. Therefore, this $\Psi[t_{xyz}]$ is a relativistically invariant concept. The restriction that the surface must be space-like makes no trouble, since the property that a surface is space-like or time-like does not depend on a special choice of the reference system. It is not necessary, from the standpoint of the relativity theory, to admit also time-like surfaces for the variable surface, as was required by Dirac and by Yukawa. Thus we consider that $\Psi[t_{xyz}]$ introduced above is already the sufficient generalization of the ordinary $\psi$-vector, and assume that the quantum-theoretical state* of the fields is represented by this functional vector.

Let $C$ denote the surface defined by the equation $t = t_{xyz}$. Then $\Psi$ is a functional of the surface $C$. We write this as $\Psi[C]$. On $C$ we take a point $P$, whose coordinates are $(xyz, t_{xyz})$, and suppose a surface $C'$ which overlaps $C$ except in a small domain about $P$. We denote the volume of the small world lying between $C$ and $C'$ by $d\omega_p$. Then we

---

*The word state is here used in the relativistic space-time meaning. Cf Dirac's book (second edition), §6

may write (25) also in the form:

$$\frac{\delta \Psi[C]}{\delta C_P} = \lim_{C' \to C} \frac{\Psi[C'] - \Psi[C]}{d\omega_P}. \tag{28}$$

Then (26) can be written in the form:

$$\left\{ H_{12}(P) + \frac{\hbar}{i} \frac{\delta}{\delta C_P} \right\} \Psi[C] = 0. \tag{29}$$

This equation (29) has now a perfect space-time form. In the first place, $H_{12}$ is a scalar according to our assumption (i); in the second place, the commutation relations between $V(P)$ and $\Lambda(P)$ contained in $H_{12}$ has the four-dimensional forms as (10), and finally the differentiation $\frac{\delta}{\delta C_P}$ is defined by (28) quite independently of any frame of reference.

A direct conclusion drawn from (29) is that $\Psi[C']$ is obtained from $\Psi[C]$ by the following infinitesimal transformation:

$$\Psi[C'] = \left\{ 1 - \frac{i}{\hbar} H_{12}(P) d\omega_P \right\} \Psi[C]. \tag{30}$$

When there exist in the space-time world two surfaces $C_1$ and $C_2$ a finite distance apart, we need only to repeat the infinitesimal transformations in order to obtain $\Psi[C_2]$ from $\Psi[C_1]$. Thus

$$\Psi[C_2] = \prod_{C_1}^{C_2} \left\{ 1 - \frac{i}{\hbar} H_{12}(P) d\omega_P \right\} \Psi[C_1]. \tag{31}$$

The meaning of this equation is as follows: We divide the world region lying between $C_1$ and $C_2$ into small elements $d\omega_p$ (it is necessary that each world element be surrounded by two space-like surfaces). We consider for each world element the infinitesimal transformation $1 - \frac{i}{\hbar} H_{12}(P) d\omega_p$. Then we take the product of these transformations, the order of the factor being taken from $C_1$ to $C_2$. This product transforms then $\Psi[C_1]$ into $\Psi[C_2]$.

The surfaces $C_1$ and $C_2$ must here be both space-like, but otherwise they may-have any form and any configuration. Thus $C_2$ does not necessarily lie afterward against $C_1$; $C_1$ and $C_2$ may even cross with each other.

The relation of the form (31) has been already introduced by Heisenberg.[7] It can be regarded as the integral form of our generalized Schrödinger equation (29).

## 4. GENERALIZED PROBABILITY AMPLITUDE

We must now find the physical meaning of the functional $\Psi[C]$. As regards this we can follow a method similar to that of Bloch in the case of ordinary many-time theory. Besides the fact that in our case there appear infinitely many time variables, one point differs from Bloch's case: in (16) the unitary operator $u$ is commutable with the coordinates $q_1, q_2, \ldots, q_N$, while our $U$ is not commutable with the field quantities $v_1(xyz)$ and $v_2(xyz)$. Noting this difference and treating the continuum infinity as the limit of a denumerable infinity by some artifice, for instance, by the procedure of Heisenberg and Pauli,[8] Bloch's consideration can be applied also here almost without any alteration. We shall give here only the results.

Let us suppose that the fields are in the state represented by a vector $\Psi[C]$. We suppose that we make measurements of a function $f(v_1, v_2, \lambda_1, \lambda_2)$ at every point on a surface $C_1$ in the space-time world. Let $P_1$ denote the variable point on $C_1$, then, if $f(P_1)$ at any two "values" of $P_1$ commute with each other, the measurements of $f$ at each of these two points do not interfere with each other. Our first conclusion says that in this case the expectation value of $f(P_1)$ is given by

$$\overline{f(P_1)} = ((\Psi[C_1], f(P_1) \Psi[C_1])) \tag{32}$$

where $f(P_1)$ means $f(V_1(P_1), \ldots$, according to our convention on page 8, and the symbol $((A, B))$ with double parentheses is the scalar product of two vectors $A$ and $B$. It is impossible in cases of continuously many degrees of freedom to represent this scalar product by an integral of the product of two functions. For this purpose we must replace the continuum infinity by an at least denumerable infinity.

More generally, we suppose a functional $F[f(P_1)]$ of the independent variable function $f(P_1)$, regarding $f(P_1)$ as a function of $P_1$. Then the expectation value of this $F$ is given by

$$\overline{F[f(P_1)]} = ((\Psi[C_1], F[f(P_1)]\Psi[C_1])). \tag{33}$$

A physically interesting $F$ is the projective operator $M[v_1'(P_1), v_2'(P_1); V_1(P_1), V_2(P_1)]$ belonging to the "eigen-value" $v_1'(P_1), v_2(P_1)$ of $V_1(P_1), V_2(P_1)$. Then its expectation value

$$\overline{M[v_1'(P_1), v_2'(P_1); V_1(P_1), V_2(P_1)]}$$
$$= ((\Psi[C_1], M[v_1'(P_1), v_2'(P_1); V_1(P_1), V_2(P_1)]\Psi[C_1])) \tag{34}$$

gives the probability that the field 1 and the field 2 have respectively the functional form $v_1'(P_1)$ and $v_2(P_1)$ on the surface $C_1$. As $C_1$ is assumed to be space-like, the measurement of the functional $M$ is possible (the measurements of $V_1(P_1)$ and $V_2(P_1)$ at all points on $C_1$ mean just the measurement of $M$).

Thus far we have made no mention of the representation of $\Psi[C]$. We use now the special representation in which $V_1(P_1)$ at all points on $C_1$ are simultaneously diagonal. It is always possible to make all $V_1(P_1)$ and $V_2(P_1)$ diagonal when the surface $C_1$ is space-like. In this representation $\Psi[C_1]$ is represented by a functional $\Psi[v_1'(P_1), v_2'(P_1); C_1]$ of the eigenvalues $v_1'(P_1)$ and $v_2'(P_1)$ of $V_1(P_1)$ and $V_2(P_1)$. The projection operator $M$ has in this representation such diagonal form that (34) is simplified as follows

$$W[v_1'(P_1), v_2'(P_1)] = \overline{M[v_1'(P_1), v_2'(P_1); V_1(P_1), V_2(P_1)]}$$
$$= |\Psi[v_1'(P_1), v_2'(P_1); C_1]|^2. \tag{35}$$

In this sense we can call $\Psi[v_1'(P_1), v_2'(P_1); C_1]$ the "generalized probability amplitude."

# 5. GENERALIZED TRANSFORMATION FUNCTIONAL

We have stated above that between $\Psi[C_1]$ and $\Psi[C_2]$ the relation (31) holds, where $C_1$ and $C_2$ are two space-like surfaces in the space-time world. We see thus that the transformation operator

$$T\,[C_2; C_1] = \prod_{C_1}^{C_2} \left(1 - \frac{i}{\hbar}H_{12}\,d\omega\right) \tag{36}$$

plays an important role. It is evident that this operator also has a space-time meaning.

Just as the special representative of the $\psi$-vector, the probability amplitude, has a distinct physical meaning, so there is a special representation in which the representative of the transformation operator $T[C_2; C_1]$ has a distinct physical meaning.

We now introduce the mixed representative of $T[C_2; C_1]$ whose rows refer to the representation in which $V_1(P_1)$ and $V_2(P_1)$ at all points on $C_1$ become diagonal and whose column refer to the representation in which $V_1(P_2)$ and $V_2(P_2)$ at all points on $C_2$ become diagonal. We denote this representation by

$$\left[v_1''(P_2),\, v_2''(P_2) \,|T\,[C_2; C_1]|\, V_1'(P_1),\, v_2'(P_1)\right], \tag{37*}$$

or simpler:

$$\left[v_1''(P_2),\, v_2''(P_2) \,\big|\, v_1'(P_1),\, v_2'(P_1)\right]. \tag{38*}$$

If we note here the relation (35), we see that we can give the matrix elements of this representation the following meaning: One measures the field quantities $V_1$ and $V_2$ at all points on $C_2$ when the fields are prepared in such a way that they have certainly the values $v_1'(P_1)$ and $v_2'(P_1)$ at all points on $C_1$. Then

$$W\left[v_1''(P_2),\, v_2''(P_2)\,;\, v_1'(P_1),\, v_2'(P_1)\right]$$
$$= |[v_1''(P_2),\, v_2''(P_2) \,|\, v_1'(P_1),\, v_2'(P_1)]|^2 \tag{39}$$

gives the probability that one obtains the result $v_1''(P_2)$ and $v_2''(P_2)$ in this measurement. In this proposition we have assumed that $C_2$ lies afterward against $C_1$.

From this physical interpretation we may regard the matrix element (37), or (38), considered as a functional of $v_1''(P_2)$, $v_2''(P_2)$ and

---

*As the matrix elements are functional of $v(P)$, we use here the square brackets.

$v'_1(P_1)$, $v'_2(P_1)$, as the generalization of the ordinary transformation function $(q''_{t_2}|q'_{t_1})$.

As a special case it may happen that $C_2$ lies apart from $C_1$ only in a portion $S_2$ and a portion $S_1$ of $C_2$ and $C_1$ respectively, the other parts of $C_1$ and $C_2$ overlapping with each other.

In this case the matrix elements of $T[C_2; C_1]$ depend only on the values of the fields on the portions $S_1$ and $S_2$ of the surfaces $C_1$ and $C_2$. In this case we need for calculating $T[C_2; C_1]$ to take the product in (36) only in the closed domain surrounded by $S_1$ and $S_2$, thus

$$T[S_2; S_1] = \prod_{S_1}^{S_2} \left(1 - \frac{i}{\hbar} H_{12}\, d\omega\right). \tag{40}$$

The matrix elements of the mixed representation of this $T$ is a functional of $v'_1(p_1)$, $v'_2(p_1)$ and $v''_1(p_2)$, $v''_2(p_2)$, where $p_1$ denotes the moving point of the portion $S_1$, and $p_2$ the moving point on the portion $S_2$. This matrix is independent on the field quantities on the other portions of the surfaces $C_1$ and $C_2$.

The matrix element of $T[S_2; S_1]$ regarded as a functional of $v'_1(p_1)$, $v'_2(p_1)$ and $v''_1(p_2)$, $v''_2(p_2)$ has the properties of g.t.f. (generalized transformation functional) of Dirac. But in defining our g.t.f. we had to restrict the surfaces $S_1$ and $S_2$ to be space-like, while Dirac has required his g.t.f. to be defined also referring to the time-like surfaces. As mentioned above, however, such a generalization as required by Dirac is superfluous so far as concerns the relativity theory.

It is to be noted that for the physical interpretation of $[v''_1(P_2), v''_2(P_2)|v'_1(P_1), v'_2(P_1)]$ it is not necessary to assume $C_2$ to lie afterward against $C_1$. Also when the inverse is the case, we can as well give the physical meaning for $W$ of (39): One measures the field quantities $V_1$ and $V_2$ at all points on $C_2$ when the fields are prepared in such a way that they would have certainly the values $v'_1(P_1)$ and $v'_2(P_1)$ at all points on $C_1$ if the fields were left alone until $C_1$ without being measured before on $C_2$. Then $W$ gives the probability that one finds the results $v''_1(P_2)$ and $v''_2(P_2)$ in this measurement on $C_2$.

# 6. CONCLUDING REMARKS

We have thus shown that the quantum theory of wave fields can be really brought into a form which reveals directly the invariance of the theory against Lorentz transformations. The reason why the ordinary formalism of the quantum field theory is so unsatisfactory is that it has been built up in a way much too analogous to the ordinary non-relativistic mechanics. In this ordinary formalism of the quantum theory of fields the theory is divided into two distinct sections: the section giving the kinematical relations between various quantities at the same instant of time, and the section determining the causal relations between quantities at different instants of time. Thus the commutation relations (1) belong to the first section and the Schrödinger equation (2) to the second.

As stated before, this way of separating the theory into two sections is very unrelativistic, since here the concept "same instant of time" plays a distinct role.

Also in our formalism the theory is divided into two sections, but now the separation is introduced in another place. One section gives the laws of behaviour of the fields when they are left alone, and the other gives the laws determining the deviation from this behaviour due to interactions. This way of separating the theory can be carried out relativistically.

Although in this way the theory can be brought into more satisfactory form, no new contents are added thereby. So, the well-known divergence difficulties of the theory are inherited also by our theory. Indeed, our fundamental equations (29) admit only catastrophic solutions, as can be seen directly in the fact that the unavoidable infinity due to non-vanishing zero-point amplitudes of the fields inheres in the operator $H_{12}(P)$. Thus, a more profound modification of the theory is required in order to remove this fundamental difficulty.

It is expected that such a modification of the theory could possibly be introduced by some revision of the concept of interaction, because we meet no such difficulty when we deal with the non-interacting fields. This revision would then have the result that in the separation of the theory into two sections, one for free fields and one for interactions, some uncertainty would be introduced. This seems to be implied by the very fact that, when we formulate the quantum field theory in a relativistically satisfactory manner, this way of separation has revealed itself as the fundamental element of the theory.

PHYSICS DEPARTMENT,
TOKYO BUNRIKA UNIVERSITY.

## REFERENCES

1. H. Yukawa, *Kagaku*, **12**, 251, 282 and 322, 1942.
2. P. A. M. Dirac, *Phys. Z. USSR.*, **3**, 64, 1933.
3. W. Pauli, *Solvay Berichte*, 1939.
4. P. A. M. Dirac, *Proc. Roy. Soc. London*, **136**, 453, 1932.
5. F. Bloch, *Phys. Z. USSR.*, **5**, 301, 1943.
6. E. Stueckelberg, *Helv. Phys. Acta*, **11**, 225, § 5, 1938.
7. W. Heisenberg, *Z. Phys.*, **110**, 251, 1938.
8. W. Heisenberg and W. Pauli, *Z. Phys.*, **56**, 1, 1929.

# Space-Time Approach to Quantum Electrodynamics

## By

## Richard Feynman

In this paper two things are done. (1) It is shown that a considerable simplification can be attained in writing down matrix elements for complex processes in electrodynamics. Further, a physical point of view is available which permits them to be written down directly for any specific problem. Being simply a restatement of conventional electrodynamics, however, the matrix elements diverge for complex processes. (2) Electrodynamics is modified by altering the interaction of electrons at short distances. All matrix elements are now finite, with the exception of those relating to problems of vacuum polarization. The latter are evaluated in a manner suggested by Pauli and Bethe, which gives finite results for these matrices also. The only effects sensitive to the modification are changes in mass and charge of the electrons. Such changes could not be directly observed. Phenomena directly observable, are insensitive to the details of the modification used (except at extreme energies). For such phenomena, a limit can be taken as the range of the modification goes to zero. The results then agree with those of Schwinger. A complete, unambiguous, and presumably consistent, method is therefore available for the calculation of all processes involving electrons and photons.

The simplification in writing the expressions results from an emphasis on the over-all space-time view resulting from a study of the solution of the equations of electrodynamics. The relation of this to the more conventional Hamiltonian point of view is discussed. It would be very difficult Co make the modification which is proposed if one insisted on having the equations in Hamiltonian form.

Reprinted with permission from the American Physical Society: Feynman, *Physical Review*, Volume 76, p. 769, 1949. ©1949, by the American Physical Society.

The methods apply as well to charges obeying the Klein-Gordon equation, and to the various meson theories of nuclear forces. Illustrative examples are given. Although a modification like that used in electrodynamics can make all matrices finite for all of the meson theories, for some of the theories it is no longer true that all directly observable phenomena are insensitive to the details of the modification used.

The actual evaluation of integrals appearing in the matrix elements may be facilitated, in the simpler cases, by methods described in the appendix.

This paper should be considered as a direct continuation of a preceding one[1] (I) in which the motion of electrons, neglecting interaction, was analyzed, by dealing directly with the *solution* of the Hamiltonian differential equations. Here the same technique is applied to include interactions and in that way to express in simple terms the solution of problems in quantum electrodynamics.

For most practical calculations in quantum electrodynamics the solution is ordinarily expressed in terms of a matrix element. The matrix is worked out as an expansion in powers of $e^2/\hbar c$, the successive terms corresponding to the inclusion of an increasing number of virtual quanta. It appears that a considerable simplification can be achieved in writing down these matrix elements for complex processes. Furthermore, each term in the expansion can be written down and understood directly from a physical point of view, similar to the space-time view in I. It is the purpose of this paper to describe how this may be done. We shall also discuss methods of handling the divergent integrals which appear in these matrix elements.

The simplification in the formulae results mainly from the fact that previous methods unnecessarily separated into individual terms processes that were closely related physically. For example, in the exchange of a quantum between two electrons there were two terms depending on which electron emitted and which absorbed the quantum. Yet, in the virtual states considered, timing relations are not significant. Only

---

[1] R. P. Feynman, Phys. Rev. **76**, 749 (1949), hereafter called I.

the order of operators in the matrix must be maintained. We have seen (I), that in addition, processes in which virtual pairs are produced can be combined with others in which only positive energy electrons are involved. Further, the effects of longitudinal and transverse waves can be combined together. The separations previously made were on an unrelativistic basis (reflected in the circumstance that apparently momentum 7but not energy is conserved in intermediate states). When the terms are combined and simplified, the relativistic invariance of the result is self-evident.

We begin by discussing the solution in space and time of the Schrödinger equation for particles interacting instantaneously. The results are immediately generalizable to delayed interactions of relativistic electrons and we represent in that way the laws of quantum electrodynamics. We can then see how the matrix element for any process can be written down directly. In particular, the self-energy expression is written down.

So far, nothing has been done other than a restatement of conventional electrodynamics in other terms. Therefore, the self-energy diverges. A modification[2] in interaction between charges is next made, and it is shown that the self-energy is made convergent and corresponds to a correction to the electron mass. After the mass correction is made, other real processes are finite and-insensitive to the "width" of the cut-off in the interaction.[3]

Unfortunately, the modification proposed is not completely satisfactory theoretically (it leads to some difficulties of conservation of energy). It does, however, seem consistent and satisfactory to define the matrix element for all real processes as the limit of that computed here as the cut-off width goes to zero. A similar technique suggested by Pauli and by Bethe can be applied to problems of vacuum polarization (resulting in a renormalization of charge) but again a strict physical basis for the rules of convergence is not known.

---

[2] For a discussion of this modification in classical physics see R. P. Feynman, Phys. Rev. **74** 939 (1948), hereafter referred to as A.

[3] A brief summary of the methods and results will be found in R. P. Feynman, Phys. Rev. **74**, 1430 (1948), hereafter referred to as B.

After mass and charge renormalization, the limit of zero cut-off width can be taken for all real processes. The results are then equivalent to those of Schwinger[4] who does not make explicit use of the convergence factors. The method of Schwinger is to identify the terms corresponding to corrections in mass and charge and, previous to their evaluation, to remove them from the expressions for real processes. This has the advantage of showing that the results can be strictly independent of particular cut-off methods.

On the other hand, many of the properties of the integrals are analyzed using formal properties of invariant propagation functions. But one of the properties is that the integrals are infinite and it is not clear to what extent this invalidates the demonstrations. A practical advantage of the present method is that ambiguities can be more easily resolved; simply by direct calculation of the otherwise divergent integrals. Nevertheless, it is not at all clear that the convergence factors do not upset the physical consistency of the theory. Although in the limit the two methods agree, neither method appears to be thoroughly satisfactory theoretically. Nevertheless, it does appear that we now have available a complete and definite method for the calculation of physical processes to any order in quantum electrodynamics.

Since we can write down the solution to any physical problem, we have a complete theory which could stand by itself. It will be theoretically incomplete, however, in two respects. First, although each term of increasing order in $e^2/\hbar c$ can be written down it would be desirable to see some way of expressing things in finite form to all orders in $e^2/\hbar c$ at once. Second, although it will be physically evident that the results obtained are equivalent to those obtained by conventional electrodynamics the mathematical proof of this is not included. Both of these limitations will be removed in a subsequent paper (see also Dyson[5]).

Briefly the genesis of this theory was this. The conventional electrodynamics was expressed in the Lagrangian form of quantum

[4] J. Schwinger, Phys. Rev. **74**, 1439 (1948), Phys. Rev. **75**, 651 (1949). A proof of this equivalence is given by F. J. Dyson, Phys. Rev. **75**, 486 (1949).

mechanics described in the Reviews of Modern Physics.[5] The motion of the field oscillators could be integrated out (as described in Section 13 of that paper), the result being an expression of the delayed interaction of the particles. Next the modification of the delta-function interaction could be made directly from the analogy to the classical case.[6] This was still not complete because the Lagrangian method had been worked out in detail only for particles obeying the non-relativistic Schrödinger equation. It was then modified in accordance with the requirements of the Dirac equation and the phenomenon of pair creation. This was made easier by the reinterpretation of the theory of holes (I). Finally for practical calculations the expressions were developed in a power series in $e^2/\hbar c$. It was apparent that each term in the series had a simple physical interpretation. Since the result was easier to understand than the derivation, it was thought best to publish the results first in this paper. Considerable time has been spent to make these first two papers as complete and as physically plausible as possible without relying on the Lagrangian method, because it is not generally familiar. It is realized that such a description cannot carry the conviction of truth which would accompany the derivation. On the other hand, in the interest of keeping simple things simple the derivation will appear in a separate paper.

The possible application of these methods to the various meson theories is discussed briefly. The formulas corresponding to a charge particle of zero spin moving in accordance with the Klein Gordon equation are also given. In an Appendix a method is given for calculating the integrals appearing in the matrix elements for the simpler processes.

The point of view which is taken here of the interaction of charges differs from the more usual point of view of field theory. Furthermore, the familiar Hamiltonian form of quantum mechanics must be

---

[5] R. P. Feynman, Rev. Mod. Phys. **20**, 367 (1948). The application to electrodynamics is described in detail by H. J. Groenewold, Koninklijke Nederlandsche Akademia van Weteschappen. Proceedings Vol. LII, **3** (226) 1949.

[6] For a discussion of this modification in classical physics see R. P. Feynman, Phys. Rev. **74** 939 (1948), hereafter referred to as A.

compared to the overall space-time view used here. The first section is, therefore, devoted to a discussion of the relations of these viewpoints.

# 1 COMPARISON WITH THE HAMILTONIAN METHOD

Electrodynamics can be looked upon in two equivalent and complementary ways. One is as the description of the behavior of a field (Maxwell's equations). The other is as a description of a direct interaction at a distance (albeit delayed in time) between charges (the solutions of Lienard and Wiechert). From the latter point of view light is considered as an interaction of the charges in the source with those in the absorber. This is an impractical point of view because many kinds of sources produce the same kind of effects. The field point of view separates these aspects into two simpler problems, production of light, and absorption of light. On the other hand, the field point of view is less practical when dealing with close collisions of particles (or their action on themselves). For here the source and absorber are not readily distinguishable, there is an intimate exchange of quanta. The fields are so closely determined by the motions of the particles that it is just as well not to separate the question into two problems but to consider the process as a direct interaction. Roughly, the field point of view is most practical for problems involving real quanta, while the interaction view is best for the discussion of the virtual quanta involved. We shall emphasize the interaction viewpoint in this paper, first because it is less familiar and therefore requires more discussion, and second because the important aspect in the problems with which we shall deal is the effect of virtual quanta.

The Hamiltonian method is not well adapted to represent the direct action at a distance between charges because that action is delayed. The Hamiltonian method represents the future as developing out of the present. If the values of a complete set of quantities are known now, their values can be computed at the next instant in time. If particles interact through a delayed interaction, however, one cannot predict the future by simply knowing the present motion of the particles. One would also have to know what the motions of the particles were in the past in view of the interaction this may have on the future motions. This is done in the Hamiltonian electrodynamics, of course, by requiring that one specify besides the present motion of the particles, the values of a host of new variables (the coordinates of the field oscillators) to keep track of that aspect of the past motions of the particles which determines their future behavior. The use of the Hamiltonian forces one to choose the field viewpoint rather than the interaction viewpoint.

In many problems, for example, the close collisions of particles, we are not interested in the precise temporal sequence of events. It is not of interest to be able to say how the situation would look at each instant of time during a collision and how it progresses from instant to instant. Such ideas are only useful for events taking a long time and for which we can readily obtain information during the intervening period. For collisions it is much easier to treat the process as a whole.[7] The Møller interaction matrix for the the collision of two electrons is not essentially more complicated than the nonrelativistic Rutherford formula, yet the mathematical machinery used to obtain the former from quantum electrodynamics is vastly more complicated than Schrödinger's equation with the $e^2/r_{12}$ interaction needed to obtain the latter. The difference is only that in the latter the action is instantaneous so that the Hamiltonian method requires no extra variables, while in the former relativistic case it is delayed and the Hamiltonian method is very cumbersome.

We shall be discussing the solutions of equations rather than the time differential equations from which they come. We shall discover that the solutions, because of the over-all space-time view that they permit, are as easy to understand when interactions are delayed as when they are instantaneous.

---

[7] This is the viewpoint of the theory of the $S$ matrix of Heisenberg.

As a further point, relativistic invariance will be self-evident. The Hamiltonian form of the equations develops the future from the instantaneous present. But for different observers in relative motion the instantaneous present is different, and corresponds to a different 3-dimensional cut of space-time. Thus the temporal analyses of different observers is different and their Hamiltonian equations are developing the process in different ways. These differences are irrelevant, however, for the solution is the same in any space time frame. By forsaking the Hamiltonian method, the wedding of relativity and quantum mechanics can be accomplished most naturally.

We illustrate these points in the next section by studying the solution of Schrödinger's equation for non-relativistic particles interacting by an instantaneous Coulomb potential (Eq. 2). When the solution is modified to include the effects of delay in the interaction and the relativistic properties of the electrons we obtain an expression of the laws of quantum electrodynamics (Eq. 4).

# 2 THE INTERACTION BETWEEN CHARGES

We study by the same methods as in I, the interaction of two particles using the same notation as I. We start by considering the non-relativistic case described by the Schrödinger equation (I, Eq. 1). The wave function at a given time is a function $\psi(\mathbf{x}_a, \mathbf{x}_b, t)$ of the coordinates $\mathbf{x}_a$ and $\mathbf{x}_b$ of each particle. Thus call $K(\mathbf{x}_a, \mathbf{x}_b, t; \mathbf{x}'_a, \mathbf{x}'_b, t')$ the amplitude that particle $a$ at $\mathbf{x}'_a$ at time $t'$ will get to $\mathbf{x}_a$ at $t$ while particle $b$ at $\mathbf{x}'_b$ at $t'$ gets to $\mathbf{x}_b$ at $t$. If the particles are free and do not interact this is

$$K(\mathbf{x}_a, \mathbf{x}_b, t; \mathbf{x}'_a, \mathbf{x}'_b, t') = K_{0a}(\mathbf{x}_a, t; \mathbf{x}'_a, t') K_{0b}(\mathbf{x}_b, t; \mathbf{x}'_b, t')$$

where $K_{0a}$ is the $K_0$ function for particle a considered as free. In *this* case we can obviously define a quantity like K, but for which the time $t$ need not be the same for particles $a$ and $b$ (likewise for $t'$); e.g.,

$$K_0(3, 4; 1, 2) = K_{0a}(3, 1) K_{0b}(4, 2) \tag{1}$$

can be thought of as the amplitude that particle $a$ goes from $\mathbf{x}_1$ at $t_1$ to $\mathbf{x}_3$ at $t_3$ and that particle $b$ goes from $\mathbf{x}_2$ at $t_2$ to $\mathbf{x}_4$ at $t_4$.

When the particles do interact, one can only define the quantity $K(3, 4; 1, 2)$ precisely if the interaction vanishes between $t_1$ and $t_2$ and also between $t_3$ and $t_4$. In a real physical system such is not the case. There is such an enormous advantage, however, to the concept that we shall continue to use it, imagining that we can neglect the effect of interactions between $t_1$ and $t_2$ and between $t_3$ and $t_4$. For practical problems this means choosing such long time intervals $t_3 - t_1$ and $t_4 - t_2$ that the extra interactions near the end points have small relative effects. As an example, in a scattering problem it may well be that the particles are so well separated initially and finally that the interaction at these times is negligible. Again energy values can be defined by the average rate of change of phase over such long time intervals that errors initially and finally can be neglected. Inasmuch as any physical problem can be defined in terms of scattering processes we do not lose much in a general theoretical sense by this approximation. If it is not made it is not easy to study interacting particles relativistically, for there is nothing significant in choosing $t_1 = t_3$ if $\mathbf{x}_1 \neq \mathbf{x}_3$ as absolute simultaneity of events at a distance cannot be defined invariantly. It is essentially to avoid this approximation that the complicated structure of the older quantum electrodynamics has been built up. We wish to describe electrodynamics as a delayed interaction between particles. If we can make the approximation of assuming a meaning to $K(3, 4; 1, 2)$ the results of this interaction can be expressed very simply.

To see how this may be done, imagine first that the interaction is simply that given by a Coulomb potential $e^2/r$ where $r$ is the distance between the particles. If this be turned on only for a very short time $\Delta t_0$ at time $t_0$ the first order correction to $K(3, 4; 1, 2)$ can be worked out exactly as was Eq. (9) of I by an obvious generalization to two particles:

$$K^{(1)}(3, 4; 1, 2) = -ie^2 \int \int K_{0a}(3, 5) K_{0b}(4, 6) r_{56}^{-1}$$

$$\times K_{0a}(5,1)K_{0b}(6,2)d^3\mathbf{x}_5 d^3\mathbf{x}_6 \Delta t_0,$$

where $t_5 = t_6 = t_0$. If now the potential were on at all times (so that strictly $K$ is not defined unless $t_4 = t_3$ and $t_1 = t_2$), the first-order effect is obtained by integrating on $t_0$, which we can write as an integral over both $t_5$ and $t_6$ if we include a delta-function $\delta(t_5 - t_6)$ to insure contribution only when $t_5 = t_6$ Hence, the first-order effect of interaction is (calling $t_5 - t_6 = t_{56}$):

$$K^{(1)}(3,4;1,2) = -ie^2 \int\int K_{0a}(3,5)K_{0b}(4,6)r_{56}^{-1} \qquad (2)$$
$$\times \delta(t_{56})K_{0a}(5,1)K_{0b}(6,2)d\tau_5 d\tau_6,$$

where $d\tau = d^3\mathbf{x}dt$.

We know, however, in classical electrodynamics, that the Coulomb potential does not act instantaneously, but is delayed by a time $r_{56}$, taking the speed of light as unity. This suggests simply replacing $r_{56}^{-1}\delta(t_{56})$ in (2) by something like $r_{56}^{-1}\delta(t_{56} - r_{56})$ to represent the delay in the effect of $b$ on $a$.

This turns out to be not quite right,[8] for when this interaction is represented by photons they must be of only positive energy, while the Fourier transform of $\delta(t_{56} - r_{56})$ contains frequencies of both signs. It should instead be replaced by $\delta_+(t_{56} - r_{56})$ where

$$\delta_+(x) = \int_0^\infty e^{-i\omega x}d\omega/\pi = \lim_{\epsilon \to 0} \frac{(\pi i)^{-1}}{x - i\epsilon} = \delta(x) + (\pi i x)^{-1}. \qquad (3)$$

This is to be averaged with $r_{56}^{-1}\delta_+(-t_{56} - r_{56})$ which arises when $t_5 < t_6$ and corresponds to $a$ emitting the quantum which $b$ receives. Since

$$(2r)^{-1}(\delta_+(t - r) + \delta_+(-t - r)) = \delta_+(t^2 - r^2),$$

this means $r_{56}^{-1}\delta(t_{56})$ is replaced by $\delta_+(s_{56}^2)$ where $s_{56}^2 = t_{56}^2 - r_{56}^2$ is the square of the relativistically invariant interval between points 5 and 6. Since in classical electrodynamics there is also an interaction through

---

[8] It and a like term for the effect of a on b, leads to a theory which, in the classical limit, exhibits interaction through half-advanced and half-retarded potentials. Classically, this is equivalent to purely retarded effects within a closed box from which no light escapes (e.g., see A, or J. A. Wheeler and R. P. Feynman, Rev. Mod. Phys. 17, 157 (1945)). Analogous theorems exist in quantum mechanics but it would lead us too far astray to discuss them now.

the vector potential, the complete interaction (see A, Eq. (I)) should be $(1 - (\mathbf{v}_5 \cdot \mathbf{v}_6)\delta_+(s_{56}^2)$, or in the relativistic case,

$$(1 - \alpha_a \cdot \alpha_b)\delta_+(s_{56}^2) = \beta_a\beta_b\gamma_{a\mu}\gamma_{b\mu}\delta_+(s_{56}^2).$$

Hence we have for electrons obeying the Dirac equation,

$$K^{(1)}(3, 4; 1, 2) = ie^2 \int \int K_{+a}(3, 5)K_{+b}(4, 6)\gamma_{a\mu}\gamma_{b\mu} \tag{4}$$
$$\times \delta_+(s_{56}^2)K_{+a}(5, 1)K_{+b}(6, 2)d\tau_5 d\tau_6,$$

where $\gamma_{a\mu}$ and $\gamma_{b\mu}$, are the Dirac matrices applying to the spinor corresponding to particles $a$ and $b$, respectively (the factor $\beta_a\beta_b$ being absorbed in the definition, I Eq. (17), of $K_+$).

This is our fundamental equation for electrodynamics. It describes the effect of exchange of one quantum (therefore first order in $e^2$) between two electrons. It will serve as a prototype enabling us to write down the corresponding quantities involving the exchange of two or more quanta between two electrons or the interaction of an electron with itself. It is a consequence of conventional electrodynamics. Relativistic invariance is clear. Since one sums over $\mu$ it contains the effects of both longitudinal and transverse waves in a relativistically symmetrical way.

We shall now interpret Eq. (4) in a manner which will permit us to write down the higher order terms. It can be understood (see Fig. 1) as saying that the amplitude for "$a$" to go from 1 to 3 and "$b$" to go from 2 to 4 is altered to first order because they can exchange a quantum. Thus, "$a$" can go to 5 (amplitude $(K_+(5, 1))$ emit a quantum (longitudinal, transverse, or scalar $\gamma_{a\mu}$) and then proceed to 3 $(K_+(3,5))$. Meantime "$b$" goes to 6 $(K_+(6, 2))$, absorbs the quantum $(\gamma_{b\mu})$ and proceeds to 4 $(K_+(4, 6))$. The quantum meanwhile proceeds from 5 to 6, which it does with amplitude $\delta_+(s_{56}^2)$. We must sum over all the possible quantum polarizations it and positions and times of emission 5, and of absorption 6. Actually if $t_5 > t_6$ it would be better to say that "$a$" absorbs and "$b$" emits but no attention need be paid to these matters, as all such alternatives are automatically contained in (4).

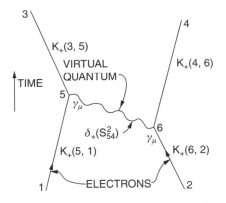

FIG. 1   The fundamental interaction Eq. (4). Exchange of one quantum between two electrons.

The correct terms of higher order in $e^2$ or involving larger numbers of electrons (interacting with themselves or in pairs) can be written down by the same kind of reasoning. They will be illustrated by examples as we proceed. In a succeeding paper they will all be deduced from conventional quantum electrodynamics.

Calculation, from (4), of the transition element between positive energy free electron states gives the Møller scattering of two electrons, when account is taken of the Pauli principle.

The exclusion principle for interacting charges is handled in exactly the same way as for noninteracting charges (I). For example, for two charges it requires only that one calculate $K(3, 4; 1, 2) - K(4, 3; 1, 2)$ to get the net amplitude for arrival of charges at 3 and 4. It is disregarded in intermediate states. The interference effects for scattering of electrons by positrons discussed by Bhabha will be seen to result directly in this formulation. The formulas are interpreted to apply to positions in the manner discussed in I.

As our primary concern will be for processes in which the quanta are virtual we shall not include here the detailed analysis of processes involving real quanta in initial or final state, and shall content ourselves by only stating the rules applying to them.[9] The result of the analysis

---

[9] Although in the expressions stemming from (4) the quanta are virtual, this is not actually a theoretical limitation. One way to deduce the correct rules for real quanta from (4) is to note that in a closed system all quanta can be considered as virtual (i.e., they

is, as expected, that they can be included by the same line of reasoning as is used in discussing the virtual processes, provided the quantities are normalized in the usual manner to represent single quanta. For example, the amplitude that an electron in going from 1 to 2 absorbs a quantum whose vector potential, suitably normalized, is $c_\mu \exp(-ik \cdot x) = C_\mu(x)$ is just the expression (I, Eq. (13)) for scattering in a potential with $\mathbf{A}$ (3) replaced by $\mathbf{C}$ (3). Each quantum interacts only once (either in emission or in absorption), terms like (I, Eq. (14)) occur only when there is more than one quantum involved. The Bose statistics of the quanta can, in all cases, be disregarded in intermediate states. The only effect of the statistics is to change the weight of initial or final states. If there are among quanta, in the initial state, some a which are identical then the weight of the state is $(1/n!)$ of what it would be if these quanta were considered as different (similarly for the final state).

# 3 THE SELF–ENERGY PROBLEM

Having a term representing the mutual interaction of a pair of charges, we must include similar terms to represent the interaction of a charge with itself. For under some circumstances what appears to be two distinct electrons may, according to I, be viewed also as a single electron (namely in case one electron was created in a pair with a positron destined to annihilate the other electron). Thus to the interaction between such electrons must correspond the possibility of the action of an electron on itself.[10]

---

have a known source and are eventually absorbed) so that in such a system the present description is complete and equivalent to the conventional one. In particular, the relation of the Einstein $A$ and $B$ coefficients can be deduced. A more practical direct deduction of the expressions for real quanta will be given in the subsequent paper. It might be noted that (4) can be rewritten as describing the action on a, $K^{(1)}(3, 1) = i \int K_+(3, 5) \times A(5) K_+(5, 1) d\tau_5$ of the potential $A_\mu(5) = e^2 \int K_+(4, 6) \delta_+(s_{56}^2) \gamma_\mu \times K_+(6, 2) d\tau_6$ arising from Maxwell's equations $-\Box^2 A_\mu = 4\pi j_\mu$ from a "current" $j_\mu(6) = e^2 K_+(4, 6) \gamma_\mu K_+(6, 2)$ produced by particle $b$ in going from 2 to 4. This is virtue of the fact that $\delta_+$ satisfies

$$-\Box_2^2 \delta_+(s_{21}^2) = 4\pi \delta(2, 1). \qquad (5)$$

---

[10] These considerations make it appear unlikely that the contention of J. A. Wheeler and R. P. Feynman, Rev. Mod. Phys. **17**, 157 (1945), that electrons do not act on themselves, will be a successful concept in quantum electrodynamics.

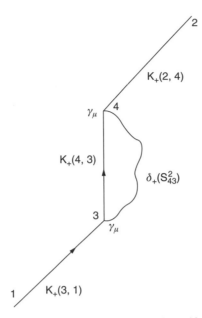

$K_+(2, 4)$

$\gamma_\mu$  4

$K_+(4, 3)$

$\delta_+(S_{43}^2)$

3
$\gamma_\mu$

1  $K_+(3, 1)$

FIG. 2   Interaction of an electron with itself, Eq. (6).

This interaction is the heart of the self energy problem. Consider to first order in $e^2$ the action of an electron on itself in an otherwise force free region. The amplitude $K(2,1)$ for a single particle to get from 1 to 2 differs from $K_+(2, 1)$ to first order in $e^2$ by a term

$$K^{(1)}(2, 1) = -ie^2 \int \int K_+(2, 4)\gamma_\mu K_+(4, 3)\gamma_\mu$$
$$\times K_+(3, 1)d\tau_3 d\tau_4 \delta_+(s_{43}^2). \tag{6}$$

It arises because the electron instead of going from 1 directly to 2, may go (Fig. 2) first to 3, $(K_+(3,1))$, emit a quantum $(\gamma_\mu)$, proceed to 4, $(K_+(4,3))$, absorb it $(\gamma_\mu)$, and finally arrive at 2 $(K_+(2, 4))$. The quantum must go from 3 to 4 $(\delta_+(s_{43}^2))$.

This is related to the self-energy of a free electron in the following manner. Suppose initially, time $t_1$, we have an electron in state $f(1)$ which we imagine to be a positive energy solution of Dirac's equation for a free particle. After a long time $t_2 - t_1$ the perturbation will alter the wave function, which can then be looked upon as a superposition of free particle solutions (actually it only contains $f$). The amplitude that $g(2)$ is contained is calculated as in (I, Eq. (21)). The diagonal element $(g = f)$ is therefore

$$\int \int \bar{f}(2)\beta K^{(1)}(2,\, 1)\beta f(1)d^3\mathbf{x}_1 d^3\mathbf{x}_2. \qquad (7)$$

The time interval $T = t_2 - t_1$ (and the spatial volume $V$ over which one integrates) must be taken very large, for the expressions are only approximate (analogous to the situation for two interacting charges).[11] This is because, for example, we are dealing incorrectly with quanta emitted just before $t_2$ which would normally be reabsorbed at times after $t_2$.

If $K^{(1)}(2,\, 1)$ from (6) is actually substituted into (7) the surface integrals can be performed as was done in obtaining I, Eq. (22) resulting in

$$-ie^2 \int \int \bar{f}(4)\gamma_\mu K_+(4,\, 3)\gamma_\mu f(3)\delta_+(s_{43}^2)d\tau_3 d\tau_4. \qquad (8)$$

Putting for $f(1)$ the plane wave $u \exp(-ip \cdot x_1)$ where $p_\mu$ is the energy $(p_4)$ and momentum of the electron $(\mathbf{p}^2 = m^2)$, and $u$ is a constant 4-index symbol, (8) becomes

$$-ie^2 \int \int (\bar{u}\gamma_\mu K_+(4,\, 3)\gamma_\mu u) \exp(ip \cdot (x_4 - x_2))\delta_+(s_{43}^2)d\tau_3 d\tau_4,$$

the integrals extending over the volume $V$ and time interval $T$. Since $K_+(4,3)$ depends only on the difference of the coordinates of 4 and 3, $x_{43\mu}$, the integral on 4 gives a result (except near the surfaces of the region) independent of 3. When integrated on 3, therefore, the result is of order $VT$. The effect is proportional to $V$, for the wave functions have been normalized to unit volume. If normalized to volume $V$, the result would simply be proportional to $T$. This is expected, for if the effect were equivalent to a change in energy $\Delta E$, the amplitude for arrival in $f$ at $t_2$ is altered by a factor $\exp(-i\Delta E(t_2 - t_1))$, or to first order by the difference $-i(\Delta E)T$.

---

[11] This is discussed in reference 5 in which it is pointed out that the concept of a wave function loses accuracy if there are delayed self-actions.

Hence, we have

$$\Delta E = e^2 \int (\bar{u}\gamma_\mu K_+(4,3)\gamma_\mu u)\exp(ip\cdot x_{43})\delta_+(s_{43}^2)d\tau_4, \quad (9)$$

integrated over all space-time $d\tau_4$. This expression will be simplified presently. In interpreting (9) we have tacitly assumed that the wave functions are normalized so that $(u^*u = (\bar{u}\gamma_4 u) = 1$. The equation may therefore be made independent of the normalization by writing the left side as $(\Delta E)(\bar{u}\gamma_4)u)$, or since $(\bar{u}\gamma_4 u) = (E/m)(\bar{u}u)$ and $m\Delta m = E\Delta E$, as $\Delta m(\bar{u}u)$ where $\Delta m$ is an equivalent change in mass of the electron. In this form invariance is obvious.

One can likewise obtain an expression for the energy shift for an electron in a hydrogen atom. Simply replace $K_+$ in (8), by $K_+^{(V)}$, the exact kernel for an electron in the potential, $\mathbf{V} = \beta e^2/r$, of the atom, and $f$ by a wave function (of space and time) for an atomic state. In general the $\Delta E$ which results is not real. The imaginary part is negative and in $\exp(-i\Delta E T)$ produces an exponentially decreasing amplitude with time. This is because we are asking for the amplitude that an atom initially with no photon in the field, will still appear after time $T$ with no photon. If the atom is in a state which can radiate, this amplitude must decay with time. The imaginary part of $\Delta E$ when calculated does indeed give the correct rate of radiation from atomic states. It is zero for the ground state and for a free electron.

In the non-relativistic region the expression for $\Delta E$ can be worked out as has been done by Bethe.[12] In the relativistic region (points 4 and 3 as close together as a Compton wave-length) the $K_+^{(V)}$ which should appear in (8) can be replaced to first order in $\mathbf{V}$ by $K_+$ plus $K_+^{(1)}$ (2,1) given in I, Eq. (13). The problem is then very similar to the radiationless scattering problem discussed below.

[12]H. A. Bethe, Phys. Rev. **72**, 339 (1947).

# 4 EXPRESSION IN MOMENTUM AND ENERGY SPACE

The evaluation of (9), as well as all the other more complicated expressions arising in these problems, is very much simplified by working in the momentum and energy variables, rather than space and time. For this we shall need the Fourier Transform of $\delta_+(s_{21}^2)$ which is

$$-\delta_+(s_{21}^2) = \pi^{-1} \int \exp(-ik \cdot x_{21})\mathbf{k}^{-2}d^4k, \tag{10}$$

which can be obtained from (3) and (5) or from I, Eq. (32) noting that $I_+(2,1)$ for $m^2 = 0$ is $\delta_+(s_{21}^2)$ from I, Eq. (34). The $\mathbf{k}^{-2}$ means $(k \cdot k)^{-1}$ or more precisely the limit as $\delta \to 0$ of $(k \cdot k + i\delta)^{-1}$. Further $d^4k$ means $(2\pi)^{-2}dk_1 dk_2 dk_3 dk_4$. If we imagine that quanta are particles of zero mass, then we can make the general rule that all poles are to be resolved by considering the masses of the particles and quanta to have infinitesimal negative imaginary parts.

Using these results we see that the self-energy (9) is the matrix element between u and u of the matrix

$$(e^2\pi i) \int \gamma_\mu(\mathbf{p} - \mathbf{k} - m)^{-1}\gamma_\mu \mathbf{k}^{-2}d^4k, \tag{11}$$

where we have used the expression (I, Eq. (31)) for the Fourier transform of $K_+$. This form for the self-energy is easier to work with than is (9).

The equation can be understood by imagining (Fig. 3) that the electron of momentum $\mathbf{p}$ emits ($\gamma_\mu$) a quantum of momentum $\mathbf{k}$, and makes its way now with momentum $\mathbf{p} - \mathbf{k}$ to the next event (factor $(\mathbf{p} - \mathbf{k} - m)^{-1}$) which is to absorb the quantum (another $\gamma_\mu$). The amplitude of propagation of quanta is $\mathbf{k}^{-2}$. (There is a factor $e^2/\pi i$ for each virtual quantum). One integrates over all quanta. The reason an electron of momentum $\mathbf{p}$ propagates as $1/(\mathbf{p} - m)$ is that this operator is the reciprocal of the Dirac equation operator, and we are simply solving this equation. Likewise light goes as $1/\mathbf{k}^2$, for this is the reciprocal D'Alembertian operator of the wave equation of light. The first $\gamma_\mu$, represents the current which generates the vector potential,

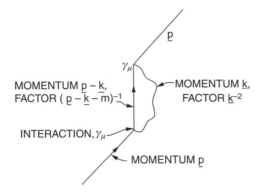

FIG. 3   Interaction of an electron with itself. Momentum space, Eq. (11).

while the second is the velocity operator by which this potential is multiplied in the Dirac equation when an external field acts on an electron.

Using the same line of reasoning, other problems may be set up directly in momentum space. For example, consider the scattering in a potential $\mathbf{A} = A_\mu \gamma_\mu$ varying in space and time as a $\exp(-iq\cdot x)$. An electron initially in state of momentum $\mathbf{p}_1 = p_{1\mu}\gamma_\mu$ will be deflected to state $\mathbf{p}_2$ where $\mathbf{p}_2 = \mathbf{p}_1 + \mathbf{q}$. The zero-order answer is simply the matrix element of a between states 1 and 2. We next ask for the first order (in $e^2$) radiative correction due to virtual radiation of one quantum. There are several ways this can happen. First for the case illustrated in Fig. 4(a), find the matrix:

$$(e^2/\pi i) \int \gamma_\mu (\mathbf{p}_2 - \mathbf{k} - m)^{-1}\mathbf{a}(\mathbf{p}_1 - \mathbf{k} - m)^{-1}\gamma_\mu \mathbf{k}^{-2} d^4 k. \quad (12)$$

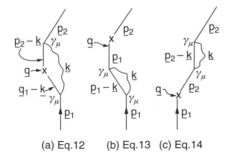

(a) Eq.12     (b) Eq.13   (c) Eq.14

FIG. 4   Radiative correction to scattering, momentum space.

For in this case, first[13] a quantum of momentum $\mathbf{k}$ is emitted $(\gamma_\mu)$, the electron then having momentum $\mathbf{p}_1 - \mathbf{k}$ and hence propagating with factor $(\mathbf{p}_1 - \mathbf{k} - m)^{-1}$. Next it is scattered by the potential (matrix $\mathbf{a}$) receiving additional momentum $\mathbf{q}$, propagating on then (factor $(\mathbf{p}_2 - \mathbf{k} - m)^{-1}$) with the new momentum until the quantum is reabsorbed $(\gamma_\mu)$. The quantum propagates from emission to absorption $(\mathbf{k}^{-2})$ and we integrate over all quanta $(d^4 k)$, and sum on polarization $\mu$. When this is integrated on $k_4$, the result can be shown to be exactly equal to the expressions (16) and (17) given in $B$ for the same process, the various terms coming from residues of the poles of the integrand (12).

Or again if the quantum is both emitted and reabsorbed before the scattering takes place one finds (Fig. 4(b))

$$(e^2/\pi i) \int \mathbf{a}(\mathbf{p}_1 - m)^{-1} \gamma_\mu (\mathbf{p}_1 - \mathbf{k} - m)^{-1} \gamma_\mu \mathbf{k}^{-2} d^4 k, \qquad (13)$$

or if both emission and absorption occur after the scattering, (Fig. 4(c))

$$(e^2/\pi i) \int \gamma_\mu (\mathbf{p}_2 - \mathbf{k} - m)^{-1} \gamma_\mu (\mathbf{p}_2 - m)^{-1} \mathbf{a} \mathbf{k}^2 d^4 k. \qquad (14)$$

These terms are discussed in detail below.

We have now achieved our simplification of the form of writing matrix elements arising from virtual processes. Processes in which a number of real quanta is given initially and finally offer no problem (assuming correct normalization). For example, consider the Compton effect (Fig. 5(a)) in which an electron in state $\mathbf{p}_1$ absorbs a quantum of momentum $\mathbf{q}_1$, polarization vector $e1\mu$ so that its interaction is $e_{1\mu}\gamma_\mu = \mathbf{e}_1$, and emits a second quantum of momentum $-\mathbf{q}_2$ polarization $\mathbf{e}_2$ to arrive in final state of momentum $\mathbf{p}_2$. The matrix for this process is $\mathbf{e}_2(\mathbf{p}_1 + \mathbf{q}_1 - m)^{-1}\mathbf{e}_1$. The total matrix for the Compton effect is, then,

$$\mathbf{e}_2(\mathbf{p}_1 + \mathbf{q}_1 - m)^{-1}\mathbf{e}_1 + \mathbf{e}_1(\mathbf{p}_1 + \mathbf{q}_2 - m)^{-1}\mathbf{e}_3, \qquad (15)$$

---

[13] First, next, etc., here refer not to the order in true time but to the succession of events along the trajectory of the electron. That is, more precisely, to the order of appearance of the matrices in the expressions.

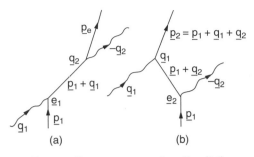

FIG. 5   Compton scattering, Eq. (15).

the second term arising because the emission of $e_2$, may also precede the absorption of $e_1$ (Fig. 5(b)). One takes matrix elements of this between initial and final electron states ($\mathbf{p}_1 + \mathbf{q}_1 = \mathbf{p}_2 - \mathbf{q}_2$), to obtain the Klein Nishina formula. Pair annihilation with emission of two quanta, etc., are given by the same matrix, positron states being those with negative time component of $\mathbf{p}$. Whether quanta are absorbed or emitted depends on whether the time component of $\mathbf{q}$ is positive or negative.

# 5 THE CONVERGENCE OF PROCESSES WITH VIRTUAL QUANTA

These expressions are, as has been indicated, no more than a re-expression of conventional quantum electrodynamics. As a consequence, many of them are meaningless. For example, the self-energy expression (9) or (11) gives an infinite result when evaluated. The infinity arises, apparently, from the coincidence of the $\delta$–function singularities in $K_+(4,3)$ and $\delta_+(s_{43}^2)$. Only at this point is it necessary to make a real departure from conventional electrodynamics, a departure other than simply rewriting expressions in a simpler form.

We desire to make a modification of quantum electrodynamics analogous to the modification of classical electrodynamics described in a previous article, $A$. There the $\delta(s_{12}^2)$ appearing in the action of interaction was replaced by $f(s_{12}^2)$ where $f(x)$ is a function of small width and great height.

The obvious corresponding modification in the quantum theory is to replace the $\delta_+(s^2)$ appearing the quantum mechanical interaction by a new function $f = (s^2)$. We can postulate that if the Fourier transform of the classical $f(s_{12}^2)$ is the integral over ail $\mathbf{k}$ of $F(\mathbf{k}^2)\exp(-ik\cdot x_{12})d^4k$, then the Fourier transform of $f_+(s^2)$ is the same integral taken over only positive frequencies $k_4$ for $t_2 > t_1$ and over only negative ones for $t_2 < t_1$ in analogy to the relation of $\delta_+(s^2)$ to $\delta(s^2)$. The function $f(s^2) = f(x \cdot x)$ can be written[14] as

$$f(x \cdot x) = (2\pi)^{-2} \int_{k_4=0}^{\infty} \int \sin(k_4|x_4|)\cos(\mathbf{K}\cdot\mathbf{x})dk_4 d^3\mathbf{K} g(k\cdot k),$$

where $g(k \cdot k)$ is $k_4^{-1}$ times the density of oscillators and may be expressed for positive $k_4$ as (A, Eq. (16))

$$g(\mathbf{k}^2) = \int_0^{\infty} (\delta(\mathbf{k}^2) - \delta(\mathbf{k}^2 - \lambda^2))G(\lambda)d\lambda,$$

where $\int_0^{\infty} G(\lambda)d\lambda = 1$ and $G$ involves values of $\lambda$ large compared to $m$. This simply means that the amplitude for propagation of quanta of momentum $\mathbf{k}$ is

$$-F_+(\mathbf{k}^3) = \pi^{-1}\int_0^{\infty} (\mathbf{k}^{-2} - (\mathbf{k}^2 - \lambda^2)^{-1}G(\lambda)d\lambda,$$

rather than $\mathbf{k}^{-2}$. That is, writing $F_+(\mathbf{k}^2) = -\pi^{-1}\mathbf{k}^{-2}C(\mathbf{k}^2)$,

$$-f_+(s_{12}^2) = \pi^{-1}\int \exp(-ik \cdot x_{12})\mathbf{k}^{-2}C(\mathbf{k}^2)d^4k. \qquad (16)$$

Every integral over an intermediate quantum which previously involved a factor $d^4k/\mathbf{k}^2$ is now supplied with a convergence factor $C(\mathbf{k}^2)$ where

$$C(\mathbf{k}^2) = \int_0^{\infty} -\lambda^2(\mathbf{k}^2 - \lambda^2)^{-1}G(\lambda)d\lambda. \qquad (17)$$

The poles are defined by replacing $\mathbf{k}^2$ by $\mathbf{k}^2 + i\delta$ in the limit $\delta \to 0$. That is $\lambda^2$ may be assumed to have an infinitesimal negative imaginary part.

---

[14]This relation is given incorrectly in $A$, equation just preceding 16.

The function $f_+(s_{12}^2)$ may still have a discontinuity in value on the light cone. This is of no influence for the Dirac electron. For a particle satisfying the Klein Gordon equation, however, the interaction involves gradients of the potential which reinstates the $\delta$ function if $f$ has discontinuities. The condition that $f$ is to have no discontinuity in value on the light cone implies $\mathbf{k}^2 C(\mathbf{k}^2)$ approaches zero as $\mathbf{k}^2$ approaches infinity. In terms of $G(\lambda)$ the condition is

$$\int_0^\infty \lambda^2 G(\lambda) d\lambda = 0. \tag{18}$$

This condition will also be used in discussing the convergence of vacuum polarization integrals.

The expression for the self-energy matrix is now

$$(e^2/\pi i) \int \gamma_\mu (\mathbf{p} - \mathbf{k} - m)^{-1} \gamma_\mu \mathbf{k}^{-2} d^4 k\, C(\mathbf{k}^2), \tag{19}$$

which, since $C(\mathbf{k}^2)$ falls off at least as rapidly as $1/\mathbf{k}^2$, converges. For practical purposes we shall suppose hereafter that $C(\mathbf{k}^2)$ is simply $-\lambda^2/(\mathbf{k}^2 - \lambda^2)$ implying that some average (with weight $G(\lambda) d\lambda$) over values of $\lambda$ may be taken afterwards. Since in all processes the quantum momentum will be contained in at least one extra factor of the form $(\mathbf{p} - \mathbf{k} - m)^{-1}$ representing propagation of an electron while that quantum is in the field, we can expect all such integrals with their convergence factors to converge and that the result of all such processes will now be finite and definite (excepting the processes with closed loops, discussed below, in which the diverging integrals are over the momenta of the electrons rather than the quanta).

The integral of (19) with $C(\mathbf{k}^2) = -\lambda^2(\mathbf{k}^2 - \lambda^2)^{-1}$ noting that $\mathbf{p}^2 = m^2$, $\lambda \gg m$ and dropping terms of order $m/\lambda$ is (see Appendix A)

$$(e^2/2\pi) \left[ 4m \left( \ln(\lambda/m) + \frac{1}{2} \right) - \mathbf{p}\left( \ln(\lambda/m) + 5/4 \right) \right]. \tag{20}$$

When applied to a state of an electron of momentum $\mathbf{p}$ satisfying $\mathbf{p}u = mu$, it gives for the change in mass (as in B, Eq. (9))

$$\Delta m = m(e^2/2\pi)\left(3\ln(\lambda/m) + \frac{3}{4}\right). \qquad (21)$$

# 6 RADIATIVE CORRECTIONS TO SCATTERING

We can now complete the discussion of the radiative corrections to scattering. In the integrals we include the convergence factor $C(\mathbf{k}^2)$, so that they converge for large $\mathbf{k}$. Integral (12) is also not convergent because of the well-known infrared catastrophy. For this reason we calculate (as discussed in B) the value of the integral assuming the photons to have a small mass $\lambda_{min} \ll m \ll \lambda$. The integral (12) becomes

$$(e^2/\pi i)\int \gamma_\mu(\mathbf{p}_2 - \mathbf{k} - m)^{-1}\mathbf{a}(\mathbf{p}_1 - \mathbf{k} - m)^{-1}$$

$$\times \gamma_\mu(\mathbf{k}^2 - \gamma_{min}^2)^{-1}d^4kC(\mathbf{k}^2 - \lambda_{min}^2),$$

which when integrated (see Appendix B) gives $(e^2/2\pi)$ times

$$\left[2\left(\ln\frac{m}{\lambda_{min}} - 1\right)\left(1 - \frac{2\theta}{\tan 2\theta}\right) + \theta\tan\theta\right.$$

$$\left. + \frac{4}{\tan 2\theta}\int_0^\theta \alpha\tan\alpha\, d\alpha\right]\mathbf{a} + \frac{1}{4m}(\mathbf{qa} - \mathbf{aq})\frac{2\theta}{\sin 2\theta} + r\mathbf{a},$$

$$(22)$$

where $(\mathbf{q}^2)^{1/2} = 2m$ and we have assumed the matrix to operate between states of momentum $\mathbf{p}_1$ and $\mathbf{p}_2 = \mathbf{p}_1 + \mathbf{q}$ and have neglected terms of order $\lambda_{min}/m$, $m/\lambda$, and $\mathbf{q}^2/\lambda^2$. Here the only dependence on the convergence factor is in the term $r\mathbf{a}$, where

$$\mathbf{r} = \ln(\lambda/m) + 9/4 - 2\ln(m/\lambda_{min}). \qquad (23)$$

As we shall see in a moment, the other terms (13), (14) give contributions which just cancel the $r\mathbf{a}$ term. The remaining terms give for small $\mathbf{q}$,

$$(e^2/4\pi) \left( \frac{1}{2m}(\mathbf{qa} - \mathbf{aq}) + \frac{4\mathbf{q}^2}{3m^2}\mathbf{a}\left( \ln \frac{m}{\lambda_{min}} - \frac{3}{8} \right) \right), \qquad (24)$$

which shows the change in magnetic moment and the Lamb shift as interpreted in more detail in B.[15]

We must now study the remaining terms (13) and (14). The integral on $\mathbf{k}$ in (13) can be performed (after multiplication by $C(\mathbf{k}^2)$) since it involves nothing but the integral (19) for the self-energy and the result is allowed to operate on the initial state $u_1$, (so that $\mathbf{p}_1 u_1 = mu_1$). Hence the factor following $\mathbf{a}(\mathbf{p}_1 - m)^{-1}$ will be just $\Delta m$. But, if one now tries to expand $1/(\mathbf{p}_1 = m) = (\mathbf{p}_1 + m)/(\mathbf{p}_1^2 - m^2)$ one obtains an infinite result, since $\mathbf{p}_1^2 = m^2$. This is, however, just what is expected physically. For the quantum can be emitted and absorbed at any time previous to the scattering. Such a process has the effect of a change in mass of the electron in the state 1. It therefore changes the energy by $\Delta E$ and the amplitude to first order in $\Delta E$ by $-i\Delta E \cdot t$ where $t$ is the time it is acting, which is infinite. That is, the major effect of this term would be canceled by the effect of change of mass $\Delta m$.

The situation can be analyzed in the following manner. We suppose that the electron approaching the scattering potential a has not been free for an infinite time, but at some time far past suffered a scattering by a potential $\mathbf{b}$. If we limit our discussion to the effects of $\Delta m$ and of the virtual radiation of one quantum between two such scatterings each of the effects will be finite, though large, and their difference is determinate. The propagation from $\mathbf{b}$ to $\mathbf{a}$ is represented by a matrix

$$\mathbf{a}(\mathbf{p}' - m)^{-1}\mathbf{b}, \qquad (25)$$

[15]That the result given in B in Eq. (19) was in error was repeatedly pointed out to the author, in private communication, by V. F. Weisskopf and J. B. French, as their calculation, completed simultaneously with the author's early in 1948, gave a different result. French has finally shown that although the expression for the radiationless scattering B, Eq. (18) or (24) above is correct, it was incorrectly joined into Bethe's non-relativistic result. He shows that the relation $\ln 2k_{max} - 1 = \ln \lambda_{min}$ used by the author should have been $2k_{max} - 5/6 = \ln \lambda_{min}$. This results in adding a term $-(1/6)$ to the logarithm in B, Eq. (19) so that the result now agrees with that of J. B. French and V. F. Weisskopf, Phys. Rev. **75**, 1240 (1949) and N. H. Kroll and W. E. Lamb, Phys. Rev. **75**, 388 (1949). The author feels unhappily responsible for the very considerable delay in the publication of French's result occasioned by this error. This footnote is appropriately numbered.

in which one is to integrate possibly over $\mathbf{p}'$ (depending on details of the situation). (If the time is long between $\mathbf{b}$ and $\mathbf{a}$, the energy is very nearly determined so that $\mathbf{p}'^2$ is very nearly $m^2$.)

We shall compare the effect on the matrix (25) of the virtual quanta and of the change of mass $\Delta m$. The effect of a virtual quantum is

$$(e^2/\pi i) \int \mathbf{a}(\mathbf{p}' - m)^{-1} \gamma_\mu (\mathbf{p}' - \mathbf{k} - m)^{-1}$$

$$\times \gamma_\mu (\mathbf{p}' - m)^{-1} \mathbf{b} k^{-2} d^4 k \, C(\mathbf{k}^2), \qquad (26)$$

while that of a change of mass can be written

$$\mathbf{a}(\mathbf{p}' - m)^{-1} \Delta m (\mathbf{p}' - m)^{-1} \mathbf{b}, \qquad (27)$$

and we are interested in the difference (26)–(27). A simple and direct method of making this comparison is just to evaluate the integral on $\mathbf{k}$ in (26) and subtract from the result the expression (27) where $\Delta m$ is given in (21). The remainder can be expressed as a multiple $-r(\mathbf{p}'^2)$ of the unperturbed amplitude (25);

$$-r(\mathbf{p}'^2) \mathbf{a}(\mathbf{p}' - m)^{-1} \mathbf{b}. \qquad (28)$$

This has the same result (to this order) as replacing the potentials $\mathbf{a}$ and $\mathbf{b}$ in (25) by $(1 - \frac{1}{2}r(\mathbf{p}'^2))\mathbf{a}$ and $(1 - \frac{1}{2}r(\mathbf{p}'^2))\mathbf{b}$. In the limit, then, as $\mathbf{p}'^2 \to m^2$ the net effect on the scattering is $-\frac{1}{2}\mathbf{ra}$ where $\mathbf{r}$, the limit of $\mathbf{r}(\mathbf{p}'^2)$ as $\mathbf{p}'^2 \to m^2$ (assuming the integrals have an infrared cut-off), turns out to be just equal to that given in (23). An equal term $-\frac{1}{2}\mathbf{ra}$ arises from virtual transitions after the scattering (14) so that the entire ra term in (22) is canceled.

The reason that $\mathbf{r}$ is just the value of (12) when $\mathbf{q}^2 = 0$ can also be seen without a direct calculation as follows: Let us call $\mathbf{p}$ the vector of length $m$ in the direction of $\mathbf{p}'$ so that if $\mathbf{p}'^2 = m(1 + \epsilon)^2$ we have $\mathbf{p}' = (1 + \epsilon 0\mathbf{p}$ and we take $\epsilon$ as very small, being of order $T^{-1}$ where $T$ is the time between the scatterings $\mathbf{b}$ and $\mathbf{a}$. Since $(\mathbf{p}' - m)^{-1} = (\mathbf{p}' + m)/(\mathbf{p}'^2 - m^2) \approx (\mathbf{p} + m)/2m^2\epsilon$, the quantity (25) is of order $\epsilon^{-1}$ or $T$. We shall compute corrections to it only to its own order ($\epsilon^{-1}$)

in the limit $\epsilon \to 0$. The term (27) can be written approximately[16] as

$$(e^2/\pi i) \int \mathbf{a}(\mathbf{p}' - m)^{-1} \gamma_\mu (\mathbf{p} - \mathbf{k} - m)^{-1}$$
$$\times \gamma_\mu (\mathbf{p}' - m)^{-1} \mathbf{b} \mathbf{k}^{-2} d^4 k C(\mathbf{k}^2),$$

using the expression (19) for $\Delta m$. The net of the two effects is therefore approximately[17]

$$-(e^2/\pi i) \int \mathbf{a}(\mathbf{p}' - m)^{-1} \gamma_\mu (\mathbf{p} - \mathbf{k} - m)^{-1} \epsilon \mathbf{p}) \mathbf{p} - \mathbf{k} - m)^{-1}$$
$$\times \gamma_\mu (\mathbf{p}' - m)^{-1} \mathbf{b} \mathbf{k}^{-2} d^4 k C(\mathbf{k}^2),$$

a term now of order $1/\epsilon$ (since $(\mathbf{p}' - m)^{-1} \approx (\mathbf{p} + m) \times (2m^2\epsilon)^{-1}$) and therefore the one desired in the limit. Comparison to (28) gives for $\mathbf{r}$ the expression

$$(\mathbf{p}_1 + m/2m) \int \gamma_\mu (\mathbf{p}_1 - \mathbf{k} - m)^{-1} (\mathbf{p}_1 m^{-1})(\mathbf{p}_1 - \mathbf{k} - m)^{-1}$$
$$\times \gamma_\mu \mathbf{k}^{-2} d^4 k C(\mathbf{k}^2). \tag{29}$$

The integral can be immediately evaluated, since it is the same as the integral (12), but with $\mathbf{q} = 0$, for $\mathbf{a}$ replaced by $\mathbf{p}_1/m$. The result is therefore $\mathbf{r} \cdot (\mathbf{p}_1/m)$ which when acting on the state $u_1$ is just $\mathbf{r}$, as $\mathbf{p}_1 u_1 = m u_1$. For the same reason the term $(\mathbf{p}_1 + m)/2m$ in (29) is effectively 1 and we are left with $-\mathbf{r}$ of (23).[18]

In more complex problems starting with a free electron the same type of term arises from the effects of a virtual emission and absorption both previous to the other processes. They, therefore, simply lead to the same factor $\mathbf{r}$ so that the expression (23) may be used directly and these renormalization integrals need not be computed afresh for each problem.

---

[16]The expression is not exact because the substitution of $\Delta m$ by the integral in (19) is valid only if $\mathbf{p}$ operates on a state such that $\mathbf{p}$ can be replaced by $m$. The error, however, is of order $\mathbf{a}(\mathbf{p}' - m)^{-1} (\mathbf{p} - m) (\mathbf{p}' - m)^{-1} \mathbf{b}$ which is $\mathbf{a}((1 + \epsilon) \mathbf{p} + m)(\mathbf{p} - m) \times ((1 + \epsilon)\mathbf{p} + m)\mathbf{p}(2\epsilon + \epsilon^2)^{-2}m^{-4}$. But since$^2 = m^2$ we have $\mathbf{p}(\mathbf{p} - m) = -m(\mathbf{p} - m) = (\mathbf{p} - m)\mathbf{p}$ so the net result is approximately $\mathbf{a}(\mathbf{p} - m)\mathbf{b}/4m^2$ and is not of order $1/\epsilon$ but smaller, so that its effect drops out in the limit.

[17]We have used, to first order, the general expansion (valid for any operators $A$, $B$)

$$(A + B0^{-1} = A^{-1} - A^{-1}BA^{-1} + A^{-1}BA^{-1}BA^{-1} - \cdots$$

with $A = \mathbf{p} - \mathbf{k} - m$ and $B = \mathbf{p}' - \mathbf{p} = \epsilon\mathbf{p}$ to expand the difference of $(\mathbf{p}' - m)^{-1}$ and $(\mathbf{p} - \mathbf{k} - m)^{-1}$.

[18]The renormalization terms appearing $B$, Eqs. (14), (15) when translated directly into the present notation do not give twice (29) but give this expression with the central $\mathbf{p}_1 m^{-1}$ factor replaced by $m\gamma_4/E_1$ where $E_1 = p_{1\mu}$, for $\mu = 4$. When integrated it therefore gives $\mathbf{ra}((\mathbf{p}_1 + m)/2m)(m\gamma_4/E_1)$ or $\mathbf{ra} - \mathbf{ra} \ (m\gamma_4/E_1)(\mathbf{p}_1 - m)/2m$. (Since $\mathbf{p}_1\gamma_4 + \gamma_4\mathbf{p}_1 = 2E_1$) which gives just $\mathbf{ra}$, since $\mathbf{p}_1 u_1 = m u_1$.

In this problem of the radiative corrections to scattering the net result is insensitive to the cut-off. This means, of course, that by a simple rearrangement of terms previous to the integration we could have avoided the use of the convergence factors completely (see for example Lewis[19]). The problem was solved in the manner here in order to illustrate how the use of such convergence factors, even when they are actually unnecessary, may facilitate analysis somewhat by removing the effort and ambiguities that may be involved in trying to rearrange the otherwise divergent terms.

The replacement of $\delta_+$ by $f_+$ given in (16), (17) is not determined by the analogy with the classical problem. In the classical limit only the real part of $\delta_+$ (i.e., just $\delta$) is easy to interpret. But by what should the imaginary part, $1/(\pi s^2)$, of $\delta_+$ be replaced? The choice we have made here (in denning, as we have, the location of the poles of (17)) is arbitrary and almost certainly incorrect. If the radiation resistance is calculated for an atom, as the imaginary part of (8), the result depends slightly on the function $f_+$. On the other hand the light radiated at very large distances from a source is independent of $f_+$. The total energy absorbed by distant absorbers will not check with the energy loss of the source. We are in a situation analogous to that in the classical theory if the entire $f$ function is made to contain only retarded contributions (see A, Appendix). One desires instead the analogue of $\langle F \rangle_{\text{ret}}$ of A. This problem is being studied.

One can say therefore, that this attempt to find a consistent modification of quantum electrodynamics is incomplete (see also the question of closed loops, below). For it could turn out that any correct form of $f_+$ which will guarantee energy conservation may at the same time not be able to make the self-energy integral finite. The desire to make the methods of simplifying the calculation of quantum electrodynamic processes more widely available has prompted this publication before an analysis of the correct form for $f_+$ is complete. One might try to take the position that, since the energy discrepancies discussed

[19] H.W. Lewis, Phys. Rev. **73**, 173 (1948).

vanish in the limit $\lambda \rightarrow \infty$, the correct physics might be considered to be that obtained by letting $\lambda \rightarrow \infty$ after mass renormalization. I have no proof of the mathematical consistency of this procedure, but the presumption is very strong that it is satisfactory. (It is also strong that a satisfactory form for $f_+$ can be found.)

## 7 THE PROBLEM OF VACUUM POLARIZATION

In the analysis of the radiative corrections to scattering one type of term was not considered. The potential which we can assume to vary as $a_\mu \exp(-iq \cdot x)$ creates a pair of electrons (see Fig. 6), momenta $\mathbf{p}_a$, $-\mathbf{p}_b$. This pair then reannihilates, emitting a quantum $\mathbf{q} = \mathbf{q}_b - \mathbf{q}_a$, which quantum scatters the original electron from state 1 to state 2. The matrix element for this process (and the others which can be obtained by rearranging the order in time of the various events) is

$$-(e^2/\pi i)(\bar{u}_2\gamma_\mu u_1) \int Sp\left[(\mathbf{p}_a + \mathbf{q} - m)^{-1}\right.$$
$$\left.\times \gamma_\mu(\mathbf{p}_a - m)^{-1}\gamma_\mu\right] d^4 p_a \mathbf{q}^{-2} C(\mathbf{q}^2) a_\nu. \tag{30}$$

This is because the potential produces the pair with amplitude proportional to $a_\nu\gamma_\nu$ the electrons of momenta $\mathbf{p}_a$, and $-(\mathbf{p}_a + \mathbf{q})$ proceed from there to annihilate, producing a quantum (factor $\gamma_\mu$) which propagates (factor $\mathbf{q}^{-2}C(\mathbf{q}^2)$) over to the other electron, by which it is absorbed (matrix element of $\gamma_\mu$, between states 1 and 2 of

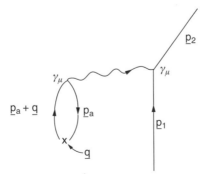

FIG. 6   Vacuum polarization effect on scattering, Eq. (30).

the original electron ($\bar{u}_2\gamma_\mu u_1$)). All momenta $\mathbf{p}_a$ and spin states of the virtual electron are admitted, which means the spur and the integral on $d^4 p_a$ are calculated.

One can imagine that the closed loop path of the positron-electron produces a current

$$4\pi J_\mu a_v,\tag{31}$$

which is the source of the quanta which act on the second electron. The quantity

$$J_{\mu v} = -(e^2/\pi i) \int Sp\left[(\mathbf{p}+\mathbf{q}-m)^{-1}\right.$$
$$\left.\times \gamma_\mu(\mathbf{p}-m)^{-1}\gamma_\mu\right] d^4 p,\tag{32}$$

is then characteristic for this problem of polarization of the vacuum.

One sees at once that $J_{\mu v}$ diverges badly. The modification of $\delta$ to $f$ alters the amplitude with which the current $j_\mu$, will affect the scattered electron, but it can do nothing to prevent the divergence of the integral (32) and of its effects.

One way to avoid such difficulties is apparent. From one point of view we are considering all routes by which a given electron can get from one region of space-time to another, i.e., from the source of electrons to the apparatus which measures them. From this point of view the closed loop path leading to (32) is unnatural. It might be assumed that the only paths of meaning are those which start from the source and work their way in a continuous path (possibly containing many time reversals) to the detector. Closed loops would be excluded. We have already found that this may be done for electrons moving in a fixed potential.

Such a suggestion must meet several questions, however. The closed loops are a consequence of the usual hole theory in electrodynamics. Among other things, they are required to keep probability conserved. The probability that no pair is produced by a potential is not unity and its deviation from unity arises from the imaginary part of $J_{\mu v}$. Again, with closed loops excluded, a pair of electrons once created cannot annihilate one another again, the scattering of light by light would be zero, etc. Although we are not experimentally sure of these phenomena, this does seem to indicate that the closed loops are necessary. To be sure, it is always possible that these matters of probability conservation, etc., will work themselves out as simply in the case of interacting particles as for those in a fixed potential. Lacking such a demonstration the presumption is that the difficulties of vacuum polarization are not so easily circumvented.[20]

An alternative procedure discussed in $B$ is to assume that the function $K_+(2, 1)$ used above is incorrect and is to be replaced by a modified function $K'_+$ having no singularity on the light cone. The effect of this is to provide a convergence factor $C(\mathbf{p}^2 - m^2)$ for every integral over electron momenta.[21] This will multiply the integrand of (32) by $C(\mathbf{p}^2 - m^2)C((\mathbf{p}+\mathbf{q})^2 - m^2)$, since the integral was originally $\delta(\mathbf{p}_a - \mathbf{p}_b + \mathbf{q})d^4p_a d^4p_b$ and both $\mathbf{p}_a$ and $\mathbf{p}_b$ get convergence factors. The integral now converges but the result is unsatisfactory.[22]

One expects the current (31) to be conserved, that is $q_\mu j_\mu = 0$ or $q_\mu J_{\mu\nu} = 0$. Also one expects no current if $a$ is a gradient, or $a_\nu = q_\nu$, times a constant. This leads to the condition $J_{\mu\nu}q_\nu = 0$ which is equivalent to $q_\mu J_{\mu\nu} = 0$ since $J_{\mu\nu}$ is symmetrical. But when the expression (32) is integrated with such convergence factors it does not satisfy this condition. By altering the kernel from $K$ to another, $K'$, which does not satisfy the Dirac equation we have lost the gauge invariance, its consequent current conservation and the general consistency of the theory.

One can see this best by calculating $J_{\mu\nu}q_\nu$ directly from (32). The expression within the spur becomes $(\mathbf{p} + \mathbf{q} - m)^{-1}\mathbf{q}(\mathbf{p} - m)^{-1}\gamma_\mu$ which can be written as the difference of two terms: $(\mathbf{p} - m)^{-1}\gamma_\mu - (\mathbf{p} + \mathbf{q} - m)^{-1}\gamma_\mu$ Each of these terms would give the same result if the integration $d^4p$ were without a convergence factor, for the first can be converted into the second by a shift of the origin of $\mathbf{p}$, namely

---

[20] It would be very interesting to calculate the Lamb shift accurately enough to be sure that the 20 megacycles expected from vacuum polarization are actually present.

[21] This technique also makes self-energy and radiationless scattering integrals finite even without the modification of $\delta_+$ to $f_+$ for the radiation (and the consequent convergence factor $C(\mathbf{k}^2)$ for the quanta). See B.

[22] Added to the terms given below (33) there is a term $\frac{1}{4}(\lambda^3 - 2\mu^2 + \frac{1}{3}\mathbf{q}^2)\delta_{\mu\nu}$ for $C(\mathbf{k}^2) = -\lambda^2(\mathbf{k}^2 - \lambda^2)^{-1}$, which is not gauge invariant. (In addition the charge renormalization has $-7/6$ added to the logarithm.)

$\mathbf{p}' = \mathbf{p} + \mathbf{q}$. This does not result in cancelation in (32) however, for the convergence factor is altered by the substitution.

A method of making (32) convergent without spoiling the gauge invariance has been found by Bethe and by Pauli. The convergence factor for light can be looked upon as the result of superposition of the effects of quanta of various masses (some contributing negatively). Likewise if we take the factor $C(\mathbf{p}^2 - m^2) = -\lambda^2(\mathbf{p}^2 - m^2 - \lambda^2)^{-1}$ so that $(\mathbf{p}^2 - m^2)^{-1}C(\mathbf{p}^2 - m^2) = (\mathbf{p}^2 - m^2)^{-1} - (\mathbf{p}^2 - m^2 - \lambda^2)^{-1}$ we are taking the difference of the result for electrons of mass $m$ and mass $(\lambda^2 + m^2)^{1/2}$. But we have taken this difference for *each* propagation between interactions with photons. They suggest instead that once created with a certain mass the electron should continue to propagate with this mass through all the potential interactions until it closes its loop. That is if the quantity (32), integrated over some finite range of $\mathbf{p}$, is called $J_{\mu\nu}(m^2)$ and the corresponding quantity over the same range of $\mathbf{p}$, but with $m$ replaced by $(m^2 + \lambda^2)^{1/2}$ is $J_{\mu\nu}(m^2 + \lambda^2)$ we should calculate

$$J_{\mu\nu}^P = \int_0^\infty \left[ J_{\mu\nu}(m^2) - J_{\mu\nu}(m^2 + \lambda^2) \right] G(\lambda)d\lambda, \qquad (32')$$

the function $G(\lambda)$ satisfying $\int_0^\infty G(\lambda)d\lambda = 1$ and $\int_0^\infty G(\lambda)\lambda^2 d\lambda = 0$. Then in the expression for $J_{\mu\nu}^P$ the range of $\mathbf{p}$ integration can be extended to infinity as the integral now converges. The result of the integration using this method is the integral on $d\lambda$ over $G(\lambda)$ of (see Appendix C)

$$J_{\mu\nu}^P = -\frac{e^2}{\pi}(q_\mu q_\nu - \delta_{\mu\nu}\mathbf{q}^2)$$

$$\times \left( -\frac{1}{3}\ln\frac{\lambda^2}{m^2} - \left[ \frac{4m^2 + 2\mathbf{q}^2}{3\mathbf{q}^2} \left( 1 - \frac{\theta}{\tan\theta} \right) - \frac{1}{9} \right] \right), \qquad (33)$$

with $\mathbf{q}^2 = 4m^2 \sin^2\theta$.

The gauge invariance is clear, since $q_\mu(q_\mu q_\nu - \mathbf{q}^2\delta_{\mu\nu}) = 0$. Operating (as it always will) on a potential of zero divergence the $(q_\mu q_\nu - \delta_{\mu\nu}\mathbf{q}^2)a_\nu$, is simply $-q^2 a_\mu$, the D'Alembertian of the potential, that is, the current producing the potential. The term $-\frac{1}{3}(\ln(\lambda^2/m^2))(q_\mu q_\nu - \mathbf{q}^2\delta_{\mu\nu})$ therefore gives a current proportional to the current producing the potential. This would have the same effect as a change in charge, so that we would have a difference $\Delta(e^2)$ between $e^2$ and the experimentally observed charge, $e^2 + \Delta(e^2)$, analogous to the difference between $m$ and the observed mass. This charge depends logarithmically on the cut-off, $\Delta(e^2)/e^2 = -(2e^2/3\pi)\ln(\lambda/m)$. After this renormalization of charge is made, no effects will be sensitive to the cut-off.

After this is done the final term remaining in (33), contains the usual effects[23] of polarization of the vacuum. It is zero for a free light quantum ($\mathbf{q}^2 = 0$). For small $\mathbf{q}^2$ it behaves as $(2/15)\mathbf{q}^2$ (adding $-\frac{1}{5}$ to the logarithm in the Lamb effect). For $\mathbf{q}^2 > (2m)^2$ it is complex, the imaginary part representing the loss in amplitude required by the fact that the probability that no quanta are produced by a potential able to produce pairs ($(\mathbf{q}^2)^{1/2} > 2m$) decreases with time. (To make the necessary analytic continuation, imagine $m$ to have a small negative imaginary part, so that $(1 - \mathbf{q}^2/4m^2 - 1)^{1/2}$ becomes $-i(\mathbf{q}^2/4m^2 - 1)^{1/2}$ as $\mathbf{q}^2$ goes from below to above $4m^2$. Then $\theta = \pi/2 + iu$ where $\sin hu = +(\mathbf{q}^2/4m^2 - 1)^{1/2}$, and $-1/\tan\theta = i\tan hu = +i(\mathbf{q}^2 - 4m^2)^{1/2}(\mathbf{q}^2)^{-1/2})$.

Closed loops containing a number of quanta or potential interactions larger than two produce no trouble. Any loop with an odd number of interactions gives zero (I, reference 9). Four or more potential interactions give integrals which are convergent even without a convergence factor as is well known. The situation is analogous to that for self-energy. Once the simple problem of a single closed loop is solved there are no further divergence difficulties for more complex processes.[24]

---

[23] E. A. Uehling, Phys. Rev. **48**, 55 (1935), R. Serber, Phys. Rev. **48**, 49 (1935).

[24] There are loops completely without external interactions. For example, a pair is created virtually along with a photon. Next they annihilate, absorbing this photon. Such loops are disregarded on the grounds that they do not interact with anything and are thereby completely unobservable. Any indirect effects they may have via the exclusion principle have already been included.

# 8 LONGITUDINAL WAVES

In the usual form of quantum electrodynamics the longitudinal and transverse waves are given separate treatment. Alternately the condition $(\partial A_\mu / \partial x_\mu)\Psi = 0$ is carried along as a supplementary condition. In the present form no such special considerations are necessary for we are dealing with the solutions of the equation $-\Box^2 A_\mu = 4\pi j_\mu$, with a current $j_\mu$, which is conserved $\partial j_\mu / \partial x_\mu = 0$. That means at least $\Box^2(\partial A_\mu / \partial x_\mu) = 0$ and in fact our solution also satisfies $\partial A_\mu / \partial x_\mu = 0$.

To show that this is the case we consider the amplitude for emission (real or virtual) of a photon and show that the divergence of this amplitude vanishes. The amplitude for emission for photons polarized in the $\mu$ direction involves matrix elements of $\gamma_\mu$. Therefore what we have to show is that the corresponding matrix elements of $q_\mu \gamma_\mu = \mathbf{q}$ vanish. For example, for a first order effect we would require the matrix element of $\mathbf{q}$ between two states $\mathbf{p}_1$ and $\mathbf{p}_2 = \mathbf{p}_1 + \mathbf{q}$. But since $\mathbf{q} = \mathbf{p}_2 - \mathbf{p}_1$ and $(\bar{u}_2 \mathbf{p}_1 u_1) = m(\bar{u}_2 u_1) = (\bar{u}_2 \mathbf{p}_2 u_1)$ the matrix element vanishes, which proves the contention in this case. It also vanishes in more complex situations (essentially because of relation (34), below) (for example, try putting $\mathbf{e}_2 = \mathbf{q}_2$ in the matrix (15) for the Compton Effect).

To prove this in general, suppose $\mathbf{a}_i$, $i = 1$ to $N$ are a set of plane wave disturbing potentials carrying momenta $\mathbf{q}_i$, (e.g., some may be emissions or absorptions of the same or different quanta) and consider a matrix for the transition from a state of momentum $\mathbf{p}_0$ to $\mathbf{p}_N$ such as $\mathbf{a}_N \Pi_{t=1}^{N-1} (\mathbf{p}_i - m)^{-1} \mathbf{a}_i$, where $\mathbf{p}_i = \mathbf{p}_{i-1} + \mathbf{q}_i$ (and in the product, terms with larger $i$ are written to the left). The most general matrix element is simply a linear combination of these. Next consider the matrix between states $\mathbf{p}_0$ and $\mathbf{p}_N + \mathbf{q}$ in a situation in which not only are the $\mathbf{a}_i$, acting but also another potential $\mathbf{a}$ $\exp(-iq \cdot x)$ where $\mathbf{a} = \mathbf{q}$. This may act previous to all $\mathbf{a}_i$ in which case it gives $\mathbf{a}_N \Pi (\mathbf{p}_i + \mathbf{q} - m)^{-1} \mathbf{a}_i (\mathbf{p}_0 + \mathbf{q} - m)^{-1} \mathbf{q}$ which is equivalent to $+\mathbf{a}_N \Pi (\mathbf{p}_i + \mathbf{q} - m)^{-1} \mathbf{a}_i$ since $+ (\mathbf{p}_0 + \mathbf{q} - m)^{-1} \mathbf{q}$ is equivalent to $(\mathbf{p}_0 + \mathbf{q} - m)^{-1} \times (\mathbf{p}_0 + \mathbf{q} - m)$ as $\mathbf{p}_0$ is equivalent to $m$ acting on the initial state. Likewise if it acts after all the potentials it gives $\mathbf{q}(\mathbf{p}_N - m)^{-1}\mathbf{a}_N \Pi(\mathbf{p}_i - m)^{-1}\mathbf{a}_i$ which is equivalent to $-\mathbf{a}_N \Pi(\mathbf{p}_i - m)^{-1}\mathbf{a}_i$ since $\mathbf{p}_N + \mathbf{q} - m$ gives zero on the final state. Or again it may act between the potential $\mathbf{a}_k$ and $\mathbf{a}_{k+1}$ for each $k$. This gives

$$\sum_{k=1}^{N-1} \mathbf{a}_N \prod_{i=k+1}^{N-1} (\mathbf{p}_i + \mathbf{q} - m)^{-1}\mathbf{a}_i(\mathbf{p}_k + \mathbf{q} - m)^{-1}$$

$$\times \mathbf{q}(\mathbf{p}_k - m)^{-1}\mathbf{a}_k \prod_{j=1}^{k-1}(\mathbf{p}_j - m)^{-1}\mathbf{a}_j.$$

However,

$$(\mathbf{p}_k + \mathbf{q} - m)^{-1}\mathbf{q}(\mathbf{p}_k - m)^{-1} = (\mathbf{p}_k - m)^{-1} - (\mathbf{p}_k + \mathbf{q} - m)^{-1},$$

(34)

so that the sum breaks into the difference of two sums, the first of which may be converted to the other by the replacement of $k$ by $k - 1$. There remain only the terms from the ends of the range of summation,

$$+\mathbf{a}_N \prod_{i=1}^{N-1} (\mathbf{p}_i - m)^{-1}\mathbf{a}_i - \mathbf{a}_N \prod_{i=1}^{N-1} (\mathbf{p}_i + \mathbf{q} - m)^{-1}\mathbf{a}_i.$$

These cancel the two terms originally discussed so that the entire effect is zero. Hence any wave emitted will satisfy $\partial A_\mu / \partial x_\mu = 0$. Likewise longitudinal waves (that is, waves for which $A_\mu = \partial\phi/\partial x_\mu$ or $\mathbf{a} = \mathbf{q}$ cannot be absorbed and will have no effect, for the matrix elements for emission and absorption are similar. (We have said little more than that a potential $A_\mu = \partial\varphi/\partial x_\mu$ has no effect on a Dirac electron since a transformation $\psi' = \exp(-i\phi)\psi$ removes it. It is also easy to see in coordinate representation using integrations by parts.)

This has a useful practical consequence in that in computing probabilities for transition for unpolarized light one can sum the squared matrix over all four directions rather than just the two special polarization vectors. Thus suppose the matrix element for some process for light polarized in direction $e_\mu$, is $e_\mu M_\mu$. If the light has wave vector

$q_\mu$, we know from the argument above that $q_\mu M_\mu = 0$. For unpolarized light progressing in the $z$ direction we would ordinarily calculate $M_x^2 + M_y^2$. But we can as well sum $M_x^2 + M_y^2 + M_z^2 - M_t^2$ for $q_\mu M_\mu$ implies $M_t = M_z$ since $q_t = q_z$ for free quanta. This shows that unpolarized light is a relativistically invariant concept, and permits some simplification in computing cross sections for such light.

Incidentally, the virtual quanta interact through terms like $\gamma_\mu \ldots \gamma_\mu \mathbf{k}^{-2} d^4 k$. Real processes correspond to poles in the formulae for virtual processes. The pole occurs when $\mathbf{k}^2 = 0$, but it looks at first as though in the sum on all four values of $\mu$, of $\gamma_\mu \ldots \gamma_\mu$ we would have four kinds of polarization instead of two. Now it is clear that only two perpendicular to $\mathbf{k}$ are effective.

The usual elimination of longitudinal and scalar virtual photons (leading to an instantaneous Coulomb potential) can of course be performed here too (although it is not particularly useful). A typical term in a virtual transition is $\gamma_\mu \ldots \gamma_\mu \mathbf{k}^{-2} d^4 k$ where the $\ldots$ represent some intervening matrices. Let us choose for the values of $\mu$, the time $t$, the direction of vector part $\mathbf{K}$, of $\mathbf{k}$, and two perpendicular directions 1, 2. We shall not change the expression for these two 1, 2 for these are represented by transverse quanta. But we must find $(\gamma_t \ldots \gamma_t) - (\gamma_\mathbf{K} \ldots \gamma_\mathbf{K})$. Now $\mathbf{k} = k_4 \gamma_t - K \gamma_\mathbf{K}$, where $K = (\mathbf{K} \cdot \mathbf{K})^{1/2}$, and we have shown above that $\mathbf{k}$ replacing the $\gamma_\mu$. gives zero.[25] Hence $K \gamma_\mathbf{K}$ is equivalent to $k_4 \gamma_t$ and

$$(\gamma_t \ldots \gamma_t) - (\gamma_\mathbf{K} \ldots \gamma_\mathbf{K}) = ((K^2 - k_4^2)/K^2)(\gamma_t \ldots \gamma_y),$$

so that on multiplying by $\mathbf{k}^{-2} d^4 k = d^4 k (k_4^2 - K^2)^{-1}$ the net effect is $-(\gamma_t \ldots \gamma_t) d^4 k / K^2$. The $\gamma_t$ means just scalar waves, that is, potentials produced by charge density. The fact that $1/K^2$ does not contain $k_4$ means that $k_4$ can be integrated first, resulting in an instantaneous interaction, and the $d^3 \mathbf{K}/K^2$ is just the momentum representation of the Coulomb potential, $1/\mathbf{r}$.

[25] A little more care is required when both $\gamma_\mu$'s act on the same particle. Define $\mathbf{x} = k_4 \gamma_t + K \gamma_\mathbf{K}$, and consider $(\mathbf{k} \ldots \mathbf{x}) + \mathbf{x} \ldots \mathbf{k})$. Exactly this term would arise if a system, acted on by potential $\mathbf{x}$ carrying momentum $-\mathbf{k}$, is disturbed by an added potential $\mathbf{k}$ of momentum $+\mathbf{k}$ (the reversed sign of the momenta in the intermediate factors in the second term $\mathbf{x} \ldots \mathbf{k}$ has no effect since we will later integrate over all $\mathbf{k}$). Hence as shown above the result is zero, but since $(\mathbf{k} \ldots \mathbf{x}) + (\mathbf{x} \ldots \mathbf{k}) = k_4^2(\gamma_t \ldots \gamma_t) - K^2(\gamma_\mathbf{K} \ldots \gamma_\mathbf{K})$ we can still conclude $(\gamma_\mathbf{K} \ldots \gamma_\mathbf{K}) = k_4^2 K^{-2}(\gamma_t \ldots \gamma_t)$.

# 9 KLEIN GORDON EQUATION

The methods may be readily extended to particles of spin zero satisfying the Klein Gordon equation,[26]

$$\Box^2 \psi - m^2 \psi = i\partial(A_\mu \psi)/\partial x_\mu + i A_\mu \partial \psi/\partial x_\mu - A_\mu A_\mu \psi. \quad (35)$$

The important kernel is now $I_+(2, 1)$ denned in (I, Eq. (32)). For a free particle, the wave function $\psi$ (20 satisfies $+\Box^2 \psi - m^2 \psi = 0$. At a point, 2, inside a space time region it is given by

$$\psi(2) = \int \left[ \psi(1)\partial I_+(2, 1)/\partial x_{1\mu} - (\partial \psi/\partial x_{1\mu}) I_+(2, 1) \right] N_\mu(1) d^3 V_1,$$

(as is readily shown by the usual method of demonstrating Green's theorem) the integral being over an entire 3-surface boundary of the region (with normal vector $N_\mu$). Only the positive frequency components of $\psi$ contribute from the surface preceding the time corresponding to 2, and only negative frequencies from the surface future to 2. These can be interpreted as electrons and positrons in direct analogy to the Dirac case.

The right-hand side of (35) can be considered as a source of new waves and a series of terms written down to represent matrix elements for processes of increasing order. There is only one new point here, the term in $A_\mu A_\mu$ by which two quanta can act at the same time. As an example, suppose three quanta or potentials, $a_\mu \exp(-iq_a \cdot x)$, $b_\mu \exp(-iq_b \cdot x)$, and $c_\mu \exp(iq_e \cdot x)$ are to act in that order on a particle of original momentum $p_{0\mu}$, so that $\mathbf{p}_a = \mathbf{p}_0 + \mathbf{q}_a$, and $\mathbf{p}_b = \mathbf{p}_a + \mathbf{q}_b$; the final momentum being $\mathbf{p}_c = \mathbf{p}_b + \mathbf{q}_c$. The matrix element is the

---

[26]The equations discussed in this section were deduced from the formulation of the Klein Gordon equation given in reference 5, Section 14. The function $\psi$ in this section has only one component and is not a spinor. An alternative formal method of making the equations valid for spin zero and also for spin 1 is (presumably) by use of the Kemmer-Duffin matrices $\beta_\mu$ satisfying the commutation relation

$$\beta_\mu \beta_\nu \beta_\sigma + \beta_\sigma \beta_\nu \beta_\mu = \delta_{\mu\nu} \beta_\sigma \delta_{\sigma\nu} \beta_\mu.$$

If we interpret a to mean $a_\mu \beta_\mu$, rather than $a_\mu \gamma_\mu$, for any $a_\mu$, all of the equations in momentum space will remain formally identical to those for the spin 1/2; with the exception of those in which a denominator $(\mathbf{p} - m)^{-1}$ has been rationalized to $(\mathbf{p} + m)(\mathbf{p}^2 - m^2)^{-1}$ since $\mathbf{p}^2$ is no longer equal to a number, $p \cdot p$. But $\mathbf{p}^3$ does equal $(p \cdot p)\mathbf{p}$ that $(\mathbf{p} - m)^{-1}$ may now be interpreted as $(mp + m^2 + \mathbf{p}^2 - p \cdot p)(p \cdot p - m^2)^{-1}$. This implies that equations in coordinate space will be valid of the function $K_+ (2,1)$ is given as $K_+(2, 1) = [(i\nabla_2 + m) - m^{-1}(\nabla_2 + \Box_2^2)]i I_+(2, 1)$ with $\nabla_2 = \beta_\mu \partial/\partial x_{2\mu}$. This is all in virtue of the fact that the many component wave function $\psi$ (5 components for spin 0, 10 for spin 1) satisfies $(i\nabla - m)\psi = \mathbf{a}\psi$ which is formally identical to the Dirac Equation. See W. Pauli, Rev. Mod. Phys. **13**, 203 (1940).

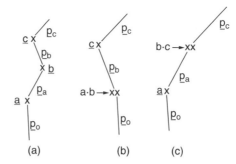

FIG. 7  Klein-Gordon particle in three potentials, Eq, (36). The coupling to the electromagnetic field is now, for example, $p_0 \cdot a + p_a \cdot a$, and a new possibility arises, (b), of simultaneous interaction with two quanta $a \cdot b$. The propagation factor is now $(p \cdot p - m^2)^{-1}$ for a particle of momentum $p_\mu$.

sum of three terms ($\mathbf{p}^2 = p_\mu p_\mu$) (illustrated in Fig. 7)

$$
\begin{aligned}
&(p_c \cdot c + p_b \cdot c)\left(\mathbf{p}_b^2 - m^2\right)^{-1}(p_b \cdot b + p_a \cdot b) \\
&\quad \times \left(\mathbf{p}_a^2 - m^2\right)^{-1}(p_a \cdot a + p_o \cdot a) \\
&\quad - (p_c \cdot c + p_b \cdot c)\left(\mathbf{p}_b^2 - m^2\right)^{-1}(b \cdot a) - (c \cdot b) \\
&\quad \times \left(\mathbf{p}_a^2 - m^2\right)^{-1}(p_a \cdot a + p_o \cdot a).
\end{aligned}
\tag{36}
$$

The first comes when each potential acts through the perturbation $i\partial(A_\mu \psi)/\partial x_\mu + i A_\mu \partial \psi/\partial x_\mu$. These gradient operators in momentum space mean respectively the momentum after and before the potential $A_\mu$ operates. The second term comes from $b_\mu$ and $a_\mu$ acting at the same instant and arises from the $A_\mu A_\mu$ term in (a). Together $b_\mu$ and $a_\mu$ carry momentum $q_{b\mu} + q_{a\mu}$ so that after $b \cdot a$ operates the momentum is $\mathbf{p}_0 + \mathbf{q}_a + \mathbf{q}_b$ or $\mathbf{p}_b$. The final term comes from $c_\mu$ and $b_\mu$ operating together in a similar manner. The term $A_\mu A_\mu$ thus permits a new type of process in which two quanta can be emitted (or absorbed, or one absorbed, one emitted) at the same time. There is no $a \cdot c$ term for the order $a$, $b$, $c$ we have assumed. In an actual problem there would be other terms like (36) but with alterations in the order in which the quanta $a$, $b$, $c$ act. In these terms $a \cdot c$ would appear.

As a further example the self-energy of a particle of momentum $p_\mu$ is

$$
\left(e^2/2\pi i m\right) \int \left[ (2p-k)_\mu \left((\mathbf{p}-k^2) - m^2\right)^{-1} \right.
$$
$$
\left. \times (2p-k)_\mu - \delta_{\mu\mu} \right] d^4 k\, \mathbf{k}^{-2} C(\mathbf{k}^2),
$$

where the $\delta_{\mu\mu}$ comes from the $A_\mu A_\mu$ term and represents the possibility of the simultaneous emission and absorption of the same virtual quantum. This integral without the $C(\mathbf{k}^2)$ diverges quadratically and would not converge if $C(\mathbf{k}^2) = -\lambda^2/(k^2 - \lambda^2)$. Since the interaction occurs through the gradients of the potential, we must use a stronger convergence factor, for example $C(\mathbf{k}^2) = \lambda^4 (k^2 - \lambda^2)^{-2}$, or in general (17) with $\int_0^\infty \lambda^2 G(\lambda) d\lambda = 0$. In this case the self-energy converges but depends quadratically on the cut-off $\lambda$ and is not necessarily small compared to $m$. The radiative corrections to scattering after mass renormalization are insensitive to the cut-off just as for the Dirac equation.

When there are several particles one can obtain Bose statistics by the rule that if two processes lead to the same state but with two electrons exchanged, their amplitudes are to be added (rather than subtracted as for Fermi statistics). In this case equivalence to the second quantization treatment of Pauli and Weisskopf should be demonstrable in a way very much like that given in $I$ (appendix) for Dirac electrons. The Bose statistics mean that the sign of contribution of a closed loop to the vacuum polarization is the opposite of what it is for the Fermi case (see I). It is ($\mathbf{p}_b = \mathbf{p}_a + \mathbf{q}$)

$$
J_{\mu v} = \frac{e^2}{2\pi i m} \int \left[ (p_{b\mu} + p_{a\mu})(p_{bv} + p_{av})(\mathbf{p}_a^2 - m^2)^{-1} \right.
$$
$$
\left. \times (\mathbf{p}_b^2 - m^2)^{-1} - \delta_{\mu v}(\mathbf{p}_a^2 - m^2)^{-1} - \delta_{\mu v}(\mathbf{p}_b^2 - m^2)^{-1} \right] d^4 p_a
$$

giving,

$$
J_{\mu v}^P = \frac{e^2}{\pi} \left( q_\mu q_v - \delta_{\mu v} \mathbf{q}^2 \right)
$$
$$
\times \left[ \frac{1}{6} \ln \frac{\lambda^2}{m^2} + \frac{1}{9} - \frac{4m^2 - \mathbf{q}^2}{3\mathbf{q}^2} \left( 1 - \frac{\theta}{\tan\theta} \right) \right],
$$

the notation as in (33). The imaginary part for $(\mathbf{q}^2)^{1/2} > 2m$ is again positive representing the loss in the probability of finding the final state to be a vacuum, associated with the possibilities of pair production. Fermi statistics would give a gain in probability (and also a charge renormalization of opposite sign to that expected).

# 10 APPLICATION TO MESON THEORIES

The theories which have been developed to describe mesons and the interaction of nucleons can be easily expressed in the language used here. Calculations, to lowest order in the interactions can be made very easily for the various theories, but agreement with experimental results is not obtained. Most likely all of our present formulations are quantitatively unsatisfactory. We shall content ourselves therefore with a brief summary of the methods which can be used.

The nucleons are usually assumed to satisfy Dirac's equation so that the factor for propagation of a nucleon of momentum $\mathbf{p}$ is $(\mathbf{p} - M)^{-1}$ where $M$ is the mass of the nucleon (which implies that nucleons can be created in pairs). The nucleon is then assumed to interact with mesons, the various theories differing in the form assumed for this interaction.

First, we consider the case of neutral mesons. The theory closest to electrodynamics is the theory of vector mesons with vector coupling. Here the factor for emission or absorption of a meson is $g\gamma_\mu$, when this meson is "polarized" in the $\mu$ direction. The factor $g$ the "mesonic charge," replaces the electric charge $e$. The amplitude for propagation of a meson of momentum $\mathbf{q}$ in intermediate states is $(\mathbf{q}^2 - \mu^2) - 1$ (rather than $\mathbf{q}^{-2}$ as it is for light) where $\mu$ is the mass of the meson. The necessary integrals are made finite by convergence factors $C(\mathbf{q}^2 - \mu^2)$ as in electrodynamics. For scalar mesons with scalar coupling the only change is that one replaces the $\gamma_\mu$ by 1 in emission and absorption. There is no longer a direction of polarization, $\mu$, to sum upon. For pseudo-scalar mesons, pseudoscalar coupling replace $\gamma_\mu$ by

$\gamma_5 = i\gamma_x\gamma_y\gamma_z\gamma_t$. For example, the self-energy matrix of a nucleon of momentum $\mathbf{p}$ in this theory is

$$(g^2/\pi i) \int \gamma_5 \left(\mathbf{p} - \mathbf{k} - M\right)^{-1} \gamma_5 d^4k \left(\mathbf{k}^2 - \mu^2\right)^{-1} C \left(\mathbf{k}^2 - \mu^2\right).$$

Other types of meson theory result from the replacement of $\gamma_\mu$, by other expressions (for example by $\frac{1}{2}(\gamma_\mu\gamma_\nu - \gamma_\nu\gamma_\mu)$ with a subsequent sum over all $\mu$ and $\nu$ for virtual mesons). Scalar mesons with vector coupling result from the replacement of $\gamma_\mu$ by $\mu^{-1}\mathbf{q}$ where $\mathbf{q}$ is the final momentum of the nucleon minus its initial momentum, that is, it is the momentum of the meson if absorbed, or the negative of the momentum of a meson emitted. As is well known, this theory with neutral mesons gives zero for all processes, as is proved by our discussion on longitudinal waves in electrodynamics. Pseudoscalar mesons with pseudo-vector coupling corresponds to $\gamma_\mu$ being replaced by $\mu^{-1}\gamma_5\mathbf{q}$ white vector mesons with tensor coupling correspond to using $(2\mu)^{-1}$ $(\gamma_\mu\mathbf{q} - \mathbf{q}\gamma_\mu)$. These extra gradients involve the danger of producing higher divergencies for real processes. For example, $\gamma_5\mathbf{q}$ gives a logarithmically divergent interaction of neutron and electron.[27] Although these divergencies can be held by strong enough convergence factors, the results then are sensitive to the method used for convergence and the size of the cut-off values of $\lambda$. For low order processes $\mu^{-1}\gamma_5\mathbf{q}$ is equivalent to the pseudoscalar interaction $2M_\mu^{-1}\gamma_5$ because if taken between free particle wave functions of the nucleon of momenta $\mathbf{p}_1$ and $\mathbf{p}_2 = \mathbf{p}_1 + \mathbf{q}$, we have

$$\left(\bar{u}_2\gamma_5\mathbf{q}u_1\right) = \left(\bar{u}_2\gamma_5 \left(\mathbf{p}_2 - \mathbf{p}_1\right) u_1\right) = - \left(\bar{u}_2\mathbf{p}_2\gamma_5 u_1\right)$$
$$- \left(\bar{u}_2\gamma_5\mathbf{p}_1 u_1\right) = -2M \left(\bar{u}_2\gamma_5 u_1\right)$$

since $\gamma_5$ anticommutes with $\mathbf{p}_2$ and $\mathbf{p}_2$ operating on the state 2 equivalent to $M$ as is 1 on the state 1. This shows that the $\gamma_5$ interaction is unusually weak in the non-relativistic limit (for example the expected value of $\gamma_5$ for a free nucleon is zero), but since $\gamma_5^2 = 1$ is not small, pseudoscalar theory gives a more important interaction in second order than it does in first. Thus the pseudoscalar coupling

[27] M. Slotnick and W. Heitler, Phys. Rev. **75**, 1645 (1949).

constant should be chosen to fit nuclear forces including these important second order processes.[28] The equivalence of pseudoscalar and pseudo-vector coupling which holds for low order processes therefore does not hold when the pseudoscalar theory is giving its most important effects. These theories will therefore give quite different results in the majority of practical problems.

In calculating the corrections to scattering of a nucleon by a neutral vector meson field ($\gamma_\mu$) due to the effects of virtual mesons, the situation is just as in electrodynamics, in that the result converges without need for a cut-off and depends only on gradients of the meson potential. With scalar (1) or pseudoscalar ($\gamma_\mu$) neutral mesons the result diverges logarithmically and so must be cut off. The part sensitive to the cut-off, however, is directly proportional to the meson potential. It may thereby be removed by a renormalization of mesonic charge $g$. After this renormalization the results depend only on gradients of the meson potential and are essentially independent of cut-off. This is in addition to the mesonic charge renormalization coming from the production of virtual nucleon pairs by a meson, analogous to the vacuum polarization in electrodynamics. But here there is a further difference from electrodynamics for scalar or pseudoscalar mesons in that the polarization also gives a term in the induced current proportional to the meson potential representing therefore an additional renormalization of the *mass of the meson* which usually depends quadratically on the cut-off.

Next consider charged mesons in the absence of an electromagnetic field. One can introduce isotopic spin operators in an obvious way. (Specifically replace the neutral $\gamma_5$, say, by $\tau_i \gamma - 5$ and sum over $i = 1, 2$, where $\tau_1 = \tau_+ + \tau_-$, $\tau_2 = i(\tau_+ - \tau_-)$ and $\tau_+$ changes neutron to proton ($\tau_+$ on proton $= 0$) and $\tau_-$ changes proton to neutron.) It is just as easy for practical problems simply to keep track of whether the particle is a proton or a neutron on a diagram drawn to help write down the matrix element. This excludes certain processes.

---

[28] H. A. Bethe, Bull. Am. Phys. Soc. **24**, 3, Z3 (Washington, 1949).

For example in the scattering of a negative meson from $\mathbf{q}_1$ to $\mathbf{q}_2$ by a neutron, the meson $\mathbf{q}_2$ must be emitted first (in order of operators, not time) for the neutron cannot absorb the negative meson $\mathbf{q}_1$ until it becomes a proton. That is, in comparison to the Klein Nishina formula (15), only the analogue of second term (see Fig. 5(b)) would appear in the scattering of negative mesons by neutrons, and only the first term (Fig. 5 (a)) in the neutron scattering of positive mesons.

The source of mesons of a given charge is not conserved, for a neutron capable of emitting negative mesons may (on emitting one, say) become a proton no longer able to do so. The proof that a perturbation $\mathbf{q}$ gives zero, discussed for longitudinal electromagnetic waves, fails. This has the consequence that vector mesons, if represented by the interaction $\gamma_\mu$, would not satisfy the condition that the divergence of the potential is zero. The interaction is to be taken[29] as $\gamma_\mu - \mu^{-2} q_\mu \mathbf{q}$ in emission and as $\gamma_\mu$ in absorption if the real emission of mesons with a non-zero divergence of potential is to be avoided. (The correction term $\mu^{-2} q_\mu \mathbf{q}$ gives zero in the neutral case.) The asymmetry in emission and absorption is only apparent, as this is clearly the same thing as subtracting from the original $\gamma_\mu \ldots \gamma_\mu$, a term $\mu^{-2} \mathbf{q} \ldots \mathbf{q}$. That is, if the term $-\mu^{-2} q_\mu \mathbf{q}$ is omitted the resulting theory describes a combination of mesons of spin one and spin zero. The spin zero mesons, coupled by vector coupling $\mathbf{q}$, are removed by subtracting the term $\mu^{-2} \mathbf{q} \ldots \mathbf{q}$.

The two extra gradients $\mathbf{q} \ldots \mathbf{q}$ make the problem of diverging integrals still more serious (for example the interaction between two

---

[29] The vector meson field potentials $\varphi_\mu$ satisfy

$$-\partial/\partial x_\nu \left(\partial\varphi_\mu/\partial x_\nu - \partial\varphi_\nu/\partial x_\mu\right) - \mu^2\varphi_\mu = -4\pi s_\mu,$$

where $s_\mu$, the source for such mesons, is the matrix element of $\gamma_\mu$ between states of neutron and proton. By taking the divergence $\partial/\partial x_\mu$ of both sides, conclude that $\partial\varphi_\nu/\partial x_\nu = 4\pi\mu^{-2}\partial s_\nu/\partial x_\nu$, so that the original equation can lie rewritten as

$$\Box^2\varphi_\mu - \mu^2\varphi_\mu = -4\pi\left(s_\mu + \mu^{-2}\partial/\partial x_\mu \left(\partial s_\nu/\partial x_\nu\right)\right).$$

The right hand side gives in momentum representation $\gamma_\mu - \mu^{-2}q_\mu q_\nu \gamma_\nu$ the left yields the $(q^2 - \mu^2)^{-1}$ and finally the interaction $s_\mu\varphi_\mu$ in the Lagrangian gives the $\gamma_\mu$ on absorption.
Proceeding in this way find generally that particles of spin one can be represented by a four-vector $u_\mu$ (which, for a free particle of momentum $q$ satisfies $q \cdot u = 0$). The propagation of virtual particles of momentum $q$ from state $\nu$ to $\mu$ is represented by multiplication by the 4-4 matrix (or tensor) $P_{\mu\nu} = (\delta_{\mu\nu} - \mu^{-2}q_\mu q_\nu) \times (q^2 - \mu^2)^{-1}$. The first-order interaction (from the Proca equation) with an electromagnetic potential $a \exp(ik \cdot x)$ corresponds to multiplication by the matrix $E_{\mu\nu} = (q_2 \cdot a + q_1 \cdot a)\delta_{\mu\nu} - q_{2\nu}a_\mu - q_{1\nu}a_\nu$, where $q_1$ and $q_2 = q_1 + k$ are the momenta before and after the interaction. Finally, two potentials $a$, $b$ may act simultaneously, with matrix $E'_{\mu\nu} = -(a \cdot b)\delta_{\mu\nu} + b_\mu a_\nu$.

protons corresponding to the exchange of two charged vector mesons depends quadratically on the cut-off if calculated in a straightforward way). One is tempted in this formulation to choose simply $\gamma_\mu \ldots \gamma_\mu$ and accept the admixture of spin zero mesons. But it appears that this leads in the conventional formalism to negative energies for the spin zero component. This shows one of the advantages of the method of second quantization of meson fields over the present formulation. There such errors of sign are obvious while here we seem to be able to write seemingly innocent expressions which can give absurd results. Pseudovector mesons with pseudovector coupling correspond to using $\gamma_5(\gamma_\mu - \mu^{-2}q_\mu \mathbf{q})$ for absorption and $\gamma_5\gamma_\mu$ for emission for both charged and neutral mesons.

In the presence of an electromagnetic field, whenever the nucleon is a proton it interacts with the field in the way described for electrons. The meson interacts in the scalar or pseudoscalar case as a particle obeying the Klein-Gordon equation. It is important here to use the method of calculation of Bethe and Pauli, that is, a virtual meson is assumed to have the same "mass" during all its interactions with the electromagnetic field. The result for mass $\mu$ and for $(\mu^2 + \lambda^2)^{1/2}$ are subtracted and the difference integrated over the function $G(\lambda)d\lambda$. A separate convergence factor is not provided for each meson propagation between electromagnetic interactions, otherwise gauge invariance is not insured. When the coupling involves a gradient, such as $\gamma - 5\mathbf{q}$ where $\mathbf{q}$ is the final minus the initial momentum of the nucleon, the vector potential $\mathbf{A}$ must be subtracted from the momentum of the proton. That is, there is an additional coupling $\pm\gamma_5\mathbf{A}$ (plus when going from proton to neutron, minus for the reverse) representing the new possibility of a simultaneous emission (or absorption) of meson and photon.

Emission of positive or absorption of negative virtual mesons are represented in the same term, the sign of the charge being determined by temporal relations as for electrons and positrons.

Calculations are very easily carried out in this way to lowest order in $g^2$ for the various theories for nucleon interaction, scattering of mesons by nucleons, meson production by nuclear collisions and by gamma-rays, nuclear magnetic moments, neutron electron scattering, etc., However, no good agreement with experiment results, when these are available, is obtaincd. Probably all of the formulations are incorrect. An uncertainty arises since the calculations are only to first order in $g^2$, and are not valid if $g^2/\hbar c$ is large.

The author is particularly indebted to Professor H. A. Bethe for his explanation of a method of obtaining finite and gauge invariant results for the problem of vacuum polarization. He is also grateful for Professor Bethe's criticisms of the manuscript, and for innumerable discussions during the development of this work. He wishes to thank Professor J. Ashkin for his careful reading of the manuscript.

# APPENDIX

In this appendix a method will be illustrated by which the simpler integrals appearing in problems in electrodynamics can be directly evaluated. The integrals arising in more complex processes lead to rather complicated functions, but the study of the relations of one integral to another and their expression in terms of simpler integrals may be facilitated by the methods given here.

As a typical problem consider the integral (12) appearing in the first order radiationless scattering problem:

$$\int \gamma_\mu \left(\mathbf{p}_2 - \mathbf{k} - m\right)^{-1} \mathbf{a} \left(\mathbf{p}_1 - \mathbf{k} - m\right)^{-1} \gamma_\mu \mathbf{k}^{-2} d^4 k C\left(\mathbf{k}^2\right), \quad (1a)$$

where we shall take $C(\mathbf{k}^2)$ to be typically $-\lambda^2(\mathbf{k}^2 - \lambda^2)^{-1}$ and $d^4 k$ means $(2\pi)^{-2} dk_1 dk_2 dk_3 dk_4$. We first rationalize the factors $(\mathbf{p} - \mathbf{k} - m)^{-1} = (\mathbf{p} - \mathbf{k} + m)((\mathbf{p} - \mathbf{k})^2 - m^2)^{-1}$ obtaining,

$$\int \gamma_\mu \left(\mathbf{p}_2 - \mathbf{k} + m\right) \mathbf{a} \left(\mathbf{p}_1 - \mathbf{k} + m\right) \gamma_\mu \mathbf{k}^{-2} d^4 k C\left(\mathbf{k}^2\right)$$
$$\times \left((\mathbf{p}_1 - \mathbf{k})^2 - m^2\right)^{-1} \left((\mathbf{p}_2 - \mathbf{k})^2 - m^2\right)^{-1}. \quad (2a)$$

The matrix expression may be simplified. It appears to be best to do so after the integrations are performed. Since $\mathbf{AB} = 2A \cdot B - \mathbf{BA}$ where

$A \cdot B = A_\mu B_\mu$ is a number commuting with all matrices, find, if $R$ is any expression, and $\mathbf{A}$ a vector, since $\gamma_\mu \mathbf{A} = -\mathbf{A}\gamma_\mu + 2A_\mu$,

$$\gamma_\mu \mathbf{A} R \gamma_\mu = -\mathbf{A}\gamma_\mu R \gamma_\mu + 2 R \mathbf{A}. \qquad (3a)$$

Expressions between two $\gamma_\mu$'s can be thereby reduced by induction. Particularly useful are

$$\gamma_\mu \gamma_\mu = 4$$

$$\gamma_\mu \mathbf{A} \gamma_\mu = -2\mathbf{A}$$

$$\gamma_\mu \mathbf{AB} \gamma_\mu = 2\,(\mathbf{AB} + \mathbf{BA}) = 4A \cdot B$$

$$\gamma_\mu \mathbf{ABC} \gamma_\mu = -2\mathbf{CBA} \qquad (4a)$$

where $\mathbf{A}, \mathbf{B}, \mathbf{C}$ are any three vector-matrices (i.e., linear combinations of the four $\gamma_\mu$s),

In order to calculate the integral in (2a) the integral may be written as the sum of three terms (since $\mathbf{k} = k_\sigma \gamma_\sigma$),

$$\gamma_\mu \left(\mathbf{p}_2 + m\right) \mathbf{a} \left(\mathbf{p}_1 + m\right) \gamma_\mu J_1 - \left[\gamma_\mu \gamma_\sigma \mathbf{a} \left(\mathbf{p}_1 + m\right) \gamma_\mu\right.$$

$$\left. + \gamma_\mu \left(\mathbf{p}_2 + m\right) \mathbf{a}\gamma_\sigma \gamma_\mu\right] J_2 + \gamma_\mu \gamma_\sigma \mathbf{a}\gamma_\tau \gamma_\mu J_3, \qquad (5a)$$

where

$$J\,(1;2;3) = \int (1; k_\sigma; k_\sigma k_\tau)\, \mathbf{k}^{-2} d^4 k\, C\left(\mathbf{k}^2\right)$$

$$\times \left(\left(\mathbf{p}_2 - \mathbf{k}\right)^2 - m^2\right)^{-1} \left(\left(\mathbf{p}_1 - \mathbf{k}\right)^2 - m^2\right)^{-1}. \qquad (6a)$$

That is for $J_1$ the $(1; k_\sigma; k_\sigma k_\tau)$ is replaced by 1, for $J_2$ by $k_\sigma$ and for $J_3$ by $k_\sigma k_\tau$.

More complex processes of the first order involve more factors like $((\mathbf{p}_3 - \mathbf{k})^2 - m^2)^{-1}$ and a corresponding increase in the number of $k$'s which may appear in the numerator, as $k_\sigma k_\tau k_\nu$ .... Higher order processes involving two or more virtual quanta involve similar integrals but with factors possibly involving $\mathbf{k} + \mathbf{k}'$ instead of just $\mathbf{k}$, and the integral extending on $\mathbf{k}^{-2} d^4 k\, C\,(\mathbf{k}^2) \mathbf{k}'^{-2} d^4 k C({'}^2)$. They can be simplified by methods analogous to those used on the first order integrals.

The factors $(\mathbf{p} - \mathbf{k})^2 - m^2$ may be written

$$\left(\mathbf{p} - \mathbf{k}\right)^2 - m^2 = \mathbf{k}^2 - 2p \cdot k - \Delta, \qquad (7a)$$

where $\Delta = m^2 - \mathbf{p}^2$, $\Delta_1 = m_1^2 - \mathbf{p}_1^2$, etc., and we can consider dealing with cases of greater generality in that the different denominators need not have the same value of the mass $m$. In our specific problem (6a) $\mathbf{p}_1^2 = m^2$ that $\Delta_1 = 0$, but we desire to work with greater generality.

Now for the factor $C(\mathbf{k}^2)/\mathbf{k}^2$ we shall use $-\lambda^2(\mathbf{k}^2 - \lambda^2)^{-1}\mathbf{k}^{-2}$. This can be written as

$$-\lambda^2/\left(\mathbf{k}^2 - \lambda^2\right)\mathbf{k}^2 = \mathbf{k}^{-2} C\left(\mathbf{k}^2\right) = -\int_0^{\lambda^2} dL\left(\mathbf{k}^2 - L\right)^{-2}. \quad (8a)$$

Thus we can replace $\mathbf{k}^{-2}C(\mathbf{k}^2)$ by $(\mathbf{k}^2 - L)^{-2}$ and at the end integrate the result with respect to $L$ from zero to $\lambda^2$. We can for many practical purposes consider $\lambda^2$ very large relative to $m^2$ or $p^2$. When the original integral converges even without the convergence factor, it will be obvious since the $L$ integration will then be convergent to infinity. If an infra-red catastrophe exists in the integral one can simply assume quanta have a small mass $\lambda_{min}$ and extend the integral on $L$ from $\lambda_{min}^2$ to $\lambda^2$, rather than from zero to $\lambda^2$.

We then have to do integrals of the form

$$\int (1; k_\sigma; k_\sigma k_\tau d^4 k(\mathbf{k}^2 - L)^{-2}(\mathbf{k}^2 - 2p_1 \cdot k - \Delta_1)^{-1}$$

$$\times (\mathbf{k}^2 - 2p_2 \cdot k - \Delta_2)^{-1}, \quad (9a)$$

where by $(1; k_\sigma; k_\sigma k_\tau)$ we mean that in the place of this symbol either 1, or $k_\sigma$ or $k_\sigma k_\tau$ may stand in different cases. In more complicated problems there may be more factors $(\mathbf{k}^2 - 2p_i \cdot k - \Delta_i)^{-1}$ or other powers of these factors (the $(\mathbf{k}^2 - L)^{-2}$ may be considered as a special case of such a factor with $\mathbf{p}_i = 0$, $\Delta_i = L$) and further factors like $k_\sigma k_\tau k_\rho \ldots$ in the numerator. The poles in all the factors are made definite by the assumption that $L$, and the $\Delta$'s have infinitesimal negative imaginary parts.

We shall do the integrals of successive complexity by induction. We start with the simplest convergent one, and show

$$\int d^4 k \left(\mathbf{k}^2 - L\right)^{-3} = (8iL)^{-1}. \quad (10a)$$

For this integral is $\int (2\pi)^{-2} dk_4 d^3\mathbf{K}(k_4^2 - \mathbf{K} \cdot \mathbf{K} - L)^{-3}$ where the vector $\mathbf{K}$, of magnitude $K = (\mathbf{K} \cdot \mathbf{K})^{1/2}$ is $k_1$, $k_2$, $k_3$. The integral on $k_4$ shows third order poles at $k_4 = + (K^2 + L)^{1/2}$ and $k_4 = -(K^2 + L)^{1/2}$. Imagining, in accordance with our definitions, that $L$ has a small negative imaginary part only the first is below the real axis. The contour can be closed by an infinite semi-circle below this axis, without change of the value of the integral since the contribution from the semi-circle vanishes in the limit. Thus the contour can be shrunk about the pole $k_4 = +(K^2 + L)^{1/2}$ and the resulting $k_4$, integral is $-2\pi i$ times the residue at this pole. Writing $k_4 = (k^2 + L)^{1/2} + \epsilon$ and expanding $(k_4^2 - K^2 - L)^{-3} = \epsilon^{-3}(\epsilon + 2(K^2 + L)^{1/2})^{-3}$ in powers of $\epsilon$, the residue, being the coefficient of the term $\epsilon^{-1}$, is seen to be $6(2(K^2 + L)^{1/2})^{-5}$ so our integral is

$$- (3i/32\pi) \int_0^\infty 4\pi K^2 dK \left(K^2 + L\right)^{-5/2} = (3/8i)(1/3L)$$

establishing (10a).

We also have $\int k_\sigma d^4 k (\mathbf{k}^2 - L)^{-3} = 0$ from the symmetry in the $k$ space. We write these results as

$$(8i) \int (1; k_\sigma) d^4 k \left(\mathbf{k}^2 - L\right)^{-3} = (1; 0) L^{-1}, \qquad (11a)$$

where in the brackets $(1; k_\sigma)$ and $(1; 0)$ corresponding entries are to he used.

Substituting $\mathbf{k} = \mathbf{k}' - \mathbf{p}$ in (11a) and calling $L - p^2 = \Delta$ shows that

$$(8i) \int (1; k_\sigma) d^4 k \left(k^2 - 2p \cdot k - \Delta\right)^{-3} = (1; p_\sigma)\left(p^2 + \Delta\right)^{-1}. \tag{12a}$$

By differentiating both sides of (12a) with respect to $\Delta$ or with respect to $p_\tau$ there follows directly

$$(24i) \int (1; k_\sigma; k_\sigma k_\tau) d^4 k \left(k^2 - 2p \cdot k - \Delta\right)^{-4}$$

$$= -\left(1; p_\sigma; p_\sigma p_\tau - \frac{1}{2}\delta_{\sigma\tau}\left(p^2 + \Delta\right)\right)\left(p^2 + \Delta\right)^{-2}. \tag{13a}$$

Further differentiations give directly successive integrals including more $k$ factors in the numerator and higher powers of $(\mathbf{k}^2 - 2p \cdot k - \Delta)$ in the denominator.

The integrals so far only contain one factor in the denominator. To obtain results for two factors we make use of the identity

$$a^{-1}b^{-1} = \int_0^1 dx\,(ax + b\,(1-x))^{-2}, \qquad (14a)$$

(suggested by some work of Schwinger's involving Gaussian integrals). This represents the product of two reciprocals as a parametric integral over one and will therefore permit integrals with two factors to be expressed in terms of one. For other powers of $a$, $b$ we make use of all of the identities, such as

$$a^{-2}b^{-1} = \int_0^1 2x\,dx\,(ax + b\,(1-x))^{-3}, \qquad (15a)$$

deducible from (14a) by successive differentiations with respect to $a$ or $b$. To perform an integral, such as

$$(8i)\int (1; k_\sigma)\,d^4k\,\left(\mathbf{k}^2 - 2p_1 \cdot k - \Delta_1\right)^{-2}\left(\mathbf{k}^2 - 2p_2 \cdot k - \Delta_2\right)^{-1}, \qquad (16a)$$

write, using (15a),

$$\left(\mathbf{k}^2 - 2p_1 \cdot k - \Delta_1\right)^{-2}\left(\mathbf{k}^2 - 2p_2 \cdot k - \Delta_2\right)^{-1}$$

$$= \int_0^1 2x\,dx\,\left(\mathbf{k}^2 - 2p_x \cdot k - \Delta_x\right)^{-3},$$

where

$$\mathbf{p}_x = x\mathbf{p}_1 + (1-x)\,\mathbf{p}_2 \quad \text{and} \quad \Delta_x = x\Delta_1 + (1-x)\,\Delta_2, \qquad (17a)$$

(note that $\Delta_x$ is *not* equal to $m^2 - \mathbf{p}_x^2$) so that the expression (16a) is $(8i)\int_0^1 2x\,dx(1; k_\sigma)d^4k(\mathbf{k}^2 - 2p_x \cdot k\Delta_x)^{-3}$ which may now be evaluated by (12a) and is

$$(16a) = \int_0^1 (1; p_{x\sigma})\,2x\,dx\,\left(\mathbf{p}_x^2 + \Delta_x\right)^{-1}, \qquad (18a)$$

where, $\mathbf{p}_x$, $\Delta_x$ are given in (17a). The integral in (18a) is elementary, being the integral of ratio of polynomials, the denominator of second degree in $x$. The general expression although readily obtained is a rather complicated combination of roots and logarithms.

Other integrals can be obtained again by parametric differentiation. For example differentiation of (16a), (18a) with respect to $\Delta_2$ or $p_{2\tau}$ gives

$$(8i) \int (1; k_\sigma; k_\sigma k_\tau) \, d^4k \left(\mathbf{k}^2 - 2p_1 \cdot k - \Delta_1\right)^{-2}$$

$$\times \left(\mathbf{k}^2 - 2p_2 \cdot k - \Delta_2\right)^{-2}$$

$$= - \int_0^1 \left(1; p_{x\sigma}; p_{x\sigma} p_{x\tau} - \frac{1}{2}\delta_{\sigma\tau} \left(x^2\mathbf{p}^2 + \Delta_x\right)\right)$$

$$\times 2x \, (1 - x) \, dx \left(\mathbf{p}_x^2 + \Delta_x\right)^{-2}, \qquad (19a)$$

again leading to elementary integrals.

As an example, consider the case that the second factor is just $(k^2 - L)^{-2}$ and in the first put $\mathbf{p}_1 = \mathbf{p}$, $\Delta_1 = \Delta$. Then $\mathbf{p}_x = x\mathbf{p}$, $\Delta_x - x\Delta + (1 - x)L$. There results

$$(8i) \int (1; k_\sigma; k_\sigma k_\tau) \, d^4k \left(\mathbf{k}^2 - L\right)^{-2} \left(\mathbf{k}^2 - 2p \cdot k - \Delta\right)^{-2}$$

$$= - \int_0^1 \left(1; xp_\sigma; x^2 p_\sigma p_\tau - \frac{1}{2}\delta_{\sigma\tau} \left(x^2\mathbf{p}^2 + \Delta_x\right)\right)$$

$$\times 2x \, (1 - x) \, dx \left(x^2\mathbf{p}^2 + \Delta_x\right)^{-2}. \qquad (20a)$$

Integrals with three factors can be reduced to those involving two by using (14a) again. They, therefore, lead to integrals with two parameters (e.g., see application to radiative correction to scattering below).

The methods of calculation given in this paper are deceptively simple when applied to the lower order processes. For processes of increasingly higher orders the complexity and difficulty increases rapidly, and these methods soon become impractical in their present form.

## A. Self-Energy

The self-energy integral (19) is

$$\left(e^2/\pi i\right) \int \gamma_\mu \left(\mathbf{p} - \mathbf{k} - m\right)^{-1} \gamma_\mu \mathbf{k}^{-2} d^4 k C\left(\mathbf{k}^2\right), \qquad (19)$$

so that it requires that we find (using the principle of (8a)) the integral on $L$ from 0 to $\lambda^2$ of

$$\int \gamma_\mu \left(\mathbf{p} - \mathbf{k} + m\right) \gamma_\mu d^4 k \left(\mathbf{k}^2 - L\right)^{-2} \left(\mathbf{k}^2 - 2p \cdot k\right)^{-1},$$

since $(\mathbf{p} - \mathbf{k})^2 - m^2 = \mathbf{k}^2 - 2p \cdot k$, as $\mathbf{p}^2 = m^2$. This is of the form (16a) with $\Delta_1 = L$, $\mathbf{p}_1 = 0$, $\Delta_2 = 0$, $\mathbf{p}_2 = \mathbf{p}$ so that (18a) gives, since $\mathbf{p}_x = (1 - x)\mathbf{p}$, $\Delta_x = xL$,

$$(8i) \int \left(1; k_\sigma\right) d^4 k \left(\mathbf{k}^2 - L\right)^{-2} \left(\mathbf{k}^2 - 2p \cdot k\right)^{-1}$$

$$= \int_0^1 \left(1; (1 - x)\, p_\sigma 02 x dx \left((1 - x)^2 m_x^2 L\right)^{-1},\right.$$

or performing the integral on $L$, as in (8),

$$(8i) \int \left(1; k_\sigma\right) d^4 k \mathbf{k}^{-2} C\left(\mathbf{k}^2\right) \left(\mathbf{k}^2 - 2p \cdot k\right)^{-1}$$

$$= \int_0^1 \left(1; (1 - x)\, p_\sigma\right) 2 dx \, \text{In} \frac{x\lambda^2 + (1 - x)^2 m^2}{(1 - x)^2 m^2}.$$

Assuming now that $\lambda^2 \gg m^2$ we neglect $(1 - x)^2 m^2$ relative to $x\lambda^2$ in the argument of the logarithm, which then becomes $(\lambda^2/m^2)(x/(1 - x)^2)$. Then since $\int_0^1 dx \ln(x(1 - x)^{-2}) = 1$ and $\int_0^1 (1 - x) dx \, \ln(x(1 - x)^{-2}) = -(1/4)$ find

$$(8i) \int \left(1; k_\sigma\right) \mathbf{k}^{-2} C\left(\mathbf{k}^2\right) d^4 k \left(\mathbf{k}^2 - 2p \cdot k\right)^{-1}$$

$$= \left(2 \ln \frac{\lambda^2}{m^2} + 2; p_\sigma \left(\ln \frac{\lambda^2}{m^2} - \frac{1}{2}\right)\right),$$

so that substitution into (19) (after the $(\mathbf{p} - \mathbf{k} - m)^{-1}$ in (19) is replaced by $(\mathbf{p} - \mathbf{k} + m)(\mathbf{k}^2 - 2p \cdot k)^{-1})$ gives

$$(19) = (e^2/8\pi)\gamma_\mu \Big[ (\mathbf{p} + m)\left(2\ln\left(\lambda^2/m^2\right) + 2\right)$$

$$-\mathbf{p}\left(\ln\left(\lambda^2/m^2\right) - \frac{1}{2}\right) \Big]\gamma_\mu$$

$$= \left(e^2/8\pi\right)\left[8m\left(\ln\left(\lambda^2/m^2\right) + 1\right) - \mathbf{p}\left(2\ln\left(\lambda^2 m^2\right) + 5\right)\right],$$

$$(20)$$

using (4a) to remove the $\gamma_\mu$'s. This agrees with Eq. (20) of the text, and gives the self-energy (21) when $\mathbf{p}$ is replaced by $m$.

## B. Corrections to Scattering

The term (12) in the radiationless scattering, after rationalizing the matrix denominators and using $p_1^2 = p_2^2 = m^2$ requires the integrals (9a), as we have discussed. This is an integral with three denominators which we do in two stages. First the factors $(\mathbf{k}^2 - 2p_1 \cdot k)$ and $(K^2 - 2p_2 \cdot k)$ are combined by a parameter $y$;

$$\left(\mathbf{k}^2 - 2p_1 \cdot k\right)^{-1}\left(\mathbf{k}^2 - 2p_2 \cdot k\right)^{-1} = \int_0^1 dy\,\left(\mathbf{k}^2 - 2p_y \cdot k\right)^{-2},$$

from (14a) where

$$p_y = yp_1 + (1-y)\,p_2. \tag{21a}$$

We therefore need the integrals

$$(8i)\int (1; k_\sigma; k_\sigma k_\tau)\,d^4k\,\left(\mathbf{k}^2 - L\right)^{-2}\left(\mathbf{k}^2 - 2p_y \cdot k\right)^{-2}, \tag{22a}$$

which we will then integrate with respect to $y$ from 0 to 1. Next we do the integrals (22a) immediately from (20a) with $p = p_y$, $\Delta = 0$:

$$(22a) = -\int_0^1 \int_0^1$$

$$\times \left(1; xp_{y\sigma}; x^2 p_{y\sigma} p_{y\tau} - \frac{1}{2}\delta_{\sigma\tau}(x^2 p_y^2 + (1-x)L)\right)$$

$$\times 2x(1-x)dx(x^2 p_y^2 + L(1-x))^{-2}dy.$$

We now turn to the integrals on $L$ as required in (8a). The first term, (1), in $(1; k_\sigma; k_{\sigma}k_\tau)$ gives no trouble for large $L$, but if $L$ is put equal to zero there results $x^{-2}p_y^{-2}$ which leads to a diverging integral on $x$ as $x \to 0$. This infra-red catastrophe is analyzed by using $\lambda_{min}^2$ for the lower limit of the $L$ integral. For the last term the upper limit of $L$ must be kept as $\lambda^2$. Assuming $\lambda_{min}^2 \ll p_y^2 \ll \lambda^2$ the $x$ integrals which remain are trivial, as in the self-energy case. One finds

$$- (8i) \int \left( \mathbf{k}^2 - \lambda_{min}^2 \right)^{-1} d^4k\, C \left( \mathbf{k}^2 - \lambda_{min}^2 \right) \left( \mathbf{k}^2 - 2p_1 \cdot k \right)^{-1}$$

$$\times \left( \mathbf{k}^2 - 2p_2 \cdot k \right)^{-1} = \int_0^1 p_y^{-2} dy \, \ln \left( p_y^2 / \lambda_{min}^2 \right) \qquad (23a)$$

$$- (8i) \int k_\sigma \mathbf{k}^{-2} d^4k\, C \left( \mathbf{k}^2 \right) \left( \mathbf{k}^2 - 2p_1 \cdot k \right)^{-1} \left( \mathbf{k}^2 - 2p_2 \cdot k \right)^{-1}$$

$$= 2 \int_0^1 p_{y\sigma} p_y^{-2} dy, \qquad (24a)$$

$$- (8i) \int k_\sigma k_\tau \mathbf{k}^{-2} d^4k\, C \left( \mathbf{k}^2 \right) \left( \mathbf{k}^2 - 2p_1 \cdot k \right)^{-1} \left( \mathbf{k}^2 - 2p_2 \cdot k \right)^{-1}$$

$$= \int_0^1 p_{y\sigma} p_{y\tau} p_y^{-2} dy - \frac{1}{2} \delta_{\sigma\tau} \int_0^1 dy \, \ln \left( \lambda^2 p_y^{-2} \right) + \frac{1}{4} \delta_{\sigma\tau}. \qquad (25a)$$

The integrals on $y$ give,

$$\int_0^1 p_y^{-2} dy \, \ln \left( p_y^2 \lambda_{min}^{-2} \right) = 4 \left( m^2 \sin 2\theta \right)^{-1}$$

$$\times \left[ \theta \ln \left( m\lambda_{min}^{-1} \right) - \int_0^\theta \alpha \tan \alpha\, d\alpha \right], \qquad (26a)$$

$$\int_0^1 p_{y\sigma} p_y^{-2} dy = \theta \left( m^2 \sin 2\theta \right)^{-1} \left( p_{1\sigma} + p_{2\sigma} \right), \qquad (27a)$$

$$\int_0^1 p_{y\sigma} p_{y\tau} p_y^{-2} dy = \theta \left( 2m^2 \sin 2\theta \right)^{-1}$$

$$\times \left( p_{1\sigma} + p_{1\tau} \left( p_{2\sigma} + p_{2\tau} \right) + \mathbf{q}^{-2} q_\sigma q_\tau \left( 1 - \theta \operatorname{ctn}\theta \right), \quad (28a) \right.$$

$$\int_0^1 dy \, \ln \left( \lambda^2 p_y^{-2} \right) = \ln \left( \lambda^2 / m^2 \right) + 2 \left( 1 - \theta \operatorname{ctn}\theta \right). \qquad (29a)$$

These integrals on $y$ were performed as follows. Since $p_2 = p_1 + \mathbf{q}$ where $\mathbf{q}$ is the momentum carried by the potential, it follows from

$\mathbf{p}_2^2 = \mathbf{p}_1^2 = m^2$ that $2p_1 \cdot q = -\mathbf{q}^2$ so that since $\mathbf{p}_y = \mathbf{p}_1 + \mathbf{q}(1 - y)$, $\mathbf{p}_y^2 = m^2 - \mathbf{q}^2 y(1 - y)$. The substitution $2y - 1 = \tan\theta$ where $\theta$ is defined by $4m^2 \sin^2 \theta = \mathbf{q}^2$ is useful for it means $\mathbf{p}_y^2 = m^2 \sec^2 \alpha / \sec^2 \theta$ and $\mathbf{p}_y^{-2} dy = (m^2 \sin 2\theta))^{-1} d\alpha$ where $\alpha$ goes from $-\theta$ to $+\theta$.

These results are substituted into the original scattering formula (2a), giving (22). It has been simplified by frequent use of the fact that $\mathbf{p}_1$ operating on the initial state is $m$ and likewise $\mathbf{p}_2$ when it appears at the left is replacable by $m$. (Thus, to simplify:

$$\gamma_\mu \mathbf{p}_2 \mathbf{a} \mathbf{p}_1 \gamma_\mu = -2\mathbf{p}_1 \mathbf{a} \mathbf{p}_2 \quad \text{by} \quad (4a),$$

$$= -2 (\mathbf{p}_2 - \mathbf{q}) \mathbf{a} (\mathbf{p}_1 + \mathbf{q}) = -2 (m - \mathbf{q}) \mathbf{a} (m + \mathbf{q}).$$

A term like $\mathbf{q}\mathbf{a}\mathbf{q} = -\mathbf{q}^2\mathbf{a} + 2(a \cdot q)\mathbf{q}$ is equivalent to just $-\mathbf{q}^2\mathbf{a}$ since $\mathbf{q} = \mathbf{p}_2 - \mathbf{p}_1 = m - m$ has zero matrix element.) The renormalization term requires the corresponding integrals for the special case $\mathbf{q} = 0$.

## C. Vacuum Polarization

The expressions (32) and (32') for $J_{\mu\nu}$ in the vacuum polarization problem require the calculation of the integral

$$J_{\mu\nu}(m^2) = \frac{e^2}{\pi i} \int Sp \left[ \gamma_\mu \left(\mathbf{p} - \frac{1}{2}\mathbf{q} + m\right) \gamma_\nu \mathbf{p} + \frac{1}{2}\mathbf{q} + m) \right] d^4 p$$

$$\times \left(\left(\mathbf{p} - \frac{1}{2}\mathbf{q}\right)^2 - m^2\right)^{-1} \left(\left(\mathbf{p} + \frac{1}{2}\mathbf{q}\right)^2 - m^2\right)^{-1},$$

$$(32)$$

where we have replaced $\mathbf{p}$ by $\mathbf{p} - \frac{1}{2}\mathbf{q}$ to simplify the calculation somewhat. We shall indicate the method of calculation by studying the integral,

$$I(m^2) = \int p_\sigma p_\tau d^4 p \left(\left(\mathbf{p} - \frac{1}{2}\mathbf{q}\right)^2 - m^2\right)^{-1} \left(\left(\mathbf{p} + \frac{1}{2}\mathbf{q}\right)^2 - m^2\right)^{-1}.$$

The factors in the denominator, $\mathbf{p}^2 - p \cdot q - m^2 + \frac{1}{4}\mathbf{q}^2$ and $\mathbf{p}^2 + p \cdot q - m^2 + \frac{1}{4}\mathbf{q}^2$ are combined as usual by (8a) but for symmetry we substitute $x = \frac{1}{2}(1 + \eta)$, $(1 - x) = \frac{1}{2}(1 - \eta)$ and integrate $\eta$ from

−1 to +1:

$$I(m^2) = \int\limits_{-1}^{+1} p_\sigma\, p_\tau d^4 p\left(\mathbf{p}^2 - \eta p \cdot q - m^2 + \frac{1}{4}\mathbf{q}^2\right)^{-2} d\eta/2. \quad (30a)$$

But the integral on $\mathbf{p}$ will not be found in our list for it is badly divergent. However, as discussed in Section 7, Eq. (34′) we do not wish $I(m^2)$ but rather $\int_0^\infty [I(m^2) - I(m^2 + \lambda^2)]G(\lambda)d\lambda$. We can calculate the difference $I(m^2) - I(m^2 + \lambda^2)$ by first calculating the derivative $I'(m^2 + L)$ of $I$ with respect to $m^2$ at $m^2 + L$ and later integrating $L$ from zero to $\lambda^2$. By differentiating (30a), with respect to $m^2$ find,

$$I'\left(m^2 + L\right) = \int\limits_{-1}^{+1} p_\sigma\, p_\tau d^4 p\left(\mathbf{p}^2 - \eta p \cdot q - m^2 - L + \frac{1}{4}\mathbf{q}^2\right)^{-3} d\eta.$$

This still diverges, but we can differentiate again to get

$$I''\left(m^2 + L\right)$$

$$= 3\int\limits_{-1}^{+1} p_\sigma\, p_\tau d^4 p\left(\mathbf{p}^2 - \eta p \cdot q - m^2 - L + \frac{1}{4}\mathbf{q}^2\right)^{-4} d\eta$$

$$= -(8i)^{-1}\int\limits_{-1}^{+1}\left(\frac{1}{4}\eta^2 q_\sigma q_\tau D^{-2} - \frac{1}{2}\delta_{\sigma r}D^{-1}\right) d\eta \quad (31a)$$

(where $D = \frac{1}{4}(\eta^2 - 1)\mathbf{q}^2 + m^2 + L$), which now converges and has been evaluated by (13a) with $\mathbf{p} = \frac{1}{2}\eta\mathbf{q}$ and $\Delta = m^2 + L - \frac{1}{4}\mathbf{q}^2$. Now to get $I'$ we may integrate $I''$ with respect to $L$ as an indefinite integral and *we may choose any convenient arbitrary constant*. This is because a constant $C$ in $I'$ will mean a term $-C\lambda^2$ in $I(m^2) - I(m^2 + \lambda^2)$ which vanishes since we will integrate the results times $G(\lambda)d\lambda$ and $\int_0^\infty \lambda^2 G(\lambda)d\lambda = 0$. This means that the logarithm appearing on integrating $L$ in (31a) presents no problem. We may take

$$I'\left(m^2 + L\right) = (8i)^{-1}\int\limits_{-1}^{+1}\left[\frac{1}{4}\eta^2 q_\sigma q_\tau D^{-1} + \frac{1}{2}\delta_{\sigma\tau}\ln D\right] d\eta + C\delta_{\sigma\tau},$$

a subsequent integral on $L$ and finally on $\eta$ presents no new problems. There results

$$-(8i) \int p_\sigma p_\tau d^4 p \left(\left(p - \frac{1}{2}q\right)^2 - m^2\right)^{-1} \left(\left(p + \frac{1}{2}q\right)^2 - m^2\right)^{-1}$$

$$= \left(q_\sigma q_\tau - \delta_{\sigma\tau}q^2\right) \left[\frac{1}{9} - \frac{4m^2 - q^2}{3q^2}\left(1 - \frac{\theta}{\tan\theta}\right) + \frac{1}{6}\ln\frac{\lambda^2}{m^2}\right]$$

$$+ \delta_{\sigma\tau}\left[(\lambda^2 + m^2)\ln\left(\lambda^2 m^{-2} + 1\right) - C'\lambda^2\right] \tag{32a}$$

where we assume $\lambda^2 \gg m^2$ and have put some terms into the arbitrary constant $C'$ which is independent of $\lambda^2$ (but in principle could depend on $\mathbf{q}^2$ and which drops out in the integral on $G(\lambda)d\lambda$. We have set $\mathbf{q}^2 = 4m^2 \sin^2\theta$.

In a very similar way the integral with $m^2$ in the numerator can be worked out. It is, of course, necessary to differentiate this $m^2$ also when calculating $I'$ and $I''$. There results

$$-(8i) \int m^2 d^4 p \left(\left(p - \frac{1}{2}q\right)^2 - m^2\right)^{-1} \left(\left(p + \frac{1}{2}q\right)^2 - m^2\right)^{-1}$$

$$= 4m^2 \left(1 - \theta \operatorname{ctn}\theta\right) - q^2/3 + 2\left(\lambda^2 + m^2\right)$$

$$\times \ln\left(\lambda^2 m^{-2} + 1\right) - C''\lambda^2), \tag{33a}$$

with another unimportant constant $C''$. The complete problem requires the further integral,

$$-(8i) \int (1; p_\sigma) d^4 p \left(\left(p - \frac{1}{2}q\right)^2 - m^2\right)^{-1} \left(\left(p + \frac{1}{2}q\right)^2 - m^2\right)^{-1}$$

$$= (1, 0) \left(4\left(1 - \theta \operatorname{ctn}\theta\right) + 2\ln\left(\lambda^2 m^{-2}\right)\right). \tag{34a}$$

The value of the integral (34a) times $m^2$ differs from (33a), of course, because the results on the right are not actually the integrals on the left, hut rather equal their actual value minus their value for $m^2 = m^2 + \lambda^2$.

Combining these quantities, as required by (32), dropping the constants $C'$, $C''$ and evaluating the spur gives (33). The spurs are evaluated in the usual way, noting that the spur of any odd number of $\gamma$ matrices vanishes and $S_p(AB) = S_p(BA)$ for arbitrary $A$, $B$. The

$S_p(1) = 4$ and we also have

$$\frac{1}{2}S_p\left[(p_1 + m_1)(p_2 - m_2)\right] = p_1 \cdot p_2 - m_1 m_2, \qquad (35a)$$

$$\frac{1}{2}S_p\left[(\mathbf{p}_1 + m_1)(\mathbf{p}_2 - m_2)(\mathbf{p}_4 - m_4)\right]$$

$$= (p_1 \cdot p_2 - m_1 m_2)(p_3 \cdot p_4 - m_3 m_4)$$

$$- (p_1 \cdot p_3 - m_1 m_3)(p_2 \cdot p_4 - m_2 m_4)$$

$$+ (P - 1 \cdot p_4 - m_1 m_4)(p_2 \cdot p_3 - m_2 m_3), \qquad (36a)$$

where $\mathbf{p}_i$, $m_i$ are arbitrary four-vectors and constants.

It is interesting that the terms of order $\lambda^2 \ln \lambda^2$ go out, so that the charge renormalization depends only logarithmically on $\lambda^2$. This is not true for some of the meson theories. Electrodynamics is suspiciously unique in the mildness of its divergence.

## D. More Complex Problems

Matrix elements for complex problems can be set up in a manner analogous to that used for the simpler cases. We give three illustrations; higher order corrections to the Møller scattering, to the Compton scattering, and the interaction of a neutron with an electromagnetic field.

For the Møller scattering, consider two electrons, one in state $u_1$ of momentum $\mathbf{p}_1$ and the other in state $u_2$ of momentum $\mathbf{p}_2$. Later they are found in states $u_3$, $\mathbf{p}_3$ and $u_4$, $\mathbf{p}_4$. This may happen (first order in $e^2/\hbar c$) because they exchange a quantum of momentum $\mathbf{q} = \mathbf{p}_1 - \mathbf{p}_3 = \mathbf{p}_4 - \mathbf{p}_2$ in the manner of Eq. (4) and Fig. 1. The matrix element for this process is proportional to (translating (4) to momentum space)

$$\left(\bar{u}_4 \gamma_\mu u_2\right)\left(\bar{u}_3 \gamma_\mu u_1\right) \mathbf{q}^{-2}. \qquad (37a)$$

We shall discuss corrections to (37a) to the next order in $e^2/\hbar c$. (There is also the possibility that it is the electron at 2 which finally arrives at 3, the electron at 1 going to 4 through the exchange of quantum of momentum $\mathbf{p}_3 - \mathbf{p}_2$. The amplitude for this process,

$(\bar{u}_4\gamma_\mu u_1)(\bar{u}_3\gamma_\mu u_2)(\mathbf{p}_3 - \mathbf{p}_2)^{-2}$ must be subtracted from (37a) in accordance with the exclusion principle. A similar situation exists to each order so that we need consider in detail only the corrections to (37a), reserving to the last the subtraction of the same terms with 3, 4 exchanged.) One reason that (37a) is modified is that two quanta may be exchanged, in the manner of Fig. 8a. The total matrix element for all exchanges of this type is

$$(e^2/\pi i) \int (\bar{u}_3\gamma_\nu(\mathbf{p}_1 - \mathbf{k} - m)^{-1}\gamma_\mu u_1)$$

$$\times(\bar{u}_4\gamma_\nu(\mathbf{p}_2 + \mathbf{k} - m)^{-1}\gamma_\mu u_2) \cdot \mathbf{k}^{-2}(\mathbf{q} - \mathbf{k})^{-2}d^4k, \qquad (38a)$$

as is clear from the figure and the general rule that electrons of momentum $\mathbf{p}$ contribute in amplitude $(\mathbf{p} - m)^{-1}$ between interactions $\gamma_\mu$ and that quanta of momentum $\mathbf{k}$ contribute $\mathbf{k}^{-2}$. In integrating on $d^4k$ and summing over $\mu$, and $\nu$, we add all alternatives of the type of Fig. 8a. If the time of absorption, $\gamma_\mu$, of the quantum $\mathbf{k}$ by electron 2 is later than the absorption, $\gamma_\mu$, of $\mathbf{q} - \mathbf{k}$, this corresponds

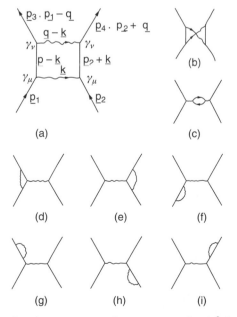

FIG. 8   The interaction between two electrons to order $(e^2/\hbar c)^2$. One adds the contribution of every figure involving two virtual quanta, Appendix D.

to the virtual state $\mathbf{p}_2 + \mathbf{k}$ being a positron (so that (38a) contains over thirty terms of the conventional method of analysis).

In integrating over all these alternatives we have considered all possible distortions of Fig. 8a which preserve the order of events along the trajectories. We have not included the possibilities corresponding to Fig. 8b, however. Their contribution is

$$\left(e^2/\pi i\right) \int \left(\bar{u}_3 \gamma_\nu \left(\mathbf{p}_1 - \mathbf{k} - m\right)^{-1} \gamma_\nu u_1\right)$$

$$\times \left(\bar{u}_4 \gamma_\mu \left(\mathbf{p}_2 + \mathbf{q} - \mathbf{K} - m\right)^{-1} \gamma_\nu u_2\right) \mathbf{k}^{-2} \left(\mathbf{q} - \mathbf{k}\right)^{-2} d^4 k, \ (39a)$$

as is readily verified by labeling the diagram. The contributions of all possible ways that an event can occur are to be added. This means that one adds with equal weight the integrals corresponding to each topologically distinct figure.

To this same order there are also the possibilities of Fig. 8d which give

$$\left(e^2/\pi i\right) \int \left(\bar{u}_3 \gamma_\nu \left(\mathbf{p}_2 - \mathbf{k} - m\right)^{-1} \gamma_\mu \left(\mathbf{p}_1 - \mathbf{k} - m\right)^{-1} \gamma_\nu u_1\right)$$

$$\times \left(\bar{u}_4 \gamma_\mu u_2\right) \mathbf{k}^{-2} \mathbf{q}^{-2} d^4 k.$$

This integral on $\mathbf{k}$ will be seen to be precisely the integral (12) for the radiative corrections to scattering, which we have worked out. The term may be combined with the renormalization terms resulting from the difference of the effects of mass change and the terms, Figs. 8f and 8g. Figures 8e, 8h, and 8i are similarly analyzed.

Finally the term Fig. 8c is clearly related to our vacuum polarization problem, and when integrated gives a term proportional to $(\bar{u}_4 \gamma_\mu u_2)(\bar{u}_3 \gamma_\nu u_1) J_{\mu\nu} \mathbf{q}^{-4}$. If the charge is renormalized the term $\ln(\lambda/m)$ in $J_{\mu\nu}$ in (33) is omitted so there is no remaining dependence on the cut-off.

The only new integrals we require are the convergent integrals (38a) and (39a). They can be simplified by rationalizing the denominators and combining them by (14a). For example (38a) involves the factors $(\mathbf{k}^2 - 2p_1 \cdot k)^{-1} (\mathbf{k}^2 + 2p_2 \cdot k)^{-1} \mathbf{k}^{-2} (\mathbf{q}^2 + \mathbf{k}^2 - 2q \cdot k)^{-2}$. The first two may be combined by (14a) with a parameter $x$, and the second pair by an expression obtained by differentation (15a) with respect to $b$ and calling the parameter $y$. There results a factor $(\mathbf{k}^2 - 2p_x \cdot k)^{-2} (\mathbf{k}^2 + y\mathbf{q}^2 - 2yq \cdot k)^{-4}$ so that the integrals on $d^4k$ now involve two factors and can be performed by the methods given earlier in the appendix. The subsequent integrals on the parameters $x$ and $y$ are complicated and have not been worked out in detail.

Working with charged mesons there is often a considerable reduction of the number of terms. For example, for the interaction between protons resulting from the exchange of two mesons only the term corresponding to Fig. 8h remains. Term 8a, for example, is impossible, for if the first proton emits a positive meson the second cannot absorb it directly for only neutrons can absorb positive mesons.

As a second example, consider the radiative correction to the Compton scattering. As seen from Eq. (15) and Fig. 5 this scattering is represented by two terms, so that we can consider the corrections to each one separately. Figure 9 shows the types of terms arising from corrections to the term of Fig. 5a. Calling $\mathbf{k}$ the momentum of the

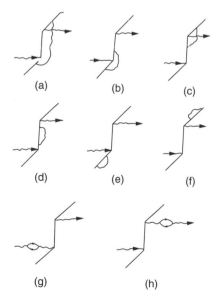

(a)  (b)  (c)

(d)  (e)  (f)

(g)  (h)

FIG. 9  Radiative correction to the Compton scattering term (a) of Fig. 5. Appendix D.

virtual quantum, Fig. 9a gives an integral

$$\int \gamma_\mu \left(\mathbf{p}_2 - \mathbf{k} - m\right)^{-1} \mathbf{e}_2 \left(\mathbf{p}_1 + \mathbf{q}_1 - \mathbf{k} - m\right)^{-1} \mathbf{e}_1$$

$$\times \left(\mathbf{p}_1 - \mathbf{k} - m\right)^{-1} \gamma_\mu \mathbf{k}^{-2} d^4 k,$$

convergent without cut-off and reducible by the methods outlined in this appendix.

The other terms are relatively easy to evaluate. Terms $b$ and $c$ of Fig. 9 are closely related to radiative corrections (although somewhat more difficult to evaluate, for one of the states is not that of a free electron, $(\mathbf{p}_1 + \mathbf{q})^2 \neq m^2$). Terms $e$, $f$, are renormalization terms. From term $d$ must be subtracted explicitly the effect of mass $\Delta m$, as analyzed in Eqs. (26) and (27) leading to (28) with $p' = \mathbf{p}_1 + \mathbf{q}$, $\mathbf{a} = \mathbf{e}_2$, $\mathbf{b} = \mathbf{e}_1$. Terms $g$, $h$ give zero since the vacuum polarization has zero effect on free light quanta, $\mathbf{q}_1^2 = 0$, $\mathbf{q}_2^2 = 0$. The total is insensitive to the cut-off $\lambda$.

The result shows an infra-red catastrophe, the largest part of the effect. When cut-off at $\lambda\min$, the effect proportional to $\ln(m/\lambda_{\min})$ goes as

$$\left(e^2/\pi\right) \ln \left(m/\lambda_{\min}\right) \left(1 - 2\theta \operatorname{ctn} 2\theta\right), \tag{40a}$$

times the uncorrected amplitude, where $(\mathbf{p}_2 - \mathbf{p}_1)^2 = 4m^2 \sin^2 \theta$. This is the same as for the radiative correction to scattering for a deflection $\mathbf{p}_2 - \mathbf{p}_1$. This is physically clear since the long wave quanta are not effected by short-lived intermediate states. The infra-red effects arise[30] from a final adjustment of the field from the asymptotic coulomb field characteristic of the electron of momentum $\mathbf{p}_1$ before the collision to that characteristic of an electron moving in a new direction $\mathbf{p}_2$ after the collision.

The complete expression for the correction is a very complicated expression involving transcendental integrals.

As a final example we consider the interaction of a neutron with an electromagnetic field in virtue of the fact that the neutron may

---

[30] F. Bloch and A. Nordsieck, Phys. Rev. **52**, 54 (1937).

emit a virtual negative meson. We choose the example of pseudoscalar mesons with pseudovector coupling. The change in amplitude due to an electromagnetic field $\mathbf{A} = \mathbf{a}\exp(-iq \cdot x)$ determines the scattering of a neutron by such a field. In the limit of small $\mathbf{q}$ it wilt vary as $\mathbf{qa} - \mathbf{aq}$ which represents the interaction of a particle possessing a magnetic moment. The first-order interaction between an electron and a neutron is given by the same calculation by considering the exchange of a quantum between the electron and the nucleon. In this case $a_{\mu}$ is $\mathbf{q}^{-2}$ times the matrix element of $\gamma_{\mu}$ between the initial and final states of the electron, the states differing in momentum by $\mathbf{q}$.

The interaction may occur because the neutron of momentum $\mathbf{p}_1$ emits a negative meson becoming a proton which proton interacts with the field and then reabsorbs the meson (Fig. 10a). The matrix for this process is ($\mathbf{p}_2 = \mathbf{p}_1 + \mathbf{q}$),

$$\int (\gamma_5 \mathbf{k} \, (\mathbf{p}_2 - \mathbf{k} - M)^{-1} \, \mathbf{a} \, (\mathbf{p}_1 - \mathbf{k} - M)^{-1} \, (\gamma_5 \mathbf{k}) \, (\mathbf{k}^2 - \mu^2)^{-1} \, d^4 k.$$

$$(41a)$$

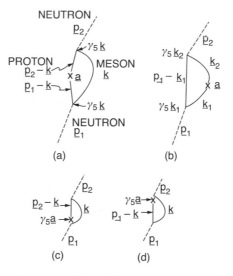

FIG. 10   According to the meson theory a neutron interacts with an electromagnetic potential $\mathbf{a}$ by first emitting a virtual charged meson. The figure illustrates the case for a pseudoscalar meson with pseudovector coupling. Appendix D.

Alternatively it may be the meson which interacts with the field. We assume that it does this in the manner of a scalar potential satisfying the Klein Gordon Eq. (35), (Fig. 10b)

$$-\int (\gamma_5 \mathbf{k}_2) \left(\mathbf{p}_1 - \mathbf{k}_1 - M\right)^{-1} (\gamma_5 \mathbf{k}_1) \left(\mathbf{k}_2^2 - \mu^2\right)^{-1}$$

$$\times (k_2 \cdot a + k_1 \cdot a) \left(\mathbf{k}_1^2 - \mu^2\right)^{-1} d^4 k_1, \tag{42a}$$

where we have put $\mathbf{k}_2 = \mathbf{k}_1 + \mathbf{q}$. The change in sign arises because the virtual meson is negative. Finally there are two terms arising from the $\gamma_5 \mathbf{a}$ part of the pseudovector coupling (Figs. 10c, 10d)

$$\int (\gamma_5 \mathbf{k}) \left(\mathbf{p}_2 - \mathbf{k} - M\right)^{-1} (\gamma_5 \mathbf{a}) \left(\mathbf{k}^2 - \mu^2\right)^{-1} d^4 k, \tag{43a}$$

and

$$\int (\gamma_5 \mathbf{a}) \left(\mathbf{p}_1 - \mathbf{k} - M\right)^{-1} (\gamma_5 \mathbf{k}) \left(\mathbf{k}^2 - \mu^2\right)^{-1} d^4 k, \tag{44a}$$

Using convergence factors in the manner discussed in the section on meson theories each integral can be evaluated and the results combined. Expanded in powers of $\mathbf{q}$ the first term gives the magnetic moment of the neutron and is insensitive to the cut-off, the next gives the scattering amplitude of stow electrons on neutrons, and depends logarithmically on the cut-off.

The expressions may be simplified and combined somewhat before integration. This makes the integrals a little easier and also shows the relation to the case of pseudoscalar coupling. For example in (41a) the final $\gamma_5 \mathbf{k}$ can be written as $\gamma_5 (\mathbf{k} - \mathbf{p}_1 + M)$ since $\mathbf{p}_1 = M$ when operating on the initial neutron state. This is $(\mathbf{p}_1 - \mathbf{k} - M)\gamma_5 + 2m\gamma_5$ since $\gamma_5$ anticommutes with $\mathbf{p}_1$ and $\mathbf{k}$. The first term cancels the $(\mathbf{p}_1 - \mathbf{k} - M)^{-1}$ and gives a term which just cancels (43a). In a like manner the leading factor $\gamma_5 \mathbf{k}$ in (41a) is written as $-2M\gamma - 5 - \gamma_5 (\mathbf{p}_2 - \mathbf{k} - M)$, the second term leading to a simpler term containing no $(\mathbf{p}_2 - k - M)^{-1}$ factor and combining with a similar one from (44a). One simplifies the $\gamma_5 \mathbf{k}_1$ and $\gamma_5 \mathbf{k}_2$ in (42a) in an analogous way. There finally results terms like (41a), (42a) but with pseudoscalar coupling $2M\gamma_5$ instead of $\gamma_5 \mathbf{k}$, no terms like (43a) or (44a) and a remainder, representing the difference in effects of pseudovector and pseudoscalar coupling. The pseudoscalar terms do not depend sensitively on the cut-off, but the difference term depends on it logarithmically. The difference term affects the electron-neutron interaction but not the magnetic moment of the neutron.

Interaction of a proton with an electromagnetic potential can be similarly analyzed. There is an effect of virtual mesons on the electromagnetic properties of the proton even in the case that the mesons are neutral. It is analogous to the radiative corrections to the scattering of electrons due to virtual photons. The sum of the magnetic moments of neutron and proton for charged mesons is the same as the proton moment calculated for the corresponding neutral mesons. In fact it is readily seen by comparing diagrams, that for arbitrary $\mathbf{q}$, the scattering matrix to *first order in the electromagnetic potential* for a proton according to neutral meson theory is equal, if the mesons were charged, to the sum of the matrix for a neutron and the matrix for a proton. This is true, for any type or mixtures of meson coupling, to all orders in the coupling (neglecting the mass difference of neutron and proton).

# THE THEORY OF POSITRONS

## BY

## RICHARD FEYNMAN

The problem of the behavior of positrons and electrons in given external potentials, neglecting their mutual interaction, is analyzed by replacing the theory of holes by a reinterpretation of the solutions of the Dirac equation. It is possible to write down a complete solution of the problem in terms of boundary conditions on the wave function, and this solution contains automatically all the possibilities of virtual (and real) pair formation and annihilation together with the ordinary scattering processes, including the correct relative signs of the various terms.

In this solution, the "negative energy states" appear in a form which may be pictured (as by Stückelberg) in space-time as waves traveling away from the external potential backwards in time. Experimentally, such a wave corresponds to a positron approaching the potential and annihilating the electron. A particle moving forward in time (electron) in a potential may be scattered forward in time (ordinary scattering) or backward (pair annihilation). When moving backward (positron) it may be scattered backward in time (positron scattering) or forward (pair production). For such a particle the amplitude for transition from an initial to a final state is analyzed to any order in the potential by considering it to undergo a sequence of such scatterings.

The amplitude for a process involving many such particles is the product of the transition amplitudes for each particle. The exclusion principle requires that antisymmetric combinations of amplitudes be chosen for those complete processes which differ only by exchange of particles. It seems that a consistent interpretation is only possible if the exclusion principle is adopted. The exclusion principle need not be taken into account in intermediate states. Vacuum problems do not arise for charges which do not interact with one another, but these are analyzed nevertheless in anticipation of application to quantum electrodynamics.

The results are also expressed in momentum-energy variables. Equivalence to the second quantization theory of holes is proved in an appendix.

# 1. INTRODUCTION

THIS is the first of a set of papers dealing with the solution of problems in quantum electrodynamics. The main principle is to deal directly with the solutions to the Hamiltonian differential equations rather than with these equations themselves. Here we treat simply the motion of electrons and positrons in given external potentials. In a second paper we consider the interactions of these particles, that is, quantum electrodynamics.

The problem of charges in a fixed potential is usually treated by the method of second quantization of the eletron field, using the ideas of the theory of holes. Instead we show that by a suitable choice and interpretation of the solutions of Dirac's equation the problem may be equally well treated in a manner which is fundamentally no more complicated than Schrödinger's method of dealing with one or more particles. The various creation and annihilation operators in the conventional electron field view are required because the number of particles is not conserved, i.e., pairs may be created or destroyed. On the other hand charge is conserved which suggests that if we follow the charge, not the particle, the results can be simplified.

In the approximation of classical relativistic theory the creation of an electron pair (electron $A$, positron $B$) might be represented by the start of two world lines from the point of creation, 1. The world lines of the positron will then continue until it annihilates another electron, $C$, at a world point 2. Between the times $t_1$ and $t_2$ there are then three world lines, before and after only one. However, the world lines of $C$, $B$, and $A$ together form one continuous line albeit the "positron part" $B$ of this continuous line is directed backwards in time. Following the charge rather than the particles corresponds to considering this continuous world line as a whole rather than breaking it up into its pieces. It is as though a bombardier flying low over a road suddenly sees three roads and it is only when two of them come together and disappear again that he realizes that he has simply passed over a long switchback in a single road.

This over-all space-time point of view leads to considerable simplification in many problems. One can take into account at the same time processes which ordinarily would have to be considered separately. For example, when considering the scattering of an electron by a potential one automatically takes into account the effects of virtual pair productions. The same equation, Dirac's, which describes the deflection of the world line of an electron in a field, can also describe the deflection (and in just as simple a manner) when it is large enough to reverse the time-sense of the world line, and thereby correspond to pair annihilation. Quantum mechanically the direction of the world lines is replaced by the direction of propagation of waves.

This view is quite different from that of the Hamiltonian method which considers the future as developing continuously from out of the past. Here we imagine the entire space-time history laid out, and that we just become aware of increasing portions of it successively. In a scattering problem this over-all view of the complete scattering process is similar to the $S$-matrix viewpoint of Heisenberg. The temporal order of events during the scattering, which is analyzed in such detail by the Hamiltonian differential equation, is irrelevant. The relation of these viewpoints will be discussed much more fully in the introduction to the second paper, in which the more complicated interactions are analyzed.

The development stemmed from the idea that in non-relativistic quantum mechanics the amplitude for a given process can be considered as the sum of an amplitude for each space-time path available.[1] In view of the fact that in classical physics positrons could be viewed

---

[1] R. P. Feynman, Rev. Mod. Phys. **20**, 367 (1948).

as electrons proceeding along world lines toward the past (reference 7) the attempt was made to remove, in the relativistic case, the restriction that the paths must proceed always in one direction in time. It was discovered that the results could be even more easily understood from a more familiar physical viewpoint, that of scattered waves. This viewpoint is the one used in this paper. After the equations were worked out physically the proof of the equivalence to the second quantization theory was found.[2]

First we discuss the relation of the Hamiltonian differential equation to its solution, using for an example the Schrödinger equation. Next we deal in an analogous way with the Dirac equation and show how the solutions may be interpreted to apply to positrons. The interpretation seems not to be consistent unless the electrons obey the exclusion principle. (Charges obeying the Klein-Gordon equations can be described in an analogous manner, but here consistency apparently requires Bose statistics.)[3] A representation in momentum and energy variables which is useful for the calculation of matrix elements is described. A proof of the equivalence of the method to the theory of holes in second quantization is given in the Appendix.

## 2. GREEN'S FUNCTION TREATMENT OF SCHRÖDINGER'S EQUATION

We begin by a brief discussion of the relation of the non-relativistic wave equation to its solution. The ideas will then be extended to relativistic particles, satisfying Dirac's equation, and finally in the succeeding paper to interacting relativistic particles, that is, quantum electrodynamics.

The Schrödinger equation

$$i\partial\psi/\partial t = H\psi, \tag{1}$$

---

[2] The equivalence of the entire procedure (including photon interactions) with the work of Schwinger and Tomonaga has been demonstrated by F. J. Dyson, Phys. Rev. **75**, 486 (1949).

[3] These are special examples of the general relation of spin and statistics deduced by W. Pauli, Phys. Rev. **58**, 716 (1940).

describes the change in the wave function $\psi$ in an infinitesimal time $\Delta t$ as due to the operation of an operator $\exp(-iH\Delta t)$. One can ask also, if $\psi(\mathbf{x}_1, t_1)$ is the wave function at $\mathbf{x}_1$ at time $t_1$, what is the wave function at time $t_2 > t_1$? It can always be written as

$$\psi(\mathbf{x}_2, t_2) = \int K(\mathbf{x}_2, t_2; \mathbf{x}_1, t_1)\, \psi(\mathbf{x}_1, t_1)\, d^3\mathbf{x}_1, \qquad (2)$$

where $K$ is a Green's function for the linear Eq. (1). (We have limited ourselves to a single particle of coordinate $\mathbf{x}$, but the equations are obviously of greater generality.) If $H$ is a constant operator having eigenvalues $E_n$, eigenfunctions $\phi_n$ so that $\psi(\mathbf{x}, t_1)$ can be expanded as $\sum_n C_n \phi_n(\mathbf{x})$, then $\psi(\mathbf{x}, t_2)=\exp(-iE_n(t_2 - t_1)) \times C_n \phi_n(\mathbf{x})$. Since $C_n = \int \phi_n^*(\mathbf{x}_1)\psi(\mathbf{x}_1, t_1)d^3\mathbf{x}_1$, one finds (where we write 1 for $\mathbf{x}_1, t_1$ and 2 for $\mathbf{x}_2, t_2$) in this case

$$K(2, 1) = \sum_n \phi_n(\mathbf{x}_2)\, \phi_n^*(\mathbf{x}_1) \exp\left(-iE_n(t_2 - t_1)\right), \qquad (3)$$

for $t_2 > t_1$. We shall find it convenient for $t_2 < t_1$ to define $K(2, 1) = 0$ (Eq. (2) is then not valid for $t_2 < t_1$). It is then readily shown that in general $K$ can be defined by that solution of

$$(i\partial/\partial t_2 - H_2)\, K(2, 1) = i\delta(2, 1), \qquad (4)$$

which is zero for $t_2 < t_1$, where $\delta(2, 1) = \delta(t_2 - t_1)\delta(x_2 - x_1) \times \delta(y_2 - y_1)\delta(z_2 - z_1)$ and the subscript 2 on $H_2$ means that the operator acts on the variables of 2 of $K(2, 1)$. When $H$ is not constant, (2) and (4) are valid but $K$ is less easy to evaluate than (3).[4]

We can call $K(2, 1)$ the total amplitude for arrival at $\mathbf{x}_2, t_2$ starting from $\mathbf{x}_1, t_1$. (It results from adding an amplitude, $\exp iS$, for each space time path between these points, where $S$ is the action along the path.[1]) The transition amplitude for finding a particle in state $\chi(\mathbf{x}_2, t_2)$ at time $t_2$, if at $t_1$ it was in $\psi(\mathbf{x}_1, t_1)$, is

$$\int \chi^*(2)K(2, 1)\psi(1)d^3\mathbf{x}_1 d^3\mathbf{x}_2. \qquad (5)$$

---

[4] For a non-relativistic free particle, where $\phi_n = \exp(i\mathbf{p}\cdot\mathbf{x})$, $E_n = \mathbf{p}^2/2m$, (3) gives, as is well known

$$K_0(2, 1) = \int \exp\left[-(i\mathbf{p}\cdot\mathbf{x}_1 - i\mathbf{p}\cdot\mathbf{x}_2) - i\mathbf{p}^2(t_2 - t_1)/2m\right]d^3p(2\pi)^{-3}$$
$$= (2\pi im^{-1}(t_2 - t_1))^{-1} \exp(\tfrac{1}{2}im(\mathbf{x}_2 - \mathbf{x}_1)^2(t_2 - t_1)^{-1})$$

for $t_2 > t_1$, and $K_0 = 0$ for $t_2 < t_1$.

A quantum mechanical system is described equally well by specifying the function $K$, or by specifying the Hamiltonian $H$ from which it results. For some purposes the specification in terms of $K$ is easier to use and visualize. We desire eventually to discuss quantum electrodynamics from this point of view.

To gain a greater familiarity with the $K$ function and the point of view it suggests, we consider a simple perturbation problem. Imagine we have a particle in a weak potential $U(\mathbf{x}, t)$, a function of position and time. We wish to calculate $K(2, 1)$ if $U$ differs from zero only for $t$ between $t_1$ and $t_2$. We shall expand $K$ in increasing powers of $U$:

$$K(2, 1) = K_0(2, 1) + K^{(1)}(2, 1) + K^{(2)}(2, 1) + \cdots. \qquad (6)$$

To zero order in $U$, $K$ is that for a free particle, $K_0(2, 1)$.[4] To study the first order correction $K^{(1)}(2, 1)$, first consider the case that $U$ differs from zero only for the infinitesimal time interval $\Delta t_3$ between some time $t_3$ and $t_3 + \Delta t_3 (t_1 < t_3 < t_2)$. Then if $\psi(1)$ is the wave function at $\mathbf{x}_1, t_1$, the wave function at $\mathbf{x}_3, t_3$ is

$$\psi(3) = \int K_0(3, 1)\, \psi(1) d^3\mathbf{x}_1, \qquad (7)$$

since from $t_1$ to $t_3$ the particle is free. For the short interval $\Delta t_3$ we solve (1) as

$$\psi(\mathbf{x}, t_3 + \Delta t_3) = \exp(-iH\Delta t_3)\psi(\mathbf{x}, t_3)$$
$$= (1 - iH_0\Delta t_3 - iU\Delta t_3)\psi(\mathbf{x}, t_3),$$

where we put $H = H_0 + U$, $H_0$ being the Hamiltonian of a free particle. Thus $\psi(\mathbf{x}, t_3 + \Delta t_3)$ differs from what it would be if the potential were zero (namely $(1 - iH_0\Delta t_3)\psi(\mathbf{x}, t_3)$) by the extra piece

$$\Delta\psi = -iU(\mathbf{x}_3, t_3) \cdot \psi(\mathbf{x}_3, t_3)\, \Delta t_3, \qquad (8)$$

which we shall call the amplitude scattered by the potential. The wave function at 2 is given by

$$\psi(\mathbf{x}_2, t_2) = \int K_0(\mathbf{x}_2, t_2; \mathbf{x}_3, t_3 + \Delta t_3)\, \psi(\mathbf{x}_3, t_3 + \Delta t_3)\, d^3\mathbf{x}_3.$$

since after $t_3 + \Delta t_3$ the particle is again free. Therefore the change in the wave function at 2 brought about by the potential is (substitute

(7) into (8) and (8) into the equation for $\psi(\mathbf{x}_2, t_2)$):

$$\Delta\psi(2) = -i \int K_0(2, 3) U(3) K_0(3, 1) \psi(1) d^3\mathbf{x}_1 d^3\mathbf{x}_3 \Delta t_3.$$

In the case that the potential exists for an extended time, it may be looked upon as a sum of effects from each interval $\Delta t_3$ so that the total effect is obtained by integrating over $t_3$ as well as $\mathbf{x}_3$. From the definition (2) of $K$ then, we find

$$K^{(1)}(2, 1) = -i \int K_0(2, 3) U(3) K_0(3, 1) \, d\tau_3, \qquad (9)$$

where the integral can now be extended over all space and time, $d\tau_3 = d^3\mathbf{x}_3 dt_3$. Automatically there will be no contribution if $t_3$ is outside the range $t_1$ to $t_2$ because of our definition, $K_0(2, 1) = 0$ for $t_2 < t_1$.

We can understand the result (6), (9) this way. We can imagine that a particle travels as a free particle from point to point, but is scattered by the potential $U$. Thus the total amplitude for arrival at 2 from 1 can be considered as the sum of the amplitudes for various alternative routes. It may go directly from 1 to 2 (amplitude $K_0(2, 1)$, giving the zero order term in (6)). Or (see Fig. 1(a)) it may go from 1 to 3 (amplitude $K_0(3, 1)$), get scattered there by the potential (scattering amplitude $-iU(3)$ per unit volume and time) and then go from 3 to 2 (amplitude $K_0(2, 3)$). This may occur for any point 3 so that summing over these alternatives gives (9).

Again, it may be scattered twice by the potential (Fig. 1(b)). It goes from 1 to 3 ($K_0(3,1)$), gets scattered there ($-iU(3)$) then proceeds to some other point, 4, in space time (amplitude $K_0(4, 3)$) is scattered again ($-iU(4)$) and then proceeds to 2 ($K_0(2, 4)$). Summing over all possible places and times for 3, 4 find that the second order contribution to the total amplitude $K^{(2)}(2, 1)$ is

$$(-i)^2 \int \int K_0(2, 4) U(4) K_0(4, 3)$$
$$\times U(3) K_0(3, 1) \, d\tau_3 d\tau_4. \qquad (10)$$

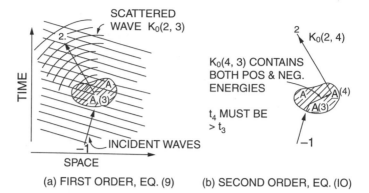

(a) FIRST ORDER, EQ. (9)    (b) SECOND ORDER, EQ. (lO)

FIG. 1  The Schrödinger (and Dirac) equation can be visualized as describing the fact that plane waves are scattered successively by a potential. Figure 1 (a) illustrates the situation in first order. $K_0(2, 3)$ is the amplitude for a free particle starting at point 3 to arrive at 2. The shaded region indicates the presence of the potential $A$ which scatters at 3 with amplitude $-iA(3)$ per $cm^3$sec. (Eq. (9)). In (b) is illustrated the second order process (Eq. (10)), the waves scattered at 3 are scattered again at 4. However, in Dirac one-electron theory $K_0(4, 3)$ would represent electrons both of positive and of negative energies proceeding from 3 to 4. This is remedied by choosing a different scattering kernel $K_+ (4, 3)$, Fig. 2.

This can be readily verified directly from (1) just as (9) was. One can in this way obviously write down any of the terms of the expansion (6).[5]

# 3. TREATMENT OF THE DIRAC EQUATION

We shall now extend the method of the last section to apply to the Dirac equation. All that would seem to be necessary in the previous equations is to consider $H$ as the Dirac Hamiltonian, $\psi$ as a symbol with four indices (for each particle). Then $K_0$ can still be defined by (3) or (4) and is now a 4–4 matrix which operating on the initial wave function, gives the final wave function. In (10), $U(3)$ can be generalized to $A_4(3) - \boldsymbol{\alpha} \cdot \mathbf{A}(3)$ where $A_4$, $\mathbf{A}$ are the scalar and vector potential (times $e$, the electron charge) and $\boldsymbol{\alpha}$ are Dirac matrices.

---

[5] We are simply solving by successive approximations an integral equation (deducible directly from (1) with $H = H_0 + U$ and (4) with $H = H_0$),

$$\psi(2) = -i \int K_0(2, 3)U(3)\psi(3)d\tau_3 + \int K_0(2, 1)\psi(1)d^3\mathbf{x}_1,$$

where the first integral extends over all space and all times $t_3$ greater than the $t_1$ appearing in the second term, and $t_2 > t_1$.

To discuss this we shall define a convenient relativistic notation. We represent four-vectors like **x**, $t$ by a symbol $x_\mu$, where $\mu = 1, 2, 3, 4$ and $x_4 = t$ is real. Thus the vector and scalar potential (times $e$) **A**, $A_4$ is $A_\mu$. The four matrices $\beta\boldsymbol{\alpha}$, $\beta$ can be considered as transforming as a four vector $\gamma_\mu$ (our $\gamma_\mu$ differs from Pauli's by a factor $i$ for $\mu = 1, 2, 3$). We use the summation convention $a_\mu b_\mu = a_4 b_4 - a_1 b_1 - a_2 b_2 - a_3 b_3 = a \cdot b$. In particular if $a_\mu$ is any four vector (but not a matrix) we write $\boldsymbol{a} = a_\mu \gamma_\mu$ so that $\boldsymbol{a}$ is a matrix associated with a vector ($\boldsymbol{a}$ will often be used in place of $a_\mu$ as a symbol for the vector). The $\gamma_\mu$ satisfy $\gamma_\mu \gamma_\nu + \gamma_\nu \gamma_\mu = 2\delta_{\mu\nu}$ where $\delta_{44} = +1$, $\delta_{11} = \delta_{22} = \delta_{33} = -1$, and the other $\delta_{\mu\nu}$ are zero. As a consequence of our summation convention $\delta_{\mu\nu} a_\nu = a_\mu$ and $\delta_{\mu\mu} = 4$. Note that $\boldsymbol{ab} + \boldsymbol{ba} = 2a \cdot b$ and that $\boldsymbol{a}^2 = a_\mu a_\mu = a \cdot a$ is a pure number. The symbol $\partial/\partial x_\mu$, will mean $\partial/\partial t$ for $\mu = 4$, and $-\partial/\partial x$, $-\partial/\partial y$, $-\partial/\partial z$ for $\mu = 1, 2, 3$. Call $\nabla = \gamma_\mu \partial/\partial x_\mu = \beta\partial/\partial t + \beta\boldsymbol{\alpha} \cdot \nabla$. We shall imagine hereafter, purely for relativistic convenience, that $\phi_n^*$ in (3) is replaced by its adjoint $\tilde{\phi}_n = \phi_n^* \beta$.

Thus the Dirac equation for a particle, mass $m$, in an external field $\boldsymbol{A} = A_\mu \gamma_\mu$ is

$$(i\nabla - m)\,\psi = \boldsymbol{A}\psi, \tag{11}$$

and Eq. (4) determining the propagation of a free particle becomes

$$(i\nabla_2 - m)\,K_+(2, 1) = i\delta(2, 1), \tag{12}$$

the index 2 on $\nabla_2$ indicating differentiation with respect to the coordinates $x_{2\mu}$ which are represented as 2 in $K_+(2, 1)$ and $\delta(2, 1)$.

The function $K_+(2, 1)$ is defined in the absence of a field. If a potential $\boldsymbol{A}$ is acting a similar function, say $K_+^{(A)}(2, 1)$ can be defined. It differs from $K_+(2, 1)$ by a first order correction given by the analogue of (9) namely

$$K_+^{(1)}(2, 1) = -i \int K_+(2, 3)\boldsymbol{A}(3)K_+(3, 1)\,d\tau_3, \tag{13}$$

representing the amplitude to go from 1 to 3 as a free particle, get scattered there by the potential (now the matrix $\boldsymbol{A}(3)$ instead of $U(3)$) and continue to 2 as free. The second order correction, analogous to

(10) is

$$K_+^{(2)}(2, 1) = - \int \int K_+(2, 4) A(4)$$
$$\times K_+(4, 3) A(3) K_+(3, 1) \, d\tau_4 d\tau_3, \tag{14}$$

and so on. In general $K_+^{(A)}$ satisfies

$$(i\nabla_2 - A(2) - m) K_+^{(A)}(2, 1) = i\delta(2, 1), \tag{15}$$

and the successive terms (13), (14) are the power series expansion of the integral equation

$$K_+^{(A)}(2, 1) = K_+(2, 1) - i \int K_+(2, 3) A(3) K_+^{(A)}(3, 1) \, d\tau_3, \tag{16}$$

which it also satisfies.

We would now expect to choose, for the special solution of (12), $K_+ = K_0$ where $K_0(2, 1)$ vanishes for $t_2 < t_1$ and for $t_2 > t_1$ is given by (3) where $\phi_n$ and $E_n$ are the eigenfunctions and energy values of a particle satisfying Dirac's equation, and $\phi_n^*$ is replaced by $\bar{\phi}_n$.

The formulas arising from this choice, however, suffer from the drawback that they apply to the one electron theory of Dirac rather than to the hole theory of the positron. For example, consider as in Fig. 1(a) an electron after being scattered by a potential in a small region 3 of space time. The one electron theory says (as does (3) with $K_+ = K_0$) that the scattered amplitude at another point 2 will proceed toward positive times with both positive and negative energies, that is with both positive and negative rates of change of phase. No wave is scattered to times previous to the time of scattering. These are just the properties of $K_0(2, 3)$.

On the other hand, according to the positron theory negative energy states are not available to the electron after the scattering. Therefore the choice $K_+ = K_0$ is unsatisfactory. But there are other solutions of (12). We shall choose the solution defining $K_+(2, 1)$ so that $K_+(2, 1)$ *for $t_2 > t_1$ is the sum of (3) over positive energy states only.* Now this new solution must satisfy (12) for all times in order that the representation be complete. It must therefore differ from the old solution $K_0$ by a solution of the homogeneous Dirac equation. It is clear from the definition that the difference $K_0 - K_+$ is the sum of (3) over all negative energy states, as long as $t_2 > t_1$. But this difference must be a solution of the homogeneous Dirac equation for all times and must therefore be represented by the same sum over negative energy states also for $t_2 < t_1$. Since $K_0 = 0$ in this case, it follows that our new kernel, $K_+(2, 1)$, *for $t_2 < t_1$ is the negative of the sum (3) over negative energy states.* That is,

$$K_+(2, 1) = \sum_{POS\ E_n} \phi_n(2)\bar{\phi}_n(1) \exp\left(-i E_n (t_2 - t_1)\right) \quad \text{for} \quad t_2 > t_1$$

$$= -\sum_{NEG\ E_n} \phi_n(2)\bar{\phi}_n(1) \exp\left(-i E_n (t_2 - t_1)\right) \quad \text{for} \quad t_2 > t_1.$$

$$(17)$$

With this choice of $K_+$ our equations such as (13) and (14) will now give results equivalent to those of the positron hole theory.

That (14), for example, is the correct second order expression for finding at 2 an electron originally at 1 according to the positron theory may be seen as follows (Fig. 2). Assume as a special example that $t_2 > t_1$ and that the potential vanishes except in interval $t_2 - t_1$ so that $t_4$ and $t_3$ both lie between $t_1$ and $t_2$.

First suppose $t_4 > t_3$ (Fig. 2(b)). Then (since $t_3 > t_1$) the electron assumed originally in a positive energy state propagates in that state (by $K_+(3, 1)$) to position 3 where it gets scattered ($A(3)$). It then proceeds to 4, which it must do as a positive energy electron. This is correctly described by (14) for $K_+(4, 3)$ contains only positive energy components in its expansion, as $t_4 > t_3$. After being scattered at 4 it then proceeds on to 2, again necessarily in a positive energy state, as $t_2 > t_4$.

In positron theory there is an additional contribution due the possibility of virtual pair production (Fig. 2(c)). A pair could be created by the potential $A$ (4) at 4, the electron of which is that found later at 2. The positron (or rather, the hole) proceeds to 3 where it annihilates the electron which has arrived there from 1.

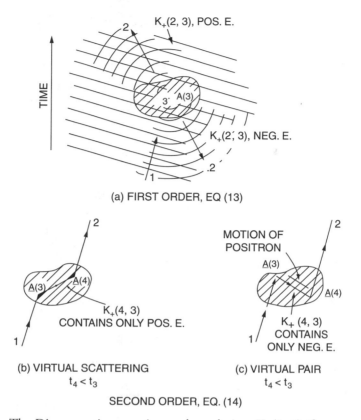

(a) FIRST ORDER, EQ (13)

(b) VIRTUAL SCATTERING
$t_4 < t_3$

(c) VIRTUAL PAIR
$t_4 < t_3$

SECOND ORDER, EQ. (14)

FIG. 2  The Dirac equation permits another solution $K_+(2, 1)$ if one considers that waves scattered by the potential can proceed backwards in time as in Fig. 2 (a). This is interpreted in the second order processes (b), (c), by noting that there is now the possibility (c) of virtual pair production at 4, the positron going to 3 to be annihilated. This can be pictured as similar to ordinary scattering (b) except that the electron is scattered backwards in time from 3 to 4. The waves scattered from 3 to 2′ in (a) represent the possibility of a positron arriving at 3 from 2′ and annihilating the electron from 1. This view is proved equivalent to hole theory: electrons traveling backwards in time are recognized as positrons.

This alternative is already included in (14) as contributions for which $t_4 < t_3$, and its study will lead us to an interpretation of $K_+(4, 3)$ for $t_4 < t_3$. The factor $K_+(2, 4)$ describes the electron (after the pair production at 4) proceeding from 4 to 2. Likewise $K_+(3, 1)$ represents the electron proceeding from 1 to 3. $K_+(4, 3)$ must therefore represent the propagation of the positron or hole from 4 to 3. That it does so is clear. The fact it in hole theory the hole proceeds in the manner of and electron of negative energy is reflected in the fact that $K_+$ (4, 3) for $t_4 < t_3$ is (minus) the sum of only negative energy components. In hole theory the real energy of these intermediate states is, of course, positive. This is true here too, since in the phases $\exp(-iE_n(t_4 - t_3))$ defining $K_+(4, 3)$ in (17), $E_n$ is negative but so is $t_4 - t_3$. That is, the contributions vary with $t_3$ as $\exp(-i|E_n|(t_3 - t_4))$ as they would if the energy of the intermediate state were $|E_n|$. The fact that the entire sum is taken as negative in computing $K_+(4, 3)$ is reflected in the fact that in hole theory the amplitude has its sign reversed in accordance with the Pauli principle and the fact that the electron arriving at 2 has been exchanged with one in the sea.[6] To this, and to higher orders, all processes involving virtual pairs are correctly described in this way.

The expressions such as (14) can still be described as a passage of the electron from 1 to 3 ($K_+(3, 1)$), scattering at 3 by $A(3)$, proceeding to 4 ($K_+(4, 3)$), scattering again, $A(4)$, arriving finally at 2. The scatterings may, however, be toward both future and past times, an electron propagating backwards in time being recognized as a positron.

This therefore suggests that negative energy components created by scattering in a potential be considered as waves propagating from the scattering point toward the past, and that such waves represent the propagation of a positron annihilating the electron in the potential.[7]

With this interpretation real pair production is also described correctly (see Fig. 3). For example in (13) if $t_1 < t_3 < t_2$ the equation gives the amplitude that if at time $t_1$ one electron is present at 1, then at time $t_2$ just one electron will be present (having been scattered at 3) and it will be at 2. On the other hand if $t_2$ is less than $t_3$, for example, if $t_2 = t_1 < t_3$, the same expression gives the amplitude that a pair, electron at 1, positron at 2 will annihilate at 3, and subsequently no particles will be present. Likewise if $t_2$ and $t_1$ exceed $t_3$ we have (minus) the amplitude for finding a single pair, electron at 2, positron

---

[6] It has often been noted that the one-electron theory apparently gives the same matrix elements for this process as does hole theory. The problem is one of interpretation, especially in a way that will also give correct results for other processes, e.g., self-energy.

[7] The idea that positrons can be represented as electrons with proper time reversed relative to true time has been discussed by the author and others, particularly by Stückelberg. E. C. C. Stückelberg, Helv. Phys. Acta **15**, 23 (1942); R. P. Feynman, Phys. Rev. **74**, 939 (1948). The fact that classically the action (proper time) increases continuously as one follows a trajectory is reflected in quantum mechanics in the fact that the phase, which is $|E_n| |t_2 - t_1|$, always increases as the particle proceeds from one scattering point to the next.

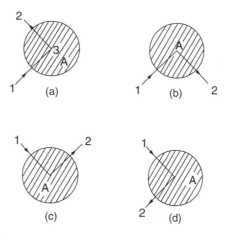

FIG. 3  Several different processes can be described by the same formula depending on the time relations of the variables $t_2$, $t_1$. Thus $P_v|K_+^{(A)}(2, 1)|^2$ is the probability that: (a) An electron at 1 will be scattered at 2 (and no other pairs form in vacuum). (b) Electron at 1 and positron at 2 annihilate leaving nothing. (c) A single pair at 1 and 2 is created from vacuum. (d) A positron at 2 is scattered to 1. ($K_+^{(A)}(2, 1)$ is the sum of the effects of scattering in the potential to all orders. $P_v$ is a normalizing constant.)

at 1 created by $A(3)$ from a vacuum. If $t_1 > t_3 > t_2$, (13) describes the scattering of a positron. All these amplitudes are relative to the amplitude that a vacuum will remain a vacuum, which is taken as unity. (This will be discussed more fully later.)

The analogue of (2) can be easily worked out.[8] It is,

$$\psi(2) = \int K_+(2, 1)N(1)\psi(1)d^3V_1, \qquad (18)$$

where $d^3V_1$ is the volume element of the closed 3-dimensional surface of a region of space time containing point 2, and $N(1)$ is $N_\mu(1)\gamma_\mu$ where $N_\mu(1)$ is the *inward* drawn unit normal to the surface at the point 1. That is, the wave function $\psi(2)$ (in this case for a free particle) is determined at any point inside a four-dimensional region if its values on the surface of that region are specified.

[8] By multiplying (12) on the right by $(-i\nabla_1 - m)$ and noting that $\nabla_1\delta(2, 1) = -\nabla_2\delta(2, 1)$ show that $K_+(2, 1)$ also satisfies $K_+(2, 1)(-i\nabla_1 - m) = i\delta(2, 1)$, where the $\nabla_1$ operates on variable 1 in $K_+(2, 1)$ but is written after that function to keep the correct order of the $\gamma$ matrices. Multiply this equation by $\psi(1)$ and Eq. (11) (with $A = 0$, calling the variables 1) by $K_+(2, 1)$, subtract and integrate over a region of space-time. The integral on the left-hand side can be transformed to an integral over the surface of the region. The right-hand side is $\psi(2)$ if the point 2 lies within the region, and is zero otherwise. (What happens when the 3-surface contains a light line and hence has no unique normal need not concern us as these points can be made to occur so far away from 2 that their contribution vanishes.)

To interpret this, consider the case that the 3-surface consists essentially of all space at some time say $t = 0$ previous to $t_2$, and of all space at the time $T > t_2$. The cylinder connecting these to complete the closure of the surface may be very distant from $x_2$ so that it gives no appreciable contribution (as $K_+(2, 1)$ decreases exponentially in space-like directions). Hence, if $\gamma_4 = \beta$, since the inward drawn normals $N$ will be $\beta$ and $-\beta$,

$$\psi(2) = \int K_+(2, 1)\beta\psi(1)d^3\mathbf{x}_1 - \int K_+(2, 1')\beta\psi(1')d^3\mathbf{x}_{1'}, \quad (19)$$

where $t_1 = 0$, $t_1' = T$. Only positive energy (electron) components in $\psi(1)$ contribute to the first integral and only negative energy (positron) components of $\psi(1')$ to the second. That is, the amplitude for finding a charge at 2 is determined both by the amplitude for finding an electron previous to the measurement and by the amplitude for finding a positron after the measurement. This might be interpreted as meaning that even in a problem involving but one charge the amplitude for finding the charge at 2 is not determined when the only thing known in the amplitude for finding an electron (or a positron) at an earlier time. There may have been no electron present initially but a pair was created in the measurement (or also by other external fields). The amplitude for this contingency is specified by the amplitude for finding a positron in the future.

We can also obtain expressions for transition amplitudes, like (5). For example if at $t = 0$ we have an electron present in a state with (positive energy) wave function $f(\mathbf{x})$, what is the amplitude for finding it at $t = T$ with the (positive energy) wave function $g(\mathbf{x})$? The amplitude for finding the electron anywhere after $t = 0$ is given by (19) with $\psi(1)$ replaced by $f(\mathbf{x})$, the second integral vanishing. Hence, the transition element to find it in state $g(\mathbf{x})$ is, in analogy to (5), just ($t_2 = T$, $t_1 = 0$)

$$\int \bar{g}(\mathbf{x}_2)\,\beta K_+(2, 1)\beta f(\mathbf{x}_1)\,d^3\mathbf{x}_1 d^3\mathbf{x}_2, \quad (20)$$

since $g^* = \bar{g}\beta$.

If a potential acts somewhere in the interval between 0 and $T$, $K_+$ is replaced by $K_+{}^{(A)}$. Thus the first order effect on the transition amplitude is, from (13),

$$-i \int \bar{g}(\mathbf{x}_2) \, \beta K_+(2, 3) A(3) K_+(3, 1) \, \beta f(\mathbf{x}_1) \, d^3\mathbf{x}_1 d^3\mathbf{x}_2. \quad (21)$$

Expressions such as this can be simplified and the 3-surface integrals, which are inconvenient for relativistic calculations, can be removed as follows. Instead of defining a state by the wave function $f(\mathbf{x})$, which it has at a given time $t_1 = 0$, we define the state by the function $f(1)$ of four variables $\mathbf{x}_1$, $t_1$ which is a solution of the free particle equation for all $t_1$ and is $f(\mathbf{x}_1)$ for $t_1 = 0$. The final state is likewise defined by a function $g(2)$ over-all space-time. Then our surface integrals can be performed since $\int K_+(3, 1)\beta f(\mathbf{x}_1)d^3\mathbf{x}_1 = f(3)$ and $\int \bar{g}(\mathbf{x}_2)\beta d^3\mathbf{x}_2 K_+(2, 3) = \bar{g}(3)$. There results

$$-i \int \bar{g}(3)A(3) f(3)d\tau_3, \quad (22)$$

the integral now being over-all space-time. The transition amplitude to second order (from (14)) is

$$-\int \int \bar{g}(2)A(2)K_+(2, 1)A(1) f(1)d\tau_1 d\tau_2, \quad (23)$$

for the particle arriving at 1 with amplitude $f(1)$ is scattered $(A(1))$, progresses to 2, $(K_+(2, 1))$, and is scattered again $(A(2))$, and we then ask for the amplitude that it is in state $g(2)$. If $g(2)$ is a negative energy state we are solving a problem of annihilation of electron in $f(1)$, positron in $g(2)$, etc.

We have been emphasizing scattering problems, but obviously the motion in a fixed potential $V$, say in a hydrogen atom, can also be dealt with. If it is first viewed as a scattering problem we can ask for the amplitude, $\phi_k(1)$, that an electron with original free wave function was scattered $k$ times in the potential $V$ either forward or backward in time to arrive at 1. Then the amplitude, after one more scattering is

$$\phi_{k+1}(2) = -i \int K_+(2, 1) V(1)\phi_k(1)d\tau_1. \quad (24)$$

An equation for the total amplitude

$$\psi(1) = \sum_{k=0}^{\infty} \phi_k(1)$$

for arriving at 1 either directly or after any number of scatterings is obtained by summing (24) over all $k$ from 0 to $\infty$;

$$\psi(2) = \phi_0(2) - i \int K_+(2, 1) V(1) \psi(1) d\tau_1. \tag{25}$$

Viewed as a steady state problem we may wish, for example, to find that initial condition $\phi_0$ (or better just the $\psi$) which leads to a periodic motion of $\psi$. This is most practically done, of course, by solving the Dirac equation,

$$(i\nabla - m)\,\psi(1) = V(1)\psi(1), \tag{26}$$

deduced from (25) by operating on both sides by $i\nabla_2 - m$, thereby eliminating the $\phi_0$, and using (12). This illustrates the relation between the points of view.

For many problems the total potential $A + V$ may be split conveniently into a fixed one, $V$, and another, $A$, considered as a perturbation. If $K_+^{(V)}$ is defined as in (16) with $V$ for $A$, expressions such as (23) are valid and useful with $K_+$ replaced by $K_+^{(V)}$ and the functions $f(1), g(2)$ replaced by solutions for all space and time of the Dirac Eq. (26) in the potential $V$ (rather than free particle wave functions).

## 4. PROBLEMS INVOLVING SEVERAL CHARGES

We wish next to consider the case that there are two (or more) distinct charges (in addition to pairs they may produce in virtual states). In a succeeding paper we discuss the interaction between such charges. Here we assume that they do not interact. In this case each particle behaves independently of the other. We can expect that if we have two particles $a$ and $b$, the amplitude that particle $a$ goes from $\mathbf{x}_1$ at $t_1$, to $\mathbf{x}_3$ at $t_3$ while $b$ goes from $\mathbf{x}_2$ at $t_2$ to $\mathbf{x}_4$ at $t_4$ is the product

$$K(3, 4; 1, 2) = K_{+a}(3, 1) K_{+b}(4, 2).$$

The symbols $a$, $b$ simply indicate that the matrices appearing in the $K_+$ apply to the Dirac four component spinors corresponding to particle $a$ or $b$ respectively (the wave function now having 16 indices). In a potential $K_{+a}$ and $K_{+b}$ become $K_{+a}^{(A)}$ and $K_{+b}^{(A)}$ where $K_{+a}^{(A)}$ is defined and calculated as for a single particle. They commute. Hereafter the $a$, $b$ can be omitted; the space time variable appearing in the kernels suffice to define on what they operate.

The particles are identical however and satisfy the exclusion principle. The principle requires only that one calculate $K(3, 4; 1, 2) - K(4, 3; 1, 2)$ to get the net amplitude for arrival of charges at 3, 4. (It is normalized assuming that when an integral is performed over points 3 and 4, for example, since the electrons represented are identical, one divides by 2.) This expression is correct for positrons also (Fig. 4). For example the amplitude that an electron and a positron found initially at $\mathbf{x}_1$ and $\mathbf{x}_4$ (say $t_1 = t_4$) are later found at $\mathbf{x}_3$ and $\mathbf{x}_2$ (with $t_2 = t_3 > t_1$) is given by the same expression

$$K_+^{(A)} (3, 1) \, K_+^{(A)} (4, 2) - K_+^{(A)} (4, 1) \, K_+^{(A)} (3, 2) \,. \tag{27}$$

The first term represents the amplitude that the electron proceeds from 1 to 3 and the positron from 4 to 2 (Fig. 4(c)), while the second term represents the interfering amplitude that the pair at 1, 4 annihilate and what is found at 3, 2 is a pair newly created in the potential. The generalization to several particles is clear. There is an additional factor $K_+^{(A)}$ for each particle, and anti-symmetric combinations are always taken.

No account need be taken of the exclusion principle in intermediate states. As an example consider again expression (14) for $t_2 > t_1$ and suppose $t_4 < t_3$ so that the situation represented (Fig. 2(c)) is that a pair is made at 4 with the electron proceeding to 2, and the positron to 3 where it annihilates the electron arriving from 1. It may be objected that if it happens that the electron created at 4 is in the same state as the one coming from 1, then the process cannot occur because of the exclusion principle and we should not have included it

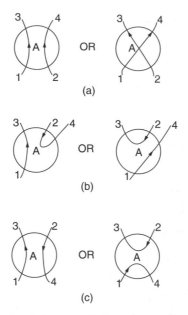

(a)

(b)

(c)

FIG. 4   Some problems involving two distinct charges (in addition to virtual pairs they may produce): $P_v | K_+^{(A)}(3, 1) K_+^{(A)}(4, 2) - K_+^{(A)}(4, 1) K_+^{(A)}(3, 2)|^2$ is the probability that: (a) Electrons at 1 and 2 are scattered to 3, 4 (and no pairs are formed). (b) Starting with an electron at 1 a single pair is formed, positron at 2, electrons at 3, 4. (c) A pair at 1, 4 is found at 3, 2, etc. The exclusion principle requires that the amplitudes for processes involving exchange of two electrons be subtracted.

in our term (14). We shall see, however, that considering the exclusion principle also requires another change which reinstates the quantity.

For we are computing amplitudes relative to the amplitude that a vacuum at $t_1$ will still be a vacuum at $t_2$. We are interested in the alteration in this amplitude due to the presence of an electron at 1. Now one process that can be visualized as occurring in the vacuum is the creation of a pair at 4 followed by a re-annihilation of the *same* pair at 3 (a process which we shall call a closed loop path). But if a real electron is present in a certain state 1, those pairs for which the electron was created in state 1 in the vacuum must now be excluded. We must therefore subtract from our relative amplitude the term corresponding to this process. But this just reinstates the quantity which it was argued should not have been included in (14), the necessary minus sign coming automatically from the definition of $K_+$. It is obviously simpler to disregard the exclusion principle completely in the intermediate states.

All the amplitudes are relative and their squares give the relative probabilities of the various phenomena. Absolute probabilities result if one multiplies each of the probabilities by $P_v$, the true probability that if one has no particles present initially there will be none finally. This quantity $P_v$ can be calculated by normalizing the relative probabilities such that the sum of the probabilities of all mutually exclusive alternatives is unity. (For example if one starts with a vacuum one can calculate the relative probability that there remains a vacuum (unity), or one pair is created, or two pairs, etc. The sum is $P_v^{-1}$.) Put in this form the theory is complete and there are no divergence problems. Real processes are completely independent of what goes on in the vacuum.

When we come, in the succeeding paper, to deal with interactions between charges, however, the situation is not so simple. There is the possibility that virtual electrons in the vacuum may interact electromagnetically with the real electrons. For that reason processes occuring in the vacuum are analyzed in the next section, in which an independent method of obtaining $P_v$ is discussed.

# 5. VACUUM PROBLEMS

An alternative way of obtaining absolute amplitudes is to multiply all amplitudes by $C_v$, the vacuum to vacuum amplitude, that is, the absolute amplitude that there be no particles both initially and finally. We can assume $C_v = 1$ if no potential is present during the interval, and otherwise we compute it as follows. It differs from unity because, for example, a pair could be created which eventually annihilates itself again. Such a path would appear as a closed loop on a space-time diagram. The sum of the amplitudes resulting from all such single closed loops we call $L$. To a first approximation $L$ is

$$L^{(1)} = -\frac{1}{2} \int \int S_p \left[ K_+(1, 2) A(1) \times K_+(1, 2) A(2) \right] d\tau_1 d\tau_2.$$

(28)

For a pair could be created say at 1, the electron and positron could both go on to 2 and there annihilate. The spur, $Sp$, is taken since one has to sum over all possible spins for the pair. The factor $\frac{1}{2}$ arises from the fact that the same loop could be considered as starting at either potential, and the minus sign results since the interactors are each $-iA$. The next order term would be[9]

$$L^{(2)} = +(i/3) \int \int \int S_p [K_+(2, 1) A(1)$$
$$\times K_+(1, 3) A(3) K_+(3, 2) A(2)] \, d\tau_1 d\tau_2 d\tau_3,$$

etc. The sum of all such terms gives $L$.[10]

In addition to these single loops we have the possibility that two independent pairs may be created and each pair may annihilate itself again. That is, there may be formed in the vacuum two closed loops, and the contribution in amplitude from this alternative is just the product of the contribution from each of the loops considered singly. The total contribution from all such pairs of loops (it is still consistent to disregard the exclusion principle for these virtual states) is $L^2/2$ for in $L^2$ we count every pair of loops twice. The total vacuum-vacuum amplitude is then

$$C_v = 1 - L + L^2/2 - L^3/6 + \cdots = \exp(-L),$$

(30)

---

[9]This term actually vanishes as can be seen as follows. In any spur the sign of all $\gamma$ matrices may be reversed. Reversing the sign of $\gamma$ in $K_+(2, 1)$ changes it to the transpose of $K_+(1, 2)$ so that the order of all factors and variables is reversed. Since the integral is taken over all $\tau_1$, $\tau_2$, and $\tau_3$ this has no effect and we are left with $(-1)^3$ from changing the sign of $A$. Thus the spur equals its negative. Loops with an odd number of potential interactors give zero. Physically this is because for each loop the electron can go around one way or in the opposite direction and we must add these amplitudes. But reversing the motion of an electron makes it behave like a positive charge thus changing the sign of each potential interaction, so that the sum is zero if the number of interactions is odd. This theorem is due to W. H. Furry, Phys. Rev. **51**, 125 (1937).

[10]A closed expression for $L$ in terms of $K_+^{(A)}$ is hard to obtain because of the factor $(1/n)$ in the $n$th term. However, the perturbation in $L$, $\Delta L$ due to a small change in potential $\Delta A$, is easy to express. The $(1/n)$ is canceled by the fact that $\Delta A$ can appear in any of the $n$ potentials. The result after summing over $n$ by (13), (14) and using (16) is

$$\Delta L = -i \int Sp \left[ \left( K_+^{(A)}(1, 1) - K_+(1, 1) \right) \Delta A(1) \right] d\tau_1.$$

(29)

. The term $K_+(1, 1)$ actually integrates to zero.

the successive terms representing the amplitude from zero, one, two, etc., loops. The fact that the contribution to $C_v$ of single loops is $-L$ is a consequence of the Pauli principle. For example, consider a situation in which two pairs of particles are created. Then these pairs later destroy themselves so that we have two loops. The electrons could, at a given time, be interchanged forming a kind of figure eight which is a single loop. The fact that the interchange must change the sign of the contribution requires that the terms in $C_v$ appear with alternate signs. (The exclusion principle is also responsible in a similar way for the fact that the amplitude for a pair creation is $-K_+$ rather than $+K_+$.) Symmetrical statistics would lead to

$$C_v = 1 + L + L^2/2 = \exp(+L).$$

The quantity $L$ has an infinite imaginary part (from $L^{(1)}$, higher orders are finite). We will discuss this in connection with vacuum polarization in the succeeding paper. This has no effect on the normalization constant for the probability that a vacuum remain vacuum is given by

$$P_v = |C_v|^2 = \exp\left( \; 2 \cdot \text{real part of } L \right),$$

from (30). This value agrees with the one calculated directly by renormalizing probabilities. The real part of $L$ appears to be positive as a consequence of the Dirac equation and properties of $K_+$ so that $P_v$ is less than one. Bose statistics gives $C_v = \exp(+L)$ and consequently a value of $P_v$ greater than unity which appears meaningless if the quantities are interpreted as we have done here. Our choice of $K_+$ apparently requires the exclusion principle.

Charges obeying the Klein-Gordon equation can be equally well treated by the methods which are discussed here for the Dirac electrons. How this is done is discussed in more detail in the succeeding paper. The real part of $L$ comes out negative for this equation so that in this case Bose statistics appear to be required for consistency.[3]

## 6. ENERGY-MOMENTUM REPRESENTATION

The practical evaluation of the matrix elements in some problems is often simplified by working with momentum and energy variables rather than space and time. This is because the function $K_+(2, 1)$ is fairly complicated but we shall find that its Fourier transform is very simple, namely $(i/4\pi^2(p - m)^{-1}$ that is

$$K_+(2, 1) = (i/4\pi^2) \int (\mathbf{p} - m)^{-1} \exp(-ip \cdot x_{21}) \, d^4p, \quad (31)$$

where $p \cdot x_{21} = p \cdot x_2 - p \cdot x_1 = p_\mu x_{2\mu} - p_\mu x_{1\mu}$, $\mathbf{p} = p_\mu \gamma_\mu$, and $d^4p$ means $(2\pi)^{-2} dp_1 dp_2 dp_3 dp_4$, the integral over all $p$. That this is true can be seen immediately from (12), for the representation of the operator $i\nabla - m$ in energy $(p_4)$ and momentum $(p_{1,2,3})$ space is $\mathbf{p} - m$ and the transform of $\delta(2, 1)$ is a constant. The reciprocal matrix $(\mathbf{p} - m)^{-1}$ can be interpreted as $(\mathbf{p} + m)(p^2 - m^2)^{-1}$ for $p^2 - m^2 = (\mathbf{p} - m)(\mathbf{p} + m)$ is a pure number not involving $\gamma$ matrices. Hence if one wishes one can write

$$K_+(2, 1) = i(i\nabla_2 + m) I_+(2, 1),$$

where

$$I_+(2, 1) = (2\pi)^{-2} \int (p^2 - m^2)^{-1} \exp(-ip \cdot x_{21}) \, d^4p, \quad (32)$$

is not a matrix operator but a function satisfying

$$\Box_2^2 I_+(2, 1) - m^2 I_+(2, 1) = \delta(2, 1), \quad (33)$$

where $-\Box_2^2 = (\nabla_2)^2 = (\partial/\partial x_{2\mu})(\partial/\partial x_{2\mu})$.

The integrals (31) and (32) are not yet completely defined for there are poles in the integrand when $p^2 - m^2 = 0$. We can define how these poles are to be evaluated by the rule that $m$ is considered to have an infinitesimal negative imaginary part. That is $m$, is replaced by $m - i\delta$ and the limit taken as $\delta \to 0$ from above. This can be seen by imagining that we calculate $K_+$ by integrating on $p_4$ first. If we call $E = +(m^2 + p_1^2 + p_2^2) + p_3^2)^{\frac{1}{2}}$ then the integrals involve $p_4$ essentially as $\int \exp(-ip_4(t_2 - t_1)) \, dp_4(p_4^2 - E^2)^{-1}$ which has poles at $p_4 = +E$ and $p_4 = -E$. The replacement of $m$ by $m - i\delta$ means that $E$ has a small negative imaginary part; the first pole is below, the second above the real axis. Now if $t_2 - t_1 > 0$ the contour can be completed around the semicircle below the real axis thus giving a residue from the $p_4 = +E$ pole, or $-(2E)^{-1} \exp(-iE(t_2 - t_1))$. If $t_2 - t_1 < 0$ the upper semicircle must be used, and $p_4 = -E$ at the pole, so that the function varies in each case as required by the other definition (17).

Other solutions of (12) result from other prescriptions. For example if $p_4$ in the factor $(p^2 - m^2)^{-1}$ is considered to have a positive imaginary part $K_+$ becomes replaced by $K_0$, the Dirac one-electron kernel, zero for $t_2 < t_1$. Explicitly the function is[11] $(\mathbf{x}, t = x_{21\mu})$

$$I_+ (\mathbf{x}, t) = - (4\pi)^- \delta \left(s^2\right) + (m/8\pi s) \, H_1^{(2)} (ms), \qquad (34)$$

where $s = +(t^2 - \mathbf{x}^2)^{\frac{1}{2}}$ for $t^2 > \mathbf{x}^2$ and $s = -i(\mathbf{x}^2 - t^2)^{\frac{1}{2}}$ for $t^2 < \mathbf{x}^2$, $H_1^{(2)}$ is the Hankel function and $\delta(s^2)$ is the Dirac delta function of $s^2$. It behaves asymptotically as $\exp(-ims)$, decaying exponentially in space-like directions.[12]

By means of such transforms the matrix elements like (22), (23) are easily worked out. A free particle wave function for an electron of momentum $p_1$ is $u_1 \exp(-ip_1 \cdot x)$ where $u_1$ is a constant spinor satisfying the Dirac equation $p_1 u_1 = mu_1$ so that $p_1^2 = m^2$. The matrix element (22) for going from a state $p_1$, $u_1$ to a state of momentum $p_2$, spinor $u_2$, is $-4\pi^2 i(\bar{u}_2 a(q) u_1)$ where we have imagined $A$ expanded in a Fourier integral

$$A(1) = \int a\left(q\right) \exp\left(-iq \cdot x_1\right) d^4q,$$

and we select the component of momentum $q = p_2 - p_1$.

The second order term (23) is the matrix element between $u_1$ and $u_2$ of

$$-4\pi^2 i \int \left(a \left(p_2 - p_1 - q\right)\right) \left(p_1 + q - m\right)^{-1} a\left(q\right) d^4q, \qquad (35)$$

---

[11] $I_+ (\mathbf{x}, t)$ is $(2i)^{-1}(D_1(\mathbf{x}, t) - iD(\mathbf{x}, t))$ where $D_1$ and $D$ are the functions defined by W. Pauli, Rev. Mod. Phys. **13**, 203 (1941).
[12] If the $-i\delta$ is kept with $m$ here too the function $I_+$ approaches zero for infinite positive and negative times. This may be useful in general analyses in avoiding complications from infinitely remote surfaces.

since the electron of momentum $p_1$ may pick up $q$ from the potential $a(q)$, propagate with momentum $p_1 + q$ (factor $(p_1 + q - m)^{-1}$) until it is scattered again by the potential, $a(p_2 - p_1 - q)$, picking up the remaining momentum, $p_2 - p_1 - q$, to bring the total to $p_2$. Since all values of $q$ are possible, one integrates over $q$.

These same matrices apply directly to positron problems, for if the time component of, say, $p_1$ is negative the state represents a positron of four-momentum $-p_1$, and we are describing pair production if $p_2$ is an electron, i.e., has positive time component, etc.

The probability of an event whose matrix element is $(\bar{u}_2 M u_1)$ is proportional to the absolute square. This may also be written $(\bar{u}_1 \bar{M} u_2)(\bar{u}_2 M u_1)$, where $\bar{M}$ is $M$ with the operators written in opposite order and explicit appearance of $i$ changed to $-i$ ($\bar{M}$ is $\beta$ times the complex conjugate transpose of $\beta M$). For many problems we are not concerned about the spin of the final state. Then we can sum the probability over the two $u_2$ corresponding to the two spin directions. This is not a complete set because $p_2$ has another eigenvalue, $-m$. To permit summing over all states we can insert the projection operator $(2m)^{-1}(p_2 + m)$ and so obtain $(2m)^{-1}(\bar{u}_1 \bar{M}(p_2 + m)Mu_1)$ for the probability of transition from $p_1$, $u_1$, to $p_2$ with arbitrary spin. If the incident state is unpolarized we can sum on its spins too, and obtain

$$(2m)^{-2} Sp \left[ \left( p_1 + m \right) \bar{M} \left( p_2 + m \right) M \right] \tag{36}$$

for (twice) the probability that an electron of arbitrary spin with momentum $p_1$ will make transition to $p_2$. The expressions are all valid for positrons when $p$'s with negative energies are inserted, and the situation interpreted in accordance with the timing relations discussed above. (We have used functions normalized to $(\bar{u}u) = 1$ instead of the conventional $(\bar{u}\beta u) = (u^*u) = 1$. On our scale $(\bar{u}\beta u) = $ energy/$m$ so the probabilities must be corrected by the appropriate factors.)

The author has many people to thank for fruitful conversations about this subject, particularly H. A. Bethe and F. J. Dyson.

# APPENDIX

## a. Deduction from Second Quantization

In this section we shall show the equivalence of this theory with the hole theory of the positron.[2] According to the theory of second quantization of the electron field in a given potential,[13] the state of this field at any time is represented by a wave function $\chi$ satisfying

$$i\partial_\chi/\partial t = H_\chi,$$

where $H = \int \Psi^*(\mathbf{x})(\boldsymbol{\alpha} \cdot (-i\boldsymbol{\nabla} - \mathbf{A}) + A_4 + m\beta)\Psi(\mathbf{x})d^3\mathbf{x}$ and $\Psi(\mathbf{x})$ is an operator annihilating an electron at position $\mathbf{x}$, while $\Psi^*(\mathbf{x})$ is the corresponding creation operator. We contemplate a situation in which at $t = 0$ we have present some electrons in states represented by ordinary spinor functions $f_1(\mathbf{x}), f_2(\mathbf{x}), \ldots$ assumed orthogonal, and some positrons. These are described as holes in the negative energy sea, the electrons which would normally fill the holes having wave functions $p_1(\mathbf{x})$, $p_2(\mathbf{x})$, $\ldots$. We ask, at time $T$ what is the amplitude that we find electrons in states $g_1(\mathbf{x}), g_2(\mathbf{x}), \ldots$ and holes at $q_1(\mathbf{x}), q_2(\mathbf{x})$, $\ldots$. If the initial and final state vectors representing this situation are $\chi_i$ and $\chi_f$ respectively, we wish to calculate the matrix element

$$R = \left(\chi_f^* \exp\left(-i \int_0^T H dt\right) \chi_i\right) = \left(\chi_f^* S \chi_i\right). \qquad (37)$$

We assume that the potential $A$ differs from zero only for times between 0 and $T$ so that a vacuum can be defined at these times. If $\chi_0$ represents the vacuum state (that is, all negative energy states filled, all positive energies empty), the amplitude for having a vacuum at time $T$, if we had one at $t = 0$, is

$$C_v = \left(\chi_0^* S \chi_0\right), \qquad (38)$$

writing $S$ for $\exp(-i \int_0^T H dt)$. Our problem is to evaluate $R$ and show that it is a simple factor times $C_v$, and that the factor involves the $K_+^{(A)}$ functions in the way discussed in the previous sections.

---

[13] See, for example, G. Wentzel, *Einführung in die Quantentheorie der Wellenfelder* (Franz Deuticke, Leipzig, 1943), Chapter V.

To do this we first express $\chi_i$ in terms of $\chi_0$. The operator

$$\Phi^* = \int \Psi^* (\mathbf{x}) \, \phi (\mathbf{x}) \, d^3\mathbf{x}, \qquad (39)$$

creates an electron with wave function $\phi(\mathbf{x})$. Likewise $\Phi = \int \phi^*(\mathbf{x}) \times \Psi(\mathbf{x}) d^3\mathbf{x}$ annihilates one with wave function $\phi(\mathbf{x})$. Hence state $\chi_i$ is $\chi_i = F_1^* F_2^* \ldots P_1 P_2 \cdots \chi_0$ while the final state is $G_1^* G_2^* \ldots \times Q_1 Q_2 \ldots \chi_0$ where $F_i$, $G_i$, $P_i$, $Q_i$ are operators defined like $\Phi$, in (39), but with $f_i, g_i, p_i, q_i$ replacing $\phi$; for the initial state would result from the vacuum if we created the electrons in $f_1, f_2 \ldots$, and annihilated those in $p_1, p_2, \ldots$. Hence we must find

$$R = \left( \chi_0^* \cdots Q_2^* Q_1^* \cdots G_2 G_1 S F_1^* F_2 \cdots P_1 P_2 \cdots \chi_0 \right). \qquad (40)$$

To simplify this we shall have to use commutation relations between a $\Phi^*$ operator and $S$. To this end consider $\exp(-i \int_0^t H dt') \Phi^* \times \exp(+i \int_0^t H dt')$ and expand this quantity in terms of $\Psi^*(\mathbf{x})$, giving $\int \Psi^*(\mathbf{x})\phi(\mathbf{x}, t)d^3\mathbf{x}$, (which defines $\phi(\mathbf{x}, t)$). Now multiply this equation by $\exp(+i \int_0^t H dt') \ldots \exp(-i \int_0^t H dt')$ and find

$$\int \Psi^* (\mathbf{x}) \, \phi (\mathbf{x}) \, d^3\mathbf{x} = \int \Psi^* (\mathbf{x}, t) \, \phi (\mathbf{x}, t) \, d^3\mathbf{x}, \qquad (41)$$

where we have defined $\Psi(\mathbf{x}, t)$ by $\Psi(\mathbf{x}, t) = \exp(+i \int_0^t H dt')\Psi(\mathbf{x}) \times \exp(-i \int_0^t H dt')$. As is well known $\Psi(\mathbf{x}, t)$ satisfies the Dirac equation, (differentiate $\Psi(\mathbf{x}, t)$ with respect to $t$ and use commutation relations of $H$ and $\Psi$)

$$i\partial\Psi(\mathbf{x}, t)/\partial t = (\boldsymbol{\alpha} \cdot (-i\nabla - \mathbf{A}) + A_4 + m\beta)\,\Psi(\mathbf{x}, t). \qquad (42)$$

Consequently $\phi(\mathbf{x}, t)$ must also satisfy the Dirac equation (differentiate (41) with respect to $t$, use (42) and integrate by parts).

That is, if $\phi(\mathbf{x}, T)$ is that solution of the Dirac equation at time $T$ which is $\phi(\mathbf{x})$ at $t = 0$, and if we define $\Phi^* = \int \Psi^*(\mathbf{x})\phi(\mathbf{x})d^3\mathbf{x}$ and $\Phi'^* = \int \Psi^*(\mathbf{x})\phi(\mathbf{x}, T)d^3\mathbf{x}$ then $\Phi'^* = S\Phi^* S^{-1}$, or

$$S\Phi^* = \Phi'^* S. \qquad (43)$$

The principle on which the proof will be based can now be illustrated by a simple example. Suppose we have just one electron initially and finally and ask for

$$r = \left( \chi_0^* GSF^* \chi_0 \right). \tag{44}$$

We might try putting $F^*$ through the operator $S$ using (43), $SF^* = F'^* S$, where $f'$ in $F'^* = \int \Psi^*(\mathbf{x}) f'(\mathbf{x}) d^3\mathbf{x}$ is the wave function at $T$ arising from $f(\mathbf{x})$ at 0. Then

$$r = \left( \chi_0^* GF'^* S \chi_0 \right) = \int g^*(\mathbf{x}) f'(\mathbf{x}) d^3\mathbf{x} \cdot C_v - \left( \chi_0^* F'^* GS \chi_0 \right), \tag{45}$$

where the second expression has been obtained by use of the definition (38) of $C_v$ and the general commutation relation

$$GF^* + F^*G = \int g^*(\mathbf{x}) f(\mathbf{x}) d^3\mathbf{x},$$

which is a consequence of the properties of $\Psi(\mathbf{x})$ (the others are $FG = -GF$ and $F^*G^* = -G^*F^*$). Now $\chi_0^* F'^*$ in the last term in (45) is the complex conjugate of $F'_{\chi_0}$. Thus if $f'$ contained only positive energy components, $F'_{\chi_0}$ would vanish and we would have reduced $r$ to a factor times $C_v$. But $f'$, as worked out here, does contain negative energy components created in the potential $\mathbf{A}$ and the method must be slightly modified.

Before putting $F^*$ through the operator we shall add to it another operator $F'''^*$ arising from a function $f''(\mathbf{x})$ containing *only negative* energy components and so chosen that the resulting $f'$ has *only positive* ones. That is we want

$$S\left( F_{pos}^* + F_{neg}''^* \right) = F_{pos}'^* S, \tag{46}$$

where the "pos" and "neg" serve as reminders of the sign of the energy components contained in the operators. This we can now use in the form

$$SF_{pos}^* = F_{pos}'^* S - SF_{neg}''^*. \tag{47}$$

In our one electron problem this substitution replaces $r$ by two terms

$$r = (\chi_0^* GF_{pos}'^* S \chi_0) - (\chi_0^* GSF_{neg}''^* \chi_0).$$

The first of these reduces to

$$r = \int g^*(\mathbf{x}) f'_{\text{pos}}(\mathbf{x}) d^3\mathbf{x} \cdot C_r,$$

as above, for $F'_{\text{pos}}\chi_0$ is now zero, while the second is zero since the creation operator $F''^*_{\text{neg}}$ gives zero when acting on the vacuum state as all negative energies are full. This is the central idea of the demonstration.

The problem presented by (46) is this: Given a function $f_{\text{pos}}(\mathbf{x})$ at time 0, to find the amount, $f''_{\text{neg}}$, of negative energy component which must be added in order that the solution of Dirac's equation at time $T$ will have only positive energy components, $f'_{\text{pos}}$. This is a boundary value problem for which the kernel $K_+^{(A)}$ is designed. We know the positive energy components initially, $f_{\text{pos}}$, and the negative ones finally (zero). The positive ones finally are therefore (using (19))

$$f'_{\text{pos}}(\mathbf{x}_2) = \int K_+^{(A)}(2, 1)\beta f_{\text{pos}}(\mathbf{x}_1) \, d^3\mathbf{x}_1, \tag{48}$$

where $t_2 = T$, $t_1 = 0$. Similarly, the negative ones initially are

$$f''_{\text{neg}}(\mathbf{x}_2) = \int K_+^{(A)}(2, 1)\beta f_{\text{pos}}(\mathbf{x}_1) \, d^3\mathbf{x}_1 - f_{\text{pos}}(\mathbf{x}_2), \tag{49}$$

where $t_2$ approaches zero from above, and $t_1 = 0$. The $f_{\text{pos}}(\mathbf{x}_2)$ is subtracted to keep in $f''_{\text{neg}}(\mathbf{x}_2)$ only those waves which return from the potential and not those arriving directly at $t_2$ from the $K_+(2, 1)$ part of $K_+^{(A)}(2, 1)$, as $t_2 \to 0$. We could also have written

$$f''_{\text{neg}}(\mathbf{x}_2) = \int \left[ K_+^{(A)}(2, 1) - K_+(2, 1) \right] \beta f_{\text{pos}}(\mathbf{x}_1) \, d^3\mathbf{x}_1. \tag{50}$$

Therefore the one-electron problem, $r = f_g^*(\mathbf{x}) f'_{\text{pos}}(\mathbf{x}) d^3\mathbf{x} \cdot C_r$, gives by (48)

$$r = C_v \int g^*(\mathbf{x}_2) K_+^{(A)}(2, 1)\beta f(\mathbf{x}_1) \, d^3\mathbf{x}_1 d^3\mathbf{x}_2,$$

as expected in accordance with the reasoning of the previous sections (i.e., (20) with $K_+^{(A)}$ replacing $K_+$).

The proof is readily extended to the more general expression $R$, (40), which can be analyzed by induction. First one replaces $F_1^*$ by a relation such as (47) obtaining two terms

$$R = \left( \chi_0^* \cdots Q_2^* Q_1^* \cdots G_2 G_1 F'^*_{1\,\text{pos}} S F_2^* \cdots P_1 P_2 \cdots \chi_0 \right)$$
$$- \left( \chi_0^* \cdots Q_2^* Q_1^* \cdots G_2 G_1 S F''^*_{1\,\text{neg}} F_2^* \cdots P_1 P_2 \cdots \chi_0 \right).$$

In the first term the order of $F'^*_{1\,\text{pos}}$ and $G_1$ is then interchanged, producing an additional term $f g_1^*(\mathbf{x}) f'_{1\,\text{pos}}(\mathbf{x}) d^3\mathbf{x}$ times an expression with one less electron in initial and final state. Next it is exchanged with $G_2$ producing an addition $- f g_2^*(\mathbf{x}) f'_{1\,\text{pos}}(\mathbf{x}) d^3\mathbf{x}$ times a similar term, etc. Finally on reaching the $Q_1^*$ with which it anticommutes it can be simply moved over to juxtaposition with $\chi_0^*$ where it gives zero. The second term is similarly handled by moving $F''^*_{1\,\text{neg}}$ through anti commuting $F_2^*$, etc., until it reaches $P_1$. Then it is exchanged with $P_1$ to produce an additional simpler term with a factor $\mp \int p_1^*(\mathbf{x}) f''_{1\,\text{neg}}(\mathbf{x}) d^3\mathbf{x}$ or $\mp \int p_1^*(\mathbf{x}_2) K_+^{(A)}(2, 1) \beta f_1(\mathbf{x}_1) d^3\mathbf{x}_1 d^3\mathbf{x}_2$ from (49), with $t_2 = t_1 = 0$ (the extra $f_1(\mathbf{x}_2)$ in (49) gives zero as it is orthogonal to $p_1(\mathbf{x}_2)$). This describes in the expected manner the annihilation of the pair, electron $f_1$, positron $p_1$. The $F''^*_{\text{neg}}$ is moved in this way successively through the $P$'s until it gives zero when acting on $\chi_0$. Thus $R$ is reduced, with the expected factors (and with alternating signs as required by the exclusion principle), to simpler terms containing two less operators which may in turn be further reduced by using $F_2^*$ in a similar manner, etc. After all the $F^*$ are used the $Q^*$'s can be reduced in a similar manner. They are moved through the $S$ in the opposite direction in such a manner as to produce a purely negative energy operator at time 0, using relations analogous to (46) to (49). After all this is done we are left simply with the expected factor times $C_v$ (assuming the net charge is the same in initial and final state.)

In this way we have written the solution to the general problem of the motion of electrons in given potentials. The factor $C_v$ is obtained by normalization. However for photon fields it is desirable to have an explicit form for $C_v$ in terms of the potentials. This is given by (30) and (29) and it is readily demonstrated that this also is correct according to second quantization.

## b. Analysis of the Vacuum Problem

We shall calculate $C_v$ from second quantization by induction considering a series of problems each containing a potential distribution more nearly like the one we wish. Suppose we know $C_v$ for a problem like the one we want and having the same potentials for time $t$ between some $t_0$ and $T$, but having potential zero for times from 0 to $t_0$. Call this $C_v(t_0)$, the corresponding Hamiltonian $Ht_0$ and the sum of contributions for all single loops, $L(t_0)$. Then for $t_0 = T$ we have zero potential at all times, no pairs can be produced, $L(T) = 0$ and $C_v(T) = 1$. For $t_0 = 0$ we have the complete problem, so that $C_v(0)$ is what is defined as $C_v$ in (38). Generally we have,

$$C_v(t_0) = \left( \chi_0^* \exp\left( -i \int_0^T Ht_0 dt \right) \chi_0 \right)$$
$$= \left( \chi_0^* \exp\left( -i \int_{t_0}^T Ht_0 dt \right) \chi_0 \right),$$

since $Ht_0$ is identical to the constant vacuum Hamiltonian $H_T$ for $t < t_0$ and $\chi_0$ is an eigenfunction of $H_T$ with an eigenvalue (energy of vacuum) which we can take as zero.

The value of $C_v(t_0 - \Delta t_0)$ arises from the Hamiltonian $Ht_0 - \Delta t_0$ which differs from $Ht_0$ just by having an extra potential during the short interval $\Delta t_0$. Hence, to first order in $\Delta t_0$, we have

$$C_v(t_0 - \Delta t_0) = \left( \chi_0^* \exp\left( -i \int_{t_0-\Delta t_0}^T Ht_0 - \Delta t_0 dt \right) \chi_0 \right)$$
$$= \left( \chi_0^* \exp\left( -i \int_{t_0}^T Ht_0 dt \right) \left[ 1 - i\Delta t_0 \int \Psi^*(\mathbf{x}) \right. \right.$$
$$\times \left. \left. \left( -\boldsymbol{\alpha} \cdot \mathbf{A}(\mathbf{x}, t_0) + A_4(\mathbf{x}, t_0) \right) \Psi(\mathbf{x}) d^3\mathbf{x} \right] \chi_0 \right);$$

we therefore obtain for the derivative of $C_v$ the expression

$$-dC_v(t_0)/dt_0 = -i\left( \chi_0^* \exp\left( -i \int_{t_0}^T Ht_0 dt \right) \right.$$
$$\times \left. \int \Psi^*(\mathbf{x}) \beta A(\mathbf{x}, t_0) \Psi(\mathbf{x}) d^3\mathbf{x}\chi_0 \right), \quad (51)$$

which will be reduced to a simple factor times $C_\nu(t_0)$ by methods analogous to those used in reducing $R$. The operator $\Psi$ can be imagined to be split into two pieces $\Psi_{\text{pos}}$ and $\Psi_{\text{neg}}$ operating on positive and negative energy states respectively. The $\Psi_{\text{pos}}$ on $\chi_0$ gives zero so we are left with two terms in the current density, $\Psi^*_{\text{pos}}\beta A\Psi_{\text{neg}}$ and $\Psi^*_{\text{neg}}\beta A\Psi_{\text{neg}}$. The latter $\Psi^*_{\text{neg}}\beta A\Psi_{\text{neg}}$ is just the expectation value of $\beta A$ taken over all negative energy states (minus $\Psi_{\text{neg}}\beta A\Psi^*_{\text{neg}}$ which gives zero acting on $\chi_0$). This is the effect of the vacuum expectation current of the electrons in the sea which we should have subtracted from our original Hamiltonian in the customary way.

The remaining term $\Psi^*_{\text{pos}}\beta A\Psi_{\text{neg}}$, or its equivalent $\Psi^*_{\text{pos}}\beta A\Psi$ can be considered as $\Psi^*(\mathbf{x})\mathbf{f}_{\text{pos}}(\mathbf{x})$ where $\mathbf{f}_{\text{pos}}(\mathbf{x})$ is written for the positive energy component of the operator $\beta A\Psi(\mathbf{x})$. Now this operator, $\Psi^*(\mathbf{x})\mathbf{f}_{\text{pos}}(\mathbf{x})$, or more precisely just the $\Psi^*(\mathbf{x})$ part of it, can be pushed through the $\exp(-i\int t_0^T H dt)$ in a manner exactly analogous to (47) when $f$ is a function. (An alternative derivation results from the consideration that the operator $\Psi(\mathbf{x}, t)$ which satisfies the Dirac equation also satisfies the linear integral equations which are equivalent to it.) That is, (51) can be written by (48), (50),

$$
-dC_\nu(t_0)/dt_0 = i\left( \chi_0^* \int\int \Psi^*(\mathbf{x}_2)\, K_+^{(A)}(2,1) \right.
$$
$$
\times \exp\left(-i\int_{t_0}^T H dt\right) A(1)\Psi(\mathbf{x}_1)\, d^3\mathbf{x}_1 d^3\mathbf{x}_2 \chi_0 \bigg)
$$
$$
+ i\left( \chi_0^* \exp\left(-i\int_{t_0}^T H dt\right) \int\int \Psi^*(\mathbf{x}_2)[K_+^{(A)}(2,1) \right.
$$
$$
\left. -K_+(2,1)]A(1)\Psi(\mathbf{x}_1)\, d^3\mathbf{x}_1 d^3\mathbf{x}_2 \chi_0 \right),
$$

where in the first term $t_2 = T$, and in the second $t_2 \to t_0 = t_1$. The $(A)$ in $K_+^{(A)}$ refers to that part of the potential $A$ after $t_0$. The first term vanishes for it involves (from the $K_+^{(A)}(2,1)$) only positive energy components of $\Psi^*$, which give zero operating into $\chi_0^*$. In the second term only negative components of $\Psi^*(\mathbf{x}_2)$ appear. If, then $\psi^*(\mathbf{x}_2)$ is interchanged in order with $\Psi(\mathbf{x}_1)$ it will give zero operating on $\chi_0$, and only the term,

$$-dC_v(t_0)/dt_0$$

$$= +i \int Sp\left[K_+^{(A)}(1, 1) - K_+(1, 1))A(1)\right] d^3\mathbf{x}_1 \cdot C_v(t_0), \quad (52)$$

will remain, from the usual commutation relation of $\Psi^*$ and $\Psi$.

The factor of $C_v(t_0)$ in (52) times $-\Delta t_0$ is, according to (29) (reference 10), just $L(t_0 - \Delta t_0) - L(t_0)$ since this difference arises from the extra potential $\Delta A = A$ during the short time interval $\Delta t_0$. Hence $-dC_v(t_0)/dt_0 = +(dL(t_0)/dt_0)C_v(t_0)$ so that integration from $t_0 = T$ to $t_0 = 0$ establishes (30).

Starting from the theory of the electromagnetic field in second quantization, a deduction of the equations for quantum electrodynamics which appear in the succeeding paper may be worked out using very similar principles. The Pauli-Weisskopf theory of the Klein-Gordon equation can apparently be analyzed in essentially the same way as that used here for Dirac electrons.

# THE RADIATION THEORIES OF TOMONAGA, SCHWINGER, AND FEYNMAN

## BY

### FREEMAN DYSON

A unified development of the subject of quantum electrodynamics is outlined, embodying the main features both of the Tomonaga-Schwinger and of the Feynman radiation theory. The theory is carried to a point further than that reached by these authors, in the discussion of higher order radiative reactions and vacuum polarization phenomena. However, the theory of these higher order processes is a program rather than a definitive theory, since no general proof of the convergence of these effects is attempted.

The chief results obtained are (a) a demonstration of the equivalence of the Feynman and Schwinger theories, and (b) a considerable simplification of the procedure involved in applying the Schwinger theory to particular problems, the simplification being the greater the more complicated the problem.

## I. INTRODUCTION

AS a result of the recent and independent discoveries of Tomonaga,[1] Schwinger,[2] and Feynman,[3] the subject of quantum electrodynamics has made two very notable advances. On

---

[1] Sin-itiro Tomonaga, Prog. Theoret. Phys. **1**, 27 (1946); Koba, Tati, and Tomonaga, Prog. Theoret. Phys. **2**, 101 198 (1947); S. Kanesawa and S. Tomonaga, Prog. Theoret. Phys. **3**, 1, 101 (1948); S. Tomonaga, Phys. Rev. **74**, 224 (1948).
[2] Julian Schwinger, Phys. Rev. **73**, 416 (1948); Phys. Rev. **74**, 1439 (1948). Several papers, giving a complete exposition of the theory, are in course of publication.
[3] R. P. Feynman, Rev. Mod. Phys. **20**, 367 (1948); Phys. Rev. **74**, 939, 1430 (1948); J. A. Wheeler and R. P. Feynman, Rev. Mod. Phys. **17**, 157 (1945). These articles describe early stages in the development of Feynman's theory, little of which is yet published.

---

Reprinted with permission from the American Physical Society: Dyson, *Physical Review*, Volume 75, p. 486, 1949. ©1949 by the American Physical Society.

the one hand, both the foundations and the applications of the theory have been simplified by being presented in a completely relativistic way; on the other, the divergence difficulties have been at least partially overcome. In the reports so far published, emphasis has naturally been placed on the second of these advances; the magnitude of the first has been somewhat obscured by the fact that the new methods have been applied to problems which were beyond the range of the older theories, so that the simplexity of the methods was hidden by the complexity of the problems. Furthermore, the theory of Feynman differs so profoundly in its formulation from that of Tomonaga and Schwinger, and so little of it has been published, that its particular advantages have not hitherto been available to users of the other formulations. The advantages of the Feynman theory are simplicity and ease of application, while those of Tomonaga-Schwinger are generality and theoretical completeness.

The present paper aims to show how the Schwinger theory can be applied to specific problems in such a way as to incorporate the ideas of Feynman. To make the paper reasonably self-contained it is necessary to outline the foundations of the theory, following the method of Tomonaga; but this paper is not intended as a substitute for the complete account of the theory shortly to be published by Schwinger. Here the emphasis will be on the application of the theory, and the major theoretical problems of gauge invariance and of the divergencies will not be considered in detail. The main results of the paper will be general formulas from which the radiative reactions on the motions of electrons can be calculated, treating the radiation interaction as a small perturbation, to any desired order of approximation. These formulas will be expressed in Schwinger's notation, but are in substance identical with results given previously by Feynman. The contribution of the present paper is thus intended to be twofold: first, to simplify the Schwinger theory for the benefit of those using it for calculations, and second, to demonstrate the equivalence of the various theories within their common domain of applicability.[1]

## II. OUTLINE OF THEORETICAL FOUNDATIONS

Relativistic quantum mechanics is a special case of non-relativistic quantum mechanics, and it is convenient to use the usual non-relativistic terminology in order to make clear the relation between the mathematical theory and the results of physical measurements. In quantum electrodynamics the dynamical variables are the electromagnetic potentials $A_\mu(\mathbf{r})$ and the spinor electron-positron field $\psi_\alpha(\mathbf{r})$; each component of each field at each point $\mathbf{r}$ of space is a separate variable. Each dynamical variable is, in the Schrödinger representation of quantum mechanics, a time-independent operator operating on the state vector $\Phi$ of the system. The nature of $\Phi$ (wave function or abstract vector) need not be specified; its essential property is that, given the $\Phi$ of a system at a particular time, the results of all measurements made on the system at that time are statistically determined. The variation of $\Phi$ with time is given by the Schrödinger equation

$$i\hbar\,[\partial/\partial t]\,\Phi = \left\{ \int H(\mathbf{r})d_\tau \right\} \Phi, \qquad (1)$$

where $H(\mathbf{r})$ is the operator representing the total energy-density of the system at the point $\mathbf{r}$. The general solution of (1) is

$$\Phi(t) = \exp\left\{ [-it/\hbar] \int H(\mathbf{r})d\tau \right\} \Phi_0, \qquad (2)$$

with $\Phi_0$ any constant state vector.

Now in a relativistic system, the most general kind of measurement is not the simultaneous measurement of field quantities at different points of space. It is also possible to measure independently field quantities at different points of space at different times, provided that the points of space-time at which the measurements are made

---

[1] After this paper was written, the author was shown a letter, published in Progress of Theoretical Physics **3**, 205 (1948) by Z. Koba and G. Takeda. The letter is dated May 22, 1948, and briefly describes a method of treatment of radiative problems, similar to the method of this paper.

lie outside each other's light cones, so that the measurements do not interfere with each other. Thus the most comprehensive general type of measurement is a measurement of field quantities at each point $\mathbf{r}$ of space at a time $t(\mathbf{r})$, the locus of the points $(\mathbf{r}, t(\mathbf{r}))$ in space-time forming a 3-dimensional surface $\sigma$ which is space-like (i.e., every pair of points on it is separated by a space-like interval). Such a measurement will be called "an observation of the system on $\sigma$." It is easy to see what the result of the measurement will be. At each point $\mathbf{r}'$ the field quantities will be measured for a state of the system with state vector $\Phi(t(\mathbf{r}'))$ given by (2). But all observable quantities at $\mathbf{r}'$ are operators which commute with the energy-density operator $H(\mathbf{r})$ at every point $\mathbf{r}$ different from $\mathbf{r}'$, and it is a general principle of quantum mechanics that if $B$ is a unitary operator commuting with $A$, then for any state $\Phi$ the results of measurements of $A$ are the same in the state $\Phi$ as in the state $B\Phi$. Therefore, the results of measurement of the field quantities at $r'$ in the state $\Phi(t(\mathbf{r}'))$ are the same as if the state of the system were

$$\Phi(\sigma) = \exp\left\{-\left[i/\hbar\right]\int t(\mathbf{r})H(\mathbf{r})d\tau\right\}\Phi_0, \tag{3}$$

which differs from $\Phi(t(\mathbf{r}'))$ only by a unitary factor commuting with these field quantities. The important fact is that the state vector $\Phi(\sigma)$ depends only on $\sigma$ and not on $\mathbf{r}'$. The conclusion reached is that observations of a system on $\sigma$ give results which are completely determined by attributing to the system the state vector $\Phi(\sigma)$ given by (3).

The Tomonaga-Schwinger form of the Schrödinger equation is a differential form of (3). Suppose the surface $\sigma$ to be deformed slightly near the point $\mathbf{r}$ into the surface $\sigma'$, the volume of space-time separating the two surfaces being $V$. Then the quotient

$$\left[\Phi\left(\sigma'\right) - \Phi(\sigma)\right]/V$$

Results of the application of the method to a calculation of the second-order radiative correction to the Klein-Nishina formula are stated. All the papers of Professor Tomonaga and his associates which have yet been published were completed before the end of 1946. The isolation of these Japanese workers has undoubtedly constituted a serious loss to theoretical physics.

tends to a limit as $V \to 0$, which we denote by $\partial\Phi/\partial\sigma(\mathbf{r})$ and call the functional derivative of $\Phi$ with respect to $\sigma$ at the point $\mathbf{r}$. From (3) it follows that

$$i\hbar c\,[\partial\Phi/\partial\sigma(\mathbf{r})] = H(\mathbf{r})\Phi, \tag{4}$$

and (3) is, in fact, the general solution of (4).

The whole meaning of an equation such as (4) depends on the physical meaning which is attached to the statement "a system has a constant state vector $\Phi_0$." In the present context, this statement means "results of measurements of field quantities at any given point of space are independent of time." This statement is plainly non-relativistic, and so (4) is, in spite of appearances, a non-relativistic equation.

The simplest way to introduce a new state vector $\Psi$ which shall be a relativistic invariant is to require that the statement "a system has a constant state vector $\Psi$" shall mean "a system consists of photons, electrons, and positrons, traveling freely through space without interaction or external disturbance." For this purpose, let

$$H(\mathbf{r}) = H_0(\mathbf{r}) + H_1(\mathbf{r}), \tag{5}$$

where $H_0$ is the energy-density of the free electromagnetic and electron fields, and $H_1$ is that of their interaction with each other and with any external disturbing forces that may be present. A system with constant $\Psi$ is, then, one whose $H_1$ is identically zero; by (3) such a system corresponds to a $\Phi$ of the form

$$\Phi(\sigma) = T(\sigma)\Phi_0, \quad T(\sigma) = \exp\left\{-[i/\hbar]\int t(\mathbf{r})H_0(\mathbf{r})d\tau\right\}. \tag{6}$$

It is therefore consistent to write generally

$$\Phi(\sigma) = T(\sigma)\Psi(\sigma), \tag{7}$$

thus defining the new state vector $\Psi$ of any system in terms of the old $\Phi$. The differential equation satisfied by $\Psi$ is obtained from (4), (5), (6), and (7) in the form

$$i\hbar c\,[\partial\Psi/\partial\sigma(\mathbf{r})] = (T(\sigma))^{-1}\,H_1(\mathbf{r})\,T(\sigma)\Psi. \tag{8}$$

Now if $q(\mathbf{r})$ is any time-independent field operator, the operator

$$q(x_0) = (T(\sigma))^{-1} q(\mathbf{r}) T(\sigma)$$

is just the corresponding time-dependent operator as usually defined in quantum electrodynamics.[4] It is a function of the point $x_0$ of space-time whose coordinates are $(\mathbf{r}, ct(\mathbf{r}))$, but is the same for all surfaces $\sigma$ passing through this point, by virtue of the commutation of $H_1(\mathbf{r})$ with $H_0(\mathbf{r}')$ for $\mathbf{r}' \neq \mathbf{r}$. Thus (8) may be written

$$i\hbar c \left[ \partial \Psi / \partial \sigma(x_0) \right] = H_1(x_0) \Psi, \tag{9}$$

where $H_1(x_0)$ is the time-dependent form of the energy-density of interaction of the two fields with each other and with external forces. The left side of (9) represents the degree of departure of the system from a system of freely traveling particles and is a relativistic invariant; $H_1(x_0)$ is also an invariant, and thus is avoided one of the most unsatisfactory features of the old theories, in which the invariant $H_1$ was added to the non-invariant $H_0$. Equation (9) is the starting point of the Tomonaga-Schwinger theory.

## III. INTRODUCTION OF PERTURBATION THEORY

Equation (9) can be solved explicitly. For this purpose it is convenient to introduce a one-parameter family of space-like surfaces filling the whole of space-time, so that one and only one member $\sigma(x)$ of the family passes through any given point $x$. Let $\sigma_0, \sigma_1, \sigma_2, \ldots$ be a sequence of surfaces of the family, starting with $\sigma_0$ and proceeding in small steps steadily into the past. By

$$\int_{\sigma_1}^{\sigma_0} H_1(x) dx$$

[4] See, for example, Gregor Wentzel, *Einführung in die Quantentheorie der Wellenfelder* (Franz Deuticke, Wien, 1943), pp. 18–26.

is denoted the integral of $H_1(x)$ over the 4-dimensional volume between the surfaces $\sigma_1$ and $\sigma_0$; similarly, by

$$\int_{-\infty}^{\sigma_0} H_1(x)dx, \quad \int_{\sigma_0}^{\infty} H_1(x)dx$$

are denoted integrals over the whole volume to the past of $\sigma_0$ and to
the future of $\sigma_0$, respectively. Consider the operator

$$U = U(\sigma_0) = \left(1 - [i/\hbar c] \int_{\sigma_1}^{\sigma_0} H_1(x)dx\right)$$

$$\times \left(1 - [i/\hbar c] \int_{\sigma_2}^{\sigma_1} H_1(x)dx\right) \cdots, \tag{10}$$

the product continuing to infinity and the surfaces $\sigma_0$, $\sigma_1$, ... being
taken in the limit infinitely close together. $U$ satisfies the differential
equation

$$i\hbar c \left[\partial U/\partial \sigma (x_0)\right] = H_1(x_0) U, \tag{11}$$

and the general solution of (9) is

$$\Psi(\sigma) = U(\sigma)\Psi_0, \tag{12}$$

with $\Psi_0$ any constant vector.

Expanding the product (10) in ascending powers of $H_1$ gives a
series

$$U = 1 + (-i/\hbar c) \int_{-\infty}^{\sigma_0} H_1(x_1)\, dx_1 + (-i/\hbar c)^2$$

$$\times \int_{-\infty}^{\sigma_0} dx_1 \int_{-\infty}^{\sigma(x_1)} H_1(x_1) H_1(x_2)\, dx_2 + \cdots. \tag{13}$$

Further, $U$ is by (10) obviously unitary, and

$$U^{-1} = \bar{U} = 1 + (i/\hbar c) \int_{-\infty}^{\sigma_0} H_1(x_1)\, dx_1 + (i/\hbar c)^2$$

$$\times \int_{-\infty}^{\sigma_0} dx_1 \int_{-\infty}^{\sigma(x_1)} H_1(x_2) H_1(x_1)\, dx_2 + \cdots. \tag{14}$$

It is not difficult to verify that $U$ is a function of $\sigma_0$ alone and is
independent of the family of surfaces of which $\sigma_0$ is one member. The
use of a finite number of terms of the series (13) and (14), neglecting

the higher terms, is the equivalent in the new theory of the use of perturbation theory in the older electrodynamics.

The operator $U(\infty)$, obtained from (10) by taking $\sigma_0$ in the infinite future, is a transformation operator transforming a state of the system in the infinite past (representing, say, converging streams of particles) into the same state in the infinite future (after the particles have interacted or been scattered into their final outgoing distribution). This operator has matrix elements corresponding only to real transitions of the system, i.e., transitions which conserve energy and momentum. It is identical with the Heisenberg $S$ matrix.[5]

## IV. ELIMINATION OF THE RADIATION INTERACTION

In most of the problem of electrodynamics, the energy-density $H_1(x_0)$ divides into two parts—

$$H_1(x_0) = H^i(x_0) + H^e(x_0), \tag{15}$$

$$H^i(x_0) = -[1/c]\, j_\mu(x_0)\, A_\mu(x_0), \tag{16}$$

the first part being the energy of interaction of the two fields with each other, and the second part the energy produced by external forces. It is usually not permissible to treat $H^e$ as a small perturbation as was done in the last section. Instead, $H^i$ alone is treated as a perturbation, the aim being to eliminate $H^i$ but to leave $H^e$ in its original place in the equation of motion of the system.

Operators $S(\sigma)$ and $S(\infty)$ are defined by replacing $H_1$ by $H^i$ in the definitions of $U(\sigma)$ and $U(\infty)$. Thus $S(\sigma)$ satisfies the equation

$$i\hbar c\,[\partial S/\partial\sigma(x_0)] = H^i(x_0)\, S. \tag{17}$$

Suppose now a new type of state vector $\Omega(\sigma)$ to be introduced by the substitution

$$\Psi(\sigma) = S(\sigma)\Omega(\sigma). \tag{18}$$

[5] Werner Heisenberg, Zeits. f. Physik **120**, 513 (1943), **120**, 673 (1943), and Zeits. f. Naturforschung **1**, 608 (1946).

By (9), (15), (17), and (18) the equation of motion for $\Omega(\sigma)$ is

$$i\hbar c \left[\partial \Omega / \partial \sigma \left(x_0\right)\right] = \left(S(\sigma)\right)^{-1} H^e \left(x_0\right) S(\sigma)\Omega. \qquad (19)$$

The elimination of the radiation interaction is hereby achieved; only the question, "How is the new state vector $\Omega(\sigma)$ to be interpreted?," remains.

It is clear from (19) that a system with a constant $\Omega$ is a system of electrons, positrons, and photons, moving under the influence of their mutual interactions, but in the absence of external fields. In a system where two or more particles are actually present, their interactions alone will, in general, cause real transitions and scattering processes to occur. For such a system it is rather "unphysical" to represent a state of motion including the effects of the interactions by a constant state vector; hence, for such a system the new representation has no simple interpretation. However, the most important systems are those in which only one particle is actually present, and its interaction with the vacuum fields gives rise only to virtual processes. In this case the particle, including the effects of all its interactions with the vacuum, appears to move as a free particle in the absence of external fields, and it is eminently reasonable to represent such a state of motion by a constant state vector. Therefore, it may be said that the operator,

$$H_T \left(x_0\right) = \left(S(\sigma)\right)^{-1} H^e \left(x_0\right) S(\sigma), \qquad (20)$$

on the right of (19) represents the interaction of a physical particle with an external field, including radiative corrections. Equation (19) describes the extent to which the motion of a single physical particle deviates, in the external field, from the motion represented by a constant state-vector, i.e., from the motion of an observed "free" particle.

If the system whose state vector is constantly $\Omega$ undergoes no real transitions with the passage of time, then the state vector $\Omega$ is called "steady." More precisely, $\Omega$ is steady if, and only if, it satisfies the equation

$$S\left(\infty\right) \Omega = \Omega. \qquad (21)$$

As a general rule, one-particle states are steady and many-particle states unsteady. There are, however, two important qualifications to this rule.

First, the interaction (20) itself will almost always cause transitions from steady to unsteady states. For example, if the initial state consists of one electron in the field of a proton, $H_T$ will have matrix elements for transitions of the electron to a new state with emission of a photon, and such transitions are important in practice. Therefore, although the interpretation of the theory is simpler for steady states, it is not possible to exclude unsteady states from consideration.

Second, if a one-particle state as hitherto defined is to be steady, the definition of $S(\sigma)$ must be modified. This is because $S(\infty)$ includes the effects of the electromagnetic self-energy of the electron, and this self-energy gives an expectation value to $S(\infty)$ which is different from unity (and indeed infinite) in a one-electron state, so that Eq. (21) cannot be satisfied. The mistake that has been made occurred in trying to represent the observed electron with its electromagnetic self-energy by a wave field with the same characteristic rest-mass as that of the "bare" electron. To correct the mistake, let $\delta m$ denote the electromagnetic mass of the electron, i.e., the difference in rest-mass between an observed and a "bare" electron. Instead of (5), the division of the energy-density $H(\mathbf{r})$ should have taken the form

$$H(\mathbf{r}) = (H_0(\mathbf{r}) + \delta mc^2 \psi^*(\mathbf{r})) \beta \psi(r)) + (H_1(\mathbf{r}) - \delta mc^2 \psi^*(\mathbf{r}) \beta \psi(\mathbf{r})).$$

The first bracket on the right here represents the energy-density of the free electromagnetic and electron fields with the observed electron restmass, and should have been used instead of $H_0(\mathbf{r})$ in the definition (6) of $T(\sigma)$. Consequently, the second bracket should have been used instead of $H_1(\mathbf{r})$ in Eq. (8).

The definition of $S(\sigma)$ has therefore to be altered by replacing $H^i(x_0)$ by[6]

$$H^1(x_0) = H^i(x_0) + H^S(x_0) = H^i(x_0) - \delta mc^2 \bar{\psi}(x_0) \psi(x_0).$$
$$(22)$$

---

[6] Here Schwinger's notation $\bar{\psi} = \psi^* \beta$ is used.

The value of $\delta m$ can be adjusted so as to cancel out the self-energy effects in $S(\infty)$ (this is only a formal adjustment since the value is actually infinite), and then Eq. (21) will be valid for one-electron states. For the photon self-energy no such adjustment is needed since, as proved by Schwinger, the photon self-energy turns out to be identically zero.

The foregoing discussion of the self-energy problem is intentionally only a sketch, but it will be found to be sufficient for practical applications of the theory. A fuller discussion of the theoretical assumptions underlying this treatment of the problem will be given by Schwing in his forthcoming papers. Moreover, it must realized that the theory as a whole cannot be put into a finally satisfactory form so long as divergencies occur in it, however skilfully these divergencies are circumvented; therefore, the present treatment should be regarded as justified by its success in applications rather than by its theoretical derivation.

The important results of the present paper up to this point are Eq. (19) and the interpretation of the state vector $\Omega$. The state vector $\Psi$ of a system can be interpreted as a wave function giving the probability amplitude of finding any particular set of occupation numbers for the various possible states of free electrons, positrons, and photons. The state vector $\Omega$ of a system with a given $\Psi$ on a given surface $\sigma$ is, crudely speaking, the $\Psi$ which the system would have had in the infinite past if it had arrived at the given $\Psi$ on $\sigma$ under the influence of the interaction $H^I(x_0)$ alone.

The definition of $\Omega$ being unsymmetrical between past and future, a new type of state vector $\Omega'$ can be defined by reversing the direction of time in the definition of $\Omega$. Thus the $\Omega'$ of a system with a given $\Psi$ on a given $\sigma$ is the $\Psi$ which the system would reach in the infinite future if it continued to move under the influence of $H^1(x_0)$ alone. More simply, $\Omega'$ can be defined by the equation

$$\Omega'(\sigma) = S(\infty)\,\Omega(\sigma). \tag{23}$$

Since S($\infty$) is a unitary operator independent of $\sigma$, the state vectors $\Omega$ and $\Omega'$ are really only the same vector in two different representations or coordinate systems. Moreover, for any steady state the two are identical by (21).

## V. FUNDAMENTAL FORMULAS OF THE SCHWINGER AND FEYNMAN THEORIES

The Schwinger theory works directly from Eqs. (19) and (20), the aim being to calculate the matrix elements of the "effective external potential energy" $H_T$ between states specified by their state vectors $\Omega$. The states considered in practice always have $\Omega$ of some very simple kind, for example, $\Omega$ representing systems in which one or two free-particle states have occupation number one and the remaining free-particle states have occupation number zero. By analogy with (13), $S(\sigma_0)$ is given by

$$S(\sigma_0) = 1 + (-i/\hbar c) \int_{-\infty}^{\sigma_0} H^I(x_1)\, dx_1 + (-i/\hbar c)^2$$

$$\times \int_{-\infty}^{\sigma_0} dx_1 \int_{-\infty}^{\sigma(x_1)} H^I(x_1)\, H^I(x_2)\, dx_2 + \cdots, \quad (24)$$

and $(S(\sigma_0))^{-1}$ by a corresponding expression analogous to (14). Substitution of these series into (20) gives at once

$$H_T(x_0) = \sum_{n=0}^{\infty} (i/\hbar c)^n \int_{-\infty}^{\sigma(x_0)} dx_1 \int_{-\infty}^{\sigma(x_1)} dx_2 \cdots \int_{-\infty}^{\sigma(x_{n-1})} dx_n$$

$$\times \left[ H^I(x_n), \left[ \cdots, \left[ H^I(x_2), \left[ H^I(x_1), H^\sigma(x_0) \right] \right] \cdots \right] \right].$$

$$(25)$$

The repeated commutators in this formula are characteristic of the Schwinger theory, and their evaluation gives rise to long and rather difficult analysis. Using the first three terms of the series, Schwinger was able to calculate the second-order radiative corrections to the equations of motion of an electron in an external field, and obtained satisfactory agreement with experimental results. In this paper the

806

development of the Schwinger theory will be carried no further; in principle the radiative corrections to the equations of motion of electrons could be calculated to any desired order of approximation from formula (25).

In the Feynman theory the basic principle is to preserve symmetry between past and future. Therefore, the matrix elements of the operator $H_T$ are evaluated in a "mixed representation;" the matrix elements are calculated between an initial state specified by its state vector $\Omega_1$ and a final state specified by its state vector $\Omega_2'$. The matrix element of $H_T$ between two such states in the Schwinger representation is

$$\Omega_2^* H_T \Omega_1 = \Omega_2'^* S(\infty) H_T \Omega_1, \tag{26}$$

and therefore the operator which replaces $H_T$ in the mixed representation is

$$H_F(x_0) = S(\infty) H_T(x_0)$$
$$= S(\infty) (S(\sigma))^{-1} H^e(x_0) S(\sigma). \tag{27}$$

Going back to the original product definition of $S(\sigma)$ analogous to (10), it is clear that $S(\infty) \times (S(\sigma))^{-1}$ is simply the operator obtained from $S(\sigma)$ by interchanging past and future. Thus,

$$R(\sigma) = S(\infty) (S(\sigma))^{-1} = 1 + (-i/\hbar c)$$
$$\times \int_\sigma^\infty H^I(x_1)\, dx_1 + (-i/\hbar c)^2 \int_\sigma^\infty dx_1$$
$$\times \int_{\sigma(x_1)}^\infty H^I(x_2) H^I(x_1)\, dx_2 + \cdots. \tag{28}$$

The physical meaning of a mixed representation of this type is not at all recondite. In fact, a mixed representation is normally used to describe such a process as bremsstrahlung of an electron in the field of a nucleus when the Born approximation is not valid; the process of bremsstrahlung is a radiative transition of the electron from a state described by aCoulomb wave function, with a plane ingoing and a spherical outgoing wave, to a state described by a Coulomb wave function with a spherical ingoing and a plane outgoing wave. The initial and final states here belong to different orthogonal systems of wave functions, and so the transition matrix elements are calculated in a mixed representation. In the Feynman theory the situation is analogous; only the roles of the radiation interaction and the external (or Coulomb) field are interchanged; the radiation interaction is used instead of the Coulomb field to modify the state vectors (wave functions) of the initial and final states, and the external field instead of the radiation interaction causes transitions between these state vectors.

In the Feynman theory there is an additional simplification. For if matrix elements are being calculated between two states, either of which is steady (and this includes all cases so far considered), the mixed representation reduces to an ordinary representation. This occurs, for example, in treating a one-particle problem such as the radiative correction to the equations of motion of an electron in an external field; the operator $H_F(x_0)$, although in general it is not even Hermitian, can in this case be considered as an effective external potential energy acting on the particle, in the ordinary sense of the words.

This section will be concluded with the derivation of the fundamental formula (31) of the Feynman theory, which is the analog of formula (25) of the Schwinger theory. If

$$F_1(x_1), \ldots, F_n(x_n)$$

are any operators defined, respectively, at the points $x_1, \ldots, x_n$ of space-time, then

$$P(F_1(x_1), \ldots, F_n(x_n)) \qquad (29)$$

will denote the product of these operators, taken in the order, reading from right to left, in which the surfaces $\sigma(x_1), \ldots, \sigma(x_n)$ occur in time. In most applications of this notation $F_i(x_i)$ will commute with $F_j(x_j)$ so long as $x_i$ and $x_j$ are outside each other's light cones; when this is the case, it is easy to see that (29) is a function of the points $x_1, \ldots, x_n$ only and is independent of the surfaces $\sigma(x_i)$. Consider now the integral

$$I_n = \int_{-\infty}^{\infty} dx_1 \cdots \int_{-\infty}^{\infty} dx_n P\left(H^e\left(x_0\right),\right.$$

$$\left. H^I(x_1), \ldots, H^I(x_n)\right).$$

Since the integrand is a symmetrical function of the points $x_1$, ..., $x_n$, the value of the integral is just $n!$ times the integral obtained by restricting the integration to sets of points $x_1$, ..., $x_n$ for which $\sigma(x_i)$ occurs after $\sigma(x_{i+1})$ for each $i$. The restricted integral can then be further divided into $(n+1)$ parts, the $j$'th part being the integral over those sets of points with the property that $\sigma(x_0)$ lies between $\sigma(x_{j-1})$ and $\sigma(x_j)$ (with obvious modifications for $j = 1$ and $j = n + 1$). Therefore,

$$I_n = n! \sum_{j=1}^{n+1} \int_{-\infty}^{\sigma(x_0)} dx_j \cdots \int_{-\infty}^{\sigma(x_n-1)} dx_n$$

$$\times \int_{\sigma(x_0)}^{\infty} dx_{j-1} \cdots \int_{\sigma(x_2)}^{\infty} dx_1 \times H^I(x_1) \cdots$$

$$H^I\left(x_{j-1}\right) H^e\left(x_0\right) H^I\left(x_j\right) \cdots H^I\left(x_n\right). \tag{30}$$

Now if the series (24) and (28) are substituted into (27), sums of integrals appear which are precisely of the form (30). Hence finally

$$H_F\left(x_0\right) = \sum_{n=0}^{\infty} (-i/\hbar c)^n \left[1/n!\right] I_n$$

$$= \sum_{n=0}^{\infty} (-i/\hbar c)^n \left[1/n!\right] \int_{-\infty}^{\infty} dx_1 \cdots \int_{-\infty}^{\infty} dx_n$$

$$\times P\left(H^e\left(x_0\right), H^I\left(x_1\right), \ldots, H^I\left(x_n\right)\right). \tag{31}$$

By this formula the notation $H_F(x_0)$ is justified, for this operator now appears as a function of the point $x_0$ alone and not of the surface $\sigma$. The further development of the Feynman theory is mainly concerned with the calculation of matrix elements of (31) between various initial and final states.

As a special case of (31) obtained by replacing $H^e$ by the unit matrix in (27),

$$S(\infty) = \sum_{n=0}^{\infty} (-i/\hbar c)^n [1/n!] \int_{-\infty}^{\infty} dx_1 \cdots \int_{-\infty}^{\infty} dx_n$$
$$\times P\left(H^I(x_1), \ldots, H^I(x_n)\right). \qquad (32)$$

# VI. CALCULATION OF MATRIX ELEMENTS

In this section the application of the foregoing theory to a general class of problems will be explained. The ultimate aim is to obtain a set of rules by which the matrix element of the operator (31) between two given states may be written down in a form suitable for numerical evaluation, immediately and automatically. The fact that such a set of rules exists is the basis of the Feynman radiation theory; the derivation in this section of the same rules from what is fundamentally the Tomonaga-Schwinger theory constitutes the proof of equivalence of the two theories.

To avoid excessive complication, the type of matrix element considered will be restricted in two ways. First, it will be assumed that the external potential energy is

$$H^e(x_0) = -[1/c]\, j_\mu(x_0)\, A^e_\mu(x_0), \qquad (33)$$

that is to say, the interaction energy of the electron-positron field with electromagnetic potentials $A^e_\mu(x_0)$ which are given numerical functions of space and time. Second, matrix elements will be considered only for transitions from a state $A$, in which just one electron and no positron or photon is present, to another state $B$ of the same character. These restrictions are not essential to the theory, and are introduced only for convenience, in order to illustrate clearly the principles involved.

The electron-positron field operator may be written

$$\psi_\alpha(x) = \sum_u \phi_{u\alpha}(x) a_u, \qquad (34)$$

where the $\phi_{u\alpha}(x)$ are spinor wave functions of free electrons and positrons, and the $a_u$ are annihilation operators of electrons and creation operators of positrons. Similarly, the adjoint operator

$$\bar{\psi}_\alpha(x) = \sum_u \bar{\phi}_{u\alpha}(x)\bar{a}_u, \tag{35}$$

where $\bar{a}_u$ are annihilation operators of positrons and creation operators of electrons. The electromagnetic field operator is

$$A_\mu(x) \sum_v \left( A_{v\mu}(x)b_v + A_{v\mu}^*(x)\bar{b}_v \right), \tag{36}$$

where $b_v$ and $\bar{b}_v$ are photon annihilation and creation operators, respectively. The charge-current 4-vector of the electron field is

$$j_\mu(x) = iec\bar{\psi}(x)\gamma_\mu\psi(x); \tag{37}$$

strictly speaking, this expression ought to be antisymmetrized to the form[7]

$$j_\mu(x) = \frac{1}{2}iec\left\{\bar{\psi}_\alpha(x)\psi_\beta(x) - \psi_\beta(x)\bar{\psi}_\alpha(x)\right\}\left(\gamma_\mu\right)_{\alpha\beta}, \tag{38}$$

but it will be seen later that this is not necessary in the present theory.

Consider the product $P$ occurring in the $n$'th integral of (31); let it be denoted by $P_n$. From (16), (22), (33), and (37) it is seen that $P_n$ is a sum of products of $(n+1)$ operators $\psi_\alpha$, $(n+1)$ operators $\bar{\psi}_\alpha$, and not more than $n$ operators $A_\mu$, multiplied by various numerical factors. By $Q_n$ may be denoted a typical product of factors $\psi_\alpha$, $\bar{\psi}_\alpha$, and $A_\mu$, not summed over the indices such as $\alpha$ and $\mu$, so that $P_n$ is a sum of terms such as $Q_n$. Then $Q_n$ will be of the form (indices omitted)

$$Q_n = \bar{\psi}(xi_0)\psi(xi_0)\bar{\psi}(xi_1)\psi(xi_1)\cdots\bar{\psi}(xi_n)\psi(xi_n)$$
$$\times A(xj_1)\cdots A(xj_m), \tag{39}$$

where $i_0, i_1, \ldots, i_n$ is some permutation of the integers $0, 1, \ldots, n$, and $j_1, \ldots, j_m$ are some, but not necessarily all, of the integers $1, \ldots, n$ in some order. Since none of the operators $\bar{\psi}$ and $\psi$ commute with each other, it is especially important to preserve the order of

[7] See Wolfgang Pauli, Rev. Mod. Phys. 13, 203 (1941), Eq. (96), p. 224.

these factors. Each factor of $Q_n$ is a sum of creation and annihilation operators by virtue of (34), (35), and (36), and so $Q_n$ itself is a sum of products of creation and annihilation operators.

Now consider under what conditions a product of creation and annihilation operators can give a non-zero matrix element for the transition $A \to B$. Clearly, one of the annihilation operators must annihilate the electron in state $A$, one of the creation operators must create the electron in state $B$, and the remaining operators must be divisible into pairs, the members of each pair respectively creating and annihilating the same particle. Creation and annihilation operators referring to different particles always commute or anticommute (the former if at least one is a photon operator, the latter if both are electron-positron operators). Therefore, if the two single operators and the various pairs of operators in the product all refer to different particles, the order of factors in the product can be altered so as to bring together the two single operators and the two members of each pair, without changing the value of the product except for a change of sign if the permutation made in the order of the electron and positron operators is odd. In the case when some of the single operators and pairs of operators refer to the same particle, it is not hard to verify that the same change in order of factors can be made, provided it is remembered that the division of the operators into pairs is no longer unique, and the change of order is to be made for each possible division into pairs and the results added together.

It follows from the above considerations that the matrix element of $Q_n$ for the transition $A \to B$ is a sum of contributions, each contribution arising from a specific way of dividing the factors of $Q_n$ into two single factors and pairs. A typical contribution of this kind will be denoted by $M$. The two factors of a pair must involve a creation and an annihilation operator for the same particle, and so must be either one $\bar{\psi}$ and one $\psi$ or two $A$; the two single factors must be one $\bar{\psi}$ and one $\psi$. The term $M$ is thus specified by fixing an integer $k$, and a permutation $r_0, r_1, \ldots, r_n$ of the integers $0, 1, \ldots, n$, and a division $(s_1, t_1), (s_2, t_2), \ldots, (s_h, t_h)$ of the integers $j_1, \ldots, j_m$ into pairs; clearly $m = 2h$ has to be an even number; the term $M$ is obtained by choosing for single factors $\bar{\psi}(x_k)$ and $\psi(x_{r_k})$, and for associated pairs of factors $(\bar{\psi}(x_i), \psi(x_{r_i}))$ for $i = 0, 1, \ldots, k-1, k+1, \ldots, n$ and $(A(x_{s_i}), A(x_{t_i}))$ for $i = 1, \ldots, h$. In evaluating the term $M$, the order of factors in $Q_n$ is first to be permuted so as to bring together the two single factors and the two members of each pair, but without altering the order of factors within each pair; the result of this process is easily seen to be

$$Q'_n = \epsilon P\left(\bar{\psi}(x_0), \psi(x_{r_0})\right) \cdots P\left(\bar{\psi}(x_n), \psi(x_{r_n})\right)$$
$$\times P\left(A(x_{s_1}), A(x_{t_1})\right) \cdots P\left(A(xs_h), A(xt_h)\right), \quad (40)$$

a factor $\epsilon$ being inserted which takes the value $\pm 1$ according to whether the permutation of $\bar{\psi}$ and $\psi$ factors between (39) and (40) is even or odd. Then in (40) each product of two associated factors (but not the two single factors) is to be independently replaced by the sum of its matrix elements for processes involving the successive creation and annihilation of the same particle.

Given a bilinear operator such as $A_\mu(x)A_\nu(y)$, the sum of its matrix elements for processes involving the successive creation and annihilation of the same particle is just what is usually called the "vacuum expectation value" of the operator, and has been calculated by Schwinger. This quantity is, in fact (note that Heaviside units are being used)

$$\langle A_\mu(x)A_\nu(y)\rangle_0 = \frac{1}{2}\hbar c\delta_{\mu\nu}\left\{D^{(1)} + iD\right\}(x-y),$$

where $D^{(1)}$ and $D$ are Schwinger's invariant $D$ functions. The definitions of these functions will not be given here, because it turns out that the vacuum expectation value of $P(A_\mu(x), A_\nu(y))$ takes an even simpler form. Namely,

$$\langle P\left(A_\mu(x), A_\nu(y)\right)\rangle_0 = \frac{1}{2}\hbar c\delta_{\mu\nu}D_F(x-y), \quad (41)$$

where $D_F$ is the type of $D$ function introduced by Feynman. $D_F(x)$ is an even function of $x$, with the integral expansion

$$D_F(x) = -\left[i/2\pi^2\right]\int_0^\infty \exp\left[i\alpha x^2\right]d\alpha, \quad (42)$$

where $x^2$ denotes the square of the invariant length of the 4-vector $x$. In a similar way it follows from Schwinger's results that

$$\langle P\left(\bar{\psi}_\alpha(x),\psi_\beta(y)\right)\rangle_0 = \frac{1}{2}\eta\left(x,y\right)S_{F\beta\alpha}\left(x-y\right), \quad (43)$$

where

$$S_{F\beta\alpha}(x) = -\left(\gamma_\mu\left(\partial/\partial x_\mu\right)+\kappa_0\right)_{\beta\alpha}\Delta_F(x), \quad (44)$$

$\kappa_0$ is the reciprocal Compton wave-length of the electron, $\eta(x,y)$ is $-1$ or $+1$ according as $\sigma(x)$ is earlier or later than $\sigma(y)$ in time, and $\Delta_F$ is a function with the integral expansion

$$\Delta_F(x) = -\left[i/2\pi^2\right]\int_0^\infty \exp\left[i\alpha x^2 - i\kappa_0^2/4\alpha\right]d\alpha. \quad (45)$$

Substituting from (41) and (44) into (40), the matrix element $M$ takes the form (still omitting the indices of the factors $\bar{\psi}$, $\psi$, and $A$ of $Q_n$)

$$M = \epsilon\prod_{i\neq k}\left(\frac{1}{2}\eta\left(x_i,x_{r_i}\right)S_F\left(x_i-x_{r_i}\right)\right)$$

$$\times\prod_j\left(\frac{1}{2}\hbar c\,D_F\left(x_{s_j}-x_{t_j}\right)\right)P\left(\bar{\psi}\left(x_k\right),\psi\left(x_{r_k}\right)\right). \quad (46)$$

The single factors $\bar{\psi}(x_k)$ and $\psi(x_{r_k})$ are conveniently left in the form of operators, since the matrix elements of these operators for effecting the transition $A\rightarrow B$ depend on the wave functions of the electron in the states $A$ and $B$. Moreover, the order of the factors $\bar{\psi}(x_k)$ and $\psi(x_{r_k})$ is immaterial since they anticommute with each other; hence it is permissible to write

$$P(\bar{\psi}(x_k),\psi(x_{r_k})) = \eta(x_k,x_{r_k})\bar{\psi}(x_k)\psi(x_{r_k}).$$

Therefore (46) may be rewritten

$$M = \epsilon'\prod_{i\neq k}\left(\frac{1}{2}S_F\left(x_i-x_{r_i}\right)\right)\prod_j\left(\frac{1}{2}\hbar c\,D_F\left(x_{s_j}-x_{t_j}\right)\right)$$

$$\times\bar{\psi}\left(x_k\right)\psi\left(x_{r_i}\right), \quad (47)$$

with

$$\epsilon' = \epsilon\prod_i\eta\left(x_i,x_{r_i}\right). \quad (48)$$

Now the product in (48) is $(-1)^p$, where $p$ is the number of occasions in the expression (40) on which the $\psi$ of a $P$ bracket occurs to the left of the $\bar{\psi}$. Referring back to the definition of $\epsilon$ after Eq. (40), it follows that $\epsilon'$ takes the value $+1$ or $-1$ according to whether the permutation of $\bar{\psi}$ and $\psi$ factors between (39) and the expression

$$\bar{\psi}\left(x_0\right) \psi\left(x_{r_0}\right) \cdots \bar{\psi}\left(x_n\right) \psi\left(x_{r_n}\right) \tag{49}$$

is even or odd. But (39) can be derived by an even permutation from the expression

$$\bar{\psi}\left(x_0\right) \psi\left(x_0\right) \cdots \bar{\psi}\left(x_n\right) \psi\left(x_n\right), \tag{50}$$

and the permutation of factors between (49) and (50) is even or odd according to whether the permutation $r_0, \ldots, r_n$ of the integers $0, \ldots, n$ is even or odd. Hence, finally, $\epsilon'$ in (47) is $+1$ or $-1$ according to whether the permutation of $r_0, \ldots, r_n$ is even or odd. It is important that $\epsilon'$ depends only on the type of matrix element $M$ considered, and not on the points $x_0, \ldots, x_n$; therefore, it can be taken outside the integrals in (31).

One result of the foregoing analysis is to justify the use of (37), instead of the more correct (38), for the charge-current operator occurring in $H^e$ and $H^i$. For it has been shown that in each matrix element such as $M$ the factors $\bar{\psi}$ and $\psi$ in (38) can be freely permuted, so that (38) can be replaced by (37), except in the case when the two factors form an associated pair. In the exceptional case, $M$ contains as a factor the vacuum expectation value of the operator $j_\mu(x_i)$ at some point $x_i$; this expectation value is zero according to the correct formula (38), though it would be infinite according to (37); thus the matrix elements in the exceptional case are always zero. The conclusion is that only those matrix elements are to be calculated for which the integer $r_i$ differs from $i$ for every $i \neq k$, and in these elements the use of formula (37) is correct.

To write down the matrix elements of (31) for the transition $A \to B$, it is only necessary to take all the products $Q_n$, replace each by the sum of the corresponding matrix elements $M$ given by (47), reassemble the terms into the form of the $P_n$ from which they were derived, and finally substitute back into the series (31). The problem of calculating the matrix elements of (31) is thus in principle solved. However, in the following section it will be shown how this solution-inprinciple can be reduced to a much simpler and more practical procedure.

# VII. GRAPHICAL REPRESENTATION OF MATRIX ELEMENTS

Let an integer $n$ and a product $P_n$ occurring in (31) be temporarily fixed. The points $x_0, x_1, \ldots, x_n$, may be represented by $(n + 1)$ points drawn on a piece of paper. A type of matrix element $M$ as described in the last section will then be represented graphically as follows. For each associated pair of factors $(\bar{\psi}(x_i), \psi(x_{r_i}))$ with $i \neq k$, draw a line with a direction marked in it from the point $x_i$ to the point $x_{r_i}$. For the single factors $\bar{\psi}(x_k), \psi(x_{r_k})$, draw directed lines leading out from $x_k$ to the edge of the diagram, and in from the edge of the diagram to $x_{r_k}$. For each pair of factors $(A(x_{s_i}), A(x_{t_i}))$, draw an undirected line joining the points $x_{s_i}$ and $x_{t_i}$. The complete set of points and lines will be called the "graph" of $M$; clearly there is a one-to-one correspondence between types of matrix element and graphs, and the exclusion of matrix elements with $r_i - i$ for $i \neq k$ corresponds to the exclusion of graphs with lines joining a point to itself. The directed lines in a graph will be called "electron lines," the undirected lines "photon lines."

Through each point of a graph pass two electron lines, and therefore the electron lines together form one open polygon containing the vertices $x_k$ and $x_{r_k}$, and possibly a number of closed polygons as well. The closed polygons will be called "closed loops," and their number denoted by $l$. Now the permutation $r_0, \ldots, r_n$ of the integers $0, \ldots,$ $n$ is clearly composed of $(l + 1)$ separate cyclic permutations. A cyclic permutation is even or odd according to whether the number of elements in it is odd or even. Hence the parity of the permutation $r_0,$ $\ldots, r_n$ is the parity of the number of even-number cycles contained in it. But the parity of the number of odd-number cycles in it is obviously the same as the parity of the total number $(n + 1)$ of elements. The total number of cycles being $(l + 1)$, the parity of the number of even-number, cycles is $(l - n)$. Since it was seen earlier that the $\epsilon'$ of Eq. (47) is determined just by the parity of the permutation $r_0, \ldots, r_n$, the above argument yields the simple formula

$$\epsilon' = (-1)^{i-n}. \tag{51}$$

This formula is one result of the present theory which can be much more easily obtained by intuitive considerations of the sort used by Feynman.

In Feynman's theory the graph corresponding to a particular matrix element is regarded, not merely as an aid to calculation, but as a picture of the physical process which gives rise to that matrix element. For example, an electron line joining $x_1$ to $x_2$ represents the possible creation of an electron at $x_1$ and its annihilation at $x_2$, together with the possible creation of a position at $x_2$ and its annihilation at $x_1$. This interpretation of a graph is obviously consistent with the methods, and in Feynman's hands has been used as the basis for the derivation of most of the results, of the present paper. For reasons of space, these ideas of Feynman will not be discussed in further detail here.

To the product $P_n$ correspond a finite number of graphs, one of which may be denoted by $G$; all possible $G$ can be enumerated without difficulty for moderate values of $n$. To each $G$ corresponds a contribution $C(G)$ to the matrix element of (31) which is being evaluated.

It may happen that the graph $G$ is disconnected, so that it can be divided into subgraphs, each of which is connected, with no line joining a point of one subgraph to a point of another. In such a case it is clear from (47) that $C(G)$ is the product of factors derived from each subgraph separately. The subgraph $G_1$ containing the point $x_0$ is called the "essential part" of $G$, the remainder $G_2$ the "inessential part." There are now two cases to be considered, according to whether the points $x_k$ and $x_{r_k}$ lie in $G_2$ or in $G_1$ (they must clearly both lie in the same subgraph). In the first case, the factor $C(G_2)$ of $C(G)$ can be seen by a comparison of (31) and (32) to be a contribution to the matrix element of the operator $S(\infty)$ for the transition $A \to B$. Now letting $G$ vary over all possible graphs with the same $G_1$ and different $G_2$, the sum of the contributions of all such $G$ is a constant $C(G_1)$ multiplied by the total matrix element of $S(\infty)$ for the transition $A \to B$. But for one-particle states the operator $S(\infty)$ is by (21) equivalent to the identity operator and gives, accordingly, a zero matrix element for the transition $A \to B$. Consequently, the disconnected $G$ for which $x_k$ and $x_{r_k}$ lie in $G_2$ give zero contribution to the matrix element of (31), and can be omitted from further consideration. When $x_k$ and $x_{r_k}$ lie in $G_1$, again the $C(G)$ may be summed over all $G$ consisting of the given $G_1$ and all possible $G_2$; but this time the connected graph $G_1$ itself is to be included in the sum. The sum of all the $C(G)$ in this case turns out to be just $C(G_1)$ multiplied by the expectation value in the vacuum of the operator $S(\infty)$. But the vacuum state, being a steady state, satisfies (21), and so the expectation value in question is equal to unity. Therefore the sum of the $C(G)$ reduces to the single term $C(G_1)$, and again the disconnected graphs may be omitted from consideration.

The elimination of disconnected graphs is, from a physical point of view, somewhat trivial, since these graphs arise merely from the fact that meaningful physical processes proceed simultaneously with totally irrelevant fluctuations of fields in the vacuum. However, similar arguments will now be used to eliminate a much more important class of graphs, namely, those involving self-energy effects. A "self-energy part" of a graph $G$ is defined as follows; it is a set of one or more vertices not including $x_0$, together with the lines joining them, which is connected with the remainder of $G$ (or with the edge of the diagram) only by two electron lines or by one or two photon lines. For definiteness it may be supposed that $G$ has a self-energy part $F$, which is connected with its surroundings only by one electron line entering $F$ at $x_1$, and another leaving $F$ at $x_2$; the case of photon lines can be treated in an entirely analogous way. The points $x_1$ and $x_2$

may or may not be identical. From $G$ a "reduced graph" $G_0$ can be obtained by omitting $F$ completely and joining the incoming line at $x_1$ with the outgoing line at $x_2$ to form a single electron line in $G_0$, the newly formed line being denoted by $\lambda$. Given $G_0$ and $\lambda$, there is conversely a well determined set $\Gamma$ of graphs $G$ which are associated with $G_0$ and $\lambda$ in this way; $G_0$ itself-is considered also to belong to $\Gamma$. It will now be shown that the sum $C(\Gamma)$ of the contributions $C(G)$ to the matrix element of (31) from all the graphs $G$ of $\Gamma$ reduces to a single term $C'(G_0)$.

Suppose, for example, that the line $\lambda$ in $G_0$ leads from a point $x_3$ to the edge of the diagram. Then $C(G_0)$ is an integral containing in the integrand the matrix element of

$$\bar{\psi}_\alpha (x_3) \tag{52}$$

for creation of an electron into the state $B$. Let the momentum-energy 4-vector of the created electron be $p$; the matrix element of (52) is of the form

$$Y_\alpha (x_3) = a_\alpha \exp\left[-i\left(p \cdot x_3\right)/\hbar\right] \tag{53}$$

with $a_\alpha$ independent of $x_3$. Now consider the sum $C(\Gamma)$. It follows from an analysis of (31) that $C(\Gamma)$ is obtained from $C(G_0)$ by replacing the operator (52) by

$$\sum_{n=0}^\infty (-i/\hbar c)^n \left[1/n!\right] \int_{-\infty}^\infty dy_1 \cdots \int_{-\infty}^\infty dy_n$$
$$\times P\left(\bar{\psi}_\alpha (x_3),\, H^I (y_1),\, \ldots,\, H^I (y_n)\right). \tag{54}$$

(This is, of course, a consequence of the special character of the graphs of $\Gamma$.) It is required to calculate the matrix element of (54) for a transition from the vacuum state $O$ to the state $B$, i.e., for the emission of an electron into state $B$. This matrix element will be denoted by $Z_\alpha$; $C(\Gamma)$ involves $\dot{Z}_\alpha$ in the same way that $C(G_0)$ involves (53). Now $Z_\alpha$ can be evaluated as a sum of terms of the same general character as (47); it will be of the form

$$Z_\alpha = \sum_i \int_{-\infty}^\infty K_i^{\alpha\beta} (y_i - x_3)\, Y_\beta (y_i)\, dy_i,$$

where the important fact is that $K_i$ is a function only of the coordinate differences between $y_i$ and $x_3$. By (53), this implies that

$$Z_\alpha = R_{\alpha\beta}(p) Y_\beta(x_3), \tag{55}$$

with $R$ independent of $x_3$. From considerations of relativistic invariance, $R$ must be of the form

$$\delta_{\beta\alpha} R_1(p^2) + (p_\mu \gamma_\mu)_{\beta\alpha} R_2(p^2),$$

where $p^2$ is the square of the invariant length of the 4-vector $p$. But since the matrix element (53) is a solution of the Dirac equation,

$$p^2 = -\hbar^2 \kappa_0^2, \quad (p_\mu \gamma_\mu)_{\beta\alpha} Y_\beta = i\hbar \kappa_0 Y_\alpha,$$

and so (55) reduces to

$$Z_\alpha = R_1 Y_\alpha(x_3),$$

with $R_1$ an absolute constant. Therefore the sum $C(\Gamma)$ is in this case just $C'(G_0)$, where $C'(G_0)$ is obtained from $C(G_0)$ by the replacement

$$\bar{\psi}(x_3) \rightarrow R_1 \bar{\psi}(x_3). \tag{56}$$

In the case when the line $\lambda$ leads into the graph $G_0$ from the edge of the diagram to the point $x_1$, it is clear that $C(\Gamma)$ will be similarly obtained from $C(G_0)$ by the replacement

$$\psi(x_3) \rightarrow R_1^* \psi(x_3). \tag{57}$$

There remains the case in which $\lambda$ leads from one vertex $x_3$ to another $x_4$ of $G_0$. In this case $C(G_0)$ contains in its integrand the function

$$\frac{1}{2}\eta(x_3, x_4) S_{F\beta\alpha}(x_3 - x_4), \tag{58}$$

which is the vacuum expectation value of the operator

$$P\left(\bar{\psi}_\alpha(x_3), \psi_\beta(x_4)\right) \tag{59}$$

according to (43). Now in analogy with (54), $C(\Gamma)$ is obtained from $C(G_0)$ by replacing (59) by

$$\sum_{n=0}^{\infty} (-i/\hbar c)^n \, [1/n!] \int_{-\infty}^{\infty} dy_1 \cdots \int_{-\infty}^{\infty} dy_n$$
$$\times P\left(\bar{\psi}_\alpha(x_3), \psi_\beta(x_4), H^I(y_1), \ldots, H^I(y_n)\right), \tag{60}$$

and the vacuum expectation value of this operator will be denoted by

$$\frac{1}{2}\eta\,(x_3, x_4)\,S'_{F\beta\alpha}\,(x_3, x_4)\,. \tag{61}$$

By the methods of Section VI, (61) can be expanded as a series of terms of the same character as (47); this expansion will not be discussed in detail here, but it is easy to see that it leads to an expression of the form (61), with $S'_F(x)$ a certain universal function of the 4-vector $x$. It will not be possible to reduce (61) to a numerical multiple of (58), as $Z_\alpha$ was in the previous case reduced to a multiple of $Y_\alpha$. Instead, there may be expected to be a series expansion of the form

$$S_{F\beta\alpha}(x) = \left(R_2 + a_1\left(\Box^2 - \kappa_0^2\right) + a_2\left(\Box^2 - \kappa_0^2\right)^2 + \cdots\right) S_{F\beta\alpha}(x)$$
$$+ \left(b_1 + b_2\left(\Box^2 - \kappa_0^2\right) + \cdots\right)$$
$$\times \left(\gamma_\mu\left[\partial/\partial x_\mu\right] - \kappa_0\right)_{\beta\gamma} S_{F\gamma\alpha}(x), \tag{62}$$

where $\Box^2$ is the Dalembertian operator and the $a$, $b$ are numerical coefficients. In this case $C(\Gamma)$ will be equal to the $C'(G_0)$ obtained from $C(G_0)$, by the replacement

$$S_F\,(x_3 - x_4) \rightarrow S'_F\,(x_3 - x_4)\,. \tag{63}$$

Applying the same methods to a graph $G$ with a self-energy part connected to its surroundings by two photon lines, the sum $C(\Gamma)$ will be obtained as a single contribution $C'(G_0)$ from the reduced graph $G_0$, $C'(G_0)$ being formed from $C(G_0)$ by the replacement

$$D_F\,(x_3 - x_4) \rightarrow D'_F\,(x_3 - x_4)\,. \tag{64}$$

The function $D'_F$ is defined by the condition that

$$\frac{1}{2}\hbar c\,\delta_{\mu\nu}\,D'_F\,(x_3 - x_4) \tag{65}$$

is the vacuum expectation value of the operator

$$\sum_{n=0}^{\infty}(-i/\hbar c)^n\,[1/n\,!]\int_{-\infty}^{\infty} dy_1 \cdots \int_{-\infty}^{\infty} dy_n$$
$$\times P\left(A_\mu\,(x_3), A_\nu\,(x_4), H^I\,(y_i), \ldots, H^I\,(y_n)\right), \tag{66}$$

and may be expanded in a series

$$D'_F(x) = \left( R_3 + c_1 \Box^2 + c_2 \left( \Box^2 \right)^2 + \cdots \right) D_F(x). \qquad (67)$$

Finally, it is not difficult to see that for graphs $G$ with self-energy parts connected to their surroundings by a single photon line, the sum $C(\Gamma)$ will be identically zero, and so such graphs may be omitted from consideration entirely.

As a result of the foregoing arguments, the contributions $C(G)$ of graphs with self-energy parts can always be replaced by modified contributions $C'(G_0)$ from a reduced graph $G_0$. A given $G$ may be reducible in more than one way to give various $G_0$, but if the process of reduction is repeated a finite number of times a $G_0$ will be obtained which is "totally reduced," contains no self-energy part, and is uniquely determined by $G$. The contribution $C'(G_0)$ of a totally reduced graph to the matrix element of (31) is now to be calculated as a sum of integrals of expressions like (47), but with a replacement (56), (57), (63), or (64) made corresponding to every line in $G_0$. This having been done, the matrix element of (31) is correctly calculated by taking into consideration each totally reduced graph once and once only.

The elimination of graphs with self-energy parts is a most important simplification of the theory. For according to (22), $H^I$ contains the subtracted part $H^S$, which will give rise to many additional terms in the expansion of (31). But if any such term is taken, say, containing the factor $H^S(x_i)$ in the integrand, every graph corresponding to that term will contain the point $x_i$ joined to the rest of the graph only by two electron lines, and this point by itself constitutes a self-energy part of the graph. Therefore, all terms involving $H^S$ are to be omitted from (31) in the calculation of matrix elements. The intuitive argument for omitting these terms is that they were only introduced in order to cancel out higher order self-energy terms arising from $H^i$, which are also to be omitted; the analysis of the foregoing paragraphs is a more precise form of this argument. In physical language, the argument can be stated still more simply; since $\delta m$ is an unobservable quantity, it cannot appear in the final description of observable phenomena.

# VIII. VACUUM POLARIZATION AND CHARGE RENORMALIZATION

The question now arises: What is the physical meaning of the new functions $D'_F$ and $S'_F$, and of the constant $R_1$? In general terms, the answer is clear. The physical processes represented by the self-energy parts of graphs have been pushed out of the calculations, but these processes do not consist entirely of unobservable interactions of single particles with their self-fields, and so cannot entirely be written off as "self-energy processes." In addition, these processes include the phenomenon of vacuum polarization, i.e., the modification of the field surrounding a charged particle by the charges which the particle induces in the vacuum. Therefore, the appearance of $D'_F$, $S'_F$, and $R_1$ in the calculations may be regarded as an explicit representation of the vacuum polarization phenomena which were implicitly contained in the processes now ignored.

In the present theory there are two kinds of vacuum polarization, one induced by the external field and the other by the quantized electron and photon fields themselves; these will be called "external" and "internal," respectively. It is only the internal polarization which is represented yet in explicit fashion by the substitutions (56), (57), (63), (64); the external will be included later.

To form a concrete picture of the function $D'_F$, it may be observed that the function $D_F(y - z)$ represents in classical electrodynamics the retarded potential of a point charge at $y$ acting upon a point charge at $z$, together with the retarded potential of the charge at $z$ acting on the charge at $y$. Therefore, $D_F$ may be spoken of loosely as "the electromagnetic interaction between two point charges." In this semiclassical picture, $D'_F$ is then the electromagnetic interaction between two point charges, including the effects of the charge-distribution which each charge induces in the vacuum.

The complete phenomenon of vacuum polarization, as hitherto understood, is included in the above picture of the function $D'_F$. There is nothing left for $S'_F$ to represent. Thus, one of the important conclusions of the present theory is that there is a second phenomenon occurring in nature, included in the term vacuum polarization as used in this paper, but additional to vacuum polarization in the usual sense of the word. The nature of the second phenomenon can best be explained by an example.

The scattering of one electron by another may be represented as caused by a potential energy (the Møller interaction) acting between them. If one electron is at $y$ and the other at $z$, then, as explained above, the effect of vacuum polarization of the usual kind is to replace a factor $D_F$ in this potential energy by $D'_F$. Now consider an analogous, but unorthodox, representation of the Compton effect, or the scattering of an electron by a photon. If the electron is at $y$ and the photon at $z$, the scattering may be again represented by a potential energy, containing now the operator $S_F(y - z)$ as a factor; the potential is an exchange potential, because after the interaction the electron must be considered to be at $z$ and the photon at $y$, but this does not detract from its usefulness. By analogy with the 4-vector charge-current density $j_\mu$ which interacts with the potential $D_F$, a spinor Compton-effect density $u_\alpha$ may be defined by the equation

$$u_\alpha(x) = A_\mu(x) \left( \gamma_\mu \right)_{\alpha\beta\psi\beta} (x),$$

and an adjoint spinor by

$$\bar{u}_\alpha(x) = \bar{\psi}_\beta(x) \left( \gamma_\mu \right)_{\beta\alpha} A_\mu(x).$$

These spinors are not directly observable quantities, but the Compton effect can be adequately described as an exchange potential, of magnitude proportional to $S_F(y - z)$, acting between the Compton-effect density at any point $y$ and the adjoint density at $z$. The second vacuum polarization phenomenon is described by a change in the form of this potential from $S_F$ to $S'_F$. Therefore, the phenomenon may be pictured in physical terms as the inducing, by a given element of Compton-effect density at a given point, of additional Compton-effect density in the vacuum around it.

In both sorts of internal vacuum polarization, the functions $D_F$ and $S_F$, in addition to being altered in shape, become multiplied by numerical (and actually divergent) factors $R_3$ and $R_2$; also the matrix elements of (31) become multiplied by numerical factors such as $R_1 R_1^*$. However, it is believed (this has been verified only for second-order terms) that all $n$'th-order matrix elements of (31) will involve these factors only in the form of a multiplier

$$\left( e R_2 R_3^{\frac{1}{2}} \right)^n ;$$

this statement includes the contributions from the higher terms of the series (62) and (67). Here $e$ is defined as the constant occurring in the fundamental interaction (16) by virtue of (37). Now the only possible experimental determination of $e$ is by means of measurements of the effects described by various matrix elements of (31), and so the directly measured quantity is not $e$ but $e R_2 R_3^{\frac{1}{2}}$. Therefore, in practice the letter $e$ is used to denote this measured quantity, and the multipliers $R$ no longer appear explicitly in the matrix elements of (31); the change in the meaning of the letter $e$ is called "charge renormalization," and is essential if $e$ is to be identified with the observed electronic charge. As a result of the renormalization, the divergent coefficients $R_1$, $R_2$, and $R_3$ in (56), (57), (62), and (67) are to be replaced by unity, and the higher coefficients $a$, $b$, and $c$ by expressions involving only the renormalized charge $e$.

The external vacuum polarization induced by the potential $A_\mu^e$ is, physically speaking, only a special case of the first sort of internal polarization; it can be treated in a precisely similar manner. Graphs describing external polarization effects are those with an "external polarization part," namely, a part including the point $x_0$ and connected with the rest of the graph by only a single photon line. Such a graph is to be "reduced" by omitting the polarization part entirely and renaming with the label $x_0$ the point at the further end of the single photon line. A discussion similar to those of Section VII leads to the conclusion that only reduced graphs need be considered in the calculation of the matrix element of (31), and that the effect of external polarization is explicitly represented if in the contributions from these graphs a replacement

$$A_\mu^e(x) \rightarrow A_\mu^{e'}(x) \tag{68}$$

is made. After a renormalization of the unit of potential, similar to the renormalization of charge, the modified potential $A_\mu^{e'}$ takes the form

$$A_\mu^{e'}(x) = \left(1 + c_1 \Box^2 + c_2 \left(\Box^2\right)^2 + \cdots\right) A_\mu^e(x), \tag{69}$$

where the coefficients are the same as in (67).

It is necessary, in order to determine the functions $D_F'$, $S_F'$, and $A_\mu^{e'}$, to go back to formulas (60) and (66). The determination of the vacuum expectation values of the operators (60) and (66) is a problem of the same kind as the original problem of the calculation of matrix elements of (31), and the various terms in the operators (60) and (66) must again be split up, represented by graphs, and analyzed in detail. However, since $D_F'$ and $S_F'$ are universal functions, this further analysis has only to be carried out once to be applicable to all problems.

It is one of the major triumphs of the Schwinger theory that it enables an unambiguous interpretation to be given to the phenomenon of vacuum polarization (at least of the first kind), and to the vacuum expectation value of an operator such as (66). In making this interpretation, profound theoretical problems arise, particularly concerned with the gauge invariance of the theory, about which nothing will be said here. For Schwinger's solution of these problems, the reader must refer to his forthcoming papers. Schwinger's argument can be transferred without essential change into the framework of the present paper.

Having overcome the difficulties of principle, Schwinger proceeded to evaluate the function $D_F'$ explicitly as far as terms of order $\alpha = (e^2/4\pi\hbar c)$ (heaviside units). In particular, he found for the coefficient $c_1$ in (67) and (69) the value $(-\alpha/15\pi\kappa_0^2)$ to this order.[8] It is hoped to publish in a sequel to the present paper a similar evaluation of the function $S_F'$; the analysis involved is too complicated to be summarized here.

---

[8] Schwinger's results agree with those of the earlier, theoretically unsatisfactory treatment of vacuum polarization. The best account of the earlier work is V. F. Weisskopf, Kgl. Danske Sels. Math.-Fys. Medd. **14**, No. 6 (1936).

## IX. SUMMARY OF RESULTS

In this section the results of the preceding pages will be summarized, so far as they relate to the performance of practical calculations. In effect, this summary will consist of a set of rules for the application of the Feynman radiation theory to a certain class of problems.

Suppose an electron to be moving in an external field with interaction energy given by (33). Then the interaction energy to be used in calculating the motion of the electron, including radiative corrections of all orders, is

$$H_E(x_0) = \sum_{n=0}^{\infty} (-i/\hbar c)^n [1/n!] J_n$$

$$= \sum_{n=0}^{\infty} (-i/\hbar c)^n [1/n!] \int_{-\infty}^{\infty} dx_1 \cdots \int_{-\infty}^{\infty} dx_n$$

$$\times P\left(H^e(x_0), H^i(x_1), \ldots, H^i(x_n)\right), \qquad (70)$$

with $H^i$ given by (16), and the $P$ notation as defined in (29).

To find the effective $n$'th-order radiative correction to the potential acting on the electron, it is necessary to calculate the matrix elements of $J_n$ for transitions from one one-electron state to another. These matrix elements can be written down most conveniently in the form of an operator $K_n$ bilinear in $\bar{\psi}$ and $\psi$, whose matrix elements for one-electron transitions are the same as those to be determined. In fact, the operator $K_n$ itself is already the matrix element to be determined if the $\bar{\psi}$ and $\psi$ contained in it are regarded as one-electron wave functions.

To write down $K_n$, the integrand $P_n$ in $J_n$ is first expressed in terms of its factors $\bar{\psi}$, $\psi$, and $A$, all suffixes being indicated explicitly, and the expression (37) used for $j_\mu$. All possible graphs $G$ with $(n+1)$ vertices are now drawn as described in Section VII, omitting disconnected graphs, graphs with self-energy parts, and graphs with external vacuum polarization parts as defined in Section VIII. It will be found that in each graph there are at each vertex two electron lines and one photon line, with the exception of $x_0$ at which there are two electron lines only; further, such graphs can exist only for even $n$. $K_n$ is the sum of a contribution $K(G)$ from each $G$.

Given $G$, $K(G)$ is obtained from $J_n$ by the following transformations. First, for each photon line joining $x$ and $y$ in $G$, replace two factors $A_\mu(x) A_\nu(y)$ in $P'_n$ (regardless of their positions) by

$$\frac{1}{2} \hbar c \delta_{\mu\nu} D'_F (x - y), \tag{71}$$

with $D'_F$ given by (67) with $R_3 = 1$, the function $D_F$ being defined by (42). Second, for each electron line joining $x$ to $y$ in $G$, replace two factors $\bar{\psi}_\alpha(x) \psi_\beta(y)$ in $P_n$ (regardless of positions) by

$$\frac{1}{2} S'_{F\beta\alpha} (x - y) \tag{72}$$

with $S'_F$ given by (62) with $R_2 = 1$, the function $S_F$ being defined by (44) and (45). Third, replace the remaining two factors $P(\bar{\psi}_\gamma(z), \psi_\delta(w))$ in $P_n$ by $\bar{\psi}_\gamma(z) \psi_\delta(w)$ in this order. Fourth, replace $A^e_\mu(x_0)$ by $A^{e'}_\mu(x_0)$ given by

$$A^{e'}_\mu(x) = A^e_\mu(x) - \left[ \alpha / 15 \pi \kappa_0^2 \right] \Box^2 A^e_\mu(x) \tag{73}$$

or, more generally, by (69). Fifth, multiply the whole by $(-1)^l$, where $l$ is the number of closed loops in $G$ as defined in Section VII.

The above rules enable $K_n$ to be written down very rapidly for small values of $n$. It should be observed that if $K_n$ is being calculated, and if it is not desired to include effects of higher order than the $n$'th, then $D'_F$, $S'_F$, and $A^{e'}_\mu$ in (71), (72), and (73) reduce to the simple functions $D_F$, $S_F$, and $A^e_\mu$. Also, the integrand in $J_n$ is a symmetrical function of $x_1, \ldots, x_n$; therefore, graphs which differ only by a relabeling of the vertices $x_1, \ldots, x_n$ give identical contributions to $K_n$ and need not be considered separately.

The extension of these rules to cover the calculation of matrix elements of (70) of a more general character than the one-electron transitions hitherto considered presents no essential difficulty. All that is necessary is to consider graphs with more than two "loose ends," representing processes in which more than one particle is involved.

This extension is not treated in the present paper, chiefly because it would lead to unpleasantly cumbersome formulas.

# X. EXAMPLE—SECOND-ORDER RADIATIVE CORRECTIONS

As an illustration of the rules of procedure of the previous section, these rules will be used for writing down the terms giving second-order radiative corrections to the motion of an electron in an external field. Let the energy of the external field be

$$- [1/c] \, j_\mu \, (x_0) \, A^e_\mu \, (x_0) \, . \tag{74}$$

Then there will be one second-order correction term

$$U = \left[\alpha/15\pi\kappa_0^2\right] [1/c] \, j_\mu \, (x_0) \, \Box^2 A^e_\mu \, (x_0)$$

arising from the substitution (73) in the zero-order term (74). This is the well-known vacuum polarization or Uehling term.[9]

The remaining second-order term arises from the second-order part $J_2$ of (70). Written in expanded form, $J_2$ is

$$J_2 = ie^3 \int_{-\infty}^{\infty} dx_1 \int_{-\infty}^{\infty} dx_2 \, P(\bar{\psi}_\alpha(x_0)(\gamma_\lambda)_{\alpha\beta}\psi_\beta(x_0)A^e_\lambda(x_0),$$
$$\times \, \bar{\psi}_\gamma(x_1)(\gamma_\mu)_{\gamma\delta}\psi_\delta(x_1)A_\mu(x_1), \, \bar{\psi}_\epsilon(x_2)(\gamma_\nu)_{\epsilon\zeta}\psi_\zeta(x_2)A_\nu(x_2)).$$

Next, all admissable graphs with the three vertices $x_0$, $x_1$, $x_2$ are to be drawn. It is easy to see that there are only two such graphs, that $G$ shown in Fig. 1, and the identical graph with $x_1$ and $x_2$ interchanged. The full lines are electron lines, the dotted line a photon line. The contribution $K(G)$ is obtained from $J_2$ by substituting according to the rules of Section IX; in this case $l = 0$, and the primes can be omitted from (71), (72), (73) since only second-order terms are required. The integrand in $K(G)$ can be reassembled into the form of a matrix product, suppressing the suffixes $\alpha, \ldots, \zeta$. Then, multiplying by a factor 2 to allow for the second graph, the complete second-order

[9] Robert Serber, Phys. Rev. **48**, 49 (1935); E. A. Uehling, Phys. Rev. 48, 55 (1935).

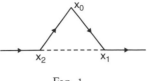

FIG. 1

correction to (74) arising from $J_2$ becomes

$$L = -i \left[ e^3/8\hbar c \right] \int_{-\infty}^{\infty} dx_1 \int_{-\infty}^{\infty} dx_2 D_F (x_1 - x_2) A_{\mu}^e (x_0)$$

$$\times \bar{\psi} (x_1) \gamma_{\nu} S_F (x_0 - x_1) \gamma_{\mu} S_F (x_2 - x_0) \gamma_{\nu} \psi (x_2).$$

This is the term which gives rise to the main part of the Lamb-Retherford line shift,[10] the anomalous magnetic moment of the electron,[11] and the anomalous hyperfine splitting of the ground state of hydrogen.[12]

The above expression $L$ is formally simpler than the corresponding expression obtained by Schwinger, but the two are easily seen to be equivalent. In particular, the above expression does not lead to any great reduction in the labor involved in a numerical calculation of the Lamb shift. Its advantage lies rather in the ease with which it can be written down.

In conclusion, the author would like to express his thanks to the Commonwealth Fund of New York for financial support, and to Professors. Schwinger and Feynman for the stimulating lectures in which they presented their respective theories.

*Notes added in proof* (To Section II). The argument of Section II is an over-simplification of the method of Tomonaga,[1] and is unsound. There is an error in the derivation of (3); derivatives occurring in $H(r)$ give rise to non-commutativity between $H(r)$ and field quantities at $r'$ when $r$ is a point on $\sigma$ infinitesimally distant from $r'$. The argument should be amended as follows. $\Phi$ is defined only for flat surfaces $t(r) = t$, and for such surfaces (3) and (6) are correct. $\Psi$ is defined for general surfaces by (12) and (10), and is verified to satisfy (9). For a

[10] W. E. Lamb and R. C. Retherford, Phys. Rev. **72**, 241 (1947).
[11] P. Kusch and H. M. Foley, Phys. Rev. **74**, 250 (1948).
[12] J. E. Nafe and E. B. Nelson, Phys. Rev. **73**, 718 (1948); Aage Bohr, Phys. Rev. **73**, 1109 (1948).

flat surface, $\Phi$ and $\Psi$ are then shown to be related by (7). Finally, since $H_1$ does not involve the derivatives in $H$, the argument leading to (3) can be correctly applied to prove that for general $\sigma$ the state-vector $\Psi(\sigma)$ will completely describe results of observations of the system on $\sigma$.

(To Section III). A covariant perturbation theory similar to that of Section III has previously been developed by E. C G. Stueckelberg, Ann. d. Phys. 21, 367 (1934); Nature, **153**, 143 (1944).

(To Section V). Schwinger's "affective potential" is not $H_r$ given by (25), but is $H'_r = Q H_T Q^{-1}$. Here $Q$ is a "square-root" of $S(\infty)$ obtained by expanding $(S(\infty))^{\frac{1}{2}}$ by the binomial theorem. The physical meaning of this is that Schwinger specifies states neither by $\Omega$ nor by $\Omega'$, but by an intermediate state-vector $\Omega'' = Q\Omega = Q^{-1}\Omega'$, whose definition is symmetrical between past and future. $H'_T$ is also symmetrical between past and future. For one-particle states, $H_T$ and $H'_T$ are identical.

Equation (32) can most simply be obtained directly from the product expansion of $S(\infty)$.

(To Section VII). Equation (62) is incorrect. The function $S'_F$ is well-behaved, but its fourier transform has a logarithmic dependence on frequency, which makes an expansion precisely of the form (62) impossible.

(To Section X). The term $L$ still contains two divergent parts. One is an "infra-red catastrophe" removable by standard methods. The other is an "ultraviolet" divergence, and has to be interpreted as an additional charge-renormalization, or, better, cancelled by part of the charge-renormalization calculated in Section VIII.

# Chapter Nine

In this section, we present two lecture series and a historical account by some of the founders of quantum theory. In 1925–26 Max Born presented a series of lectures at MIT entitled "Problems of Atomic Dynamics." This was immediately after Heisenberg had, with Born's aid, worked out the first quantum theory of the atom, and it was during this time that Schrodinger published his wave theory of quantum mechanics. These lectures present the atomic theory right as it was being worked out. Born starts by introducing the classical theory of mechanics and then shows how the Bohr model of the atom was able to explain the hydrogen atom. He explains why Bohr's model was unsuccessful for other elements including helium. Finally, he presents the matrix method for the new quantum theory, and shows how to derive fundamental commutation relationships upon which the uncertainty principle is founded. During the course of the lecture series insights by other researchers were being discovered. For example, Pauli's introduction of the spin quantum number and Schrodinger's and Dirac's formulations of quantum theory all happened over the course of this series, and it is interesting to see the evolution of Born's ideas as the new quantum theory took shape.

The second work in this chapter is a historical account by the physicist George Gamow of his experiences working with the founders of quantum theory. In the late 1920s Gamow received a fellowship to study at Niels Bohr's institute in Copenhagen. There he worked with many of quantum theory's greatest pioneers, including, of course, Bohr himself. In "Thirty Years that Shook Physics," he gives an insider's perspective on the historical development of quantum theory. He shares numerous anecdotes about his experiences with these great thinkers, providing a refreshingly human face to this esoteric but fascinating field.

The last work in this section is a series of four lectures delivered by Paul Dirac at Yeshiva University outlining developments in quantum theory. In the first lecture, Dirac starts with the Hamiltonian formulation of classical mechanics and shows how to develop quantum mechanics by implementing a quantization principle. In the second lecture, Dirac generalizes the method in the first lecture by showing how to obtain a quantum field theory from a classical field theory. Dirac is ultimately concerned with obtaining a quantum theory that can incorporate Einstein's theory of general relativity. From general relativity we know that gravity curves space, and thus Dirac wanted to know if it is possible to derive a relativistic quantum theory on curved surfaces. In the third lecture, Dirac examines this question and decides that in general it is not possible to obtain relativistic quantum theory on a curved surface. Finally, he shows that it is possible to develop a relativistic quantum theory on a flat surface. Thus, we can obtain a quantum theory which is consistent with special but not general relativity. The search for a quantum theory that is consistent with general relativity or a quantum gravity theory is still an unsolved problem in physics. Finding this theory is, perhaps, the primary goal of theoretical physics. Currently the most favored type of quantum gravity theory is known as string theory, but it still remains to be shown if string theory is an accurate description of reality. It will be fascinating to see in the coming years which quantum gravity theory best describes our universe, because once this theory is found, we will for the first time have a fundamental understanding of all known physical laws.

# Problems of Atomic Dynamics

## By
## Max Born

## PREFACE

The lectures which constitute this book are given just as they were presented at the Massachusetts Institute of Technology from November 14, 1925, to January 22, 1926, without any amplification. They do not purport to be a text-book—for of these we have enough—but rather an exposition of the present status of research in those regions of physics in which I myself have made investigations, and of which I therefore believe that I can take a comprehensive view. In the short time that was at my disposal, I could neither seek for completeness nor consider minutiæ. It was my purpose to present methods, objects of investigation, and the most important results. I have avoided references and have only occasionally named individual authors. I take this occasion to ask the pardon of all those colleagues whose names I have omitted to mention.

The lectures on the theory of lattices are essentially an abstract of certain sections of my book, *Atomtheorie des festen Zustandes*, and of subsequent works on this topic. In the same manner, the earlier lectures on the structure of the atom are closely related to my book *Atommechanik*, but I soon made the transition to a different point of view. At the time I began this course of lectures, Heisenberg's first paper on the new quantum theory had just appeared. Here his masterly treatment gave the quantum theory an entirely new turn. The paper of Jordan and myself, in which we recognized the matrix calculus as the proper formulation for Heisenberg's ideas, was in press, and the

This work was originally published in 1926 by the Massachusetts Institute of Technology.

manuscript of a third paper by the three of us was almost completed. Though the results contained in this third paper left no doubt in my mind as to the superiority of the new methods to the old, I could not bring myself to plunge directly into the new quantum mechanics. To do this would not only be to deny to Bohr's great achievement its due need of credit, but even more to deprive the reader of the natural and marvelous development of an idea. I have consequently begun by presenting the Bohr theory as an application of classical mechanics, but have emphasized more than is usual its weaknesses and conceptual difficulties. It is perhaps superfluous to state that this is only done to establish the necessity of a new conception, and is not intended as a hostile criticism of Bohr's immortal work. As the course proceeded, further achievements of the new method came to my notice. I was able to introduce some of these into the lectures. Pauli's theory of the hydrogen atom is a case in point. Of others, such as the treatment of the theory of aperiodic processes in terms of a general calculus of operators, developed by N. Wiener and myself, I was able to give a sketch. These sections are not so much a report on scientific results as an enumeration of the problems which seem of most interest to us theoretical physicists.

I wrote the original text in German. It was then translated into English by Dr. W. P. Allis and Mr. Hans Müller, and read through by me. Mr. F. W. Sears revised the second part; finally Dr. M. S. Vallarta went carefully through the complete text in order to verify the formulas and make the English idiomatic. I hereby express my sincere thanks to all these gentlemen, who have spent much labor on this work of revision, and have sacrificed much valuable time, as well as to Assistant Dean H. E. Lobdell, who has taken great pains in the supervision of the work of publication.

I feel it as a great honor that this book appears as a publication of the Massachusetts Institute of Technology. For this I wish to express my thanks to President S. W. Stratton and Professor C. L. Norton, the Head of the Department of Physics. To Professor Paul Heymans, who has not merely extended to my wife and myself the hospitality of his house, but also shared his office with me during the three months of my stay at the Institute, I wish to express my gratitude in visible form by the dedication of this little book.

MAX BORN

MASSACHUSETTS INSTITUTE OF TECHNOLOGY
*January*, 1926

# Series I
## The Structure of the Atom

### LECTURE 1

Comparison between the classical continuum theory and the quantum theory—Chief experimental results on the structure of the atom—General principles of the quantum theory—Examples.

Physics today is everywhere based on the theory of atoms. Through experimental and theoretical researches we have reached the conviction that matter is not infinitely divisible, but that there exist ultimate units of matter which cannot be further divided. However, it is not the atoms of the chemists that we feel authorized in calling "indivisible"; on the contrary they are very complicated structures composed of smaller elements. These are, from the point of view of recent investigations, the atoms of electricity, the (negative) electrons and the (positive) protons. It is conceivable that at a later epoch science will change its point of view and penetrate to still smaller elements; in this case the philosophical significance of atomistics could no longer be valued as highly. The last units would not be anything absolute, but only a measure of the present status of science. But I do not think that is so; I believe that we can hope that we have not to do with an endless chain of divisions, but that we are near the end of a finite chain, perhaps we have even attained it. The reasons that can be given for this optimism lie less in the experimental evidence for the reality of atoms, protons and electrons, which the new physics has furnished, than in the special character of the laws which govern the interactions of elementary electric particles. These laws have indeed properties which permit us to conclude that we are near their final formulation.

Such an assertion may seem too bold, because all philosophies of all ages have taught that human knowledge is incomplete, that each goal of knowledge is attained only at the cost of new puzzles.

Up to the present, in physics as in other sciences, every result that our age has proclaimed as absolute has had to fall after a few years, decades or centuries, because new investigations have brought new knowledge and we have become uscd to consider the true laws of nature as unattainable ideals to which the so-called laws of physics are only successive approximations. Now, when I say that certain formulations of the laws of the atomistics of today have a character which is in a certain sense final, this does not fit in with our scheme of successive approximations and it becomes necessary that I offer an explanation. This special character that the atom possesses is the appearance of *whole numbers*. We pretend not only that in any body, for instance a piece of metal, there exist a certain finite number of atoms or electrons, but further that the properties of a single atom and the processes which occur during the interaction of several atoms are capable of being described by whole numbers. This is the substance of the *quantum theory*, the fundamental significance of which is based not only in its practical application but above all in its philosophical consequences considered here. To illustrate this idea we consider a small body free to move in a straight line. According to the usual ideas it can be at any time at any point. To fix this point we give the coördinate $x$ measured from a point 0.

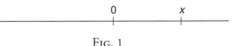

FIG. 1

But the accuracy of this indication depends entirely on experimental means of observation. If $x$ can vary continuously, a more exact measurement may give us another decimal. For the processes in the atom, however, conditions seem to be different. We may compare them with the behavior of this body if we consider it infinitely small and allow it to occupy only certain discrete pointe which we shall number 1, 2, 3, .... The coördinate $x$ can therefore only take the values 1, 2, 3, ... but not, for instance, $\frac{1}{2}$ or 3.7. This is in fact the behavior of the so-called quantum numbers by means of which we describe today the state of atoms. Should this process be always satisfactory we evidently stand before a new state of our knowledge. If the value of $x$ should be exactly a whole number, then a determination of such a number once made could not be altered. If it had been determined that $x$ is certainly not equal to 1, nor to 3 or 4, nor to any greater number, then there remains for $x$ only the value 2, and a more accurate measurement cannot change anything. We have therefore definite elements in the statements of laws, and there seems to exist a tendency that laws obtain this essential final character when expressed as relations between whole numbers. I therefore do not exaggerate when I say that the year 1900, when Planck first stated his theory of quanta, marked the beginning of an entirely new conception of nature.

The theory of matter, as treated up to the present, still falls very short of this extreme view. To emphasize this standpoint, we consider again the body in the straight line with the coördinate $x$. Then the usual quantum theory corresponds somewhat to the condition that $x$ be allowed to take all possible continuous values, but that then the integral values of $x$ should be selected as stationary states through the so-called quantum conditions. This conception is altogether too unsatisfactory. For this reason we have sought at Göttingen to find a new formulation of the quantum theory in which only these integral values of $x$ occur and intermediate fractional values have no meaning. This theory has been verified in the sense that certain fundamental difficulties that existed in the old quantum theory are not encountered in the new. On the other hand, calculations are rather complicated. Therefore I shall not begin my course of lectures with this new theory, but start with a short survey of the old theory. Let me remind you of the most important experimental investigations of the structure of atoms.

The first of these is the conception developed by Lenard and Rutherford, that the atom is composed of a positive *nucleus* surrounded by negative *electrons*. The simplest atom, that of hydrogen, consists of one electron revolving around the simplest nucleus, a *pro-ton*, each having the same charge $e = 4.77 \times 10^{-10}$ e.s.u., but different masses, the ratio of the masses being 1:1830. The nuclei of the other atoms are complex structures built up of protons and electrons, as shown by radioactive phenomena, but in these lectures we shall not discuss the structure of these nuclei, but treat them as masspoints with a charge which is an integral multiple $Z$ of the charge $e$ given above. This number $Z$ is known as the atomic number and determines the position of the element in the periodic system. In the neutral atoms the number of electrons is also $Z$; in negative ions the number of electrons is greater than $Z$, in positive ions it is less.

The forces binding the electrons to the nucleus are certainly of electrical nature. This has been proved by the experiments of Lenard on the scattering of cathode rays and the experiments of Rutherford and his students on the scattering of $\alpha$-rays, in which it was shown that Coulomb's law of force holds for distances of the order of magnitude involved in this theory.

But the supposition of purely electrical forces leads to difficulties. There is a mathematical theorem which states that a system of electric charges cannot be in stable equilibrium; therefore Rutherford was compelled to assume that the electrons move around the nucleus in such a way that the centrifugal force balances the resultant of the electrical forces. But if electromagnetic laws can be applied to such a system, it must radiate energy until the electrons fall into the nucleus. A second difficulty arises from the kinetic theory of gases. We know that every molecule or atom of a gas under normal conditions collides with other molecules or atoms about 100,000,000 times per second. If the ordinary laws of mechanics held, there would be expected a slight change of the electron orbits at each collision and these changes would accumulate so that after one second the system would be materially altered. But we know that every molecule has a definite set of properties. It is therefore necessary to find a principle of stability which evidently cannot be derived from the ordinary laws of mechanics.

Niels Bohr has given this principle by applying the rules of quantum theory to atomic systems. These rules were developed by Max Planck in the study of the laws of heat radiation. He proved that it is impossible to explain the spectral distribution of the energy radiated by a black body if we make the ordinary assumption that energy can be divided into infinitely small parts; but it may be explained if we assume that the energy exists in quanta of finite size, $h\nu$, where $\nu$ is the frequency of the radiation and $h$ is a constant, $h = 6.54 \times 10^{-27}$ erg. sec. This remarkable idea has been of the greatest fruit-fulness in the development of physics, for it has been shown that the constant $h$ and the quantum $h\nu$ play important rôles in many phenomena. In the photoelectric effect the kinetic energy of the photoelectron is given by $mv^2/2 = h\nu$, where $\nu$ is the frequency of the incident radiation. This equation, proposed by Einstein, was proved experimentally by Millikan and others and gave the first direct evidence of the existence of the quantum. It was followed by many other experiments of a similar kind of which I will mention only one group: that investigating the relation between the kinetic energy of an electron and the frequency of the light emitted as the result of the collision between this electron and an atom, first tested by Franck and Hertz and later developed by Compton, Foote, Mohler and many other American physicists.

All these experiments show that the production of radiation of a certain frequency requires a certain amount of kinetic energy. Niels Bohr has made the assumption that this law holds not only between kinetic energy and radiation, but between all kinds of energy and radiation. In this way he found a very simple interpretation of the fact that isolated atoms, as in a rarefied gas, emit a line spectrum, that is, a set of monochromatic light waves. He assumes that Einstein's law can be applied to the emission of a line in the spectrum in such a way that while the system loses a finite amount of internal energy $W_1 - W_2$, the frequency $\nu$ of the emitted light is connected with this loss of energy by the equation

$$W_1 - W_2 = h\nu. \tag{1}$$

To explain the whole system of lines of the atom Bohr postulated the existence of a system of so-called "stationary states" in which the atom can exist without loss of energy by radiation, while keeping its total energy contents $W_1, W_2 \ldots$. The frequency of every spectral line appears now as the difference of two terms $W_1/h$ and $W_2/h$, in perfect agreement with the well-known optical fact formulated in the Ritz combination principle. At the same time this hypothesis solves the difficulty previously mentioned concerning the stability of atomic systems, for the energy necessary to change an atom from one stationary state to another is large, larger than that available at ordinary temperatures as a consequence of thermal agitation, therefore the atom remains unaltered.

These assumptions are in direct contradiction to classical dynamics, but lacking any knowledge of the exact laws of the new theory we use the classical laws as far as possible and then seek to alter them, when they lead nowhere. The chief problem becomes the determination of the stationary states and their energies; but first we show, after Einstein, that Bohr's principle suffices to give a very simple derivation of Planck's formula for *black-body radiation*.

Consider two stationary states $W_1$ and $W_2$ ($W_1 > W_2$). In statistical equilibrium they may exist in the amounts $N_1$ and $N_2$. Then by Boltzmann's principle

$$\frac{N_2}{N_1} = \frac{e^{-W_2/kT}}{e^{-W_1/kT}} = e^{\frac{W_1-W_2}{kT}}$$

and using Bohr's frequency condition (1):

$$\frac{N_2}{N_1} = e^{\frac{h\nu}{kT}}$$

In the classical theory, the interaction of atomic systems and radiation is made up of three processes:

1. If the atom is in a state of higher energy it loses energy spontaneously by outward radiation.

2. The external radiation field adds or subtracts energy to or from the atom depending on the phase and amplitude of the waves of which it consists. We call these processes:

   (a) positive absorption if the atom gains energy,

   (b) negative absorption if the atom loses energy through the action of the external field.

In the last two cases the contribution of these processes to the change of energy is proportional to the energy density $\rho_v$.

In analogy to these we assume for the quantic interaction three corresponding processes. The following transitions occur between the two energy levels $W_1$ and $W_2$:

1. Spontaneous decrease in energy through changes from $W_1$ to $W_2$. The frequency with which these transitions occur is proportional to the number of systems, $N_1$, which are in the initial state $W_1$ and is also dependent on the final state. For the number of these transitions we therefore write,

$$A_{12} N_1.$$

2a. Increases in energy due to the radiation field (transitions from $W_2$ to $W_1$). We place, likewise, for the number of such transitions

$$B_{21} N_{2\rho_v}.$$

2b. Decreases in energy due to the radiation field (transitions from $W_1$ to $W_2$). The number of such transitions is

$$B_{12} N_{1\rho_v}.$$

For statistical equilibrium between the states $W_1$ and $W_2$ it is required that

$$A_{12} N_1 = (B_{21} N_2 - B_{12} N_1)\, \rho_v,$$

from which

$$\rho_v = \frac{A_{12}}{B_{21} \frac{N_2}{N_1} - B_{12}} = \frac{A_{12}}{B_{21} e^{\frac{hv}{kT}} - B_{12}}. \tag{2}$$

It is natural to suppose that the classical laws are limiting cases of the quantum laws. Here the limiting case is that of high temperatures, where $h\nu$ is small compared with $kT$. Under such conditions Equation (2) should go over into the classical law of Rayleigh and Jeans,

$$\rho_\nu = \frac{8\pi}{c^3}\nu^2 kT.$$

For large values of $T$ (2) has the form

$$\rho_\nu = \frac{A_{12}}{B_{21} - B_{12} + B_{21}\frac{h\nu}{kT} + \cdots}.$$

These two expressions become identical if

$$B_{12} = B_{21},$$

and

$$\frac{A_{12}}{B_{12}} = \frac{8\pi}{c^3}\nu^3 h.$$

Inserting these values in (2) we obtain Planck's radiation formula

$$\rho_\nu = \frac{8\pi h}{c^3}\frac{\nu^3}{e^{\frac{h\nu}{kT}} - 1}.$$

We see that the validity of Planck's formula is quite independent of the determination of the stationary states.

We shall now consider the problem of the *determination of the stationary states*. The simplest model of a radiating system is the *harmonic oscillator*, the equation of motion of which is

$$m\ddot{q} + \kappa q = 0$$

where $q$ is the distance of the moving point from the position of equilibrium, $m$ its mass, and $\kappa$ a constant which is connected with the natural frequency $\nu_0$ by the relation

$$\kappa = m\left(2\pi\nu_0\right)^2.$$

The motion of a point obeying this equation is very closely related to the motion of the field vector in a monochromatic light wave. An immediate assumption is that the frequency of such a linear oscillator is the same as the frequency of the emitted light; then it follows from Bohr's frequency condition (1) that the energies of the stationary states of the oscillator must differ by $h\nu_0$, that is they are, for an appropriate choice of additive constants,

$$W_0 = 0, \quad W_1 = h\nu_0, \quad W_2 = 2h\nu_0 \cdots, \quad W_n = nh\nu_0 \cdots .$$

In the case of one degree of freedom the motion is completely determined by the energy, so in this simple example the stationary states are completely known,

$$q = \sqrt{\frac{W}{2\pi m \nu_0^2}} \cos\left(2\pi \nu_0 t + \delta\right).$$

From the system of energy levels it is possible to derive the complete system of spectral lines by taking all possible differences,

$$\nu = \frac{1}{h}\left(nh\nu_0 - kh\nu_0\right) = \nu_0\left(n - k\right).$$

We see that Bohr's principle gives for the frequency of the emitted light not only the fundamental frequency, $\nu_0$, as in the classical theory, but also the overtones $\nu_0(n - k)$. But in this simple case of the linear oscillator we should expect that both theories give exactly the same result; therefore we need a new principle to eliminate superfluous overtones. Bohr has supplied this under the name of the *Principle of Correspondence*. He makes the assumption we have already used once, that the quantum laws must go over into the classical laws in the limiting case. If the oscillator has a very large amount of energy, that is if $n$ is large, the difference between two neighboring energy-levels is small compared with their absolute values, and approximately the series of values of $W_n$ may be looked upon as varying continuously, as in the classical theory. Therefore Bohr assumes that the classical theory remains approximately valid in this limiting case. Then the emitted light may be calculated classically, and the light vector is proportional to the electric moment of the vibrating system. In the case of one coördinate $q$, this moment is $eq$, where $e$ is the charge of the moving point, and for the oscillator we have,

$$eq = e\sqrt{\frac{W}{2\pi m \nu_0^2}} \cos\left(2\pi \nu_0 t + \delta\right).$$

In the general case the electric moment is a Fourier series with an infinite number of terms of the same form. The squares of the coefficients of the terms of the series will be a measure of the intensity of light emitted at the corresponding frequency of the overtone; this measure must also hold approximately in the quantum theory. In this way we get a rough estimate of the intensities of spectral lines even for small quantum numbers. But in one case we may expect this rule to give the exact results, namely, when one Fourier coefficient is identically zero. Then we may assume that the corresponding frequency is not emitted at all and that the corresponding transition does not occur. In our case we see that the Fourier series has only one term, corresponding to the transition $n - k = 1$. In this way Bohr's correspondence principle reduces the number of frequencies to that of the classical theory. It seems quite trivial in this example, but we shall see that in other cases it gives valuable information concerning possible transitions.

We go over now to more complicated systems, at first those of one degree of freedom but of any energy-function. In a system of this kind we have in general, if we exclude orbits which go to infinity, periodic motions, so that the coördinate $q$ can be expanded in a Fourier series of the time $t$. It might be thought that it would be possible to determine the stationary states, as in the case of the oscillator, by making the energies integral multiples of $h\nu$, where $T = 1/\nu$ is the period of the motion, i.e., the time of a complete revolution. We shall see that this is not possible on account of another principle, the last to be mentioned in this introduction: the *adiabatic hypothesis* of Ehrenfest.

We consider the action of an external force on the atomic system. There are two limiting cases: constant forces and oscillating forces of high frequency. We know that in the second case classical mechanics cannot be applied, for the action of light consists in the production of transitions or quantum jumps which cannot be described by classical theory. But suppose that we investigate the action of a force changing slowly relatively to the inner motions. From the assumption of stationary states it follows that this force can have either no effect or a finite effect resulting in a quantum jump; it is natural to suppose that the latter case will be more and more improbable as the rate of change of the force diminishes. So we see that quantities which are suitable for fixing the stationary states must have the property of not being altered by slowly changing forces. Ehrenfest calls them "adiabatic invariants," from analogy to similar quantities in thermodynamics. The question is whether such quantities can be found at all in classical mechanics. In the case of a linear oscillator it is not the energy which has this property, because the frequency $\nu$ is not constant under the influence of slowly changing forces, but it can be proved that the quotient $W/\nu$ is an adiabatic invariant. Indeed our fixation of the stationary states of the oscillator can be formulated by giving to this quotient the discrete values $0h$, $1h$, $2h$ ... etc. One of the aims of these lectures will be to find adiabatic invariants for every atomic system; we shall show that they exist not only for simple periodic systems, but also for the larger class of so-called multipleperiodic systems.

Periodic properties are closely connected with the laws of the quantum theory. A process which can be resolved into rotations and oscillations falls in its domain. Therefore, it will be our first problem to study systematically the most general systems having periodic properties.

# LECTURE 2

General introduction to mechanics—Canonical equations and canonical transformations.

The laws of mechanics can be best condensed, by means of the Principle of Least Action, in Hamilton's formula,

$$\int_{t_1}^{t_2} L\, dt \text{ is stationary.} \tag{1}$$

In this formula, the so-called Lagrangian function $L$ depends on the coördinates and the components of the velocities and the extremal is understood to be taken by comparison of all motions leading from one given point $Q_1$ at the time $t_1$ to another given point $Q_2$ at the time $t_2$. This formulation has the advantage that it is independent of the system of coördinates. In the following we shall always operate with generalized independent coördinates $q_1, q_2 \ldots$. Using the variational principle we obtain, from the variation of equation (1), the equations of motion in the form of Euler-Lagrange

$$\frac{d}{dt}\frac{\partial L}{\partial \dot{q}_k} - \frac{\partial L}{\partial q_k} = 0. \tag{2}$$

We shall now give the expression for $L$ in three important cases:

1. In the mechanics of Galileo and Newton,

$$L = T - U$$

where $T$ is the kinetic and $U$ the potential energy. Denoting the velocity vectors by $\mathbf{v}_k$ and the masses by $m_k$ we have

$$T = \frac{1}{2}\sum_k m_k \mathbf{v}_k^2$$

and Equations (2) take the Newtonian form,

$$\frac{d}{dt}(m_k \mathbf{v}_k) = \mathbf{F}_k \tag{3}$$

where the components of the forces $\mathbf{F}_k$ are obtained from $U$ by differentiation.

$$F_{kx} = -\frac{\partial U}{\partial x_k}\cdots.$$

2. In Einstein's relativistic mechanics we have

$$L = T^\times - U$$

where

$$T^\times = \sum_k m_k^0 c^2 \left( 1 - \sqrt{1 - (v_k/c)^2} \right) \qquad (4)$$

and $m_k^0$ is the rest-mass, $v_k$ the magnitude of the velocity, and $c$ the velocity of light. This $T^\times$ is *different* from the kinetic energy

$$T = \sum_k m_k^0 c^2 \left( \frac{1}{\sqrt{1 - (v_k/c)^2}} - 1 \right). \qquad (5)$$

In this case the equations of motion can also be written in the form (3) if the mass depends on the velocity according to the law:

$$m_k = \frac{m_k^0}{\sqrt{1 - \left(\frac{v_k}{c}\right)^2}}. \qquad (6)$$

3. If magnetic forces are acting on the system we have,

$$L = T - U - \frac{1}{c} \sum_k e_k \mathbf{A}_k \cdot \mathbf{v}_k \qquad (7)$$

where $e_k$ is the charge of the particle and $\mathbf{A}_k$ the vector potential of the magnetic field for that configuration of the system.

The equations of motion are of the second order with respect to time and it is often convenient to transform them into twice as many equations of the first order. This has been done by Hamilton in the following symmetrical way:

Introducing as unknown functions besides the coördinates the momenta

$$p_k = \frac{\partial L}{\partial \dot{q}_k} \qquad (8)$$

and using instead of $L(q_1, \dot{q}_1, q_2, \dot{q}_2 \ldots)$ the function

$$H(q_1, p_1, q_2, p_2 \cdots) = \sum_k p_k \dot{q}_k - L \qquad (9)$$

the Principle of Least Action can be written

$$\int_{t_1}^{t_2} \left[ \sum_k p_k \dot{q}_k - H\left(q_1, p_1 \cdots\right) \right] dt \text{ is stationary} \qquad (10)$$

and the equations of Euler-Lagrange take the symmetrical form

$$\begin{cases} \dot{q}_k = \dfrac{\partial H}{\partial p_k} \\[2ex] \dot{p}_k = -\dfrac{\partial H}{\partial q_k} \end{cases} \qquad (11)$$

Equations (11) are also true if the Hamilton function $H$ depends explicitly on the time $t$. If this is *not* the case we find

$$\frac{dH}{dt} = \sum_k \left( \frac{\partial H}{\partial q_k} \dot{q}_k + \frac{\partial H}{\partial p_k} \dot{p}_k \right) = 0$$

or

$$H = \text{const.} \qquad (12)$$

We shall now discuss the physical meaning of $H$ in the same three cases considered above:

1. In the mechanics of Galileo and Newton, where $T$ is a homogeneous quadratic function of the components of the velocities we have, according to Euler's theorem,

$$2T = \sum_k \frac{\partial T}{\partial \dot{q}_k} \dot{q}_k = \sum_k \frac{\partial L}{\partial \dot{q}_k} \dot{q}_k = \sum_k p_k \dot{q}_k$$

and therefore, according to (9) and since $L = T - U$,

$$H = T + U.$$

$H$ is therefore the total energy and (12) is the law of conservation of energy. This holds only for "inertial systems" and not for accelerated systems of coördinates. In such a system, for instance in a rotating system, $H$ is constant, but does not represent the energy.

2. In reltitivistic mechanics we find by a simple calculation

$$H = \sum_k m_k^0 c^2 \left( \frac{1}{\sqrt{1 - \left(\frac{v_k}{c}\right)}} - 1 \right) + U = T + U$$

whence $H$ is also the total energy.

If it is desired to express this in terms of the momenta, we find, combining the momenta corresponding to the components of velocity with a vector momentum

$$\mathbf{P}_k = m_k \mathbf{v}_k = \frac{m_k^0 \mathbf{v}_k}{\sqrt{1 - \left(\frac{v_k}{c}\right)^2}}$$

and by elimination of $v_k$

$$H = \sum_k m_k^0 c^2 \left( \sqrt{1 + \frac{\mathbf{p}_k^2}{\left(m_k^0\right)^2 c^2}} - 1 \right) + U. \tag{13}$$

3. In a magnetic field we do not have a simple proportionality between velocity and momentum, but

$$\mathbf{p}_k = m_k \mathbf{v}_k - \frac{e_k}{c} \mathbf{A}_k$$

and even in this case $H$ is the total energy

$$H = T + U.$$

Introducing the momenta we find a rather complicated expression

$$H = \sum_k \left( \frac{\mathbf{p}_k^2}{2m_k} + \frac{e_k}{c m_k} \mathbf{A}_k \cdot \mathbf{p}_k + \frac{e_k^2}{2m_k c^2} \mathbf{A}_k \cdot \mathbf{A}_k \right) + U. \tag{14}$$

Before beginning with the general integration of the canonical equations, we shall consider some simple examples. If the Hamiltonian function, $H$, is independent of one coördinate, for example of $q_1$,

$$H = H(p_1, q_2, p_2 \cdots t),$$

we get from the canonical equations

$$\dot{p}_1 = 0$$

therefore,

$$p_1 = \text{const.},$$

and so we have found an integral of these equations. Such is the case, for example, if $q_1$ is the angle of rotation about an axis passing through the center of gravity of a solid body, therefore the coördinate is called a "cyclic" variable. In this case it is easily shown that $p_1$ is the moment of momentum of the system about the axis.

It may happen that $H$ is independent of all the $q_k$'s

$$H(p_1 p_2 \cdots t)$$

then the canonical equations are completely integrated by the formulas

$$\dot{p}_k = 0 \qquad\qquad p_k = \alpha_k$$
$$\dot{q}_k = \frac{\partial H}{\partial p_k} = \omega_k \qquad q_k = \omega_k t + \beta_k \tag{15}$$

where the $\omega_k$'s are characteristic constants of the system, $\alpha_k$ and $\beta_k$ constants of integration.

We see that the mechanical problem is solved if we can find such coördinates that $H$ depends only on the momenta. This is the method of integration which we shall use in the following. The difficulty now is that variables of this type cannot be found by means of a simple point transformation of $q_k$, but only by a simultaneous transformation of $q_k$ and $p_k$.

We shall now find all the transformations of $p_k$ and $q_k$ which do not change the form of the canonical equations. Such transformations are called "canonical transformations." This condition is evidently fulfilled if the Principle of Least Action (1) does not change its form by a transformation

$$p_k = p_k(\bar{q}_1, \bar{q}_2 \cdots \bar{p}_1, \bar{p}_2, \cdots t)$$

and

$$q_k = q_k(\bar{q}_1, \bar{q}_2, \cdots \bar{p}_1, \bar{p}_2, \cdots t);$$

in other words, if the sum

$$\sum_k p_k \dot{q}_k - H(q_1, p_1 \cdots t)$$

differs from the corresponding expression in the new coördinates by a quantity which is a total differential of the time. We must set, therefore,

$$\left[ \sum_k p_k \dot{q}_k - H(q_1, p_1 \cdots t) \right] - \left[ \sum_k \bar{p}_k \dot{\bar{q}}_k - H(\bar{q}_1, \bar{p}_1 \cdots t) \right] = \frac{dV}{dt}. \tag{16}$$

This equation is easily satisfied. Let us choose for $V$ an arbitrary function of the new and old coördinates and of the time

$$V(q_1, \bar{q}_1, \cdots t).$$

We obtain by comparison of the coefficients of $\dot{q}_k$ and $\dot{\bar{q}}_k$:

$$\begin{cases} p_k = \dfrac{\partial}{\partial q_k} V(q_1, \bar{q}_1, \cdots t) \\[2mm] \bar{p}_k = -\dfrac{\partial}{\partial \bar{q}_k} V(q_1, \bar{q}_1, \cdots t) \\[2mm] H = \bar{H} - \dfrac{\partial}{\partial t} V(q_1, \bar{q}_1, \cdots t). \end{cases} \tag{17}$$

Expressing $\bar{q}_k$, $\bar{p}_k$ in terms of $q_k$, $p_k$ we obtain the desired equations of transformation. But we can give to these canonical transformations several other forms, by using, instead of $q_k$, $\bar{q}_k$, other independent variables. There are in all four such combinations possible from which we select the common case where $q_k$, $\bar{p}_k$ are used as independent variables. To do this we write instead of $V$

$$V - \sum_k \bar{p}_k \bar{q}_k$$

which is evidently, like $V$, an arbitrary function, and consider here $V$ as a function of $\bar{q}_k$, $\bar{p}_k$. Then we obtain

$$\left[ \sum_k p_k \dot{q}_k - H(q_1, p_1 \cdots t) \right] - \left[ -\sum_k \bar{q}_k \dot{\bar{p}}_k - \bar{H}(\bar{q}_1, \bar{p}_1 \cdots t) \right]$$

$$= \frac{d}{dt} V(q_1, \bar{p}_1 \cdots t)$$

and therefore, by comparison of the coefficients,

$$
\begin{cases}
p_k = \dfrac{\partial}{\partial q_k} V(q_1, \bar{p}_1 \dots t) \\[2em]
\bar{q}_k = \dfrac{\partial}{\partial \bar{p}_k} V(q_1, \bar{p}_1 \dots t) \\[2em]
H = \bar{H} - \dfrac{\partial}{\partial t} V(q_1, \bar{p}_1 \dots t).
\end{cases}
\tag{18}
$$

We illustrate this equation by a few simple examples:

The function

$$
V = q_1 \bar{p}_1 + q_2 \bar{p}_2
$$

gives the identical transformation

$$
q_1 = \bar{q}_1, \quad p_1 = \bar{p}_1, \quad q_2 = \bar{q}_2, \quad p_2 = \bar{p}_2.
$$

The function

$$
V = q_1 \bar{p}_1 \pm q_1 \bar{p}_2 + q_2 \bar{p}_2
$$

gives

$$
\begin{cases}
q_1 = \bar{q}_1 & p_1 = \bar{p}_1 \pm \bar{p}_2 \\
q_2 = \bar{q}_2 \pm \bar{q}_1 & p_2 = \bar{p}_2.
\end{cases}
$$

For three pairs of variables the function

$$
V = q_1(\bar{p}_1 + \bar{p}_2 + \bar{p}_3) + q_2(\bar{p}_1 + \bar{p}_3) + q_3 \bar{p}_3
$$

gives the transformation

$$
\begin{array}{ll}
q_1 = \bar{q}_1 & p_1 = \bar{p}_1 + \bar{p}_2 + \bar{p}_3 \\
q_2 = \bar{q}_2 - \bar{q}_1 & p_2 = \bar{p}_2 + \bar{p}_3 \\
q_3 = \bar{q}_3 - \bar{q}_2 & p_3 = \bar{p}_3.
\end{array}
$$

In these examples the coördinates and impulses are transformed among themselves. The general condition is that $V$ shall be a linear function of $q$ and $\bar{p}$

$$
V = \sum_{i,k} \alpha_{ik} q_i \bar{p}_k + \sum_k \beta_k q_k + \sum_k \gamma_k \bar{p}_k.
$$

Then we have

$$p_i = \sum_k \alpha_{ik} \bar{p}_k + \beta_k$$

$$\bar{q}_i = \sum_k \alpha_{ki} q_k + \gamma_i.$$

If $\beta_i$ and $\gamma_i$ vanish we have

$$\sum_k p_k q_k = \sum_{k,l} \alpha_{kl} \bar{p}_l q_k = \sum_l \bar{q}_l \bar{p}_l.$$

This transformation is linear, homogeneous and contragredient. To this group belongs the case of orthogonal transformations, for instance the rotation of rectangular coördinates.

We obtain a point transformation, that is a transformation of the $q_k$'s among themselves, when $V$ is linear in $\bar{p}$:

$$V = \sum_k f_k(q_1, q_2 \dots) \bar{p}_k + g(q_1, q_2 \dots)$$

that is

$$\begin{cases} p_k = \sum_l \dfrac{\partial f_1}{\partial q_k} \bar{p}_l + \dfrac{\partial g}{\partial q_k} \\ \bar{q}_k = f_k(q_1, q_2 \dots) \end{cases}$$

and we have corresponding relations for the momenta.

As an example we shall give the transformation of rectangular coördinates into polar coördinates. Here we place,

$$-V = p_x r \, \cos \phi \, \sin \theta + p_y r \, \sin \phi \, \sin \theta + p_z r \, \cos \theta.$$

Then we have

$$\begin{cases} x = r \, \cos \phi \, \sin \theta & p_r = p_x \cos \phi \sin \theta + p_y \sin \phi \sin \theta + p_z \cos \theta \\ y = r \, \sin \phi \, \sin \theta & p_\phi = -p_x r \, \sin \phi \sin \theta + p_y r \, \cos \phi \sin \theta \\ z = r \, \cos \theta & p_\theta = p_x r \, \cos \phi \cos \theta + p_y r \, \sin \phi \cos \theta \\ & \quad - p_z r \, \sin \theta. \end{cases}$$

and the expression $p_x^2 + p_y^2 + p_z^2$ is transformed into

$$p_r^2 + \frac{1}{r^2} p_\theta^2 + \frac{1}{r^2 \sin^2 \theta} p_\phi^2.$$

As an example of the first form given to the canonical transformation, where $V$ depends on $q$ and $\bar{q}$, we choose

$$V = \frac{c}{2} q^2 \cot \bar{q}.$$

Then we have

$$p = cq \cot \bar{q}$$

$$\bar{p} = \frac{c}{2} q^2 \frac{1}{\sin^2 \bar{q}}$$

or

$$q = \sqrt{\frac{2\bar{p}}{c}} \sin \bar{q}$$

$$p = \sqrt{2c\bar{p}} \cos \bar{q}.$$

Hence the expression

$$\frac{1}{2} \left( p^2 + c^2 q^2 \right)$$

is transformed into $c\bar{p}$.

This example can be used to explain how the canonical transformations are employed in the integration of the equations of motion. For this we consider the *harmonic oscillator* in which

$$T = \frac{m}{2} \dot{q}^2 \qquad U = \frac{\kappa}{2} q^2.$$

Therefore

$$H = \frac{p^2}{2m} + \frac{\kappa}{2} q^2 = \frac{1}{2m} \left( p^2 + m\kappa q^2 \right).$$

If in the last transformation given we place $c^2 = m_\kappa$, $H$ is transformed to $c\bar{p}/m$. This is the solution of the problem. For now $\bar{q} = \phi$ is a cyclic variable and we have

$$\bar{p} = \alpha$$

$$\bar{q} = \phi = \omega t + \beta, \qquad \omega = \frac{\partial H}{\partial \bar{p}} = \frac{c}{m} = \sqrt{\frac{\kappa}{m}}.$$

In the original coördinates the motion is represented by

$$q = \sqrt{\frac{2\alpha}{m\omega}} \sin (\omega t + \beta)$$

$$H = \omega \alpha.$$

# LECTURE 3

The Hamilton-Jacobi partial differential equation—Action and angle variables—The quantum conditions.

In the same way we can now consider the most general case. Let us suppose that $H$ does not depend explicitly on $t$. We shall denote constant momenta by $\alpha_k$, the new variables which are linear functions of the time by $\phi_k$, the number of degrees of freedom by $f$. Then we have to determine a function

$$S\left(q_1, \ q_2 \ldots q_f, \ \alpha_1, \ \alpha_2 \ldots \alpha_f\right)$$

so that, by the transformation,

$$
\begin{cases}
p_k = \dfrac{\partial}{\partial q_k} S\left(q_1, q_2 \ldots q_f, \alpha_1, \alpha_2 \ldots \alpha_f\right) \\[2mm]
\phi_k = \dfrac{\partial}{\partial \alpha_k} S\left(q_1, q_2 \ldots q_f, \alpha_1, \alpha_2 \ldots \alpha_f\right)
\end{cases}
\tag{1}
$$

$H$ becomes a function depending only on the $\alpha_k$'s,

$$W\left(\alpha_1, a_2 \ldots \alpha_f\right).$$

Replacing $p_k$ by its value in

$$H\left(q_1, q_2 \ldots p_1, \ p_2 \ldots\right)$$

we obtain the condition

$$H\left(q_1, q_2 \ldots q_f, \ \frac{\partial S}{\partial q_1}, \ \frac{\partial S}{\partial q_2} \ldots \frac{\partial S}{\partial q_f}\right) = W\left(\alpha_1, \alpha_2 \ldots \alpha_f\right). \tag{2}$$

This expression can be looked upon as a partial differential equation for the determination of $S$. The problem is now to determine a so-called complete integral of this equation, that is, an integral which depends of $f-1$ arbitrary constants $\alpha_2 \ldots \alpha_f$, where $\alpha_1$ is to be identified with $W$, or, if no particular constant $\alpha_1$ is to be privileged in this manner, then we must find an integral which depends on $f$ constants $\alpha_1 \ldots \alpha_f$, among which there exists a relation

$$W = W\left(\alpha_1 \ldots \alpha_f\right).$$

857

The motion is then represented by

$$\phi_k = \omega_k t + \beta_k, \qquad \omega_k = \frac{\partial W}{\partial \alpha_k} \tag{3}$$

We shall call Equation (2) the *Hamilton-Jacobi differential equation* and *S* the *action-function*. An important property of *S* is the following: We have

$$dS = \sum_k \frac{\partial S}{\partial q_k} dq_k = \sum p_k dq_k.$$

Therefore *S* is a line integral, taken along the orbit, from a fixed point $Q_0$ to a moving point $Q$.

$$S = \int_{Q_0}^{Q} \sum_k p_k dq_k. \tag{4}$$

In Galilean-Newtonian mechanics this has a simple significance, because in this case,

$$2T = \sum_k p_k \dot{q}_k$$

and we have

$$S = 2 \int_{t_0}^{t} T \, dt = 2 \bar{T}(t - t_0) \tag{5}$$

where $\bar{T}$ is the time average of $T$.

We have seen already that the quantum theory is closely related to the periodic properties of the motion. In fact, Bohr's theory permits the definition of stationary states only for such motions as can be decomposed by harmonic analysis into periodic components. The astronomers call this class of motions "conditioned periodic." We prefer to call them "multiple periodic." These motions are defined in the following way: It is possible to introduce instead of variables $q_k$, $p_k$ new variables $w_k$, $I_k$ by means of the canonic transformation

$$p_k = \frac{\partial}{\partial q_k} S \left( q_1, I_1, q_2, I_2 \ldots q_f, I_f \right)$$

$$w_k = \frac{\partial}{\partial I_k} S \left( q_1, I_1, q_2, I_2, \ldots q_f, I_f \right)$$

which satisfy the following conditions:

(*A*) The position of the system depends periodically on $w_k$, with the fundamental period 1. That is, if the $q_k$'s are uniquely determined by the position of the system, then they can be expanded in a Fourier series:

$$q_k = \sum_{\tau} C_{\tau}^{(k)} e^{2\pi i (w\tau)}$$

where $\tau$ represents a number of integers $\tau_1, \tau_2 \ldots \tau_f$ and we place

$$(w\tau) = w_1 \tau_1 + w_2 \tau_2 + \cdots + w_f \tau_f.$$

If one of the $q_k$'s is an angle, it is not uniquely determined by the position of the system, but only within a multiple of a constant, as for instance $2\pi$. Then the above condition of periodicity is also true except for a multiple of that constant.

(*B*) Hamilton's function can be transformed into a function $W$ which depends only on the $I_k$'s.

It follows that the $I$'s are constants and the $w$'s are linear functions of the time $t$,

$$w_k = v_k t + \beta_k.$$

The $q$'s can therefore be represented by trigonometric series in $t$ with the frequencies

$$v_1 \tau_1 + v_2 \tau_2 + \cdots + v_f \tau_f$$

where, according to the results obtained above,

$$v_k = \frac{\partial W}{\partial I_k}.$$

$w_k$, $I_k$ are not yet uniquely determined by these conditions. For instance we can set

$$\bar{w}_k = w_k + f\left(I_1 \ldots I_f\right)$$

and

$$\bar{I}_k = I_k + C_k.$$

These form a canonic transformation, which is evidently compatible with the conditions (A) and (B). In order to exclude this indetermination we further set the condition:

(C) The function $S^\times = S - \sum_k w_k I_k$

shall be periodic in $w_k$ with the period 1:

$$S^\times = \sum_\tau C_\tau^\times e^{2\pi i (w\tau)}.$$

The canonic transformation in question can also be expressed by means of the function $S^\times$ as follows:

$$p_k = \frac{\partial}{\partial q_k} S^\times \left( q_1 \ldots q_f, w_1 \ldots w_f \right)$$

$$I_k = -\frac{\partial}{\partial w_k} S^\times \left( q_1 \ldots q_f, w_1 \ldots w_f \right).$$

Then indeed we can prove rigorously that $w_k$, $I_k$, which are called *angle* and *action variables*, are essentially uniquely determined by the conditions (A), (B) and (C). "Essentially" expresses the following: If we make a canonic transformation of the form

$$w_k = \sum_l c_{kl} \bar{w}_l$$

$$I_k = \sum_l c_{lk} \bar{I}_l$$

where the $c$'s are whole numbers and the determinant $|c_{kl}| = \pm 1$, all the conditions (A), (B), (C) are still satisfied. Aside from this indetermination, however, $w_k$, $I_k$ are really uniquely determined in all cases when the mechanical system is *not degenerate*, that is when there is no identical relation in $v_k$ of the form

$$v_1 \tau_1 + v_2 \tau_2 + \cdots + v_f \tau_f = 0$$

with the $\tau$'s whole numbers.

This theorem was first given by Burgers but his proof is not sufficient. A rigorous proof can be found in my book "Atommechanik"; this proof was given by my associate, F. Hund. This arbitrariness in the determination of the $I_k$'s, whereby the latter are determined except for a whole-number transformation of determinant $\pm 1$, is of essential importance for the applications of the quantum theory, for it is just these quantities that are equated to multiples of Planck's constant $h$; i.e.,

$$I_1 = n_1 h, \quad I_2 = n_2 h, \ldots \quad I_f = n_f h,$$

and from these equations it follows that also the $\bar{I}_k$'s are multiples of $h$.

# LECTURE 4

Adiabatic invariants—The principle of correspondence.

In order to justify this method of quantization, it must be shown in the first place that the $I$'s are adiabatic invariants. The general proof of this theorem was first outlined by Burgers and also by Krutkow; later more rigorous proofs were given by von Laue, Dirac and also by Jordan and myself. I shall not give here these rather complicated considerations, but shall only explain the significance of $I$ and its adiabatic invariance using the example of the harmonic resonator. Using the Hamiltonian function,

$$H = \frac{1}{2m}\left(p^2 + m\kappa q^2\right),$$

and then applying a canonic transformation, we have found above a solution of the problem of the resonator, which, although not quite satisfying the conditions $(A)$, $(B)$, $(C)$, is easy to transform into one which fulfils these conditions. It is only necessary to place

$$\phi = 2\pi w, \quad \alpha = \frac{I}{2\pi}.$$

Then the transformation is

$$q = \sqrt{\frac{I}{\pi m\omega}}\,\sin 2\pi w, \quad p = \sqrt{\frac{I\,m\omega}{\pi}}\,\cos 2\pi w$$

and the energy-function becomes

$$H = W = \omega\alpha = \frac{\omega}{2\pi}I = \nu I$$

where

$$\omega = 2\pi\nu$$

and also

$$w = \nu t + \delta, \quad \nu = \frac{dW}{dI}.$$

As $q$ is periodic in $w$ with the period 1, and as $H$ depends only on $I$, therefore the conditions $(A)$ and $(B)$ are fulfilled. To see whether condition ($C$) is also satisfied, we must only remember that the canonic transformation was found through the function

$$V = \frac{m\omega}{2}q^2 \cot 2\pi w$$

and then through the formulas,

$$p = \frac{\partial V}{\partial q}, \quad I = \frac{\partial V}{\partial w}.$$

This $V$ is therefore identical with the $S^\times$ introduced above. It can be written in the form

$$V = S^\times = \frac{I}{2\pi\, m\omega} \sin 2\pi w \cos 2\pi w$$

and, since it is periodic, the condition ($C$) is also fulfilled.

The quantum condition

$$I = nh$$

gives therefore the energy levels,

$$W = nh\nu$$

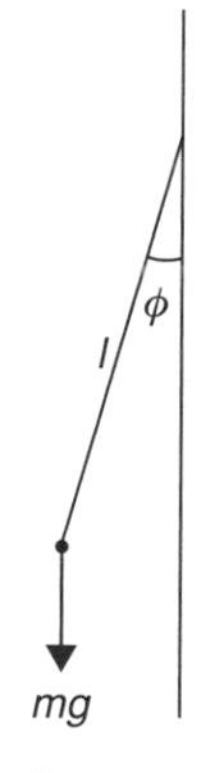

FIG. 2

in agreement with Planck's hypothesis. In order to verify that $I = W/\nu$ is really an adiabatic invariant, we represent the resonator by a pendulum swinging with small amplitude. Let $m$ be the mass of the bob, $l$ the length of the wire and $g$ the acceleration of gravity. Suppose now that the length $l$ is changed very slowly: the problem is to calculate how $W$ and $\nu$ vary. The forces which stretch the wire for any value of the angle $\phi$ are the component of gravity $mg \cos \phi = mg(1 - \phi^2/2)$ and the centrifugal force $ml\dot\phi^2$. The work done in shortening the wire is therefore

$$A = -mg \int (1 - \phi^2/2)\, dl - ml \int \dot\phi^2 dl. \qquad (1)$$

If the process of shortening is slow enough and has no period comparable with that of the pendulum, then it is permissible to introduce a mean amplitude, and we may write

$$dA = -mg \left(1 - \frac{\bar\phi^2}{2}\right) dl - ml\bar{\dot\phi}^2 dl$$

where the dash denotes average over a period. The work done is now split up in two parts: $-mgdl$ is the work done in lifting the bob, and

$$dW = \left(\frac{mg}{2}\bar\phi^2 - ml\bar{\dot\phi}^2\right) dl.$$

is the increase in the energy of oscillation. Now we know that for harmonic oscillations:

$$\frac{W}{2} = \frac{m}{2}l^2 \bar{\dot\phi}^2 = \frac{m}{2}gl\bar\phi^2$$

and hence

$$dW = -\frac{W}{2l}dl.$$

Now, since $v$ is proportional to $1/\sqrt{l}$, therefore $\frac{dv}{v} = -\frac{dl}{2l}$ and

$$\frac{dW}{W} = \frac{dv}{v}$$

whence, by integration,

$$\frac{W}{v} = \text{constant} \qquad (2)$$

which proves our theorem. The general proof of adiabatic invariance consists essentially of quite analogous considerations.

As another important example let us consider the rotator, that is, a body which can be rotated about an axis. If $A$ is the moment of inertia with respect to the axis and $\phi$ the angle of rotation, then we have

$$H = \frac{A}{2}\dot\phi^2$$

whence it follows, for the momentum $p$ corresponding to $\phi$

$$p = A\dot{\phi}.$$

$p$ is the angular momentum and we have

$$H = \frac{p^2}{2A}.$$

$\phi$ is therefore a cyclic variable and

$$p = \text{constant}.$$

If we set $\phi = 2\pi w$, the position of the system is a periodie function of $w$ of period 1. The canonic transformation

$$(\phi, p) \rightarrow (w, I)$$

is evidently characterized by the function $S = \phi I/2\pi$ and has the form

$$p = \frac{\partial S}{\partial \phi} = \frac{I}{2\pi}, \quad w = \frac{\partial S}{\partial I} = \frac{\phi}{2\pi},$$

whence it follows that $S^\times = S - wI = 0$ is a periodic function. Finally, we obtain

$$H = W = \frac{I^2}{8\pi^2 A}.$$

The conditions (A), (B), (C) are fulfilled and we have to set

$$I = h\nu,$$

which gives the energy levels

$$W = \frac{h^2}{8\pi^2 A} n^2. \tag{3}$$

This model is applied to the explanation of the band spectra of molecules. If a molecule rotates about a fixed axis, the emitted frequencies, according to Bohr, are given by the relation $\nu = \frac{1}{h}(W_m - W_n) = \frac{h}{8\pi^2 A}(m^2 - n^2)$, but, as in the case of the oscillator, the number of different frequencies given by this formula is too large. Out of these frequencies we must choose certain ones by the Principle of Correspondence. For this purpose we consider a component of the electric moment of the rotator. Evidently, in this case the motion is also given by a simple harmonic oscillation, whence we conclude, as above, that there are no other jumps of $n$ than those where $n$ changes by $\pm 1$, i.e., that $n - m = \pm 1$. Introducing this restriction, we obtain for the emitted frequencies (placing $m - n = 1$, or $m = n + 1$):

$$\nu = \frac{h}{8\pi^2 A}\left((n+1)^2 - n^2\right) = \frac{h}{8\pi^2 A}(2n+1)$$

$$\nu = \frac{h}{4\pi^2 A}\left(n + \frac{1}{2}\right).$$

The rotation frequency of the rotator itself is given by

$$\nu_0 = \frac{dW}{dI} = \frac{I}{4\pi^2 A} = \frac{nh}{4\pi^2 A}.$$

Therefore, as $n$ increases, the relative difference between the rotation and the emitted frequency becomes smaller. In both cases we have an equidistant series of frequencies and, indeed, the band spectrum emitted by a rotating molecule appears as a first approximation to consist of such a series.

We shall not go further into this problem, but instead will now consider the general relation which exists, according to the Principle of Correspondence, between the frequencies and the intensities of the spectral lines calculated classically and the corresponding quantities calculated according to the quantum theory. We consider the electric moment of the system having a Fourier expansion analogous to that of the coördinates

$$\mathbf{M} = \sum_k e_k \mathbf{r}_k = \sum_\tau \mathbf{C}_\tau e^{2\pi i(w\tau)} = \sum_\tau \mathbf{C}_\tau e^{2\pi i[(\nu\tau)t + (\delta_\tau)]} \qquad (4)$$

The frequencies can be written

$$\nu_{cl} = (\nu\tau) = \sum_k \nu_k \tau_k = \sum_k \tau_k \frac{\partial W}{\partial I_k}. \qquad (5)$$

Let a stationary state be determined by

$$I_k^{(1)} = n_k^{(1)} h$$

and another by

$$I_k^{(2)} = n_k^{(2)} h;$$

then we can consider in the $I_k$-space of $f$ dimensions the two points connected by the straight line

$$I_k = I_k^{(1)} + \tau_k \lambda; \quad 0 \leq \lambda \leq h$$

where

$$\tau_k = n_k^{(2)} - n_k^{(1)}.$$

Then

$$\frac{d I_k}{d\lambda} = \tau_k$$

and

$$v_{cl} = \sum_k \frac{\partial W}{\partial I_k} \frac{d I_k}{dt} = \frac{d W}{d\lambda}.$$

On the other hand the quantum frequency is

$$v_{qu} = \frac{W_1 - W_2}{h} \tag{6}$$

and the relation between the frequencies in the classical and in the quantum theory is the same as that between derivative and difference-ratio. It is also possible to consider the quantum-theory frequencies as the straight-line mean of the classical frequencies, as follows,

$$v_{qu} = \frac{1}{h} \int dW = \frac{1}{h} \int_0^h \frac{d W}{d\lambda} d\lambda = \frac{1}{h} \int_0^h v_{cl} d\lambda. \tag{7}$$

If the changes of the quantum numbers are small compared with the numbers themselves, the two expressions for $v_{qu}$ and $v_{cl}$ respectively differ very little. As to the intensities, we expect that they vary approximately in the same way as the quantities $|\mathbf{C}_\tau|^2$, where $\mathbf{C}_\tau$ is a function of the $I_k$'s and of $\tau_k = n_k^{(1)} - n_k^{(2)} = (I_k^{(1)} - I_k^{(2)})/h$. It is seen that this statement has a definite meaning only if $n_k$ is large, because only in this case is it immaterial whether we place in $\mathbf{C}_\tau(I) = \mathbf{C}_{n^{(1)} - n^{(2)}}(n)$ for $n$ the initial value $n^{(1)}$ or the final value $n^{(2)}$. On the other hand, this statement has a unique meaning if $\mathbf{C}_\tau(I)$ is identically zero for all $I$'s, for then we expect that a jump of $\tau$ does not occur. In other cases the difficulty has been evaded by taking a suitable mean of $\mathbf{C}_\tau(I)$ over the values of $I$ between the initial and the final states. By this method,

Kramers has succeeded in representing satisfactorily the results of observations in certain cases. It is not satisfactory in principle that we should not find in the quantum theory, in the form here presented, a unique determination of the intensities. This is one of the main reasons which led us to formulate our new quantum theory, where this difficulty is overcome.

# LECTURE 5

Degenerate systems—Secular perturbations—The quantum integrals.

We now say a few words about the case, left aside so far, of degeneration, that is, that in which there exist identical relations in the $I_k$'s of the form

$$(\nu\tau) = \nu_1\tau_1 + \nu_2\tau_2 + \cdots + \nu_n\tau_n = 0. \tag{1}$$

Then our theorem of uniqueness no longer holds, and it is no longer possible to formulate the quantum conditions in the form

$$I_k = n_k h.$$

Such is the case, for instance, in the harmonic oscillator of two degrees of freedom.

$$H = \frac{1}{2m}\left(p_x^2 + p_y^2\right) + \frac{m}{2}\left(\omega_x^2 x^2 + \omega_y^2 y^2\right).$$

The solution of the equations of motion can be written down immediately, because the two coördinates are separable. We obtain

$$x = \sqrt{\frac{I_x}{\pi\omega_x m}}\,\sin 2\pi w_x, \quad p_x = \sqrt{\frac{\omega_x m I_x}{\pi}}\,\cos 2\pi w_x$$

$$y = \sqrt{\frac{I_y}{\pi\omega_y m}}\,\sin 2\pi w_y, \quad p_y = \sqrt{\frac{\omega_y m I_y}{\pi}}\,\cos 2\pi w_y.$$

Here $I_x, w_x; I_y, w_y$ are two conjugate pairs of action and angle variables. If now $w_x$ and $w_y$ are *not* commensurable the motion given by placing (Fig. 3)

$$w_x = \omega_x t + \delta_x, \quad w_y = \omega_y t + \delta_y$$

is a so-called Lissajous figure, in which the path comes as near as desired to any point within a rectangle. But if a relation of the form

$$\tau_x \omega_x + \tau_y \omega_y = 0$$

exists, for instance if (Fig. 4.)

$$\omega_x = \omega_y = \omega_0 = 2\pi\nu_0,$$

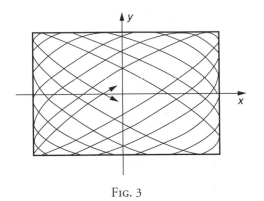

FIG. 3

then the orbit is simple periodic (ellipse). We can now rotate the system of coördinates arbitrarily without changing the form of the solution. But in doing so the sides of the rectangle change continuously and so do the magnitudes $\sqrt{I_z}$, $\sqrt{I_y}$, which differ from them only by the constant factor $1/\sqrt{\pi\omega m}$. It is therefore impossible to place $I_x$ and $I_y$ proportional to whole numbers $n_x$, $n_y$. The diagonal of the rectangle, however, that is the square root of the quantity

$$(\sqrt{I_x})^2 + (\sqrt{I_y})^2 = I_x + I_y = I,$$

remains invariant for such a rotation. We can therefore set

$$I_x + I_y = nh$$

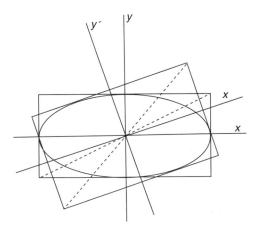

FIG. 4

whereby the total energy

$$W = \frac{\omega_0}{2\pi} \left( I_x + I_y \right) = \nu_0 I$$

is uniquely determined. $W = nh\nu_0$ has therefore exactly the same value as for the linear oscillator. We can describe this behavior as follows: If we introduce, instead of $I_x$ and $I_y$, two new variables, $I_x + I_y = I$ and $I_x - I_y = I'$, then two new conjugate angle variables, $w$, $w'$, correspond to the latter, with the frequencies,

$$\nu = \frac{dW}{dI} = \nu_0, \quad \nu' = \frac{dW}{dI'} = 0,$$

and we can only quantize the variable $I$ which alone appears in $W$ and which therefore alone corresponds to a frequency different from zero.

The following rule holds in general: In cases of degeneration it is always possible to attain, by means of a linear whole-number transformation of determinant $\pm 1$, that $H = W$ shall depend only on a number $s$ of $I$-variables, among which there are no commensurability relations. We call such variables $I_\alpha$. To these variables correspond $s$ frequencies $\nu_\alpha$ different from zero, while the other $f - s$ frequencies $\nu_\rho$ vanish. Only such variables $I_\alpha$ are to be equated to multiples of $h$. Bohr calls $s$ the *degree of periodicity* of the system.

Evidently we can increase the degree of periodicity of a system by introducing perturbing forces, for instance by placing the system in an electric or in a magnetic field. Then the original energy-function, which we shall call $H_0$, is increased by an additional energy which we shall call "perturbation energy" and denote by $\lambda H_1$, where $\lambda$ is a measure of the magnitude of the additional energy. If the perturbation is small, that is, if $\lambda$ is small, then there is a simple process whereby the new motions which are to be added to the system which was originally degenerate can be calculated. The influence of the perturbation energy is to change slightly all the magnitudes $w$, $I$, but the influence is different for these two kinds of variables. Those angle variables $w_\rho$, which belong to the zero frequencies of the unperturbed system, and which therefore were constant, now change slowly with frequencies which are proportional to $\lambda$. The other angular variables $w_\alpha$ will only undergo small variations of their frequencies. If we take the $w^0$, $I^0$'s of the unperturbed system as initial variables for the perturbation problem, then we have

$$H = H_0\left(I_\alpha^0\right) + \lambda H_1(I_\alpha^0, w_\alpha^0; I_\rho^0, w_\rho^0) \tag{2}$$

and $w_\alpha^0$ will complete its period many times during a period of $w_\rho^0$. Therefore an approximation can be made by taking an average over $w_\alpha^0$

$$\bar{H} = H_0\left(I_\alpha^0\right) + \lambda \bar{H}_1(I_\alpha^0; I_\rho^0, w_\rho^0). \tag{3}$$

This function can be considered as the energy-function of a new problem of motion for the formerly degenerate variables $I_\rho^0$, $w_\rho^0$. It is required to solve the equations of motion

$$\dot{w}_\rho^0 = \lambda \frac{\partial \bar{H}_1}{\partial I_\rho^0}, \quad \dot{I}_\rho^0 = -\lambda \frac{\partial \bar{H}_1}{\partial w_\rho^0}, \tag{4}$$

that is, to find a canonical substitution

$$\left(I_\rho^0, w_\rho^0\right) \rightarrow \left(I_\rho, w_\rho\right)$$

such that $\bar{H}_1$ is transformed into a function $W_1$ which depends only on $I$, i.e., ($I_\alpha$ is written for $I_\alpha^0$).

$$H = W_0\left(I_\alpha\right) + \lambda W_1(I_\alpha, I_\rho).$$

The perturbation frequencies are then

$$\nu_\alpha = \frac{\partial W_0}{\partial I_\alpha} + \lambda \frac{\partial W_1}{\partial I_\alpha}, \quad \nu_\rho = \lambda \frac{\partial W_1}{\partial I_\rho}.$$

The name "secular perturbations" is given in celestial mechanics to the slow motions of frequencies $\nu_\rho$.

We shall only touch upon the question of how the angle and action variables can be actually found in given cases. The method of separation of variables is often used: It is applicable if it is possible to find canonic variables $p_k, q_k$ for which the Hamilton-Jacobi differential equation can be solved by setting

$$S = S_1(q_1) + S_2(q_2) + \cdots + S_f(q_f). \tag{5}$$

Then

$$p_k = \frac{\partial S_k}{\partial q_k} \tag{6}$$

is a function of $q_k$ only and we can show that the integrals taken over a period

$$I_k = \int_0 p_k dq_k = \int_0 \frac{\partial S_k}{\partial q_k} dq_k \tag{7}$$

are the action variables. The functions $S_k(qk)$ depend also on these constants $I_k$. Applying a canonic transformation, in which the $I_k$'s enter as new action variables, the corresponding new angle variables are defined by

$$w_k = \frac{\partial S}{\partial I_k} = \sum_l \frac{\partial S_l}{\partial I_k}. \tag{8}$$

Since $S$ satisfies the Hamilton-Jacobi differential equation, $H$ is transformed into a function $W(I_1 \cdots I_f)$ and the condition $(A)$ is satisfied.

If any coördinate $q_k$ varies once between its limits, while the other coördinates $q_k$ are kept constant, the change of any variable $w_k$ is

$$\Delta_h w_k = \int_0 \frac{\partial w_k}{\partial q_h} dq_h.$$

Now,

$$\frac{\partial w_k}{\partial q_h} = \sum_l \frac{\partial^2 S_l}{\partial I_k \partial q_h} = \frac{\partial}{\partial I_k} \sum_l \frac{\partial S_l}{\partial q_h} = \frac{\partial}{\partial I_k} \frac{\partial S_h}{\partial q_h}$$

whence

$$\Delta_h w_k = \frac{\partial}{\partial I_k} \int_0 \frac{\partial S_h}{\partial q_h} dq_h = \frac{\partial I_k}{\partial I_k} = \begin{cases} 1 \text{ for } h = k. \\ 0 \text{ for } h \neq k. \end{cases} \tag{9}$$

If any point $q_1^0 \ldots q_f^0$ in $q$-space, to which corresponds the point $w_1^0 \ldots w_f^0$ in $w$-space, describes a closed curve, then the point $w$ need not return to its original position, but the end point is given by an expression of the form $w_k^0 + (\tau_1 w_1^0 + \cdots + \tau_f w_f^0)$ where the $\tau$'s are whole numbers. The $q$'s are therefore periodic functions of the $w$'s, with the fundamental period 1. Condition $(B)$ is thus satisfied.

According to the definition of $I_k$, $S$ increases by the amount $I_k$ every time that $q_k$ varies over a cycle, the other variables being held constant. As $w_k$ increases by 1 at the same time, the function $S^\times = S - \sum_k w_k I_k$ remains unchanged. Therefore, it is periodic, and the condition $(C)$ is satisfied, whence it is proved that $w$, $I$ are the angle and action variables.

Many authors introduce the quanta by this integral definition, but it appears to me, as to Bohr, better to define them generally by the properties of periodicity, that is, by the three conditions $(A)$, $(B)$, $(C)$.

# LECTURE 6

Bohr's theory of the hydrogen atom—Relativity effect and fine structure—Stark and Zeeman effects.

After these general considerations we now take up the applications to the theory of atomic structure. As you know it was with the *hydrogen atom* that Bohr first developed his ideas. We have in this case one nucleus and one electron, that is, a two-body problem which can be reduced, as you know, to a onebody problem: the motion of a point around a fixed center of attraction. If $r$, $\phi$, and $\theta$ are the polar coördinates of the electron relative to the nucleus, and if we place

$$\frac{1}{\mu} = \frac{1}{M} + \frac{1}{m}$$

where $M$ is the mass of the nucleus, and $m$ that of the electron, we have

$$H = \frac{\mu}{2}\left(\dot{r}^2 + r^2\dot{\theta}^2 + r^2\dot{\phi}^2\sin^2\theta\right) + U(r).$$

The potential energy of the Coulomb force between a nucleus carrying a $Z$-fold charge and an electron is

$$U(r) = -\frac{Ze^2}{r}$$

but we shall also consider general central forces with an arbitrary function $U(r)$.

Introducing the momenta we obtain

$$H = \frac{1}{2\mu}\left(p_r^2 + \frac{1}{r^2}p_\theta^2 + \frac{p_\phi^2}{r^2\sin^2\theta}\right) + U(r). \tag{1}$$

The corresponding Hamilton-Jacobi differential equation can be easily solved by separation of variables. In the case of Coulomb's law

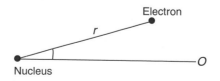

Electron

$r$

Nucleus

$O$

FIG. 5

875

we obtain the well-known Keplerian motions; of these only periodic orbits, i.e., ellipses, come into consideration in the quantum theory. It is seen at once that the motion is doubly degenerate for it has three degrees of freedom but is only simple-periodic. There is therefore only one action quantity $I$ and one quantum condition. Calculation shows that $I$ is related to the major axis $a$ of the ellipse by the formula

$$a = \frac{I^2}{4\pi^2 \mu e^2 Z}$$

and for the energy we obtain

$$W = \frac{2\pi^2 \mu c^4 Z^2}{I^2}. \tag{2}$$

Referred to a system of axes directed along the axes of the ellipse, the motion is represented by simple Fourier series

$$\frac{x}{a} = -\frac{3}{2}\varepsilon + \sum_{\tau=1}^{\infty} C_\tau(\varepsilon)\cos 2\pi w\tau$$

$$\frac{y}{a} = \sum_{\tau=1}^{\infty} D_\tau(\varepsilon)\sin 2\pi w\tau \tag{3}$$

the coefficients of which are continuous functions of the eccentricity $\varepsilon$. The angle variable $w$ is, except for the factor $2\pi$, the "mean anomaly" of astronomers.

These were the starting formulas for Bohr's theory of the hydrogen atom. Placing

$$I = nh$$

and

$$R = \frac{2\pi^2 \mu c^4}{h^3} \tag{4}$$

he found

$$W = -\frac{RhZ^2}{n^2} \tag{5}$$

and obtained for the frequencies of the emitted light

$$\nu = \frac{1}{h}(W_1 - W_2) = RZ^2\left(\frac{1}{n_2^2} - \frac{1}{n_1^2}\right). \tag{6}$$

For the hydrogen atom $Z = 1$ and this formula gives in fact all the known lines of hydrogen, in particular the Balmer series ($n_2 = 2$),

$$v = R_H \left( \frac{1}{4} - \frac{1}{n_1^2} \right) \quad n_1 = 3, \ 4, \ 5 \cdots .$$

The formula gives not only the dependence on $n_1$ but, what is more important, the correct value of $R_H$. To calculate the latter we have to replace $\mu$ by the expression

$$\mu = \frac{mM}{m + M} = m \frac{1}{1 + \dfrac{m}{M}} .$$

We may therefore write

$$R_H = R_\infty \frac{1}{1 + \dfrac{m}{M}} , \quad R_\infty = \frac{2\pi^2 m e^4}{h^3} = 3.28 \times 10^{15} \ \text{sec.}^{-1}$$

in which $e$, $m$ and $h$ are replaced by the best experimental values. Neglecting the small fraction $m/M$, which is about $1/1830$, we obtain, dividing by the velocity of light, $c = 3 \times 10^{10}$ cm./sec.,

$$\frac{R_H}{c} = \frac{3.28 \times 10^{15}}{c} = 1.09 \times 10^5 \ \text{cm.}^{-1}$$

while spectroscopic measurements give $109678$ cm.$^{-1}$.

The series given by $n_2 = 1$, $n_2 = 2$, $n_2 = 3$, $n_2 = 4$, $n_2 = 5$, have also been measured (by Lyman, Paschen, Brackett). Moreover Bohr was justified in maintaining that the series which is obtained by putting $Z = 2$, and which had until then been ascribed to hydrogen, must belong to ionized helium,

$$v = 4R_{H_e} \left( \frac{1}{n_2^2} - \frac{1}{n_1^2} \right) = R_{H_e} \left( \frac{1}{\left( \dfrac{n_2}{2} \right)^2} - \frac{1}{\left( \dfrac{n_2}{2} \right)^2} \right) .$$

The fraction $m/M$ is now four times smaller than for the H-atom, because the He-atom is four times heavier. Therefore, the lines for same $n_1$ and $n_2$ do not exactly coincide with the hydrogen lines. This separation is observed experimentally and now we are certain that the spectrum is that of ionized helium, to be sure the most beautiful result of Bohr's theory.

Bohr's theory of all other spectra may be briefly described as an attempt to consider them as modifications of the hydrogen spectrum. Two lines of attack are to be distinguished here; the first is to calculate the influence of secondary effects on the hydrogen atom: The dependence of mass on velocity is taken into consideration and gives the fine structure of the lines, then the influence of external electric and magnetic fields (Stark and Zeeman effects). The second line of attack leads to the study of other atoms and, together with it, to a theoretical systematic study of relations among the atoms and of the periodic system of the elements. Let me speak about the first line of attack.

Sommerfeld was the first to point out and carry through the idea that the variation of mass demanded by the *theory of relativity* must have an effect on the spectrum. He replaced the classical energy-function by the relativistic one given in the first lecture:

$$H = m_0 c^2 \left[ \sqrt{1 + \frac{p^2}{m_0 c^2}} - 1 \right] - \frac{e^2 Z}{r}. \tag{7}$$

On account of the smallness of the effect it is sufficient to take into account the first term in the expansion in powers of $\frac{p^2}{m_0 c^2}$ and write:

$$H = H_0 + H_1$$

where $H_0$ is the classical energy-function and

$$H_1 = -\frac{1}{8 m_0^3 c^2} \left( p_x^2 + p_y^2 + p_z^2 \right)^2$$

is the perturbation function.

The law of areas holds also in relativistic mechanics. Therefore the orbit is plane, but the plane orbit is no longer a simple periodic ellipse, but is transformed into a "rose-shaped" figure. The motion can also be described as an elliptic motion the major axis of which is rotating uniformly. The law of precession of the perihelion is found, following the method of secular perturbations, by taking the average

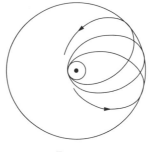

FIG. 6

of the function $H_1$ over the unperturbed motion

$$\bar{H}_1 = -\frac{RhZ^2}{n^2}\frac{\alpha^2 Z^2}{n^2}\left(\frac{I}{I'} - \frac{3}{4}\right) \tag{8}$$

where

$$\alpha = \frac{2\pi e^2}{hc} = 7.29 \times 10^{-3}$$

is a numerical constant, $I'$ is the angle variable conjugate to the azimuth $w'$ of the major axis and depends on the eccentricity $\varepsilon$ by the simple relation

$$I' = I\sqrt{1 - \varepsilon^2}.$$

Since $w'$ does not appear in $H_1$, therefore it is a cyclic variable and we have the new quantum condition

$$I' = kh. \tag{9}$$

$k$ is called the azimuthal quantum number to distinguish it from the main quantum number $n$. $k$ is always less than or equal to $n$. The total energy becomes

$$W = -\frac{RhZ^2}{n^2}\left[1 + \frac{\alpha^2 Z^2}{n^2}\left(\frac{n}{k} - \frac{3}{4}\right)\right]. \tag{10}$$

This formula expresses that every term of the unperturbed spectrum is separated into a number of terms which correspond to the values $k = 1, 2 \cdots n$. From this arises a splitting of the spectral lines,

$$\nu = \frac{1}{h}\left[W(n_1, k_1) - W(n_2, k_2)\right]$$

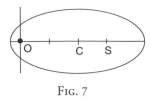

FIG. 7

in such a way that $k$ is changed only by $\pm 1$, for the rotation of the perihelion determined by $I' = kh$ is a simple harmonic motion. This split of spectral lines was predicted by Sommerfeld and experimentally verified for hydrogen and ionised helium, not only for the number of lines but also for the absolute value of the separation.* Kramers has also calculated the intensity of the lines by means of the principle of correspondence and found good agreement with observations.

The influence of an external electric field, i.e., the *Stark effect*, can be treated in quite an analogous way. The perturbation energy is in this case,

$$H_1 = eEz \tag{11}$$

where $z$ is the coördinate of the electron along the $z$-axis taken parallel to the field $E$. It is therefore simply required to calculate the average of $z$. This depends not only on the position of the major axis of the ellipse in the orbital plane, but also on the orientation of this plane in space. It can be shown, however, that the problem of the secular perturbation can be reduced to one of one degree of freedom. The calculation gives for the energy,

$$W = -\frac{RhZ^2}{n^2} \pm \frac{3Eh^2}{8\pi^2 \mu e Z} n n_e \tag{12}$$

where $n_e$ is a new quantum number which varies between $-(n - 1)$ and $(n - 1)$. The motion itself can be described in the following way. If we calculate the electric "center of gravity" $S$ of the electron in its orbit, that is, the average value of its coördinates over one revolution, we find that it is on the major axis at a distance $\frac{3}{2}a\varepsilon$ from the nucleus $O$ in the direction towards the aphelion. On account of

---

*For hydrogen the quantitative results are not yet quite certain.

secular perturbations, this point moves with simple harmonic motion in a plane perpendicular to the field $E$, whence it follows that $n_e$ can only change by $\pm 1$. The split of spectral lines is completely determined hereby, in good agreement with the results of experiment, also as regards the intensities calculated by Kramers.

For the *Zeeman effect* the calculation is still simpler and more-over can be carried out for atoms with an arbitrary number of electrons. The expression given earlier for the energy in a magnetic field is, neglecting terms containing the square of the field strength (Equation (14), Lecture 2):

$$H = H_0 + \frac{e}{c\mu} \sum A \cdot p \tag{13}$$

where $H_0$ is the energy of the unperturbed system. The Vector potential of a homogeneous field is

$$A = \frac{1}{2} H \times r.$$

Therefore

$$\sum A \cdot p = \frac{1}{2} \sum H \times r \cdot p = \frac{1}{2} H \sum r \times p = \frac{1}{2} |H| P_\phi$$

where $P_\phi$ is the component of angular momentum $\mathbf{P} = \sum r \times p$ parallel to the field. For the unperturbed system the angular momentum is constant in magnitude $|\mathbf{P}|$ and direction. It is easy to see that $2\pi |\mathbf{P}|$ is an action integral. We place therefore

$$2\pi |\mathbf{P}| = jh. \tag{14}$$

The components of $\mathbf{P}$ are also constant, but they are evidently conjugate to degenerate angle variables. In the magnetic field the degeneration of the angle $\phi$, which fixes the position of the plane determined by the field and the angular momentum with respect to a fixed plane parallel to the direction of the field, is removed and the system precesses around the direction of the field. $P_\phi$ is conjugate to $\phi$, as easily seen. We have therefore the new quantum condition

$$2\pi P_\phi = mh. \tag{15}$$

If $\alpha$ is the angle between the angular momentum and the direction of the field, then evidently

$$\cos \alpha = \frac{P_\phi}{|\mathbf{P}|} = \frac{m}{j}.$$

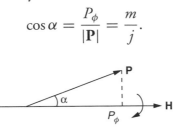

FIG. 8

The axis of angular momentum can therefore only take $2j + 1$ different directions ($m = -j, \cdots + j$) with respect to the direction of the field axis. We shall call this result, following Sommerfeld, "directional quantization."

The energy is

$$H = W_0 \pm \frac{eh}{4\pi\mu c} |\mathbf{H}| \, m \tag{16}$$

whence the number of revolutions of the axis of angular momentum, the so-called "Larmor frequency," is

$$v_m = \frac{\partial H}{\partial \, 2\pi P_\phi} = \frac{1}{h} \frac{\partial H}{\partial m} = \frac{e \, |\mathbf{H}|}{4\pi\mu c} = 4.70 \times 10^{-5} \, |\mathbf{H}| \, \mathrm{cm.}^{-1}. \tag{17}$$

Precession does not influence the components of the motion of the electrons in the direction parallel to the field. Therefore there is no additional term in the $z$-component of the electric moment and light oscillating parallel to $z$ corresponds to jumps for which $m$ does not change. The components of motion perpendicular to the field are altered however by simple rotations in one direction or the other, hence the emitted light must be decomposed into two circularly polarized waves in opposite directions, to which correspond the jumps

$$m \to m \pm 1.$$

We obtain therefore the classical Zeeman triplet without any change. This contradicts experiment, however, for in most cases spectral lines are split up in a much more complicated way. Bohr's theory in its present form gives no explanation of this more complicated effect. According to it we should expect in all cases and for every atom normal Larmor precession and the normal spectral triplet. At this point many attempts have been made to change the theory. Starting from Sommerfeld's researches, Landé has succeeded in decomposing the observed Zeeman separation of most spectral lines into terms and discovered their relation to the periodic system of the elements. Heisenberg, Pauli and many others have investigated this problem further. The essential result of all these investigations is that the so-called "abnormal" Zeeman effect—which is, however, certainly the normal case—finds no place in the semi-classic theory which we have developed here.

The positive result is that the Zeeman effect is closely connected with the construction of atoms out of electrons describing orbits to which correspond fixed quantum numbers. We shall now treat this problem of the arrangement of electronic orbits in the atom, following the method of Bohr, who considered the series spectra as modifications of the hydrogen spectrum.

# LECTURE 7

Attempts towards a theory of the helium atom and reasons for their failure—Bohr's semi-empirical theory of the structure of higher atoms—The optical electron and the Rydberg-Ritz formula for spectral series—The classification of series—The main quantum numbers of the alkali atoms in the unexcited state.

The most obvious way of finding an exact theory of atomic structure would be to consider successively the simplest atoms, helium, lithium, etc., following hydrogen in the series of the elements. This has been tried, but even the first step from the hydrogen to the helium atom proved unsuccessful. The helium atom is an instance of the three-body problem: one nucleus and two electrons. It is well known that the three-body problem has greatly perplexed astronomers and that it has not been possible to represent the motion by analytical expressions (series) which really permit a survey of the motion at all times. In the case of atomic structure, conditions are even less favorable, for in celestial mechanics there is at least the advantage that the attraction towards the central body is much greater than the other attractions on account of the preponderant mass of the sun, so that all these other attractions can be looked upon as small "perturbations." In atomic mechanics, however, all the attractions and repulsions of electric charges are of the same order of magnitude. On the other hand, the atomic problem has an advantage of a different kind, precisely on account of the postulate of the quantum theory, that only certain "stationary" orbits come into consideration. It has been shown that the quantum conditions allow only very simple types of orbits, because they exclude certain librations (oscillations).

Based on this result, attempts have been made to find the stationary orbits for the helium atom, and calculate its energy levels. The line of attack has been along two directions: Some investigators have considered the normal state of the helium atom (Bohr, Kramers, van Vleck), others the excited state, in which one electron is in the nearest orbit to the nucleus and the other revolves in a very distant orbit (van Vleck, Born and Heisenberg). Both calculations give incorrect results: The calculated energy of the normal state does not agree with experimental results (ionization energy of the normal helium atom), and the calculated term system for the excited states is different from that observed, qualitatively as well as quantitatively.

After all, no other result could be expected, for the validity of the frequency condition is sufficient to show conclusively that in the realm of atomic processes the laws of classical theories (geometry, kinematics or mechanics, electrodynamics) are not right. That in certain simple cases, as for a single electron, they give partially correct results is, in fact, more astonishing than that they fail in the more complicated cases of several electrons. This failure of the theory in the case of interactions among several electrons is evidently connected with the following fact: We know that electrons react quite unclassically to light waves, because the latter produce quantum jumps. In a system made up of several electrons, each electron is in the oscillating field due to all other electrons and the periods of these fields are of the same order of magnitude as those of light waves, therefore we have no reason to expect that the electron should react classically to this oscillating field. This point of view gives grounds for understanding why we obtain, by the classical theory, correct results in many cases of the one-electron problem.

Appreciating these difficulties, Bohr has given up the attempt to construct a truly deductive theory, and, instead, has endeavored, with the greatest success, to discover, by interpretation of facts, above all of the facts relating to the spectra and the chemical and magnetic properties of the atoms, something about the arrangement of the electrons. The starting point was the observation of the fact that the spectra of certain atoms are of a type quite similar to the hydrogen spectrum. The lines, or better, the terms form series quite similar to the series of terms $\frac{R}{n^2}$ of the H-atom. Rydberg, for instance, showed that in many cases expressions of the form $\frac{R}{(n+\delta)^2}$ with $\delta =$ constant, are sufficient for expressing the terms. This is the case for the alkali metals, for part of the lines of Cu, Ag, Au, and for other similar cases, all of which share chemical properties which indicate the easy detachment of one electron. From this Bohr concludes that all these spectra, as that of hydrogen, are produced by the jumps of *one* electron, the "optical electron." This electron, however, does not move around a simple nucleus, but around a *core* consisting of the nucleus and all the remaining electrons. If the optical electron is more and more strongly excited—that is, brought up to levels of higher energy—the state of total separation called ionization is gradually reached and then the core is left as an "ion." This argument agrees with the results of the chemists, as formulated by Lewis, Langmuir and Kossel. According to these theories the ions of the alkali metals have the same structure as the atoms of the neighboring inert gases: The latter are the most stable closed electron configurations.

Now it can be shown that the orbit of the optical electron in the lower stationary states must penetrate into the core, for otherwise the terms would differ but little from those of the H-atom. Moreover, the radii of the ions are fairly well known from the theory of electrolytes and of polar crystals and it can be estimated, by a method to be given shortly, that the orbit of the optical electron must go through the core (Schrödinger, Bohr).

In postulating such "penetrating orbits," a step is taken which is incompatible with ordinary mechanics, for following our quantization rules we must assume that the orbit of the optical electron is exactly periodic, but this cannot be understood from the standpoint of mechanics because of the intensive interaction with the inner electrons. It would be necessary to assume that the whole electronic structure is rigorously periodic in every quantized state and it is very questionable whether there exist such solutions of the mechanical equations. If in spite of this we wish to describe the paths of the optical electron within the realm of our theory and more or less approximately, this may be done following Bohr and in a purely formal way, by replacing the action of the core on the electron by a central force and neglecting altogether the reaction of the electron on the atomic core. Then the conservation of energy is always satisfied for the electron alone and we have to do, as hitherto, with a one-body problem. Conservation of angular momentum also holds and the orbit is plane.

We can show, following Bohr, that the terms must be approximately expressible by formulas of Rydberg type or more accurately of Ritz type, assuming that the core is small compared with the size of the orbit of the optical electron. The outer part of the orbit differs only slightly from a Keplerian ellipse, while the inner part is an oval of small radius of curvature, because there the electron comes in a region of strong nuclear attraction (Fig. 9).

If we replace the outer part of the orbit by an ellipse, then its energy becomes

$$W = -\frac{RhZ^{\times 2}}{n^{\times 2}}. \tag{1}$$

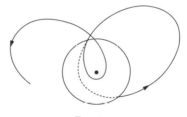

FIG. 9

In this formula $n^{\times}$ depends on the aphelion distance $2\,a^{\times}$, in the same way as was given above for other quantized orbits, i.e.

$$a^{\times} = \frac{h^2}{4\pi^2 \mu e^2 Z^{\times}} n^{\times 2} \tag{2}$$

and $Z^{\times}$ is the "effective nuclear charge," that is the difference between the charge of the nucleus and of the screening electrons of the ion. $n^{\times}$ need not be an integer, for it is not the main quantum number of the whole orbit; if we call the latter $n$ then the frequency of the motion from one aphelion to the next is given by

$$\nu = \frac{\partial W}{\partial I} = \frac{1}{h}\frac{\partial W}{\partial n}. \tag{3}$$

On account of our assumption that the core is small, the time of revolution $\frac{1}{\nu}$ the whole orbit differs slightly from the time of revolution $\frac{1}{\nu^\times}$ of the ellipse replacing the actual path. The latter is given by

$$\nu^\times = \frac{1}{h}\frac{\partial W}{\partial n^\times} = \frac{2\,RZ^{\times 2}}{n^{\times 3}}. \tag{4}$$

We therefore place

$$\frac{1}{\nu} = \frac{1}{\nu^\times} + b \quad \text{or} \quad \frac{\nu^\times}{\nu} = 1 + b\nu^\times = 1 + \frac{2\,RbZ^{\times 2}}{n^{\times 3}}$$

and consider $b$ to be approximately constant; then we have

$$\frac{dn}{dn^\times} = \frac{\nu^\times}{\nu} = 1 + \frac{2\,RbZ^\times}{n^{\times 3}}$$

which integrated gives

$$n = n^\times - \delta_1 - \frac{\delta_2}{n^{\times 2}}$$

and solving approximately for $n^\times$

$$n^\times = n + \delta_1 + \frac{\delta_2}{n^2} + \cdots, \tag{5}$$

where $\delta_1$ is an integration constant. $\delta_2 = RbZ^{\times 2}$ is determined by the mechanical system. Of course $\delta_1$ depends also on the second quantum number of the system, for the motion in a central field is double-periodic; one of the periods is the one which has been already considered, namely the motion of the electron from perihelion to perihelion with the main quantum number $n$; the second is that of the revolution of the perihelion itself with the quantum number $k$; $hk$ is therefore the angular momentum of the electron and as the rotation of the perihelion is simple-periodic therefore $k$ can change only by $\pm 1$, as in the case of the relativity correction for the H-atom. $\delta_1$ is a function of $k$; this can also be approximately found by means of relatively simple considerations.

The expression which we have found in this way for the value of the term

$$\frac{W}{h} = -\frac{RZ^{\times 2}}{(n + \delta_1(k) + \delta_1/n^2 + \cdots)^2} \tag{6}$$

agrees exactly with the term formulas found empirically by Rydberg ($\delta_1$-term) and Ritz ($\delta_1$- and $\delta_2$-terms).

Since $k$ can have different values, every atom has several series of terms. In fact we should expect, on account of the selection rule for $k$ ($k \rightarrow k \pm 1$), that the latter can be so classified that a term of a series can be combined only with the terms of neighboring series. This indeed is the case. It is usual to classify the terms in series according to the following scheme:

$$1s \quad 2s \quad 3s \quad 4s \quad 5s \cdots$$
$$2p \quad 3p \quad 4p \quad 5p \cdots$$
$$3d \quad 4d \quad 5d \cdots$$
$$4f \quad 5f \cdots$$

where an $s$-term can be combined only with a $p$-term, a $p$-term with $s$- and $d$-terms, etc. From this we conclude with Sommerfeld that the following correspondence holds:

$$s \quad p \quad d \quad f$$
$$k = 1 \quad 2 \quad 3 \quad 4$$

We shall now proceed to determine the main quantum numbers for all the observed terms. For this purpose we must above all determine whether the path in question is a penetrating orbit. We calculate from the observed term $\frac{W}{h}$ the effective quantum number $n^\times$ in accordance with the formula

$$n^\times = Z^\times \sqrt{\frac{Rh}{W}}.$$

The aphelion distance then is known, that is the major axis $2\,a^\times$ of the ellipse replacing the path. Moreover the parameter $2\,P$ of this ellipse is known. This parameter depends on the value of $k$ as shown by the formula

$$P = \frac{h^2}{4\pi^2 \mu e^2 Z^\times} k^2.$$

Therefore, the whole equivalent ellipse is known approximately and it is possible to determine whether it penetrates into the atomic core, the size of which is known from the ionic volume. If in this way the conclusion is reached that the path is wholly outside the core, then the Rydberg correction $\delta_1$ is small, that is $n^\times$ is nearly an integer. If such is the case then $n$ can be chosen as the next integer to $n^\times$. In fact all the terms corresponding to exterior paths $d(k = 3)$, $f(k = 4) \ldots$ behave in this manner.

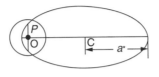

FIG. 10

On the other hand the $s$-terms ($k = 1$) and the $p$-terms ($k = 2$) correspond in general to penetrating orbits. Here $n^\times$ departs considerably from integral values. $\delta_1(k)$ is then quite large, frequently larger than 1 or 2. The actual determination of $\delta^1$ requires approximation formulas, for the derivation of which fairly rough assumptions are sufficient. In every case the main quantum number $n$ can be determined with fair certainty.

The main quantum number of the normal state is thus of the greatest interest. The most important result can be stated as follows: For every alkali atom (hydrogen included) the main quantum number of the normal state of the optical electron is increased by 1:

$$\begin{array}{ccccccc} \text{H} & \text{Li} & \text{Na} & \text{K} & \text{Rb} & \text{Cs} \cdots \\ n = 1 & 2 & 3 & 4 & 5 & 6 \ldots \end{array}$$

# LECTURE 8

Bohr's principle of successive building of atoms—Arc and spark spectra—X-ray spectra—Bohr's table of the completed numbers of electrons in the stationary states.

Bohr's construction of the periodic system is based on the supposition that every atom can be derived by the addition of one electron to an ion which is constructed essentially as the previous atom and has the same number of electrons. On this depends the possibility of deriving the structure of one atom from that of the previous one. It is first assumed that the core of the second atom has the same structure as that of the first atom and then, on the basis of a simple estimate of the Rydberg constants, it is seen whether the spectrum is not in contradiction with their value. In many cases we also know the spark spectrum, that is the spectrum of the ionized atom, which is considered as produced by one optical electron rotating around a core of a structure similar to that of the second previous atom having the same number of electrons. We understand from this the so-called "spectroscopic displacement law" given by Sommerfeld and Kossel. The structure of the spectrum of a neutral atom (often called "arc spectrum" because of the most convenient means for its production) resembles the first spark spectrum of the next higher atom, the second spark spectrum of the following atom and so on; except that Rydberg's constant $R$ must be replaced by $4 R$, $9 R$, ... or generally $Z^{\times 2} R$. We have already used the simplest example of this rule, where the correspondence of the spectra is quite exact, when we spoke of the spectra of H, $He^+$, $Li^{++}$, ... together by introducing an arbitrary nuclear charge $Z$.

An electronic configuration once formed is buried more and more deeply inside the atom as it proceeds along the periodic system of the elements. Now the X-ray spectra furnish means of examining the inner parts of the atom. The production of these spectra depends, according to Kossel, on the following process: As all the quantum orbits are, so to speak, full, it is impossible for an electron to jump from one orbit to another. It is necessary that an electron be previously removed by supplying energy (electron impact or absorption of X-rays). Then other electrons may fall from higher orbits into the gaps left free and in this way the emission lines of the X-ray spectrum are produced. According to whether the removed electron had the main quantum number $n = 1, 2, 3 \ldots$ we name the line emitted when this electron is replaced, a $K, L, M \ldots$ line; and according to the origin of the substituting electron we indicate the line by indices $K_\alpha$, $K_\beta \ldots L_\alpha$, $L_\beta \ldots$ or by new quantum numbers. The correctness of this conception can be tested by observing that for the X-ray lines, the Ritz combination principle must hold. Of course the energy values on the differences of which the frequencies depend are directly given by the so-called *absorption limits*. In the spectrum of absorption of an atom there must exist sharp limits or "edges" which separate the frequencies the energy-quanta $h\nu$ of which are greater or less than the work necessary to remove to infinity the electron describing the orbit which is responsible for the absorption. In this way the system of X-ray terms is determined as exactly as that of the optical spectra.

If we consider the X-ray terms as functions of the atomic numbers $Z$ we obtain in general smooth curves, first discovered by Moseley and Darwin. Only at places where any irregularity in the introduction of electrons occurs are there slight kinks. In this way we can verify the arrangement of the electrons derived from the study of optical spectra. The chief result of this discussion of observed spectra is the following: It is not at all true that all electrons are first introduced in the orbits $n = 1$, then in those for which $n = 2$, $n = 3$, and so on, but on the contrary it is possible that, with electrons already in the orbits $n = 4$, new electrons of higher azimuthal quantum number $k$ fill up an inner shell, for instance $n = 3$. This can be deduced partly from spectroscopic, partly from chemical evidence. If two neighboring elements differ only in that the number of inner electrons, for instance those for which $n = 3$, differ by one, while the number of outer electrons, i.e. $n = 4$, remains constant (for instance, two), then we should expect that these elements are chemically very similar. We have such groups of similar elements in the fourth period, in the group Sc, Ti, . . . Ni, which have the common property of paramagnetism or ferromagnetism. Even more remarkable are the rare earths which are similar in every respect. This is seen in the following presentation of the periodic system of the elements:

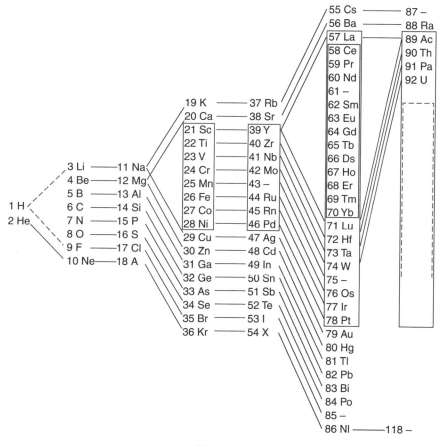

FIG. 11

As a result of all these considerations we give Bohr's table for the arrangement of the electrons:

| | $1_1$ | $2_1\,2_2$ | $3_1\,3_2\,3_3$ | $4_1\,4_2\,4_3\,4_4$ | $5_1\,5_2\,5_3\,5_4\,5_5$ | $6_1\,6_2\,6_3\,6_4\,6_5\,6_6$ | $7_1\,7_2$ |
|---|---|---|---|---|---|---|---|
| 1 H | 1 | | | | | | |
| 2 He | 2 | | | | | | |
| 3 Li | 2 | 1 | | | | | |
| 4 Be | 2 | 2 | | | | | |
| 5 B | 2 | 2 1 | | | | | |
| 6 C | 2 | 2 (2) | | | | | |
| -- | - | -- | | | | | |
| 10 Ne | 2 | 8 | | | | | |
| 11 Na | 2 | 8 | 1 | | | | |
| 12 Mg | 2 | 8 | 2 | | | | |
| 13 Al | 2 | 8 | 2 1 | | | | |
| 14 Si | 2 | 8 | 2 (2) | | | | |
| -- | - | -- | -- | | | | |
| 18 A | 2 | 8 | 8 | | | | |
| 19 K | 2 | 8 | 8 | 1 | | | |
| 20 Ca | 2 | 8 | 8 | 2 | | | |
| 21 Sc | 2 | 8 | 8 1 | (2) | | | |
| 22 Ti | 2 | 8 | 8 2 | (2) | | | |
| -- | - | | -- | - | | | |
| 29 Cu | 2 | 8 | 18 | 1 | | | |
| 30 Zn | 2 | 8 | 18 | 2 | | | |
| 31 Ga | 2 | 8 | 18 | 2 1 | | | |
| -- | - | -- | -- | -- | | | |
| 36 Kr | 2 | 8 | 18 | 8 | | | |
| 37 Rb | 2 | 8 | 18 | 8 | 1 | | |
| 38 Sr | 2 | 8 | 18 | 8 | 2 | | |
| 39 Y | 2 | 8 | 18 | 8 1 | (2) | | |
| 40 Zr | 2 | 8 | 18 | 8 2 | (2) | | |
| -- | - | -- | -- | -- | - | | |
| 47 Ag | 2 | 8 | 18 | 18 | 1 | | |
| 48 Cd | 2 | 8 | 18 | 18 | 2 | | |
| 49 In | 2 | 8 | 18 | 18 | 2 1 | | |
| -- | - | -- | -- | -- | -- | | |
| 54 X | 8 | 8 | 18 | 18 | 8 | | |
| 55 Cs | 2 | 8 | 18 | 18 | 8 | 1 | |
| 56 Ba | 2 | 8 | 18 | 18 | 8 | 2 | |
| 57 La | 2 | 8 | 18 | 18 | 8 1 | (2) | |
| 58 Ce | 2 | 8 | 18 | 18 1 | 8 1 | (2) | |
| 59 Pr | 2 | 8 | 18 | 18 2 | 8 1 | (2) | |
| -- | - | -- | -- | -- | -- | | |
| 71 Cp | 2 | 8 | 18 | 32 | 8 1 | (2) | |
| 72 Hf | 2 | 8 | 18 | 32 | 8 2 | (2) | |
| -- | - | -- | -- | -- | -- | - | |
| 79 Au | 2 | 8 | 18 | 32 | 18 | 1 | |
| 80 Hg | 2 | 8 | 18 | 32 | 18 | 2 | |
| 81 Tl | 2 | 8 | 18 | 32 | 18 | 2 1 | |
| -- | - | -- | -- | -- | -- | -- | |
| 86 Nt | 2 | 8 | 18 | 32 | 18 | 8 | |
| 87 – | 2 | 8 | 18 | 32 | 18 | 8 | 1 |
| 88 Ra | 2 | 8 | 18 | 32 | 18 | 8 | 2 |
| 89 Ac | 2 | 8 | 18 | 32 | 18 | 8 1 | (2) |
| 90 Th | 2 | 8 | 18 | 32 | 18 | 8 2 | (2) |
| -- | - | -- | -- | -- | -- | -- | - |
| -- | - | -- | -- | -- | -- | -- | -- |
| 118 – | 2 | 8 | 18 | 32 | 32 | 18 | 8 |

Fig. 12

We see here how the electronic shell $n = 3$, $k = 1, 2$, which was completed with 8 electrons with Sc ($Z = 21$) begins to increase again for $n = 3$, $k = 3$. The same occurs with Y ($Z = 39$) for $n = 4$, and with La ($Z = 57$) for $n = 5$.

The X-ray spectra confirm the assumption that internal changes begin with these elements. There are obvious kinks in the curves

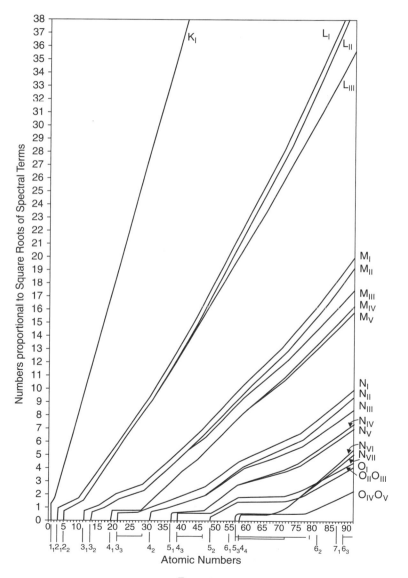

FIG. 13

expressing the relation between X-ray terms and atomic number, which otherwise are quite smooth, for the elements of atomic number $Z = 21, 39, 57$ (Fig. 13).

We may therefore assume that Bohr's arrangement of electronic shells is correct, at any rate as far as the numbers $n$, $k$ are concerned.

We know nothing about the dynamic mechanism which results in these simple laws. Above all it cannot be explained mechanically why a certain group with a certain main quantum number $n$ is "filled" by a certain limited number of electrons, first 2, then 8, then 18, or why the sub-groups defined by $k$ can also take only a definite number of electrons.

# LECTURE 9

Sommerfeld's inner quantum numbers—Attempts toward their interpretation by means of the atomic angular momentum—Breakdown of the classical theory—Formal interpretation of spectral regularities—Stoner's definition of subgroups in the periodic system—Pauli's introduction of four quantum numbers for the electron—Pauli's principle of unequal quantum numbers—Report on the development of the formal theory.

The view of the atom just described has recently led much further in the investigation of the so-called multiplets. Many spectral lines which we have considered here as though they were simple are in fact multiple. For instance, the $D$-line of sodium is double. Sommerfeld first resolved these lines into terms by introducing a new *inner quantum number j* and giving a selection rule for this number. The possibility of a third quantum number of the optical electron is indicated by the fact that it has *three* degrees of freedom: It need only be supposed that the core is not spherically symmetrical but only symmetrical about an axis. Then the optical electron no longer moves in a central field and therefore the orbit is no longer plane, but, to a first approximation, the motion can be described thus: Assume that the orbit is plane for a single revolution and has the angular momentum $k$. Then this orbit together with the axis of the atomic core, regarded as a rigid system, is endowed with a precession of angular momentum $R$ around the total momentum $J$ considered fixed in space. $K, R, J$, as easily shown, are action variables conjugate to the corresponding angles of rotation. We place, therefore,

$$K = kh, \quad R = rh, \quad J = jh,$$

where $k$ is the azimuthal quantum number, already introduced above, of the optical electron in its orbit. The quantum number $r$ characterizes the constitution of the core, for, given $r$ and $k$, $j$ cannot have any value, but only those between $|k - r|$ and $|k + r|$. Also, $j$, as precessional momentum, can only make jumps

$$j \to j \begin{array}{l} \nearrow j - 1 \\ \phantom{j} j \\ \searrow j + 1 \end{array}$$

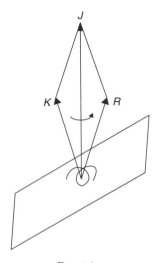

FIG. 14

where the jumps $j \to j \pm 1$ correspond to oscillations of the electric moment perpendicular to the $J$-axis and $j \to j$ corresponds to oscillations parallel to the $J$-axis.

We have thus the possibility of explaining multiplets, and the selection rule found empirically by Sommerfeld for the inner quantum number agrees with that found theoretically; but the number of components for given $k$ and $r$ is not verified by experiments. For instance, we are inclined to ascribe to the inert gases, which are certainly highly symmetrical, the angular momentum zero and hence the same to the core of the alkali atoms. But then they should show no separation. If we assume for the inert gases $r = 1$, the values of $j$ lie between $k - 1$ and $k + 1$, therefore $j = k - 1$, $j = k$ and $j = k + 1$. But the alkali atoms have *no* triplets. In the $s$-states ($k = 1$) they have single lines and in all other states ($k = 2, 3 \ldots$) doublets.

This constitutes a violation of Bohr's principle of selection. The number of possible states of a system consisting of an ion to which an electron has been added is *not* equal to the product of the number of the states of the ion by the number of possible electronic orbits, but one less. Bohr calls this a "non-mechanical constraint" and has repeatedly emphasized that this is a most important deviation from mechanical laws. The diffculty has been overcome formally by introducing half-quantum numbers $\ldots -\frac{3}{2}, -\frac{1}{2}, +\frac{1}{2}, +\frac{3}{2}, \ldots$ and a cabalistic rule.

If a mechanically "logical" system leads to the quantum numbers

$$-3, -2, -1, 0, +1, +2, +3$$

then, by this rule, we replace this series by the one given below:

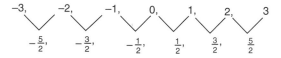

in which the number of terms is one less. In this way the doublets of the alkalies are "explained." The three positions of the core of inert gas type $(r = 1)$ with respect to the electronic orbits give only 2 $j$-values,

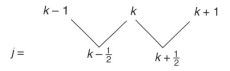

The absolute value of $j$ is of course still arbitrary. Instead of $j = k - \frac{1}{2}, k + \frac{1}{2}$ we can write $j = k - 1, k$, or choose any other normalization which is convenient for the purpose on hand. Only the *number* of possible values of $j$ is important.

The anomalous Zeeman effect is quite similar. Here the classical theory gives for an atom with the total angular momentum $\frac{hj}{2\pi}$, $2j + 1$ orientations in a magnetic field, namely all the values of the magnetic quantum number $m$ between $-j$ and $j$. In fact, however, only $2j$ terms exist, corresponding to the scheme,

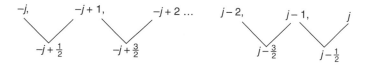

In this way important advances have been made in the systematization of spectra. We shall now review these briefly.

Stoner has given an important generalization of Bohr's theory of the periodic system. He recognized that the electrons in the closed inert gas configurations can be arranged in the following groups:

$$n \quad \textit{Number of entering electrons}$$

He......1    $2 = 2$
Ne......2    $8 = 2 + 2 + 4$
 A......3    $8 = 2 + 2 + 4$
Kr......4    $18 = 2 + 2 + 4 + 4 + 6$
Xe......5    $18 = 2 + 2 + 4 + 4 + 6$
Em......6    $\underline{32 = 2 + 2 + 4 + 4 + 6 + 6 + 8}$
         $j = 1 \quad \underbrace{1 \quad 2} \quad \underbrace{2 \quad 3} \quad \underbrace{3 \quad 4}$
         $k = 1 \qquad 2 \qquad\quad 3 \qquad\quad 4$

By dividing the completed number of electrons in each group by 2, the numbers labeled "quantum number $j$" are obtained. For every $j > 1$, two neighboring values of $j$ are combined to form a larger group with a "quantum number $k$" equal to the larger $j$. In this way a one-to-one correspondence is established between quantum numbers and electrons. Stoner was able to show that this development of Bohr's scheme leads to the explanation of many properties of atoms, especially of their spectra in the optical and X-ray regions.

To give only one example: The ionized carbon atom $C^+$ has a doublet spectrum, which has been analyzed by Fowler. The lines of this spectrum which correspond to jumps to the fundamental orbit give information about this orbit, that is about the normal state of the singly ionized carbon atom $C^+$. Bohr had originally attributed to the carbon atom four equivalent electrons with $n = 2, k = 1$ because chemical facts seemed to demand this equivalence. That would leave to the $C^+$-ion three equivalent electrons $n = 2, k = 1$. Stoner's scheme evidently gives for the carbon atom two electrons $n = 2, k = 1$ and two electrons $n = 2, k = 2$, of which the former are more strongly attached because they belong to ellipses. The ion $C^+$ therefore has two electrons $n = 2, k = 1$ and one electron $n = 2, k - 2$. The jumps of the latter give rise to the spectrum of $C^+$. The fundamental orbit must hence be shown by its combinations to be a *p*-term ($k = 2$). In fact observations verify this conclusion and not Bohr's original hypothesis. In the table on p. 57 this result has already been taken account of. We cannot enter into further details here.

Pauli showed Stoner's arrangement to be a consequence of a very general principle. He started from the assumption that spectra behave as though *four* quantum numbers belonged to each electron, that is, that beside the three numbers *n*, *k*, *j* used so far there exists still a fourth *m* which determines the magnetic separation. Until now *j* has been interpreted as the resultant of the angular momentum of the electron and the core. Pauli abandons this idea and ascribes all four quantum numbers to *one* electron. The difficulty is that the electron, according to our ordinary ideas, has only three degrees of freedom. It will be seen later that the newest development of the quantum theory seems to lead to a fourth degree of freedom, i. e. an axial rotation, for the electron. Let us forego for the moment a physical explanation and describe briefly Pauli's method. His quantum numbers are normalized somewhat differently from those introduced above. He employs *n* and *k* in the usual way, denoting however the latter by $k_1$, and using instead of *j* a number $k_2$ which can *always* have exactly two values $k_2 = k_1 - 1$ and $k_2 = k_1$. Each *single* electron behaves therefore as the optical electron of the alkalies and gives a doublet. The same must hold also for the magnetic quantum number *m*. For the alkalies *m* takes, according to observation, 2 $k_2$ different values—this can be shown from our cabalistic rule, if $k_2$ is interpreted as the total momentum. Therefore, all terms belonging to a given $k_1$ take in all $2(k_1 - 1) + 2k_1 = 2(2k_1 - 1)$ values. Pauli next observed that Bohr's method of building up atoms by successive steps can be kept in this way. The number of possible states is simply the sum of those of the core and of those of the newly-entering electron (permanence of quantum numbers). If we pass, for instance, from an alkali atom to the neighboring alkaline earth, then the doublet system of the first becomes a system of single and triplet terms. In the singlet system a state with given *n*, $k_1$ is decomposed into $1 \times (2k_1 - 1)$

terms, in the triplet system into $3(2k_1 - 1)$ terms. This has been interpreted up to the present as meaning that, in strong fields, there corresponds to the optical electron in spite of mechanics, $2k_1 - 1$ orientations in every case, while the core is oriented in the single terms along one direction, in the triplet terms along three directions. The latter contradicts the principle of permanence because the free alkali atom can have, in the unexcited state ($s$-state $k_1 = 1$), only two such orientations. The totality of the $4(2k_1 - 1)$ states of the atom can be interpreted as meaning that the core, as in the free alkali atom, can have two states and the optical electron, as in the alkalies, $2(2k_1 - 1)$ states. A corresponding explanation can be given in general, but we shall pass over all these details and consider now the connection between Pauli's ideas and Stoner's classification of the periodic system.

Pauli found, that the latter is equivalent to the following general principle: *It never happens that any two electrons in the atom have the same four quantum numbers $n$, $k_1$, $k_2$, $m$.* If $n$, $k_1$, $k_2$ are given, the number of possible values of $m$, as was seen above, is $2k_2$. Therefore, the greatest number of "equivalent" electrons, that is having the same $n$, $k_1$, $k_2$, is also $2k_2$, otherwise $m$ would be equal for two of these electrons. If the quantum number $k_2$ is identified with Stoner's $j$, which also takes for every $k$ (or $k_1$) the values $k$, $k - 1$, then Stoner's classification is shown to be a consequence of Pauli's principle. The object of the theory is therefore to understand Pauli's principle, that is either to derive it from the laws of quantum mechanics or to show that it belongs to indemonstrable basic postulates.

Further developments can be briefly described in the following way: According to Pauli electrons aggregate into systems while keeping their own quantum numbers. The energy of such a system, whether a core, a group of outer electrons or a complete atom, depends, however, only on a certain resultant of the quantum numbers of the individual electrons. A distinction must therefore be made between the quantum numbers of the individual electrons and the resultant quantum numbers of the electron groups. If this group is identical with the whole atom, the resulting quantum numbers fix the terms which determine the spectrum. The rules governing the formation of this resultant are mainly of empirical origin. The following is alone mentioned as a theoretical guiding principle. Paschen and Back have discovered that in strong magnetic fields the Zeeman components of a multiplet are displaced with respect to each other so that they correspond to the normal separation (Larmor frequency). This can be interpreted theoretically by assuming that the individual electrons move practically independently of one another in fields where the magnetic energy is much larger than that due to the interaction between the electrons. The electrons therefore precess with normal Larmor frequency. From this follows that in strong fields the magnetic quantum numbers behave additively and the construction of the resultant is referred to this idea.

The development of this conception was very much aided by the investigations of Russell and Saunders. It was already known, according to Götze, that lines appear which correspond to combinations of $p$-terms with other $p'$-terms, therefore having the same azimuthal quantum number. Bohr accounted for this by assuming a simultaneous jump of two electrons, whereby the simple harmonic character of the motion, from which we derive the selection rule, $k \rightarrow k \pm 1$, is lost. Russell and Saunders found that there exist negative $p'$-terms, which hence correspond to a state of the atom of higher energy than required for ionization. They were able to explain their observations by introducing a resultant quantum number for the electrons jumping simultaneously and then treating this system with respect to the core in the same way as the optical electron with respect to the core of the alkali atoms. This method was developed systematically by Heisenberg and applied to the practical interpretation of numerous spectra by Hund. The latter succeeded in analyzing completely, among others, the series of the magnetic atoms beginning with scandium and ending with the group iron, cobalt, nickel, deducing not only the completed numbers of the electron groups in the normal states, but also interpreting in a rough way the character of the spectra.

With this we have come to the limit which can be attained by the development of Bohr's fundamental ideas. There is material a plenty.

It is now time for the theorist to take the initiative again and lay the foundations of a real dynamics of atoms. Heisenberg found a short time ago the key to the gate, closed for such a long time, which kept us from the realm of atomic laws. In his brief paper, the leading physical ideas are clearly stated, but only exemplified on account of the lack of appropriate mathematical equipment. The required machinery Jordan and I have discovered in the matrix calculus. Shortly afterwards, as I learned later, Dirac also found an algorithm which is equivalent to ours, but without noticing its identity with the usual mathematical theory of matrices.

# LECTURE 10

Introduction to the new quantum theory—Representation of a coördinate by a matrix—The elementary rules of matrix calculus.

In seeking a line of attack for the remodelment of the theory, it must be borne in mind that weak palliatives cannot overcome the staggering difficulties so far encountered, but that the change must reach its very foundations. It is necessary to search for a general principle, a philosophical idea, which has proved successful in other similar cases. We look back to the time before the advent of the theory of relativity, when the electrodynamics of moving bodies was in difficulties similar to those of the atomic theory of today. Then Einstein found a way out of the difficulty by noting that the existing theory operated with a conception which did not correspond to any observable phenomenon in the physical world, the conception of simultaneity. He showed that it is fundamentally impossible to establish the simultaneity of two events occurring in different localities, but rather that a new definition, prescribing a definite method of measurement is required. Einstein gave a method of measurement adapting itself to the structure of the laws of propagation of light and of electromagnetic phenomena in general. Its success justified the method and with it the initial principle involved: *The true laws of nature are relations between magnitudes which must be fundamentally observable.* If magnitudes lacking this property occur in our theories, it is a symptom of something defective. The development of the theory of relativity has shown the fertility of this idea, for the attempt to state the laws of nature in invariant form, independently of the system of coördinates, is nothing but the expression of the desire of avoiding magnitudes which are not observable. A similar situation exists in other branches of physics.

In the case of atomic theory, we have certainly introduced, as fundamental constituents, magnitudes of very doubtful observability, as, for instance, the position, velocity and period of the electron. What we really want to calculate by means of our theory and can be observed experimentally, are the energy levels and the emitted light frequencies derivable from them. The mean radius of the atom (atomic volume) is also an observable quantity which can be determined by the methods of the kinetic theory of gases or other analogous methods. On the other hand, no one has been able to give a method for the determination of the period of an electron in its orbit or even the position of the electron at a given instant. There seems to be no hope that this will ever become possible, for in order to determine lengths or times, measuring rods and clocks are required. The latter, however, consist themselves of atoms and therefore break down in the realm of atomic dimensions. It is necessary to see clearly the following points: All measurements of magnitudes of atomic order depend on indirect conclusions; but the latter carry weight only when their train of thought is consistent with itself and corresponds to a certain region of our experience. But this is precisely not the case for atomic structures such as we have considered so far. I have already called attention to the points where the theory fails.

At this stage it appears justified to give up altogether the description of atoms by means of such quantities as "coördinates of the electrons" at a given time, and instead utilize such magnitudes as are really observable. To the latter belong, besides the energy levels which are directly measureable by electron impacts and the frequencies which are derivable from them and which are also directly measurable, the intensity and the polarization of the emitted waves. We therefore take from now on the point of view that the *elementary waves* are the primary data for the description of atomic processes; all other quantities are to be derived from them. That this standpoint offers more possibilities than the assumption of electronic motions is best understood by considering the Compton effect.

If an X-ray wave of frequency $v$ impinges on free or loosely bound electrons, it transmits to the latter impacts in every direction. At the same time a secondary X-radiation is emitted having a frequency $v'$ dependent on the azimuth. According to Compton and Debye this can be quantitatively explained if the energies $hv$, $hv'$ and the momenta $\frac{hv}{c}$, $\frac{hv'}{c}$ are ascribed to the waves and then the laws of conservation of energy and of impulse are applied to the light-quanta and the electrons. But if the process is considered from the point of view of the wave theory, then the change of frequency must be interpreted as a Doppler effect. A calculation of the velocity of the wave-center gives then extremely large values in the direction of the primary X-ray and not in that of the electron. We have therefore struck upon a case in which motion of the electron and motion of the wave-center do not coincide. In the classical theory, where the emitted waves are determined by the harmonic components of the electronic motion, this is of course absolutely unexplainable. We therefore stand before a new fact which forces us to decide whether the electronic motion or the wave shall be looked upon as the primary act. After all theories which postulate the motion have proved unsatisfactory we investigate if this is also the case for the waves.

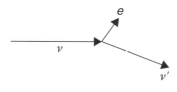

FIG. 15

To begin with consider processes which in the classical theory would correspond to a one-dimensional motion given by a Fourier series for the coördinate $q$.

$$q(t) = \sum_r q_r e^{2\pi i \nu r t}. \tag{1}$$

We consider now, not the motion $q(t)$, but the set of all the elementary oscillations

$$q_r e^{2\pi i \nu r t}$$

and try to change them so that they are suitable for the representation, not of the higher harmonics of the motion, but of the real waves of an atom.

Frequencies are *not* therefore in general harmonics $(\nu \tau)$, but can be expressed according to Ritz's combination principle as differences of every pair of terms of the series

$$\frac{W_1}{h}, \frac{W_2}{h}, \frac{W_3}{h} \cdots.$$

We therefore write

$$v(nm) = \frac{1}{h}(W_n - W_m). \tag{2}$$

To every jump $n \to m$ corresponds an amplitude and a phase that we denote by the complex amplitude,

$$q(nm) = |q(nm)|e^{i\delta(nm)}. \tag{3}$$

The set of all possible oscillations is best expressed by ordering them in a square array of terms,

$$\left\| \begin{array}{ll} q(11)e^{2\pi iv(11)t} & q(12)e^{2\pi iv(12)t} \cdots \cdots \cdots \\ q = (21)e^{2\pi iv(21)t} & q(22)e^{2\pi iv(22)t} \cdots \cdots \cdots \\ \cdots \cdots \cdots \cdots \cdots \cdots \cdots \cdots \cdots \cdots \cdots \cdots \cdots \end{array} \right\|$$

We abbreviate this to

$$q = (q(nm)e^{2\pi iv(nm)t}) = (|q(nm)|e^{2\pi iv(nm)t+i\delta(nm)}). \tag{4}$$

In order that this array correspond to a real Fourier series $q(t)$ the condition $\delta(mn) = -\delta(nm)$ must be added, or its equivalent, that $q(nm)$ be transformed into its conjugate $q^*(nm)$ by an interchange of $m$ and $n$, i.e.,

$$q(mn) = q^*(nm) \tag{5}$$

because for a real Fourier series the corresponding relation $C_{-r} = C_r^*$ holds.

The manifold of elementary oscillations is thus naturally represented by a two-dimensional array, while the manifold of harmonics of a motion is represented by a one-dimensional series,

$$C_1 e^{2\pi ivt}, C_2 e^{2\pi i2vt}, C_3 e^{2\pi i3vt \cdots}.$$

It is for this reason that, in the theory presented so far, it was necessary to consider simultaneously a whole series of motions, i.e., the stationary states, which are distinguished by another index, i.e., the quantum number $n$, whereby $C$ and $v$ become functions of $n$. The array found in this way has neither the correct frequencies nor a simple and unique correspondence to the jumps.

We must now find the laws determining the amplitudes $q(nm)$ and the frequencies $v(nm)$. For this purpose we utilize the principle of making the new laws as similar as possible to those of classical mechanics, for the fact that the classical theory of conditioned-periodic motions is in a position to account qualitatively for many quantum phenomena shows that the essential point is not the overthrow of mechanics, but rather a change from classical geometry and kinematics to the new method of representation by means of elementary waves.

As simplest example in classical mechanics we consider the oscillator. We are already familiar with the fact that everything follows once the potential energy $\frac{k}{2}q^2(t)$ is known. The potential energy can also be expressed in terms of elmentary waves, because the square of a Fourier series is also a Fourier series

$$q^2(t) = \left( \sum_\tau C_\tau e^{2\pi i v \tau t} \right)^2 = \sum_\tau D_\tau e^{2\pi i v \tau t} \qquad (6)$$

where

$$D_\tau = \sum_\sigma C_\sigma C_{\tau-\sigma}.$$

The set of quantities $D_\tau$ represents therefore the function $q^2(t)$ in quite the same way as the set $C_\tau$ represents the function $q(t)$. This can be translated into our square array, as follows: We ask, is it possible to find a multiplication rule for $q(nm)$ by which, out of every array $q$ we can construct a new array which we shall write symbolically $q^2$, but in which no new frequencies appear? The latter condition is essential and corresponds to the theorem of classical theory, that the square of a Fourier series, or the product of two such series with the same fundamental frequency, is also a Fourier series with the same fundamental frequency.

This question can be answered by looking upon the square array from the point of view of the mathematician, considering it as a *matrix* and applying the known rule for the multiplication of matrices. The product of the two matrices

$$a = (a(nm)), \quad b = (b(nm))$$

is defined by the matrix

$$c = (c(nm)) = \left( \sum_k a(nk)b(km) \right) = ab. \tag{7}$$

If we apply this rule to our array of elementary waves $q$ and multiply it by another array $p$ which has the *same* frequencies $v(nm)$ we obtain

$$qp = \left( \sum_k q(nk)e^{2\pi iv(nk)t} p(km)e^{2\pi iv(km)t} \right),$$

but we have

$$v(nk) + v(km) = \frac{1}{h}(W_n - W_k) + \frac{1}{h}(W_k - W_m)$$

$$= \frac{1}{h}(W_n - W_m) = v(nm).$$

Therefore

$$qp = \left( \sum_k q(nk)p(km)e^{2\pi iv(nm)t} \right); \tag{8}$$

that is, the symbolic product has the same frequencies as its factors. This formula is a profound generalization of the rule for obtaining the Fourier coefficients of the product of two Fourier series. We see that the multiplication rule of matrices is very closely connected with Ritz's principle of combination.

We now give the fundamental rules of matrix calculus. Addition and subtraction are performed by carrying out the required operation on each element:

$$a \pm b = (a(mn) \pm b(mn)). \tag{9}$$

The notation can be simplified further by dropping the factors $e^{2\pi ivt}$. The matrix $q = (q(nm))$ represents therefore one coördinate.

The derivative of a matrix with respect to time is the matrix

$$\dot{q} = (2\pi iv(nm)q(nm)) \tag{10}$$

where again the exponential factor is dropped. The operation of differentiation can also be expressed in terms of multiplication of matrices. For this purpose we introduce the unit matrix

$$1 = \begin{Vmatrix} 1 & 0 & 0\cdots \\ 0 & 1 & 0\cdots \\ 0 & 0 & 1\cdots \\ \cdots\cdots\cdots \end{Vmatrix} = (\delta_{nm}), \qquad (11)$$

where

$$\delta_{nm} = \begin{cases} 1 & \text{if } n = m \\ 0 & \text{if } n \neq m. \end{cases}$$

From this we form a diagonal matrix

$$W = (W(nm)) = (W_n\delta_{nm}) = \begin{Vmatrix} W_1 & 0 & 0 & 0\cdots \\ 0 & W_2 & 0 & 0\cdots \\ 0 & 0 & W_3 & 0\cdots \\ \cdots\cdots\cdots\cdots\cdots \end{Vmatrix} \qquad (12)$$

We now multiply this matrix by the matrix $(q(nm))$. In this connection we note an extremely important theorem in the development of the theory, i.e., that the multiplication of matrices is not commutative. We have

$$Wq = \left(\sum_k W(nk)q(km)\right) = \left(W_n \sum_k \delta_{nk}q(km)\right) = (W_n q(nm)),$$

but

$$qW = \left(\sum_k q(nk)W(km)\right) = \left(\sum_k q(nk)W_k\delta_{km}\right) = (W_m q(nm)).$$

If we take the difference,

$$Wq - qW = ((W_n - W_m)q(nm)), \qquad (13)$$

we see that, from Ritz's combination principle,

$$\nu(nm) = \frac{1}{h}(W_n - W_m),$$

follows the formula,

$$\dot{q} = \frac{2\pi i}{h}(Wq - qW). \qquad (14)$$

# LECTURE 11

The commutation rule and its justification by a correspondence consideration—
Matrix functions and their differentiation with respect to matrix arguments.

We shall now try to translate classical mechanics, as slightly altered as possible, into matrix form. To each coördinate matrix $q$ corresponds a momentum matrix $p$. We form out of these matrices, by matrix addition and multiplication, in some cases repeated an infinite number of times, the Hamiltonian function $H$ and try to establish the analogue of the canonical differential equations. Here we again encounter the difficulty that products are now non-commutative; $qp$ is not in general equal to $pq$. At this point the quantum theory makes its appearance. I maintain that the condition

$$pq - qp = \frac{h}{2\pi i} \tag{1}$$

must be introduced, whereby Planck's constant $h$ is bound up closely with the foundations of the theory. This relation can be made plausible by showing that, in the case of large quantum numbers, it becomes identical with the quantum condition for periodic systems. This limiting case can be described more accurately as follows: We consider large values of $m$ and $n$, and assume that all $q(mn)$, $p(mn)$ are vanishingly small except if $|m - n| = \tau$ is small compared with $m$ and $n$. For simplicity we consider only the case where $p = \mu \dot{q}$, therefore

$$p(mn) = 2\pi i \mu \nu(mn) q(mn).$$

Let us consider especially the diagonal elements of our quantum condition (1)

$$\sum_k (p(nk)q(kn) - q(nk)p(kn)) = \frac{h}{2\pi i} \tag{2}$$

or

$$\sum_k \nu(nk)|q(nk)|^2 = -\frac{h}{8\pi^2 \mu}$$

for which we can write

$$\sum_{\tau>0}(v(n, n+\tau)|q(n, n+\tau)|^2 + v(n, n-\tau)|q(n, n-\tau)|^2) = -\frac{h}{8\pi^2\mu}$$

or, since $v(mn) = -v(nm)$,

$$\sum_{\tau>0}(v(n+\tau, n)|q(n+\tau, n)|^2 - v(n, n-\tau)|q(n, n-\tau)|^2) = \frac{h}{8\pi^2\mu}.$$

If we place

$$f_\tau(n) = v(n, n-\tau)|q(n, n-\tau)|^2.$$

we may write

$$\sum_{\tau>0}\tau.\frac{f_\tau(n+\tau) - f_\tau(n)}{\tau} = \frac{h}{8\pi^2\mu}.$$

If we pass to the limit $n \gg \tau$, we obtain the classical formulas. Placing $nh = I$, we obtain

$$v(n, n-\tau) = \tau\frac{W(n) - W(n-\tau)}{\tau h} \rightarrow \tau\frac{dW}{dI} = \tau v \qquad (3)$$

which is the classical frequency of $\tau$th harmonic. Further, the corresponding amplitude is

$$q(n, n-\tau) \rightarrow q_\tau(I).$$

Therefore

$$f_\tau(n) \rightarrow f_\tau(I) = v\tau|q_\tau(I)|^2$$

and

$$\frac{1}{8\pi^2\mu} = \sum_{\tau>0}\tau\frac{f_\tau(n+\tau) - f_\tau(n)}{h\tau} \rightarrow \sum_{\tau>0}\tau\frac{\partial}{\partial I}f_\tau(I). \qquad (4)$$

This formula, however, is the quantum condition of Bohr's theory

$$\int_0 pdq = I = hn,$$

for if we set

$$q(t) = \sum_\tau q_\tau e^{2\pi iv\tau t}$$

913

we obtain

$$I = \mu \int_0^{\frac{1}{v}} \dot{q}^2 dt = -\mu(2\pi)^2 \int_0^{\frac{1}{v}} v\sigma q_\tau q_\sigma e^{2\pi i v(\tau+\sigma)^t} dt$$

$$= 4\pi^2 \mu \cdot 2 \sum_{\tau>0} \tau^2 v q_\tau q_{-\tau} = 8\pi^2 \mu \sum_{\tau>0} \tau \cdot v\tau \, |q_\tau|^2$$

and differentiating with respect to $I$

$$\frac{1}{8\pi^2\mu} = \sum_{\tau>0} \tau \frac{\partial}{\partial I} v\tau \, |q_\tau|^2 = \sum_{\tau>0} \tau \frac{\partial}{\partial I} f_\tau \, (I) \tag{5}$$

in agreement with the limit given above.

These correspondence considerations justify in a certain sense the diagonal elements of the fundamental relation (1). In order to approach as closely as possible to commutativity it is reasonable to set all elements except those on the diagonal equal to zero. Owing to this commutation law, calculations with matrices become determinate. We can therefore construct functions of $p$ and $q$ by repeated multiplications and additions.

We have for instance the energy-function of the harmonic oscillator (Mass $= \mu$):

$$H = \frac{1}{2\mu}p^2 + \frac{\kappa}{2}q^2. \tag{6}$$

To form the canonical equations we must first introduce the operation of differentiation. The derivative of a matrix function $f(x)$ with respect to the argument-matrix $x$ is defined by

$$\frac{df}{dx} = \lim_{\alpha \to 0} \frac{f(x+\alpha) - f(x)}{\alpha} \tag{7}$$

where $\alpha(mn)$ is the product of the unit matrix by a number $\alpha$

$$\alpha(mn) = \alpha\delta_{mn}.$$

The multiplication by such a matrix, or its reciprocal

$$\alpha^{-1}(mn) = \frac{1}{\alpha}\delta_{mn}$$

is commutative and therefore our definition has a unique meaning. We have, for instance,

$$\frac{dx}{dx}(mn) = \lim_{\alpha \to 0} \frac{1}{\alpha}[x(mn) + \alpha\delta_{mn} - x(mn)] = \delta_{mn},$$

that is

$$\frac{dx}{dx} = 1.$$

Similarly

$$\frac{dx^2}{dx}(mn) = \lim_{\alpha \to 0} \frac{1}{\alpha}\left[\sum_k (x_{mk} + \alpha\delta_{mk})(x_{kn} + \alpha\delta_{kn}) - \sum_k x_{mk}x_{kn}\right]$$

$$= 2x_{mn},$$

that is

$$\frac{dx^2}{dx} = 2x.$$

The product rule

$$\frac{d}{dx}(\phi\psi) = \phi\frac{d\psi}{dx} + \frac{d\phi}{dx}\psi \tag{8}$$

is proved as in ordinary calculus:

$$\frac{d}{dx}(\phi\psi) = \lim_{\alpha \to 0} \frac{1}{\alpha}[\phi(x + \alpha)\psi(x + \alpha) - \phi(x)\psi(x)]$$

$$= \lim_{\alpha \to 0} \frac{1}{\alpha}[\phi(x + \alpha)\psi(x + \alpha) - \phi(x + \alpha)\psi(x)$$

$$+ \phi(x + \alpha)\psi(x) - \phi(x)\psi(x)]$$

$$= \phi\frac{d\psi}{dx} + \frac{d\phi}{dx}\psi$$

where it should be observed that the order $\phi$, $\psi$ must be conserved. From this we deduce at once

$$\frac{dx^n}{dx} = nx^{n-1},$$

whence it follows that all the rules of ordinary differential calculus hold. The partial derivative of a matrix function of several argument-matrices $f(x_1, x_2 \ldots)$ with respect to one of them, say $x_1$, is obtained by applying our definition of differentiation to $x_1$ only, while $x_2, x_3 \ldots$ are held constant.

# LECTURE 12

The canonical equations of mechanics—Proof of the conservation of energy and of the "frequency condition"—Canonical transformations—The analogue of the Hamilton-Jacobi differential equation.

We can now write the *canonical equations*

$$\left.\begin{array}{l} \dot{q} = \dfrac{\partial H}{\partial p} \\[2mm] \dot{p} = -\dfrac{\partial H}{\partial q} \end{array}\right\} \tag{1}$$

They form in reality an infinite number of equations for an infinite number of unknowns, for the matrices on the right and left-hand sides must be equal element by element.

To establish the law of conservation of energy we need the following lemmas: Let $f(qp)$ be any matrix function of $p$ and $q$. Then

$$\left.\begin{array}{l} fq - qf = \dfrac{h}{2\pi i}\dfrac{\partial f}{\partial p} \\[3mm] pf - fp = \dfrac{h}{2\pi i}\dfrac{\partial f}{\partial q} \end{array}\right\} \tag{2}$$

To prove these relations we first assume that they are true for any two given functions $\phi$ and $\psi$ and show that they are also true for $\phi + \psi$ and $\phi\psi$. For $\phi + \psi$ this is trivial, for $\phi\psi$ a simple calculation gives

$$\phi\psi q - q\phi\psi = \phi\left(\psi q - q\psi\right) + \left(\phi q - q\phi\right)\psi$$

$$= \frac{h}{2\pi i}\left(\phi\frac{\partial\psi}{\partial p} + \frac{\partial\phi}{\partial p}\psi\right) = \frac{h}{2\pi i}\frac{\partial}{\partial p}\phi\psi$$

and an analogous relation for $p\phi\psi - \phi\psi p$. But our relations hold for $f = p$ and $f = q$ and therefore hold for every function, as functions have already been defined by repeated application of the elementary operations.

By Equations (14), Lecture 10, and (2), this Lecture, we may write the canonical equations (1)

$$\left.\begin{array}{l} Wq - qW = Hq - qH \\ Wp - pW = Hp - pH \end{array}\right\} \qquad (3)$$

or

$$(W - H)\,q - q\,(W - H) = 0$$
$$(W - H)\,p - p\,(W - H) = 0.$$

$W - H$ is therefore commutable with $p$ and $q$, hence also with any function of $p$ and $q$, in particular with $H(pq)$. We thus have

$$(W - H)\,H - H\,(W - H) = 0$$

or

$$WH - HW = 0.$$

From this follows, by Equation (14), Lecture 10,

$$\dot{H} = 0 \qquad (4)$$

which proves the *conservation of energy*. $H$ is thus seen to be a diagonal matrix

$$H(nm) = \left\{\begin{array}{ll} H_n & \text{for } n = m \\ 0 & \text{for } n \neq m \end{array}\right\} \qquad (5)$$

For the elements, the first of Equations (3) can be written,

$$q\,(nm)\,(W_n - W_m) = q\,(nm)\,(H_n - H_m)\,.$$

Therefore,

$$H_n - H_m = W_n - W_m = h\nu(nm) \qquad (6)$$

whence *Bohr's frequency condition* follows as a consequence of our postulates. By a suitable choice of an arbitrary constant we can place

$$H_n = W_m \qquad (7)$$

and this gives to the Ritz combination principle the more precise meaning of the Einstein-Bohr frequency condition.

The whole proof can also be reversed. We know that the principle of conservation of energy and the frequency condition are correct. If, therefore, the energy-function $H$ is given as an analytic function of any two variables $P$, $Q$, then, provided that

$$PQ - QP = \frac{h}{2\pi i},$$

the canonical equations

$$\dot{Q} = \frac{\partial H}{\partial P}, \qquad \dot{P} = -\frac{\partial H}{\partial Q}$$

hold. This is true because the expressions $HP - PH$ and $HQ - QH$ can always, as we have shown, be interpreted in two ways, either as partial derivatives of $H$ or, as $H$ is constant, as derivatives of $Q$ or $P$ with respect to time. Therefore, we understand by a *canonic transformation* $pq \to PQ$ one for which

$$pq - qp = PQ - QP = \frac{h}{2\pi i} \tag{8}$$

for then the canonical equations hold for $p$, $q$ as well as for $P$, $Q$.

A general transformation which satisfies this condition is

$$\left.\begin{array}{l} P = SpS^{-1} \\ Q = SqS^{-1} \end{array}\right\} \tag{9}$$

where $S$ is any arbitrary matrix. Probably, this is the most general canonic transformation. It has the simple property that for any function $f(PQ)$ the relation

$$f(PQ) = Sf(pq)S^{-1} \tag{10}$$

holds, where $f(pq)$ is formed from $f(PQ)$ by replacing $P$ by $p$ and $Q$ by $q$ without changing the form of the function. We shall show that if this theorem is true for two functions $\phi$, $\psi$, it is also true for $\phi + \psi$ and $\phi\psi$. For $\phi + \psi$ it is evident. For $\phi\psi$ we have

$$\phi(PQ)\psi(PQ) = S\phi(pq)S^{-1}S\psi(pq)S^{-1} = S\phi(pq)\psi(pq)S^{-1}.$$

As the proposition holds for $f = p$ or $f = q$, it holds in general for all analytic functions.

The importance of the canonic transformations is based on the following theorem: If any pair of variables $p_0$, $q_0$ is given which satisfies the condition

$$p_0 q_0 - q_0 p_0 = \frac{h}{2\pi i}$$

we can reduce the problem of integrating the canonical equations for an energy function $H(pq)$ to the following one: A function $S$ is to be determined, such that

$$H(pq) = SH(p_0 q_0) S^{-1} = W \qquad (11)$$

becomes a diagonal matrix. Then the solution of the canonical equations has the form

$$p = Sp_0 S^{-1}, \quad q = Sq_0 S^{-1}.$$

We have therefore a complete analogue of *Hamilton-Jacobi's differential equation.* $S$ corresponds to the action-function.

# LECTURE 13

The example of the harmonic oscillator—Perturbation theory.

Let us now illustrate these abstract considerations by an example. For this purpose we choose the harmonic oscillator, for which

$$H = \frac{p^2}{2\mu} + \frac{\kappa}{2}q^2. \tag{1}$$

The canonical equations

$$\dot{q} = \frac{p}{\mu}, \quad \dot{p} = -\kappa q \tag{2}$$

give by elimination of $p$ and placing $\frac{\kappa}{\mu} = (2\pi \nu_0)^2$

$$\ddot{q} + (2\pi \nu_0)^2 q = 0 \tag{3}$$

or, more explicitly,

$$\left[\nu^2(nm) - \nu_0^2\right] q(nm) = 0. \tag{4}$$

To this is added the commutation relation which gives

$$\sum_k \left[\nu(nk) - \nu(km)\right] q(nk)q(km) = \left\{ \begin{array}{ll} -\dfrac{h}{4\pi^2\mu} & \text{if} \quad n = m \\ 0 & \text{if} \quad n \neq m \end{array} \right\} \tag{5}$$

There follows from the equation of motion that $q(nm)$ can differ from zero only if

$$\nu(nm) = \frac{1}{h}(W_n - W_m) = \pm\nu_0. \tag{6}$$

In the row $m$ of the matrix there are therefore at most two non-vanishing elements, i.e., those for which

$$W_n = W_m + h\nu_0 \quad \text{or} \quad W_n = W_m + h\nu_0.$$

Evidently the order of the elements in the diagonal of a matrix is of no importance. If we perform the same permutation on the rows and columns, all matrix equations are unaltered. We can therefore choose $W_m = W_0$ arbitrarily and denote the "neighboring values" $W_0 + h\nu_0$ and $W_0 - h\nu_0$ of $W_0$ by the symbols $W_1$ and $W_{-1}$. Each of these has again neighboring values which differ from it by $h\nu_0$, etc. In this way we obtain an arithmetical series of energy levels,

$$W_n = W_0 \pm nh\nu_0. \tag{7}$$

The diagonal elements of the commutation relation (5) give

$$\frac{h}{8\pi^2\mu} = -\sum_k \nu(nk)\, |q(nk)|^2$$

$$= \nu_0 \left[ |q(n, n+1)|^2 - |q(n, n-1)|^2 \right]. \tag{8}$$

Whence it follows that $|q(n, n+1)|^2$ also form an arithmetical series with the difference $h/8\,\pi^2\mu\nu_0$. Since all these terms are positive, the series must stop somewhere. We have therefore

$$|q(1, 0)|^2 = \frac{h}{8\pi^2\mu\nu_0}$$

$$|q(n+1, n)|^2 = |q(n, n-1)|^2 + \frac{h}{8\pi^2\mu\nu_0}, \quad n = 1, 2, 3 \ldots$$

therefore

$$|q(n+1, n)|^2 = (n+1)\frac{h}{8\pi^2\mu\nu_0}. \tag{9}$$

It is apparent at once that all other elements of the matrix $pq - qp$ are actually zero. We verify further the conservation of energy:

$$H(nm) = 4\pi^2\mu \sum_k \left[ \nu_0^2 - \nu(nk)\nu(km) \right] q(nk)q(km). \tag{10}$$

This vanishes for $n \neq m$ and we have

$$H(nn) = 4\pi^2\mu\nu_0^2 \left[ |q(n+1, n)|^2 + |q(n, n-1)|^2 \right]$$

$$= h\nu_0 \frac{1}{2}(2n+1) = h\nu_0 \left( n + \frac{1}{2} \right). \tag{11}$$

The quantity $W_0$ introduced above has therefore the value $\frac{1}{2}h\nu_0$. The energy at absolute zero, which has been considered already by Planck and Nernst in statistical problems of the quantum theory, appears here quite naturally.

The formula for the complex amplitudes

$$q(n+1, n)\, e^{2\pi i\nu_0 t} = \sqrt{\frac{h}{8\pi^2\mu\nu_0}(n+1)}\, e^{i(2\pi\nu_0 t + \phi_n)} \tag{12}$$

involves arbitrary phases $\phi_n$ which are of great importance for the statistical behavior of the resonator. Besides, Equation (12) goes over into the classical formula

$$q(t) = \sqrt{\frac{I}{8\pi^2 \mu \nu_0}} e^{i(2\pi \nu_0 t + \phi)}, \quad I = hn \qquad (13)$$

for large values of $n$.

The theory of the harmonic oscillator can be used as a starting point for the calculation of more general systems, if we consider these as derived from the former by variation of one of its parameters. The required process can be developed in a way closely analogous to the classical perturbation theory.

We assume the energy given as a power series in the parameter $\lambda$,

$$H = H_0(pq) + \lambda H_1(pq) + \lambda^2 H_2(pq) + \cdots . \qquad (14)$$

Let the mechanical problem defined by $H_0(pq)$ be solved. We know the solution $p_0, q_0$ which satisfies the condition

$$p_0 q_0 - q_0 p_0 = \frac{h}{2\pi i}$$

and for which $H_0(p_0 q_0)$ becomes a diagonal matrix $W^0$. Now we try to determine a transformation $S$ such that if

$$p = S p_0 S^{-1}, \quad q = S q_0 S^{-1}, \qquad (15)$$

$H(pq)$ is transformed into a diagonal matrix $W$. This means that $S$ satisfies the Hamilton-Jacobi equation

$$H(pq) = S H(p_0 q_0) S^{-1} = W. \qquad (16)$$

To solve this equation we place

$$\left. \begin{array}{l} W = W^0 + \lambda W^{(1)} + \lambda^2 W^{(2)} + \cdots \\ S = 1 + \lambda S_1 + \lambda^2 S_2 + \cdots . \end{array} \right\} \qquad (17)$$

Then we have

$$S^{-1} = 1 - \lambda S_1 + \lambda^2 \left( S_1^2 - S_2 \right) - \cdots + \cdots .$$

Substituting in Equation (16),

$$\left(1 + \lambda S_1 + \lambda^2 S_2 + \cdots\right)\left(H_0\left(p_0 q_0\right) + \lambda H_1\left(p_0 q_0\right) + \lambda^2 H_2\left(p_0 q_0\right) + \cdots\right)$$
$$\left(1 - \lambda S_1 + \lambda^2\left(S_1^2 - S_2\right) + \cdots\right) = W^0 + \lambda W^{(1)} + \lambda^2 W^{(2)} + \cdots$$

and equating the coefficients of like powers of $\lambda$ we obtain the following system of approximate equations:

$$\left.\begin{array}{c} H_0\left(p_0 q_0\right) = W^0 \\ S_1 H_0 - H_0 S_1 + H_1 = W^{(1)} \\ S_2 H_0 - H_0 S_2 + H_0 S_1^2 - S_1 H_0 S_1 + S_1 H_1 - H_1 S_1 + H_2 = W^{(2)} \\ \cdots\cdots\cdots\cdots\cdots\cdots\cdots\cdots\cdots\cdots\cdots\cdots\cdots\cdots\cdots\cdots \\ S_r H_0 - H_0 S_r + F_r\left(H_0, \cdots H_r, S_0, \cdots S_{r-1}\right) = W^{(r)} \end{array}\right\}$$

$$(18)$$

where $H_0, H_1 \ldots$ are to be considered as functions of $p_0, q_0$.

The first equation is satisfied. The others can be solved successively in a way quite analogous to that used in classical theory: The average of the energy is first formed in order to fix the energy constant, for

$$S_r H_0 - H_0 S_r = -\left(W^0 S_r - S_r W^0\right)$$

has no diagonal terms. There follows, in general,

$$W^{(r)} = \bar{F}_r, \quad \text{i.e.} \quad W_n^{(r)} = F_r(nn).$$

We have, further,

$$W_n^0 S_r(mn) - W_m^0 S_r(mn) + F_r(mn) = 0, \, m \neq n$$

or

$$S_r(mn) = \frac{F_r(mn)}{h\nu_0(mn)}\left(1 - \delta_{mn}\right) \tag{19}$$

where $\nu_0(mn)$ are the frequencies of the unperturbed motion.

This solution satisfies the condition

$$S\tilde{S}^* = 1 \tag{20}$$

where the symbol $\sim$ denotes transposition of the rows and columns and the symbol* the substitution of conjugate complex quantities. As $S$ is only obtained by successive calculation of the approximations

$S_1, S_2 \ldots$ this relation can be proved only by successive steps. We shall restrict ourselves to the first step. If we must have

$$S\bar{S}^* = (1 + \lambda S_1 + \cdots)(1 + \lambda \bar{S}_1^* + \cdots) = 1$$

then

$$S_1 + \bar{S}_1^* = 0,$$

but our general formula (19) gives

$$S_1(mn) = \frac{H_1(mn)}{h\nu_0(mn)}(1 - \delta_{mn}),$$

therefore

$$\bar{S}_1^*(mn) = S_1^*(nm) = \frac{H_1^*(nm)}{h\nu_0(nm)}(1 - \delta_{mn}), \tag{21}$$

As $H_1$ is an Hermitian matrix, that is since

$$H_1^*(nm) = H_1(mn),$$

it follows that

$$\bar{S}_1^*(mn) = \frac{H_1(mn)}{-h\nu_0(mn)}(1 - \delta_{mn}) = -S_1(mn).$$

The importance of the relation $S\bar{S}^* = 1$ arises from the fact that the Hermitianness of the matrices $p, q$ is a consequence of this relation. The rule

$$(\bar{a}b) = \bar{b}\bar{a} \tag{22}$$

holds, as can be easily deduced from the definition of the products:

$$\sum_k a(nk)b(km) = \sum_k \bar{b}(mk)\bar{a}(kn).$$

From this follows that

$$q^* = S^* q_0^* (S^*)^{-1} = \bar{S}^{-1} \bar{q}_0 \bar{S} = \bar{q} \tag{23}$$

and similarly for $p$.

If we place,

$$\left. \begin{aligned} q = q_0 + \lambda q_1 + \cdots &= (1 + \lambda S_1 + \cdots) q_0 (1 - \lambda S_1 + \cdots) \\ p = p_0 + \lambda p_1 + \cdots &= (1 + \lambda S_1 + \cdots) p_0 (1 - \lambda S_1 + \cdots) \end{aligned} \right\}$$

then we have, as a first approximation,

$$
\left.\begin{array}{l}
q_1 = S_1 q_0 - q_0 S_1, \\
p_1 = S_1 p_0 - p_0 S_1.
\end{array}\right\}
$$

Or more explicitly,

$$
\left.\begin{array}{l}
q_1(mn) = \dfrac{1}{h} \sum_{k}{}' \left( \dfrac{H_1(mk)q_0(kn)}{v_0(mk)} - \dfrac{q_0(mk)H_0(kn)}{v_0(kn)} \right) \\[3mm]
p_1(mn) = \dfrac{1}{h} \sum_{k}{}' \left( \dfrac{H_1(mk)p_0(kn)}{v_0(mk)} - \dfrac{p_0(mk)H_0(kn)}{v_0(kn)} \right).
\end{array}\right\} \quad (24)
$$

For the energy we obtain, as a second approximation,

$$
W^{(2)} = \overline{H_0 S_1^2} - \overline{S_1 H_0 S_1} + \overline{S_1 H_1} - \overline{H_1 S_1} + \overline{H_2},
$$

or

$$
\left.\begin{array}{l}
W_n^{(2)} = \sum_{k}{}' \left( W_n^0 S_1(nk)S_1(kn) - S_1(nk)S_1(kn)W_k^0 \right. \\[2mm]
\qquad\qquad \left. + S_1(nk)H_1(kn) - H_1(nk)S_1(kn) \right) + H_2(nn) \\[3mm]
W_n^{(2)} = H_2(nn) + \dfrac{1}{h} \sum_{k}{}' \dfrac{H_1(nk)H_1(kn)}{v_0(nk)}.
\end{array}\right\} \quad (25)
$$

# LECTURE 14

The meaning of external forces in the quantum theory and corresponding perturbation formulas—Their application to the theory of dispersion.

Before discussing the significance of these formulas, we consider the more general case where the Hamiltonian function contains the time $t$ explicitly. This can be easily taken account of formally by introducing in $H(t, p, q)$ instead of $t$ a new coördinate $q^0$, to which corresponds a momentum $p^0$, and considering the Hamiltonian function

$$H^\times = H\left(q^0, p, q\right) + p^0, \tag{1}$$

To $q^0$, $p^0$ correspond the canonical equations,

$$\dot{q}^0 = \frac{\partial H^\times}{\partial p^0} = 1, \; \dot{p}^0 = -\frac{\partial H^\times}{\partial q^0} = -\frac{\partial H}{\partial t} \tag{2}$$

of which the first says that $q^0$ is the time and the second defines $p^0$.

A closer consideration leads to an important difficulty. The introduction of a function $H$ depending explicitly on $t$ has evidently the physical meaning that the reaction of the system $A$ in question on other systems $B$ which act on $A$ is so small that it can be neglected, and that the quantities depending on these external systems $B$ can be considered to be the same functions of time as they would be without the presence of $A$. In classical theory, where the interactions of two systems depend only on their instantaneous motion, the condition for this is that the coupling energy be small. But in the quantum theory this is not obviously so. Here the reaction depends, as our perturbation formulas show, not only on the instantaneous state of the system, but on all the states of the system together, for the products occurring in the formulas contain sums over all the states. The perturbation of the system $A$, due to a motion of the system $B$ given as a function of the time can be taken care of only as long as approximations are restricted to those for which the quantities belonging to $B$ enter only linearly in the perturbation function $H_1$. Higher approximations have no meaning even in the case of a weak coupling. But if the assumption is made that the system $A$ under consideration is negligible energetically compared with the external systems $B$, then going over to higher approximations can also be justified in the quantum theory.

We shall restrict ourselves here to the first approximation $q_1$, $p_1$. We consider the special case where the system defined by $H_0$ is acted upon by an electric field $\mathbf{E}$. Then the perturbation function is, to a first approximation,

$$H_1 = eq_0 E. \tag{3}$$

According to what has been said above, $\mathbf{E}$ can be looked upon as a function of the time. If in particular we are considering a monochromatic light wave of the frequency $\nu$

$$E = E_0 \cos 2\pi \nu t,$$

therefore

$$H_1 = e E_0 q_0 \cos 2\pi \nu t = \frac{1}{2} e E_0 q_0 \left( e^{2\pi i \nu t} + e^{-2\pi i \nu t} \right)$$

then we get for the perturbation of the coördinates,

$$q_1(mn) = \frac{E_0 e}{2h} \sum_k \left( \frac{q_0(mk)q_0(kn)}{\nu_0(mk) + \nu} - \frac{q_0(mk)q_0(kn)}{\nu_0(kn) + \nu} \right)$$

or, as $p_1 = \mu \dot{q}_1$,

$$q_1(mn) = \frac{E_0 e}{2h \cdot 2\pi i \mu} \sum_k \frac{q_0(mk)p_0(kn) - p_0(mk)q_0(kn)}{(\nu_0(mk) + \nu)(\nu_0(kn) + \nu)}. \tag{4}$$

For the diagonal terms we have, in particular,

$$q_1(nn) = -\frac{E_0 e}{2h \cdot 2\pi i \mu} \sum_k \frac{q_0(nk)p_0(kn) - p_0(nk)q_0(kn)}{\nu_0^2(nk) - \nu^2}. \tag{5}$$

The polarization produced by the field $\mathbf{E}$ is obtained by multiplying $q_1$ by the charge $e$ and then the index of refraction can be calculated by well-known methods.

This formula for $q_1(nn)$ contains Kramers' theory of dispersion, which was found by considerations of correspondence. To understand its meaning we recall the relation between the theory of dispersion and the quantum theory of multiple-periodic systems. When a light wave acts on such a system, the electronic orbits perform oscillations. The resonance points of these forced oscillations lie evidently where the Fourier analysis of the orbits leads to a harmonic overtone. Debye attempted to calculate the dispersion formula for the hydrogen molecule using the model shown in Fig. 16, and Sommerfeld extended this process to more general molecular models with electrons arranged in rings. If they found a fairly good agreement with measurements of refractive indices it was only because the range of measurements lay very far from the characteristic resonance points. The incorrectness of the formula follows already from the fact that some of the resonance points have imaginary proper frequencies, which is always a sign of instability of the motion. It is rendered more evident by the fact that the resonance points have no relation to the frequencies which the system would emit according to the quantum theory. It is quite clear, however, that the frequencies actually emitted must determine essentially the resonance or dispersion curve, and not the higher harmonics of the stationary motion which are not optically observable.

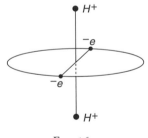

FIG. 16

The first step towards a rational change of the theory of dispersion in this sense was made by Ladenburg. His dispersion formula consists essentially of those terms in the above expression (5) for $q_1(nn)$ for which $n < k$ and which, therefore, correspond to "upward jumps," that is, to absorption processes. Ladenburg also discovered a relation between the numerator of the dispersion formula $|q_0(nk)|^2 \nu_0(kn)$ and the transition probabilities between the states $n$ and $k$ which appear in Einstein's derivation of Planck's formula.

Kramers has given the complete expression for $q_1(nn)$ in which the emission terms $(n > k)$ are also taken into account. Since

$$v_0(kn) = -v_0(nk)$$

these terms give "negative" contributions to the dispersion. Kramers' formula has the great advantage of reducing in the limit to the classical formula for the influence of an alternating field on a multiple-periodic system. It therefore satisfies the principle of correspondence.

The case of a constant electric field (Stark effect), as represented by our original formula, was used by Pauli to estimate the intensity of the spectral lines of the mercury atom which do not appear in its natural state (for which $q_0(nm) = 0$) and which are first excited by the field ($q_1(nm) \neq 0$).

Kramers' work suggested to me that quite generally the perturbation energy cannot depend on the classical frequencies of the unperturbed system, but rather on the quantic frequencies and this has been recently confirmed by Schrödinger's considerations on the actual structure of certain line spectra, e.g. aluminum. By correspondence considerations I arrived at the expression (25) Lecture 13 for the perturbation energy $W^{(2)}$. By similar considerations Heisenberg and Kramers also found and discussed the expression $q_1(nm)$ [Eq. (4), Lecture 14] for a light wave. They correspond to the phenomenon that light of frequency $v$ is not only scattered as light of the same frequency, as in the classical theory, but also as light of other colors belonging to the combination frequencies $v \pm v_0(nk)$. This phenomenon had been already postulated by Smekal from considerations on light quanta.

Consider finally the limiting case of very high frequencies of the exciting light

$$v \gg |v_0(mk)|, \quad v \gg |v(kn)|.$$

We then obtain

$$q_1 = -\frac{E_0 e}{2h \cdot 2\pi i v^2 \mu}(p_0 q_0 - q_0 p_0),$$

and since

$$p_0 q_0 - q_0 p_0 = \frac{h}{2\pi i}$$

therefore

$$q_1 = \frac{E_0 e}{8\pi^2 \nu^2 \mu}. \tag{6}$$

Compare this with the excitation of a free electron by the same electric field $E_0 \cos 2\pi \nu t$. Here we must only take that part $\frac{1}{2} E_0 e^{2\pi i \nu t}$ corresponding to an element of the matrix. We have the differential equation

$$\mu \ddot{q}_1 = \frac{1}{2} e E_0 e^{2\pi i \nu t}$$

the solution of which is

$$q_1 = \frac{E_0 e}{8\pi^2 \nu^2 \mu}.$$

Our quantic commutation rule can therefore be interpreted as the condition that, for sufficiently high frequencies, the electron behave as in classical theory, where the scattered light of frequency $\nu$ has the correct intensity and the scattered light of the combination frequencies vanishes. Starting from this condition Kuhn and Thomas found a formula equivalent to the commutation rule, as already stated above, and they and Reiche applied it to dispersion problems.

# LECTURE 15

Systems o$_\perp$ more than one degree of freedom—The commutation rules—The analogue of the Hamilton-Jacobi theory—Degenerate systems.

We now consider systems of $f$ degrees of freedom. By an immediate generalization they may be represented by $2f$-dimensional matrices

$$\left. \begin{array}{l} q_k = \Big( q_k \Big( n_1, n_2 \cdots n_f; m_1, m_2 \cdots m_f \Big) \Big) \\[2mm] p_k = \Big( p_k \Big( n_1, n_2 \cdots n_f; m_1, m_2 \cdots m_f \Big) \Big) \end{array} \right\} \tag{1}$$

This representation is sometimes very convenient and clear, but not at all necessary. We can always imagine the matrix written in two dimensions. Then, as already shown for one degree of freedom, the expression of the stationary states as given by the arrangement of the rows is quite immaterial, as contrasted with the older theory. We can therefore always transform a $2f$-dimensional matrix into a two-dimensional one. We can for instance write a 4-dimensional matrix $q(n_1, n_2; m_1, m_2)$ as follows:

$$q = \begin{Vmatrix} q\,(1, 1; 1, 1)\, q\,(1, 1; 1, 2) \cdots q\,(1, 1; 2, 1)\, q\,(1, 1; 2, 2) \cdots \\ q\,(1, 2; 1, 1)\, q\,(1, 2; 1, 2) \cdots q\,(1, 2; 2, 1)\, q\,(1, 2; 2, 2) \cdots \\ \cdots\cdots\cdots\cdots\cdots\cdots\cdots\cdots\cdots\cdots\cdots\cdots\cdots \\ q\,(2, 1; 1, 1)\, q\,(2, 1; 1, 2) \cdots q\,(2, 1; 2, 1)\, q\,(2, 1; 2, 2) \cdots \\ q\,(2, 2; 1, 1)\, q\,(2, 2; 1, 2) \cdots q\,(2, 2; 2, 1)\, q\,(2, 2; 2, 2) \cdots \\ \cdots\cdots\cdots\cdots\cdots\cdots\cdots\cdots\cdots\cdots\cdots\cdots\cdots \end{Vmatrix}$$

The definitions of addition and of multiplication are quite independent of the order of the indices. The rules of matrix calculus are therefore applicable as before. We can therefore define a Hamiltonian function

$$H \left( q_1 \cdots q_f, p \cdots p_f \right)$$

and have the equations of motion:

$$\dot{q}_k = \frac{\partial H}{\partial p_k}, \quad \dot{p}_k = -\frac{\partial H}{\partial q_k}. \tag{2}$$

The *quantic commutation rules* are fundamental. We make the following immediate generalization:

$$\left.\begin{aligned} p_k q_l - q_l p_k &= \frac{h}{2\pi i}\delta_{kl} \\[2mm] p_k p_l - p_l p_k &= 0 \\[1mm] q_k q_l - q_l q_k &= 0 \end{aligned}\right\} \tag{3}$$

Whence follows, as before, for any arbitrary function $f(q_1 \ldots q_f, p_1, \ldots p_f)$,

$$\left.\begin{aligned} p_k f - f p_k &= \frac{h}{2\pi i}\frac{\partial f}{\partial q_k} \\[2mm] f q_k - q_k f &= \frac{h}{2\pi i}\frac{\partial f}{\partial p_k} \end{aligned}\right\} \tag{4}$$

Therefore the proof of the conservation of energy and of the frequency condition remains the same as above, as does the concept of canonic transformations

$$p_k = S p_k^0 S^{-1}, \; q_k = S q_k^0 S^{-1} \tag{5}$$

and the Hamilton-Jacobi equation,

$$H(pq) = S H (p_0 q_0) S^{-1} = W. \tag{6}$$

The large number of commutation rules gives rise to the question whether $p_k$, $q_k$ can be at all determined so as to satisfy all the conditions. It is easily seen that all the condition equations are not independent. From the canonical equations of motion alone follows, for instance,

$$\frac{d}{dt} \sum_k (p_k q_k - q_k p_k) = 0.$$

The general proof of the possibility of satisfying all the conditions can be given by means of the theory of perturbations, starting from an unperturbed system with the energy-function

$$H_0(pq) = \sum_{k=1}^{f} H^{(k)} (p_k q_k)$$

which therefore consists of $f$ uncoupled systems. Let the motions of these be represented by two-dimensional matrices $q_k^0$, $p_k^0$. If these $f$ uncoupled systems are considered formally as a single system of $f$ degrees of freedom, $q_k^0$, $p_k^0$ are to be represented by 2 $f$-dimensional matrices for which the relations

$$\left.\begin{array}{l} q_k^0 \left( n_1 \cdots n_f, m_1 \cdots m_f \right) = \delta_k q_k^0 \left( n_k m_k \right) \\[2mm] p_k^0 \left( n_1 \cdots n_f, m_1 \cdots m_f \right) = \delta_k q_k^0 \left( n_k m_k \right) \end{array}\right\}$$

hold where

$$\left.\begin{array}{l} \delta_k = 1 \text{ if } n_j = m_j \text{ for every } j \text{ except } j = k \\[2mm] \delta_k = 0 \text{ if any } j \neq k, n_j \neq m_j. \end{array}\right\}$$

From these there follow in the first place the relations

$$\left.\begin{array}{l} p_k^0 q_l^0 - q_l^0 p_k^0 = 0 \quad \text{for } l \neq k \\[2mm] p_k^0 p_l^0 - p_l^0 p_k^0 = 0 \\[2mm] q_k^0 q_l^0 - q_l^0 q_k^0 = 0 \end{array}\right\} \tag{7}$$

Next the relation originally postulated for the 2-dimensional matrices,

$$p_k^0 q_k^0 - q_k^0 p_k^0 = \frac{h}{2\pi i} \tag{8}$$

is also correct for 2 $f$-dimensional matrices.

If now the Hamiltonian function for the coupled system is

$$H = H_0 + \lambda H_1 + \lambda^2 H_2 + \cdots, \tag{9}$$

then we have shown that a solution of the unperturbed system exists which satisfies all the commutation relations. If we further suppose that the system $H_0$ is not degenerate, that is, that in the diagonal matrix $W^0$, which results from $H_0$ by the introduction of $q_k^0$, $p_k^0$, no two diagonal elements are equal, then we can find the motion of the perturbed system by the method of successive approximations discussed above. We place

$$q_k = S q_k^0 S^{-1}, \quad p_k = S p_k^0 S^{-1}$$
$$S = 1 + \lambda S_1 + \lambda^2 S_2 + \cdots$$

where $S$ is determined by the equation

$$S H \left( p^0 q^0 \right) S^{-1} = W.$$

The commutation relations and the equations of motion are then evidently satisfied and the desired proof is thus now complete.

The commutation relations are also invariant with respect to linear orthogonal transformations of $q_k$ and $p_k$. For if we place

$$q'_k = \sum_l a_{kl} q_l \qquad \sum_l a_{kl} a_{jl} = \delta_{kj}$$
$$p'_k = \sum_l a_{kl} p_l \tag{10}$$

Then

$$p'_k q'_l - q'_l p'_k = \sum_{ij} a_{ki} a_{lj}(p_i q_j - q_j p_i) = \frac{h}{2\pi i} \sum_j a_{kj} a_{lj} = \frac{h}{2\pi i} \delta_{kl}$$

and similarly for the other relations. Therefore, if we postulate that our fundamental relations hold for one cartesian system, they hold for any such system.

Proceeding systematically, we have now to study degenerate systems, that is, systems such that several of the $W_n$-values are equal, therefore several of the frequencies $v(nm)$ are zero. The constancy of the energy $\dot{H} = 0$ can still be deduced from the equations of motion and the commutation relations, but it *no longer* follows in general from $\dot{H} = 0$ that $H$ is a diagonal matrix and therefore the proof of the frequency theorem cannot be carried through. The equations of motion and the commutation relations alone are here insufficient for a unique determination of the properties of the system and a further restriction of the fundamental equations is necessary. It is obvious that this restriction must be as follows: *As fundamental equations, the commutation relations and*

$$H = W = diagonal\ matrix \tag{11}$$

*shall hold.* Then the validity of the frequency condition is also assured for degenerate systems.

Although, except in singular cases, the energy is uniquely determined by these conditions, the coördinates $q_k$ are *not* uniquely determined. In non-degenerate systems, as was seen already in the example of the harmonic oscillator, only certain phase constants are arbitrary;

one for each stationary state. In degenerate systems a much greater indetermination exists, which evidently is related to a sort of lability which allows arbitrarily small external perturbations to produce finite changes in the coördinates. But it can be shown that even then those properties of the system on which the polarization of the emitted light depends vary only continuously, a fact which Heisenberg has called "spectroscopic stability." We shall not discuss this question further.

# LECTURE 16

Conservation of angular momentum—Axial symmetrical systems and the quantization of the axial component of angular momentum.

The applications of the basic principles, which have been considered so far, suppose that several specially simple systems which are used as starting points in the calculus of perturbations are completely known. For this purpose we have until now studied the particular example of the harmonic oscillator. We must now develop general methods for the direct integration of the fundamental equations. These methods are the same as those used in classical mechanics, i.e., general properties of the energy-function $H$ are used to find integrals. The conservation of energy has been thus derived as a consequence of the property of $H$ not to depend explicitly on the time. The conservation of momentum and moment of momentum will now be developed, making the same hypotheses on $H$ as in ordinary mechanics. The integration method is quite similar to that used in the derivation of the conservation of energy. The equations of motion, considered for the elements of the matrices, form an infinite system for an infinite number of variables: in general each equation contains an infinite number of variables. To begin with, a function $A(pq)$, constant according to the fundamental equations, and therefore a diagonal matrix for non-degenerate systems is determined. If $\phi(pq)$ is any function, the difference

$$\phi A - A\phi = \psi$$

can be calculated from our commutation rules. But as $A$ is a diagonal matrix each of the equations for the elements contains only one of the elements of $\phi$ and $\psi$, besides two diagonal terms of $A$.

In Galilean-Newtonian mechanics, as well as in Einstein's ("relativistic") mechanics:

$$H = H'(p) + H''(q). \tag{1}$$

The components of momentum are

$$
\left.
\begin{aligned}
p_x &= \sum_{k=1}^{3} p_{kx} \\
p_y &= \sum_{k=1}^{3} p_{ky} \\
p_z &= \sum_{k=1}^{3} p_{kz}
\end{aligned}
\right\}
\tag{2}
$$

and the components of moment of momentum

$$
\left.
\begin{aligned}
M_x &= \sum_{k=1}^{3} \left(q_k p_{kz} - p_{ky} q_{kz}\right) \\
M_y &= \sum_{k=1}^{3} \left(q_{kz} p_{kx} - p_{kx} q_{kx}\right) \\
M_z &= \sum_{k=1}^{3} \left(q_{kx} p_{ky} - p_{kx} q_{ky}\right)
\end{aligned}
\right\}
\tag{3}
$$

If derivatives with respect to time are now taken and it is noted that, because of our assumption on $H$, all $\dot{p}_{kx}\ldots$ etc. depend only on $q_{kx}\ldots$ etc., and all $\dot{q}_{kx}$ etc. on $p_{kx}\ldots$ etc., it is seen that all these derivatives have the form $\phi(q) + \psi(p)$. Now since all the $q$'s and all the $p$'s are interchangeable among themselves, these expressions all vanish under the same conditions as in classical mechanics. The theorems on uniform motion of the center of gravity and on the conservation of angular momentum (law of areas) therefore hold exactly as in classical theory.

Let us now build up the expression

$$
\begin{aligned}
M_x M_y - M_y M_x &= \sum_{k,l} \big[\left(q_{ky} p_{kz} - p_{ky} q_{kz}\right)\left(q_{lz} p_{lx} - p_{lz} q_{lx}\right) \\
&\quad - \left(q_{kz} p_{kx} - p_{kz} q_{kx}\right)\left(q_{ly} p_{lz} - p_{ly} q_{lz}\right)\big] \\
&= \sum_{k,l} \big[q_{ky} p_{lx}\left(p_{kz} p_{lz} - q_{lz} p_{kz}\right) \\
&\quad + p_{ky} q_{lx}\left(q_{kz} p_{lz} - p_{lz} q_{lz}\right)\big] \\
&= -\frac{h}{2\pi i} \sum_{k} \left(q_{kx} p_{ky} - p_{kx} q_{ky}\right)
\end{aligned}
$$

therefore

$$M_x M_y - M_y M_x = -M_z \epsilon, \quad \epsilon = \frac{h}{2\pi i} \tag{4}$$

whence it is seen that the law of areas, as in classical mechanics, holds either only for one or for all three axes.

It will now be assumed that the system consists only of discrete energy levels, further that it is not degenerate and that the law of areas holds for one of the momenta, for instance $\bar{M}_x = 0$. This is e.g. the case if the external forces acting on the atom are symmetrical with respect to the $z$-axis. Then $M_z$ is a diagonal matrix and the individual elements $M_{xn}$ are to be interpreted as the angular momenta of the atom around the $z$-axis for the corresponding individual states.

From the definition of $M_x$, $M_y$, $M_z$, and the commutation rules follow the matrix equations

$$\left. \begin{aligned} q_{lx} M_z - M_z q_{lx} &= +\epsilon q_{ly} \\ q_{ly} M_z - M_z q_{ly} &= -\epsilon q_{lx} \\ q_{lz} M_z - M_z q_{lz} &= 0 \end{aligned} \right\} \tag{5}$$

As

$$M_z(nm) = \delta_{nm} M_{zn},$$

these expressions can be rewritten

$$\left. \begin{aligned} q_{lx}(nm)\,(M_{zn} - M_{zm}) &= +\epsilon q_{ly}(nm) \\ q_{ly}(nm)\,(M_{zn} - M_{zm}) &= -\epsilon q_{lx}(nm) \\ q_{lz}(nm)\,(M_{zn} - M_{zm}) &= 0 \end{aligned} \right\} \tag{6}$$

Equations (6) express, in the ordinary language of Bohr's theory, the following: For a quantum jump in which the angular momentum $M_{zn}$ changes, $q_{lz}(nm) = 0$ and the plane of oscillation of the emitted light wave is therefore perpendicular to the $z$-axis. For jumps in which $M_{zn}$ does not change $q_{lx}(nm) = 0$, $q_{ly}(nm) = 0$ and the emitted light therefore vibrates parallel to the $z$-axis. Moreover, in the former case,

$$\left[ (M_{zn} - M_{zm})^2 - \frac{h^2}{4\pi^2} \right] q_{l\eta}(nm) = 0, \quad \eta = x, y. \tag{7}$$

That is: for every quantum jump $M_{zn}$ changes by 0 or by $\pm\frac{h}{2\pi}$. In the former case the emitted light is linearly polarized parallel to the $z$-axis, in the latter case it is circularly polarized around this axis. $M_{zn}$ can therefore be represented by

$$M_{zn} = \frac{h}{2\pi}(n_1 - C), \quad n_1 = \cdots - 2, -1, 0, 1, 2 \cdots. \quad (8)$$

If states existed the angular momentum of which did not find a place in this series, there could be no jumps or interactions between them and those belonging to the above series.

From these results it is seen that the index $n$ can be split up into two components, one of which is the number $n_1$ which has already been introduced, while the other, $n_2$, numbers the different $n$'s with the same $n_1$. Our matrices become four-dimensional and the "polarization rules" already derived are equivalent to the following expressions:

$$\left.\begin{array}{c} q_{lx}(nm) = \delta_{1,|n_1-m_1|}q_{lx}(nm) \\ q_{ly}(nm) = \delta_{1,|n_1-m_1|}q_{ly}(nm) \\ q_{lz}(nm) = \delta_{n_1,m_1}q_{lz}(nm) \\ q_{lx}(n_1, n_2; n_1 \pm 1, m_2) \mp iq_{ly}(n_1,n_2; n_1 \pm 1, m_2) = 0 \end{array}\right\} \quad (9)$$

All these relations hold if $q_{lx}, q_{ly}, q_{lz}$ are replaced by $p_{lx}, p_{ly}, p_{lz}$ or by $M_x, M_y, M_z$.

In particular we note that

$$\left.\begin{array}{c} M_x(nm) = \delta_{1,|n_1-m_1|}M_x(nm) \\ M_y(nm) = \delta_{1,|n_1-m_1|}M_y(nm) \\ M_x(n_1, n_2; n_1 \pm 1, m_2) \mp iM_z(n_1,n_2; n_1 \pm 1, m_2) = 0 \end{array}\right\}$$

Further we need the following derived commutation relations: If,

$$q_l^2 = q_l^2 = q_{lx}^2 + q_{ly}^2 + q_{lz}^2, \quad \mathbf{M}^2 = M^2 = M_x^2 = M_y^2 = M_z^2,$$

then simple calculations give

$$\left.\begin{array}{c} q_l^2 M_z - M_z q_l^2 = 0 \\ M^2 M_z - M_z M^2 = 0 \end{array}\right\} \quad (10)$$

which means that $q_l^2$ and $M^2$ are diagonal matrices with respect to the quantum number $n_1$.

The two components $M_x$, $M_y$ may also be constant, but can never be diagonal matrices. For from

$$M_y M_z - M_z M_y = -\epsilon M_x$$

or

$$M_y(nm)\,(M_{zn} - M_{zm}) = -\epsilon M_x(nm)$$

it would follow that, for $M_y(nm) = M_{yn}\delta_{nm}$, $M_x$ would vanish identically, and therefore that $M_y$, $M_z$ would also vanish identically. Such a system with a constant vector $\mathbf{M}$, for instance a system moving freely in space, is hence necessarily degenerate.

Consider now a system the energy function of which is

$$H = H_0 + \lambda H_1 + \cdots$$

under the following assumptions: *For $\lambda = 0$ the law of areas holds for all three directions. For $\lambda \neq 0$ the system is not degenerate, but $M_z$ is constant. The energy $H_0$ does not depend on $n_1$.* A system of this kind is, for instance, an atom in an axially symmetrical field of strength proportional to $\lambda$. This investigation leads also to definite information about the degenerate system with the energy-function $H_0$, for every property of the perturbed system which is independent of $\lambda$ or the choice of the privileged direction $z$ must remain valid for $\lambda = 0$.

According to our hypothesis that for $\lambda = 0$ the law of areas along all three directions holds, $\dot{M}_x$, $\dot{M}_y$ and therefore also $d/dt(M^2)$ have no terms without $\lambda$. Therefore

$$\left.\begin{aligned}
v_0(nm)M_x^0(nm) &= 0\\
v_0(nm)M_y^0(nm) &= 0\\
v_0(nm)(M^0)^2(nm) &= 0.
\end{aligned}\right\} \tag{11}$$

As it has been further supposed that $H_0 = W^0$ is independent of the quantum number $n_1$, we have

$$v_0\,(n_1, n_2; m_1, n_2) = W_{m2}^0 - W_{m2}^0 = 0$$
$$v_0\,(n_1, n_2; m_1, m_2) = W_{n2}^0 - W_{m2}^0 \neq 0 \text{ for } n_2 \neq m_2,$$

whence

$$\left.\begin{array}{c} M_x^0(nm) = \delta_{n_2 m_2} M_x^0(nm) \\ M_y^0(nm) = \delta_{n_2 m_2} M_y^0(nm) \\ (M^0)^2(nm) = \delta_{n_2 m_2}(M^0)^2(nm) \end{array}\right\} \tag{12}$$

It was shown earlier that $M^2$ is quite in general a diagonal matrix with respect to $n_1$; it is now shown that $(M^0)^2$ is a diagonal matrix with respect to both quantum numbers $n_1$, $n_2$. The same holds for $(M_x^0)^2 + (M_y^0)^2(M^0)^2 - M_z^2$. Now

$$\left(M_x^0\right)^2 (n_1 n_2 m_1 m_2) = \sum_{k_1 k_2} M_x^0 (n_1 n_2 k_1 k_2)\, M_x^0 (k_1 k_2 m_1 m_2)$$

$$= \delta_{n_2 m_2} \sum_{k_1} M_x^0 (n_1 n_2 k_1 n_2)\, M_x^0 (k_1 n_2 m_1 n_2)$$

and

$$\left((M_x^0)^2 + (M_y^0)^2\right) (n_1 n_2 m_1 n_2)$$

$$= \delta_{n_1 m_1} \delta_{n_2 m_2} \sum_{k_1} \left\{ M_x^0 (n_1 n_2 k_1 n_2)\, M_x^0 (k_1 n_2 n_1 n_2) \right.$$

$$\left. + M_y^0 (n_1 n_2 k_1 n_2)\, M_y^0 (k_1 n_2 n_1 n_2) \right\}$$

$$= \delta_{n_1 m_1} \delta_{n_2 m_2} \sum_{kl} \left[ \left| M_x^0 (n_1 n_2 k_1 n_2) \right|^2 + \left| M_y^0 (n_1 n_2 k_1 n_2) \right|^2 \right]$$

$$\tag{13}$$

The diagonal terms of $(M^0)^2 - M_z^2$ are therefore always positive; and since $(M^0)^2$ does not depend on $n_1$, the number of possible values of $M_{zn1}^2 = (\frac{h}{2\pi})^2(n_1 + C)^2$ for a given $n_2$, therefore for a given $(M_{n_2}^0)^2$, is finite. In other words the number of values $n_1$ for a given $n_2$ is finite. Hence the sum

$$\sum_{k_1 k_2} M_x^0 (n_1, n_2; k_1, k_2)\, M_y^0 (k_1, k_2; m_1, m_2)$$

$$= \delta_{n_2 m_2} \sum_{k_1} M_x^0 (n_1, n_2; k_1, n_2)\, M_y^0 (k_1, n_2; m_1, n_2)$$

has only a finite number of terms. This sum is an element of $M_x^0 M_y^0$. If we now form in the same way $M_y^0 M_x^0$ and sum the equations

$$-\epsilon M_z^0 = M_x^0 M_y^0 - M_y^0 M_x^0$$

over $n_1$ for fixed $n_2$, this sum is zero on the right-hand side because in general, for *finite* matrices, the diagonal sum of $ab$ is equal to that of $ba$:

$$\sum_n \left( \sum_k a(nk)b(kn) \right) = \sum_n \left( \sum_k b(nk)a(kn) \right).$$

Therefore

$$\sum_{n_1} M_z = \frac{h}{2\pi} \sum_{n_1} (n_1 + C) = 0. \tag{14}$$

This holds for every complete series of $n_1$. Therefore the possible values of $n_1$ which go with a fixed $n_2$ always form a symmetrical series with respect to the origin. Hence $(n_1 + C)$ runs through a finite series of whole numbers ... $-2, -1, 0, 1, 2$ ... or of "half-numbers" $\ldots - \frac{3}{2}, -\frac{1}{2}, +\frac{1}{2} + \frac{3}{2}, +, \ldots$.

In the literature $m$ (magnetic quantum number) is used in place of $n_1 + C$. It has therefore been shown that the quantum number $m$ defined by the diagonal term of $M_z = \frac{h}{2\pi}m$ is either a whole or a half number and that the selection rule

$$m \to \begin{cases} m+1 \\ m \\ m-1 \end{cases} \tag{15}$$

holds.

This result does not seem to lead much further than that which was obtained from the classical theory of multiple-periodic systems, but it must be borne in mind that in classical theory certain orbits frequently had to be ruled out by additional excluding rules. For instance, in the theory of the hydrogen atom, orbits leading to a collision between the electron and the nucleus were excluded. In the present theory no such additional rules are necessary, a fact which must be regarded as an essential step forward. To this must be added the full justification of half and whole quantum numbers, which so far could not be explained theoretically, while the empirical facts necessarily led to the introduction of the former, as already shown.

# LECTURE 17

Free systems as limiting cases of axially symmetrical systems—Quantization of the total angular momentum—Comparison with the theory of directional quantization—Intensities of the Zeeman components of a spectral line—Remarks on the theory of Zeeman separation.

The detailed presentation of the derivations in the preceding lecture are, I believe, sufficient to show the method clearly. From now on I shall mainly emphasize results. Proceeding along the same line of reasoning we arrive at a new quantum number $j$ which determines, in the limit $\lambda \to 0$, the diagonal terms of $M^2$, as follows,

$$\text{Diagonal terms of } M^2 = \left(\frac{h}{2\pi}\right)^2 j\,(j+1)\,. \tag{1}$$

Further $j$ is always equal to the maximum value of the quantum number $m$ and therefore is a whole or a half number. The selection rule

$$j \to \begin{cases} j+1 \\ j \\ j-1 \end{cases} \tag{2}$$

holds. The proof is quite similar to that in classical theory. In the latter a new rectangular system of coördinates is introduced whose $z$-axis coincides, for $\lambda = 0$, with the fixed direction of the angular momentum. Considerations concerning the total angular momentum are quite similar, for this system, to those for $M_z$ in the case of axial symmetry. A linear combination of the coördinate matrices is formed which corresponds formally to a rotation of the system of coördinates into the desired position ($z$-axis parallel to the angular momentum). The equations obtained from these expressions have a finite number of matrix elements of a type similar to that obtained previously for the coördinates themselves except that $M^2$ occurs instead of $M_z$. We find from $M_z$ and $M$ by means of the identity

$$\left(M_x + iM_y\right)\left(M_x - iM_y\right) = M_x^2 + M_y^2 - i\left(M_x M_y - M_y M_x\right)$$
$$= M^2 - M_z^2 + i\epsilon M_z$$

and the above relations for $M_x$ and $M_y$ [Equations (3), (4), Lecture (16)]

$$\left.\begin{aligned}
\left(M_x + i M_y\right) (j, m-1; j, m) &= \frac{1}{2}\frac{h}{2\pi}\sqrt{j(j+1) - m(m-1)} \\
\left(M_x - i M_y\right) (j, m; j, m-1) &= \frac{1}{2}\frac{h}{2\pi}\sqrt{j(j+1) - m(m-1)}
\end{aligned}\right\}$$
(3)

It is also possible to express explicitly the coördinates $q_{lx}, q_{ly}, q_{lz}$ in terms of the quantum numbers $m, j$. The result can be most clearly stated if written separately for the three possible jumps of $j$.

$$j \to j \left\{\begin{aligned}
&\left(q_{lx} + i q_{ly}\right)(j, m-1; j, m) \\
&\quad = A\frac{1}{2}\sqrt{j(j+1) - m(m-1)} \\
&\left(q_{lx} - i q_{ly}\right)(j, m; j, m-1) \\
&\quad = A\frac{1}{2}\sqrt{j(j+1) - m(m-1)} \\
&q_{lz}(j, m) = Am
\end{aligned}\right\}$$
(4)

$$j \to j-1 \left\{\begin{aligned}
&\left(q_{lx} + i q_{ly}\right)(j, m-1; j-1, m) \\
&\quad = B\frac{1}{2}\sqrt{(j-m)(j-m+1)} \\
&\left(q_{lx} - i q_{ly}\right)(j, m; j-1, m-1) \\
&\quad = -B\frac{1}{2}\sqrt{(j+m)(j+m-1)} \\
&q_{lz}(j, m; j-1, m) = B\sqrt{j^2 - m^2}
\end{aligned}\right\}$$
(5)

$$j \to j+1 \left\{\begin{aligned}
&\left(q_{lx} + i q_{ly}\right)(j, m; j+1, m+1) \\
&\quad = C\frac{1}{2}\sqrt{(j+m+2)(j+m+1)} \\
&\left(q_{lx} - i q_{ly}\right)(j, m; j+1, m-1) \\
&\quad = C\frac{1}{2}\sqrt{(j-m+2)(j-m+1)} \\
&q_{lz}(j, m; j+1, m) = C\sqrt{(j+1)^2 - m^2}
\end{aligned}\right\}$$
(6)

where $A, B, C$, depend in some way on the other quantum numbers of the system.

These expressions, as the perturbation and dispersion formulas given above (Lecture 14), had been found previously by considerations of correspondence before they were derived by the methods of our theory. This is seen most easily from formula (4) for $j \to j$ by going over to the limiting case of large quantum numbers $m$, $j$. Then 1 can be neglected compared with $m$ and $j$ and we find for the ratio of the intensities of the two circular vibrations and the linear oscillation:

$$\left| q_{lx} + i q_{ly} \right|^2 : \left| q_{lx} - i q_{ly} \right|^2 : \left| q_{lz} \right|^2$$

$$= \frac{1}{4} \left( j^2 - m^2 \right) : \frac{1}{4} \left( j^2 - m^2 \right) : m^2$$

$$= \frac{1}{4} \left( M^2 - M_z^2 \right) : \frac{1}{4} \left( M^2 - M_z^2 \right) : M_z^2$$

$$= \frac{1}{4} \left( M_x^2 + M_y^2 \right) : \frac{1}{4} \left( M_x^2 - M_y^2 \right) : M_z^2 \tag{7}$$

where $M$, $M_x$, $M_y$, $M_z$ denote the total angular momentum and its components in the quantized state in question $m$, $j$. These formulas can, however, be obtained classically as follows:

Consider the motion of the electrons as represented by the motion of their electric center of gravity $S$. The motion of $S$ is decomposed into a linear component parallel to the angular momentum $\mathbf{M}$ and two circular components rotating in opposite directions perpendicular to $\mathbf{M}$; the first component corresponds alone to the jump $j \to j$, the two others correspond to jumps $j \to j \pm 1$. The linear oscillation parallel to $\mathbf{M}$ is given by

$$\left. \begin{array}{l} q_x = a \, \sin \omega t \, \cos \phi \, \sin \theta \\ q_y = a \, \sin \omega t \, \sin \phi \, \sin \theta \\ q_z = a \, \sin \omega t \, \cos \theta \end{array} \right\}$$

where $\phi$, $\theta$ are the polar coördinates of the direction of $\mathbf{M}$ in a fixed system of coördinates. The motion in the $xy$-plane can be decomposed into two rotations in opposite directions,

$$\left. \begin{array}{l} q_x = q_x' + q_x'' \\ q_y = q_y' + q_y'' \end{array} \right\}$$

where

$$q'_x = \frac{a}{2} \sin \theta \sin (\omega t + \phi), \quad q''_x = \frac{a}{2} \sin \theta \sin (\omega t - \phi)$$
$$q'_y = -\frac{a}{2} \sin \theta \cos (\omega t + \phi), \quad q''_y = \frac{a}{2} \sin \theta \cos (\omega t - \phi)$$

To these correspond the jumps $m \rightarrow m \pm 1$, while the component $q_z$ corresponds to jumps $m \rightarrow m$. The intensities are proportional to

$$q'^2_x + q'^2_y : q''^2_x + q''^2_y : q^2_z = \frac{a^2}{4} \sin^2 \theta : \frac{a^2}{4} \sin^2 \theta : a^2 \cos^2 \theta. \quad (8)$$

If a weak outer field is now established in the $z$-direction, the whole atom rotates slowly around this direction. The circular frequencies of the two circular component oscillations are thereby changed slightly as well as their intensities, but in the limiting case of an infinitely weak field these changes can be neglected. Then we have

$$M_z = M \cos \theta, \ \sqrt{M_x^2 + M_y^2} = M \sin \theta .$$

If these are introduced in the above relations the formula (8) given above is shown to be the limiting value of the rigorous formula (4), obtained through the new quantum theory. The cases $j \rightarrow j \pm 1$ can be interpreted in a precisely similar manner.

Historically, in fact, the opposite has been done. Starting from the classical motion, intensity formulas have been found, which should be correct for large quantum numbers, but which need a correction for smaller $m$ and $j$. This correction has been found in a number of ways. Goudsmit and Kronig have used the so-called "boundary principle," that is the requirement that the intensities must vanish when either of the states disappears. It was noted above that $j$ is the maximum value of $m$; therefore, the intensities of all jumps for which $j$ remains unchanged but to changes vanish if we place $m = j + 1$. Our formula is seen to satisfy this condition.

The first incentive to the investigation of these intensity laws was the experimental researches on the relative brightness of the components of the Zeeman effect. These, under the direction of Ornstein, were carried out by Moll, Burgers, Dorgelo and others. These investigators at Utrecht first found empirical whole-number laws for the intensities in the Zeeman effect and gave simple rules for their calculation. The theory then developed as outlined above.

Our formulas fit exactly the case of an atom in a weak magnetic field. The number and position of the components into which a line is split up by the field cannot as yet be calculated theoretically. We shall return to this question later. But, if we consider the system of split-up lines given experimentally, we can read off from it the value of $j$. Thus, starting from the middle, the Zeeman components vibrating parallel to the field ($m \to m$) are assigned half or whole numbers and the largest value of $m$ is equal to $j$. We have therefrom the corresponding values of $m$ and $j$ for each line and hence can calculate the relative intensities by our formula. Comparisons with observations have in all cases verified the theory.

As already stated, the actual magnetic split of spectral lines is not given by the present theory as it stands. For if linear terms in the field strength, such as

$$\frac{e}{2\mu c} \, |\mathbf{H}| \, M_z$$

are alone taken into account in the magnetic additional terms of the energy, as usual, then we obtain, on account of our formula for $M_z$, the equidistant term sequence

$$\frac{e}{4\pi \mu c} \, |\mathbf{H}| \, m = \nu_m m$$

with the *normal* separation $\nu_m$ corresponding to the classical Larmor precession. Experimentally, however, the separation is found to be $g\nu_m$; the numbers $g$ have been determined empirically by Landé as functions of the quantum numbers characterizing the corresponding spectral lines. All attempts to derive these $g$-formulas from classical models led to similar formulas, but never the correct ones. In these formulas the square of the angular momentum $M^2$ enters, among other quantities. The former, of course, has always been replaced so far by $(jh/2\pi)^2$, but Landé's empirical formulas always require the expression $j(j + 1)$ instead of $j^2$. Our new quantum theory gives in fact $M^2 = (h/2\pi)^2 j(j+1)$, a circumstance which encourages us to further researches.

We have now to turn our attention to a new point which distinguishes the new from the classical theory and may result in 'rendering Larmor's theorem *invalid*: This new point is the neglect of the terms in $\mathbf{H}^2$ entering in the expression for the magnetic energy. These can certainly be neglected in the classical theory for orbits of small dimensions but can *not* be neglected for orbits of large dimensions or hyperbolic paths. In the limiting case of a free electron the period of revolution is just twice the normal Larmor precession. In quantum mechanics all these orbits, the distant as well as the near ones, are so intimately connected due to the peculiar kinematics and geometry that the justification of neglecting $\mathbf{H}^2$ is no longer evident, for the probability of transition even from the unexcited state to that of a free electron is always considerable. For the oscillator we have certainly the normal Zeeman effect. For the nuclear atom, however, it is possible that the intimate connection between inner and outer orbits may lead to different results. There are, on the other hand, powerful arguments against such an explanation, particularly the intimate connection between the anomalous Zeeman effect and the multiplet structure of spectral lines. A new physical concept seems to be required here. Such an idea has been formulated by Uhlenbeck and Goudsmit, but here I can only indicate it. Pauli, from the study of multiplets, has been led to attribute to each electron not three quantum numbers, as would correspond to its number of degrees of freedom, but *four*. Until now this has been considered as something purely formal, to be eliminated if possible. Uhlenbeck and Goudsmit, however, take this hypothesis earnestly. They attribute to the electron a proper rotation and a corresponding magnetic field determined by the fourth quantum number. Preliminary calculations by Heisenberg and Jordan have shown that this idea forms a basis for an exact theory of the abnormal Zeeman effect, but I am unable to give further details on this at present.

# LECTURE 18

Pauli's theory of the hydrogen atom.

We now come to the crucial question for the whole new theory: Is it able to account for the properties of the hydrogen atom? Let us recall that the explanation of the hydrogen spectrum (Balmer's formula) was the first great success of Bohr's theory and has since remained its keynote. If the new theory failed here it would have to be abandoned in spite of its many conceptual advantages, but, as Pauli has shown, it stands the test successfully. I can give here only the fundamental ideas and results of this development, as yet unpublished.

In the classical theory of Keplerian motion it is customary to operate with polar coördinates. This process fails here because it does not seem possible to consider angular variables as matrices. Pauli avoided this difficulty by retaining rectangular coördinates and introducing an additional coördinate, the radius vector $r$, which is related to $x$, $y$, $z$, by the relation

$$r^2 = x^2 + y^2 + z^2.$$

The process will first be explained using the classical model. We have the energy function

$$H = \frac{1}{2\mu}\mathbf{p}^2 - \frac{Ze^2}{r} \tag{1}$$

and the equations of motion,

$$\dot{\mathbf{r}} = \frac{1}{\mu}\mathbf{p}, \quad \dot{\mathbf{p}} = -\frac{Ze^2\mathbf{r}}{r^3}. \tag{2}$$

From these follows that the angular momentum

$$\mathbf{M} = \mathbf{r} \times \mathbf{p} \tag{3}$$

is constant with respect to time. There follows, further, by using the relation

$$\mathbf{M} \times \mathbf{r} = (\mathbf{r} \times \mathbf{p}) \times \mathbf{r} = \mathbf{p}r^2 - (\mathbf{p} \cdot \mathbf{r})\mathbf{r}$$

that the vector

$$\mathbf{A} = \frac{1}{Ze^2\mu}\mathbf{M} \times \mathbf{p} + \frac{\mathbf{r}}{r} \qquad (4)$$

is also constant with respect to time. Placing $M = |\mathbf{M}|$ we obtain at once

$$\mathbf{A} \cdot \mathbf{r} = -\frac{1}{Ze^2\mu}M^2 + r,$$

which is the equation of a conic. If we take the $xy$-plane in the plane of this curve and the $x$-axis in the direction of $\mathbf{A}$, that is

$$\left.\begin{array}{ll} x = r\cos\phi & A_x = |\mathbf{A}| = A \\ y = r\sin\phi & A_y = 0 \\ z = 0 & A_z = 0 \end{array}\right\}$$

we obtain

$$Ar\cos\phi = -\frac{M^2}{Ze^2\mu} + r$$

or

$$r = \frac{M^2}{Ze^2\mu}\left(\frac{1}{1 - A\cos\phi}\right).$$

$A$ is therefore the eccentricity and we find for the energy

$$W\frac{2}{Z^2e^4\mu}M^2 = A^2 - 1. \qquad (5)$$

This calculation can be repeated, with only slight changes, in matrix mechanics.

The matrices $x, y, z, r$ are commutative among themselves, as also the momentum matrices $p_x, p_y, p_z, p_r$. The following are also commutative:

$$x \quad \text{with } p_y, p_z \ldots$$
$$p_x \text{ with } y, z \ldots$$

but,

$$p_x x - x p_x = \frac{h}{2\pi i}, \text{ etc.}$$

The energy and the equations of motion are the same as above. From the latter there follows at once, as has been shown generally before, that the angular momentum is constant in time.

It can be shown further that the vector

$$\mathbf{A} = \frac{1}{Ze^2\mu} \frac{1}{2} \left( \mathbf{M} \times \mathbf{p} + \mathbf{p} \times \mathbf{M} \right) + \frac{\mathbf{r}}{r}$$

is constant in time. To prove this a longer calculation is necessary and secondary commutation relations are needed, as for instance

$$y M_z - M_z y = -\frac{h}{2\pi i} x, \qquad M_y z - z M_y = \frac{h}{2\pi i}$$

$$\cdots\cdots\cdots\cdots\cdots \qquad\qquad \cdots\cdots\cdots\cdots\cdots$$

$$p_x r - r p_x = \frac{h}{2\pi i} \frac{x}{r}, \cdots.$$

$$p_x \frac{x}{r} - \frac{x}{r} p_x = \frac{h}{2\pi i} \frac{y^2 - z^2}{r^3}, \cdots p_x \frac{y}{r} - \frac{y}{r} p_x = -\frac{h}{2\pi i} \frac{xy}{r^3}.$$

Derivatives with respect to time are transformed by means of the formula

$$\frac{d}{dt} \frac{x}{r} = \frac{2\pi i}{h} \left( W \frac{x}{r} - \frac{x}{r} W \right) = \frac{2\pi i}{h} \frac{1}{2\mu} \left( p^2 \frac{x}{r} - \frac{x}{r} p^2 \right).$$

The problem is now to find the constant vectors $\mathbf{M}$ and $\mathbf{A}$. For these the following commutation relations hold:

$$\left.\begin{aligned}
M_x M_y - M_y M_x &= -\frac{h}{2\pi i} M_z \cdots \\
A_x M_y - M_y A_x &= -\frac{h}{2\pi i} A_z \cdots \\
A_x A_y - A_y A_x &= -\frac{h}{2\pi i} A_z \cdots
\end{aligned}\right\} \qquad (6)$$

$$A_x A_y - A_y A_x = \frac{h}{2\pi i} \frac{2}{\mu Z^2 e^4} W M_z \cdots \qquad (7)$$

Finally, the following equation is found

$$\left( M^2 + \frac{h^2}{4\pi^2} \right) \frac{2}{\mu Z^2 e^4} W = A^2 - 1. \qquad (8)$$

This equation differs from the corresponding classical equation only by the term $\frac{h^2}{4\pi^2}$ added to $M^2$. This is just one important characteristic of the new theory.

In the solution of these equations $W$ is always a diagonal matrix, but the constant components of the vector matrices $\mathbf{p}$, $\mathbf{A}$ are not diagonal matrices, as was shown in general above. According to our previous results, the requirement that, besides $W$, $p_z$ and $p^2$ be diagonal matrices has a definite meaning, i.e., the addition of a weak axially symmetrical perturbing field of force the energy of which depends on $p_z$ and $p$.

The same method used above (Lecture 16) can now be applied to determine the vector **P**. Then Equations (6) are exactly the same as Equations (4) and (5), Lecture 16, except that the coördinates $q_{lx}$, $q_{ly}$, $q_{lz}$ are replaced by $A_x$, $A_y$, $A_z$. Instead of the quantum numbers $n_1$, $n_2$ we write the usual symbols $k$ and $m$, where $k$ determines the total angular momentum, denoted by $j$ above, and $m$ its $z$-component. We then have

$$p_z(k, m; k, m) = \frac{h}{2\pi} m, \qquad p^2 = \frac{h^2}{4\pi^2} k(k+1)$$

$$|p_x(k, m; k, m \pm 1)|^2 = |p_y(k, m; k, m \pm 1)|^2$$

$$= \frac{1}{4} \frac{h^2}{4\pi^2} [k(k+1) - m(m+1)] \qquad (9)$$

where $m$ runs through a complete series of half or whole numbers from $-k$ to $+k$. Further we obtain for $A_x$, $A_y$, $A_z$ expressions quite similar to those found before for $q_{lx}$, $q_{ly}$, $q_{lz}$, for example the following one:

$$|A_x(k+1, m; k, m \pm 1)|^2 = |A_y(k+1, m; k, m \pm 1)|^2$$

$$= \frac{1}{4} C(k+1, k)(k \mp m)(k \mp m + 1)$$

$$|A_z(k+1, m; k, m)|^2 = C(k+1, k)\left((k+1)^2 - m^2\right).$$

Consider in Equations (7) a given $W$ and the smallest possible $k$. Then a closer discussion shows that the equations in question can be satisfied if $m$ is zero and only zero, hence $k_{min} = 0$. Herein is contained the *integrality* of $k$ and $m$.

Formula (7) gives further for the functions $C(k+1, k) = C(k, k+1)$ the equation,

$$(2k - 1) C(k, k - 1) - (2k + 3) C(k+1, k) = \frac{|W|}{Rh}, k \neq 0.$$

$W$ is here assumed to be *negative*, i.e., ellipse, not hyperbola; $R$ is Rydberg's constant. As solution of this equation we obtain

$$C\,(k+1,k) = \frac{|W|}{Rh}\frac{(k_m - k)\,(k_m + k + 2)}{(2k+1)\,(2k+3)}$$

where $k_m$ is the maximum value of $k$ for a given $|W|$. We now have the components of **A** and, therefore, also the value of

$$A^2 = A_x^2 + A_y^2 + A_z^2,$$

that is,

$$\begin{aligned}
A^2\,(k, m; k, m) &= (k+1)\,(2k+3)\,C\,(k+1,k) \\
&\quad + k\,(2k-1)\,C\,(k,k-1) \\
&= \frac{|W|}{Rh}\left[k_m^2 + 2k_m - k\,(k+1)\right].
\end{aligned} \tag{10}$$

There follows finally from Equation (8):

$$1 = \frac{|W|}{Rh}\,(k_m + 1)^2.$$

If we write $n = k_m + 1$, then $n$ corresponds to the main quantum number of Bohr's theory and takes the values $1, 2, 3, \ldots$. For a given $n$, $k$ has the values $k = 0, 1, 2, \ldots n - 1$. We have thus found Balmer's formula

$$W = -\frac{Rh}{n^2}(n = 1, 2, 3 \ldots) \tag{11}$$

and have shown at the same time how each term is split up on removal of the degeneration by the addition of weak perturbing forces. This split is given by $k = 0, 1, 2, \ldots n - 1; m = -k, -k + 1, \ldots k - 1, k$.

A characteristic trait of the new theory is that the value $k = n$ does not occur. It follows in particular that in the unexcited state $n = 1, k = 0$ and, therefore, $m = 0$; in other words, the normal state is not magnetic. This result must be revised, however, if the rotating magnetic electron of Uhlenbeck and Goudsmit is accepted.

Pauli has succeeded in deriving in a similar way the Stark effect for the hydrogen atom. In this case also no additional conditions need be imposed. The same holds in the case where an electric and a magnetic field act in arbitrary directions (crossed fields). Just here the classical theory of multipleperiodic systems encountered great difficulties, for the frequencies of the overtones are made up of two fundamental periods (the electric frequency $v_c$ and the magnetic Larmor frequency $v_m$) and, therefore, protracted commensurabilities appear when the fields are varied, that is equations of the form

$$\tau_1 v_e + \tau_2 v_m = 0$$

with integral $\tau_1$, $\tau_2$. This means that arbitrarily small adiabatic changes, of the electric field, for instance, will produce degeneration. The validity of Ehrenfest's adiabatic hypothesis is no longer certain and, therefore, the quantum rules become doubtful. All these difficulties disappear in the new theory.

Pauli has also attacked the theory of fine structure (relativistic change of mass), without yet quite attaining his goal.

# LECTURE 19

Connection with the theory of Hermitian forms—Aperiodic motions and continuous spectra.

Let us now inquire how aperiodic motions, such as hyperbolic orbits in the hydrogen atom, can be treated in the new theory. It is to be expected *a priori* that there is no essential difference in the treatment of periodic and aperiodic processes because the postulate of periodicity does not appear explicitly in the fundamental equations. The notion of matrix can be generalized at once so as to permit the representation of aperiodic processes. The indices $n$, $m$ have only to be considered as continuous variables and the matrix product defined by the integral

$$pq = \left( \int p(nk)q(km)dk \right),$$

but difficulties appear at once if we attempt to generalize the notion of unit matrix to include these continuous matrices. This must be done because the unit matrix enters in the commutation relation

$$pq - qp = \frac{h}{2\pi i} 1. \tag{1}$$

That function $f(nm)$ is to be taken as unit matrix which vanishes for $n \neq m$ and becomes infinite for $n = m$ in such a way that the integrals

$$\int \int (nk)dk \quad \text{and} \quad \int \int (kn)dk$$

become unity, for then

$$qf = \left( \int q(nk)f(km)dk \right) = (q(nk)) = q$$

and at the same time $f_q = q$. It is clear that operating with such unusual functions is not convenient. In circumventing this difficulty a way indicated by entirely different lines of reasoning has been followed.

In classical mechanics the known theory of oscillation of a system is intimately related to the theory of quadratic forms. Oscillations occur when the potential energy is a "definite" quadratic form of the variables, i.e., one which does not change its sign. For two variables $x$, $y$, for instance,

$$U = \frac{1}{2}\left(a_{11}x^2 + 2a_{12}xy + a_{22}y^2\right).$$

Oscillations are obtained in the simplest way by transforming this form to a sum of squares by means of the linear transformation,

$$x = b_{11}\xi + b_{12}\eta$$
$$y = b_{21}\xi + b_{22}\eta.$$

We now try to effect this transformation in such a way that the kinetic energy $T = \frac{m}{2}(\dot{x}^2 + \dot{y}^2)$, which is already a sum of squares, retains this characteristic and is transformed into

$$T = \frac{m}{2}\left(\dot{\xi}^2 + \dot{\eta}^2\right).$$

As the velocities are transformed in the same way as the coordinates we have the condition that the linear transformation must leave the quantity $x^2 + y^2$ invariant, i.e.,

$$x^2 + y^2 = \xi^2 + \eta^2.$$

Such transformations are called "orthogonal." They correspond geometrically to *rotations* of the coördinate system around the origin in the $xy$-plane, because for such a rotation the distance $r$ or $r^2 = x^2 + y^2$ is in fact invariant. Now an equation of the form

$$a_{11}x^2 + 2a_{12}xy + a_{22}y^2 = 2U = \text{const.}$$

with a definite left-hand side represents an ellipse with the origin at the center. This ellipse has two principal axes $a$, $b$. If these are chosen as $\xi\eta$-axes the equation of the ellipse $\frac{\xi^2}{a^2} + \frac{\eta^2}{b^2}$ becomes

$$2U = \kappa_1\xi^2 + \kappa_2\eta^2,$$

where $a^2 = 2U/\kappa_1$ and $b^2 = 2U/\kappa_2$, and we have the desired expression. The equations of motion are now

$$m\ddot{\xi} + \kappa_1\xi = 0, \quad m\ddot{\eta} + \kappa_2\eta = 0$$

and, therefore, the frequencies are

$$\nu_1 = \frac{1}{2\pi}\sqrt{\frac{\kappa_1}{m}} = \frac{1}{2\pi a}\sqrt{\frac{2U}{m}}$$

$$\nu_2 = \frac{1}{2\pi}\sqrt{\frac{\kappa_2}{m}} = \frac{1}{2\pi b}\sqrt{\frac{2U}{m}}.$$

Similar relations hold for any arbitrary number of degrees of freedom.

Formerly, in order to interpret line spectra, attempts were made to construct mechanical systems which should have just the observed lines as proper frequencies, but none of them gave rise to useful results; in other words, none led to oscillating systems built out of known elementary particles (protons and electrons) governed by known laws or reasonable modifications of them and having these frequencies.

In our new theory the relation between the principal axes of the quadratic form and the frequency enters again, except that instead of the observed frequencies the values of the *terms* or *energy levels* occur. These appear as the reciprocal axes of a certain Hermitian form. The frequencies appear later as differences between terms.

To each matrix $a = (a(nm))$ corresponds a bilinear form

$$A(xy) = \sum_{m,n} a(nm)x_n J_m \tag{2}$$

of two systems of variables. If the matrix is Hermitian,

$$\tilde{a} = a^*, \quad a(mn) = a^*(nm), \tag{3}$$

where the symbol $\sim$ indicates interchange of rows and columns and the symbol $*$ change to conjugate complex quantities, then the form $A$ assumes real values if we place for the variables $y_n$ the values conjugate to $x_n$:

$$A(xx^*) = \sum_{n,m} a(nm)x_n x_m^* \text{ is real.} \tag{4}$$

Let us recall the rule, easily proved, that $(\tilde{a}b) = \tilde{b}\tilde{a}$. Applying a linear transformation to $x_n$,

$$x_n = \sum_l v(ln)y_l \tag{5}$$

with the complex matrix $v = (v(ln))$ the bilinear form $A$ is transformed into

$$A(xx^*) = B(yy^*) = \sum_{n,m} b(nm)y_n y_m^*$$

where

$$b(nm) = \sum_{k,l} v(nk)a(kl)v^*(ml)$$

or in matrix notation

$$b = va\bar{v}^*. \tag{6}$$

The matrix $b$ is said to be the transform of $a$. The matrix $b$ is again Hermitian, for

$$\bar{b} = v^*\bar{a}\bar{v} = v^*a^*\bar{v} = b^*. \tag{7}$$

The matrix $v$ is said to be orthogonal if the corresponding transformation leaves the Hermitian unit form

$$E(xx^*) = \sum_n x_n x_n^*$$

invariant. According to the result just obtained this is true when, and only when,

$$v\bar{v}^* = 1 \quad \text{or} \quad \bar{v}^* = v^{-1}. \tag{8}$$

For a finite number of variables the same theorems hold in general for Hermitian forms as for real quadratic forms. Here also there always exists an orthogonal principal-axis transformation by which $A$ becomes a sum of squares,

$$A(xx^*) = \sum_n W_n y_n y_n^*.$$

For matrices this means that there exists a matrix $v$ for which

$$v\bar{v}^* = 1 \text{ and } va\bar{v}^* = vav^{-1} = W \tag{9}$$

where $W = (W_n \delta_{mn})$ is a diagonal matrix.

A similar theorem exists for infinite matrices in all cases so far investigated. It may happen that, in the right-hand side of these equations, $n$ takes, besides a discrete series of values, a continuous series, to each of which correspond integral components in our formulas. The quantities $W_n$ are called "characteristic values," their totality constitutes the "mathematical" spectrum of the form, consisting of a "point" spectrum and an "interval" spectrum. This spectrum is, as pointed out before and as will be shown presently, identical with the "term spectrum" of physics, while the "frequency spectrum" is obtained from the former by difference relations.

The transformation along the principal axes gives at once the solution of the dynamical problem which can be formulated as follows: Let any system of coördinates and momenta $q_k^0$, $p_k^0$ be given satisfying the commutation relations, for instance those of a system of uncoupled resonators. A transformation $(q_k^0 p_k^0) \rightarrow (q_k p_k)$ must be found leaving the commutation relations (1) invariant and transforming the energy into a diagonal matrix. According to the theorem above an orthogonal matrix $S$ for which

$$S \bar{S}^* = 1 \quad S^* S = 1$$

exists such that by the transformation,

$$\left. \begin{array}{l} p_k = S p_k^0 \bar{S}^* = S p_k^0 S^{-1} \\ q_k = S q_k^0 \bar{S}^* = S q_k^0 S^{-1}, \end{array} \right\} \tag{10}$$

(1) the Hermitian character of $p_k^0$, $q_k^0$ is conserved for $p_k$, $q_k$,
(2) the commutation relations remain invariant,
(3) the energy is transformed into a diagonal matrix

$$H(pq) = SH(p^0 q^0)S^{-1} = W. \tag{11}$$

It is important to add that the transforming matrix and the series of $W$-values may have continuous parts. This has been shown by Hilbert and Hellinger for a certain class of infinite matrices belonging to the so-called "bounded forms." The same must be expected *a priori* of our matrices which in general do not satisfy the condition of bounded forms. A continuous series of energy values $W$ or of terms $\frac{W}{h}$ is thus obtained. Accordingly there are evidently three kinds of elements in the coördinate matrices:

(1) Those for which both $m$ and $n$ belong to the discrete series of values of $W$. These correspond to jumps between periodic orbits and give the line spectrum.

(2) Those for which $n$ belongs to the discrete and $m$ to the continuous series of values of $W$ or conversely. These correspond to jumps between periodic and aperiodic orbits and give those known continuous spectra which exist beyond the limits of line series.

(3) Those for which both $n$ and $m$ belong to the continuous series of $W$-values. These correspond to jumps between two aperiodic orbits and give the continuous spectrum in the proper sense.

The actual mathematical calculation of the continuous spectrum on the basis of this theory is, however, impossible partly on account of the intricacy of the calculations and more particularly because of difficulties of convergence. The integrals are improper or altogether divergent. This is related to the fact that aperiodic motions approach uniform rectilinear motion asymptotically in the limit of infinite distance. This motion has evidently no period and represents the case of greatest singularity. It is not amenable to matrix representation, even if continuous matrices are mustered for the purpose.

# LECTURE 20

Substitution of the matrix calculus by the general operational calculus for improved treatment of aperiodic motions—Concluding remarks.

In the case of aperiodic straight-line motion, another procedure must therefore be adopted, which Wiener and I have recently developed. Only an outline of the fundamental ideas can be given here. An Hermitian form can be associated with every matrix, as already shown; likewise a linear transformation of the form used above,

$$x_n = \sum_t v(ln)y_l. \tag{1}$$

Then the product of two matrices corresponds to the successive application of two such transformations:

$$x_n = \sum_k q(nk)y_k, \quad y_k = \sum p(km)z_m.$$

These together give

$$x_n = \sum_m qp(nm)z_m, \tag{2}$$

where

$$qp(nm) = \sum_k q(nk)p(km).$$

As seen, the matrix enters here not as a "quantity" or "system of quantities," but as an *operator* which, from an infinite system of quantities $y_1, y_2 \ldots$, yields another system $x_1, x_2 \ldots$. The precise physical significance of these quantities is still very obscure. A calculus of operators can, therefore, be substituted for the matrix calculus and this method becomes fruitful if applied in the following way: An infinite system of quantities $x_1, x_2 \ldots$ may define a function with a continuous range of arguments; for instance these quantities may be taken as the coefficients of a Fourier series. It is advantageous to operate with this function instead of with the coefficients, because we then have the whole machinery of the calculus at our disposal, and differential or integral equations replace infinite sets of simultaneous equations in

an infinite number of variables. These equations possess solutions under certain conditions even when the original representation in series
collapses. Of course Fourier series will not be used here, but general
trigonometric series of the form

$$x(t) = \sum_n x_n e^{\frac{2\pi i}{b} W_n t}. \tag{3}$$

The coefficients $x_n$ are determined from the function $x(t)$ by taking
averages,

$$x_n = \lim_{T \to \infty} \frac{1}{2T} \int_{-T}^{T} x(s) e^{-\frac{2\pi i}{b} W_{n^s}} ds. \tag{4}$$

Instead of the matrix $q = (q(mn))$ we make use of the function of
two variables,

$$q(t, s) = \sum_{mn} q_{mn} e^{\frac{2\pi i}{b}(W_m t - W_n s)} \tag{5}$$

and of the derived "average operator,"

$$q = \left( \lim_{T \to \infty} \frac{1}{2T} \int_{-T}^{T} q(t, s) ds \cdots \right). \tag{6}$$

It is then easy to show that operator products, formed by the successive
application of operators, correspond to matrix products. An explicit
representation of operators is not, however, necessary. It is sufficient to
consider linear operators in general, that is such operators for which
the simple formula

$$q(x(t) + y(t)) = qx(t) + qy(t)$$

holds. Thus multiplication by a function of $t$, differentiation and
integration with respect to $t$, for example, are all operators. Of special
importance is the differential operator $D = d/dt$.

Under certain conditions a matrix can be associated with an operator. The sequence of energy levels of this matrix is ordered not with
respect to indices $m$, $n$, but relatively to the energy values themselves.
The elements of the matrix which correspond to the operator $q$ are
defined by

$$q(V, W) = \lim_{T \to \infty} \frac{1}{2T} \int_{-T}^{T} e^{-\frac{2\pi i}{b} Vt} q e^{\frac{2\pi i}{b} Wt} dt. \tag{7}$$

In many cases this matrix does not exist although the sum of elements in a row

$$q(t, W) = e^{-\frac{2\pi i}{h}Wt} q\, e^{\frac{2\pi i}{h}Wt} \qquad (8)$$

does. For the operator $D$ for instance,

$$q(V, W) = \lim_{T \to \infty} \frac{1}{2T} \int_{-T}^{T} \frac{2\pi i}{h} W e^{\frac{2\pi i}{h}(W-V)t}$$

$$dt = \begin{cases} \dfrac{2\pi i}{h} W & \text{if} \quad V = W \\ 0 & \text{otherwise,} \end{cases}$$

$q(V, W)$ therefore does not exist as a continuous function. If $W$ has discrete values, $q(V, W)$ is the diagonal matrix $(W_n \delta_{nm})$. The sum of the elements in a row

$$q(t, W) = \frac{2\pi i}{h} W$$

however, always exists. From this example it is seen how the operational method permits a treatment of singular cases where the matrix representation breaks down.

A more exhaustive treatment of the method is as yet unwarranted. Suffice to say that it has been possible to show that, in the case of the harmonic oscillator, the operational calculus gives the same result as the matrix calculus. Moreover, treatment of uniform rectilinear motion is possible, a case where the matrix calculus breaks down completely. Investigations on the theorems of angular momentum, on hyperbolic orbits in the hydrogen atom and on similar problems are in progress.

In closing I should like to add a few general remarks. The first concerns the question of whether it is possible to visualize the laws of physics as formulated in this new manner and whether the processes in the atom can be conceived to exist in space and time. A definite answer will only be possible when we can see all the consequences of the new theory, perhaps only when new principles have been discovered. But it already seems certain that the usual conceptions of space and time are not rigorously compatible with the character of the new laws.

Consider for instance the hydrogen atom. The classical theory not only gives the orbits of the electron but seems to assign a meaning to the position of the electron at each instant. In the new theory the energy and moment of momentum of a state can be given, but it appears to be impossible to give any further description of this state as a geometrical orbit and even more impossible to fix the position of the electron at any instant. Space points and time points in the ordinary sense do not exist. These conceptions can only be introduced subsequently in limiting cases.

On the other hand it seems to me that we have a right to use the terms "orbit" or even "ellipse," "hyperbola," etc. in the new theory, if we agree to interpret them rationally and to understand by them the quantum processes which go over in the limit to the orbits, ellipses, hyperbolas, etc. of the classical theory. This not only gives a convenient terminology but expresses the following fact: The world of our imagination is narrower and more special in its logical structure than the world of physical things. Our imagination is restricted to a limiting case of possible physical processes. This philosophical point of view is not new: it has always been the guiding thought of physicists since Copernicus, and it came so clearly to the fore in the theory of relativity that philosophy was compelled to take a definite stand towards it. In the quantum theory this guiding principle assumes an even more predominant rôle, but in this case it is supported by such an enormous weight of evidence that a flat denial seems much more difficult than it was when the theory of relativity came up for consideration.

Only a further extension of the theory, which in all likelihood will be very laborious, will show whether the principles given above are really sufficient to explain atomic structure. Even if we are inclined to put faith in this possibility, it must be remembered that this is only the first step toward the solution of the riddles of the quantum theory. Our theory gives the *possible* states of the system, but no indication of whether a system is in a given state. It gives at most the probability of the jumps. However, the statement that a system is at a given time and place in a certain state probably has a meaning, which the present state of our theory does not allow us to formulate. This is also the case with regard to the problem of light-quanta. Here the Compton effect and the related experiments of Bothe and Geiger, Compton and Simon have shown that both the energy and the momentum of light travel as a projectile from atom to atom. But the existence of interference, that is the fact that light added to light can produce darkness, is just as certain. It cannot yet be seen how these two views can be reconciled or whether a matrix representation of the electromagnetic field will lead further. An attempt to treat the statistics of cavity radiation by the new method has resulted in the elimination of serious contradictions in the classical theory. Many puzzling questions remain which fall outside the scope of these lectures.

In the further development of the new quantum theory, the physicist cannot dispense with the aid of the mathematician. The close alliance between mathematics and physics which has reigned during the best periods of both sciences will, I hope, return and banish the mystic cloud in which physics has of late been enshrouded. The activity of the mathematician must however not carry him as far as in the theory of relativity, where the clarity of his reasoning has come to be hidden by the erection of a structure of pure speculation so vast that it is impossible to view it in its entirety. A single crystal can be clear, nevertheless a mass of fragments of this crystal is opaque. Even the theoretical physicist must be guided by the ideal of the closest possible contact with the world of facts. Only then do the formulas live and beget new life.

# EXCERPTS FROM THIRTY YEARS THAT SHOOK PHYSICS: THE STORY OF QUANTUM THEORY (CHAPTERS I AND IV)

By

GEORGE GAMOW

Illustrations by the Author

Courtesy of Dover Publications

# CHAPTER I

## M. PLANCK AND LIGHT QUANTA

The roots of Max Planck's revolutionary statement that light can be emitted and absorbed only in the form of certain discrete energy packages goes back to much earlier studies of Ludwig Boltzmann, James Clerk Maxwell, Josiah Willard Gibbs, and others on the statistical description of the thermal properties of material bodies. The Kinetic Theory of Heat considered heat to be the result of random motion of the numerous individual molecules of which all material bodies are formed. Since it would be impossible (and also purposeless) to follow the motion of each single individual molecule participating in thermal motion, the mathematical description of heat phenomena must necessarily use statistical method. Just as the government economist does not bother to know exactly how many acres are seeded by farmer John Doe or how many pigs he has, a physicist does not care about the position or velocity of a particular molecule of a gas which is formed by a very large number of individual molecules. All that counts here, and what is important for the economy of a country or the observed

967

macroscopic behavior of a gas, are the averages taken over a large number of farmers or molecules.

One of the basic laws of *Statistical Mechanics*, which is the study of the average values of physical properties for very large assemblies of individual particles involved in random motion, is the so-called *Equipartition Theorem*, which can be derived mathematically from the Newtonian laws of Mechanics. It states that: *The total energy contained in the assembly of a large number of individual particles exchanging energy among themselves through mutual collisions is shared equally (on the average) by all the particles.* If all particles are identical, as for example in a pure gas such as oxygen or neon, all particles will have on the average equal velocities and equal kinetic energies. Writing $E$ for the total energy available in the system, and $N$ for the total number of particles, we can say that the average energy per particle is $E/N$. If we have a collection of several kinds of particles, as in a mixture of two or more different gases, the more massive molecules will have the lesser velocities, so that their kinetic energies (proportional to the mass and the square of the velocity) will be on the average the same as those of the lighter molecules.

Consider, for example, a mixture of hydrogen and oxygen. Oxygen molecules, which are 16 times more massive than those of hydrogen, will have average velocity $\sqrt{16} = 4$ times smaller than the latter.*

While the equipartition law governs the *average distribution of energy* among the members of a large assembly of particles, the velocities and energies of individual particles may deviate from the averages, a phenomenon known as *statistical fluctuations*. The fluctuations can also be treated mathematically, resulting in curves showing the relative number of particles having velocities greater or less than the average for any given temperature. These curves, first calculated by J. Clerk Maxwell and carrying his name, are shown in Fig. 1 for three different temperatures of the gas. The use of the statistical method in the study of thermal motion of molecules was very successful in explaining the

---

*Since kinetic energy is the product of [mass] × [velocity]$^2$, this product will remain the same if the mass increases by a factor 16 and velocity decreases by a factor 4. In fact, $4^2 = 16$!

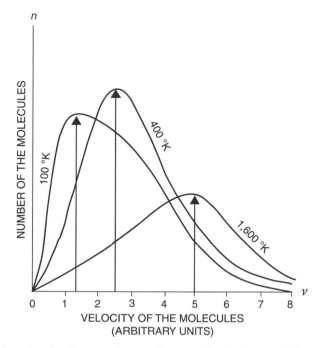

FIG. 1   Maxwell's distribution: the number of molecules having different velocities v is plotted against the velocities for three different temperatures, 100°, 400°, and 1600°K. Since the number of molecules in the container remains constant, the areas under the three curves are the same. The average velocities of the molecules increase proportionally to the square root of the absolute temperature.

thermal properties of material bodies, especially in the case of gases; in application to gases the theory is much simplified by the fact that gaseous molecules fly freely through space instead of being packed closely together as in liquids and solids.

## Statistical Mechanics and Thermal Radiation

Toward the end of the nineteenth century Lord Rayleigh and Sir James Jeans attempted to extend the statistical method, so helpful in understanding thermal properties of material bodies, to the problems of thermal radiation. All heated material bodies emit electromagnetic waves of different wavelengths. When the temperature is comparatively low-the boiling point of water, for example-the predominant wavelength of the emitted radiation is rather large. These waves do not affect the retina of our eye (that is, they are invisible) but are absorbed by our skin, giving the sensation of warmth, and one speaks therefore of heat or *infrared radiation*. When the temperature rises to about 600°C (characteristic of the heating units of an electric range) a faint red light is seen. At 2000°C (as in the filament of an electric bulb) a bright white light which contains all the wavelengths of the *visible radiation spectrum* from red to violet is emitted. At the still higher temperature of an electric arc, 4000°C, a considerable amount of invisible *ultraviolet radiation* is emitted, the intensity of which rapidly increases as the temperature rises still higher. At each given temperature there is one predominant vibration frequency for which the intensity is the highest, and as the temperature rises this predominant frequency becomes higher and higher. The situation is represented graphically in Fig. 2, which gives the distribution of intensity in the spectra corresponding to three different temperatures.

Comparing the curves in Figs. 1 and 2, we notice a remarkable qualitative similarity. While in the first case the increase of temperature moves the maximum of the curve to higher molecular velocities, in the second case the maximum moves to higher radiation frequencies. This similarity prompted Rayleigh and Jeans to apply to thermal radiation the same Equipartition Principle that had turned out to be so successful in the case of gas; that is, to assume that the total available energy of radiation is distributed equally among all possible vibration frequencies. This attempt led, however, to catastrophic results! The trouble was that, in spite of all similarities between a gas formed by individual molecules and thermal radiation formed by electromagnetic vibrations, there exists one drastic difference: while the number of gas molecules in a given enclosure is always finite even though usually very large, the number of possible electromagnetic vibrations in the same enclosure is always infinite. To understand this statement, one must remember that the wavemotion pattern in a cubical enclosure, let us say, is formed by the superposition of various standing waves having their nodes on the walls of the enclosure.

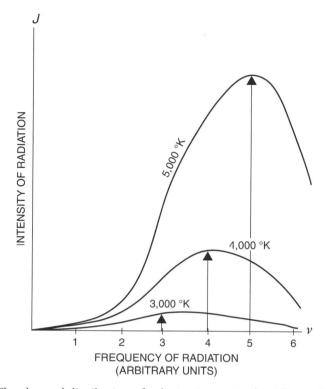

FIG. 2  The observed distribution of radiation intensities for different frequencies $\nu$ is plotted against the frequencies. Since the radiation energy content per unit volume increases as the fourth power of the absolute temperature $T$, the areas under the curves increase. The frequency corresponding to maximum intensity increases proportionally to the absolute temperature.

The situation can be visualized more easily in a simpler case of one-dimensional wave motion, as of a string fastened at its two ends. Since the ends of the string cannot move, the only possible vibrations are those shown in Fig. 3 and correspond in musical terminology to the fundamental tone and various overtones of the vibrating string. There may be one half-wave on the entire length of the string, two half-waves, three half-waves, ten half-waves, ... a hundred, a thousand, a million, a billion ... any number of half-waves. The corresponding vibration frequencies of various overtones will be double, triple ... tenfold, a hundredfold, a millionfold, a billionfold ... etc., of the basic tone.

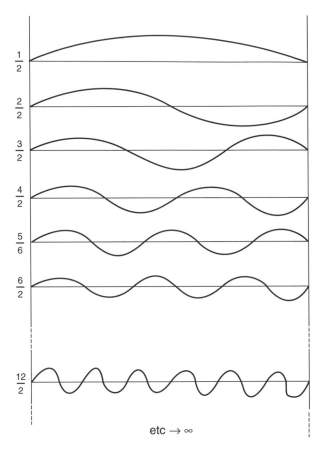

FIG. 3 The basic tone and higher overtones in the case of the one-dimensional continuum–for example, a violin string.

In the case of standing waves within a three-dimensional container, such as a cube, the situation will be similar though somewhat more complicated, leading to unlimited numbers of different vibrations with shorter and shorter wavelengths and correspondingly higher and higher frequencies. Thus, if $E$ is the total amount of radiant energy available in the container, the Equipartition Principle will lead to the conclusion that each individual vibration will be allotted $E/\infty$, an infinitely small amount of energy! The paradoxicalness of this conclusion is evident, but we can point it even more sharply by the following discussion.

Suppose we have a cubical container, known as "Jeans' cube," the inner walls of which are made of ideal mirrors reflecting 100 per cent of the light falling on them. Of course, such mirrors do not exist and cannot be manufactured; even the best mirror absorbs a small fraction of the incident light. But we can use the notion of such ideal mirrors in theoretical discussions as the limiting case of very good mirrors. Such reasoning, whereby one *thinks* what would be the result of an experiment in which ideal mirrors, frictionless surfaces, weightless bars, etc., are employed, is known as a "thought experiment" (*Gedankenexperiment* is the original term), and is often used in various branches - of theoretical physics. If we make in a wall of Jeans' cube a small window and shine in some light, closing the ideal shutter after that operation, the light will stay in for an indefinite time, being reflected to and fro from the ideal mirror walls. When we open the shutter sometime later we will observe a flash of the escaping light. The situation here is identical in principle to pumping some gas into a closed container and letting it out again later. Hydrogen gas in a glass container can stay indefinitely, representing an ideal case. But hydrogen will not stay long in a container made of palladium metal, since hydrogen molecules are known to diffuse rather easily through this material. Nor can one use a glass container for keeping hydrofluoric acid, which reacts chemically with glass walls. Thus, Jeans' cube with the ideal mirror walls is after all not such a fantastic thing!

There is, however, a difference between the gas and the radiation enclosed in a container. Since the molecules are not mathematical points but have certain finite diameters, they undergo numerous mutual collisions in which their energy can be exchanged. Thus, if we inject into a container some hot gas and some cool gas, mutual collisions between the molecules will rapidly slow down the fast ones and speed up the slow ones, resulting in even distribution of energy in accordance to the Equipartition Principle. In the case of an ideal gas formed by point-molecules, which of course does not exist in nature, mutual collisions would be absent and the hot fraction of the gas would remain hot while the cool fraction would remain cool. The exchange of energy between the molecules of an ideal gas can be stimulated, however, by introducing into the container one or several particles with finite though small diameters (Brownian particles). Colliding with them, fast point-sized molecules will communicate to them their energy, which will be communicated in turn to the other slower point-sized molecules.

In the case of light waves the situation is different, since two light beams crossing each other's path do not affect each other's propagation in any way.* Thus, to procure the exchange of energy between the standing waves of different lengths, we must introduce into the container small bodies that can absorb and re-emit all possible wavelengths, thus permitting energy exchange among all possible vibrations. Ordinary black bodies, such as charcoal, have this property, at least in the visible part of the spectrum, and we may imagine "ideal black bodies" which behave in the same way for all possible wavelengths. Placing into Jeans' cube a few particles of ideal coal dust, we will solve our energy-exchange problem.

Now let us perform a thought experiment, injecting into an originally empty Jeans' cube a certain amount of radiation of a given wavelength-let us say some red light. Immediately after injection, the interior of the cube will contain only red standing waves extending from wall to wall, while all other modes of vibrations will be absent. It is as if one strikes on a grand piano one single key. If, as it is in practice, there is only very weak energy exchange among different strings of the instrument, the tone will continue to sound until all the energy communicated to the string will be dissipated by damping. If, however, there is a leak of energy among the strings through the armature to which they are attached, other strings will begin to vibrate too until, according to the Equipartition Theorem, all 88 strings will have energy equal to 1/88 of the total energy communicated.

---

*To avoid an objection on the part of those readers who know much more than necessary for understanding this discussion, the author hastens to state that, according to modern quantum electrodynamics, some scattering of light by light must be expected because of virtual electron-pair formation. But Jeans and Planck did not know this.

FIG. 4  A piano with an unlimited number of keys extending into the ultrasonic region all the way to infinite frequencies. The equipartition law would require all the energy supplied by a musician to one of the low-frequency keys to travel all the way into the ultrasonic region out of the audible range!

But if a piano is to represent a fairly good analogy of the Jeans' cube, it must have many more keys extending beyond any limit to the right into the ultrasonic region (Fig. 4). Thus the energy communicated to one string in an audible region would travel to the right into the region of higher pitches and be lost in the infinitely far regions of the ultrasonic vibrations, and a piece of music played on such a piano would turn into a sharp shrill. Similarly *the energy of red light injected into Jeans' cube would turn into blue, violet, ultraviolet, X-rays, γ-rays, and so on without any limit.* It would be foolhardy to sit in front of a fireplace since the red light coming from the friendly glowing cinders would quickly turn into dangerous high-frequency radiation of fission products!

The runaway of energy into the high-pitch region does not represent any real danger to concert pianists, not only because the keyboard is limited on the right, but mostly because, as was mentioned before, the vibration of each string is damped too fast to permit a transfer of even a small part of energy to a neighboring string. In the case of radiant energy, however, the situation is much more serious, and, if the Equipartition Law should hold in that case, the open door of a boiler would be an excellent source of X- and $\gamma$-rays. Clearly something must be wrong with the arguments of nineteenth-century physics, and some drastic changes must be made to avoid the Ultraviolet Catastrophe, which is expected theoretically but never occurs in reality.

## MAX PLANCK AND THE QUANTUM OF ENERGY

The problem of radiation-thermodynamics was solved by Max Planck, who was a 100 per cent classical physicist (for which he cannot be blamed). It was he who originated what is now known as *modern physics*. At the turn of the century, at the December 14, 1900 meeting of the German Physical Society, Planck presented his ideas on the subject, which were so unusual and so grotesque that he himself could hardly believe them, even though they caused intense excitement in the audience and in the entire world of physics.

Max Planck was born in Kiel, in 1858, and later moved with his family to Munich. He attended Maximilian Gymnasium (high school) in Munich and, after graduation, entered the University of Munich, where he studied physics for three years. The following year he spent at the University of Berlin, where he came in contact with the great physicists of that time, Herman von Helmholtz, Gustav Kirchhoff, and Rudolph Clausius, and learned much about the theory of heat, technically known as thermodynamics. Returning to Munich, he presented a doctoral thesis on the Second Law of Thermodynamics, receiving his Ph.D. degree in 1879, and then became an instructor at that university. Six years later he accepted the position of associate professor in Kiel. In 1889 he moved to the University of Berlin as an associate professor, becoming a full professor in 1892. The latter position was, at that time, the highest academic position in Germany, and Planck kept it until his retirement at the age of seventy. After retirement he continued his activities and delivered public speeches until his death at the age of almost ninety. Two of his last papers (*A Scientific Autobiography* and *The Notion of Causality in Physics*) were published in 1947, the year he died.

Planck was a typical German professor of his time, serious and probably pedantic, but not without a warm human feeling, which is evidenced in his correspondence with Arnold Sommerfeld who, following the work of Niels Bohr, was applying the Quantum Theory to the structure of the atom. Referring to the quantum as Planck's notion, Sommerfeld in a letter to him wrote:

> You cultivate the virgin soil,
> Where picking flowers was *my* only toil.

and to this answered Planck:

> *You* picked flowers–well, so have *I*.
> Let them be, then, combined;
> Let us exchange our flowers fair,
> And in the brightest wreath them bind.*

For his scientific achievements Max Planck received many academic honors. He became a member of the Prussian Academy of Sciences in 1894, and was elected a foreign member of the Royal Society of London in 1926. Although he made no contribution to the science of astronomy, one of the newly discovered asteroids was called Planckiana in his honor.

Throughout all his long life Max Planck was interested almost exclusively in the problems of thermodynamics, and the many papers he published were important enough to earn him the honorable position of full professor in Berlin at the age of thirty-four. But the real outburst in his scientific work, the discovery of the *quantum of energy*, for which, in 1918, he was awarded the Nobel Prize, came rather late in life, at the age of forty-two. Forty-two years is not so late in the life of a man in the usual run of occupations or professions, but it usually happens that the most important work of a theoretical physicist is done at the age of about twenty-five, when he has had time to learn enough of the existing theories but while his mind is still agile enough to conceive new, bold revolutionary ideas. For

---

* *Scientific Autobiography* by M. Planck. Translated by F. Gaynor. New York: Philosophical Library (1949).

example, Isaac Newton conceived the Law of Universal Gravity at the age of twenty-three; Albert Einstein created his Theory of Relativity at the age of twenty-six; and Niels Bohr published his Theory of the Atomic Structure at the age of twenty-seven. In his small way, the author of this book also published his most important work, on natural and artificial transformations of the atomic nucleus, when he was twenty-four. In his lecture Planck stated that according to his rather complicated calculations the paradoxical conclusions obtained by Rayleigh and Jeans could be remedied and the danger of the Ultraviolet Catastrophe avoided *if one postulates that the energy of electromagnetic waves (including light waves) can exist only in the form of certain discrete packages, or quanta, the energy content of each package being directly proportional to the corresponding frequency.*

Theoretical considerations in the field of statistical physics are notoriously difficult, but by inspecting the graph in Fig. 5 one can get some notion of how Planck's postulate "discourages" radiant energy from leaking into the limitless high-frequency region of the spectrum.

In this graph the frequencies possible within a "one-dimensional" Jeans' cube are plotted on the abscissa axes and marked 1, 2, 3, 4, etc.; on the ordinate axes are plotted the vibration energies that can be allotted to each possible frequency. According to classical physics any value of energy (that is, any point on the vertical lines drawn through 1, 2, 3, etc.) is permitted, the distribution resulting statistically in the Equipartition of Energy among all possible frequencies. On the other hand, Planck's postulate permits only a discrete set of energy values, equal to one, two, three, etc., energy packages corresponding to the given frequency. Since the energy contained in each package is assumed to be proportional to the frequency, we obtain the permitted energy values shown by large black dots in the diagram. The higher the frequency, the smaller is the number of possible energy values below any given limit, a fact which restricts the capacity of the high-frequency vibrations to take up more additional energy. As a result, the amount of energy that can be taken by high-frequency vibrations

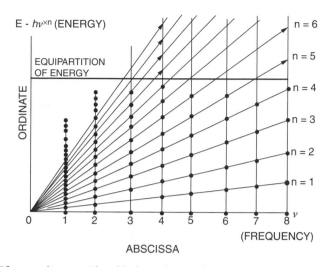

FIG. 5 If, according to Planck's hypothesis, the energy corresponding to each frequency $v$ must be an integer of the quantity hv, the situation is quite different from that shown in the previous diagram. For example, for $v = 4$ there are eight possible vibration states, whereas for $v = 8$ there are only four. This restriction reduces the number of possible vibrations at high frequencies and cancels Jeans' paradox.

becomes finite in spite of their infinite number, and everything is dandy.

It has been said that there are "lies, white lies, and statistics," but in the case of Planck's calculations the statistics turned out to be well-nigh true. He had obtained for energy distribution in thermal radiation spectrum a theoretical formula that stood in perfect agreement with the observation shown in Fig. 2.

While the Rayleigh-Jeans formula shoots sky high, demanding an infinite amount of total energy, Planck's formula comes down at high frequencies and its shape stands in perfect agreement with the observed curves. Planck's assumption that the energy content of a radiation quantum is proportional to the frequency can be written as:

$$E = hv$$

where $v$ (the Greek letter *nu*) is the frequency and $h$ is a universal constant known as *Planck's Constant*, or the *quantum constant*. In order to make Planck's theoretical curves agree with the observed ones, one has to ascribe to $h$ a certain numerical value, which is found to be $6.77 \times 10^{-27}$ in the centimeter-gram-second unit system.*

The numerical smallness of that value makes quantum theory of no importance for the large-scale phenomena which we encounter in everyday life, and it emerges only in the study of the processes occurring on the atomic scale.

## LIGHT QUANTA AND THE PHOTOELECTRIC EFFECT

Having let the spirit of quantum out of the bottle, Max Planck was himself scared to death of it and preferred to believe the packages of energy arise not from the properties of the light waves themselves but rather from the internal properties of atoms which can emit and absorb radiation only in certain discrete quantities. Radiation is like butter, which can be bought or returned to the grocery store only in quarter-pound packages, although the butter as such can exist in any desired amount (not less, though, than one molecule!). Only five years after the original Planck proposal, the light quantum was established as a physical entity existing independently of the mechanism of its emission or absorption by atoms. This step was taken by Albert Einstein in an article published in 1905, the year of his first article on the Theory of Relativity. Einstein indicated that the existence of light quanta rushing freely through space represents a necessary condition for explaining empirical laws of the photoelectric effect; that is, the emission of electrons from the metallic surfaces irradiated by violet or ultraviolet rays.

An elementary arrangement for demonstrating photoelectric effect, shown in Fig. 6a, consists of a negatively charged ordinary electroscope with a clean metal plate $P$ attached to it. When a light from an electric arc $A$, which is rich in violet and ultraviolet rays, falls on the plate, one observes that the leaves $L$ of the electroscope collapse as the electroscope discharges. That negative particles (electrons) are discharged from the metal plate was demonstrated repeatedly, by the

---

*The physical dimension of the quantum constant $h$ is a product of energy and time, or /erg · sec/ in c.g.s. units, and is known in classical mechanics as *action*; action appears in many important considerations, such as the Hamilton's Principle of Least Action.

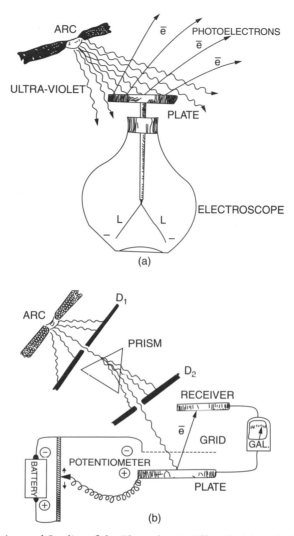

FIG. 6 Experimental Studies of the Photoelectric Effect. In (a) a primitive method for a demonstration of the photoelectric effect is illustrated. The ultraviolet radiation emitted by an electric arc ejects the electrons from a metal plate attached to an electroscope. The negatively charged leaves $L$, which have been repelling each other, lose charge and collapse. In (b) the modern method is shown. Ultraviolet radiation from an electric arc passes through a prism allowing only one selected frequency to fall on the plate. Turning the prism, one can select a monochromatic light and direct it to the plate. The energy of photoelectrons is measured by their ability to get through from plate to receiver, moving against the electric force produced by a potentiometer between plate and grid.

American physicist Robert Millikan (1868–1953) among others. If a glass plate, which absorbs ultraviolet radiation, is interposed between the arc and the metal plate, electrons are not given off, conclusive evidence that the action of the rays causes the emission. A more elaborate arrangement used for the detailed study of the laws of photoelectric effect is shown schematically in Fig. 6b. It consists of:

1. A quartz or fluoride prism (transparent for ultraviolet) and a slit permitting the selection of a monochromatic radiation of desired wavelength.
2. A set of rotating discs with triangular openings of various sizes, permitting a change in the intensity of radiation.
3. An evacuated container somewhat similar to the electron tubes used in radio sets. A variable electric potential is applied between the plate $P$, from which the photoelectrons are emitted, and the grid $G$. If the grid is charged negatively, and potential difference between the grid and the plate is equal to or larger than the kinetic energy of photoelectrons expressed in electron volts, no current will flow through the system. In the opposite case there will be a current, and its strength can be measured by the galvanometer $GM$. Using this arrangement, one can measure the number and the kinetic energy of electrons ejected by the incident light of any given intensity and wavelength (or frequency).

The study of photoelectric effect in different metals resulted in two simple laws:

I. *For light of a given frequency but varying intensity the energy of photoelectrons remains constant while their number increases in direct proportion to the intensity of light* (Fig. 7a).
II. *For varying frequency of light no photoelectrons are emitted until that frequency exceeds a certain limit $v_0$, which is different for different metals. Beyond that frequency threshold the energy of photoelectrons increases linearly, being proportional to the difference between the*

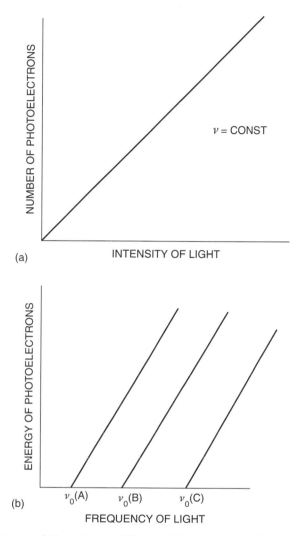

FIG. 7   The Laws of Photoelectric Effect. In (a) the number of electrons is plotted as the function of the intensity of the incident monochromatic light. In (b) the energy of photoelectrons is shown as the function of the frequency of the incident monochromatic light for three different metals: A, B and C.

*frequency of the incident light and the critical frequency $\nu_0$ of the metal* (Fig. 7b).

These well-established facts could not be explained on the basis of the classical theory of light; in some points they even contradicted it. Light is known to be short electromagnetic waves, and the increase of intensity of light must mean an increase of the oscillating electric and magnetic forces propagating through space. Since the electrons apparently are ejected from the metal by the action of electric force, their energy should increase with the increase of light intensity, instead of remaining constant as it does. Also, in the classical electromagnetic theory of light, there was no reason to expect a linear dependence of the energy of photoelectrons on the frequency of the incident light.

Using Planck's idea of light quanta and assuming the reality of their existence as independent energy packages flying through space, Einstein was able to give a perfect explanation of both empirical laws of photoelectric effect. He visualized the elementary act of the photoelectric effect as the result of a collision between a single incident light quantum and one of the conductivity electrons carrying electric current in the metal. In this collision the light quantum vanishes, giving its entire energy to the conductivity electron at the metallic surface. But, in order to cross the surface and to get into the free space, the electron must spend a certain amount of energy disengaging itself from the attraction of metallic ions. This energy, known by the somewhat misleading name of "work function," is different for different metals and is usually denoted by a symbol $W$. Thus, the kinetic energy $K$ with which a photoelectron gets out of the metal is:

$$K = h\left(\nu - \nu_0\right) = h\nu - W$$

where $\nu_0$ is the critical frequency of light below which the photoelectric effect does not occur. This picture explains at once the two laws derived from experiment. If the frequency of the incident light is kept constant, the energy content of each quantum remains the same, and the increase of light intensity results only in the corresponding increase of the number of light quanta. Thus more photoelectrons are ejected, each of them with the same energy as before. The formula giving $K$ as the function of $\nu$ explains the empirical graphs shown in Fig. 7b, predicting that the slope of the line should be the same for all metals having a numerical value equal to $h$. This consequence of

Einstein's picture of photoelectric effect stands in complete agreement with experiment and leaves no doubt of the reality of light quanta.

## THE COMPTON EFFECT

An important experiment proving the reality of light quanta was performed in 1923 by an American physicist, *Arthur Compton*, who wanted to study a collision of light quantum with an electron moving freely through space. The ideal situation would be to observe such collisions by sending a beam of light through an electron beam. Unfortunately, the number of electrons in even the most intense electron beams available is so small that one would have to wait for centuries for a single collision. Compton solved the difficulty by using X-rays, the quanta of which carry very large amounts of energy because of the very high frequency involved. As compared with the energy carried by each X-ray quantum, the energy with which electrons are bound in the atoms of light elements can be disregarded and one can regard them (the electrons) as being unbound and quite free. Considering a free collision between light quantum and an electron in the same way as one considers a collision between two elastic balls, one would expect that the energy, and hence the frequency, of scattered X-rays would

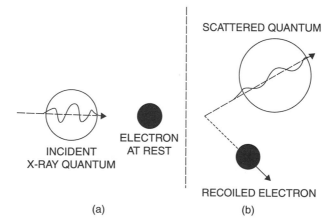

INCIDENT
X-RAY QUANTUM

ELECTRON
AT REST

SCATTERED QUANTUM

RECOILED ELECTRON

(a)                    (b)

FIG. 8 Compton scattering of X-rays. Notice that after the collision the wavelength of X-ray quantum increases because of loss of energy to the electron.

decrease with the increasing scattering angle. Compton's experiments (Fig. 8) stood in complete agreement with this theoretical prediction, and with the formula derived on the basis of conservation of energy and mechanical momentum in the collision of two elastic spheres. This agreement gave additional confirmation of the existence of light quanta.

## CHAPTER IV

## L. DE BROGLIE AND PILOT WAVES

Louis Victor, Duc de Broglie, born in Dieppe in 1892, who became
the Prince de Broglie on the death of an elder brother, had a rather
unusual scientific career. As a student at the Sorbonne he decided
to devote his life to medieval history, but the onset of World War I
induced him to enlist in the French Army. Being an educated man,
he got a position in one of the field radio-communication units, a
novelty at that time, and turned his interest from Gothic cathedrals to
electromagnetic waves. In 1925 he presented a doctoral thesis which
contained such revolutionary ideas concerning a modification of the
Bohr original theory of atomic structure that most physicists were
rather skeptical; some wit, in fact, dubbed de Broglie's theory "*la
ComéAdie Française.*"

Having worked with radio waves during the war, and being a
connoisseur of chamber music, de Broglie chose to look at an atom as
some kind of musical instrument which, depending on the way it is

constructed, can emit a certain basic tone and a sequence of overtones. Since by that time Bohr's electronic orbits were fairly well established as characterizing different quantum states of an atom, he chose them as a basic pattern for his wave scheme. He imagined that each electron moving along a given orbit is accompanied by some mysterious pilot waves (now known as de Broglie waves) spreading out all along the orbit. The first quantum orbit carried only one wave, the second two waves, the third three, etc. Thus the length of the first wave must be equal to the length $2\pi r_1$ of the first quantum orbit, the length of the second wave must be equal to one-half of the length of the second orbit, $\frac{1}{2} \cdot 2\pi r_2$, etc. In general, the $n$th quantum orbit carries $n$ waves with the length $\frac{1}{n} 2\pi r_n$ each.

As we have seen in Chapter II, the radius of the $n$th orbit in Bohr's atom is

$$r_n = \frac{1}{4\pi^2} \frac{h^2}{me^2} n^2$$

From the equality of the centrifugal force due to the orbital motion, and the electrostatic attraction beween the charged particles, we obtain:

$$\frac{mv_n^2}{r_n} = \frac{e^2}{r_n^2}$$

or

$$e^2 = mv_n^2 r_n$$

Substituting this value of $e^2$ into the original formula, we get

$$r_n = \frac{1}{4\pi^2} \frac{h^2 n^2}{m} \cdot \frac{1}{mv_n^2 r_n}$$

or

$$(2\pi r_n)^2 = \frac{h^2 n^2}{m^2 v^2}$$

Extracting the square root from both sides of this equation we finally obtain:

$$2\pi r_n = n \cdot \frac{h}{mv_n}$$

Thus, if the length λ of the wave accompanying an electron is equal to Planck's constant $h$ divided by the mechanical momentum $mv$ of the particle, then

$$\lambda = \frac{h}{mv}$$

and de Broglie could satisfy his desire to introduce waves of such a nature that *1, 2, 3, etc., of them would fit exactly into the 1st, 2nd, 3rd of Bohr's quantum orbits* (Fig. 19). The result given is mathematically equivalent to Bohr's original quantum condition and brings in nothing physically new-nothing, that is, but *an idea*: the motion of the electrons along Bohr's quantum orbits is accompanied by mysterious waves of the lengths determined by the mass and the velocity of the moving particles. If these waves represented some kind of physical reality, they should also accompany particles moving freely through space, in which case their existence or non-existence could be checked by direct experiment. In fact, if the motion of electrons is always guided by de Broglie waves, a beam of electrons under proper conditions should

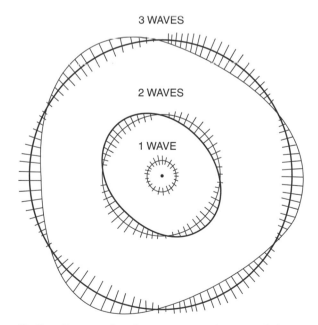

FIG. 19   De Broglie waves fitted to quantum orbits in Bohr's atom model.

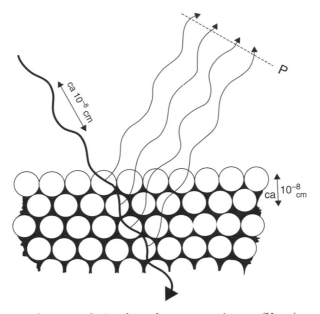

FIG. 20   An incident wave, be it a short electromagnetic wave (X-ray) or a de Broglie wave associated with a beam of fast electrons, produces wavelets as it passes through the successive layers of a crystal lattice. Depending on the angle of incidence, dark and light interference fringes appear. (P is the phase plane.)

show diffraction phenomena similar to those characteristic of beams of light. Electron beams accelerated by electron tensions of several kilovolts (which are commonly used in laboratory experiments) should, according to de Broglie's formula, be accompanied by pilot waves of about $10^{-8}$ cm wavelength, which is comparable to the wavelength of ordinary X-rays. This wavelength is too short to show a diffraction in ordinary optical gratings and should be studied with the technique of standard X-ray spectroscopy. In this method the incident beam is reflected from the surface of a crystal, and the neighboring crystalline layers, located about $10^{-8}$ cm apart, have the function of the more widely separated lines in opitical diffraction gratings (Fig. 20). This experiment was carried out simultaneously and independently by Sir George Thomson (son of Sir J. J. Thomson) in England, and G. Davisson and L. H. Germer in the United States, who used a crystal arrangement similar to that of Bragg and Bragg, but substituted for the beam of X-rays a beam of electrons moving at a given velocity. In the experiments a characteristic diffraction pattern appeared on the screen (or photographic plate) that was placed in the way of the reflected beam, and the diffraction bands widened or narrowed when the velocity of incident electrons was increased or decreased. The measured wavelength coincided exactly in all cases with that given by the de Broglie formula. Thus the de Broglie waves became an indisputable physical reality, although nobody understood what they were.

Later on a German physicist, Otto Stern, proved the existence of the diffraction phenomena in the case of atomic beams. Since atoms are thousands of times more massive than electrons, their de Broglie waves were expected to be correspondingly shorter for the same velocity. To make atomic de Broglie waves of a length comparable with the distances between the crystalline layers (about $10^{-8}$ cm), Stern decided to use the thermal motion of atoms, since he could regulate the velocity simply by changing the temperature of the gas. The source consisted of a ceramic cylinder heated by an electric wire wound around it. At one end of the closed cylinder was a tiny hole through which the atoms escaped at their thermal velocity into a much larger evacuated container, and in their flight through space they hit a crystal placed in their way. The atoms reflected in different directions stuck to metal plates cooled by liquid air, and the number of atoms on the different plates was counted by a complicated method of chemical microanalysis. Plotting the number of atoms scattered in different directions against the scattering angle, Stern obtained again a perfect diffraction pattern corresponding exactly to the wavelength calculated from de Broglie's formula. And the bands became wider or thinner when the temperature of the cylinder was changed.

When in the late twenties I was working at Cambridge University with Rutherford, I decided to spend Christmas vacation in Paris (where I had never been before) and wrote to de Broglie, saying that I would like very much to meet him and to discuss some problems of the Quantum Theory. He answered that the University would be closed but that he would be glad to see me in his home. He lived in a magnificent family mansion in the fashionable Parisian suburb Neuillysur-Seine. The door was opened by an impressive-looking butler.

*"Je veux voir Professeur de Broglie."*

*"Vous voulez dire, Monsieur le Duc de Broglie," retorted the butler.*

"O.K., *le Duc de Broglie," said I, and was let into the house.*

De Broglie, wearing a silk house coat, met me in his sumptuously furnished study, and we started talking physics. He did not speak any English; my French was rather poor. But somehow, partly by using my broken French and partly by writing formulas on paper, I managed to convey to him what I wanted to say and to understand his comments. Less than a year later, de Broglie came to London to deliver a lecture at the Royal Society, and I was, of course, in the audience. He delivered a brilliant lecture, in perfect English, with only a slight French accent. Then I understood another of his principles: when foreigners come to France they must speak French.

A number of years later when I was planning a trip to Europe and de Broglie asked me to deliver a special lecture in the institute of Henri Poincaré, of which he was a director, I decided to come well prepared. I planned to write the lecture down in my (still) poor French on board the liner crossing the Atlantic, have somebody in Paris correct the text, and use it as notes at the lecture. But, as everybody knows, all good resolutions collapse on an ocean voyage offering many distractions, and I had to face the audience in the Sorbonne completely unprepared. The lecture went through somewhat stumblingly, but my French held, and everybody understood what I had to say. After the lecture I told de Broglie that I was sorry that I did not carry out my original plan of having the corrected French notes. *"Mon Dieu!"* he exclaimed, "it is lucky that you didn't.

De Broglie told me about a lecture delivered by the noted British physicist R. H. Fowler. It is well known that since English is the best language in the world, the English are of the opinion that all foreigners should learn it, thus freeing themselves from the need to learn anyone else's language. Since the lecture in the Sorbonne had to be in French, Fowler had prepared the complete English text of his lecture, and he sent it well in advance to de Broglie, who had personally translated it into French. Thus Fowler lectured in French, using the typewritten French text. De Broglie said that after the lecture a group of students came to him, "*Monsieur le Professeur*," they said, "we are greatly puzzled. We expected that Professor Fowler would lecture in English, and we all know enough English to be able to understand. But he did not speak English but some other language and we cannot figure out what language it was." "*Et parfois!*" added de Broglie, "I had to tell them that Professor Fowler was lecturing in French!"

## SCHRÖDINGER'S WAVE EQUATION

Creating the revolutionary idea that the motion of atomic particles is guided by some mysterious pilot waves, de Broglie was too slow to develop a strict mathematical theory of this phenomenon, and, in 1926, about a year after his publication, there appeared an article by an Austrian physicist, Erwin Schrödinger, who wrote a general equation for de Broglie waves and proved its validity for all kinds of electron motion. While de Broglie's model of the atom resembled more an unusual stringed instrument, or rather a set of vibrating concentric metal rings of different diameters, Schrödinger's model was a closer analogy to wind instruments; in his atom the vibrations occur throughout the entire space surrounding the atomic nucleus.

Consider a flat metal disc something like a cymbal fastened in the center (Fig. 21a). If one strikes it, it will begin to vibrate with its rim moving periodically up and down (Fig. 21b). There exist also more complicated kinds of vibrations (overtones) like the pattern shown in Fig. 21c where the center of the plate and all the points located along the circle between the center and circumference (marked by heavy line in the figure) are at rest, so that, when the material bulges up within that circle the material outside the circle moves down, and vice versa. The motionless points and lines of a vibrating elastic surface are called the *nodal* points and lines; one can extend Fig. 21c by drawing higher

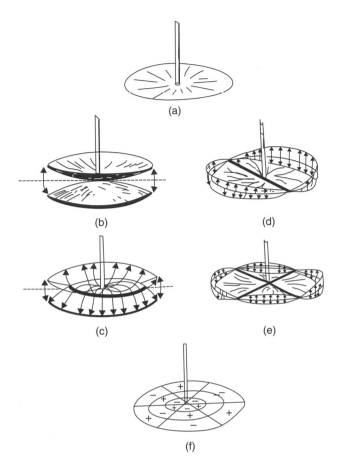

FIG. 21  Various vibration modes of an elastic disc fastened in the center: (a) state of rest; (b) nodal point in the center; (c) one circular nodal line; (d) one radial nodal line; (e) two radial nodal lines; (f) three radial and two circular nodal lines.

overtones which correspond to two or more nodal circles around the central nodal point.

Besides such "radial" vibrations there also exist the "azimuthal vibrations" in which the nodal lines are straight lines passing through the center as shown in Fig. 21d and e, where arrows indicate whether the membrane lifts or sinks in respect to the equilibrium horizontal position. Of course, the radial and azimuthal vibrations can exist simultaneously in a given membrane. The resulting complex state of motion should be described by two integers $n_r$ and $n_\phi$, giving the numbers of radial and azimuthal nodal lines.

Next in complexity are the three-dimensional vibrations such as, for example, the sound waves in the air filling a rigid metal sphere. In this case it becomes necessary to introduce the third kind of nodal lines and also the third integer $n_1$ giving their number.

This kind of vibrations was studied in theoretical acoustics many years ago, and, in particular, Hermann von Helmholtz in the last century made detailed studies of the vibrations of air enclosed in rigid metal spheres (Helmholtz resonators). He drilled a little hole in the sphere, to let in sound from the outside, and used a siren which emitted a pure tone, the pitch of which could be changed continuously by changing the rotation speed of the siren's disc. When the frequency of the siren's sound coincided with one of the possible vibrations of air in the sphere, resonance was observed. These experiments stood in perfect agreement with the mathematical solutions of the wave equation for sound, which is too complicated to be discussed in this book.

The equation written by Schrödinger for de Broglie's waves is very similar to the well-known wave equations for the propagation of sound and light (that is, electromagnetic) waves, except that for a few years there remained the mystery of just *what was vibrating*. We will return to this question in the next chapter.

When an electron moves around a proton in a hydrogen atom the situation is somewhat similar to the vibration of gas within a rigid spherical enclosure. But whereas for Helmholtz vibrators there is a rigid wall preventing the gas from expanding beyond it, the atomic electron is subject to electric attraction of the central nucleus which slows down the motion when the electron travels farther and farther from the center, and stops it when it goes beyond the limit permitted by its kinetic energy. The situation in both cases is shown graphically in Fig. 22. In the figure on the left the "potential hole" (that is, the lowering of potential energy in the neighborhood of a certain point) resembles a cylindrical well; the figure on the right looks more like a funnel-shaped hole in the ground. The horizontal lines represent the quantized energy levels, the lowest of them corresponding to the

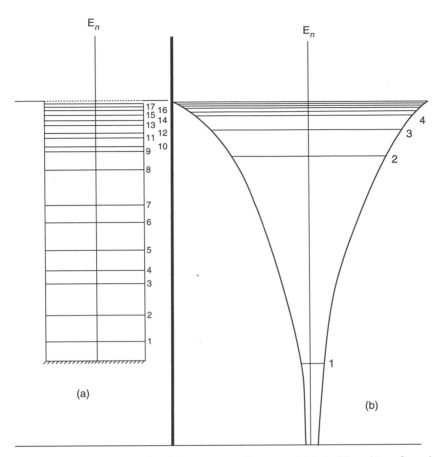

FIG. 22   Quantum energy levels in a rectangular potential hole (a) and in a funnel-shaped potential hole (b).

lowest energy the particle can have. Comparing Fig. 22b with Fig. 12 of Chapter II, we find that the levels of the hydrogen atom calculated on the basis of Schrödinger's equation are identical with those obtained from Bohr's old theory of quantum orbits. But the physical aspect is quite different. Instead of sharp circular and elliptic orbits along which point-shaped electrons run, we have now a full-bodied atom represented by multishaped vibrations of something which in the early years of wave mechanics was called, for lack of a better name, a $\psi$-function (Greek letter *psi*).

It must be remarked here that the rectangular well potential distribution shown in Fig. 22a turned out to be very useful for a description of proton and neutron motion within the atomic nucleus, and later was used successfully by Maria Goeppert Mayer and, independently, by Hans Jensen for an understanding of the energy levels within the atomic nuclei and the origin of $\gamma$-ray spectra of radioactive nuclear species.

The frequencies of different $\psi$-vibration modes do not correspond to the frequencies of the light wave emitted by the atom, but to the energy values of the different quantum states divided by $h$. Thus, the emission of a spectral line necessitated the excitation of two vibration modes, say $\psi_m$ and $\psi_n$, with the resulting composite frequency

$$\nu_{m,n} = \frac{E_m}{h} - \frac{E_n}{h} = \frac{E_m - E_n}{h}$$

which is the same as Bohr's expression for the frequency of light quantum resulting from the transition of the atomic electron from the energy level $E_m$ to the lower energy level $E_n$.

## APPLYING WAVE MECHANICS

Apart from giving a more rational foundation to Bohr's original idea of quantum orbits, and removing some discrepancies, wave mechanics could explain some phenomena well beyond the reach of the old Quantum Theory. As was mentioned in Chapter II, the author of the present book and, independently, a team consisting of Ronald Gurney and Edward Condon successfully applied Schrödinger's wave equation to the explanation of the emission of $\alpha$-particles by radioactive elements, and their penetration into the nuclei of other lighter elements with the resulting transformation of elements. To understand this rather complicated phenomenon, we will compare an atomic nucleus to a fortress surrounded on all sides by high walls; in nuclear physics the analogy of the fortress walls is known as a *potential barrier*. Due to the fact that both the atomic nucleus and the $\alpha$-particle carry a positive electric charge, there exists a strong repulsive Coulomb force* acting on the $\alpha$-particle approaching a nucleus. Under the action of that force an $\alpha$-particle shot at the nucleus may be stopped and thrown back before it comes into direct contact with the nucleus. On the other hand, $\alpha$-particles that are inside the various nuclei as constituent parts of them are prevented from escaping by very strong attractive nuclear forces (analogous to the cohesion forces in ordinary liquids), but these nuclear forces act only when the particles are closely packed, being in direct contact with one another. The combination of these two forces forms a potential barrier preventing the inside particles from getting out and the outside particles from getting in, unless their kinetic energy is high enough to climb over the top of the potential barrier.

Rutherford found experimentally that the $\alpha$-particles emitted by various radioactive elements, such as uranium and radium, have much smaller kinetic energy than that needed to get out over the top of the barrier. It was also known that when $\alpha$-particles are shot at the nuclei from outside with less kinetic energy than needed to reach the top of the potential barrier they often penetrate into the nuclei, producing artificial nuclear transformations. According to the basic principles of classical mechanics, both phenomena were absolutely impossible, so that no spontaneous nuclear decay resulting in the emission of $\alpha$-particles, and no artificial nuclear transformations under the influence of $\alpha$-bombardment could possibly exist. And yet both were experimentally observed!

If one looks on the situation from the point of view of wave mechanics, it appears quite different, since the motion of the particles is governed by de Broglie's pilot waves. To understand how wave mechanics explains these classically impossible events, one should remember that wave mechanics stands in the same relation to the classical Newtonian mechanics as wave optics to the old geometrical optics. According to Snell's Law, a light ray falling on a glass surface

---

*During the early studies of electric phenomena, the French physicist Charles de Coulomb found that forces acting between charged particles are proportional to the product of their electric charges and inversely proportional to the square of the distance between them. This is known as Coulomb's Law.

at a certain incidence angle $i$ (Fig. 23a) is refracted at a smaller angle $r$, satisfying the condition $sin\ i/sin\ r = n$ where $n$ is the refractive index of glass. If we reverse the situation (Fig. 23b), and let a light ray propagating through glass exit into the air, the angle of refraction will be larger than that of incidence and we will have $sin\ i/sin\ r = 1/n$. Thus a light ray falling on the interface between the glass and air at an angle of incidence greater than a certain critical value will not enter into the air at all but will be totally reflected back into the glass. According to the wave theory of light the situation is different. Light waves undergoing total internal reflection are not reflected from the mathematical boundary between the two substances, but penetrate into the second medium (in this case air) to the depth of several wavelengths $\lambda$ and then are thrown back into the original medium (Fig. 23c). Therefore, if we place another plate of glass a few wavelengths away (a few microns, in the case of visible light), some amount of light coming into the air will reach the surface of that glass and continue to propagate in the original direction (Fig. 23d). The theory of this phenomenon can be found in the books on optics published a century ago and represents a standard demonstration in many university courses on optics.

Similarly, de Broglie waves which guide the motion of $\alpha$-particles and other atomic projectiles can penetrate through the regions of space which are prohibited to particles by classical Newtonian mechanics, and $\alpha$-particles, protons, etc., can cross the potential barriers whose height is greater than the energy of the incident particle. But the probability of penetration is of physical importance only for particles of atomic mass, and for barriers not more than $10^{-12}$ or $10^{-13}$ cm wide. Let us take, for example, a uranium nucleus which emits an $\alpha$-particle after an interval of about $10^{10}$ years. An $\alpha$-particle imprisoned within the uranium potential barrier hits the barrier wall some $10^{21}$ times per second, which means that the chance of escape after a simple hit is one out of $10^{10} \times 3 \cdot 10^7 \times 10^{21} \cong 3 \cdot 10^{38}$ hits (here $3 \cdot 10^7$ is the number of seconds in a year). Similarly, the chances that an atomic projectile will enter the nucleus are very small

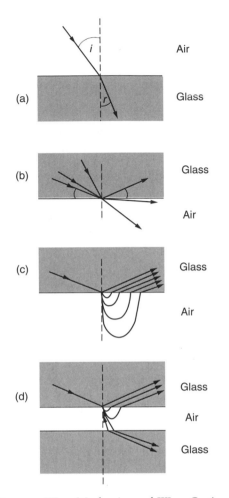

FIG. 23   Analogy Between Wave Mechanics and Wave Optics. In (a) we have the familiar picture of refraction of light entering from the rarer into the denser medium. In (b) we have the reverse case when the light entering from the denser into the rarer medium can be completely reflected from the interface if the angle of incidence exceeds a certain critical value. According to the wave theory of light, the reflection takes place not on the mathematical surface separating the two media, but within a certain layer several wavelengths thick. Thus, if the second layer of the denser medium is placed a few wavelengths beyond the first layer, a fraction of the incident light will not be totally reflected but will penetrate into the second dense layer propagating in the original direction. Similarly, according to wave mechanics, some particles can penetrate through the regions prohibited by classical mechanics, where the potential is higher than the original kinetic energy of the particles.

for each individual hit, but may become considerable if a very large number of nuclear collisions are involved. It was shown in 1929 by Fritz Houtermans and Robert Atkinson that the nuclear collisions caused by intensive thermal motion, known as *thermonuclear reactions*, are responsible for the production of energy in the Sun and stars. Physicists are working hard now to produce the so-called "controlled thermonuclear reactions" which would supply us with cheap, inexhaustible, and harmless sources of nuclear energy. All this would have been impossible if Newton's classical mechanics had not been replaced by de Broglie-Schrödinger wave mechanics.

# EXCERPTS FROM LECTURES ON QUANTUM MECHANICS

By

PAUL A.M. DIRAC

DR. DIRAC

*Lecture No. 1*

## THE HAMILTONIAN METHOD

I'm very happy to be here at Yeshiva and to have this chance to talk to you about some mathematical methods that I have been working on for a number of years. I would like first to describe in a few words the general object of these methods.

In atomic theory we have to deal with various fields. There are some fields which are very familiar, like the electromagnetic and the gravitational fields; but in recent times we have a number of other fields also to concern ourselves with, because according to the general ideas of De Broglie and Schrödinger every particle is associated with waves and these waves may be considered as a field. So we have in atomic physics the general problem of setting up a theory of various fields in interaction with each other. We need a theory conforming to the principles of quantum mechanics, but it is quite a difficult matter to get such a theory.

One can get a much simpler theory if one goes over to the corresponding classical mechanics, which is the form which quantum mechanics takes when one makes Planck's constant $\hbar$ tend to zero. It is very much easier to visualize what one is doing in terms of classical mechanics. It will be mainly about classical mechanics that I shall be talking in these lectures.

Now you may think that that is really not good enough, because classical mechanics is not good enough to describe Nature. Nature is described by quantum mechanics. Why should one, therefore, bother so much about classical mechanics? Well, the quantum field theories are, as I said, quite difficult and so far, people have been able to build up quantum field theories only for fairly simple kinds of fields with simple interactions between them. It is quite possible that these simple fields with the simple interactions between them are not adequate for a description of Nature. The successes which we get with quantum field theories are rather limited. One is continually running into difficulties and one would like to broaden one's basis and have some possibility of bringing more general fields into account. For example, one would like to take into account the possibility that Maxwell's equations are not accurately valid. When one goes to distances very close to the charges that are producing the fields, one may have to modify Maxwell's field theory so as to make it into a nonlinear electrodynamics. This is only one example of the kind of generalization which it is profitable to consider in our present state of ignorance of the basic ideas, the basic forces and the basic character of the fields of atomic theory.

In order to be able to start on this problem of dealing with more general fields, we must go over the classical theory. Now, if we can put the classical theory into the Hamiltonian form, then we can always apply certain standard rules so as to get a first approximation to a quantum theory. My talks will be mainly concerned with this problem of putting a general classical theory into the Hamiltonian form. When one has done that, one is well launched onto the path of getting an accurate quantum theory. One has, in any case, a first approximation.

Of course, this work is to be considered as a preliminary piece of work. The final conclusion of this piece of work must be to set up an accurate quantum theory, and that involves quite serious difficulties, difficulties of a fundamental character which people have been worrying over for quite a number of years. Some people are so much impressed by the difficulties of passing over from Hamiltonian classical mechanics to quantum mechanics that they think that maybe the whole method of working from Hamiltonian classical theory is a bad method. Particularly in the last few years people have been trying to set up alternative methods for getting quantum field theories. They have made quite considerable progress on these lines. They have obtained a number of conditions which have to be satisfied. Still I feel that these alternative methods, although they go quite a long way towards accounting for experimental results, will not lead to a final solution to the problem. I feel that there will always be something missing from them which we can only get by working from a Hamiltonian, or maybe from some generalization of the concept of a Hamiltonian. So I take the point of view that the Hamiltonian is really very important for quantum theory.

In fact, without using Hamiltonian methods one cannot solve some of the simplest problems in quantum theory, for example the problem of getting the Balmer formula for hydrogen, which was the very beginning of quantum mechanics. A Hamiltonian comes in therefore in very elementary ways and it seems to me that it is really quite essential to work from a Hamiltonian; so I want to talk to you about how far one can develop Hamiltonian methods.

I would like to begin in an elementary way and I take as my starting point an action principle. That is to say, I assume that there is an action integral which depends on the motion, such that, when one varies the motion, and puts down the conditions for the action integral to be stationary, one gets the equations of motion. The method of starting from an action principle has the one great advantage, that one can easily make the theory conform to the principle of relativity. We need our atomic theory to conform to relativity because in general we are dealing with particles moving with high velocities.

If we want to bring in the gravitational field, then we have to make our theory conform to the general principle of relativity, which means working with a space-time which is not flat. Now the gravitational field is not very important in atomic physics, because gravitational forces are extremely weak compared with the other kinds of forces which are present in atomic processes, and for practical purposes one can neglect the gravitational field. People have in recent years worked to some extent on bringing the gravitational field into the quantum theory, but I think that the main object of this work was the hope that bringing in the gravitational field might help to solve some of the difficulties. As far as one can see at present, that hope is not realized, and bringing in the gravitational field seems to add to the difficulties rather than remove them. So that there is not very much point at present in bringing gravitational fields into atomic theory. However, the methods which I am going to describe are powerful mathematical methods which would be available whether one brings in the gravitational field or not.

We start off with an action integral which I denote by

$$I = \int L\, dt. \tag{1-1}$$

It is expressed as a time integral, the integrand $L$ being the Lagrangian. So with an action principle we have a Lagrangian. We have to consider how to pass from that Lagrangian to a Hamiltonian. When we have got the Hamiltonian, we have made the first step toward getting a quantum theory.

You might wonder whether one could not take the Hamiltonian as the starting point and short-circuit this work of beginning with an action integral, getting a Lagrangian from it and passing from the Lagrangian to the Hamiltonian. The objection to trying to make this short-circuit is that it is not at all easy to formulate the conditions for a theory to be relativistic in terms of the Hamiltonian. In terms of the action integral, it is very easy to formulate the conditions for the theory to be relativistic: one simply has to require that the action integral shall be invariant. One can easily construct innumerable examples of action integrals which are invariant. They will automatically lead to equations of motion agreeing with relativity, and any developments from this action integral will therefore also be in agreement with relativity.

When we have the Hamiltonian, we can apply a standard method which gives us a first approximation to a quantum theory, and if we are lucky we might be able to go on and get an accurate quantum theory. You might again wonder whether one could not short-circuit that work to some extent. Could one not perhaps pass directly from the Lagrangian to the quantum theory, and short-circuit altogether the Hamiltonian? Well, for some simple examples one *can* do that. For some of the simple fields which are used in physics the Lagrangian is quadratic in the velocities, and is like the Lagrangian which one has in the non-relativistic dynamics of particles. For these examples for which the Lagrangian is quadratic in the velocities, people have devised some methods for passing directly from the Lagrangian to the quantum theory. Still, this limitation of the Lagrangian's being quadratic in the velocities is quite a severe one. I want to avoid this limitation and to work with a Lagrangian which can be quite a general function of the velocities. To get a general formalism which will be applicable, for example, to the non-linear electrodynamics which I mentioned previously, I don't think one can in any way short-circuit the route of starting with an action integral, getting a Lagrangian, passing from the Langrangian to the Hamiltonian, and then passing from the Hamiltonian to the quantum theory. That is the route which I want to discuss in this course of lectures.

In order to express things in a simple way to begin with, I would like to start with a dynamical theory involving only a finite number of degrees of freedom, such as you are familiar with in particle dynamics. It is then merely a formal matter to pass from this finite number of degrees of freedom to the infinite number of degrees of freedom which we need for a field theory.

Starting with a finite number of degrees of freedom, we have dynamical coordinates which I denote by $q$. The general one is $q_n$, $n = 1, \ldots, N$, $N$ being the number of degrees of freedom. Then we have the velocities $dq_n/dt = \dot{q}_n$. The Lagrangian is a function $L = L(q, \dot{q})$ of the coordinates and the velocities.

You may be a little disturbed at this stage by the importance that the time variable plays in the formalism. We have a time variable $t$ occurring already as soon as we introduce the Lagrangian. It occurs again in the velocities, and all the work of passing from Lagrangian to Hamiltonian involves one particular time variable. From the relativistic point of view we are thus singling out one particular observer and making our whole formalism refer to the time for this observer. That, of course, is not really very pleasant to a relativist, who would like to treat all observers on the same footing. However, it is a feature of the present formalism which I do not see how one can avoid if one wants to keep to the generality of allowing the Lagrangian to be *any* function of the coordinates and velocities. We can be sure that the contents of the theory are relativistic, even though the form of the equations is not manifestly relativistic on account of the appearance of one particular time in a dominant place in the theory.

Let us now develop this Lagrangian dynamics and pass over to Hamiltonian dynamics, following as closely as we can the ideas which one learns about as soon as one deals with dynamics from the point of view of working with general coordinates. We have the Lagrangian equations of motion which follow from the variation of the action integral:

$$\frac{d}{dt}\left(\frac{\partial L}{\partial \dot{q}_n}\right) = \frac{\partial L}{\partial q_n}. \tag{1-2}$$

To go over to the Hamiltonian formalism, we introduce the momentum variables $p_n$, which are defined by

$$p_n = \frac{\partial L}{\partial \dot{q}_n}. \tag{1-3}$$

Now in the usual dynamical theory, one makes the assumption that the momenta are independent functions of the velocities, but that assumption is too restrictive for the applications which we are going to make. We want to allow for the possibility of these momenta *not* being independent functions of the velocities. In that case, there exist certain relations connecting the momentum variables, of the type $\phi(q, p) = 0$.

There may be several independent relations of this type, and if there are, we distinguish them one from another by a suffix $m = 1, \ldots, M$, so we have

$$\phi_m (q, p) = 0. \tag{1-4}$$

The $q$'s and the $p$'s are the dynamical variables of the Hamiltonian theory. They are connected by the relations (1-4), which are called the *primary constraints* of the Hamiltonian formalism. This terminology is due to Bergmann, and I think it is a good one.

Let us now consider the quantity $p_n \dot{q}_n - L$. (Whenever there is a repeated suffix I assume a summation over all values of that suffix.) Let us make variations in the variables $q$ and $\dot{q}$, in the coordinates and the velocities. These variations will cause variations to occur in the momentum variables $p$. As a result of these variations,

$$\delta\,(p_n\dot{q}_n - L) = \delta p_n\dot{q}_n + p_n\delta\dot{q}_n - \left(\frac{\partial L}{\partial q_n}\right)\delta q_n - \left(\frac{\partial L}{\partial \dot{q}_n}\right)\delta \dot{q}_n$$
$$= \delta p_n\dot{q}_n - \left(\frac{\partial L}{\partial q_n}\right)\delta q_n$$

(1-5)

by (1-3). Now you see that the variation of this quantity $p_n\dot{q}_n - L$ involves only the variation of the $q$'s and that of the $p$'s. It does not involve the variation of the velocities. That means that $p_n\dot{q}_n L -$ can be expressed in terms of the $q$'s and the $p$'s, independent of the velocities. Expressed in this way, it is called the *Hamiltonian H*.

However, the Hamiltonian defined in this way is not uniquely determined, because we may add to it any linear combination of the $\phi$'s, which are zero. Thus, we could go over to another Hamiltonian

$$H^* = H + c_m\phi_m,$$

(1-6)

where the quantities $c_m$ are coefficients which can be any function of the $q$'s and the $p$'s. $H^*$ is then just as good as $H$; our theory cannot distinguish between $H$ and $H^*$. The Hamiltonian is not uniquely determined.

We have seen in (1-5) that

$$\delta H = \dot{q}_n\delta p_n - \left(\frac{\partial L}{\partial q_n}\right)\delta q_n.$$

This equation holds for any variation of the $q$'s and the $p$'s subject to the condition that the constraints (1-4) are preserved. The $q$'s and the $p$'s cannot be varied independently because they are restricted by (1-4), but for any variation of the $q$'s and the $p$'s which preserves these conditions, we have this equation holding. From the general method of the calculus of variations applied to a variational equation with constraints of this kind, we deduce

$$\dot{q}_n = \frac{\partial H}{\partial p_n} + u_m \frac{\partial \phi_m}{\partial p_n}$$

and

$$-\frac{\partial L}{\partial q_n} = \frac{\partial H}{\partial q_n} + u_m \frac{\partial \phi_m}{\partial q_n}$$

(1-7)

or

$$p_n = -\frac{\partial H}{\partial q_n} - u_m \frac{\partial \phi_m}{\partial q_n},$$

(1-8)

with the help of (1-2) and (1-3), where the $u_m$ are unknown coefficients. We have here the *Hamiltonian equations of motion*, describing how the variables $q$ and $p$ vary in time, but these equations involve unknown coefficients $u_m$.

It is convenient to introduce a certain formalism which enables one to write these equations briefly, namely the Poisson bracket formalism. It consists of the following: If we have two functions of the $q$'s and the $p$'s, say $f(q, p)$ and $g(q, p)$, they have a *Poisson bracket* $[f, g]$ which is defined by

$$[f, g] = \frac{\partial f}{\partial q_n} \frac{\partial g}{\partial p_n} - \frac{\partial f}{\partial p_n} \frac{\partial g}{\partial q_n}.$$

(1-9)

The Poisson brackets have certain properties which follow from their definition, namely $[f, g]$ is antisymmetric in $f$ and $g$:

$$[f, g] = -[g, f];$$

(1-10)

it is linear in either member:

$$[f_1 + f_2, g] = [f_1, g] + [f_2, g], \text{ etc.;}$$

(1-11)

and we have the product law,

$$[f_1 f_2, g] = f_1 [f_2, g] + [f_1, g] f_2.$$

(1-12)

Finally, there is the relationship, known as the *Jacobi Identity*, connecting three quantities:

$$[f, [g, h]] + [g, [h, f]] + [h, [f, g]] = 0.$$

(1-13)

With the help of the Poisson bracket, one can rewrite the equations of motion. For any function $g$ of the $q$'s and the $p$'s, we have

$$\dot{g} = \frac{\partial g}{\partial q_n}\dot{q}_n + \frac{\partial g}{\partial p_n}\dot{p}_n. \tag{1-14}$$

If we substitute for $\dot{q}_n$ and $\dot{p}_n$ their values given by (1-7) and (1-8), we find that (1-14) is just

$$\dot{g} = [g, H] + u_m [g, \phi_m]. \tag{1-15}$$

The equations of motion are thus all written concisely in the Poisson bracket formalism.

We can write them in a still more concise formalism if we extend the notion of Poisson bracket somewhat. As I have defined Poisson brackets, they have a meaning only for quantities $f$ and $g$ which can be expressed in terms of the $q$'s and the $p$'s. Something more general, such as a general velocity variable which is not expressible in terms of the $q$'s and $p$'s, does not have a Poisson bracket with another quantity. Let us extend the meaning of Poisson brackets and suppose that they exist for any two quantities and that they satisfy the laws (1-10), (1-11), (1-12), and (1-13), but are otherwise undetermined when the quantities are not functions of the $q$'s and $p$'s.

Then we may write (1-15) as

$$\dot{g} = [g, H + u_m\phi_m]. \tag{1-16}$$

Here you see the coefficients $u$ occurring in one of the members of a Poisson bracket. The coefficients $u_m$ are *not* functions of the $q$'s and the $p$'s, so that we cannot use the definition (1-9) for determining the Poisson bracket in (1-16). However, we can proceed to work out this Poisson bracket using the laws (1-10), (1-11), (1-12), and (1-13). Using the summation law (1-11) we have:

$$[g, H + u_m\phi_m] = [g, H] + [g, u_m\phi_m] \tag{1-17}$$

and using the product law (1-12),

$$[g, u_m\phi_m] = [g, u_m]\phi_m + u_m[g, \phi_m]. \tag{1-18}$$

The last bracket in (1-18) is well-defined, for $g$ and $\phi_m$ are both functions of the $q$'s and the $p$'s. The Poisson bracket $[g, u_m]$ is *not* defined, but it is multiplied by something that vanishes, $\phi_m$. So the first term on the right of (1-18) vanishes. The result is that

$$[g, H + u_m\phi_m] = [g, H] + u_m[g, \phi_m], \qquad (1\text{-}19)$$

making (1-16) agree with (1-15).

There is something that we have to be careful about in working with the Poisson bracket formalism: We have the constraints (1-4), but must not use one of these constraints *before* working out a Poisson bracket. If we did, we would get a wrong result. So we take it as a rule that Poisson brackets must all be worked out before we make use of the constraint equations. To remind us of this rule in the formalism, I write the constraints (1-4) as equations with a different equality sign $\approx$ from the usual. Thus they are written

$$\phi_m \approx 0. \qquad (1\text{-}20)$$

I call such equations weak equations, to distinguish them from the usual or strong equations.

One can make use of (1-20) only after one has worked out all the Poisson brackets which one is interested in. Subject to this rule, the Poisson bracket (1-19) is quite definite, and we have the possibility of writing our equations of motion (1-16) in a very concise form:

$$\dot{g} \approx [g, H_T] \qquad (1\text{-}21)$$

with a Hamiltonian I call the *total* Hamiltonian,

$$H_T = H + u_m\phi_m. \qquad (1\text{-}22)$$

Now let us examine the consequences of these equations of motion. In the first place, there will be some consistency conditions. We have the quantities $\phi$ which have to be zero throughout all time. We can apply the equation of motion (1-21) or (1-15) taking $g$ to be one of the $\phi$'s. We know that $\dot{g}$ must be zero for consistency, and so we get some consistency conditions. Let us see what they are like. Putting $g = \phi_m$ and $\dot{g} = 0$ in (1-15), we have:

$$[\phi_m, H] + u_{m'}[\phi_m, \phi_{m'}] \approx 0. \qquad (1\text{-}23)$$

We have here a number of consistency conditions, one for each value of $m$. We must examine these conditions to see what they lead to. It is possible for them to lead directly to an inconsistency. They might lead to the inconsistency $1 = 0$. If that happens, it would mean that our original Lagrangian is such that the Lagrangian equations of motion are inconsistent. One can easily construct an example with just one degree of freedom. If we take $L = q$ then the Lagrangian equation of motion (1-2) gives immediately $1 = 0$. So you see, we cannot take the Lagrangian to be completely arbitrary. We must impose on it the condition that the Lagrangian equations of motion do not involve an inconsistency. With this restriction the equations (1-23) can be divided into three kinds.

One kind of equation reduces to $0 = 0$, i.e. it is identically satisfied, with the help of the primary constraints.

Another kind of equation reduces to an equation independent of the $u$'s, thus involving only the $q$'s and the $p$'s. Such an equation must be independent of the primary constraints, otherwise it is of the first kind. Thus it is of the form

$$\chi\,(q,\,p) = 0. \tag{1-24}$$

Finally, an equation in (1-23) may not reduce in either of these ways; it then imposes a condition on the $u$'s.

The first kind we do not have to bother about any more. Each equation of the second kind means that we have another constraint on the Hamiltonian variables. Constraints which turn up in this way are called *secondary constraints*. They differ from the primary constraints in that the primary constraints are consequences merely of the equations (1-3) that define the momentum variables, while for the secondary constraints, one has to make use of the Lagrangian equations of motion as well.

If we have a secondary constraint turning up in our theory, then we get yet another consistency condition, because we can work out $\dot{\chi}$ according to the equation of motion (1-15) and we require that

$\dot{\chi} \approx 0$. So we get another equation

$$[\chi, H] + u_m [\chi, \phi_m] \approx 0. \qquad (1\text{-}25)$$

This equation has to be treated on the same footing as (1-23). One must again see which of the three kinds it is. If it is of the second kind, then we have to push the process one stage further because we have a further secondary constraint. We carry on like that until we have exhausted all the consistency conditions, and the final result will be that we are left with a number of secondary constraints of the type (1-24) together with a number of conditions on the coefficients $u$ of the type (1-23).

The secondary constraints will for many purposes be treated on the same footing as the primary constraints. It is convenient to use the notation for them:

$$\phi_k \approx 0, \; k = M+1, \ldots, M+K, \qquad (1\text{-}26)$$

where $K$ is the total number of secondary constraints. They ought to be written as weak equations in the same way as primary constraints, as they are also equations which one must not make use of before one works out Poisson brackets. So all the constraints together may be written as

$$\phi_j \approx 0, \; j = 1, \ldots, M+K \equiv J. \qquad (1\text{-}27)$$

Let us now go over to the remaining equations of the third kind. We have to see what conditions they impose on the coefficients $u$. These equations are

$$[\phi_j, H] + u_m [\phi_j, \phi_m] \approx 0 \qquad (1\text{-}28)$$

where $m$ is summed from 1 to $M$ and $j$ takes on any of the values from 1 to $J$. We have these equations involving conditions on the coefficients $u$, insofar as they do not reduce merely to the constraint equations.

Let us look at these equations from the following point of view. Let us suppose that the $u$'s are unknowns and that we have in (1-28) a number of non-homogeneous linear equations in these unknowns $u$, with coefficients which are functions of the $q$'s and the $p$'s. Let us look for a solution of these equations, which gives us the $u$'s as functions of the $q$'s and the $p$'s, say

$$u_m = U_m(q, p). \tag{1-29}$$

There must exist a solution of this type, because if there were none it would mean that the Lagrangian equations of motion are inconsistent, and we are excluding that case.

The solution is not unique. If we have one solution, we may add to it any solution $V_m(q, p)$ of the homogeneous equations associated with (1-28):

$$V_m\left[\phi_j, \phi_m\right] = 0, \tag{1-30}$$

and that will give us another solution of the inhomogeneous equations (1-28). We want the most general solution of (1-28) and that means that we must consider *all* the independent solutions of (1-30), which we may denote by $V_{am}(q, p), a = 1, \ldots, A$. The general solution of (1-28) is then

$$u_m = U_m + v_a V_{am}, \tag{1-31}$$

in terms of coefficients $v_a$ which can be arbitrary.

Let us substitute these expressions for $u$ into the total Hamiltonian of the theory (1-22). That will give us the total Hamiltonian

$$H_T = H + U_m \phi_m + v_a V_{am} \phi_m. \tag{1-32}$$

We can write this as

$$H_T = H' + v_a \phi_a, \tag{1-33}$$

where

$$H' = H + U_m \phi_m \tag{1-33'}$$

and

$$\phi_a = V_{am} \phi_m. \tag{1-34}$$

In terms of this total Hamiltonian (1-33) we still have the equations of motion (1-21).

As a result of carrying out this analysis, we have satisfied all the consistency requirements of the theory and we still have arbitrary coefficients $v$. The number of the coefficients $v$ will usually be less than the number of coefficients $u$. The $u$'s are not arbitrary but have to satisfy consistency conditions, while the $v$'s are arbitrary coefficients. We may take the $v$'s to be arbitrary functions of the time and we have still satisfied all the requirements of our dynamical theory.

This provides a difference of the generalized Hamiltonian formalism from what one is familiar with in elementary dynamics. We have arbitrary functions of the time occurring in the general solution of the equations of motion with given initial conditions. These arbitrary functions of the time must mean that we are using a mathematical framework containing arbitrary features, for example, a coordinate system which we can choose in some arbitrary way, or the gauge in electrodynamics. As a result of this arbitrariness in the mathematical framework, the dynamical variables at future times are not completely determined by the initial dynamical variables, and this shows itself up through arbitrary functions appearing in the general solution.

We require some terminology which will enable one to appreciate the relationships between the quantities which occur in the formalism. I find the following terminology useful. I define any dynamical variable, $R$, a function of the $q$'s and the $p$'s, to be *first-class* if it has zero Poisson brackets with all the $\phi$'s:

$$[R, \phi_j] \approx 0, \qquad j = 1, \ldots, J. \qquad (1\text{-}35)$$

It is sufficient if these conditions hold weakly. Otherwise $R$ is *second-class*. If $R$ is *first-class*, then $[R, \phi_j]$ has to be strongly equal to some linear function of the $\phi$'s, as anything that is weakly zero in the present theory is strongly equal to some linear function of the $\phi$'s. The $\phi$'s are, by definition, the only independent quantities which are weakly zero. So we have the strong equations

$$[R, \phi_j] = r_{jj'}\phi_{j'}. \qquad (1\text{-}36)$$

Before going on, I would like to prove a

*Theorem*: the Poisson bracket of two first-class quantities is also first-class. *Proof*: Let $R$, $S$ be first-class: then in addition to (1-36), we have

$$[S, \phi_j] = s_{jj'} \phi_{j'}. \qquad (1\text{-}36')$$

Let us form $[[R, S], \phi_j]$. We can work out this Poisson bracket using Jacobi's identity (1-13)

$$\begin{aligned}
[[R, S], \phi_j] &= [[R, \phi_j], S] - [[S, \phi_j], R] \\
&= [r_{jj'}\phi'_j, S] - [s_{jj'}\phi_{j'}, R] \\
&= r_{jj'}[\phi_{j'}, S] - [r_{jj'}, S]\phi_{j'} - s_{jj'}[\phi_{j'}, R] \\
&\quad - [s_{jj'}, R]\phi_{j'} \\
&\approx 0
\end{aligned}$$

by (1-36), (1-36)$'$, the product law (1-12), and (1-20). The whole thing vanishes weakly. We have proved therefore that $[R, S]$ is first-class.

We have altogether four different kinds of constraints. We can divide constraints into first-class and second-class, which is quite independent of the division into primary and secondary.

I would like you to notice that $H'$ given by (1-33)$'$ and the $\phi_a$ given by (1-34) are first-class. Forming the Poisson bracket of $\phi_a$ with $\phi_j$ we get, by (1-34), $V_{am}[\phi_m, \phi_j]$ plus terms that vanish weakly. Since the $V_{am}$ are defined to satisfy (1-30), $\phi_a$ is first-class. Similarly (1-28) with $U_m$ for $u_m$ shows that $H'$ is first-class. Thus (1-33) gives the total Hamiltonian in terms of a first-class Hamiltonian $H'$ together with some first-class $\phi$'s.

Any linear combination of the $\phi$'s is of course another constraint, and if we take a linear combination of the primary constraints we get another primary constraint. So each $\phi_a$ is a primary constraint; and it is first-class. So the final situation is that we have the total Hamiltonian expressed as the sum of a first-class Hamiltonian plus a linear combination of the primary, first-class constraints.

The number of independent arbitrary functions of the time occurring in the general solution of the equations of motion is equal to the number of values which the suffix $a$ takes on. That is equal to the number of independent primary first-class constraints, because all the independent primary first-class constraints are included in the sum (1-33).

That gives you then the general situation. We have deduced it by just starting from the Lagrangian equations of motion, passing to the Hamiltonian and working out consistency conditions.

From the practical point of view one can tell from the general transformation properties of the action integral what arbitrary functions of the time will occur in the general solution of the equations of motion. To each of these functions of the time there must correspond some primary first-class constraint. So we can tell which primary first-class constraints we are going to have without going through all the detailed calculation of working out Poisson brackets; in practical applications of this theory we can obviously save a lot of work by using that method.

I would like to go on a bit more and develop one further point of the theory. Let us try to get a physical understanding of the situation where we start with given initial variables and get a solution of the equations of motion containing arbitrary functions. The initial variables which we need are the $q$'s and the $p$'s. We don't need to be given initial values for the coefficients $v$. These initial conditions describe what physicists would call the *initial physical state* of the system. The physical state is determined only by the $q$'s and the $p$'s and not by the coefficients $v$.

Now the initial state must determine the state at later times. But the $q$'s and the $p$'s at later times are not uniquely determined by the initial state because we have the arbitrary functions $v$ coming in. That means that the state does not uniquely determine a set of $q$'s and $p$'s, even though a set of $q$'s and $p$'s uniquely determines a state. There must be several choices of $q$'s and $p$'s which correspond to the same state. So we have the problem of looking for all the sets of $q$'s and $p$'s that correspond to one particular physical state.

All those values for the $q$'s and $p$'s at a certain time which can evolve from one initial state must correspond to the same physical state at that time. Let us take particular initial values for the $q$'s and the $p$'s at time $t = 0$, and consider what the $q$'s and the $p$'s are after a short time interval $\delta t$. For a general dynamical variable $g$, with initial value $g_0$, its value at time $\delta t$ is

$$
\begin{aligned}
g\,(\delta t) &= g_0 + \dot{g}\delta t \\
&= g_0 + [g,\,H_T]\,\delta t \qquad\qquad\qquad (1\text{-}37) \\
&= g_0 + \delta t\,\{[g,\,H'] + v_a\,[g,\,\phi_a]\}\,.
\end{aligned}
$$

The coefficients $v$ are completely arbitrary and at our disposal. Suppose we take different values, $v'$, for these coefficients. That would give a different $g(\delta t)$, the difference being

$$
\Delta g\,(\delta t) = \delta t \left(v_a - v'_a\right) [g,\,\phi_a]\,. \qquad (1\text{-}38)
$$

We may write this as

$$
\Delta g\,(\delta t) = \varepsilon_a\,[g,\,\phi_a]\,, \qquad (1\text{-}39)
$$

where

$$
\varepsilon_a = \delta t \left(v_a - v'_a\right) \qquad (1\text{-}40)
$$

is a small arbitrary number, small because of the coefficient $\delta t$ and arbitrary because the $v$'s and the $v'$'s are arbitrary. We can change all our Hamiltonian variables in accordance with the rule (1-39) and the new Hamiltonian variables will describe the same state. This change in the Hamiltonian variables consists in applying an infinitesimal contact transformation with a generating function $\varepsilon_a\phi_a$. We come to the conclusion that the $\phi_a$'s, which appeared in the theory in the first place as the primary first-class constraints, have this meaning: *as generating functions of infinitesimal contact transformations, they lead to changes in the $q$'s and the $p$'s that do not affect the physical state.*

However, that is not the end of the story. We can go on further in the same direction. Suppose we apply two of these contact transformations in succession. Apply first a contact transformation with generating function $\varepsilon_a\phi_a$ and then apply a second contact transformation with generating function $\gamma_{a'}\phi_{a'}$, where the gamma's are some new small coefficients. We get finally

$$
g' = g_0 + \varepsilon_a\,[g,\,\phi_a] + \gamma_{a'}\,[g + \varepsilon_a\,[g,\,\phi_a],\,\phi_{a'}]\,. \qquad (1\text{-}41)
$$

(I retain the second order terms involving products $\varepsilon\gamma$, but I neglect the second order terms involving $\varepsilon^2$ or involving $\gamma^2$. This is legitimate and sufficient. I do that because I do not want to write down more than I really need for getting the desired result.) If we apply the two transformations in succession in the reverse order, w get finally

$$g'' = g_0 + \gamma_{a'}[g, \phi_{a'}] + \varepsilon_a[g + \gamma_{a'}[g, \phi_{a'}], \phi_a]. \qquad (1\text{-}42)$$

Now let us subtract these two. The difference is

$$\Delta g = \varepsilon_a \gamma_{a'}\{[[g, \phi_a], \phi_{a'}] - [[g, \phi_{a'}], \phi_a]\}. \qquad (1\text{-}43)$$

By Jacobi's identity (1-13) this reduces to

$$\Delta g = \varepsilon_a \gamma_{a'}[g, [\phi_a, \phi_{a'}]]. \qquad (1\text{-}44)$$

This $\Delta g$ must also correspond to a change in the $q$'s and the $p$'s which does not involve any change in the physical state, because it is made up by processes which individually don't involve any change in the physical state. Thus we see that we can use

$$[\phi_a, \phi_{a'}] \qquad (1\text{-}45)$$

as a generating function of an infinitesimal contact transformation and it will still cause no change in the physical state.

Now the $\phi_a$ are first-class: their Poisson brackets are weakly zero, and therefore strongly equal to some linear function of the $\phi$'s. This linear function of the $\phi$'s must be first-class because of the theorem I proved a little while back, that the Poisson bracket of two first-class quantities is first-class. So we see that the transformations which we get this way, corresponding to no change in the physical state, are transformations for which the generating function is a first-class constraint. The only way these transformations are more general than the ones we had before is that the generating functions which we had before are restricted to be first-class primary constraints. Those that we get now could be first-class secondary constraints. The result of this calculation is to show that we might have a first-class secondary constraint as a generating function of an infinitesimal contact transformation which leads to a change in the $q$'s and the $p$'s without changing the state.

For the sake of completeness, there is a little bit of further work one ought to do which shows that a Poisson bracket $[H', \phi_a]$ of the first-class Hamiltonian $H'$ with a first-class $\phi$ is again a linear function of first-class constraints. This can also be shown to be a possible generator for infinitesimal contact transformations which do not change the state.

The final result is that those transformations of the dynamical variables which do not change physical states are infinitesimal contact transformations in which the generating function is a primary first-class constraint or possibly a secondary first-class constraint. A good many of the secondary first-class constraints do turn up by the process (1-45) or as $[H', \phi_a]$. I think it may be that all the first-class secondary constraints should be included among the transformations which don't change the physical state, but I haven't been able to prove it. Also, I haven't found any example for which there exist first-class secondary constraints which do generate a change in the physical state.

DR. DIRAC
*Lecture No. 2*

# THE PROBLEM OF QUANTIZATION

We were led to the idea that there are certain changes in the $p$'s and $q$'s that do not correspond to a change of state, and which have as generators first-class secondary constraints. That suggests that one should generalize the equations of motion in order to allow as variation of a dynamical variable $g$ with the time not only any variation given by (1-21), but also any variation which does not correspond to a change of state. So we should consider a more general equation of motion

$$\dot{g} = [g, H_E] \qquad (2\text{-}1)$$

with an extended Hamiltonian $H_E$, consisting of the previous Hamiltonian, $H_T$, plus all those generators which do not change the state, with arbitrary coefficients:

$$H_E = H_T + v'_{a'}\phi_{a'}. \qquad (2\text{-}2)$$

Those generators $\phi_{a'}$, which are not included already in $H_T$ will be the first-class secondary constraints. The presence of these further terms in the Hamiltonian will give further changes in $g$, but these further changes in $g$ do not correspond to any change of state and so they should certainly be included, even though we did not arrive at these further changes of $g$ by direct work from the Lagrangian.

That, then, is the general Hamiltonian theory. The theory as I have developed it applies to a finite number of degrees of freedom but we can easily extend it to the case of an infinite number of degrees of freedom. Our suffix denoting the degree of freedom is $n = 1, \ldots, N$; we may easily make $N$ infinite. We may further generalize it by allowing the number of degrees of freedom to be continuously infinite. That is to say, we may have as our $q$'s and $p$'s variables $q_x, p_x$ where $x$ is a suffix which can take on all values in a continuous range. If we work with this continuous $x$, then we have to change all our sums over $n$ in the previous work into integrals. The previous work can all be taken over directly with this change.

There is just one equation which we will have to think of a bit differently, the equation which defines the momentum variables,

$$p_n = \frac{dL}{d\dot{q}_n}. \tag{1-3}$$

If $n$ takes on a continuous range of values, we have to understand by this partial differentiation a process of partial functional differentiation that can be made precise in this way: We vary the velocities by $\delta\dot{q}_x$ in the Lagrangian and then put

$$\delta L = \int p_x \delta\dot{q}_x. \tag{2-3}$$

The coefficient of $\delta\dot{q}_x$ occurring in the integrand in $\delta L$ is defined to be $p_x$.

After giving this general abstract theory, I think it would be a help if I gave a simple example as illustration. I will take as an example just the electromagnetic field of Maxwell, which is defined in terms of potentials $A_\mu$. The dynamical coordinates now consist of the potentials for all points of space at a certain time. That is to say, the dynamical coordinates consist of $A_{\mu x}$, where the suffix $x$ stands for the three coordinates $x^1$, $x^2$, $x^3$ of a point in three-dimensional space at a certain time $x^0$ (not the four $x$'s which one is used to in relativity). We shall have then as the dynamical velocities the time derivatives of the dynamical coordinates, and I shall denote these by a suffix 0 preceded by a comma.

Any suffix with a comma before it denotes differentiation according to the general scheme

$$\xi_{,\mu} = \frac{d\xi}{dx^\mu}. \tag{2-4}$$

We are dealing with special relativity so that we can raise and lower these suffixes according to the rules of special relativity: we have a change in sign if we raise or lower a suffix 1, 2, or 3 but no change of sign when we raise or lower the suffix 0.

We have as our Lagrangian for the Maxwell electrodynamics, if we work in Heaviside units,

$$L = -\frac{1}{4} \int F_{\mu\nu} F^{\mu\nu} d^3x. \tag{2-5}$$

Here $d^3x$ means $dx^1\, dx^2\, dx^3$, the integration is over three-dimensional space, and $F_{\mu\nu}$ means the field quantities defined in terms of the potentials by

$$F_{\mu\nu} = A_{\nu,\mu} - A_{\mu,\nu}. \tag{2-6}$$

This $L$ is the Lagrangian because its time integral is the action integral of the Maxwell field.

Let us now take this Lagrangian and apply the rules of our formalism for passing to the Hamiltonian. We first of all have to introduce the momenta. We do that by varying the velocities in the Lagrangian. If we vary the velocities, we have

$$\begin{aligned}
\delta L &= -\frac{1}{2} \int F^{\mu\nu}\, \delta F_{\mu\nu}\, d^3x \\
&= \int F^{\mu 0}\, \delta A_{\mu,0}\, d^3x.
\end{aligned} \tag{2-7}$$

Now the momenta $B^\mu$ are defined by

$$\delta L - \int B^\mu\, \delta A_{\mu 0}\, d^3x \tag{2-8}$$

and these momenta will satisfy the basic Poisson bracket relations

$$\left[A_{\mu x}, B^\nu_{x'}\right] = g^\nu_\mu\, \delta^3\left(x - x'\right); \qquad \mu, \nu = 0, 1, 2, 3. \tag{2-9}$$

In this formula $A$ is taken at a point $x$ in three-dimensional space and $B$ is taken at a point $x'$ in the three-dimensional space. $g^\nu_\mu$ is just the Kronecker delta function. $\delta^3(x - x')$ is the three-dimensional delta function of $x - x'$.

We compare the two expressions (2-7) and (2-8) for $\delta L$ and that gives us

$$B^\mu = F^{\mu 0}. \tag{2-10}$$

Now $F^{\mu\nu}$ is anti-symmetrical

$$F^{\mu\nu} = -F^{\nu\mu}. \tag{2-11}$$

So if we put $\mu = 0$, in (2-10) we get zero. Thus $B_x^0$ is equal to zero. This is a primary constraint. I write it as a weak equation:

$$B_x^0 \approx 0. \tag{2-12}$$

The other three momenta $B^r (r = 1, 2, 3)$ are just equal to the components of the electric field.

I should remind you that (2-12) is not just one primary constraint: there is a whole threefold infinity of primary constraints because there is the suffix $x$ which stands for some point in three-dimensional space; and each value for $x$ will give us a different primary constraint.

Let us now introduce the Hamiltonian. We define that in the usual way by

$$
\begin{aligned}
H &= \int B^\mu A_{\mu,0} d^3 x - L \\
&= \int \left( F^{r0} A_{r,0} + \frac{1}{4} F^{rs} F_{rs} + \frac{1}{2} F^{r0} F_{r0} \right) d^3 x \\
&= \int \left( \frac{1}{4} F^{rs} F_{rs} - \frac{1}{2} F^{r0} F_{r0} + F^{r0} A_{0,r} \right) d^3 x \\
&= \int \left( \frac{1}{4} F^{rs} F_{rs} + \frac{1}{2} B^r B^r - A_0 B^r_{,r} \right) d^3 x.
\end{aligned}
\tag{2-13}
$$

I've done a partial integration of the last term in (2-13) to get it in this form. Now here we have an expression for the Hamiltonian which does not involve any velocities. It involves only dynamical coordinates and momenta. It is true that $F_{rs}$ involves partial differentiations of the potentials, but it involves partial differentiations only with respect to $x^1$, $x^2$, $x^3$. That does not bring in any velocities. These partial derivatives are functions of the dynamical coordinates.

We can now work out the consistency conditions by using the primary constraints (2-12). Since they have to remain satisfied at all times, $[B^0, H]$ has to be zero. This leads to the equation

$$B^r_{,r} \approx 0. \tag{2-14}$$

This is again a constraint because there are no velocities occurring in it. This is a secondary constraint, which appears in the Maxwell theory in this way. If we proceed further to examine the consistency relations, we must work out

$$[B^r_{,r}, H] = 0. \qquad (2\text{-}15)$$

We find that this reduces to $0 = 0$. It does not give us anything new, but is automatically satisfied. We have therefore obtained all the constraints in our problem. (2-12) gives the primary constraints. (2-14) gives the secondary constraints.

We now have to look to see whether they are first-class or second-class, and we easily see that they are all first-class. The $B_0$ are momenta variables. They all have zero Poisson brackets with each other. $B^r_{,r}$ and $B_0$ also have zero Poisson brackets with each other. And $B^r_{,rx}$ and $B^r_{,rx'}$ also have zero Poisson brackets with each other. All these quantities are therefore first-class constraints. There are no second-class constraints occurring in the Maxwell electrodynamics.

The expression (2-13) for $H$ is first-class, so this $H$ can be taken as the $H'$ of (1-33). Let us now see what the total Hamiltonian is:

$$H_T = \int \left( \frac{1}{4} F^{rs} F_{rs} + \frac{1}{2} B_r B_r \right) d^3x - \int A_0 B^r_{,r} \, d^3x$$
$$+ \int v_x B^0 d^3x. \qquad (2\text{-}16)$$

This $v_x$ is an arbitrary coefficient for each point in three-dimensional space. We have just added on the primary first-class constraints with arbitrary coefficients, which is what we must do according to the rules to get the total Hamiltonian.

In terms of the total Hamiltonian we have the equation of motion in the standard form

$$\dot{g} \approx [g, H_T]. \qquad (1\text{-}21)$$

The $g$ which we have here may be any field quantity at some point $x$ in three-dimensional space, or may also be a function of field quantities at *different* points in three-dimensional space. It could, for example, be an integral over three-dimensional space. This $g$ can be perfectly generally any function of the $q$'s and the $p$'s throughout three-dimensional space.

It is permissible to take $g = A_0$ and then we get

$$A_{0,0} = v, \qquad (2\text{-}17)$$

because $A_0$ has zero Poisson brackets with everything except the $B_0$ occurring in the last term of (2-16). This gives us a meaning for the arbitrary coefficient $v_x$ occurring in the total Hamiltonian. It is the time derivative of $A_0$.

Now to get the most general motion which is physically permissible, we ought to pass over to the extended Hamiltonian. To do this we add on the first-class secondary constraints with arbitrary coefficients $u_x$. This gives the extended Hamiltonian:

$$H_E = H_T + \int u_x B^r_{,r} \, d^3x. \qquad (2\text{-}18)$$

Bringing in this extra term into the Hamiltonian allows a more general motion. It gives more variation of the $q$'s and the $p$'s, of the nature of a gauge transformation. When this additional variation of the $q$'s and the $p$'s is brought in, it leads to a further set of $q$'s and $p$'s which must correspond to the same state.

That is the result of working out, according to our rules, the Hamiltonian form of the Maxwell theory. When we've got to this stage, we see that there is a certain simplification which is possible. This simplification comes about because the variables $A_0$, $B_0$ are not of any physical significance. Let us see what the equations of motion tell us about $A_0$ and $B_0$. $B_0 = 0$ all the time. That is not of interest. $A_0$ is something whose time derivative is quite arbitrary. That again is something which is not of interest. The variables $A_0$ and $B_0$ are therefore not of interest at all. We can drop them out from the theory and that will lead to a simplified Hamiltonian formalism where we have fewer degrees of freedom, but still retain all the degrees of freedom which are physically of interest.

In order to carry out this discard of the variables $A_0$ and $B_0$, we drop out the term $v_x B^0$ from the Hamiltonian. This term merely has the effect of allowing $A_0$ to vary arbitrarily. The term $-A_0 B^r_{,r}$ in $H_T$ can be combined with the $u_x B^r_{,r}$ in the extended Hamiltonian. The coefficient $u_x$ is an arbitrary coefficient in any case. When we combine these two terms, we just have this $u_x$ replaced by $u'_x = u_x - A_0$ which is equally arbitrary. So that we get a new Hamiltonian

$$H = \int \left( \frac{1}{4} F^{rs} F_{rs} + \frac{1}{2} B_r B_r \right) d^3x + \int u'_x B^r_{,r} \, d^3x. \qquad (2\text{-}19)$$

This Hamiltonian is sufficient to give the equations of motion for all the variables which are of physical interest. The variables $A_0$, $B_0$ no longer appear in it. This is the Hamiltonian for the Maxwell theory in its simplest form.

Now the usual Hamiltonian which people work with in quantum electrodynamics is not quite the same as that. The usual one is based on a theory which was originally set up by Fermi. Fermi's theory involves putting this restriction on the potentials:

$$A^\mu_{,\mu} = 0. \qquad (2\text{-}20)$$

It is quite permissible to bring in this restriction on the gauge. The Hamiltonian theory which I have given here does not involve this restriction, so that it allows a completely general gauge. It's thus a somewhat different formalism from the Fermi formalism. It's a formalism which displays the full transforming power of the Maxwell theory, which we get when we have completely general changes of gauge. This Maxwell theory gives us an illustration of the general ideas of primary and secondary constraints.

I would like now to go back to general theory and to consider the problem of quantizing the Hamiltonian theory. To discuss this question of quantization, let us first take the case when there are no second-class constraints, when all the constraints are first-class. We make our dynamical coordinates and momenta, the $q$'s and $p$'s, into operators satisfying commutation relations which correspond to the Poisson bracket relations of the classical theory. That is quite straightforward. Then we set up a Schrödinger equation

$$i\hbar \frac{d\psi}{dt} = H'\psi. \qquad (2\text{-}21)$$

$\psi$ is the wave function on which the $q$'s and the $p$'a operate. $H'$ is the first-class Hamiltonian of our theory.

We further impose certain supplementary conditions on the wave function, namely:

$$\phi_j \psi = 0. \qquad (2\text{-}22)$$

Each of our constraints thus leads to a supplementary condition on the wave function. (The constraints, remember, are now all first-class.)

The first thing we have to do now is to see whether these equations for $\psi$ are consistent with one another. Let us take two of the supplementary conditions and see whether they are consistent. Let us take (2-22) and

$$\phi_{j'} \psi = 0. \qquad (2\text{-}22')$$

If we multiply (2-22) by $\phi_{j'}$, we get

$$\phi_{j'} \phi_j \psi = 0. \qquad (2\text{-}23)$$

If we multiply (2-22)' by $\phi_j$, we get

$$\phi_j \phi_{j'} \psi = 0. \qquad (2\text{-}23')$$

If we now substract these two equations, we get:

$$[\phi_j, \phi_{j'}]\,\psi = 0. \qquad (2\text{-}24)$$

This further condition on $\psi$ is necessary for consistency. Now we don't want to have any fresh conditions on $\psi$. We want all the conditions on $\psi$ to be included among (2-22). That means to say, we want to have (2-24) a consequence of (2-22) which means we require

$$[\phi_j, \phi_{j'}] = c_{jj'j''}\phi_{j''}. \qquad (2\text{-}25)$$

If (2-25) *does* hold, then (2-24) is a consequence of (2-22) and is not a new condition on the wave function.

Now we know that the $\phi$'s are all first-class in the classical theory, and that means that the Poisson bracket of any two of the $\phi$'s is a linear combination of the $\phi$'s in the classical theory. When we go over to the quantum theory, we must havea similar equation holding for the commutator, but it does not necessarily follow that the coefficients $c$ are all on the left. We need to have these coefficients all on the left, because the $c$'s will in general be functions of the coordinates and momenta and will not commute with the $\phi$'s in the quantum theory, and (2-24) will be a consequence of (2-22) only provided the $c$'s are all on the left.

When we set up the quantities $\phi$ in the quantum theory, there may be some arbitrariness coming in. The corresponding classical expressions may involve quantities which don't commute in the quantum theory and then we have to decide on the order in which to put the factors in the quantum theory. We have to try to arrange the order of these factors so that we have (2-25) holding with all the coefficients on the left. If we can do that, then we have the supplementary conditions all consistent with each other. If we cannot do it, then we are out of luck and we cannot make an accurate quantum theory. In any case we have a first approximation to the quantum theory, because our equations would be all right if we look at them only to the order of accuracy of Planck's constant $\hbar$ and neglect quantities of order $\hbar^2$.

I have just discussed the requirements for the supplementary conditions to be consistent with one another. There is a similar discussion needed in order to check that the supplementary conditions shall be consistent with the Schrödinger equation. If we start with a $\psi$ satisfying the supplementary conditions (2-24) and let that $\psi$ vary with the time in accordance with the Schrödinger equation, then after a lapse of a short interval of time will our $\psi$ still satisfy the supplementary conditions? We can work out the requirement for that to be the case and we get

$$\left[\phi_j, H\right]\psi = 0, \tag{2-26}$$

which means that $[\phi_j, H]$ must be some linear function of the $\phi$'s:

$$\left[\phi_j, H\right] = b_{jj'}\phi_{j'}, \tag{2-27}$$

if we are not to get a new supplementary condition. Again we have anequation which we know is all right in the classical theory. $\phi_j$, and

$H$ are both first-class, so their Poisson bracket vanishes weakly. The Poisson bracket is thus strongly equal to some linear function of the $\phi$'s in the classical theory. Again we have to try to arrange things so that in the corresponding quantum equation we have all our coefficients on the left. That is necessary to get an accurate quantum theory, and we need a bit of luck, in general, in order to be able to bring it about.

Let us now consider how to quantize a Hamiltonian theory in which there are second-class constraints. Let us think of this question first in terms of a simple example. We might take as the simplest example of two second-class constraints

$$q_1 \approx 0 \quad \text{and} \quad p_1 \approx 0. \tag{2-28}$$

If we have these two constraints appearing in the theory, then their Poisson bracket is not zero, so they are second-class. What can we do with them when we go over to the quantum theory? We cannot impose (2-28) as supplementary conditions on the wave function as we did with the first-class constraints. If we try to put $q_1\psi = 0$, $p_1\psi = 0$, then we should immediately get a contradiction because we should have $(q_1 p_1 - p_1 q_1)\psi = i\hbar\psi = 0$. So that won't do. We must adopt some different plan.

Now in this simple case it's pretty obvious what the plan must be. The variables $q_1$ and $p_1$ are not of interest if they are both restricted to be zero. So the degree of freedom 1 is not of any importance. We can just discard the degree of freedom 1 and work with the other degrees of freedom. That means a different definition for a Poisson bracket. We should have to work with a definition of a Poisson bracket in the classical theory

$$[\xi, \eta] = \frac{\partial\xi}{\partial q_n}\frac{\partial\eta}{\partial p_n} - \frac{\partial\xi}{\partial p_n}\frac{\partial\eta}{\partial q_n} \quad \text{summed over } n = 2, \ldots N.$$

$$\tag{2-29}$$

This would be sufficient because it would deal with all the variables which are of physical interest. Then we could just take $q_1$ and $p_1$ as identically zero. There's no contradiction involved there, and we can pass over to the quantum theory, setting it up in terms only of the degrees of freedom $n = 2, \ldots, N$.

In this simple case it is fairly obvious what we have to do to build up a quantum theory. Let us try now to generalize it. Suppose we have $P_1 \approx 0, q_1 \approx f(q_r, P_r), r = 2, \ldots, N$, so $f$ is any function of all the other $q$'s and $p$'s. We could drop out the number 1 degree of freedom if we substitute $f(q_r, p_r)$ for $q_1$ in the Hamiltonian and in all the other constraints. Again we can forget about the number 1 degree of freedom and simply work with the other degrees of freedom and pass over to a quantum theory in these other degrees of freedom. Again we should have to work with the (2-29) kind of Poisson bracket, referring only to the other degrees of freedom.

That is the idea which one uses for quantizing a theory which involves second-class constraints. The existence of second-class constraints means that there are some degrees of freedom which are not physically important. We have to pick out these degrees of freedom and set up new Poisson brackets referring only to the other degrees of freedom which *are* of physical importance. Then in terms of those new Poisson brackets we can pass over to the quantum theory. I would like to discuss a general procedure for carrying that out.

For the present, we are going back to the classical theory. We have a number of constraints $\phi_j \approx 0$, some of them first-class, some second-class. We can replace these constraints by independent linear combinations of them, which will do just as well as the original constraints. We try to arrange to take the linear combinations in such a way as to have as many constraints as possible brought into the first class. There may then be some left in the second class which we just cannot bring into the first class by taking linear combinations of them. Those which are left in the second class I will call $\chi_s, s = 1, \ldots, S$. $S$ is the number of second-class constraints which are such that no linear combination of them is first-class.

We take these surviving second-class constraints and we form all their Poisson brackets with each other and arrange these Poisson brackets as a determinant $\Delta$:

$$\Delta = \begin{vmatrix} 0 & [\chi_1, \chi_2] & [\chi_1, \chi_3] & \cdots & [\chi_1, \chi_s] \\ [\chi_2, \chi_1] & 0 & [\chi_2, \chi_3] & \cdots & [\chi_2, \chi_s] \\ \vdots & \vdots & \vdots & & \vdots \\ [\chi_s, \chi_1] & [\chi_s, \chi_2] & [\chi_s, \chi_3] & \cdots & \vdots \end{vmatrix}$$

I would like now to prove a

*Theorem*: The determinant $\Delta$ does not vanish, not even weakly.

*Proof*: Assume that the determinant *does* vanish. I'm going to show that we get a contradiction. If the determinant vanishes, then it is of some rank $T < S$. Now let us set up the determinant $A$:

$$A = \begin{vmatrix} \chi_1 & 0 & [\chi_1, \chi_2] & \cdots & [\chi_1, \chi_T] \\ \chi_2 & [\chi_2, \chi_1] & 0 & & [\chi_2, \chi_T] \\ \vdots & \vdots & \vdots & & \vdots \\ \chi_{T+1} & [\chi_{T+1}, \chi_1] & [\chi_{T+1}, \chi_2] & \cdots & [\chi_{T+1}, \chi_T] \end{vmatrix}$$

$A$ has $T + 1$ rows and columns. $T + 1$ might equal $S$ or might be less than $S$. If we expand $A$ in terms of the elements of its first column, we will get each of these elements multiplied into one of the sub-determinants of $\Delta$. Now I don't want all of these sub-determinants to vanish. It might so happen that they do all vanish. And in that case, I would choose the $\chi$'s which are referred to among the rows and columns of $A$ in a different way. There must always be some way of choosing the $\chi$'s which occur in $A$ so that the sub-determinants don't all vanish, because $\Delta$ is of rank $T$. So we choose the $\chi$'s m such a way that the coefficients of the elements in the first column are not all zero.

Now I will show that $A$ has zero Poisson brackets with any of the $\phi$'s. If we form the Poisson bracket of $\phi$ with a determinant, we get the result by forming the Poisson bracket of $\phi$ with the first column of the determinant, adding on the result of forming the Poisson bracket of $\phi$ with the second column of the determinant, and so on. Thus

$$[\phi, A] = \begin{vmatrix} [\phi, \chi_1] & 0 & \cdots \\ [\phi, \chi_2] & [\chi_2, \chi_1] & \cdots \\ \vdots & \vdots & \\ [\phi, \chi_{T+1}] & [\chi_{T+1}, \chi_1] & \cdots \end{vmatrix}$$

$$+ \begin{vmatrix} \chi_1 & 0 & \cdots \\ \chi_2 & [\phi, [\chi_2, \chi_1]] & \cdots \\ \vdots & \vdots & \\ \chi_{T+1} & [\phi, [\chi_{T+1}, \chi_1]] & \cdots \end{vmatrix}$$

$$+ \begin{vmatrix} \chi_1 & 0 & [\phi, [\chi_1, \chi_2]] & \cdots \\ \chi_2 & [\chi_2, \chi_1] & 0 & \cdots \\ \vdots & \vdots & \vdots & \\ \chi_{T+1} & [\chi_{T+1}, \chi_1] & [\phi, [\chi_{T+1}, \chi_2]] & \cdots \end{vmatrix} + \cdots.$$

This looks rather complicated, but one can easily see that every one of these determinants vanishes. In the first place, the first determinant on the right vanishes: if $\phi$ is first class, then the first column vanishes; if $\phi$ is second class, then $\phi$ is one of the $\chi$'s and we have a determinant which is a part of the determinant $\Delta$ with $T + 1$ rows and columns. But $\Delta$ is assumed to be of rank $T$, so that any part of it with $T + 1$ rows and columns vanishes. Now, the second determinant on the right vanishes weakly because the first column vanishes weakly. Similarly all the other determinants vanish weakly. The result is that the whole right-hand side vanishes weakly. Thus $A$ is a quantity whose Poisson bracket with every one of the $\phi$'s vanishes weakly.

Also, we can expand the determinant $A$ in terms of the elements of the first column, and get $A$ as a linear combination of the $\chi$'s. So we have the result that a certain linear combination of the $\chi$'s has zero Poisson brackets with all the $\phi$'s. That means that this linear combination of the $\chi$'s is first class. That contradicts our assumption that we have put as many $\chi$'s as possible into the first class. That proves the theorem.

Incidentally, we see that the number of surviving $\chi$'s, which cannot be brought into the first class, must be even, because the determinant

$\Delta$ is antisymmetrical. Any antisymmetrical determinant with an odd number of rows and columns vanishes. This one doesn't vanish and therefore must have an even number of rows and columns.

Because this determinant, $\Delta$, doesn't vanish, we can bring in the reciprocal $c_{ss'}$ of the matrix whose determinant is $\Delta$. We define the matrix $c_{ss'}$ by

$$c_{ss'}[\chi_{s'}, \chi_{s''}] = \delta_{ss''}. \tag{2-30}$$

We now define new Poisson brackets in accordance with this formalism: any two quantities $\xi$, $\eta$ have a new Poisson bracket defined by

$$[\xi, \eta]^* = [\xi, \eta] - [\xi, \chi_s]\, c_{ss'}\, [\chi_{s'}, \eta]. \tag{2-31}$$

It is easy to check that new Poisson brackets defined in this way satisfy the laws which Poisson brackets usually satisfy: $[\xi, \eta]^*$ is antisymmetrical between $\xi$ and $\eta$, is linear in $\xi$, is linear in $\eta$, satisfies the product law $[\xi_1 \xi_2, \eta]^* = \xi_1 [\xi_2 \eta]^* + [\xi_1, \eta]^* \xi_2$, and obeys the Jacobi identity $[[\xi, \eta]^*, \zeta]^* + [[\eta, \zeta]^* \xi]^* + [[\zeta, \xi]^*, \eta]^* = 0$. I don't know of any neat way of proving the Jacobi identity for the new Poisson brackets. If one just substitutes according to the definition and works it out in a complicated way, one does find that all the terms cancel out and that the left-hand side equals zero. I think there ought to be some neat way of proving it, but I haven't been able to find it. The straightforward method I have given in the *Canadian Journal of Mathematics*, **2**, 147 (1950). The problem has been dealt with by Bergmann, *Physical Review*, **98**, 531 (1955).

Now let us see what we can do with these new Poisson brackets. First of all, I would like you to notice that the equations of motion are as valid for the new Poisson brackets as for the original ones.

$$[g, H_T]^* = [g, H_T] - (g, \chi_s)\, c_{ss'}\, [\chi_{s'}, H_T]$$
$$\approx [g, H_T]$$

because the terms $[\chi_{s'}, H_T]$ all vanish weakly on account of $H_T$ being first-class. Thus we can write

$$\dot{g} \approx [g, H_T]^*.$$

Now if we take any function $\xi$ whatever of the $q$'s and $p$'s, and form its new Poisson bracket with one of the $\chi$'s, say $\chi_{s''}$, we have

$$
\begin{aligned}
[\xi, \chi_{s''}]^* &= [\xi, \chi_{s''}] - [\xi, \chi_s] \, c_{ss'} \, [\chi_{s'}, \chi_{s''}] \\
&= [\xi, \chi_{s''}] - [\xi, \chi_s] \, \delta_{ss''} \quad\quad\quad (2\text{-}30) \\
&= 0.
\end{aligned}
$$

Thus we can put the $\chi$'s equal to 0 before working out new Poisson brackets. This means that the equation

$$
\chi_s = 0 \quad\quad\quad\quad\quad\quad (2\text{-}32)
$$

may be considered as a strong equation.

We modify our classical theory in this way, bringing in these new Poisson brackets, and this prepares the ground for passing to the quantum theory. We pass over to the quantum theory by taking the commutation relations to correspond to die new Poisson bracket relations and taking the strong equations (2-32) to be equations between operators in the quantum theory. The remaining weak equations, which are all first class, become again supplementary conditions on the wave functions. The situation is then reduced to the previous case where there were only first-class $\phi$'s. We have again, therefore, a method of quantizing our general classical Hamiltonian theory. Of course, we again need a bit of luck in order to arrange that the coefficients are all on the left in the consistency conditions.

That gives the general method of quantization. You notice that when we have passed over to the quantum theory, the distinction between primary constraints and secondary constraints ceases to be of any importance. The distinction between primary and secondary constraints is not a very fundamental one. It depends very much on the original Lagrangian which we start off with. Once we have gone over to the Hamiltonian formalism, we can really forget about the distinction between primary and secondary constraints. The distinction between first-class and second-class constraints is very important. We must put as many as possible into the first class and bring in new Poisson brackets which enable us to treat the surviving second-class constraints as strong.

*Lecture No. 3*

# QUANTIZATION ON CURVED SURFACES

We started off with a classical action principle. We took our action integral to be Lorentz-invariant. This action gives us a Lagrangian. We then passed from the Lagrangian to the Hamiltonian, and then to the quantum theory by following through certain rules. The result is that, starting with a classical field theory, described by an action principle, we end up with a quantum field theory. Now you might think that that finishes our work, but there is one important problem still to be considered: whether our quantum field theory obtained in this way is a relativistic theory. For the purposes of discussion, we may confine ourselves to special relativity. We have then to consider whether our quantum theory is in agreement with special relativity.

We started from an action principle and we required that our action should be Lorentz-invariant. That is sufficient to ensure that our classical theory shall be relativistic. The equations of motion that follow from a Lorentz invariant action principle must be relativistic equations. It is true that when we put these equations of motion into the Hamiltonian form, we are disturbing the four-dimensional symmetry. We are expressing our equations in the form

$$\dot{g} \approx [g, H_T]. \qquad (1\text{-}21)$$

The dot here means *dg/dt* and refers to one absolute time, so that the classical equations of motion in the Hamiltonian form are not manifestly relativistic, but we know that they must be relativistic in content because they follow from relativistic assumptions.

However, when we pass over to the quantum theory we are making new assumptions. The expression for $H_T$ which we have in the classical theory does not uniquely determine the quantum Hamiltonian. We have to decide questions about the order in which to put non-commuting factors in the quantum theory. We have something at our disposal in choosing this order, and so we are making new assumptions. These new assumptions may disturb the relativistic invariance of the theory, so that the quantum field theory obtained by this method is not necessarily in agreement with relativity. We now have to face the problem of seeing how we can ensure that our quantum theory shall be a relativistic theory.

For that purpose we have to go back to first principles. It is no longer sufficient to consider just one time variable referring to one particular observer; we have to consider different observers moving relatively to one another. We must set up a quantum theory which applies equally to any of these observers, that is, to any time axis. To get a theory involving all the different time axes, we should first get the corresponding *classical* theory and then pass from this classical theory to the quantum theory by the standard rules.

I would like to go back to the beginning of our Hamiltonian development and consider a special case. We started our development by taking a Lagrangian $L$, which is a function of dynamical coordinates and velocities $q$, $\dot{q}$, introducing the momenta, then introducing the Hamiltonian. Let us take the special case when $L$ is homogeneous of the first degree in the $\dot{q}$'s. Then Euler's Theorem tells us that

$$\dot{q}_n \frac{\partial L}{\partial \dot{q}_n} = L. \tag{3-1}$$

That just tells us that $p_n \dot{q}_n - L = 0$. Thus we get in this special case a Hamiltonian that is zero.

We necessarily get primary constraints in this case. There must certainly be one primary constraint, because the $p$'s are homogeneous functions of degree zero in the velocities. The $p$'s are thus functions only of the ratios of the velocities. The number of $p$'s is equal to $N$, the number of degrees of freedom, and the number of ratios of the velocities is $N - 1$. $N$ functions of $N - 1$ ratios of the velocities cannot be independent. There must be at least one function of the $p$'s and $q$'s which is equal to zero; there must be at least one primary constraint. There may very well be more than one. One can also see that, if we are to have any motion at all with a zero Hamiltonian, we must have at least one primary first-class constraint.

We have the expression (1-33) for the total Hamiltonian

$$H_T = H' + v_a \phi_a.$$

$H'$ must be a first-class Hamiltonian, and as 0 is certainly a first-class quantity we may take $H' = 0$. Our total Hamiltonian is now built up entirely from the primary first-class constraints with arbitrary coefficients:

$$H_T = v_a \phi_a, \tag{3-2}$$

showing that there must be at least one primary first-class constraint if we are to have any motion at all.

Our equations of motion now read like this:

$$\dot{g} \approx v_a \, [g, \phi_a] \,.$$

We can see that the $\dot{g}$'s may all be multiplied by a factor because, since the coefficients $v$ are arbitrary, we may multiply them all by a factor. If we multiply all the $dg/dt$'s by a factor, it means that we have a different time scale. So we have now Hamiltonian equations of motion in which the time scale is arbitrary. We could introduce another time variable $\tau$ instead of $t$ and use $\tau$ to give us equations of motion

$$\frac{dg}{d\tau} \approx v'_a \, [g, \phi_a] \,. \tag{3-3}$$

So we have now a Hamiltonian scheme of equations of motion in which there is no absolute time variable. Any variable increasing monotonically with $t$ could be used as time and the equations of motion would be of the same form. Thus the characteristic of a Hamiltonian theory where the Hamiltonian $H'$ is zero and where every Hamiltonian is weakly equal to zero, is that there is no absolute time.

We may look at the question also from the point of view of the action principle. If $I$ is the action integral, then

$$I = \int L(q, \dot{q}) \, dt = \int L \left( q, \frac{dq}{d\tau} \right) d\tau, \qquad (3\text{-}4)$$

because $L$ is homogeneous of the first degree in the $dq/dt$. So we can express the action integral with respect to $\tau$ in the same form as with respect to $t$. That shows that the equations of motion which follow from the action principle must be invariant under the passage from $t$ to $\tau$. The equations of motion do not refer to any absolute time.

We have thus a special form of Hamiltonian theory, but in fact this form is not really so special because, starting with any Hamiltonian, it is always permissible to take the time variable as an extra coordinate and bring the theory into a form in which the Hamiltonian is weakly equal to zero. The general rule for doing this is the following: we take $t$ and put it equal to another dynamical coordinate $q_0$. We set up a new Lagrangian

$$L^* = \frac{dq_0}{d\tau} L \left( q, \frac{dq/d\tau}{dq_0/d\tau} \right)$$

$$= L^* \left( q_k, \frac{dq_k}{d\tau} \right), \qquad k = 0, 1, 2, \ldots, N \qquad (3\text{-}5)$$

$L^*$ involves one more degree of freedom than the original $L$. $L^*$ is *not* equal to L but

$$\int L^* d\tau = \int L \, dt.$$

Thus the action is the same whether it refers to $L^*$ and $\tau$ or to $L$ and $t$. So for any dynamical system we can treat the time as an extra coordinate $q_0$ and then pass to a new Lagrangian $L^*$, involving one extra degree of freedom and homogeneous of the first degree in the velocities. $L^*$ gives us a Hamiltonian which is weakly equal to zero.

This special case of the Hamiltonian formalism where the Hamiltonian is weakly equal to zero is what we need for a relativistic theory, because in a relativistic theory we don't want to have one particular time playing a special role; we want to have the possibility of various

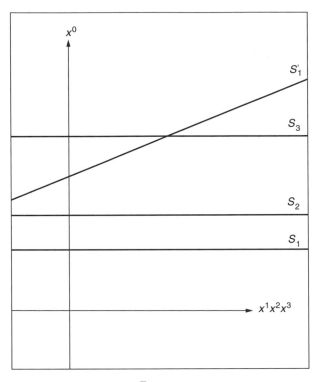

FIG. 1

times $\tau$ which are all on the same footing. Let us see in detail how we can apply this idea.

We want to consider states at specified times with respect to different observers. Now if we set up a space-time picture as in Fig. 1, the state at a certain time refers to the physical conditions on a three-dimensional flat space-like surface $S_1$ which is orthogonal to the time axis. The state at different times will refer to physical conditions on different surfaces $S_2, S_3, \ldots$ Now we want to bring in other time axes referring to different observers and the state, with respect to the other time axes, will involve physical conditions on other flat space-like surfaces like $S'_1$. We want to have a Hamiltonian theory which will enable us to pass from the state, $S_1$ say, to the state $S'_1$. Starting off with given initial conditions on the surface $S'_1$ and applying the equations of motion, we must be able to pass over to the physical conditions on the surface $S_1'$. There must thus be four freedoms in the motion of a state, one freedom corresponding to the movement of the surface parallel to itself, then three more freedoms corresponding to a general change of direction of this flat surface. That means that there will be four arbitrary functions occurring in the solution of the equations of motion which we are trying to get. So we need a Hamiltonian theory with (at least) four primary first-class constraints.

There may be other primary first-class constraints if there are other kinds of freedom in the motion, for example, if we have the possibility of the gauge transformations of electrodynamics. To simplify the discussion, I will ignore this possibility of other first-class primary constraints, and consider only the ones which arise from the requirements of relativity.

We could proceed to set up our theory referring to these flat space-like surfaces which can move with the four freedoms, but I would like first to consider a more general theory in which we consider a state to be defined on an arbitrary curved space-like surface, such as $S$ of Fig. 2. This represents a three-dimensional surface in space-time which has the property of being everywhere space-like, that is to say, the normal to the surface must lie within the light-cone. We may set up a Hamiltonian theory which tells us how the physical conditions vary when we go from one of the curved space-like surfaces to a neighboring one.

Now, bringing in the curved surfaces means bringing in something which is not necessary from the point of view of special relativity. If we wanted to bring in general relativity and gravitational fields, then it would be essential to work with these curved surfaces, but for special relativity, the curved surfaces are not essential. However, I like to bring them in at this stage, even for the discussion of a theory in special relativity, because I find it easier to explain the basic ideas with reference to these curved surfaces than with reference to the flat surfaces. The reason is that with these curved surfaces we can make

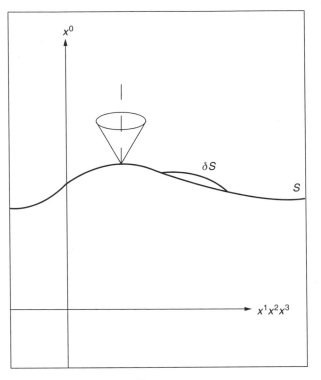

FIG. 2

local deformations of the surface like $\delta S$ in Fig. 2, and discuss the equations of motion with respect to these local deformations of the surface.

One way of proceeding now would be to refer our action integral to a set of curved surfaces, like $S$, take the amount of action between two neighboring curved surfaces, divide it by some parameter $\delta\tau$ expressing the distance between these two surfaces, take this amount of action as our Lagrangian, then apply our standard method of passing from the Lagrangian to the Hamiltonian. Our Lagrangian would necessarily be homogeneous of the first degree in the velocities with respect to the time parameter $\tau$ which specifies the passage from one of these space-like surfaces to a neighboring one, and it would lead to a Hamiltonian theory for which the Hamiltonian is weakly equal to zero.

However, I don't want to go through all the work of following through in detail what we get from an action principle. I want to short-circuit that work and discuss the form of the final Hamiltonian theory which results. We can get quite a lot of information about the form of this Hamiltonian theory just from our knowledge that there must be freedom for the space-like surface to move arbitrarily so long as it remains always space-like. This freedom of motion of the space-like surface must correspond to first-class primary constraints in the Hamiltonian, there being one primary first-class constraint for each type of elementary motion of the surface which can be set up. I shall develop the theory from that point of view.

First of all we have to introduce suitable dynamical variables. Let us describe a point on the space-like surface $S$ by three curvilinear coordinates $(x^1, x^2, x^3) = (x^r)$. In order to fix the position of this space-like surface in space-time, we introduce another set of coordinates $y_\Lambda (\Lambda = 0, 1, 2, 3)$, which we may take to be rectilinear, orthogonal coordinates in special relativity. (I use a capital suffix for referring to the $y$ coordinate system and a small suffix such as $r$ for referring to the $x$ coordinate system.) The four functions $y_\Lambda$, of $x^r$, will specify the surface $S$ in space time and will also specify its parameterization, i.e. the system of coordinates $x^1, x^2, x^3$.

We can use these $y_\Lambda$ as dynamical coordinates, $q$'s. If we form

$$y_{\Lambda,r} = \frac{\partial y_\Lambda}{\partial x^r}, \qquad (r = 1, 2, 3) \qquad (3\text{-}6)$$

this is a function of the $q$'s, the dynamical coordinates.

$$\dot{y}_\Lambda = \frac{\partial y_\Lambda}{\partial \tau}, \qquad (3\text{-}7)$$

$\tau$ being the parameter changing from one surface to the neighboring surface, will be a velocity, a $\dot{q}$. Thus $y_\Lambda$ are the dynamical coordinates needed for describing the surface and $\dot{y}_\Lambda$ are the velocities.

We shall need to introduce momentum variables $w_\Lambda$ conjugate to these dynamical coordinates. The momentum variables will be connected with the coordinates by the Poisson bracket relations

$$[y_{\Lambda x}, w_{\Gamma x'}] = g_{\Lambda\Gamma}\delta^3 (x - x'). \qquad (3\text{-}8)$$

We shall need other variables for describing any physical fields which occur in the problem. If we are dealing with a scalar field $V$, then $V(x)$ for all values of $x^1$, $x^2$, $x^3$ will provide us with further dynamical coordinates, $q$'s. $V_{,r}$ will be functions of the $q$'s. $\partial V/\partial \tau$ will be a velocity. The derivative of $V$ in any direction is expressible of terms of $\partial V/\partial \tau$ of and $V_{,r}$ and so is expressible in terms of the dynamical coordinates and velocities. The Lagrangian will involve these $V$'s differentiated in general directions and is thus a function of the dynamical coordinates and velocities. For each $V$, we shall need a conjugate momentum $U$, satisfying the Poisson bracket conditions

$$\left[ V(x), U(x') \right] = \delta^3 (x - x'). \tag{3-9}$$

That is how one would treat a scalar field. There is a similar method for vector, tensor, or spinor fields, just bringing in the necessary additional suffixes. I need not go into that.

Now let us see what the Hamiltonian will be like. The Hamiltonian has to be a linear function of primary first-class constraints of the type (3-2). First of all I shall put down what the primary first-class constraints are like. There must be primary first-class constraints which allow for arbitrary deformations of the surface. They must involve the variables $w$ to which are conjugate to the $y$'s, in order to make the $y$'s vary, and they will involve other field quantities. We can express them in the form

$$w_\Lambda + K_\Lambda \approx 0, \tag{3-10}$$

where $K_\Lambda$ is some function of the Hamiltonian variables, the $q$'s and $p$'s, not involving the $w$'s.

Now we can assert that the Hamiltonian is just an arbitrary linear function of all the quantities (3-10):

$$H_T = \int c^\Lambda (w_\Lambda + K_\Lambda)\, d^3x. \tag{3-11}$$

This is integrated over the three $x$'s which specify a point on the surface. The $c$'s are arbitrary functions of the three $x$'s and the time.

The general equation of motion is of course $\dot{g} \approx [g, H_T]$, We can get a meaning for the coefficient $c_\Lambda$ by taking this equation of motion and applying it for $g$ equal to one of the $y$ variables. For $g = y_\Lambda$ at some particular point $x^1$, $x^2$, $x^3$ we get

$$\dot{y}_\Lambda = \left[ y_\Lambda, \int c'^\Gamma \left( w'_\Gamma + K'_\Gamma \right) d^3 x' \right]$$

$$= \int c'^\Gamma [y_\Lambda, w'_\Gamma + K'_\Gamma] d^3 x. \qquad (3\text{-}12)$$

Here the attached to a field quantity $c^\Gamma$, $w_\Gamma$, or $K_\Gamma$ denotes the value of that quantity at the point $X^{1'}$, $X^{2'}$, $X^{3'}$. $y_\Lambda$ has zero Poisson brackets with $K'_\Gamma$ because $K'_\Gamma$ is independent of the $w$'s, so we just have to take into account the Poisson bracket of $y_\Lambda$ with $w'_\Gamma = w_\Gamma(x')$. This gives us the delta function and so

$$\dot{y}_\Lambda = c_\Lambda. \qquad (3\text{-}13)$$

Thus the coefficients $c_\Lambda$ turn out to be the velocity variables which tell us how our surface varies with the parameter $\tau$. We can get an arbitrary variation of the surface with $\tau$ by choosing these $c_\Lambda$ in an arbitrary way.

This tells us what the Hamiltonian is like for a field theory expressed with respect to states on curvilinear surfaces.

We can make a deeper analysis of this Hamiltonian by resolving the vectors which occur in it into components which are normal and tangential to the surface. If we have any vector whatever, $\xi_\Lambda$, we can obtain from $\xi_\Lambda$ a *normal component*

$$\xi_\perp = \xi_\Lambda l^\Lambda,$$

where $l^\Lambda$ is the unit normal vector, and *tangential components* (referred to the $x$ coordinate system)

$$\xi_r = \xi_\Lambda y^\Lambda_{,r}.$$

The $l$ are determined by the $y^\Lambda_{,r}$ and are thus functions of the dynamical coordinates. Any vector can be resolved in this way into a part normal to the surface and a part tangential to the surface. We have the scalar product law

$$\xi_\Lambda \eta^\Lambda = \xi_\perp \eta_\perp + \gamma^{rs} \xi_r \eta_s, \qquad (3\text{-}14)$$

where $\gamma_{rs}\ dx^r\ dx^s$ is the metric in the surface referred to the $x$-coordinate system. $\gamma^{rs}$ is the reciprocal matrix of the $\gamma_{rs}$. ($r$, $s = 1$, 2, 3).

We can use this scalar product law (3-14) to express our total Hamiltonian in terms of the tangential and normal components of $w$ and $K$:

$$H_T = \int \dot{y}^\Lambda \left(w_\Lambda + K_\Lambda\right) d^3x$$

$$= \int \left(\dot{y}_\perp \left(w_\perp + K_\perp\right) + \gamma^{rs} \dot{y}_r \left(w_s + K_s\right)\right) d^3x. \quad (3\text{-}15)$$

Here $\dot{y} = \dot{y}^\Lambda l_\Lambda$ and $\dot{y}_r = \dot{y}^\Lambda y_{\Lambda, r}$.

We shall need the Poisson bracket relationships between the normal and tangential terms in (3-15). I will first write down the Poisson bracket relations for the different components of $w$. We have of course

$$\left[w_\Lambda, w'_\Gamma\right] = 0, \quad (3\text{-}16)$$

referred to the external coordinates $y$; but when we resolve our $w$'s into normal and tangential components, they will no longer have zero Poisson brackets with each other. The Poisson brackets can easily be worked out by straightforward arguments. I don't want to go through the details of that work. I will just mention that the details can be found in a paper of mine (*Canadian Journal of Mathematics*, **3**, 1 (1951)). The results are

$$\left[w_r, w'_s\right] = w_s \delta_{,r} \left(x - x'\right) + w'_r \delta_{,s} \left(x - x'\right), \quad (3\text{-}17)$$

$$\left[w_\perp, w'_r\right] = w'_\perp \delta_{,r} \left(x - x'\right), \quad (3\text{-}18)$$

$$\left[w_\perp, w'_\perp\right] = -2w^r \delta_{,r} \left(x - x'\right) - w^r_{,r} \delta \left(x - x'\right), \quad (3\text{-}19)$$

Now we know that

$$\left[w_\mu + K_\mu, w'_\nu + K'_\nu\right] \approx 0 \quad \text{for} \quad \mu, \nu = r, s \text{ or } \perp. \quad (3\text{-}20)$$

We can infer that

$$\left[w_r + K_r, w'_s + K'_s\right] = (w_s + K_s)\delta_{,r} \left(x - x'\right)$$
$$+ (w'_r + K'_r)\delta_{,s} \left(x - x'\right), \quad (3\text{-}21)$$

$$\left[w_\perp + K_\perp, w'_r + K'_r\right] = (w'_\perp + K'_\perp)\delta_{,r} \left(x - x'\right), \quad (3\text{-}22)$$

$$[w_\perp + K_\perp, w'_\perp + K'_r] = -2(w^r + K^r)\delta_{,r}(x - x')$$
$$- (w^r + K^r)_{,r}\delta(x - x'). \quad (3\text{-}23)$$

These results could be worked out directly from the definitions of the normal and tangential components of the $w$'s, but they can be inferred more simply by the following argument. Since $w_r + K_r$, $w_\perp + K_\perp$ are all first class, their Poisson brackets are zero weakly. Thus $[w_r + K'_r, w'_s + K'_s]$, $[W_\perp + K_\perp, w'_r + K'_r]$ and $[w_\perp + K_\perp, w'_\perp + K'_\perp]$ must all be weakly equal to zero. We can now infer what they are equal to strongly. We have to put on the right-hand side in each of (3-21,) (3-22,) and (3-23) a quantity which is weakly equal to zero and which is therefore built up from $w_r + K_r$ and $w_\perp + K_\perp$ with certain coefficients. Further, we can see what these coefficients are by working out what terms containing $w$ there are on the right-hand sides. Terms containing $w$'s can arise only from taking the Poisson bracket of a $w$ with a $w$, according to (3-17), (3-18), and (3-19). Taking a Poisson bracket $[w, K']$ will not lead to anything involving $w$, because it means taking the Poisson bracket of a $w$ momentum with some functions of dynamical coordinates and momenta other than $w$'s, and that won't involve the $w$ momentum variables. Similarly the Poisson bracket of a $K$ with a $K$ won't involve any $w$ variables. Thus the only to variables which occur on the right side of (3-21) will be the ones which occur on the right side of (3-17). We have to put certain further terms in the right side of (3-21) in order that the total expression shall be weakly equal to zero. It is then quite clear what we should put here, namely, $(w_s + K_s)\delta_{,r}(x - x') + (w'_r + K'_r)\delta_{,s}(x - x')$. We do the same with the right sides of (3-22), (3-23).

The next thing to notice is that the terms $w_s + K_s$ in the Hamiltonian (3-15) correspond to a motion in which we change the system of coordinates in the curved surface but do not have the surface moving. It corresponds to each point in the surface moving tangentially to the surface.

Let us put $\dot{y}_\perp = 0$, which means that we are taking no motion of the surface perpendicular to itself but are merely making a change of the coordinates of the surface, and then we have equations of motion of the type

$$\dot{g} = \int \gamma^{rs} \dot{y}_r \, [g, w_s + K_s] \, d^3x. \qquad (3\text{-}24)$$

This must be the equation of motion which tells us how $g$ varies when we change the system of coordinates in the surface without moving the surface itself. Now this change in $g$ must be a trivial one, which can be inferred merely from the geometrical nature of the dynamical variable $g$. If $g$ is a scalar, then we know how that changes when we change the system of coordinates $x^1$, $x^2$, $x^3$. If it is a component of a vector or a tensor there will be a rather more complicated change for $g$, but still we can work it out; similarly if $g$ is a spinor. In every case, this change of $g$ is a trivial one. That means that $K_s$, can be determined from geometrical arguments only.

I will give one or two examples of that. For a scalar field $V$ with a conjugate momentum $U$, there is a term

$$V_{,r} U \qquad (3\text{-}25)$$

in $K_r$. For a vector field, say a three-vector $A_s$, with conjugate $B^s$, there is a term

$$A_{s,r} B^s - (A_r B^s)_{,s} \qquad (3\text{-}26)$$

in $K_r$; and so on for tensors, with something rather more complicated for spinors. The first term in (3-26) is the change in $A_s$ coming from the translation associated with the change in the system of coordinates, and the second is the change in the $A_s$ arising from the rotation associated with the change in the system of coordinates. There is no such rotation term coming in in the case (3-25) of the scalar.

We can obtain the total $K_r$ by adding the contribution needed for all the different kinds of fields which are present in the problem. The result is that we can work out this tangential component of $K$ just from geometrical arguments. One can see in this way that the tangential component of $K$ is something which is not of real physical importance, it is just concerned with the mathematical technique. The quantity which is of real physical importance is the normal component of $K$ in (3-15). This normal component of $K$ added on to the normal component of $w$ gives us the first-class constraint which is associated with a motion of the surface normal to itself. That is something which is of dynamical importance.

The problem of getting a Hamiltonian field theory on these curved surfaces involves finding the expressions $K$ to satisfy the required Poisson bracket relations (3-21), (3-22), and (3-23). The tangential part of $K$ can be worked out from geometrical arguments as I discussed, and when we have worked it out we should find of course that it satisfies the first Poisson bracket relation (3-21). The second Poisson bracket relation (3-22) involves $K_\perp$ linearly and this Poisson bracket relation would be satisfied by any quantity $K_\perp$ which satisfies the condition of being a scalar density. This Poisson bracket relation really tells us that if the quantity $K_\perp$ varies suitably under a change of coordinate system $X^1$, $X^2$, $X^3$, this Poisson bracket relation will be fulfilled. The difficult relation to fulfill is the third one, which is quadratic in $K_\perp$. So the problem of setting up a Hamiltonian field theory on curved space-like surfaces is reduced to the problem of finding a normal component of $K$ which is a scalar density and which satisfies the Poisson bracket relationship (3-23).

One way of finding such a normal component of $K$ is to work from a Lorentzinvariant action principle. We might obtain all the components of $K$ by working from the action principle. If we did that, the tangential part of $K$ which we get would not necessarily be the same as that built up from terms like (3-25) and (3-26), because it might differ by a contact transformation. But one could eliminate such a contact transformation by rewriting the action principle, adding to it a perfect differential term. This doesn't affect the equations of motion. By such a change of the action principle, one can arrange that the tangential part of $K$ given by the action principle agrees precisely with the value which is obtained by the simple application of geometrical arguments. We are then able to find the normal component of $K$ by working with our general method of passing from the action principle to the Hamiltonian. If the action principle is relativistic, then the normal component of $K$ obtained in this way would have to satisfy the condition (3-23).

We can now discuss the passage to the quantum theory. Quantization involves making the quantities $w$ and the variables which enter in $K$ into operators. We have to be careful now how we define the tangential and the normal components of $w$, and I choose this way to define them:

$$w_r = y_{\Lambda,r} w^{\Lambda}, \qquad (3\text{-}27)$$

putting the momentum variable $w$ on the right. (In the quantum theory, you see, the result is different, depending on whether we put the $w$ on the right or the left.) Similarly,

$$w_{\perp} = l_{\Lambda} w^{\Lambda}. \qquad (3\text{-}28)$$

Then these quantities are well defined.

Now in the quantum theory we have the weak equations $w_r + K_r \approx 0$ and $w_{\perp} + K_{\perp} \approx 0$, which provide us with supplementary conditions on the wave function:

$$(w_r + K_r)\,\psi = 0, \qquad (3\text{-}29)$$

$$(w_{\perp} + K_{\perp})\,\psi = 0, \qquad (3\text{-}30)$$

corresponding to (2-22). We require that these supplementary conditions be consistent. According to (2-25), we must arrange that in the commutation relations (3-21), (3-22), and (3-23) the coefficients on the right-hand sides stand before (on the left of) the constraints.

In the case of (3-21), the tangential components, the conditions fit if we choose the order of the factors in $K_r$ so that the momentum variables are always on the right. We have now in (3-21) a number of quantities, linear in the momentum variables with the momentum variables on the right, and the commutator of any two such quantities will again be linear in the momentum variables with the momentum variables on the right. Thus we shall always have the momentum variables on the right and we shall always have our factors occurring in the order in which we want them to.

Now we have the problem of bringing in $K_\perp$, which cannot be disposed of so simply. $K_\perp$ will usually involve the product of non-commuting factors and we have to arrange the order of those factors so that (3-22) and (3-23) shall be satisfied with the coefficients occurring on the left in every term on the right-hand side. The equation (3-22) is again a fairly simple one to dispose of. If we simply take $K_\perp$ to be a scalar density, that is all that is needed, because we have $w_\perp + K_\perp$ occurring on the right-hand side without any coefficients which don't commute with it; the only coefficient is the delta function, which is a number.

But the relationship (3-23) is the troublesome one. For the purposes of the quantum theory, I ought to write out the right-hand side here rather more explicitly:

$$\left[w_\perp + K_\perp, w'_\perp + K'_\perp\right] = -2\gamma^{rs}\left(w_s + K_s\right)\delta_{,r}\left(x - x'\right)$$
$$- \left(\gamma^{rs}\left(w_s = K_s\right)\right)_{,r}\delta\left(x - x'\right). \quad (3\text{-}31)$$

I've written this out with the coefficients $\gamma^{rs}$ occurring on the left, and that is how we need to have these coefficients in the quantum theory.

The problem of setting up a quantum field theory on general curved surfaces involves finding $K_\perp$ so that this Poisson bracket relationship (3-31) holds with the coefficients $\gamma^{rs}$ occurring on the left. If we do satisfy (3-31), then the supplementary conditions (3-30) are consistent with each other, and we already have (3-29) consistent with each other and (3-30) consistent with (3-29).

There we have formulated the conditions for our quantum theory to be relativistic. We need a bit of luck to be able to satisfy the conditions. We cannot always satisfy them. There is one general rule which is of importance, which tells us that when we've got a $K_\perp$ satisfying these conditions and certain other conditions, we can easily construct other $K_\perp$'s to satisfy the conditions. Let us suppose that we have a solution in which $K_\perp$ involves only undifferentiated momentum variables together with dynamical coordinates which may be differentiated. There are a number of simple fields for which $K_\perp$ does satisfy the Poisson bracket relations (3-22) and (3-23) and does have this simple character. Then we may add to $K_\perp$ any function of the undifferentiated $q$'s. That is to say, we take a new $K_\perp$,

$$K_\perp^* = K_\perp + \phi(q).$$

Then we see that adding on this $\phi$ to $K_\perp$ can affect the right-hand side of (3-31) only by bringing in a multiple of the delta function. We cannot get any differentiations of the delta function coming in, because the extra terms come from Poisson brackets of $\phi(q)$ with undifferentiated momentum variables. So that the only effect on the right-hand side of adding the term $\phi$ to $K_\perp$ can be adding on a multiple of the delta function. But the right-hand side has to be anti-symmetrical between $x$ and $x'$, because the left-hand side is obviously antisymmetrical between $x$ and $x'$. That prevents us from just adding a multiple of the delta function to the right-hand side of (3-31), so that it is *not* altered at all. Thus if the original $K_\perp$ satisfies the Poisson bracket relation (3-31), then the new one will also satisfy it.

There is a further factor which has to be taken into account to complete the proof. $\phi$ may also involve $\Gamma = \sqrt{-\det g_{rs}}$. One finds that $[w_\perp, \Gamma']$ involves $\delta(x - x')$ undifferentiated (one just has to work this out) and thus we can bring $\Gamma$ into $\phi$ without disturbing the argument. In fact, we have to bring in $\Gamma$ in order to preserve the validity of (3-22), which requires that $K_\perp^*$ and $K_\perp$ shall be scalar densities. We must thus bring in a suitable power of $\Gamma$ to make $\phi$ a scalar density.

This is the method which is usually used in practice for bringing in interaction between fields without disturbing the relativistic character of the theory. For various simple fields the conditions turn out to be satisfied. We have the necessary bit of luck, and we can bring in interaction between fields of the simple character described and the conditions for the quantum theory to be relativistic are preserved.

There are some examples for which we *don't* have the necessary luck and we just cannot arrange the factors in $K_\perp$ to get (3-31) holding with the coefficients on the left, and then we do not know how to quantize the theory with states on curved surfaces. But actually, we are trying to do rather more than is necessary when we try to set up our quantum theory with states on curved surfaces. For the purposes of getting a theory in agreement with special relativity, it would be quite sufficient to have our states defined only on flat surfaces. That will involve some conditions on $K_\perp$ which are less stringent than those which I have formulated here. And it may be that we can satisfy these less stringent conditions without being able to satisfy those which I have formulated here.

An example for that is provided by the Born-Infeld electrodynamics, which is a modification of the Maxwell electrodynamics based on a different action integral, an action integral which is in agreement with the Maxwell one for weak fields, but differs from it for strong fields. This Born-Infeld electrodynamics leads to a classical $K_\perp$ which involves square roots. It is of such a nature that it doesn't seem possible to fulfill the conditions which are necessary for building up a relativistic quantum theory on curved surfaces. However, it does seem to be possible to build up a relativistic quantum theory on flat surfaces, for which the conditions are less stringent.

DR. DIRAC

*Lecture No. 4*

# QUANTIZATION ON FLAT SURFACES

We have been working with states on general space-like curved surfaces in space-time. I will just summarize the results that we obtained concerning the conditions for a quantum field theory, formulated in terms of these states, to be relativistic. We introduce variables to describe the surface, consisting of the four coordinates $Y^\Lambda$ of each point $x^r = (x^1, x^2, x^3)$ on the surface. The $x$'s form a curvilinear system of coordinates on the surface. Then the $y$'s are treated as dynamical coordinates and there are momenta conjugate to them, $w_\Lambda(x)$, again functions of the $x$'s. And then we get a number of primary first-class constraints appearing in the Hamiltonian formalism, of the nature

$$w_\Lambda + K_\Lambda \approx 0. \tag{3-10}$$

The $K$'s are independent of the $w$'s, but may be functions of any of the other Hamiltonian variables. The $K$'s will involve the physical fields which are present. We analyze these constraints by resolving them into components tangential to the surface and normal to the surface. The tangential components are

$$w_r + K_r \approx 0, \tag{4-1}$$

and the normal components is

$$w_\perp + K_\perp \approx 0. \tag{4-2}$$

With this analysis, we find that the $K_r$ can be worked out just from geometrical considerations. The $K_r$ should be looked upon as something rather trivial, associated with transformations in which the coordinates of the surface are varied, but the surface itself doesn't move. The first-class constraints (4-2) are associated with the motion of the surface normal to itself and are the important ones physically.

Certain Poisson bracket relations (3-21), (3-22), and (3-23) have to be fulfilled for consistency. Some of the Poisson bracket relations involve merely the $K_r$, and they are automatically satisfied when the $K_r$ are chosen in accordance with the geometrical requirements. Some of the consistency conditions are linear in $K_\perp$ and they are automatically satisfied provided we choose $K_\perp$ to be a scalar density. Then finally we have the consistency conditions which are quadratic in the $K_\perp$ and those are the important ones, the ones which cannot be satisfied by trivial arguments.

These important consistency conditions can be satisfied in the classical theory if we work from a Lorentz-invariant action principle and calculate the $K_\perp$ by following the standard rules of passing from the action principle to the Hamiltonian. The problem of getting a relativistic quantum theory then reduces to the problem of suitably choosing the non-commuting factors which occur in the quantum $K_\perp$ in such a way that the quantum consistency conditions are fulfilled, which means that the commutator of two of the quantities (4-2) at two points in space $x^1$, $x^2$, $x^3$ has to be a linear combination of the constraints with coefficients occurring on the left. These quantum consistency conditions will usually be quite difficult to satisfy. It turns out that one can satisfy them with certain simple examples, but with more complicated examples it doesn't seem to be possible to satisfy them. That leads to the conclusion that one cannot set up a quantum theory for these more general fields with the states defined on general curved surfaces.

I might mention that the quantities $K$ have a simple physical meaning. $K_r$ can be interpreted as the momentum density, $K_\perp$ as the energy density; so the momentum density, expressed in terms of Hamiltonian variables, is something which is always easy to work out just from the geometrical nature of the problem and the energy density is the important quantity which one has to choose correctly (satisfying certain commutation relations) in order to satisfy the requirements of relativity.

If we cannot set up a quantum theory with states on general curved surfaces, it might still be possible to set it up with states defined only on flat surfaces.

We can get the corresponding classical theory simply by imposing conditions which make our previous curved surface into a flat surface. The conditions will be the following: The surface is specified by $Y_\Lambda(x)$; in order to make the surface flat, we require that these functions shall be in the form

$$y_\Lambda(x) = a_\Lambda + b_{\Lambda r} x^r, \qquad (4\text{-}3)$$

where the $a$'s and $b$'s are independent of the $x$'s. This will result in the surface being flat, and in the system of coordinates $x^r$ being rectilinear. At present we are not imposing the conditions that the $x^r$ coordinate system shall be orthogonal: I shall bring that in a little later. We are thus working with general, oblique, rectilinear axes $x^r$.

We now have our surface fixed by quantities $a_\Lambda$, $b_{\Lambda r}$ and these quantities will appear as the dynamical variables needed to fix the surface. We have far fewer of them than previously. In fact, we have only $4 + 12 = 16$ variables here. We have these 16 dynamical coordinates to fix the surface instead of the previous $Y_\Lambda(x)$, which meant $4 \cdot \infty^3$ dynamical coordinates.

When we restrict the surface in this way, we may look upon the restriction as bringing a number of constraints into our Hamiltonian formalism, constraints which express the $4 \cdot \infty^3$ $y$ coordinates in terms of 16 coordinates. These constraints will be second-class. Their presence means a reduction in the number of effective degrees of freedom for the surface from $4 \cdot \infty^3$ to 16, a very big reduction!

In a previous lecture I gave the general technique for dealing with second-class constraints. The reduction in the number of effective degrees of freedom leads to a new definition of Poisson brackets. This general technique is not needed in our present case, where conditions are sufficiently simple for one to be able to use a more direct method. In fact, we can work out directly what effective momentum variables remain in the theory when we have reduced the number of effective degrees of freedom for the surface.

With our dynamical coordinates restricted in this way, we have of course the velocities restricted by the equation

$$\dot{y}_\Lambda = \dot{a}_\Lambda + b_{\Lambda r} x^r. \tag{4-4}$$

The dot refers to differentiation with respect to some parameter $\tau$. As $\tau$ varies, this flat surface varies, moving parallel to itself and also changing its direction. The surface thus moves with a four-fold freedom, and this motion is expressed by our taking $a_\Lambda$, $b_{\Lambda r}$ to be functions of the parameter $\tau$.

The total Hamiltonian is now

$$H_T = \int \dot{y}^\Lambda \left( w_\Lambda + K_\Lambda \right) d^3 x$$
$$= \dot{a}^\Lambda \int (w_\Lambda + K_\Lambda) d^3 x + b_r^\Lambda \int x^r (w_\Lambda + K_\Lambda) d^3 x. \tag{4-5}$$

(I have taken the quantities $\dot{a}^\Lambda$, $b^\Lambda$. outside the integral signs, because they are independent of the $x$ variables.) (4-5) involves the $\omega$ variables only through the combinations $\int w_\Lambda d^3 x$ and $\int x^r w_\Lambda d^3 x$. We have here 16 combinations of the $w$'s, which will be the new momentum variables conjugate to the 16 variables $a$, $b$ which are now needed to describe the surface.

We can again express $H_T$ in terms of the normal and tangential components of these quantities:

$$H_T = \dot{a}^\Lambda l_\Lambda \int (w_\perp + K_\perp) d^3 x + \dot{a}^\Lambda b_{\Lambda r} \int (w^r + K^r) d^3 x$$
$$+ b_r^\Lambda l_\Lambda \int x^r (w_\perp + K_\perp) d^3 x + b_{\Lambda r} b_s^\Lambda \int x^r (w^s + K^s) d^3 x. \tag{4-6}$$

Let us now bring in the condition that the $x^r$ coordinate system is orthogonal. That means

$$b_{\Lambda r} b_s^\Lambda = g_{rs} = -\delta_{rs}. \tag{4-7}$$

Differentiating (4-7) with respect to $\tau$, we get

$$b_{\Lambda r}b_s^\Lambda + b_{\Lambda s}b_r^\Lambda = 0 \tag{4-8}$$

(I have been raising the $\Lambda$ suffixes quite freely because the $\Lambda$ coordinate system is just the coordinate system of special relativity.) This equation tells us that $b_{\Lambda r}b_s^\Lambda$ is antisymmetric between $r$ and $s$. So the last term in (4-6) is equal to

$$\frac{1}{2}b_{\Lambda r}b_s^\Lambda \int \{x^r(w^s + K^s) - x^3(w^r + K^r)\}d^3x.$$

Now you see that we don't have so many linear combinations of the $w$'s occurring in the $H_T$ as before. The only linear combinations of the $w$'s which survive are the following ones:

$$P_\perp \equiv \int w_\perp d^3x, \tag{4-9}$$

$$P_r \equiv \int w_r d^3x, \tag{4-10}$$

and also

$$M_{r\perp} \equiv \int x^r w_\perp d^3x, \tag{4-11}$$

and

$$M_{rs} \equiv \int (x_r w_s - x_s w_r)d^3x. \tag{4-12}$$

(We can raise and lower the suffixes $r$ quite freely now because they refer to rectilinear orthogonal axes.) These are the momentum variables which are conjugate to the variables needed to fix the surface when the surface is restricted to be a flat one referred to rectilinear orthogonal coordinates.

The whole set of momentum variables included in (4-9), (4-10), (4-11), and (4-12) can be written as $P_\mu$ and $M_{\mu\nu} = -M_{\nu\mu}$, where the suffixes $\mu$ and $\nu$ take on 4 values, a value 0 associated with the normal component, and 1, 2, 3 associated with the three $x$'s. $\mu$, $\nu$ are small suffixes referring to the $x$ coordinate system, to distinguish them from the capital suffixes $\Lambda$ referring to the fixed $y$ coordinate system.

So now our momentum variables are reduced to just 10 in number, and associated with these 10 momentum variables we have 10 primary first-class constraints, which we may write

$$P_\mu + p_\mu \approx 0, \tag{4-13}$$

$$M_{\mu\nu} + m_{\mu\nu} \approx 0, \tag{4-14}$$

where

$$p_\perp \equiv \int K_\perp d^3x, \tag{4-15}$$

$$p_r \equiv \int K_r d^3x, \tag{4-16}$$

$$m_{r\perp} \equiv \int x_r K_\perp d^3x, \tag{4-17}$$

and

$$m_{rs} \equiv \int (x_r K_s - x_s K_r) d^3x. \tag{4-18}$$

We have now 10 primary first-class constraints associated with a motion of the flat surface. In Lecture (3) I said that we would need 4 primary first-class constraints (3-10) to allow for the general motion of a flat surface. We see now that 4 is not really adequate. The 4 has to be increased to 10, because 4 elementary motions of the surface normal to itself and changing its direction would not form a group; in order to have these elementary motions forming a group, we have to extend the 4 to 10, the extra 6 members of the group including the translations and rotations of the surface, which motions affect merely the system of coordinates in the surface without affecting the surface as a whole. In this way we are brought to a Hamiltonian theory involving 10 primary first-class constraints.

We have now to discuss the consistency conditions, the conditions in terms of Poisson bracket relations which are necessary for all the constraints to be first-class. Let us first discuss the Poisson bracket relations between the momentum variables $P_\mu$, $M_{\mu\nu}$. We are given these momentum variables in terms of the to variables (4-9) to (4-12), and we know the Poisson bracket relations (3-17), (3-18), and (3-19) between the $w$ variables, so we can calculate the Poisson bracket relations between the $P$ and $M$ variables. It is not really necessary to go through all this work to determine the Poisson bracket relations between the $P$ and $M$ variables. It is sufficient to realize that these variables just correspond to the operators of translation and rotation in four-dimensional flat space-time, and thus their Poisson bracket relations must just correspond to the commutation relations between the operators of translation and rotation. In either way we get the following Poisson bracket relations:

$$[P_\mu, P_\nu] = 0 \tag{4-19}$$

which expresses that the various translations commute;

$$[P_\mu, M_{\rho\sigma}] = g_{\mu\rho}P_\sigma - g_{\mu\sigma}P_\rho; \tag{4-20}$$

and

$$[M_{\mu\nu}, M_{\rho\sigma}] = -g_{\mu\rho}M_{\nu\sigma} + g_{\nu\rho}M_{\mu\sigma} + g_{\mu\sigma}M_{\nu\rho} - g_{\nu\sigma}M_{\mu\rho}. \tag{4-21}$$

Let us now consider the requirements for the equations (4-13) and (4-14) to be first-class. The Poisson bracket of any two of them must be something which vanishes weakly and must therefore be a linear combination of them. So we are led to these Poisson bracket relations:

$$[P_\mu + p_\mu, P_\nu + p_\nu] = 0, \tag{4-22}$$

$$[P_\mu + p_\mu, M_{\mu\sigma} + m_{\mu\sigma}] = g_{\mu\rho}(P_\sigma + p_\sigma) - g_{\mu\sigma}(P_\rho + p_\rho), \tag{4-23}$$

and

$$\begin{aligned}[M_{\mu\nu} + m_{\mu\nu}, M_{\rho\sigma} + m_{\rho\sigma}] = &-g_{\mu\rho}(M_{\nu\sigma} + m_{\nu\sigma}) + g_{\nu\rho}(M_{\mu\sigma} + m_{\mu\sigma}) \\ &+ g_{\mu\sigma}(M_{\nu\rho} + m_{\nu\rho}) \\ &- g_{\nu\sigma}(M_{\mu\rho} + m_{\mu\rho}). \end{aligned} \tag{4-24}$$

The argument for getting these relations is that, on the right-hand sides we had to put something which is weakly equal to zero in each case, and we know the terms on the right-hand sides which involve the momentum variables $P$, $M$ because these terms come only from the Poisson brackets of momenta with momenta and so are given by (4-19), (4-20), and (4-21). (I have already used the same argument in the curvilinear case for (3-21), (3-22), and (3-23), so there is no need to go into detail here. For example, see how (4-23) comes about. The terms involving P are just the same as in (4-20). They come from the

Poisson bracket of $P$ and $M$. The remaining terms are filled in in order to make the total expression weakly equal to zero.) (4-22), (4-23), and (4-24) are the requirements for consistency.

We can make a further simplification, which we could not do in the case of curvilinear coordinates, in this way: Let us suppose that our basic field quantities are chosen to refer only to the $x$ coordinate system. They are field quantities at specific points $x$ in the surface, and we can choose them so as to be quite independent of the $y$ coordinate system. Then the quantities $K_\perp$, $K_r$ will be quite independent of the $y$ coordinate system, and that means that they will have zero Poisson brackets with the variables $P$, $M$. We then have a zero Poisson bracket between each of the variables, $p$, $m$ and each of the $P$, $M$.

This condition follows with the natural choice of dynamical variables to describe the physical fields which are present. We cannot do the corresponding simplification when we are working with the curved surfaces, because the $g_{rs}$ variables that fix the metric will enter into the quantities $K_\perp$, $K_r$. The result is that we cannot set them up in a form which does not refer at all to the $y$ coordinate system, because the $y$ coordinates enter into the $g_{rs}$ variables. However, with the flat surfaces, we can make this simplification, and that results in equations (4-22), (4-23), and (4-24) simplifying to

$$\left[ p_\mu, p_\nu \right] = 0; \tag{4-25}$$

$$\left[ p_\mu, m_{\rho\sigma} \right] = g_{\mu\rho} p_\sigma - g_{\mu\sigma} p_\rho; \tag{4-26}$$

and

$$\left[ m_{\mu\nu}, m_{\rho\sigma} \right] = -g_{\mu\rho} m_{\nu\sigma} + g_{\nu\rho} m_{\mu\sigma} + g_{\mu\sigma} m_{\nu\rho} - g_{\nu\sigma} m_{\mu\rho}. \tag{4-27}$$

$P$ and $M$ have disappeared from these equations, so the consistency conditions now involve only the field variables, and not the variables, which are introduced for describing the surface. In fact, these conditions merely say that the $p$, $m$ shall satisfy Poisson bracket relations corresponding to the operators of translation and rotation in flat space-time. The problem of setting up a relativistic field theory now reduces to finding the quantities $p$, $m$ to satisfy the Poisson bracket relations (4-25), (4-26), and (4-27).

These quantities, remember, are defined in terms of $K_\perp$ and $K_r$, the energy density and the momentum density. The expression for the momentum density is just the same as in curvilinear coordinates. It is determined by geometrical arguments only. Our problem reduces to finding the energy density $K_\perp$ leading to $p$'s and $m$'s such that the Poisson bracket relations (4-25), 4-26), and (4-27) are fulfilled.

If we work from a Lorentz-invariant action integral and deduce $K_\perp$ from it by standard Hamiltonian methods, $K_\perp$ will automatically satisfy these requirements in the classical theory. The problem of getting a relativistic quantum theory then reduces the problem of suitably choosing the order of factors which occur in $K_\perp$ so as to satisfy the equations (4-25), (4-26), and (4-27) also in the quantum theory, where the Poisson bracket becomes a commutator and the $p$, $m$ involve non-commuting quantities.

Let us look at (4-25), (4-26), and (4-27) and substitute for $p$ and $m$ their values in terms of $K$'s. Then you see that some of these conditions will be independent of $K_\perp$. These are automatically satisfied when we choose $K_r$ properly, in accordance with the geometrical requirements. Some of the conditions are linear in $K_\perp$. These will be satisfied by taking $K_\perp$ to be any three-dimensional scalar density in the space of the $x$'s. So that there is no problem in satisfying the conditions which are linear in $K_\perp$. The awkward ones to satisfy are the ones which are quadratic in $K_\perp$. They are the following:

$$\left[\int x_r K_\perp d^3x, \int K'_\perp d^3x'\right] = \int K_r d^3x. \qquad (4\text{-}28)$$

(This equation comes from (4-26) where we put $\mu = \perp$, $\rho = r$, and $\sigma = \perp$.)

$$\left[\int x_r K_\perp d^3x, \int x'_s K'_\perp d^3x'\right] = -\int (x_r K_s - x_r K_r)d^3x \qquad (4\text{-}29)$$

(from (4-27) where we take $\nu = \perp$ and $\sigma = \perp$). So the problem of getting a relativistic quantum field theory now reduces to the problem of finding an energy density $K_\perp$ which satisfies the conditions (4-28) and (4-29) when we take into account non-commutation of the factors.

We can analyze these conditions a little more when we take into account that the Poisson bracket connecting $K_\perp$ at one point and $K'_\perp$ at another point will be a sum of terms involving delta functions and derivatives of delta functions:

$$\left[K_\perp, K'_\perp\right] = a\delta + 2b_r\delta_{,r} + c_{rs}\delta_{,rs} + \cdots \cdot \quad (4\text{-}30)$$

(This delta is the three-dimensional delta function involving the three coordinates $x$ and the three coordinates $x'$ of the first and second points.) Here $a = a(x)$, $b = b(x)$, $c = c(x)$, ... One could have the coefficients involving also $x'$, but then one could replace them by coefficients involving $x$ only at the expense of making some changes in the earlier coefficients in the series. There is no fundamental dissymmetry between $x$ and $x'$, only a dissymmetry in regard to the way the equation is written.

(4-30) is the general relationship connecting the energy density at two points. Now for many examples, including all the more usual fields, derivatives of the delta function higher than the second do not occur. Let us examine this case further.

Assume derivatives higher that the second do not occur. That means that the series (4-30) stops at the third term. In this special case we can get quite a bit of information about the coefficients $a$, $b$, $c$ by making use of the condition that the Poisson bracket (4-30) is anti-symmetrical between the two points $x$ and $x'$. Interchanging $x$ and $x'$ in (4-30), we get

$$\left[K'_\perp, K_\perp\right] = a'\delta - 2b'_r\delta_{,r} + c'_{rs}\delta_{,rs}$$
$$= a'\delta - 2(b'_r\delta)_{,r} + (c'_{rs}\delta)_{,rs}$$

(since $\partial b_r(x')/\partial x^r = 0$, etc.)

$$= a\delta - 2(b_r\delta)_{,r} + (c_{rs}\delta)_{,rs}$$
$$= (a - 2b_{r,r} + c_{rs,rs})\delta + (-2b_r + 2c_{rs,r})\delta_{,r} + c_{rs}\delta_{,rs}. \quad (4\text{-}31)$$

The expression (4-31) must equal minus the expression (4-30) identically. In order that the coefficients of $\delta_{,rs}$ shall agree we must have

$$c_{rs} = 0. \tag{4-32}$$

This then makes the coefficients of $\delta_{,r}$ agree. Finally, in order that the coefficients of $\delta$ shall agree, we must have

$$a = b_{r,r}. \tag{4-33}$$

This gives us the equation

$$\left[K_\perp, K'_\perp\right] = 2b_r\delta_{,r} + b_{r,r}\delta. \tag{4-34}$$

Let us now substitute in (4-28) and (4-29). They become:

$$\int K_r d^3x = \int\int x_r(2b_s\delta_{,s} + b_{s,s}\delta)d^3x d^3x'$$

$$= \int x_r b_{s,s} d^3x$$

$$= \int b_r d^3x. \tag{4-35}$$

(Note that $x_{r,s} = \partial x_r/\partial x^3 = -\delta_{rs}$.)

$$-\int (x_r K_s - x_s K_r)\, d^3x = \int\int x_r x'_s(2b_t\delta_{,t} + b_{t,t}\delta)d^3x\, d^3x'$$

$$= \int (-2x_r b_s + x_r x_s b_{t,t})d^3x$$

$$= \int (-x_r b_s + x_s b_r)d^3x. \tag{4-36}$$

This is what our consistency conditions reduce to, and we see that they are satisfied by taking $b_r = K_r$. This is not quite the most general solution; more generally we could have

$$b_r = K_r + \theta_{rs,s} \tag{4-37}$$

for any quantity $\theta_{rs}$ satisfying the condition that

$$\int (\theta_{rs} - \theta_{sr})d^3x = 0. \tag{4-38}$$

Thus $\theta$ can have any symmetrical part and its anti-symmetrical part must be a divergence.

That gives the general requirement for a field theory to be relativistic. We have to find the energy density $K_\perp$ satisfying the Poisson bracket relation (4-34) where $b_r$ is connected with the momentum density by (4-37). If we work out the energy density from a Lorentz-invariant action then this condition will certainly be satisfied in the classical theory. It might not be satisfied in the quantum theory because the order of the factors might be wrong. It is only when one can choose the order of the factors in the energy density so as to make (4-34), (4-37) hold accurately that we have a relativistic quantum theory. The conditions which we have here for a quantum theory to be relativistic are less stringent than the ones which we obtained when we had states defined on general curved surfaces.

I would like to illustrate that by taking the example of Born-Infeld electrodynamics. This is an electrodynamics which is in agreement with Maxwell electrodynamics for weak fields but differs from it for strong fields. (We now refer the electromagnetic field quantities to some absolute unit defined in terms of the charge of the electron and classical radius of the electron, so that we can talk of strong fields and weak fields.) The general equations of the Born-Infeld electrodynamics follow from the action principle:

$$I = \int \sqrt{-\det \left(g_{\mu\nu} + F_{\mu\nu}\right)}\, d^4x. \qquad (4\text{-}39)$$

We may use curvilinear coordinates at this stage. $g_{\mu\nu}$ gives the metric referred to these curvilinear coordinates and $F_{\mu\nu}$ gives the electromagnetic field referred to the absolute unit.

We can pass from this action integral to a Hamiltonian by using the general procedure. The result is to give us a Hamiltonian in which we have, in addition to the variables needed to describe the surface, the dynamical coordinates $A_r$, $r = 1, 2, 3$. $A_0$ turns out to be an unimportant variable just like in the Maxwell field. The conjugate momenta $D^r$ to the $A_r$ are the components of the electric induction, and satisfy the Poisson bracket relations

$$\left[A_r, D'^s\right] = g_r^s \delta \left(x - x'\right). \qquad (4\text{-}40)$$

It turns out that in the Hamiltonian we only have $A$ occurring through its curl, namely through the field quantities:

$$B^r = \frac{1}{2}\varepsilon^{rst} F_{st} = \varepsilon^{rst} A_{t,s}. \tag{4-41}$$

$\varepsilon^{rst} = 1$ when $(rst) = (1, 2, 3)$ and is anti-symmetrical between the suffixes. The commutation relation between $B$ and $D$ is

$$\left[ B^r, D'^s \right] = \varepsilon^{rst} \delta_{,t}(x - x'). \tag{4-42}$$

The momentum density now has the value

$$K_r = F_{rs} D^s. \tag{4-43}$$

This is just the same as in the Maxwell theory. It is in agreement with the general principle that the momentum density depends only on geometrical arguments, i.e. on the geometrical character of the fields we are using, and the action principle doesn't matter.

The energy density now has the value

$$K_\perp = \left\{ \Gamma^2 - \gamma_{rs} \left( D^r D^s + B^r B^s \right) - \gamma^{rs} F_{rt} F_{su} D^t D^u \right\}^{1/2} \tag{4-44}$$

Here $\gamma_{rs}$ is the metric in the three-dimensional surface and

$$-\Gamma^2 = \det \gamma_{rs}. \tag{4-45}$$

If we work with curved surfaces we require $K_\perp$ to satisfy the Poisson bracket relation (3-31). In the classical theory it must do so because it is deduced from a Lorentz-invariant action integral. But we cannot get it to satisfy the required commutation relationship in the quantum theory. The expression for $K_\perp$ has a square root occurring in it, which makes it very awkward to work with. It seems to be quite hopeless to try to get the commutation relations correctly fulfilled with the coefficients $\gamma^{rs}$ occurring on the left. So it does not seem to be possible to get a Born-Infeld quantum electrodynamics with the state defined on general curved surfaces.

Let us, however, go over to flat surfaces. For that purpose, we need to work out the Poisson bracket relationship (4-34). Now we know that conditions are all right in the classical theory. Classically we must therefore have the Poisson bracket relationship:

$$\left[ K_\perp, K'_\perp \right] = 2K_r \delta_{,r} + K_{r,r} \delta. \tag{4-46}$$

We can see without going into detailed calculations that this must hold also in the quantum theory, because $K_r$ is built up entirely from the quantities $D^s$ and $B^t$. When we work things out in the quantum theory, we shall have the $D$'s and the $B$'s occurring in a certain order, but the $D$'s and $B$'s all commute with each other when we take them at the same point. We see that from (4-42). If we put the $x' = x$ we get

$$[B^r, D^s] = \varepsilon^{rst}\delta_{,t}(0) = 0 \qquad (4\text{-}47)$$

(the derivative of the delta function with the argument 0 is to be taken as zero). Thus we are not bothered by the non-commutation of the $D$'s and the $B$'s that occur in $K_r$. We must therefore get the classical expression, so that the consistency conditions *are* fulfilled.

So for the Born-Infeld electrodynamics, the consistency conditions for the quantum theory on flat surfaces are fulfilled, while they are not fulfilled on curved surfaces. Physically that means that we can set up the basic equations for a quantum theory of the Born-Infeld electrodynamics agreeing with special relativity, but we should have difficulties if we wanted to have this quantum theory agreeing with general relativity.

That completes the discussion of the consistency requirements for the quantum theory to be relativistic. However, even if we have satisfied these consistency requirements, we have not yet disposed of all the difficulties. There are some quite formidable difficulties still lying ahead of us. If we were dealing with a system involving only a finite number of degrees of freedom, then we should have disposed of all of the difficulties, and it would be a straightforward matter to solve the differential equations on $\psi$. But with field theory, we have an infinite number of degrees of freedom, and this infinity may lead to trouble. It usually does lead to trouble.

We have to solve equations in which the unknown, the wave function $\psi$, involves an infinite number of variables. The usual method that people have for solving this kind of equation is to use perturbation methods in which the wave function is expanded in powers of some small parameter, and one tries to get a solution step by step. But one usually runs into the difficulty that after a certain stage the equations lead to divergent integrals.

People have done a great deal of work on this problem. They have found methods for handling these divergent integrals which seem to be tolerable to physicists even though they cannot be justified mathematically, and they have built up the renormalization technique, which allows one to disregard the infinities in the case of certain kinds of field theory.

So, even when we have formally satisfied the consistency requirements, we still have the difficulty that we may not know how to get solutions of the wave equation satisfying the required supplementary conditions. If we can get such solutions, there remains the further problem of introducing scalar products for these solutions, which means considering these solutions as the vectors in a Hilbert space. It is necessary to introduce these scalar products before we can get a physical interpretation for our wave function in terms of the standard rules for the physical interpretation of quantum mechanics. It is necessary that we should have scalar products for the wave functions which satisfy the supplementary conditions, but we do not need to worry about scalar products for general wave functions which do not satisfy the supplementary conditions. There may be no way of defining scalar products for these general wave functions, but that would not matter at all. The physical interpretation for quantum mechanics requires that scalar products exist only for wave functions satisfying all the supplementary conditions.

You see that there are quite formidable difficulties in getting the Hamiltonian theory to work, in connection with quantum mechanics. So far as concerns classical mechanics, the method seems to be fairly complete and we know exactly what the situation is; but for quantum mechanics we have only really started on the problem. There are the difficulties of finding solutions even when the supplementary conditions are formally consistent, and possibly also the difficulty of introducing scalar products of the solutions.

The difficulties are quite serious, and they have led some physicists to challenge the whole Hamiltonian method. A good many physicists are now working on the problem of trying to set up a quantum field theory independently of any Hamiltonian. Their general method is to introduce quantities which are of physical importance, then to bring in accepted general principles in order to impose conditions on these quantities; and their hope is that ultimately they will get enough conditions imposed on these quantities of physical importance to be able to calculate them. They are still very far from achieving that end, and my own belief is that it will not be possible to dispense entirely with the Hamiltonian method. The Hamiltonian method dominates mechanics from the classical point of view. It may be that our method of passing from classical mechanics to quantum mechanics is not yet correct. I still think that in any future quantum theory there will have to be something corresponding to Hamiltonian theory, even if it is not in the same form as at present.

I have given the treatment of the Hamiltonian method as far as it has yet been developed. It is quite a general and powerful method which can be adapted to a variety of problems. It can be adapted to problems where singularities (point or surface) occur in the field. The general idea governing this development of the Hamiltonian theory is to find an action $I$ which involves certain parameters $q$, such that when we vary the $q$'s, $\delta I$ is linear in the $\delta q$'s. It is indispensable that we should have $\delta I$ linear in the $\delta q$'s in order that we may apply the treatment described in these lectures.

The way to bring about linearity when we have singularities is to work in terms of curvilinear coordinates, and not to vary any equations which determine the position of a singular point or a singular surface. For example, if we are dealing with a singular surface specified by an equation $f(x) = 0$, then we must have a variation principle in which $f(x)$ is not varied. If we allow $f(x)$ to vary, if we treat $f$ itself as providing some of the $q$'s, then we do not have $\delta I$ linear in the $\delta q$'s. But we can keep $f(x)$ fixed with respect to some curvilinear coordinate system $x$ and we can vary the surface by varying the curvilinear coordinate system without varying the function $f$. Then the general method which I have discussed here works very well in the classical theory. When we go over to quantization we have the difficulties arising which I have discussed.

# ACKNOWLEDGMENTS

This book would not have been possible without the help of a number of talented people who made different contributions at various stages of the book's development. Among those deserving special thanks are Joel Allred, David Goldberg, Leonard Mlodinow, and Karen Pelaez.